D0926444

Energy, Power, and Transportation Technology

Len S. Litowitz
Professor, Industry and Technology
Millersville University
Millersville, Pennsylvania

Ryan A. Brown
Associate Instructor
Indiana University
Indianapolis, Indiana

Contributing Photographer
Jack Klasey
Kankakee, Illinois

Publisher
The Goodheart-Willcox Company, Inc.
Tinley Park, Illinois
www.g-w.com

Manufactured in the United States of America.

Library of Congress Catalog Card Number 2005046305

ISBN-13: 978-1-59070-221-5

ISBN-10: 1-59070-221-2

3 4 5 6 7 8 9 10—07—10 09 08 07

Library of Congress Cataloging-in-Publication Data

Litowitz, Len S.
Energy, power, and transportation technology / by Len S. Litowitz, Ryan A. Brown; contributing photographer Jack Klasey.

p. cm.
ISBN 1-59070-221-2
1. Power resources. 2. Transportation. I. Brown, Ryan, A. II. Title.
TJ163.2.L58 2007
621.042—dc22 2005046305

Introduction

Our lives are affected in many ways by advancements in technology. *Energy, Power, and Transportation Technology* provides a comprehensive study of the basic elements of energy, power, and transportation and how they affect the world we live in. This textbook covers the resources, processes, and systems used in these industries.

Energy, Power, and Transportation Technology is organized into 27 chapters. The first chapter introduces the concepts of energy, power, and transportation and their relationships to one another. The next section of chapters discusses energy and its various sources. The following section explains power and different types of power systems. The next section deals with transportation systems and the vehicular systems that make them work. The final section takes a look at the effects of energy, power, and transportation on the environment and gives you a glimpse into the future of these technological industries.

Each chapter begins with a list of objectives, divided into Basic Concepts, Intermediate Concepts, and Advanced Concepts. New terms related to energy, power, and transportation appear in ***bold italics***. These terms are defined in a running glossary in the margins. At the end of each chapter, a list of these terms is given so you can review them. Also given at the end of each chapter are Test Your Knowledge questions, which will help you review the material in the chapter. Several end-of-chapter activities are also provided so you can apply the knowledge you gained in the chapter.

As you use this textbook, you will find many boxed features. These features are designed to give you additional information about energy, power, and transportation technology and how it relates to the world around you. The four types of features are the following:

- **Career Connections.** These provide additional information about careers related to energy, power, and transportation.
- **Curricular Connections.** These provide additional information to emphasize how math, science, and social studies are connected to the world of energy, power, and transportation.
- **Technology Links.** These provide additional information about how energy, power, and transportation relate to and work with other types of technology, such as agriculture, communication, construction, manufacturing, and medicine.
- **Tech Extensions.** These provide additional information about interesting aspects of energy, power, and transportation technology.

A sound understanding of energy, power, and transportation technology is essential for making wise choices. With the knowledge you will gain in this course, you will be prepared to take an active and responsible part in our technological world. You are now ready to embark on this challenging study of energy, power, and transportation!

About the Authors

Dr. Len S. Litowitz is a full professor in the Department of Industry and Technology at Millersville University of Pennsylvania, where he also serves as program coordinator for the technology education teacher preparation program. He has taught courses in energy, power, transportation, and automation technologies to students of various age levels for approximately 25 years. Dr. Litowitz has published and presented extensively in the field of technology education, served as a consultant to numerous school districts, and taught summer workshops at several colleges and universities throughout the country. Additionally, Dr. Litowitz has studied the Design and Technology curriculum in the United Kingdom and served terms on the Board of Directors for the International Technology Education Association (ITEA) and the Technology Student Association (TSA). In 2002, he completed a term as President of the Technology Education Association of Pennsylvania (TEAP). Dr. Litowitz resides in Conestoga, Pennsylvania with his wife, Evanna Morris; daughter, Lindsay Rose; son, Evan; and cat, Jingle Bell.

Mr. Ryan A. Brown is an associate instructor at Indiana University. He is currently pursuing a Doctor of Philosophy degree in curriculum studies, with a minor in educational policy. Previously, he has taught technology education at the secondary level. He taught a variety of courses, including transportation systems, design process, and fundamentals of engineering. Mr. Brown coauthored *Technology: Design and Applications* with Dr. R. Thomas Wright. He has also written titles in both the *HITS* and *KITS* series for the ITEA, as well as in the *Activity!* series for the Center for Implementing Technology Education. Mr. Brown's educational background includes a bachelor's degree and master's degree from Ball State University. He has been recognized with the Young Professionals Award from the Technology Educators of Indiana. The Foundation for Technology Education also named him a Maley Spirit of Excellence Outstanding Graduate Student. Mr. Brown and his wife, Heather, reside in Indianapolis, Indiana.

Brief Contents

Table of Contents

Features

Career Connections

Curricular Connections

Math

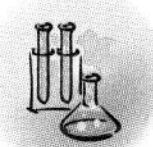

Science

Social Studies

Technology Links

Agriculture

Communication

Construction

Manufacturing

Medicine

Tech Extensions

Acknowledgments

The authors and publisher wish to thank the following companies, organizations, and individuals for their generous contributions of photographic images, artwork, and resource material.

Adrienne Levatino
Airbus
American Coal Foundation
America West Airlines
Amtrak
Autodesk, Inc.
Baldor Electric
Balluff
Battery Council International
Bayliner Marine Corporation
Bell Helicopter Textron
Bill Lyons, Spire Solar Chicago
The Boeing Company
Bosch Automation Technology
Boston Gear
Briggs and Stratton Corporation
British Columbia Transit
British Information Services
Burlington Northern Santa Fe
Canadian Nuclear Association
Canadian Solar Industries Association
Carnival Cruise Lines
Caterpillar, Inc.
Cessna Aircraft Company
Chuck Meyers, Office of Surface Mining, Department of the Interior
Cleveland Gear
The Coastal Corporation
CSIA
CSX Creative Services
DaimlerChrysler
Darrell Hunt
Deere & Company
Delta Queen Steamboat Company
DG Flugzeugbau, GmbH
Edison Electric Institute
Estes
Eurotunnel
Exelon Energy
Exide
Fisher Controls International, Inc.
Ford Motor Company
Freightliner Corporation
French Technology Press Office
Fruehauf Trailer Operations
Garmin International
GM-Cadillac Motor Car Division
GM-Hughes
The Goodyear Tire & Rubber Company
Go-Power Corporation
Greyhound
Hearth and Home Technologies—Quadra-Fire
Hobart Brothers Company
Howard Bud Smith
International Atomic Energy Agency
John Wright, Jr.
Joint European Torus/European Fusion Development Agreement (JET/EFDA)
Joseph M. McCade
Kawasaki
Kenneth P. DeLucca
Kerr-McGee Corporation
Kohler Company
Library of Congress
Lincoln-Mercury
Lockheed Missiles and Space Company, Inc.
Mack Truck Company
Mazda
Mercedes-Benz
Miller Electric Manufacturing Company
Monarch Instruments
Monnier, Inc.
Murphy/Jahn, Architects
National Aeronautics and Space Administration (NASA)
National Aeronautics and Space Administration's Jet Propulsion Laboratory
National Aeronautics and Space Administration's Johnson Space Center
National Aeronautics and Space Administration's Marshall Space Flight Center
National Oceanic and Atmospheric Administration
National Park Service, Natchez Trace Parkway
Nissan
Norfolk Southern Corporation
OMC
P&H Construction Equipment
P&H Mining Equipment
Polaris
Port of Long Beach
Quaker State Oil Refining Corporation
Randy Montoya
Rupp Industries, Inc.
Saab
Saturn Corporation
Scaled Composites
Scott Haas
Sears, Roebuck and Co.
Sharon A. Brusic
Shell Energy
Shell Oil Company
Shell Wind Energy
Siemens
Siemens VDO
SI Handling Systems, Inc.
Sikorsky Helicopter
Smithsonian Institution
Solar Power Corporation
Steve Olewinski
Subaru
Susan Hunt
Tecumseh Products Company
Terex/Cedarapids
Transrapid International, Inc., GmbH
Turbo Power, United Technologies Corporation
Union Electric Company
United Airlines
United Parcel Service
Uniweld
U.S. Air Force
U.S. Army
U.S. Bureau of Reclamation
U.S. Coast Guard
U.S. Department of Agriculture
U.S. Department of Energy (DOE)
U.S. Energy Information Administration
U.S. Navy
U.S. Postal Service
Van Hool
Wisconsin Department of Tourism
Wisconsin Department of Transportation

1

Energy, Power, and Transportation Technologies

Basic Concepts

- List the various forms of energy and power.
- Define *energy*, *power*, and *transportation*.
- Name the various modes of transportation.
- Identify a technological system.

Intermediate Concepts

- Discuss the importance of the study of energy, power, and transportation.
- Explain various factors affecting technological development.
- Describe the elements of a technological system.
- Recognize the inputs of a technological system.
- Give examples of the components of transportation systems that are common to all modes of transportation.

Advanced Concepts

- Differentiate between the scientific method and the technological method of problem solving.
- Summarize the technological method of problem solving.

The study of energy, power, and transportation technologies is, in many ways, the study of the evolution of society. The quality of energy consumed, the power produced, and the transportation modes of choice are linked to quality of life, at least as defined by the standards of Western civilization. This book is about the energy resources that fuel our society, the power we produce from energy for comfort and convenience, and the methods by which we transport people and products from place to place. The transportation sector is one of the largest energy-consuming sectors in the United States. This is based in part on America's dependence on automobiles. See **Figure 1-1.**

Figure 1-1. In the United States, the transportation sector—especially private automobiles—is one of the largest consumers of energy resources.

Throughout this book, you will learn about the technological aspects of energy, power, and transportation. You will also learn about other aspects of these technologies. Technology does not exist in a vacuum. There are a host of other factors that come into play when considering the development or expansion of a particular energy, power, or transportation technology. These factors include economic considerations, environmental concerns, and even political and social influences. You will learn about technological trade-offs. ***Trade-offs*** are situations in which technological developments solve one problem, only to create other problems. Such is the nature of technology. All citizens need to know how to make good decisions about the development of energy, power, and transportation technologies for present and future generations. The decisions we make now about these technologies will undoubtedly have a profound impact on more than just our nation. They will also affect our entire planet in the not so distant future. Good decisions can only come from a well-informed citizenry.

Trade-off: A situation in which a technological development solves one problem, only to create other problems.

Energy

Energy is defined as the ability to do work. Without energy resources, there would be no ability to produce power. Without power, there would be no modern conveniences. Electricity and contemporary forms of transportation make many conveniences possible. Some of what you will learn about energy may surprise you. For instance, did you know that almost all the energy utilized worldwide is in the form of fossil fuels? Well, unfortunately, this is true. See **Figure 1-2.**

Energy: The ability to do work.

Solar, nuclear, tidal, and wind energy are some alternative types of energy. They are in various stages of use or development. These sources offer some inherent advantages over the continued extensive use of fossil fuels. They all, however, offer some inherent disadvantages as well.

Curricular Connection

Social Studies: Society's Impact on Technology

Using available science and technology, many items have been developed that are intended to make life easier or help society reach a particular goal. After widespread use, sometimes the side effects of the application or handling of these items proves hazardous to the people involved or those exposed. In such cases, the government often bans the use of the substance or develops guidelines that alter its use for the sake of public safety. Asbestos is one such substance.

Asbestos is a naturally occurring mineral that was widely used during the 1800s for insulation on machinery and related items that reached high operating temperatures, such as boilers and steam pipes. It was also commonly used in roofing and flooring materials. The mineral was an extremely effective insulator that was strong and long lasting. Asbestos has also been used in brake pads of various transportation vehicles.

In the early 1900s, it was discovered that inhaling airborne fibers of asbestos, created while mining and working with the substance, caused a condition called asbestosis. This condition affects the airways and air sacs in both lungs and impairs breathing. People diagnosed with asbestosis are also at a greater risk of developing lung cancer.

While the benefits of using asbestos in energy conservation and transportation are great, the effect of exposure on the workers and general public are dire. The trade-offs for taking advantage of this effective material are the health and well-being of the society in which it is used. Other examples of items developed through technology that proved harmful to society include the following:

- Foam insulation.
- Lead paint.
- DDT (a pesticide).
- Agent Orange (an herbicide).

Society obviously impacts the development of technology. It is imperative that new products are proven acceptable, in terms of effectiveness and safety. Regardless of the benefits of any technology, if the side effects are harmful, the technology will not be used in society.

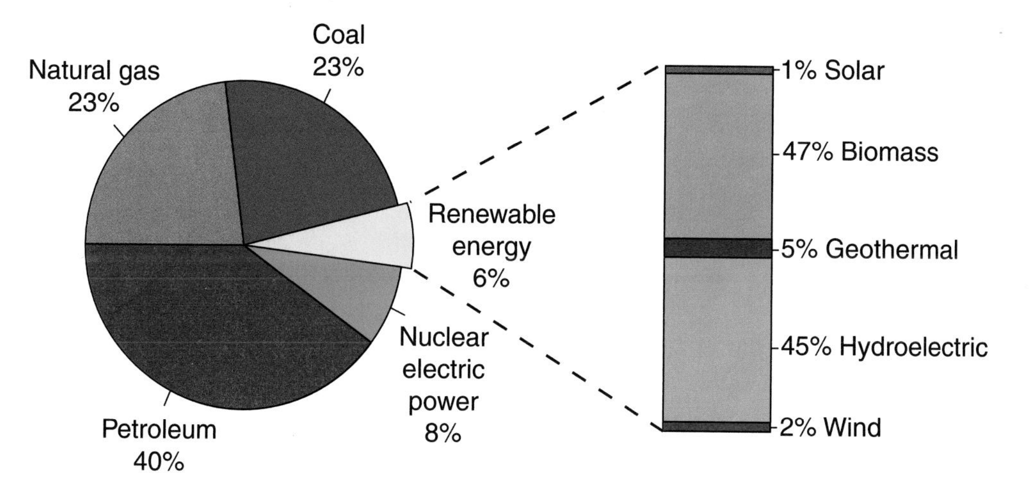

Figure 1-2. Fossil fuels are the largest source of energy used in the United States. In a recent year, renewable energy accounted for only about 6% of the total, as shown in this graph. (U.S. Energy Information Administration)

Energy is divided into the categories of nonrenewable energy resources and renewable and inexhaustible energy resources. Nonrenewable energy resources include the fossil fuels and uranium. Renewable and inexhaustible energy resources include wind and solar energy. You will learn about these resources and specific ways to help conserve energy. Several examples of ways in which energy can be conserved include reducing heat loss from structures, using energy-efficient appliances, and wisely selecting means of transportation.

Power

Power: The rate at which work is performed or energy is expended.

Power is defined as the rate at which work is performed or energy is expended. The amount of energy available in a given quantity of material is rather useless on its own. It is usually more beneficial if the energy can be converted to useful power to perform work. There are three main forms of power. They are electrical power, mechanical power, and fluid power. See **Figure 1-3.** All three forms of power are used extensively throughout society.

We see the results of electrical power, or electricity, everywhere. It is, perhaps, the most versatile of the three forms of power. Electricity can be used to power many things. It is used extensively for lighting, heat, tools, and appliances. Even within an automobile where electricity is not primarily responsible for motive power, its uses are critical to the function of the vehicle. Imagine a car without headlights, a charging system, and all the dashboard gauges powered by electricity.

Mechanical power predates electrical power. It is the easiest form of power to conceptualize because the components of mechanical power are all visual in nature. Some of these components are gears, levers, cables, and shafts. Mechanical power is very useful for performing work. Electrical power and fluid power are often converted back into mechanical power for end use. For instance, an electric drill uses electricity as a power source. The electricity must be converted to mechanical power, however, to perform the task of drilling a hole. Earth-moving equipment, like a

Figure 1-3. The three forms of power.

Electrical Power

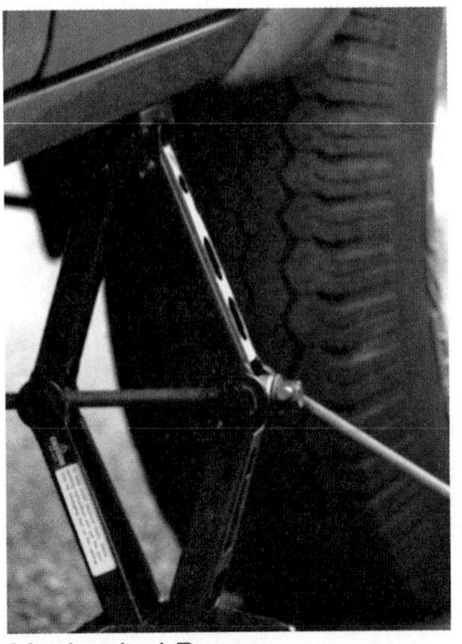

Mechanical Power

Fluid Power

backhoe, uses fluid power to provide strength. The fluid power must be converted back to mechanical power, however, to scoop the earth.

Fluid power is not as common as the other two forms of power. Its applications, however, are critical to many industries. Fluid power is used widely in the manufacturing, construction, and transportation industries. This is because fluid power is ideal for pure strength applications. The wing flaps on an airplane can be adjusted by fluid power, even though they are meeting massive air resistance when the plane is in flight. The braking and steering systems on most modern cars are assisted by fluid power. Most heavy construction equipment involves the extensive use of fluid power.

Control technology and automation also affect our lives in many ways. Automation relies on input sensors and output devices. The sensors often send signals to a computer programmed for decision making. Output devices include motors, lights, buzzers, and relays. These fundamental concepts of automation are discussed in more detail later in this book.

There are many devices that perform conversions from one form of power to another or from a form of power into a form of energy. Some, however, are used much more extensively than others. For instance, the ability to convert electricity into visible light allows us to see when it is dark out. The lightbulb is certainly one of the most popular conversion devices ever invented. Speakers convert electricity into sound. Other converters are used for convenience and automation. One of the most practical and popular sensors is the thermostat. It allows the home to remain at a relatively constant temperature without constant adjustments to the furnace. Perhaps you have a photoelectric eye that turns on a night-light. This device can convert visible light to an electrical signal.

Technology Link

Agriculture: Irrigation Pumps

Agriculture often relies on power technology to ensure successful crops. Irrigation systems can be used to support agriculture in dry or unpredictable climates. These systems provide artificial watering to maintain proper plant growth.

A sprinkler irrigation system produces artificial rain to water the crops. A pump forces water from a lake, a river, a reservoir, or an underground aquifer into the main distribution lines. The pressurized water flows through the main lines to a series of lateral pipes. At the end of each pipe is a sprinkler head. The water enters the sprinkler heads, which spray water onto the land. Valves between the main and lateral lines control the water flow.

The fluid power used to pump water from the water source to the farming fields is crucial to agricultural technology. Without these irrigation systems, it would be difficult to distribute water to areas where it is needed. Agricultural water pumps allow water to reach much more land than would ever be possible without them, and they make farming a more efficient enterprise.

The internal combustion engine is a conversion device. It has many applications in today's society. Small gas engines power everything from weed whackers and snowblowers to generators and irrigation pumps. Larger internal combustion engines form the backbone of a large segment of the transportation sector of our economy.

Transportation

Transportation: The movement of people or products from one place to another.

Intermodal: A transportation system that combines various modes of transportation to move people and products.

Transportation is defined as the movement of people or products from one place to another. Most people take transportation for granted. They usually do not think about how it affects their lives. Transportation is, however, essential in nearly everything you do. See **Figure 1-4.** Think of what your life would be like if you had no transportation. There are various modes of transportation. They are land, water, air, and space. ***Intermodal*** transportation systems also move many people and products. These systems combine various modes of transportation. You will learn about the vehicular systems common to all forms of transportation. These include guidance, control, and suspension systems.

Environmental Impacts and Looking Ahead to the Future

It is an unfortunate reality that the same use and development of energy, power, and transportation technologies that improve our quality of life also pollute our environment. In fact, the consumption of coal for the production of electricity and the consumption of gasoline to power the transportation sector of our economy are responsible for much of our pollution. Throughout this book, environmental concerns are emphasized. We will take a specific look at environmental impacts of energy, power, and transportation technologies and what is being done to improve the

Figure 1-4. We depend on several modes of transportation in our daily lives. The four main modes of transportation are land, water, air, and space.

Career Connection

NASCAR Race Car Mechanics

Many professions are necessary for energy, power, and transportation technologies to function properly. These technologies make our lives easier, safer, and more convenient. Sometimes, these technologies also make our lives more fun! NASCAR race car mechanics play an important role in the sport of racing, and they help provide us with entertainment.

NASCAR race car mechanics often travel to racing events, so their schedules are constantly changing. Usually, they spend a few days every week working in the shop on cars they are preparing for upcoming events. Some mechanics have specializations, such as working on the brakes or engines of the cars. A brake specialist, for example, might be responsible for the brakes, the hydraulics, and the cooling systems. He also could be in charge of restocking the trucks with all the brakes that will be needed at the next event.

After traveling to the race track, the mechanics check in at the track and have their cars inspected. The drivers practice and qualify their cars. The brake specialist helps make changes to the suspension and brakes during the practice sessions. Additional preparations are made to the cars after the practice sessions are complete. On the day of the race, the mechanics finish preparing the cars and help set up the pit area. See **Figure 1-A.** After the race, they help load the cars back onto the transporter and head back to the shop to start working on the cars for the next race.

One advantage of this job is that it is always changing. The cars are different from week to week, as are the tracks and racing locations. Some disadvantages, however, are the long season and the lack of time off.

To become a NASCAR race car mechanic, at least a high school diploma is required, though most positions require some sort of engineering degree. An enthusiasm for cars and experience working on them are also essential to get into this profession. Race car mechanics can make anywhere from $40,000 to $80,000 a year.

Figure 1-A. NASCAR mechanics have to work quickly at the pit stop during a race. (Sears, Roebuck and Co.)

environmental impacts of such technologies. The future of energy, power, and transportation technologies is also worth investigating. Some new research on power production and transportation techniques will be discussed. We will also look at ideals necessary for future technological developments in energy, power, and transportation technologies.

Studying Technology and Science

Science: The body of knowledge related to the natural world and its phenomena.

Technology: The body of knowledge related to the human-made world. The technological world includes human-made products and their impacts.

Scientific method: Methodology that pursues new knowledge by the collection of data through observation and experimentation to test a hypothesis.

The terms *science* and *technology* are frequently used hand in hand. To many people, there is little difference between science and technology. Other people associate technology with computers. Neither of these explanations of technology, however, is correct. A better way to explain the difference between the study of science and the study of technology is to consider their definitions.

- ***Science*** is the body of knowledge related to the natural world and its phenomena.
- ***Technology*** is the body of knowledge related to the human-made world. The technological world includes human-made products and their impacts.

Another means of differentiating technology from science is to review the methodology used by scientists and technologists. Scientists use the ***scientific method*** of inquiry. This method pursues new knowledge by the collection of data through observation and experimentation to test a hypothesis. Technologists use the ***technological method*** of problem solving to yield new products through a process of researching, testing, and refining. The two methods are compared in **Figure 1-5.**

The Scientific Method

State the Problem
Gather Information
Formulate a Hypothesis
Test the Hypothesis
Record and Analyze the Data
State a Conclusion

The Technological Method

Identify the Problem/Analyze the Situation
Clarify and Specify the Problem
Research and Investigate—Gather Information
Brainstorm Alternative Solutions
Choose the Best Solution
Model and Prototype a Solution
Test and Evaluate
Refine the Solution
Observe/Analyze/Synthesize

Figure 1-5. A comparison of the scientific method and the technological method.

The technological method is probably not as ingrained in you as the scientific method is. It is, however, a method that is well known to American industry. This method is used extensively to test and refine new products prior to large-scale manufacturing. It involves a design loop. This loop allows for continuous evaluation and modification.

Technological method: Method of problem solving that yields new products through a process of researching, testing, and refining.

Technological Systems

You are surrounded by various technological systems. Technology is the application of knowledge and creative thinking that changes resources to meet human needs. This usually results in a more convenient lifestyle. A ***system*** involves a combination of related parts. These parts work together to accomplish a desired result. A familiar system is the circulatory system located within your body. You may also know about our system of government. Both of these systems have separate, but related, parts working together. When you combine the two terms, *technology* and *system*, the result is a technological system. See **Figure 1-6.** Note that the technological systems model is sometimes referred to as the universal systems model because it can be applied to virtually any technology.

System: A combination of related parts that work together to accomplish a desired result.

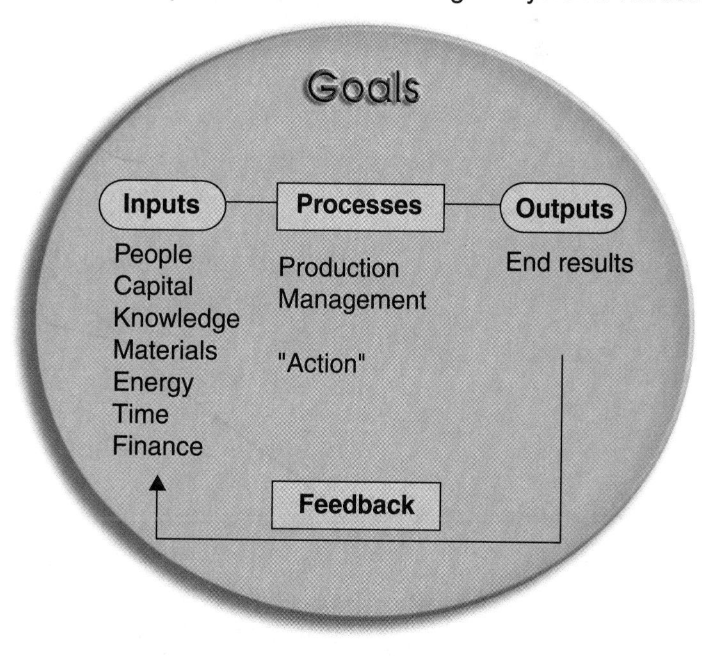

Figure 1-6. Inputs, processes, outputs, and feedback are the vital components of the technological systems model.

All ***technological systems*** consist of inputs, processes, and outputs. Technological systems are designed to meet human needs and wants. In the technological systems model, these needs and wants can be expressed as goals. One goal of construction technology is to build habitats. The primary goal of manufacturing technology is to convert raw materials into useful products. The main goal of transportation technology is to move people and products from place to place. In order to determine how well the goal is being met, all technological systems require feedback. ***Feedback*** provides information on how the system is performing or has performed. It is usually provided in multiple forms. An aircraft pilot, for example, looks at the instrument panel for feedback. Several gauges provide information about the altitude and speed of the aircraft. This lets the pilot know the performance of the aircraft. In a similar manner, power plant operators monitor electrical output and coolant temperature.

Technological system: The inputs, processes, and outputs designed to meet human needs and wants.

Feedback: Information on how a system is performing or has performed.

System Inputs

The ***inputs*** of a system are the resources needed to begin the system. These resources are needed to make the system operate. There are several different types of resources, or inputs, that must be considered for a technological system to function properly. See **Figure 1-7.** These resources are the following:

Input: A resource needed to begin a system and make it operate.

Figure 1-7. These inputs are necessary for any technological system to operate properly. (Ford Motor Company)

- **People.** Humans are a very important resource for any technological system. People design, create, and maintain systems.
- **Capital.** Capital is all the tangible items needed and used within a technological system. Storage buildings, roads, computers, and vehicles are all examples of capital within a transportation system, and power plants and repair trucks are parts of an energy and power system.
- **Knowledge.** Through experience and application, people acquire knowledge and skills necessary to reach the goals of technological systems. For example, a train engineer must have the knowledge to read the instruments in order for the train to stay on schedule and function properly.
- **Materials.** Materials are major resources used to make any technological system function. Some materials can be used as they are, while others are changed by the system.
- **Energy.** All energy, power, and transportation systems operate from some source of energy. Sources include humans, animals, the wind, the sun, and petroleum.
- **Time.** Time is the duration of any activity. Scheduling and time are essential considerations for any technological system.
- **Finance.** Finance is money. Money is used to buy the resources used to begin a system.

System Processes

Process: The portion of the technological systems model in which the desired goal is in sight. The inputs are being changed in this step.

Output: The end result of a system.

All the inputs feed into the process portion of the technological systems model. During a ***process***, the desired goal is in sight. The inputs you started with are now being changed. This change will produce an ***output***, or result. The inputs are combined to make the system run. For the system to run properly, the system must be managed. People manage systems. For transportation to be efficient, management is a necessity. The job of a manager is to plan, organize, and control the system. For example, people who manage bus lines route the schedules of the buses so people can plan their trips in advance. Railroads and trucklines also need to be managed. The movement of people and cargo begins once the transportation process is started. Plant engineers monitor and manage power distribution grids so enough electricity is available to the grid during varying demand periods. Suppliers manage resource recovery and processing so enough energy is available to a specific location, in order to meet the needs of the population.

System Outputs

The outputs of a system are the end results. Inputs and processes assist in reaching the output. Technological systems are designed to meet human needs and wants. Therefore, it makes sense that many of the outputs are desired outputs, such as tangible products. Without inputs, processes, and outputs, there would be no system. An example of a desired output from the transportation industry would be a safe and successful trip from point A to point B in a timely fashion. See **Figure 1-8.** An example of a desired output from the energy and power industry would be safe and successful power generation without interruption.

Sometimes, in the course of producing desired outputs, some undesirable outputs also occur. Can you think of any undesirable outputs from the transportation or energy and power industries? One unintended consequence common to both industries is various forms of pollution. Technology often creates air, water, visual, and noise pollution. The burning of fossil fuels puts hydrocarbons and other pollutants into the atmosphere. Aircraft cause noise pollution, especially for people living near airports. Can you think of other undesirable outputs of transportation? Sometimes, undesired consequences are inevitable. Hydroelectric dams produce inexpensive electricity without the concern of fossil fuel pollutants. They also change the ecology of rivers, however, thus creating a different form of environmental pollution.

Figure 1-8. Completing a journey from one airport to another is an example of achieving a desired output.

Goals of a System

The success of all technological systems involves desirable goals. The goal could be to be the best package delivery service in the world. This is a particular system goal of a company. There are also societal and personal goals that apply to any technological system. To arrive at a destination comfortably and in style is a personal goal. Societal goals are much broader than system goals. They might include the following:

- Maintaining a low rate of unemployment.
- Maintaining a high standard of living.
- Maintaining technological superiority.
- Maintaining national safety.

Society puts restrictions on technological systems. It does this for good reason. Cars often need to pass exhaust inspections. Workers at nuclear plants are limited to a certain amount of yearly radiation exposure. A nuclear plant is limited to a certain amount of radiation emission. American culture generally sets high expectations for technological systems. We like our products to function properly and safely. Also, we want them produced inexpensively and quickly. The goals of a technological system and the goals of society are sometimes aligned. They are, however, sometimes opposed. Because of this, technological systems are constantly being adjusted. For example, a car that gets 60 miles per gallon, but is regarded as a safety hazard while on the road, would probably not be accepted by society. A nuclear power plant that could produce cheap electricity by eliminating the costs of an environmental containment structure and a security staff would also probably not be accepted.

Seven Major Technological Systems

The components of a technological system were briefly described in the previous section. Each of those components can be applied to any of the major technological systems. There are seven major technological systems. See **Figure 1-9.**

Most technological systems are of an industrial nature. Systems like manufacturing and communication are designed to serve very large numbers of people. There are seven major technological systems that can be used to classify different technologies.

- ***Agricultural technologies*** include systems that produce outputs by growing plants and raising animals. The outputs of agricultural technologies are typically foods and fibers.
- ***Communication technologies*** are systems associated with the dissemination of information and ideas. Products of communication technologies include schematics, advertisements, Web pages, and media messages.
- ***Construction technologies*** include systems associated with the creation of structures for residential, commercial, industrial, and civil use.

Agricultural technology: Systems that produce outputs by growing plants and animals. The outputs are typically foods and fibers.

Communication technology: Systems associated with the dissemination of information and ideas. Products include schematics, advertisements, Web pages, and media messages.

Construction technology: Systems associated with the creation of structures for residential, commercial, industrial, and civil use.

Agricultural Technologies

Energy and Power Technologies

Communication Technologies

Manufacturing Technologies

Construction Technologies

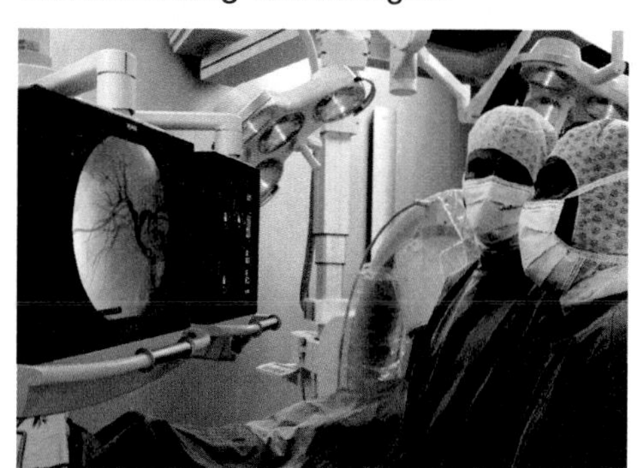

Medical Technologies

Transportation Technologies

Figure 1-9. The seven major types of technological systems.

- ***Energy and power technologies*** are systems that gather energy and convert it to useful power for the use and benefit of society. Products of energy and power technologies include electricity for household use and the engines that produce mechanical power for automobiles.
- ***Manufacturing technologies*** include systems that transform raw materials into useful products in a central location. These products must be marketed, transported, and distributed for end use.
- ***Medical technologies*** include systems used to maintain health and treat injuries and illnesses. The end product of medical technologies is a healthier society.
- ***Transportation technologies*** include systems designed to move people and products from one place to another. Systems involved with transportation technologies include propulsion, guidance, control, suspension, structure, and support.

This book focuses on two of these major technological systems: energy and power systems and transportation systems.

Energy and power technology: Systems that gather energy and convert it to useful power for the use and benefit of society. Products include electricity for household use and the engines that produce mechanical power for automobiles.

Manufacturing technology: Systems that transform raw materials into useful products in a central location. These products must be marketed, transported, and distributed for end use.

Medical technology: Systems used to maintain health and treat injuries and illnesses. The end product of medical technologies is a healthier society.

Transportation technology: Systems designed to move people and products from one place to another.

Summary

Technology is associated with the study of the human-made world. Science is associated with the phenomena of the natural world. All technological systems rely on a model known as the technological systems model. This model describes a series of inputs, processes, and outputs to accomplish goals. It can be applied to any major technological system. Feedback is used to constantly adjust the performance of the technological system. When addressing a technological problem, it is often helpful to make use of the technological method. This method of inquiry differs from the scientific method in that a useful product is usually the desired end result. Steps in the technological method include gathering information and researching, brainstorming alternative solutions, prototyping the best solution, testing and refining the solution, and synthesizing the entire process. Energy, power, and transportation technologies are part of a larger family of technological systems. This group of systems also includes agricultural, communication, construction, manufacturing, and medical technologies.

Key Words

All the following words have been used in this chapter. Do you know their meanings?

agricultural technology
communication technology
construction technology
energy
energy and power technology
feedback
input
intermodal
manufacturing technology
medical technology
output
power
process
science
scientific method
system
technological method
technological system
technology
trade-off
transportation
transportation technology

Test Your Knowledge

Write your answers on a separate sheet of paper. Do not write in this book.

1. *True or False?* The quality of energy, power, and transportation systems available to a culture greatly affects that society's quality of life.
2. The development of any particular technology is influenced by many factors besides technical know-how. Identify and describe three of these factors in one sentence each.
3. _____ is defined as the ability to do work.

4. The majority of all energy converted to power comes from _____.
5. _____ is defined as the rate of doing work.
6. Power is produced in these three forms: _____, _____, and _____.
7. Modes of transportation include air, land, _____, and _____.
8. Name two vehicular systems found in all forms of transportation systems.
9. *True or False?* Intermodal transportation is a synonym for land transportation.
10. The study of _____ is concerned with the natural world and its phenomena.
11. The study of _____ is concerned with the human-made world and its impacts and consequences.
12. Explain how the technological method differs from the scientific method.
13. Identify and briefly describe the steps in the technological problem-solving method. Use one sentence per step.
14. *True or False?* A transportation system is a technological system.
15. Gauges on an airplane and quarterly sales reports from a manufacturing firm are similar in that they both provide _____ in the universal systems model.
16. Time, capital, and finance are all considered _____ in the universal systems model.
17. *True or False?* A finished product is an output of a manufacturing system.
18. Power generation is a desired result of any power plant. What is an example of an undesired result?
19. *True or False?* Undesired consequences of particular technologies are sometimes inevitable.
20. Transportation is one of seven major technological systems. Energy and power is another. _____, _____, _____, _____, and _____ are also major technological systems.

Activities

1. Choose topics on alternative energy resources to research. Identify current uses for these forms of energy. Try to determine why these forms of energy are not used more extensively.
2. Invite a representative from a bus company to address your class on how the company sets up routes to transport people around town.
3. Analyze the desired and undesired results of using the automobile as a primary source of transportation.

2

An Introduction to Energy

Basic Concepts

- Identify types of energy surrounding us.
- Differentiate among renewable, nonrenewable, and inexhaustible energy sources.
- Explain the difference between potential and kinetic energy.
- Name and describe the six forms of energy.

Intermediate Concepts

- Describe the history of energy consumption in the United States.
- Name various sectors of society associated with energy consumption.
- Summarize the present energy consumption trends in the United States and worldwide.
- Discuss the concept of efficiency.
- Define the law of entropy.

Advanced Concepts

- Recognize various factors that influence the exploration and development of different energy resources.
- Give examples of reasons for growth in the demand for energy and power.

Energy is the ability to do work. It is one of the driving forces behind a modern, technologically advanced economy. The study of energy is important because serious decisions about the development of energy resources and associated technologies will have to be made in your lifetime. Issues surrounding energy resources are often complex. In addition to technical factors, factors such as politics, economics, and environmental concerns must be considered when making decisions about the development and use of particular energy technologies.

Energy: The Ability to Do Work

Often, we notice the work produced from energy and miss the energy being used. Work causes change. Wind that causes a windmill to operate a pump and pump water is an example of the effects of work. We often forget, however, that it is the energy produced by the wind that causes the work. Energy is at work all around us. The sun heats and lights the earth. See **Figure 2-1.** We, as humans, exert energy. Walking and bicycling are examples of work being done with the energy within us. See **Figure 2-2.** Energy affects our lives in many ways. It is easy to see some effects of energy. Still, we often fail to notice energy's many influences on our world. These influences range from impacts on our environment to economic and political considerations.

Figure 2-1. The sun provides the earth with huge amounts of energy.

Figure 2-2. Human beings convert energy sources, such as food, into energy used in work and play. (Wisconsin Department of Tourism)

For instance, you can see the sunlight and feel the sun's heat. You can feel the wind. The effects of big gusts of wind and tornadoes can be seen. You can see an airplane as it soars through the air. It is harder, however, to see the effects of energy on plants, lakes, and forests within our environment. See **Figure 2-3.**

Renewable energy source: A resource that can be replaced when needed.

Energy is used in many ways around us, but what is its source? It comes from many sources. These sources can be organized into the following groups:

- Renewable energy resources.
- Nonrenewable energy resources.
- Inexhaustible energy resources.

Renewable energy sources are those resources that can be replaced when needed. Food is one example that supplies energy for humans. Wood, cornstalks, and sugarcane are all examples of fuel sources that can be harvested and regenerated in a relatively short period of time.

Nonrenewable energy sources are those that cannot be replaced once used. Fossil fuels, such as coal, oil, and natural gas, are good examples of nonrenewable energy sources. Since it takes hundreds of thousands of years to produce the natural resources that constitute fossil fuels, these sources are considered nonrenewable. Uranium, which is mined from the ground much like coal and used as fuel for nuclear power plants, is also considered a nonrenewable energy source.

Inexhaustible energy sources are those that will never run out. At least, they will last for the next several million years! The sun, the wind, and waves are the most obvious examples of inexhaustible energy sources. These different sources of energy will be discussed in much greater detail in the following chapters.

Figure 2-3. Acid rain has damaged forests in the northeastern United States and eastern Canada. This scientist is gathering samples in a Massachusetts wetland to monitor water acidity. (National Oceanic and Atmospheric Administration)

Types of Energy

You have just been introduced to the different groups of energy resources. All energy formed from these sources can be classified into two types. Energy is either in the form of potential energy or kinetic energy.

Nonrenewable energy source: A resource that cannot be replaced once used.

Potential Energy

Potential energy is energy waiting to happen. For instance, a gallon of gasoline can produce a tremendous amount of mechanical power when properly consumed within an automobile engine. It is only potential energy, however, while waiting in a gas tank until it is burned to produce power. Similarly, water behind a hydroelectric dam is only potential energy until it is used to spin a turbine for producing electricity. It then becomes kinetic energy.

Inexhaustible energy source: An energy source that will never run out.

Potential energy: Energy waiting to happen.

Career Connection

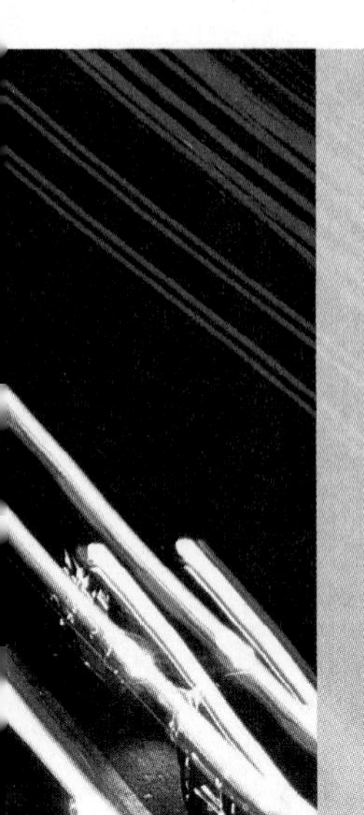

Geologists

A geologist uses scientific methods to study the earth's materials and their relationships to one another and our world. An understanding of the world's natural history and resources is an important tool in this profession. This information is needed for future planning.

In many cases, a geologist spends most of his time doing research. This work helps to gather environmental data, and it is also beneficial for predicting natural disasters. A geologist communicates her research to others in the field either with technical papers or with demonstrations using drafting software.

A geologist must be skilled in several areas, including math and science. Geologists must also have good observational skills. Verbal and written communication skills are necessary to convey data and findings. A master's degree is typical for workers in this field. The yearly salary may range from $32,000 to $79,000.

Kinetic energy: Energy in motion.

Light energy: Energy visible to the eye.

Heat energy: Energy with a longer wavelength than light energy. It is generally not visible to the eye, but it can be measured in terms of temperature. Also referred to as *infrared energy.*

Mechanical energy: Energy produced by mechanical devices, such as gears, pulleys, levers, or more complex devices, such as internal combustion engines.

Chemical energy: The potential energy locked within a substance.

Kinetic Energy

Kinetic energy is often defined as energy in motion. The water that spins a turbine in a hydroelectric plant to produce electricity is an example of kinetic energy. The wind that can power a windmill or wind generator and the radiant energy from the sun are also examples of kinetic energy.

Forms of Energy

As we just discussed, energy is either potential or kinetic. Potential energy and kinetic energy are related to the form in which the energy is found. All the energy around us comes in different forms. These forms are as follows:

- ***Light energy*** is energy visible to the eye. It represents a very small portion of all radiant energy, which collectively is known as the electromagnetic spectrum.
- ***Heat energy***, also referred to as *infrared energy*, has a longer wavelength than light energy. This longer wavelength does not allow heat to pass through certain materials, like glass, as readily as light does. Heat energy is generally not visible to the eye, but it can be measured in terms of temperature.
- ***Mechanical energy*** is energy produced by mechanical devices, such as gears, pulleys, levers, or more complex devices, like internal combustion engines.
- ***Chemical energy*** is the term used to describe the potential energy locked within a substance. For instance, 50 lbs. of red oak might be capable of producing the same amount of heat energy as 15 lbs. of high-grade coal or 1 gallon of refined heating oil.
- ***Electrical energy*** is the energy associated with the flow of electrons. Electricity is used extensively in contemporary society to power the majority of our convenience appliances.

- ***Nuclear energy*** is the term associated with the power of the atom. It was initially harnessed during the 1940s. The initial use of nuclear energy was for war, but with the conclusion of World War II, peacetime uses for nuclear energy, such as power generation, emerged.

All six of these forms of energy are used to aid us in our everyday lives. They are used to do work for us. These forms of energy will be discussed in more detail in other chapters.

Measuring Energy

The most basic unit of heat energy is known as the ***British thermal unit (Btu)***. The Btu is a very small amount of energy. It is often compared to the amount of energy given off by one wood-stem kitchen match, if the match is burned completely. The Btu is an often-referenced energy unit because all forms of energy can be related to the amount of Btu they can produce. **Figure 2-4** shows how much energy it would take to produce 100,000 Btu with the use of various energy sources. Since the Btu represents such a small amount of energy, the term *heating unit* is often used when discussing energy for structural heating. One ***heating unit*** is equivalent to 100,000 Btu.

When measuring the ***energy consumption***, or use of energy resources, of a large city, country, or continent, a term known as the *quad* is often used. A ***quad*** is an accepted abbreviation for 1 quadrillion Btu. (One quadrillion is a one followed by fifteen zeros). Homes and communities do not consume entire quads of energy, making the term a more conceptual reference point for extremely large-scale energy consumption. **Figure 2-5**

Electrical energy: The energy associated with the flow of electrons.

Nuclear energy: The power of the atom.

British thermal unit (Btu): The most basic unit of heat energy. The Btu is a very small amount of energy. It is often compared to the amount of energy given off by one wood-stem kitchen match, if the match is burned completely.

Heating unit: The equivalent of 100,000 Btu.

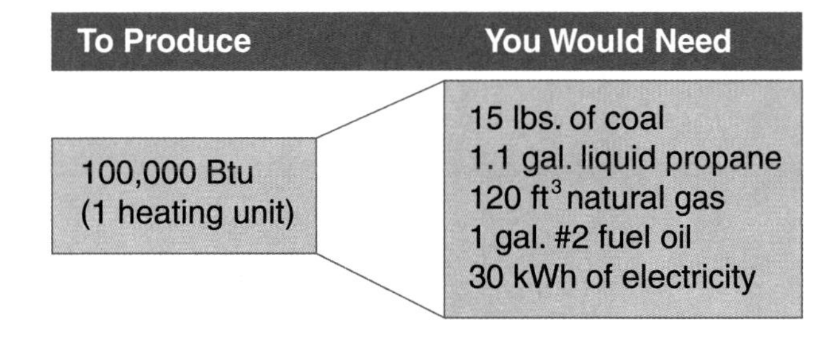

To Produce
You Would Need
100,000 Btu
(1 heating unit)
15 lbs. of coal
1.1 gal. liquid propane
120 ft³ natural gas
1 gal. #2 fuel oil
30 kWh of electricity

Figure 2-4. Quantity of fuel necessary to produce 1 heating unit (100,000 Btu).

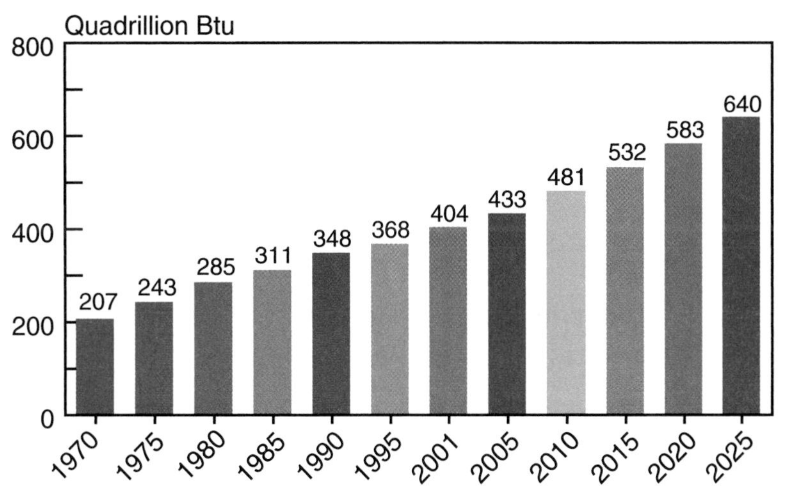

Figure 2-5. Actual and projected world energy consumption from 1970 to 2025, expressed in quadrillions of British thermal units (Btu). (U.S. Energy Information Administration)

Energy consumption: The use of energy resources.

shows the calculated and projected world energy consumption from 1970–2025, expressed in quads.

A Brief History of Energy Consumption in the United States

Energy sources in early America primarily consisted of wind, water, and wood. Factories were constructed near waterways to make use of flowing water by converting it to mechanical power with the use of a waterwheel. Farmers made use of the wind to create mechanical power for grinding flour and pumping water. Wood was burned as a source of heat.

By the 1800s, technology was demanding the use of a better heat source. The development of the steam engine led to the steam locomotive and the steamship. Coal replaced wood as the main energy source. Approximately the same amount of energy is produced by 15 lbs. of coal and 20–50 lbs. of wood. It is easy to see why coal would have been necessary to power a large steam engine, such as that of a steam locomotive or ship.

The internal combustion engine had been perfected by 1900, to the point that it was being used to power the first automobiles. See **Figure 2-6.** Americans soon began to depend on their cars. Pound for pound, gasoline contains much more energy than coal.

Figure 2-6. The introduction of the automobile in the early 1900s began the shift from coal to petroleum as the primary energy source. (Ford Motor Company)

There have been times when the supply of energy has not met the demand. In 1960, a group of nations committed to the strength and success of the oil market formed the ***Organization of Petroleum Exporting Countries (OPEC)***. The members of the OPEC decided to restrict the amount of crude oil they would sell to the United States, as a means of penalizing Americans for consuming too much energy. This type of restriction of trade for political means is known as an ***embargo***. The resulting effects of the embargo were devastating to the American economy. One way America responded to these crises was to produce smaller four-cylinder cars that offered better gas mileage than the larger engines popular in the 1960s and 1970s. Another response was an increased emphasis on ***energy conservation***, which involves making better use of the available supplies of energy.

Quad: An accepted abbreviation for 1 quadrillion Btu.

Organization of Petroleum Exporting Countries (OPEC): A group of nations committed to the strength and success of the oil market.

Embargo: Restriction of trade for political means.

Energy conservation: Making better use of the available supplies of energy.

By the early 1980s, America had begun to shift emphasis from an industrial society to a service society. Industries heavily dependent on energy, such as the steel industry, had begun to leave the United States. New jobs were formed in the service sectors of the economy, and the development of the personal computer ushered the nation into what many have termed the *information age*. Many start-up companies that focused on alternative energies, such as those that installed wind turbines and solar collectors, were eliminated.

It is important to recognize that the majority of all energy consumed in the United States is consumed in the industrial and transportation sectors of our economy. See **Figure 2-7.** Although alternative energy sources are expected to play a greater role in the world energy mix in coming decades, fossil fuels are anticipated to remain the dominant sources of energy for the foreseeable future. See **Figure 2-8.** World energy consumption on the whole is also expected to rise sharply in the future. The Energy Information Administration of the U.S. Department of Energy (DOE) estimates that world energy consumption will increase by over 50% by 2025.

As a society advances, so does its producing capability. Likewise, its energy consumption advances. The vast majority of all energy consumed in America and worldwide comes from fossil fuels. Although fossil fuels are nonrenewable, polluting, and sometimes economically volatile, we use far more fossil fuels to create power than other more environmentally friendly forms of energy. Fossil fuels simply yield more energy per volume than many other forms of energy. The use of fossil fuels is responsible for the creation of millions of jobs. It also helps the United States engage in a world economy.

Figure 2-7. Together, the industrial and transportation sectors account for 60% of the total energy consumption in the United States. (U.S. Energy Information Administration)

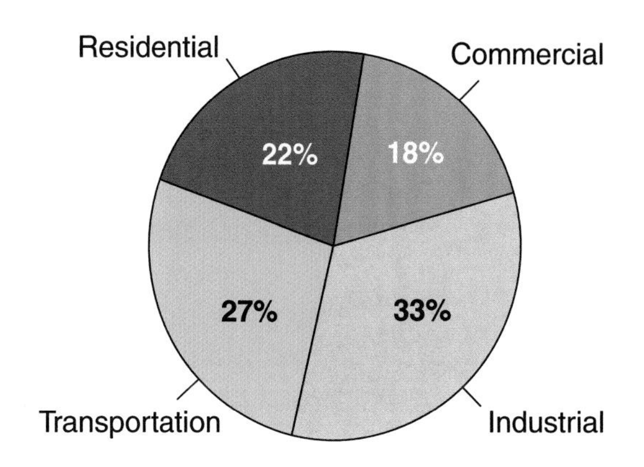

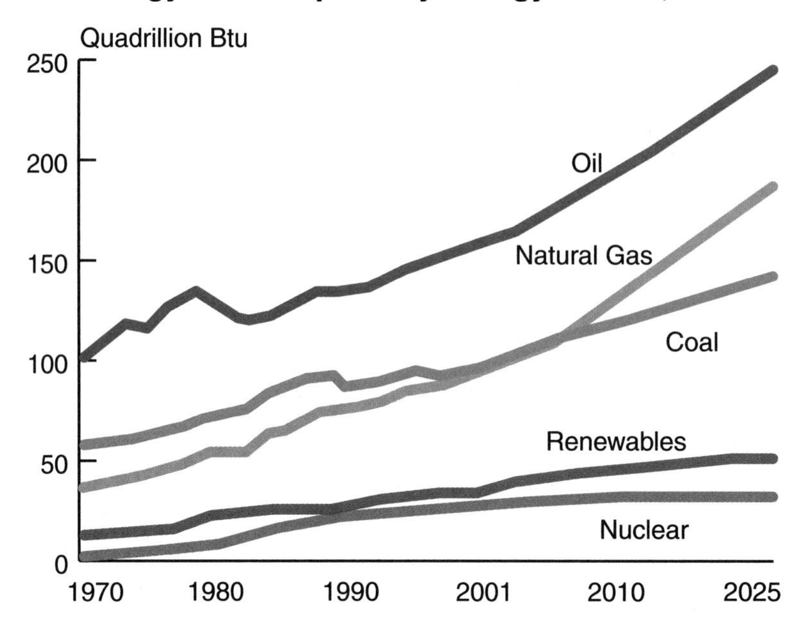

Figure 2-8. World energy consumption is projected to rise in every category except nuclear energy through 2025. (U.S. Energy Information Administration)

Energy Conversion

Energy conversion: The changing of one form of energy into another.

Technological advances, such as the development of the electric motor, the home heating furnace, and solar collectors, allow for energy conversion. ***Energy conversion*** is the changing of one form of energy into another. For example, an electric motor converts electrical energy to mechanical energy. A furnace converts a potential energy source, like heating oil or natural gas, into a kinetic energy source in the form of heat. A solar collector converts sunlight to heat.

Efficiency: The extent to which an energy form is usefully converted into another form of energy.

Efficiency is a term used to measure the extent to which an energy form is usefully converted to another form of energy. For instance, for many years, the internal combustion engine could only convert about 27% of the gasoline consumed into useful mechanical energy. Technological advancements over the past 30 years, however, have led to computer-controlled ignition, precise fuel injection, spark plug improvements, and more efficient engine designs. All of these factors translate to greater engine efficiency, which is now above 30% for many automobile engines. Improving the efficiency of energy conversion devices, such as engines, furnaces, and generators, is one of the primary ways to improve energy, power, and transportation technologies.

Entropy: A measure of the unavailable energy in a closed system.

Some conversion devices, such as some home heating furnaces, have efficiency ratings in the 90% range. This means that, out of all the possible Btu contained in a gallon of heating oil, 90% or more are converted into heat. Other conversion devices, such as the internal combustion engine, are nowhere near as efficient. One reason is the law of entropy. ***Entropy*** is a measure of the unavailable energy in a closed system. The law of entropy states that whenever an energy form is converted from one form to another, some loss will occur. **Figure 2-9** shows the approximate efficiencies for various conversion devices.

Figure 2-9. Energy efficiencies for common converters.

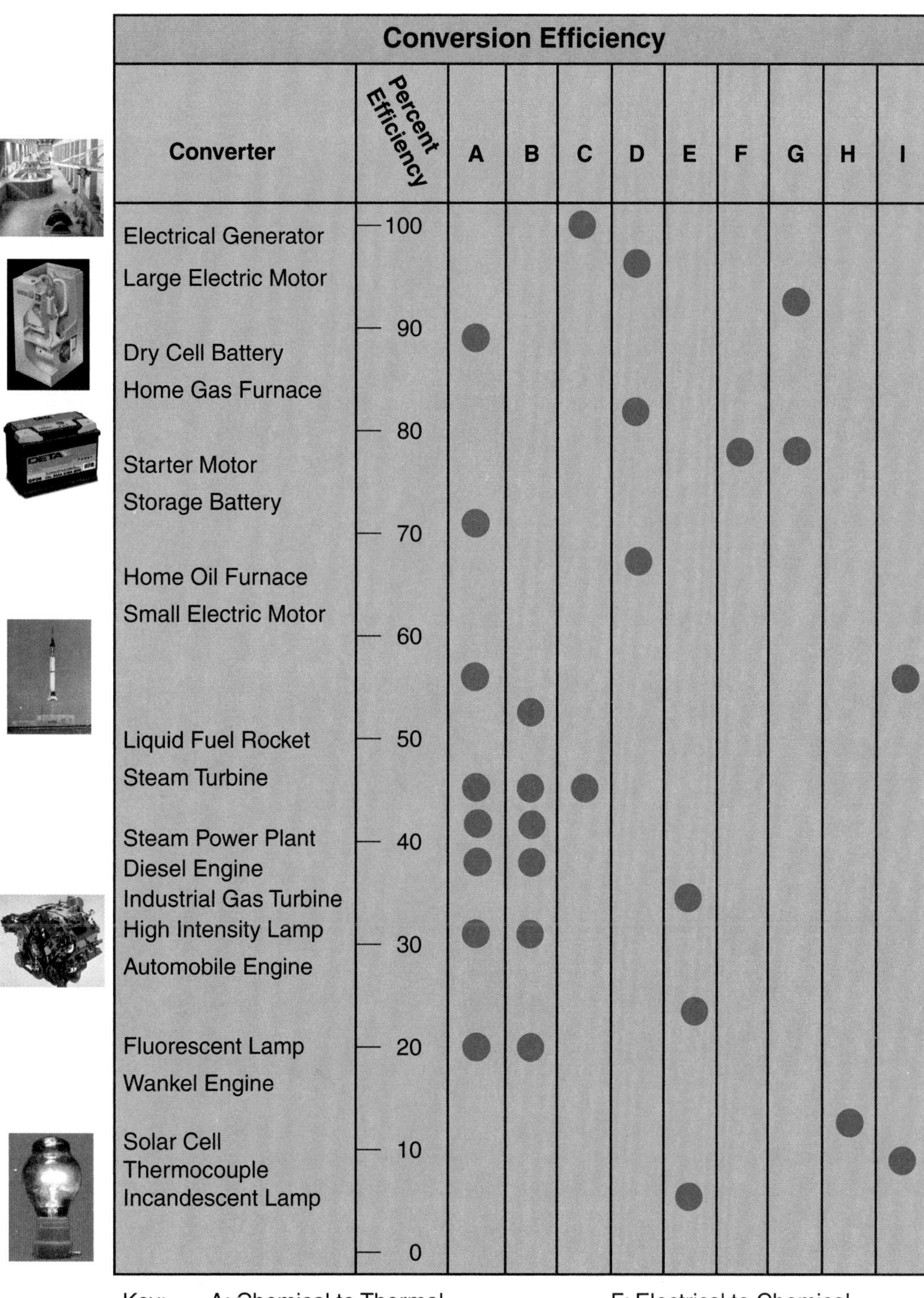

Key:

A: Chemical to Thermal
B: Thermal to Mechanical
C: Mechanical to Electrical
D: Electrical to Mechanical
E: Electrical to Radiant
F: Electrical to Chemical
G: Chemical to Electrical
H: Radiant to Electrical
I: Thermal to Electrical

Curricular Connection

Math: Efficiency

Efficiency is a measurement of input to output. It is calculated as follows:

$$\frac{\text{output}}{\text{input}} \times 100 = \%\ \text{efficiency}$$

When discussing energy, efficiency measures the extent to which an energy form is usefully converted to another form of energy.

A furnace for a building consumes oil at 1.5 gal/hr. At this rate, the potential energy of the oil is 150,000 Btu/hr. The furnace produces, however, only about 130,000 Btu/hr because it is not completely efficient. What is the efficiency of the furnace?

To solve this problem, we use the efficiency formula. The following calculation tells us the efficiency of the furnace:

$$\frac{130{,}000\ \text{Btu/hr}}{150{,}000\ \text{Btu/hr}} \times 100 = 87\%\ \text{efficiency}$$

This means about 87% of the fuel consumed by the furnace is being converted into usable heat.

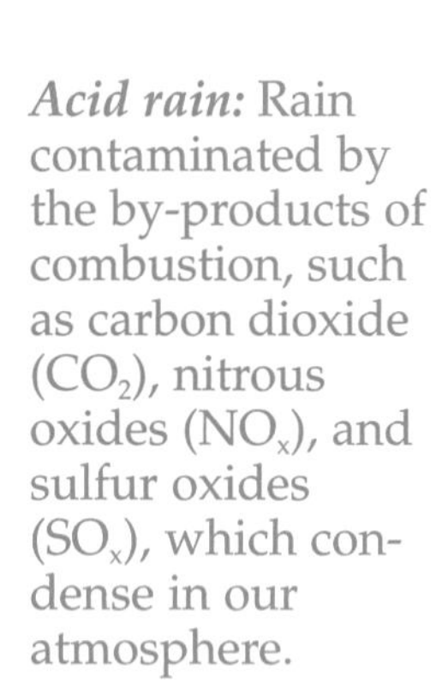

Acid rain: Rain contaminated by the by-products of combustion, such as carbon dioxide (CO_2), nitrous oxides (NO_x), and sulfur oxides (SO_x), which condense in our atmosphere.

Greenhouse effect: The situation caused by a layer of greenhouse gases surrounding our planet, produced by the burning of fossil fuels. This layer does not allow the heat produced by the sun to escape the earth's atmosphere as easily as it once did.

Energy and the Environment

The consumption of fossil fuels leads to environmental problems, such as acid rain and the greenhouse effect. ***Acid rain*** occurs when by-products of combustion, such as carbon dioxide (CO_2), nitrous oxides (NO_x), and sulfur oxides (SO_x), condense in our atmosphere, only to come back down to earth with rain. The effects of acid rain can be devastating to forests, ponds, and lakes, killing fish and altering entire ecosystems. Canada has been at frequent odds with the United States over acid rain that is created in the United States but drifts over the boundary water of Canada, polluting Canadian waters.

The ***greenhouse effect*** is said to occur because of gases produced by the burning of fossil fuels as well. When sunlight strikes the earth, a portion is reradiated back off the earth and into the atmosphere. The layer of greenhouse gases surrounding our planet is not allowing the heat produced by the sun to escape the earth's atmosphere as easily as it once did. The net result is that a partial trap occurs. This leads to a phenomenon known as ***global warming***. The effects of global warming are not widely understood, but at a minimum, global warming could cause the melting of ice caps, which could alter shorelines, and the changing of weather patterns, which could alter agricultural productivity.

Problems such as acid rain and the greenhouse effect are expected to increase. Developing nations are expected to consume more fuel to develop their economies. As the use of fossil fuels increases, so do harmful emissions.

Agriculture: Biodiesel

Are you looking for a clean-burning alternative fuel produced from domestic resources? Biodiesel contains no petroleum, but it can be blended with conventional diesel fuel to create a biodiesel blend that can be used in diesel engines without any major modifications. The fuel itself is created from fatty acids common in vegetable oils or animal fats. Such waste oil, a by-product of cooking, is routinely collected at restaurants. Restaurants usually give the oil away, as disposal is a nuisance. If biodiesel becomes popular, however, this could change.

Biodiesel can be used directly as a fuel, but it is often mixed with diesel fuel. This mix creates by-products that are considerably healthier for the environment than those created when pure diesel fuel is burned. The biodiesel actually reduces the amount of unburned hydrocarbons and nitrogen oxides released into the air during engine exhaust. It also virtually eliminates sulfur oxides and sulfates common in acid rain. In fact, biodiesel is the only alternative fuel to have successfully met the testing requirements established for health effects in the Clean Air Act. Now for the best part—fuel economy, horsepower (hp), and torque do not suffer when using biodiesel or a biodiesel blend.

This type of fuel is hardly a new invention. Most people do not know that the original diesel engine, as patented by Rudolph Diesel in 1893, was designed to run on peanut oil! Only the discovery of inexpensive crude oil put biodiesel on the shelf at that time. Now, as gasoline prices have risen dramatically, the country may need to take a serious look at alternative fuels and methods of easing our dependency on foreign oil. Biodiesel may enjoy a newfound resurgence. Your car might someday be powered by the waste oil from last week's breakfast!

Fighting Back: Conservation and Recycling

Recycling saves money, and it also saves energy. In fact, one of the reasons recycling works from a financial standpoint is precisely because of the energy saved to produce a recycled product. Consider these facts about recycling:

- Aluminum can recycling helps to save 95% of the energy required to produce new aluminum products.
- Recycling one plastic bottle can save enough energy to light a 60-watt bulb for 6 hours.
- If your school recycles 1 ton of paper this year, it will save the following:
  - 6953 gallons of water
  - 463 gallons of oil
  - 587 lbs. of air pollution
  - 4077 kilowatt-hours (kWh) of electricity
- Every day, Americans buy 62 million newspapers and throw out 44 million. This is the equivalent of dumping 500,000 trees into a landfill every week.
- Reducing your home's waste newsprint, cardboard, glass, and metal can reduce CO_2 emissions by 850 lbs. per year.

Recycling is one of the best ways to conserve energy, as the energy to create new paper, plastic, or glass bottles out of recycled materials is far less than what it takes using new raw materials. See **Figure 2-10.**

Global warming: An increase in the average temperature of the earth's atmosphere, possibly resulting in the melting of ice caps, which could alter shorelines, and the changing of weather patterns, which could alter agricultural productivity.

Figure 2-10. A scientist at Argonne National Laboratory trains interns in methods of characterizing and segregating materials in the solid waste stream. They will put their training to use in setting up a community recycling project. (U.S. Department of Energy)

Figure 2-11. An energy guide is provided on new appliances to help consumers identify the most energy-efficient equipment. This clothes washer has guides for both the United States and Canada.

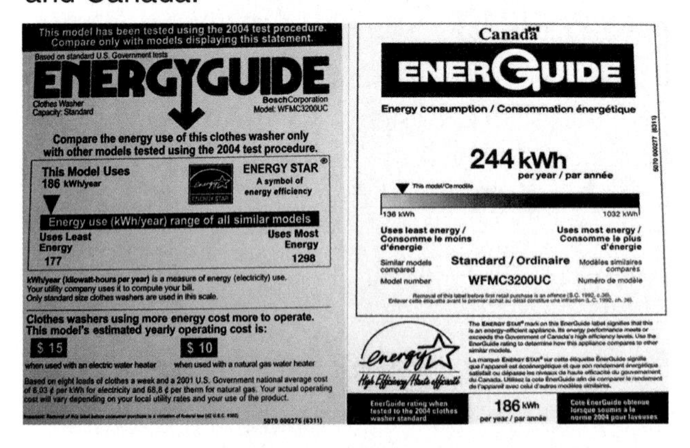

Another popular means of saving energy is through conservation measures. For instance, does anyone in your house turn the thermostat down at night, or do you leave it at the same temperature throughout the night, even though everyone is sleeping? Reducing the heat during the night is one simple means of conserving energy. Another is to shop for energy-efficient appliances. **Figure 2-11** shows an energy guide for a new appliance. The guide will indicate the estimated energy consumption for this particular appliance, in comparison to similar appliances by other manufacturers. More efficient appliances earn higher ratings.

Tech Extension

Infiltration

Infiltration is the term used to describe cold air forcing its way into a home or other building during the heating season through cracks and other openings, such as exhaust vents, chimneys, and cracks around exterior doors and windows. Hot air can also infiltrate during the air conditioning season. Excessive infiltration wastes heating or cooling energy. It also drives up energy costs for the building owner. As a building ages, it tends to settle, creating gaps and cracks that allow outside air to infiltrate the structure. Additionally, older buildings usually have doors and windows that are not as efficient at preventing infiltration as modern doors and windows, which typically seal extremely well. Reducing infiltration is an important part of energy conservation.

Summary

Energy is the ability to do work. It is important to study because it affects our lives in so many ways. Energy affects our ability to function in a technological world. Many sources of energy come from the natural forces within our world. The sun, wind, and water are examples of inexhaustible energy sources. Other energy sources are renewable or nonrenewable. Energy is classified as potential or kinetic energy. Potential energy is at rest, and kinetic energy is in motion. There are six different forms of energy. They are heat, light, mechanical, chemical, electrical, and nuclear.

Most of the energy consumed in the world today is in the form of fossil fuels. These fossil fuels provide good energy content, or Btu, per volume, making them desirable for use. They are also responsible, however, for environmental destruction, including the production of greenhouse effect gases and pollutants that form acid rain. Additionally, there are political implications to using fossil fuels, since some of them are imported from other countries. At present, no one form of energy can solve all our energy needs. Fossil fuels and alternatives all offer inherent advantages and disadvantages. Converting energy from one form to another—for example, light to heat or heat to mechanical—will result in some loss, due to entropy.

It is important to understand energy and where it comes from. A knowledge of energy helps us understand energy, power, and transportation systems. Energy is the backbone of all power and transportation systems.

Key Words

All the following words have been used in this chapter. Do you know their meanings?

acid rain
British thermal unit (Btu)
chemical energy
efficiency
electrical energy
embargo
energy conservation
energy consumption
energy conversion
entropy
global warming
greenhouse effect
heat energy
heating unit
inexhaustible energy source
kinetic energy
light energy
mechanical energy
nonrenewable energy source
nuclear energy
Organization of Petroleum Exporting Countries (OPEC)
potential energy
quad
renewable energy source

Test Your Knowledge

Write your answers on a separate sheet of paper. Do not write in this book.

1. Define *energy*.
2. List some ways in which energy affects your life.
3. What are the differences between renewable, nonrenewable, and inexhaustible resources?
4. *True or False?* Energy waiting to happen is known as potential energy.
5. List the six forms of energy.
6. *True or False?* A quad is a means of measuring how much energy you consume in your home every month.
7. Name three sources of energy.
8. Why has the demand for energy and power grown throughout history?
9. In three sentences, discuss the history of energy consumption in the United States.
10. Explain what an embargo is.
11. Which two sectors of the economy are responsible for consuming the most energy?
12. *True or False?* The majority of all energy consumed in America and worldwide is in the form of fossil fuels.
13. Why are alternative energy sources, like wind generators, not more popular?
14. Summarize the concept of efficiency.
15. *True or False?* When energy is converted from one form to another, some loss will occur.

Activities

1. Build and demonstrate a device that changes a renewable source of energy into a form that will do work.
2. Collect pictures from old magazines showing examples of potential energy in the world around you.
3. Construct a device that demonstrates potential energy being converted into kinetic energy.
4. Start a new recycling program in your school as a means of saving energy.
5. Brainstorm about the advantages and disadvantages of various conventional and alternative energy sources.

Harnessing the energy of the wind is a rapidly growing technology. These wind turbines are part of a 63-turbine "wind farm" in Illinois that generates enough power for 15,000 homes per year. The column on the right is the base of one of the turbines, which require only a small area of ground, allowing farmers to continue producing crops on the land. (Steve Olewinski)

3

Nonrenewable Sources of Energy

Basic Concepts

- Name three nonrenewable sources of energy.
- List the characteristics of the different types of coal.
- State what synfuels are and why they may be important in the future.
- Define the Organization of Petroleum Exporting Countries (OPEC) and the relationship of the United States with the OPEC.
- Identify basic environmental concerns associated with the use of fossil fuels.

Intermediate Concepts

- Recognize the major components of a coal-fired generating plant.
- Explain the purpose of a fractionating tower.
- Give examples of some positive and negative impacts of importing oil.

Advanced Concepts

- Discuss various influences, including economic, environmental, technological, and political influences, with regard to the use and continued development of fossil fuels.

Nonrenewable energy sources are those sources you cannot replace once they have been used. Examples of nonrenewable energy sources are coal, oil, natural gas, and uranium. Over 90% of our energy needs are met by using these few sources of energy. America's heavy dependence on the automobile as a primary means of transportation is a major reason we are so heavily dependent on nonrenewable resources, especially oil. Nonrenewable energy sources can be divided up into two different groups: fossil fuels and uranium. Uranium will be covered in Chapter 4.

Fossil Fuels

Natural gas: Gas usually found within the vicinity of petroleum reserves.

The three primary fossil fuels are coal, oil, and natural gases. These come from the ground. Fossil fuels are formed by the decays of plants and animals. These decays have been buried in the ground for millions of years. The remains of such plants and animals have gradually built up in layers over the years. Over time, soil is piled on top of the deposits. Fossil fuels burn easily. See **Figure 3-1.** One main use for them is to heat homes and other buildings.

Figure 3-1. Coal burns easily and continues to be used for home heating, although it has been replaced in many areas by natural gas and heating oil. (Quaker State Oil Refining Corporation)

Fossil fuels come in the form of solids, such as coal. They can also be in the form of liquids or semisolids, which include petroleum and tar. Another form of fossil fuel is a gas. Gases are usually found within the vicinity of petroleum reserves. This type of gas is called ***natural gas***. Since fossil fuels are nonrenewable, they are decreasing in our environment. In a developed nation like the United States, it takes great quantities of fossil fuel to maintain our technologically advanced economy. Because of the increases in demand brought about by an increasing population and an increased standard of living, more fuel is needed. Due to the millions of years it takes for fossil fuels to form, we cannot replace them. Fossil fuels will one day run so low that it will no longer be economically viable to locate, recover, and refine them. The need to develop alternative energy sources is important for this reason.

Coal

The forming of coal in the earth began about 500 million years ago. It took approximately 85 million years for plant and animal decay to form into coal. Coal can be a soft or rocklike material. It is black or brown, depending on its age. The older the coal is, the denser it is, and the blacker it becomes. Coal is combustible and frequently used to generate electricity by power companies, to generate heat for industrial processes, and to heat some homes and buildings. See **Figure 3-2.**

Figure 3-2. A large number of electrical generating plants burn coal to generate steam that drives the turbines.

Removing coal from the ground is called *mining*. Coal, once mined, can be burned immediately as a fuel. There are two mining methods: deep mining and surface mining. ***Deep mining*** uses shafts and special machinery to remove the coal from deep below the earth's surface. See **Figure 3-3.** This type of mining can be dangerous due to cave-ins and machinery accidents.

Figure 3-3. Deep mining involves following coal seams deep underground, breaking up the coal into manageable-sized pieces, and bringing it to the surface. The operation is highly mechanized, but mining continues to be one of the most dangerous occupations. (American Coal Foundation)

Deep mining: A mining operation that uses shafts and special machinery to remove the coal from deep below the earth's surface.

Strip mining: The mining or removing of coal close to the earth's surface. It is done mainly with the use of large pieces of machinery, such as mechanical shovels and bulldozers. Also called *surface mining*.

Land reclamation: The restoration of land to a usable condition after strip mining has taken place.

Overburden: When strip mining, the soil remaining after the topsoil is removed.

Surface mining, or ***strip mining,*** is the mining or removing of coal close to the earth's surface. It is done mainly with the use of large pieces of machinery, such as mechanical shovels and bulldozers. See **Figure 3-4.** Surface mining is much safer and less expensive than deep mining.

The one major disadvantage to surface mining is the condition it leaves the land in after the coal has been gathered. In some instances, mining companies are required to restore the land to usable condition after strip mining has taken place. This restoration is referred to as ***land reclamation.*** Strip mining is the preferred method of retrieving coal as long as the coal seam is less than 60′ deep. When strip mining, the topsoil is removed and stored in piles for later use during reclamation. Next, the remaining soil is removed down to the top of the coal seam. This remaining soil is referred to as the ***overburden.***

Once mined, the coal is typically transported by rail, but it can also be transported by barge or truck to power plants and other destinations for end use. See **Figure 3-5.** The coal at power plants is typically stored in a huge pile, since power plants are required to have a 60-day supply of coal on hand. The 60-day supply is necessary in case of labor strikes or natural disasters that could prevent coal shipments for a period of time. At the power plant, the coal is fed into mills that grind it into a fine powder. In this way, the coal produces the greatest amount of energy and the least

Figure 3-4. Coal located near the surface of the earth is mined by stripping away the dirt and rock above the coal seam. The exposed coal can then be broken up and shipped to electrical generating plants or other end users. Very large equipment, such as this shovel and truck, is used to mine with maximum efficiency. (P&H Mining Equipment)

Figure 3-5. Coal is typically shipped from mines to end users by railcar or on large barges, such as this one. Coal-fired electrical generating plants are often located along waterways to take advantage of lower-cost barge shipment.

amount of waste when injected into the boiler. In the boiler, the burning coal is used to heat water into steam. The steam drives a turbine. The turbine spins a generator, which produces the electricity. The electricity is then transformed into very high-voltage electricity, which is suitable for long-distance distribution. It is sent through transmission lines to various communities. See **Figure 3-6.**

Types of coal

Peat is a step in the formation of coal more than a type of coal. It is formed from water and the decomposition of organic materials. When dried, this material can be used as a fuel source for home heating. Some pellet stoves and coal stoves are designed for use with peat. Since peat takes so much less time to form than coal, it is actually considered to be a renewable form of energy. It is simply mentioned here as the first step in the formation of coal.

Peat: The first step in the formation of coal. It is formed from water and the decomposition of organic materials.

Lignite coal is the next youngest type of coal. This material contains some woody decomposition that can be recognized as peat, but it has a higher energy content than peat. It too contains a large amount of moisture and is brownish in color, as opposed to the black substance typically recognized as coal. Lignite has enough energy content per volume that it can be used for the production of electricity or as an industrial fuel.

Figure 3-6. A schematic of a large coal-fired electric generating plant.

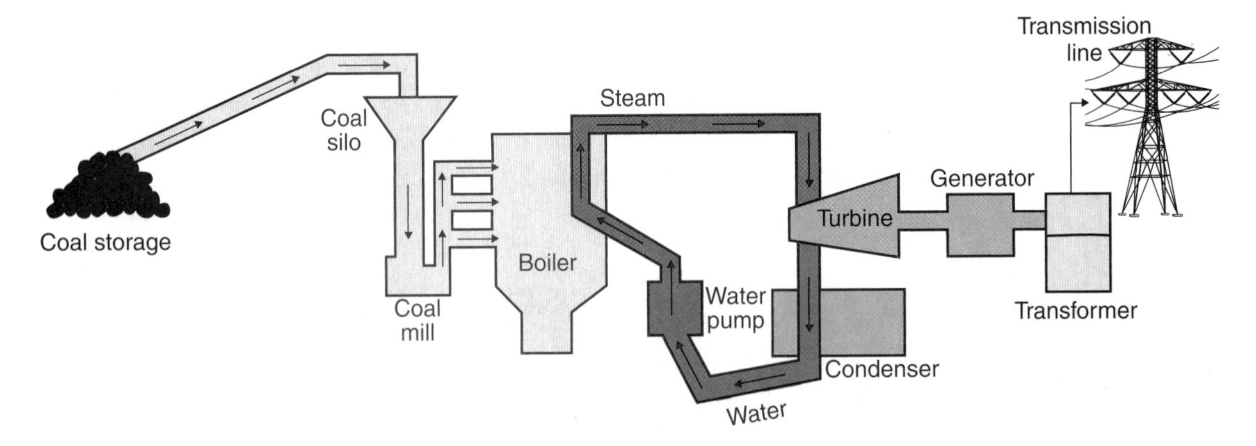

Subbituminous coal is estimated to be at least 200 million years old. It contains greater energy content per volume than lignite and is frequently used as an industrial fuel, for space heating, and for the generation of electricity. This coal is dull black in color and is categorized as a "soft" coal, a name used to describe the physical characteristics of peat, lignite, and subbituminous coals.

Bituminous coal has a high carbon content. It is denser and blacker than most other forms of coal. This type of coal is principally used for the production of electricity.

Anthracite coal is hard and brittle. It appears shiny on the surface and has a high carbon content. Because it burns cleaner than other forms of coal, it is often used for home heating. It does not have as much energy content per volume as bituminous coal. **Figure 3-7** shows the various types of coal and compares their carbon content and energy potential.

Lignite coal: A type of coal containing some woody decomposition that can be recognized as peat, but it has a higher energy content than peat. It contains a large amount of moisture and is brownish in color, as opposed to the black substance typically recognized as coal.

Subbituminous coal: Coal that contains greater energy content per volume than lignite and is frequently used as an industrial fuel, for space heating, and for the generation of electricity. This coal is dull black in color and is categorized as a "soft" coal.

Bituminous coal: Coal with a high carbon content. It is denser and blacker than most other forms of coal. This type of coal is principally used for the production of electricity.

Anthracite coal: Coal that is hard and brittle. It appears shiny on the surface and has a high carbon content. Because it burns cleaner than other forms of coal, it is often used for home heating. It does not have as much energy content per volume as bituminous coal.

Figure 3-7. There are several different types of coal. They differ in carbon percentage and energy content.

Types of Coal		
Rank	**Carbon (%)**	**Energy Content (Btu/lb)**
Peat	< 30	1000–4000
Lignite	30	5000–7000
Subbituminous	40	8000–10,000
Bituminous	50–70	11,000–15,000
Anthracite	90	14,000

Supply and demand for coal

World coal reserves are concentrated in the United States, the former Soviet Union, and China. Coal is used extensively for the production of electricity throughout the world. The use of anthracite coal primarily for home heating is expected to decline in future decades. Coal is a difficult and labor-intensive fuel, compared to oil and natural gas.

The worldwide demand for coal has been in a period of slow growth since the late 1980s. This trend is expected to continue, with growth of worldwide coal consumption estimated at about 1.5% per year. More than half of all coal consumed worldwide is for power generation. Virtually all projected growth for the use of coal is for power generation, since other fuels are cleaner to heat with and easier to distribute in small quantities.

Coal can also be converted to a synthetic gas or liquefied fuel. These fuels are typically referred to as *synfuels*. They could be used to supplement oil and natural gas supplies, if these supplies become scarce.

Coal and the environment

Strip mining can be devastating to the land. Fortunately, reclamation techniques often convert a strip-mined plot to an area suitable for farming, once strip mining has been completed. See **Figure 3-8.**

Figure 3-8. Reclamation efforts by coal-mining companies are designed to restore strip-mined land to farming or other uses. A—Crops planted on restored land at a former strip mine in Ohio. B—Large-scale reforestation is becoming widely practiced. This project is at a former mine in the state of Washington. C—Mining land can be reclaimed for recreational use, such as this golf course in Pennsylvania. (Chuck Meyers, Office of Surface Mining, Department of the Interior)

A

B

C

Burning coal can be a dirty process. This is particularly true when the coal is burned in small quantities, such as in a coal stove in a residential setting. When large amounts of coal are consumed in one area, the process can be much more effectively monitored and treated, from an environmental standpoint. Pollution control techniques have improved drastically in recent decades. Devices such as bag filters and electrostatic precipitators filter solid particles out of the exhaust gases before they are discharged into the atmosphere.

A ***bag filter*** works essentially the same way as a bag on a vacuum cleaner. All solid particles in the waste stream are trapped in the bag filter prior to the hot waste gases exiting through the smokestack of the power plant. An ***electrostatic precipitator*** works by positively charging the waste particles and attracting them to a negatively charged electrode. The solid particles are then washed off the electrode and collected. This solid waste by-product, known as ***fly ash***, historically had no use. It was stored in fly ash pools, which were large holes often dug near the power plant. More recently, fly ash has been used as traction grit in the northern states, where the roads become covered by snow and ice. It can also be used as an additive in concrete and as an absorbent material for oil spills.

Bag filter: A device that works like a bag on a vacuum cleaner, trapping all solid particles in the waste stream prior to the hot waste gases exiting through the smokestack of a power plant.

Electrostatic precipitator: A device that works by positively charging waste particles and attracting them to a negatively charged electrode. The solid particles are then washed off the electrode and collected.

Fly ash: A solid waste by-product produced by burning coal.

Petroleum: Oil.

Crude oil: Oil in its natural state.

Oil

Oil is also known as ***petroleum***. Like coal, petroleum is a fossil fuel. In its natural state, petroleum is called ***crude oil***. As crude oil comes from the ground, it is a mixture of semisolids, liquids, and gases. Oil has many uses. The biggest users of oil are transportation systems. Some form of oil fuels almost all transportation vehicles. Like coal, oil is found underground. To get oil out of the ground, an operation known as drilling is necessary. See **Figure 3-9.**

Before drilling for oil, there are several tests that need to be done on the surface and underlying rock. Certain drill bits are selected, depending on the method used to drill. Large drill bits are used to cut through layers of the earth. Drilling can occur as far as 3–5 miles below the earth's surface.

Figure 3-9. Oil is extracted from formations beneath the land and oceans by drilling into the formations and pumping the oil to the surface for transport by ship or pipeline. This is a large floating platform anchored over a well in England's North Sea oil field. (Kerr-McGee Corporation)

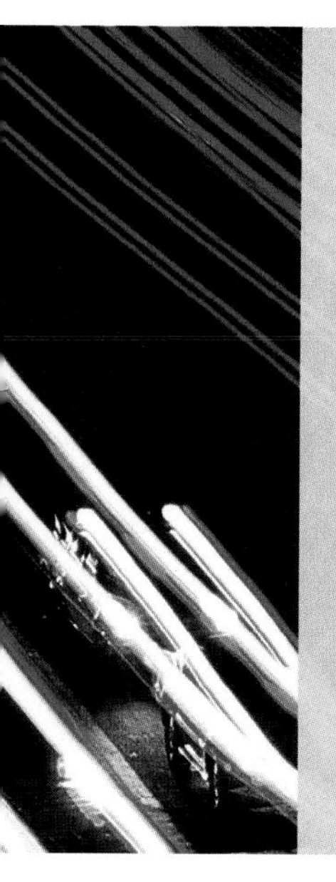

Career Connection

Petroleum Engineers

Petroleum is an exhaustible energy resource, and it is considered to be nonrenewable because it will take millions of years to replenish our supply once we have used it all. It must be drilled and refined in order to be useful. Petroleum engineers are responsible for overseeing the process of gathering and testing the oil.

The work of petroleum engineers entails designing the equipment used to drill and refine the oil. The tasks of these engineers also include researching to find the best locations for oil wells. Once the oil is drilled, petroleum engineers sample it and test its quality. They also work on making the process as efficient as possible while staying within environmental safety limits.

A petroleum engineer must be skilled in math and physics and possess an understanding of the properties of oil. In addition, leadership skills are also necessary to guide workers through the oil-refining process. To be a qualified petroleum engineer, a bachelor's degree is required. The yearly salary may range from $49,000 to $134,000.

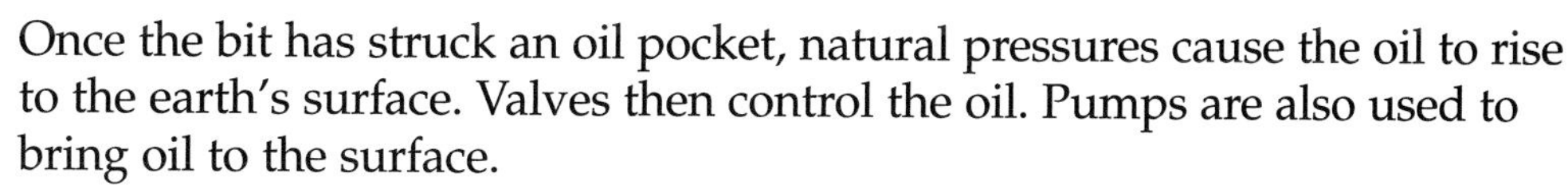

Once the bit has struck an oil pocket, natural pressures cause the oil to rise to the earth's surface. Valves then control the oil. Pumps are also used to bring oil to the surface.

Once the oil is brought to the surface, it is transported by pipeline to a refinery. See **Figure 3-10.** A refinery is a place that turns crude oil into products such as gasoline, diesel fuel, kerosene, and lubricating oil. In a large tower, called a ***fractionating tower***, crude oil is separated into various

Fractionating tower: A large tower in which crude oil is separated into various products.

Figure 3-10. Refineries break down crude oil into gasoline, diesel fuel, heating oil, and other petroleum products.

products. The crude oil travels through a pipe into a furnace at the bottom of the tower. The furnace heats the crude oil until it boils. The vapors rise in the tower, and the temperature in the tower varies with the elevation. Some "fractions" condense at higher temperatures, while some condense at lower temperatures. Trays at the various levels collect the condensed fractions and drain them out of pipes. See **Figure 3-11.** The different products are then transported by way of pipeline to storage tanks or another refinery.

Oil shale: 40–50 million-year-old sedimentary rock containing an oily substance.

Kerogen: An oily substance contained by 40–50 million-year-old sedimentary rock.

Oil shale and tar sands

In a region known as the Green River formation, where Utah, Wyoming, and Colorado join, there exists more shale oil than there are proven crude oil reserves in the United States. Recovering the oil is not easy. Like other oils, shale oil is the result of decaying vegetation that is millions of years old. ***Oil shale*** is a 40–50 million-year-old sedimentary rock containing an oily substance. This substance is called ***kerogen.*** The oil is trapped in the rock and is generally crushed and heated to a minimum

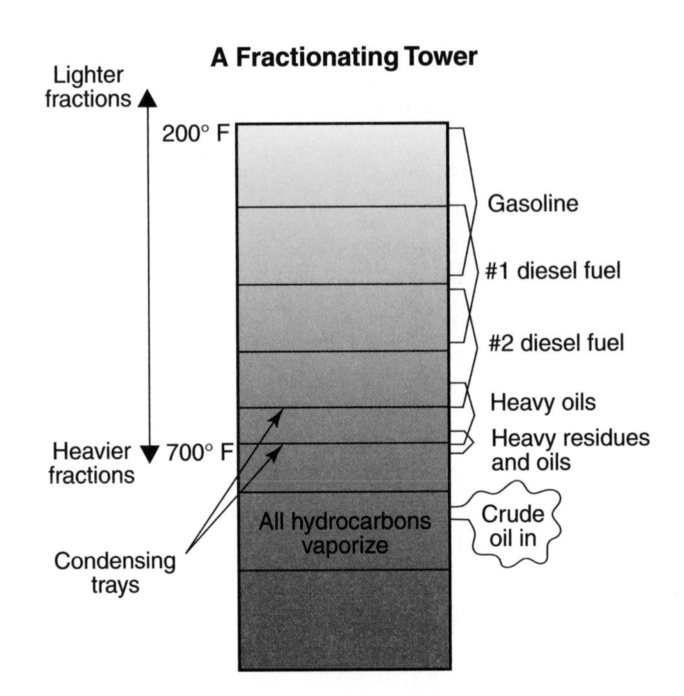

Figure 3-11. The fractionating tower is at the heart of the refining process. The oil is heated, causing all the hydrocarbon compounds to vaporize. Different "fractions" of the oil vapors, such as gasoline or diesel oil, condense out as they rise in the column.

of 800°F, so it can become liquefied for separation. This process is known in the industry as *retorting*. The yield from oil shale is only about 30 gallons of oil for every ton of rock that is excavated and crushed. Additionally, the conventional refining process requires large amounts of water in an area of the country where water is not plentiful. Shale oil is known to be present, but the economics of recovery are not yet favorable, as was proven by several attempts in the 1980s to develop pilot sites in the Green River formation.

Tech Extension

The In Situ Process

One experimental retorting process, known as the *in situ* (meaning "in place") process, may have some potential. This process calls for extraction of oil, while leaving the rock in place. See **Figure 3-A.** Imagine a tunnel that runs down one shaft, across the bottom of a plot of land known to be rich in shale oil, and then up another shaft on the other side. In the two vertical ends of the shaft, electrodes used to microwave the ground in between are inserted. The horizontal shaft below serves as a collection point, where heated oil is pumped to the surface. Other more conventional means of generating heat in the rock, such as injecting steam, are also used to retort the shale oil from the rock without removing the rock.

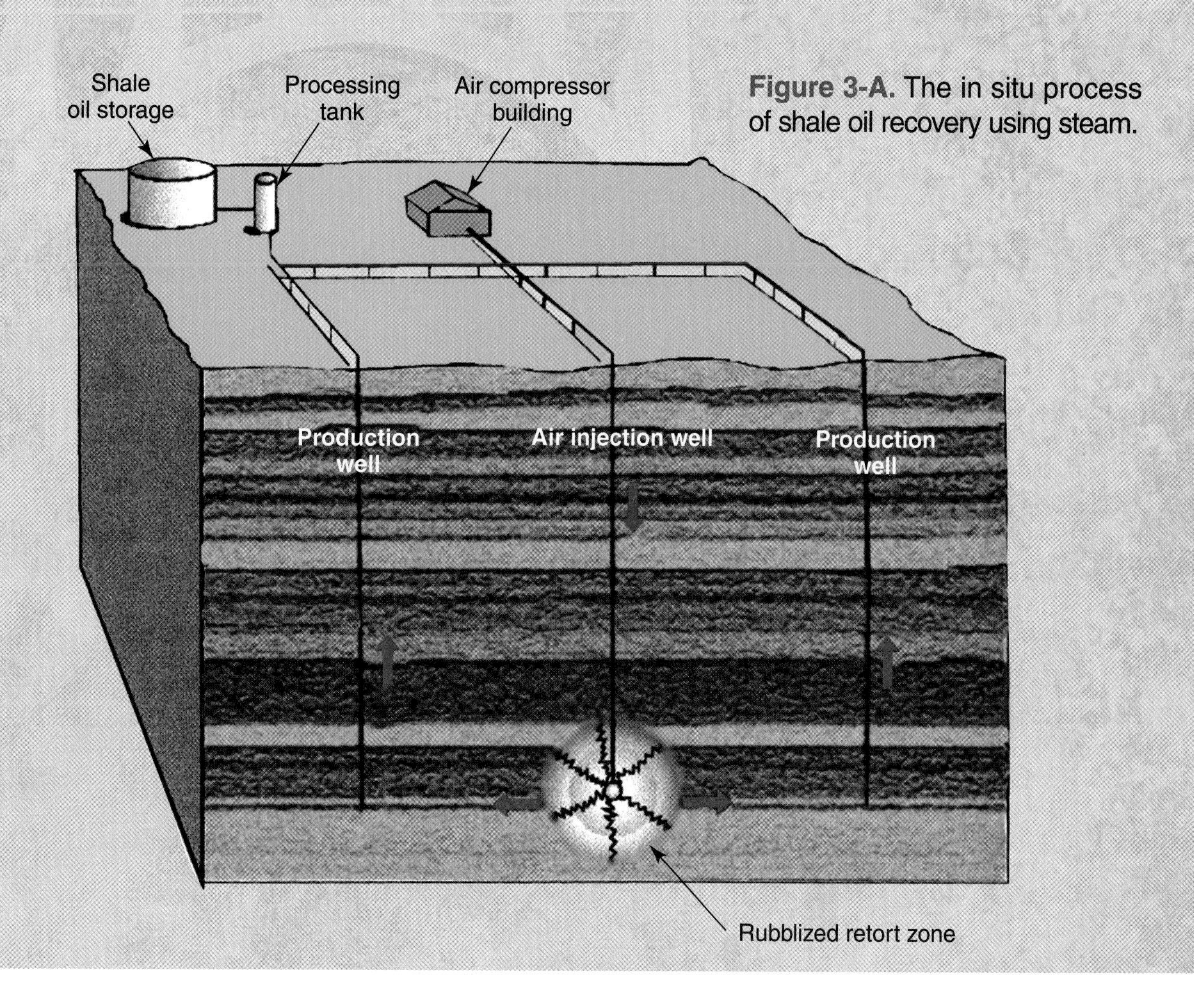

Figure 3-A. The in situ process of shale oil recovery using steam.

Tar sand: A source of crude oil. The sand is mined and mixed with hot water or steam to extract the thick oil known as bitumen.

Another important source of crude oil for the future may lie in ***tar sands***. These tar sands are similar to oil shale in that the oil cannot be retrieved through conventional means, such as well drilling. The largest reserves of tar sands in the world are located in Alberta, Canada. Again, extracting the oil from the sand is not easy. At present, the sands are mined and mixed with hot water or steam to extract the thick oil known as *bitumen*. The United States has some limited proven reserves of tar sands.

Supply and demand for oil

Over the past 25 years, oil prices have been the most volatile of all fossil fuels. This is due in large part to the instability of the Middle East, where much of the oil is produced, and the fact that a significant percentage of the oil used in the United States and elsewhere is imported. Since the United States imports large quantities of oil, the United States is subject to the discretions of other countries with regard to oil more than any other energy source.

The United States is vulnerable when it comes to imported oil. Oil prices have been significantly affected by labor strikes, wars, and the desire of the Organization of Petroleum Exporting Countries (OPEC) leaders to reduce production in order to keep the price of imported oil high. Limits to long-term oil price escalation may be on the horizon, however. They include the introduction of new fuels as a gasoline substitute, increased fuel efficiency leading to greater fuel economy, and new non-OPEC world markets for oil production, particularly in the former Soviet Union, where many deep water reserves of oil have recently been discovered. Projections for domestic oil consumption indicate it will continue to rise at a rate faster than any other fuel, with the exception of natural gas. This is due, in large part, to the fact that the United States uses oil so extensively as a fuel in the various transportation industries. See **Figure 3-12.**

Oil and the environment

Oil is a relatively safe fuel to transport and work with, even when it is refined into gasoline. For all the oil consumed on a daily basis, relatively

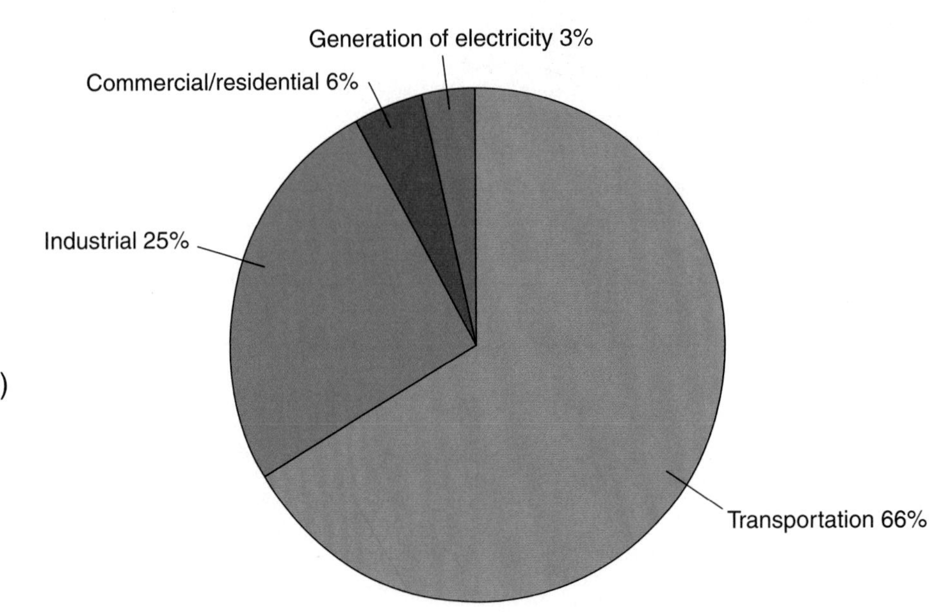

Figure 3-12. U.S. oil consumption. Two-thirds of all oil used in the year 2003 was consumed by the transportation sector, reflecting the nation's heavy dependence on vehicles powered by internal combustion engines. (U.S. Energy Information Administration)

Technology Link

Construction: The Trans-Alaska Pipeline

In 1968, the largest oil reserve on American soil was discovered on the North Slope of Alaska. This reserve is estimated to contain about 10–20 billion barrels of oil. An 800-mile long, 48″ diameter pipeline was completed in 1977 to carry oil from the North Slope to the Gulf of Alaska, where it can be shipped to refineries. See **Figure 3-B.** Construction of the pipeline cost more than $8 billion. This was the largest privately funded construction project of its time. Construction took more than three years and crossed over three mountain ranges. Thirty-one lives were lost while constructing the pipeline. The estimated 1.2 million barrels of oil per day that flow through the Trans-Alaska Pipeline represent approximately 20% of all domestic oil production in the United States. More recently, another huge oil reserve has been detected in the Arctic National Wildlife Refuge (ANWR). Drilling within the refuge would allow oil companies to make use of the already existing Trans-Alaska Pipeline to transport oil down to the shipping terminal, but there is opposition to the plan from many environmentalists. The refuge was created to protect the land from development, and it is home to many animals and birds. On the other hand, continued reliance on foreign oil and the increased cost of oil are reasons proponents argue in favor of opening up at least a small part of the refuge for drilling and continued exploration.

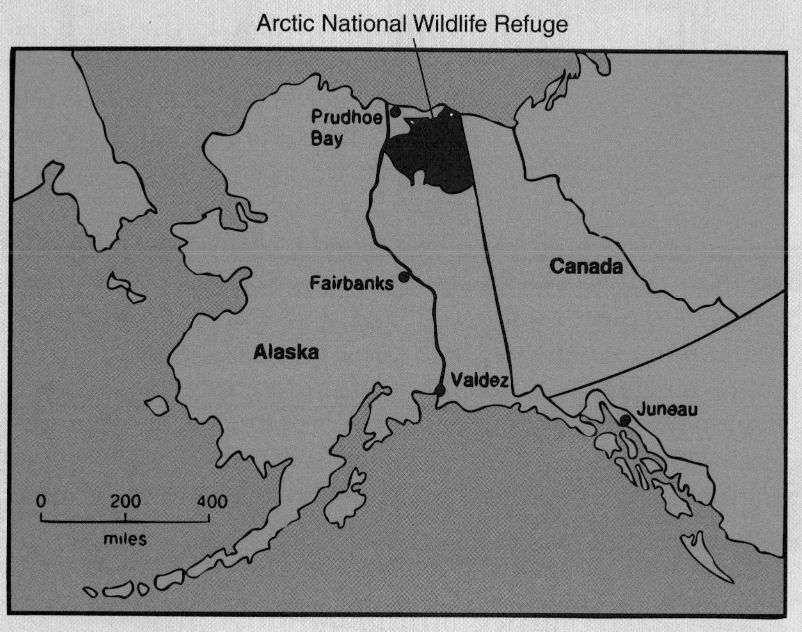

Figure 3-B. The Trans-Alaska Pipeline stretches 800 miles from the North Slope to the Gulf of Alaska.

few accidents occur with the fuel. It is possible for a fuel tank to explode if it is punctured in a certain way, but this is not common. Yet, oil is a fossil fuel, and like all fossil fuels, it has environmental drawbacks. Consumption of oil results in the production of carbon dioxide (CO_2), carbon monoxide (CO), nitrogen oxides (NO_x), and sulfur oxides (SO_x). Nitrogen oxides are major contributors to smog and respiratory problems. Sulfur oxides are other contributors to smog and the principal contributors to acid rain, which is harmful to the ecology of lakes and streams. Another environmental concern about oil is how it is transported. Since so much oil is imported worldwide, shipping of oil is common. Occasionally, mishaps occur that create environmental disasters, such as the *Exxon Valdez* accident in 1989.

Natural Gas

The gaseous portion of petroleum is called *natural gas*. Natural gas is always found wherever oil deposits are discovered. See **Figure 3-13.** This gas comprises ethane, propane, methane, and butane. These are all combustible gases. Natural gas is primarily composed of methane. Of the gases found within natural gas, butane offers the greatest British thermal unit (Btu) content per volume. Natural gas is the cleanest burning and presently the least expensive of the fossil fuels, and it has become one of the major sources of energy because of these characteristics.

Figure 3-13. Natural gas is always found above crude oil deposits.

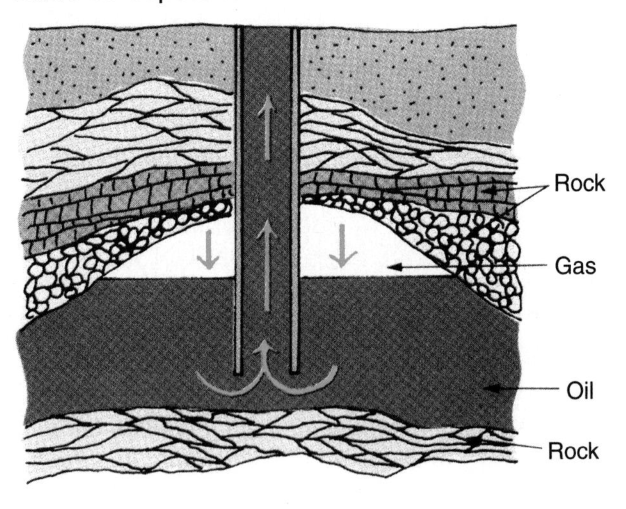

Removing gas from the ground is done in much the same way as recovering oil. Wells are drilled into the ground. The gas is removed and transported by way of pipeline to a processing plant. At the plant, all the impurities, like dirt, moisture, and sulfur, are removed. The gas is then transported to homes and industrial establishments for use.

Gas can be easily transported through pipelines in its natural state, helping to keep the cost of transporting the fuel to populated areas quite low. It is usually stored underground. During the warm months, when not much gas is needed for heating purposes, the gas from the pipelines is stored in underground reservoirs. These reservoirs are called ***aquifers***. An aquifer is a rock formation underground. These reservoirs hold large quantities of water. When the gas is pumped in under pressure, it pushes the water down farther into the ground. See **Figure 3-14.**

Aquifer: An underground rock formation that acts as a reservoir for large quantities of water. It is used to store gas from pipelines.

Liquefied natural gas (LNG): Gas that has been placed under pressure at very low temperatures, allowing transportation by railroad tankers, truck tankers, and ships.

The transporting of gas in pipelines to more rural areas can get rather expensive. Building pipelines to transport gas everywhere is difficult. To solve these problems, gas has been placed under pressure at very low temperatures to create ***liquefied natural gas (LNG)***. LNG provides a safe and viable alternative to gas in its natural state for many rural communities in which installing natural gas lines would not be economically possible. It can be delivered by truck and stored in tanks on-site at rural farms, businesses, and homes. See **Figure 3-15.** With liquefied gas, transportation can also occur by using railroad tankers and ships. Much more liquefied gas can be stored in a holding tank than gas in its natural state. When

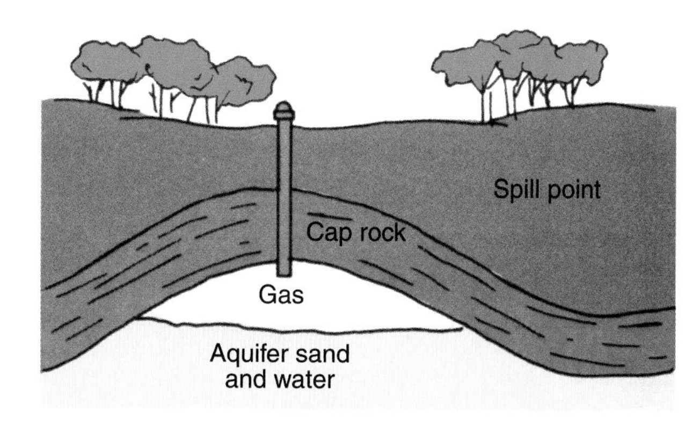

Figure 3-14. Natural underground aquifer formations covered with an impervious cap rock layer can be used for storage. The gas is pumped into the formation under pressure, displacing the water.

Figure 3-15. Homes in rural areas away from pipelines and distribution systems rely on liquefied natural gas (LNG) stored in tanks. A—A storage tank outside a home. B—A specially designed small tanker truck used to fill storage tanks at customers' homes or businesses.

A

B

purchasing gas, it is sold by the cubic foot (ft^3). A ***therm*** is equivalent to 100 ft^3 of natural gas and is typically used instead of the ft^3, since the ft^3 is such a small quantity. The therm is abbreviated as ccf, representing 100 ft^3.

Therm: The equivalent of 100 cf (cubic feet) of natural gas.

Supply and demand for natural gas

Natural gas is expected to be the fastest-growing component of world energy consumption over the next 25 years. The developing nations are anticipated to require the most growth in natural gas energy consumption. In the United States, natural gas is used extensively in the industrial, commercial, and residential sectors of our economy. It is an ideal fuel for structural heating, cooking, and many industrial processes. One sector of our economy that natural gas is not used for extensively at this time is the transportation sector. See **Figure 3-16.** Efforts are also underway to replace more polluting energy sources with natural gas in major urban areas, such as Beijing and Shanghai, China. Natural gas reserves have improved since the 1970s, but like most energy resources, the distribution is not evenly spread throughout the world. Over 70% of the current proven reserves are located in the Middle East and the former Soviet Union nations. Russia and Iran alone account for almost half of the world's proven reserves of natural gas.

Figure 3-16. U.S. natural gas consumption by end-use sector. (U.S. Energy Information Administration)

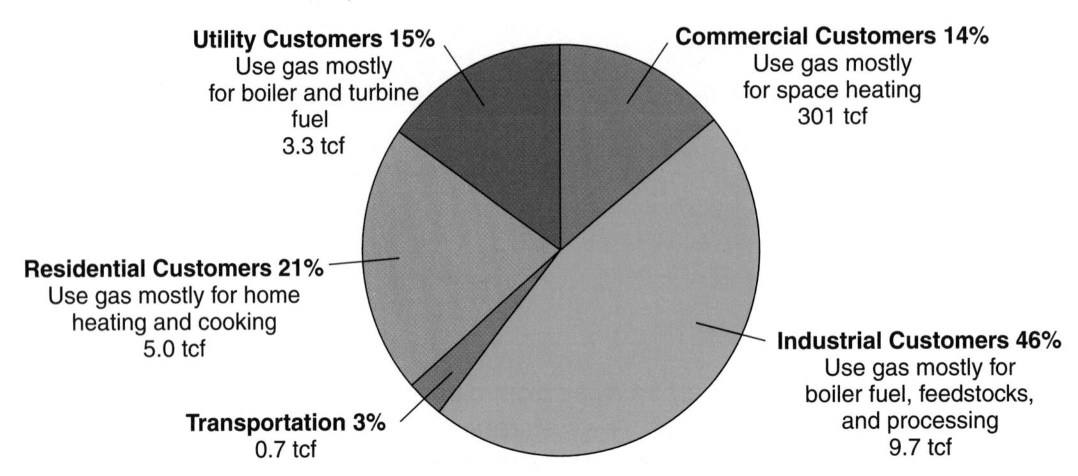

Natural gas and the environment

Natural gas is the cleanest burning of all the fossil fuels. It may also be the most dangerous of all the fossil fuels. When a natural gas leak occurs within a home, the carbon in the fuel mixes into the bloodstream and works to put people and animals into a deep sleep. This is known as carbon monoxide poisoning.

Acid Rain and the Greenhouse Effect

Two of the environmental problems most closely associated with the burning of fossil fuels are the creation of acid rain and greenhouse effect gases. *Acid rain* is a broad term often used to describe the primary ways acid comes back to the ground from our atmosphere. A more accurate term used among those in the field is *acid deposition*. Wet deposition is related to acidic rain, fog, and snow. Sulfur dioxide (SO_2) and nitrogen oxides created by the burning of fossil fuels linger in the atmosphere until mixing with rain, fog, or snow and returning to earth. As the acidic water flows over and through the ground, it affects plants and animals in many ways. The types of vegetation and the buffering capacity of the soil to filter out the acid all factor into how much damage acid rain can cause. Fish in ponds suffering from acid rain are particularly susceptible. About half of all acid rain is created from dry deposition, or gases and particles, that falls back to earth on trees and structures until washed off by the next rainfall. This can create a particularly concentrated dose of acid rain. Winds can also blow the dry deposition to other places.

The primary culprit in the production of acid rain is the electric utility industry. Scientists have confirmed that over 65% of all sulfur dioxide and about 25% of all nitrogen oxides in the environment are results of burning coal and other fossil fuels for electric power generation. The sulfur dioxide and nitrogen oxides react with water, oxygen, and other chemicals in the atmosphere to form acidic compounds. Sunlight increases the rate at which these chemical reactions occur, and the result is mild sulfuric acid and nitric acid that can be damaging to entire ecological systems and even buildings. Acid rain is monitored using the pH scale for measuring acidity. See **Figure 3-17.** The lower the pH of a substance, the more acidic it is. Pure water has a pH of 7.0, but normal rain is slightly more acidic from absorbing CO_2 as it falls to earth. The most acidic rain on record in the United States had a pH of approximately 4.3. Some other countries with government-owned electrical generating facilities are much worse. These plants are often constructed with far less environmental pollution control devices than those required in the United States. The ***Environmental Protection Agency (EPA)*** supports the monitoring of acid rain in this country.

Environmental Protection Agency (EPA): An organization that supports the monitoring of acid rain in the United States.

The *greenhouse effect* is a term used to describe the gradual warming of the earth's lower atmosphere. Some elements of the greenhouse effect occur naturally and have occurred long before human inhabitance of the earth. The greenhouse effect is important because, without it, the planet would not be warm enough for humans to live. Certainly, forest fires, volcanoes, and other naturally occurring events have been pumping gases from combustion into the atmosphere since the formation of the earth.

Figure 3-17. Efforts to clean up emissions from fossil fuel–burning power plants and other sources contributing to acid rain have been effective. The lower the number and the more orange the color, the more acidic the reading. A—The situation in 1994 shows very high acidity in large areas of the United States. B—In 2003, the readings were generally lower in all areas. (National Oceanic and Atmospheric Administration)

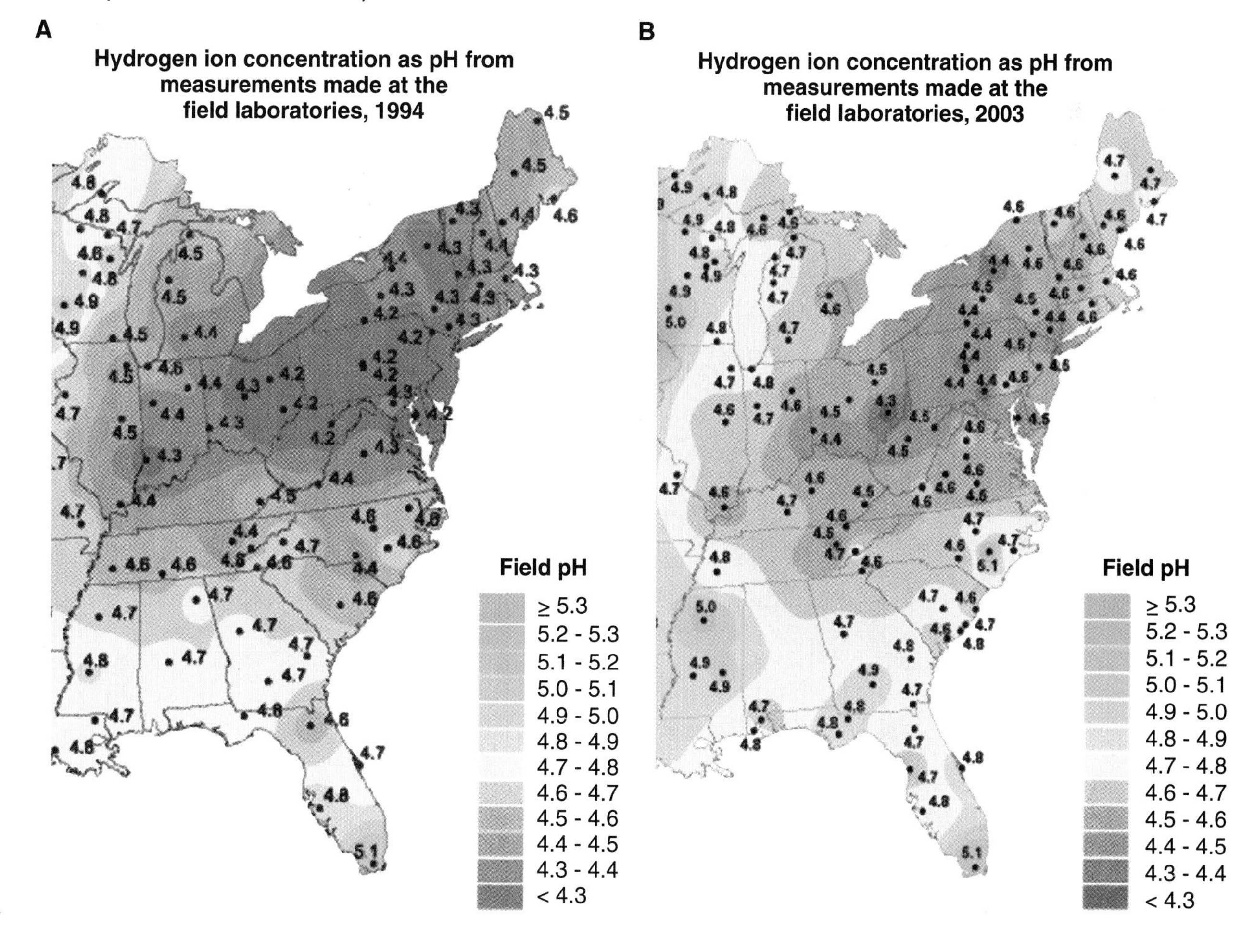

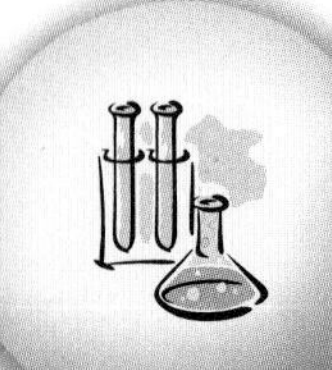

Curricular Connection

Science: The Greenhouse Effect's Impact on the Polar Ice Cap

With the rise in average global temperatures, the polar ice cap is slowly melting. The polar ice cap reflects sunlight back into space, rather than allowing the earth to absorb the heat. This process is called *albedo*, and it helps maintain the earth's temperature.

As the polar ice cap melts and is reduced in size, it cannot reflect as much sunlight back into space. This causes more heat to be retained by the earth's surface. This adds to the gases and particles involved in the greenhouse effect and global warming. The problem seems to be an unchangeable cycle, but scientists continually research the ways we can slow or halt the progression of global warming.

There is growing evidence, however, that since the mid-1800s, when the use of fossil fuels started to become widespread, there has been a marked increase in greenhouse gases. The greenhouse effect occurs naturally. As sunlight strikes the earth's surface, some of it is reflected back toward the atmosphere. Some light is also converted to heat used to warm the land and seas of the planet. Ultimately, this heat is also given up to the atmosphere, as these bodies cool. Greenhouse gases in the atmosphere serve to blanket the planet, holding the heat in our near atmosphere. See **Figure 3-18.** The concern is that, through the consumption of fossil fuels, we are creating too many greenhouse gases, including carbon, hydrogen, and even oxygen. These gases, which can form as water vapor, CO_2, methane, and ozone greenhouse particles, such as dust and soot, are helping to hold too much heat in the earth's atmosphere. The earth's average temperature is known to have increased by almost 1°F since the late 1800s. The cause of this increase has not yet been proven. There is likely a relationship, however, between CO_2 levels in the atmosphere and the continued consumption of fossil fuels. Continuing fossil fuel consumption at present rates could lead to a global warming effect that could alter weather and vegetation patterns throughout much of the earth. Other potential consequences of global warming could be the melting of icebergs and polar ice enough to literally change coastlines. Some estimates predict that permanent damage from global warming may begin to appear in the near future.

Figure 3-18. The greenhouse effect holds in heat, creating global warming, which can cause serious environmental damage to the planet.

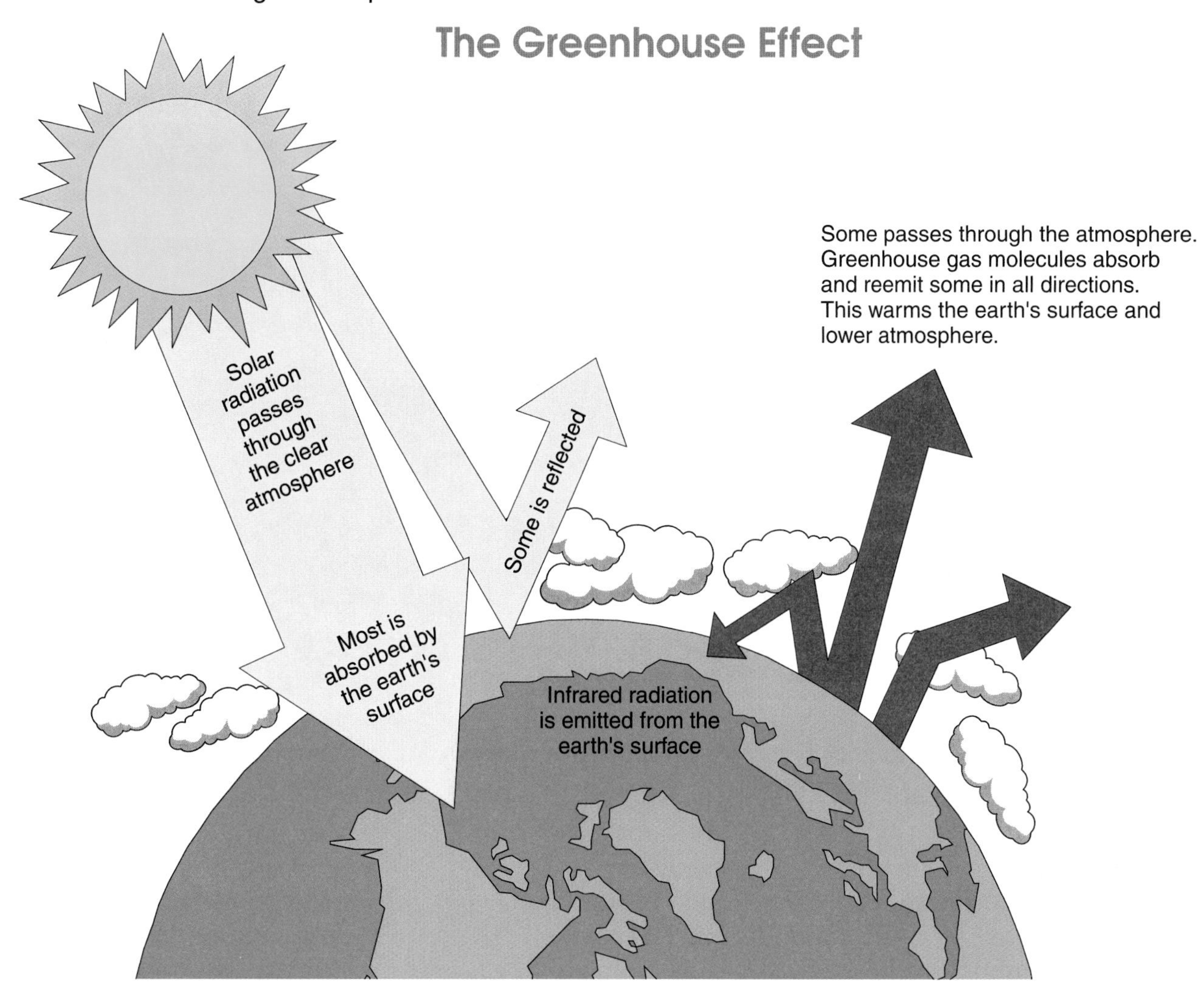
The Greenhouse Effect
Solar radiation passes through the clear atmosphere
Some is reflected
Most is absorbed by the earth's surface
Infrared radiation is emitted from the earth's surface
Some passes through the atmosphere. Greenhouse gas molecules absorb and reemit some in all directions. This warms the earth's surface and lower atmosphere.

Summary

Fossil fuels are responsible for about 90% of all energy production worldwide. They were formed over millions of years, as organic material decomposed. The consumption of these nonrenewable sources of energy creates environmental consequences, such as greenhouse effect gases and acid rain. Additional environmental hazards include the threat of oil spills, such as the *Exxon Valdez* oil spill, which occurred in 1989. Some fossil fuels, such as oil, must be imported from other countries, which can lead to economic and political consequences. Other fossil fuels, such as coal, are plentiful in the United States and can be converted to gas or liquefied petroleum products. This is unlikely to happen unless prices of conventional fuels rise dramatically. Fossil fuels such as tar sands and shale oil contain much energy that, to date, has largely gone untapped. This is because of the difficulty in retrieving these resources. Despite all the misgivings about the use of fossil fuels, these fuels are highly likely to remain the major sources of energy worldwide for several decades, until some combination of economics, technology, politics, and environmental concerns usher in sources to replace them.

Key Words

The following words have been used in this chapter. Do you know their meanings?

anthracite coal
aquifer
bag filter
bituminous coal
crude oil
deep mining
electrostatic precipitator
Environmental Protection Agency (EPA)
fly ash
fractionating tower
kerogen
land reclamation
lignite coal
liquefied natural gas (LNG)
natural gas
oil shale
overburden
peat
petroleum
strip mining
subbituminous coal
tar sand
therm

Test Your Knowledge

Write your answers on a separate sheet of paper. Do not write in this book.

1. *True or False?* Fossil fuels are responsible for about 30% of all energy consumed in the United States.
2. *True or False?* The primary use for oil in the United States is for home heating.
3. Explain why the use of fossil fuels remains so prevalent today.
4. *True or False?* Natural gas is often found where oil is found.
5. *True or False?* Coal is primarily used in the production of electricity.

6. *True or False?* The United States is fortunate to have massive coal reserves.
7. The two methods of mining are known as _____ and _____ mining.
8. Summarize the major components of a coal-fired generating plant.

Matching questions: For Questions 9 through 13, match the phrases on the left with the correct term on the right.

9. _____ Formed from water and the decomposition of organic materials.
10. _____ Very moist and brownish in color.
11. _____ Dull black and "soft."
12. _____ Very dense and black, with a high carbon content.
13. _____ Shiny, brittle, and often used for heating homes.

A. Anthracite.
B. Bituminous.
C. Lignite.
D. Peat.
E. Subbituminous.

14. Write a brief justification for the use of coal liquefaction.
15. Means of limiting emissions from a coal-fired power plant include _____ and _____.
16. A(n) _____ tower is used to distill crude oil into various products.
17. What is the relationship between the United States and the Organization of Petroleum Exporting Countries (OPEC)?
18. Discuss both the positive and negative impacts of importing oil.
19. *True or False?* Tar sands and shale oil are major contributors to the world energy production, at present.
20. _____ has the greatest British thermal unit (Btu) content of all the hydrocarbons that comprise natural gas.
21. Natural gas is transported primarily by:
    A. tanker ship.
    B. rail.
    C. pipeline.
    D. truck.
22. *True or False?* Natural gas is measured by the ccf or therm.
23. Which of the following fossil fuels is most abundant in the United States?
    A. Coal.
    B. Oil.
    C. Natural gas.
    D. Tar sands.
24. List and describe at least four concerns about the use of fossil fuels.
25. *True or False?* The pH scale is often used to measure the energy content of coal.

26. Which of the following measurements would indicate the greatest acidity?
    A. 3.2.
    B. 4.6.
    C. 7.0.
    D. 8.3.
27. *True or False?* One positive outcome of the consumption of fossil fuels is that of global warming.
28. Problems associated with the consumption of fossil fuels include:
    A. acid rain.
    B. the greenhouse effect.
    C. global warming.
    D. all of the above.

Activities

1. Visit a power plant powered by a fossil fuel.
2. Calculate how much energy of a given source would be required to heat a specific space.
3. Write a paper about environmental concerns associated with burning fossil fuels.

4

Nuclear Energy

Basic Concepts

- Describe the process of how electrical power is generated using nuclear fission as an energy source.
- Discuss the history of the nuclear power industry in the United States.
- List the environmental concerns associated with nuclear power.
- Identify how spent nuclear fuel is stored today and state plans for future storage.

Intermediate Concepts

- Describe how nuclear fuel is produced.
- Explain the major differences between the types of nuclear reactor designs in commercial operation in the United States today.
- Summarize why it is necessary to continue to explore nuclear fusion as an energy source.

Advanced Concepts

- Analyze the trade-offs associated with the use of nuclear power, in comparison to generating power via other means.
- Demonstrate how breeder reacting could multiply the energy potential of proven uranium reserves.

Nuclear energy, sometimes referred to as *atomic energy*, is the energy released from an atom in a nuclear reaction or through the radioactive decay of materials. There are two primary methods of creating nuclear reactions. One method involves fusing two smaller atoms together to produce one larger atom and a tremendous amount of energy. This method, known as ***nuclear fusion***, is only in experimental stages and is not presently in use to produce commercial power. The other method involves splitting a larger atom to produce two smaller atoms and a tremendous amount of energy. This process is known

Nuclear fusion: The combining of two nuclei into a larger nucleus. The large nucleus weighs less than the two smaller nuclei that formed it. The result of this process yields a large energy release.

Nuclear fission. The process of splitting a larger atom to produce two smaller atoms and a tremendous amount of energy.

as ***nuclear fission***. It is used extensively throughout the United States and many other countries to produce heat for steam to make electricity. The entire process relies on an element in nature that is suitable to fission. This element is known as uranium.

Uranium: The Fuel for Nuclear Fission

Uranium is a nonrenewable energy source thought to be a substance in volcanic ash. This energy source probably spewed from a volcano onto the earth's surface millions of years ago. Rains dissolved the uranium out of the ash. The uranium was then carried back into the ground, where it has hardened into ore over the years. It is located near the surface of the earth. Therefore, it is mined in much the same way as coal. Like fossil fuels, uranium supplies will one day run out. Uranium is the most common type of nuclear fuel. It is used to fuel large nuclear power plants. Uranium can generate large amounts of energy in the form of heat. The heat produced is used to boil water, thus creating steam. The steam is then used to drive turbine-powered generators, which produce electricity. See **Figure 4-1.**

Figure 4-1. Nuclear reactors generate heat to produce steam that, in turn, drives generators to produce electricity. A—This nuclear power station is one of more than 450 in operation or under construction worldwide. B—A simplified diagram of nuclear reactor operation. (Canadian Nuclear Association)

A

B

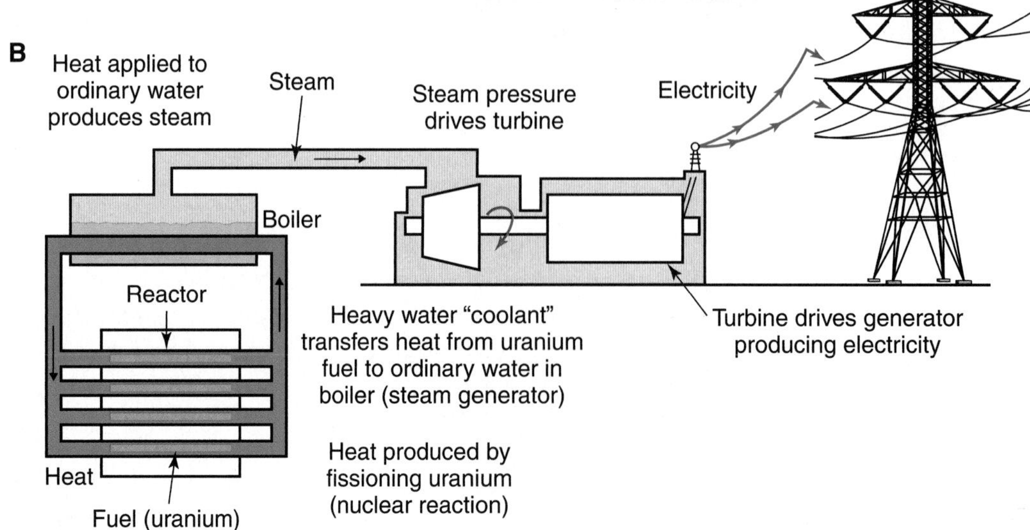

Uranium is an atom with a large and heavy nucleus. A ***nucleus*** is the center portion of an atom containing the protons and neutrons. ***Protons*** are positively charged particles. ***Neutrons*** are uncharged particles. Therefore, a lot of energy is bound inside the nucleus and can be used as nuclear fuel. Uranium atoms are bombarded by neutrons. This causes the nuclei to split apart. This splitting is known as *nuclear fission*. Lots of energy is released at the time of the splitting. When one nucleus splits, it releases energy and neutrons, which in turn, split other nuclei. A chain reaction thus occurs. This reaction produces huge amounts of energy in the form of heat. All of this reaction takes place inside a strong, closed container called a *reactor.* A nuclear reactor is to a nuclear power plant what a furnace is to a home. It produces the heat used to generate power at the plant. See **Figure 4-2.**

Nucleus: The center portion of an atom containing the protons and neutrons.

Proton: A positively charged atomic particle.

Neutron: An uncharged atomic particle.

Figure 4-2. The pressure vessel where nuclear fission takes place, generating the heat to turn water to steam. The control rods are used to regulate the amount of nuclear activity taking place and, thus, the amount of heat generated. (Edison Electric Institute)

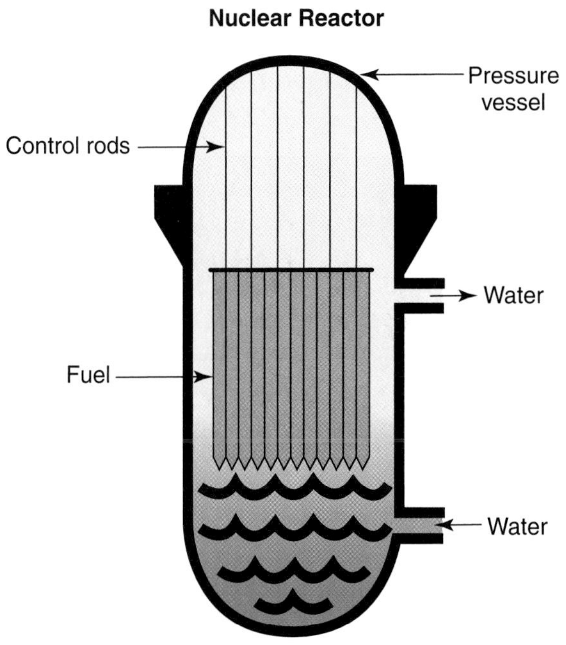

Nuclear Reactor

Pressure vessel

Control rods

Water

Fuel

Water

The fission of uranium is the only developed source of nuclear power in commercial use in the United States. The use of nuclear power produces about 20% of all the electricity produced in the United States per year. Nuclear power in the form of electricity is produced at a nuclear power plant. The amount of energy produced by uranium is incredible. In terms of pure British thermal units (Btu) per volume, no other fuel source comes close. One uranium pellet (which is about the size of your fingertip) can produce as much energy as 1780 pounds of coal, 149 gallons of oil, or 157 gallons of regular gasoline. See **Figure 4-3.**

There are some major concerns about using uranium to produce electricity. Uranium is a very hazardous substance. When the atoms of uranium decay, they give off atomic particles. This reaction produces ***radioactivity,*** which is harmful to humans and other living things. Research is being done, especially at the Argonne National Laboratory, to make spent nuclear fuel less hazardous. See **Figure 4-4.**

Figure 4-3. The amounts of various fuels needed to generate the same amount of energy.

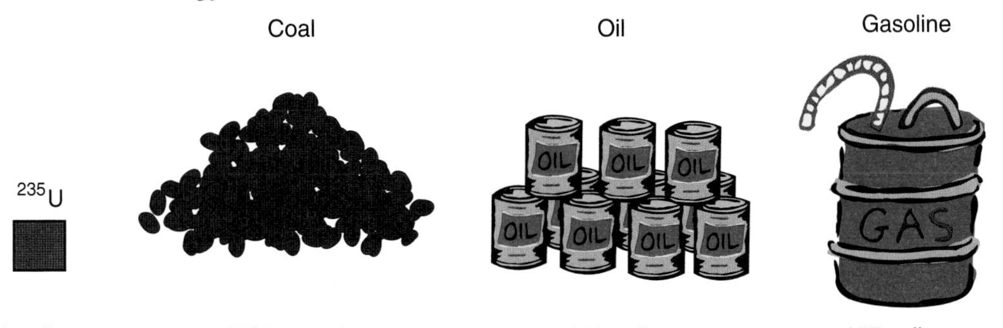

Radioactivity: A property of some atoms, such as those of uranium decay, in which they give off atomic particles. The particles emitted are harmful to humans and other living things.

Figure 4-4. Researchers are working on reducing the risks of storing spent nuclear fuel. The electrorefiner equipment in the background separates spent fuel into long-lived and short-lived fission products. The long-lived products, such as plutonium, can be recycled to create new fuel. Short-lived (and lower radiation level) products can be stored with less costly methods than the original spent fuel. The research is a joint project of Argonne National Laboratory and the Central Research Institute of Japan's electric and atomic power industries. (U.S. Department of Energy)

Technology Link

Medicine: Health Risks of Radiation Exposure

Nuclear power plants contain many harmful elements. Power plant workers risk exposure to these elements every day. Radiation from these materials can cause cancer or possible genetic defects.

The materials found in a nuclear power plant include both uranium and plutonium. Exposure to these radioactive elements can result in, among other things, lung cancer and kidney damage. Workers in power plants come in contact with radioactive materials as seldom as possible, but the health risks still remain.

Medical technology is necessary to combat these health risks. Modern medicine offers options to prevent and treat disorders related to radiation exposure. The advances in medical technology allow for the diagnosis and treatment of radiation-related diseases.

While nuclear power plant workers are subject to radiation exposure, a properly functioning nuclear power plant emits almost no excess radiation. Plant workers wear dosimeters on their clothing, much like name badges. These devices closely monitor exposure levels. The workers are also checked for radioactivity as they enter and leave work. The instrumentation is so accurate that plant workers who have eaten bananas (a natural source high in potassium) can trip the alarm when entering the power plant. The bottom line is that people are exposed to more radiation on one long plane ride than most plant workers are exposed to in a year of on-site work.

Curricular Connection

Social Studies: The History of Nuclear Power

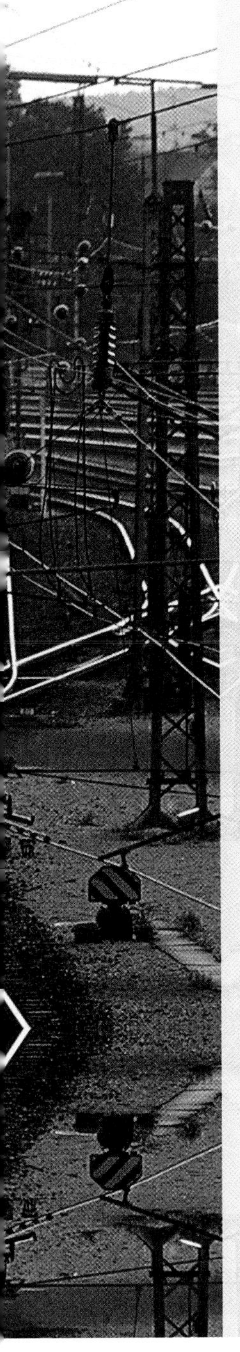

- During World War II, German and American scientists began working on nuclear weapons.
- In 1945, the American bomber, *Enola Gay*, dropped a nuclear bomb on Hiroshima, Japan. Three days later, another nuclear bomb was dropped on Nagasaki, Japan. These bombings ended the war, as Japan surrendered shortly afterward.
- The Atomic Energy Commission began investigating peaceful uses of atomic energy in 1947.
- President Eisenhower began a research and development program in 1953, entitled "Atoms for Peace."
- During the mid- to late 1950s, the government and industry cooperated and developed nuclear power plants from developmental reactors that had shown promise. See **Figure 4-A.**
- By the 1960s, utility companies were ordering nuclear power plants as an alternative to fossil fuel plants, since nuclear power plants could produce electricity at much less expense than conventional power plants.
- Many nuclear power plants were constructed in the United States during the 1970s. By the late 1970s, however, the tide was beginning to turn against nuclear power because of cost overruns on construction and environmental concerns about nuclear waste.
- In 1979, America's worst peacetime nuclear disaster occurred when a reactor at the Three Mile Island (TMI) nuclear plant near Harrisburg, Pennsylvania reached excessively high temperatures through a series of faulty readings and operator errors. Not one new order for a nuclear power plant has been placed in the United States since then.
- A much more severe accident, the Chernobyl accident, occurred at a nuclear power plant in 1986, in Kiev, a city in the former Soviet Union. A poorly designed experiment led to a large quantity of radiation being released into the environment. Many people were killed, and the radiation spread throughout much of Europe. Radiation from the accident was even detected in the United States several days later.
- In recent years, several of the more than 110 nuclear power plants that once operated in the United States have closed down permanently.
- Most recently, utility companies are again expressing interest in nuclear power, as fossil fuel costs have risen dramatically.

Figure 4-A. Dresden I, a commercial nuclear reactor, opened in 1959 in Morris, Illinois, west of Chicago. The unit was retired from active service in 1978, although two later units remain in operation at the same site. The spherical object is the containment structure of the Dresden I reactor. (Exelon Energy)

The Fission Process

Uranium 235 (U235): An element whose atoms can be split more easily than most others, making it suitable for refining into nuclear fuel.

Fission is the term associated with the splitting of atoms. The nucleus of any atom can be split if it is smashed hard enough with a neutron. In the 1930s, however, it was discovered that one element, ***uranium 235 (U235)***, is split more easily than most others. The *235* in U235 is the atomic weight, which is the mass of the atom, relative to other atoms. The atomic weight of any element is approximately equal to the number of protons and neutrons in its nucleus. The result of the fission process of U235 yields several products, including a large amount of energy, small amounts of barium and krypton, and two additional neutrons. The mass of the various products does not quite equal the mass of the U235 atom. The cause of this imbalance is that some mass has been directly converted to energy. Additionally, the two free neutrons that are liberated from the nucleus can be used to split other atoms. This is how a chain reaction occurs. See **Figure 4-5.**

The Terminology and Measurement of Radiation

Half-life: The time it takes for half the atoms present in an unstable element to transform into a new element.

Half-life is a term most often associated with the radioactive by-products of nuclear fission, although it can be used when describing anything radioactive. As radioactive elements decay, they will eventually form other elements. A half-life is the time it takes for half the atoms present in an unstable element to transform into a new element.

After a period of one half-life, only half of the original atoms will remain. As a substance reaches each half-life, it becomes more stable and emits less radiation. Not all substances decay at the same rate as others, however. Half-lives of substances can range from a few seconds to thousands of years! **Figure 4-6** lists the half-lives of various elements, including U235, which is often used for fuel in nuclear power plants.

After a period of one half-life, the radioactivity of an element is decreased by 50%. The radioactivity is reduced to 25% of the original level after two half-lives. After three half-lives, the radioactivity would be reduced to 12.5% of the original state. When an element has reached its tenth half-life, it will emit less than .1% of the original radiation emitted.

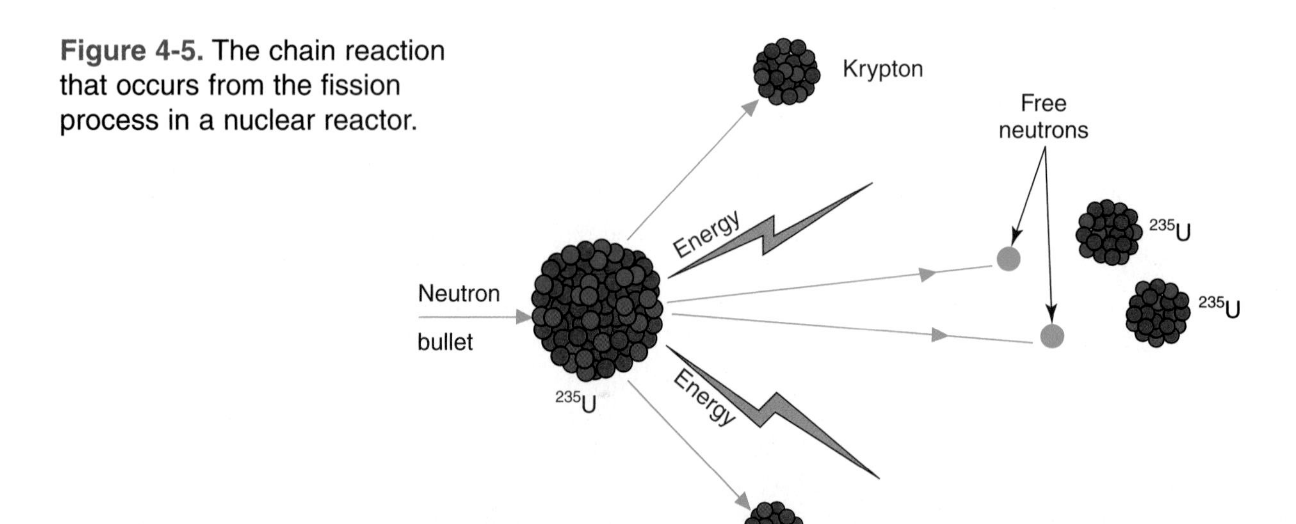

Figure 4-5. The chain reaction that occurs from the fission process in a nuclear reactor.

Element	Half-Life	Primary Use
Technetium 99	6 hours	Medical imaging
Xenon 133	2.3 days	Lung ventilation, blood flow studies
Iodine 131	8 days	Diagnosis and treatment of thyroid problems
Strontium 89	54 days	Treatment of bone pain
Cobalt 60	5.2 years	Treatment of cancerous tumors
Plutonium 239	24,400 years	Nuclear power, nuclear weapons
Uranium 235	704 million years	Nuclear power, nuclear weapons
Uranium 238	4.5 billion years	Nuclear power, nuclear weapons

Figure 4-6. The half-lives of various radioactive elements.

The concept of the half-life is very important to the nuclear power industry, since some of the elements with the greatest half-lives are associated with nuclear power. Elements such as U235 exist in nature. Other elements, such as certain plutonium isotopes, are human-made. This leads to ethical questions about the need to produce such elements, only to require that they be isolated from humanity for thousands and thousands of years. Nuclear research has yielded some drastic technological improvements for the medical industry, however. Medicine makes use of some radioactive treatments that have been designed to destroy tumors and then lose their radioactive potency quickly, before damaging surrounding healthy tissue.

How a Nuclear Reactor Works

There are two basic types of fission reactors in use in the United States today. Both are referred to as *light water reactors* because they use ordinary water instead of "heavy water" within the reactor core. *Heavy water* is a term used to describe water made up of oxygen and deuterium (a heavy isotope of hydrogen). It is used as a moderator within the reactor core to slow the rate of fission in many foreign reactors. The two types of reactors are boiling water reactors (BWRs) and pressurized water reactors (PWRs).

Boiling Water Reactors (BWRs)

The ***boiling water reactor (BWR)*** is the simplest of the reactors to describe. See **Figure 4-7.** Water surrounds the nuclear fuel core within the reactor. The ***control rods*** sit between the fuel rods. The fuel rods are assemblies that hold the fuel pellets of uranium in a specific configuration. Placement of the fuel rods is important, since they must be appropriately spaced for fission to occur at the right pace and so control rods can be inserted and retracted out of them to control the fission process. It is the job of the control rods to absorb stray neutrons. As long as they are properly inserted, fission cannot occur. When the control rods are retracted, the fission process begins to occur. A tremendous amount of heat is produced. This heat converts the surrounding water to steam. The force of the expanding steam spins a turbine to produce electricity, much the same way a turbine and generator are propelled with steam created in other

Boiling water reactor (BWR): A type of fission reactor in which water surrounds the nuclear fuel core within the reactor. Control rods sit between the fuel rods and absorb stray neutrons. When the control rods are retracted, the fission process begins to occur, and a tremendous amount of heat is produced. This heat converts the surrounding water to steam. The force of the expanding steam spins a turbine to produce electricity.

Control rod: Part of a fission reactor that sits between the fuel rods and absorb stray neutrons. When the control rods are retracted, the fission process begins to occur.

Figure 4-7. The operation of a boiling water reactor (BWR).

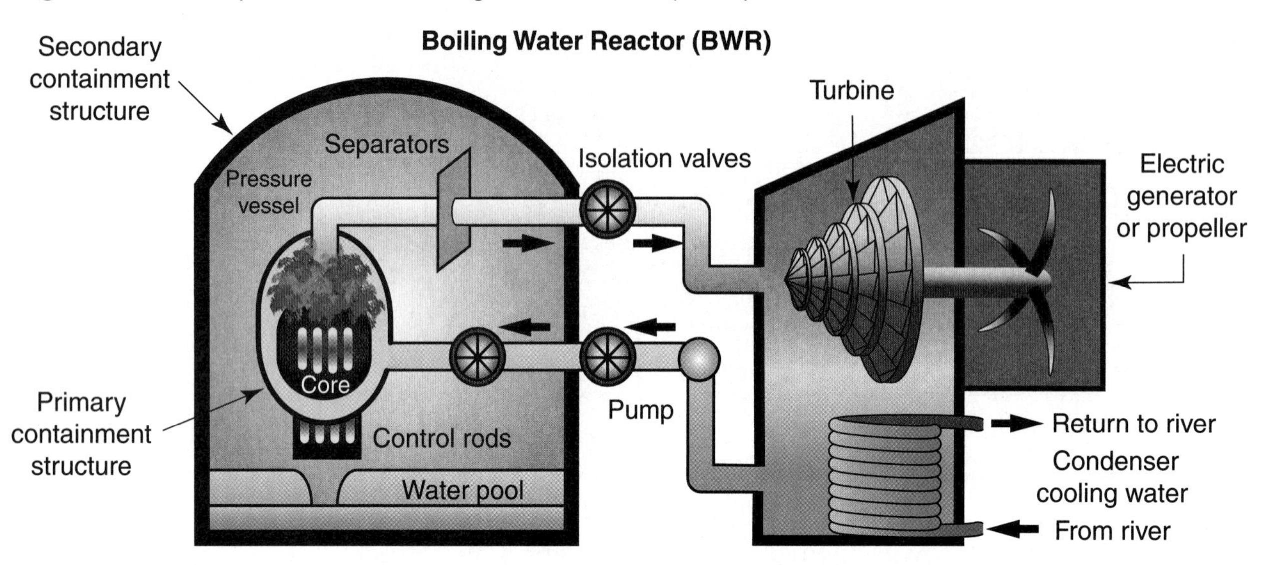

ways, such as by burning coal. The spent steam is condensed back into a liquid when it is surrounded by cooler water in the condenser of the reactor. It is then pumped back to the reactor to be heated again.

The condenser is responsible for cooling the water in the steam loop. It often gets this water from a nearby river or lake and returns it to the river or lake. Cooling towers are used to reduce the temperature of the condensing water, since returning the water to the river or lake at too high a temperature could damage the ecosystem of the water source. These towers are massive structures. They actually have nothing to do, however, with the radioactive part of a nuclear power plant. In fact, cooling towers may be used at fossil fuel power plants for the very same purpose. The coolant is typically water. Some reactor designs, however, use liquid sodium or gases to dissipate heat.

The control rods are used to slow the rate of the chain reaction when fission is occurring. They are made of neutron-absorbing elements, such as graphite, cadmium, and carbon. When fully inserted, the control rods should be capable of stopping the chain reaction known as the fission process.

Pressurized water reactor (PWR): A reactor that works similarly to a boiling water reactor (BWR), except it makes use of a heat exchanger known as a steam generator. A PWR can operate at higher pressures and temperatures than a BWR. Unlike the BWR, the steam generator in a PWR allows the turbine to remain free of radioactive contamination.

Primary loop: The part of a pressurized water reactor (PWR) in which the water is heated. It surrounds the reactor core.

Secondary loop: The part of a pressurized water reactor (PWR) in which steam is created.

Pressurized Water Reactors (PWRs)

A ***pressurized water reactor (PWR)*** works similar to a BWR, except it makes use of a heat exchanger known as a steam generator. The PWR is slightly more complicated than the BWR. See **Figure 4-8.** The PWR can operate at higher pressures and temperatures than the BWR, since the water heated in the ***primary loop*** surrounding the reactor core does not come into direct contact with the turbine. The purpose of the primary loop is to collect heat from the reactor core and transfer the heat to a steam generator. Inside the steam generator, the heat from the primary loop is transferred to the water in the ***secondary loop***, thereby creating steam in the secondary loop. The steam in the secondary loop is used to drive the turbine that spins the generator. The generator is a big heat exchanger that allows for the transfer of heat from the primary loop to the secondary loop, while keeping the water in each loop isolated. This is

Figure 4-8. The operation of a pressurized water reactor (PWR).

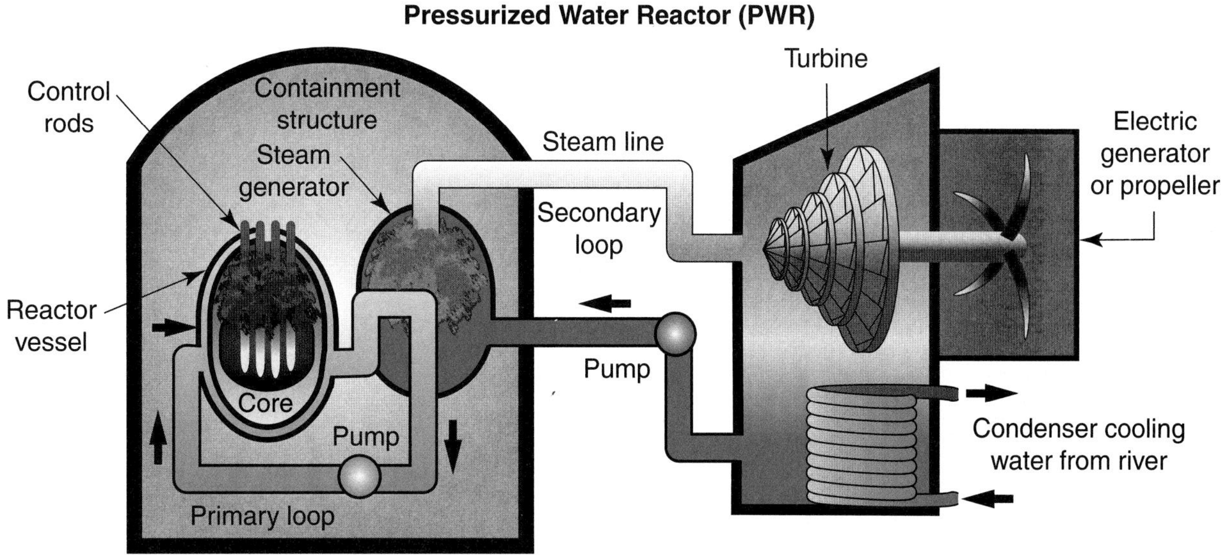

important, since the water in the primary loop is radioactive, but the water in the secondary loop is not. Unlike the BWR, the steam generator in a PWR allows the turbine to remain free of radioactive contamination. This can be beneficial from a maintenance standpoint.

Supply and Demand for Nuclear Power

Today, approximately 100 operating power plants supply about 20% of all electricity produced in the United States. Regionally, the New England states rely on nuclear power the most. Nuclear power accounts for more than one-third of all electricity produced throughout New England. The western states have fewer nuclear power plants by far. See **Figure 4-9.**

Worldwide, nuclear power is expected to share a shrinking percentage of the world's electricity production over the next 20 years, even though generating capacity is increasing. Six new reactors began generating power in 1992. Four were located in China, one in South Korea, and another in the Czech Republic. As of 2004, there are 439 nuclear power reactors in operation around the world and several more under construction. See **Figure 4-10.** More than 19 countries worldwide rely on nuclear power for at least 20% of their electricity. Lithuania leads the way, with about 80% of all electrical generating capacity coming from nuclear power. France produces about 78% of its generating capacity from nuclear fission. Slovakia produces about 58% of its generating capacity from nuclear power. Even though there have not been orders for new nuclear power plants in the United States for some time, the economics of nuclear plant construction is more favorable in other countries. In certain countries that lack their own natural resources and have access to inexpensive labor markets, nuclear power is more appealing.

Nuclear power may also be a more desirable option to those countries that wish to increase capacity while reducing carbon dioxide (CO_2) emissions to reach targets set in the ***Kyoto Protocol***. This set of rules was devised in 1997, at a meeting of more than 160 nations in Kyoto, Japan.

Kyoto Protocol: Targets set in 1997 by countries wishing to increase capacity while reducing carbon dioxide (CO_2) emissions.

Figure 4-9. The distribution of the operating nuclear power plants in the United States. Only 2 are less than 10 years old.

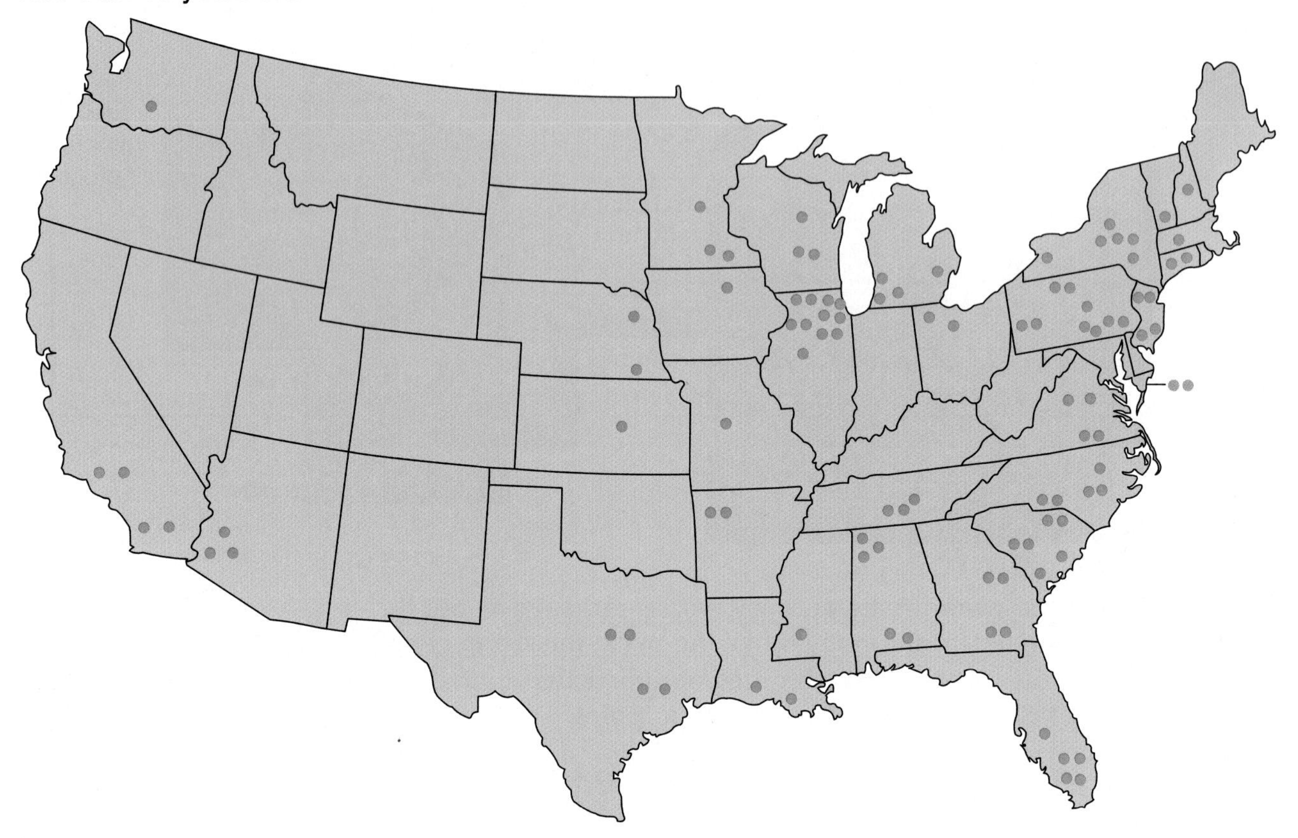

Figure 4-10. Worldwide nuclear power plant distribution in 2003. Europe and North America account for two-thirds of the existing or under-construction facilities. Asia accounts for most of the remaining one-third. (The graph was developed from the International Atomic Energy Agency data.)

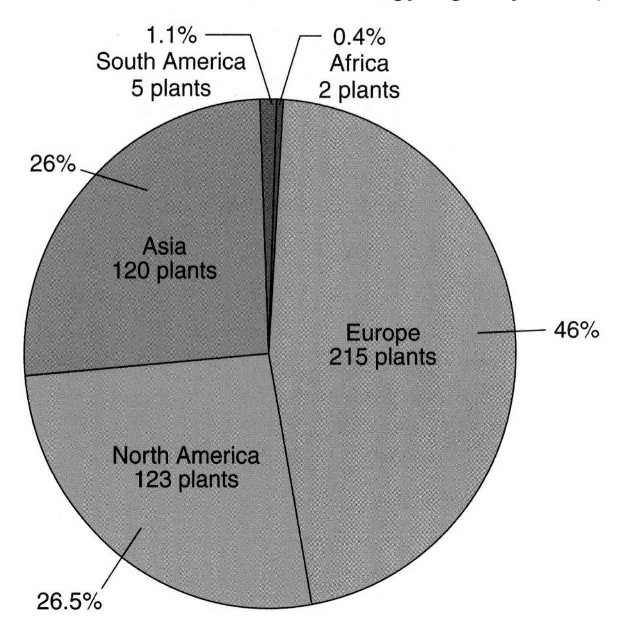

Tech Extension

The Nuclear Fuel Cycle

The production of nuclear fuel begins with mining in a similar fashion to the way other resources, such as coal and metal ores, are recovered. See **Figure 4-B.** Most uranium mining operations in the United States exist in the western states of Colorado, Wyoming, Utah, and New Mexico. Some uranium mining is also performed in Texas. Worldwide, Russia and China appear to have an abundance of proven uranium reserves.

After the uranium-bearing ore has been extracted from the ground, uranium mills separate the uranium from other elements and refine it to an oxide known as yellowcake. The yellowcake is then shipped to a fuel enrichment plant, where it will be enriched to about 3–4% pure uranium 235 (U235), for use as a fuel in nuclear power plants. The fuel enrichment process is extremely technical in nature and not easily performed.

Once enriched, the fuel is manufactured into fuel pellets and loaded into fuel rod assemblies that space the fuel correctly for optimal fission and control. A fuel rod assembly can be expected to power a nuclear power plant for approximately 3–4.5 years before being removed for storage. Each year, approximately one-third of the fuel rods within a nuclear reactor are removed and replaced with fresh fuel rods. The oldest fuel rods are removed. The remaining fuel rods are reconfigured within the reactor core for maximum efficiency. The removed fuel rods are stored on-site in storage pools at the power plant to await permanent storage. Some low-level waste is sent to low-level storage facilities. Someday, it may be possible to reprocess nuclear waste into less harmful products.

Figure 4-B. The nuclear fuel cycle extends from mining uranium ore to storing spent nuclear fuel.

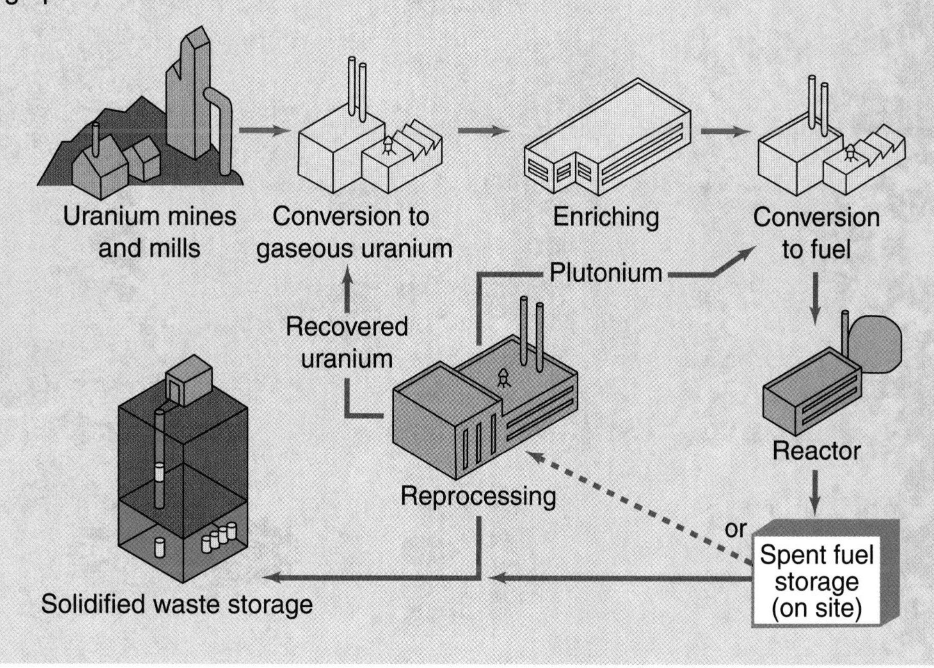

These nations negotiated binding limitations on greenhouse gasses for developed nations. The outcome of the meeting was that developed nations agreed to limit their greenhouse gas emissions, relative to the levels emitted in 1990. The United States viewed the proposed limitations as too restrictive and did not sign the Kyoto Protocol. In spite of the controversies surrounding nuclear power, many scientists see the long-term use of more nuclear power as inevitable, both domestically and abroad.

Nuclear Power and the Environment

There are many aspects of nuclear power to consider when it comes to protecting the environment. The first that comes to mind is the storage of nuclear waste. As described earlier, almost all high-level waste is stored in on-site pools at the various power plants. These pools were never intended to serve as permanent storage facilities. Initially, operators of the power plants intended to chemically separate the unused uranium and plutonium taken from spent fuel rods and then recycle it as fuel for another fuel rod assembly. Reprocessing, however, concentrates plutonium into a form that is capable of being used to produce nuclear weapons, so President Jimmy Carter banned the process in 1977. Since the fuel could no longer be reprocessed, storage pools began to fill up. The storage pools at several nuclear power plants have now been filled to capacity, with no means of expanding them or reconfiguring the spent fuel to fit more in. If the spent fuel is placed too close together, chain reactions could occur. The solution has been the development of temporary aboveground nuclear waste storage casks.

What will become of all this nuclear waste? The federal government is supposed to take it from the utilities for permanent storage. Congress promised this in 1982, when it passed the ***Nuclear Waste Policy Act***. Since that time, utilities have charged their customers a one-cent surcharge for every kilowatt-hour (kWh) generated using nuclear power. The revenue generated is intended to help solve the long-term storage issue. The multibillion-dollar fund known as the ***Nuclear Waste Fund*** is used for the development of a permanent waste disposal site.

Nuclear Waste Policy Act: An act passed by Congress in 1982 promising that the federal government is to take nuclear waste from the utilities for permanent storage.

Nuclear Waste Fund: A multibillion-dollar fund used for the development of a permanent nuclear waste disposal site.

Storage of Nuclear Waste

For decades, scientists have been recommending that nuclear waste be stored in stable rock formations deep within the earth's surface. The most promising site appears to be the government-owned Yucca Mountain facility in southern Nevada. The area sits adjacent to the Nevada Test Site, which served as a testing ground for early nuclear bombs. The land consists of porous rock that acts as a natural barrier to radiation. The location is seismically stable. Additionally, the water table in this unpopulated area is extremely deep. The idea for permanent storage would be to drill a shaft approximately 1000′ below the surface and then branch out in several directions with tunnels. An aboveground facility would accept the waste in transport casks and place it in permanent storage casks for underground disposal. The ***Yucca Mountain storage facility*** has been delayed several times. The opening date was originally scheduled for 1985. It was postponed until 1989, then 1998, and then 2003. Most recently, the scheduled opening date has been billed as 2010, but this is subject to

Yucca Mountain storage facility: A government-owned facility in southern Nevada that is a planned site for permanent storage of nuclear waste. The waste would be stored in stable rock formations deep within the earth's surface.

change. Recent world events like the 9/11 attacks on the World Trade Center Towers in New York City have called for renewed vigor in creating a permanent nuclear waste storage facility that would be difficult to sabotage.

Transporting Nuclear Waste

If nuclear waste is to be stored in a long-term storage facility, or if it is to be reprocessed, chances are it must be shipped from its current location to another location, often thousands of miles away. Spent fuel is protected for shipment against accidents and sabotage by the design of the ***shipping cask*** created to contain the fuel. See **Figure 4-11.** The tests these casks go through are extraordinary. A prototype cask is typically designed for testing. The casks are dropped onto the ground out of airplanes and helicopters, run into by diesel locomotives, and doused with jet fuel and ignited. All this is done to ensure the cask cannot be breached, even in a worst-case scenario during transport from one facility to another.

Shipping cask: A container designed to ship spent fuel from one facility to another.

Recycling Nuclear Waste

With nuclear waste piling up in storage pools at power plants, spent fuels that have half-lives of thousands of years, and the dangers of nuclear proliferation, much attention has recently been focused on the possibility of recycling nuclear waste, rather than burying it. While funding for research remains critical, some progress has been made that would indicate that nuclear waste, even high-level waste, can be rendered harmless through processing techniques. Of course, one primary obstacle to recycling or reprocessing of nuclear waste may be more economic than technological. When all is said and done, it may simply be less expensive to bury the waste than it could be to recycle it. Burial of nuclear waste may prevail over other options that have less long-term risk, but are simply more expensive.

Figure 4-11. The test of a cask design for use in shipping spent nuclear fuel to long-term storage facilities, such as Yucca Mountain. The cask is shown in a receiving facility at the Hanford Nuclear Facility in the state of Washington. (U.S. Department of Energy)

Nuclear Reactor Safety

Nuclear reactors have had a good track record in terms of environmental protection, particularly in the United States, where the reactor vessel is located within an environmental containment shield. Contamination from an explosion within the reactor vessel, regardless of how the explosion is caused, should not be capable of breaching the environmental containment shield. This is considerably different from the way some other countries constructed nuclear power plants. Many of the power plants in the former Soviet Union, including Chernobyl, were constructed within industrial buildings without any extraordinary effort to create an environmental shield. The reactor vessel was assumed to provide enough environmental protection. When a steam explosion demolished the reactor vessel and much of the surrounding building, there was simply nothing left to contain the release of radiation into the atmosphere. The results were that radiation fallout was found throughout much of Europe. See **Figure 4-12.**

All commercial reactors constructed within the United States must contain an environmental shield that has been engineered to withstand plane crashes, extreme weather conditions, and the possibility of explosion from within the reactor vessel. Since the reactor fuel is not configured like a nuclear bomb, it is doubtful a nuclear explosion could occur. A steam explosion, such as that which demolished the Chernobyl plant, is much more likely. Regardless of the means of explosion, what is most important is that the environment is protected. In the case of the ***Three Mile Island (TMI) accident***, a small portion of the reactor core was melted when the reactor coolant was accidentally shut off. Emergency cooling water flooded the reactor core. Due to human error, however, the core was allowed to overheat for a period of time that caused irreparable damage. While only a small portion of the core was damaged, the image of a nuclear meltdown that destroyed the nuclear reactor and ultimately breached the containment vessel became etched in the minds of many Americans. Following the TMI accident, the entire nuclear power industry was overhauled to rely on more passive safety technology that is not capable of being manually overridden. This passive safety technology is regarded as being much more failsafe, as it does not rely on human intervention to protect the power plant. An example of a passive technology

Three Mile Island (TMI) accident: A nuclear disaster occurring in 1979 near Harrisburg, Pennsylvania. The Unit 2 reactor reached excessively high temperatures through a series of faulty readings and operator errors. Eventually, a small piece of the reactor core melted, rendering the reactor unusable before the situation was brought under control.

Figure 4-12. Radiation fallout from the Chernobyl nuclear power plant disaster.

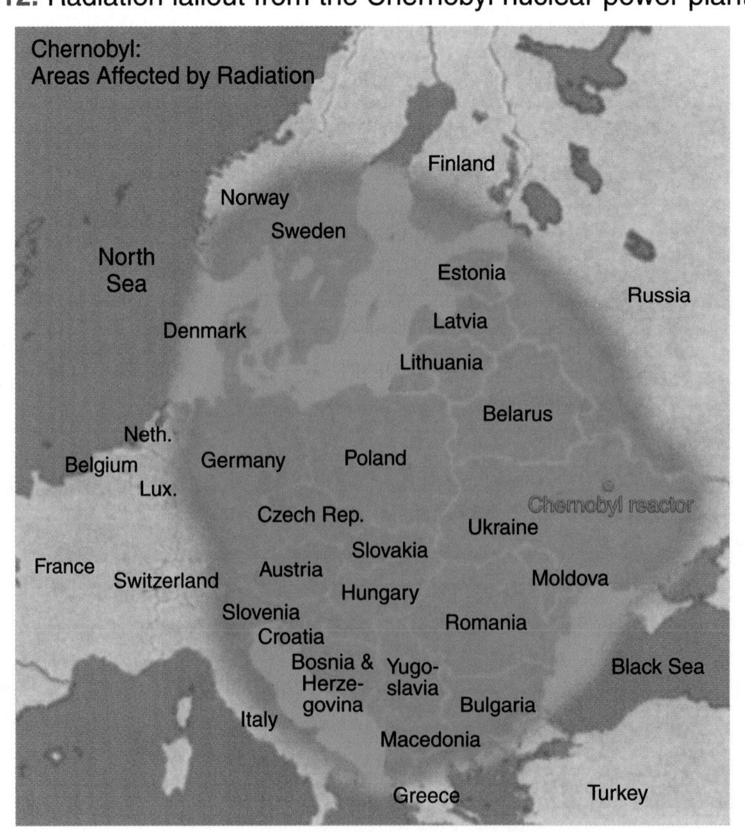

would be a weld that is designed to fail if the reactor reaches a certain temperature. When the weld fails, it allows emergency cooling water to flood the reactor core without the aid of pumps, sensors, or human intervention. As a result of improved safety procedures and techniques, a detailed risk assessment of the probability of a major accident at a nuclear power plant involving death is estimated at 1 in 1 million. See **Figure 4-13.**

Breeder Reacting

Breeder reactor technology was born in America. It has been pursued for development, however, in other countries more than in the United States. ***Breeder reacting*** is a process that could extend the life of our proven uranium reserves by hundreds or perhaps even thousands of

Breeder reacting: Creating nuclear fuel from a substance that is not fissionable.

Figure 4-13. Estimates of health damage from a nuclear reactor accident.

Estimates of Frequency of and Damage to the Public from Three Types of Major Accidents to Nuclear Reactors					
Accident Type	**Frequency (Chance per Reactor–Year)**	**Health Effects Within One Year**		**Total First Year's Health Effects Plus Delayed Health Effects**	
		Deaths	**Illnesses**	**Cancer Deaths**	**Genetic Defects (for All Generations)**
Core meltdown	1 in 20,000	Negligible	Negligible	3	75
Plus above-ground breach of containment	1 in 1,000,000	1	300	5000	3800
Plus adverse weather conditions and population density	1 in 1,000,000,000	3000	45,000	45,000 (~1500/y)	30,000

Career Connection

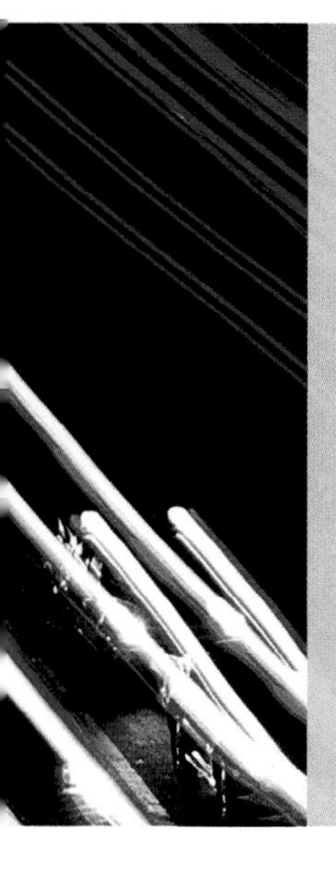

Nuclear Engineers

Nuclear power generates about 20% of the electricity in the United States. Because of this, engineers are needed to design, operate, and maintain the equipment used in nuclear power plants and reactors. The work done by nuclear engineers may also be put to use in other areas, however, varying from the military to medical technology.

Much of a nuclear engineer's time may be spent either developing or modifying technology. This work may involve conducting research, developing formulas, and performing experiments. The task must also stay within the limits of safety regulations.

A nuclear engineer must have a broad range of skills. Proficiency is required not only in science and math, but also in reading comprehension and writing. A nuclear engineer must be able to read technical journals and prepare reports, proposals, and instructional materials. A bachelor's degree is required for work in this field. The yearly salary may range from $58,000 to $111,000.

years. The technology of breeder reacting has been around since the early 1950s. In fact, the first electricity produced from a nuclear reactor was from a breeder reactor in 1951.

A breeder reactor is specifically designed to create nuclear fuel from a substance that is not fissionable. Imagine filling up your gas tank with 5 gallons of gas and 10 gallons of water and then driving around for a while. When you return home and check your tank, you have 14 gallons of gas! This is literally what breeder reacting can do. When uranium is mined out of the ground, about 99% of all uranium found is ***uranium 238 (U238)***, a nonfissionable element. Only the remaining 1% of uranium is U235, which is suitable for refining into nuclear fuel. A breeder reactor can create a fissionable fuel known as ***plutonium 239 (Pu-239)*** from U238. See **Figure 4-14.** This Pu-239 can then be used as fuel in other commercial nuclear power plants in the way that U235 is presently used. Thus, breeder reacting could have tremendous potential as an energy producer.

Uranium 238 (U238): A type of uranium that is a nonfissionable element. About 99% of all uranium mined is this type.

Plutonium 239 (P-239): A fissionable fuel created from uranium 238 (U238) by a breeder reactor.

Figure 4-14. Converting uranium 238 (U238) to plutonium in a breeder reactor.

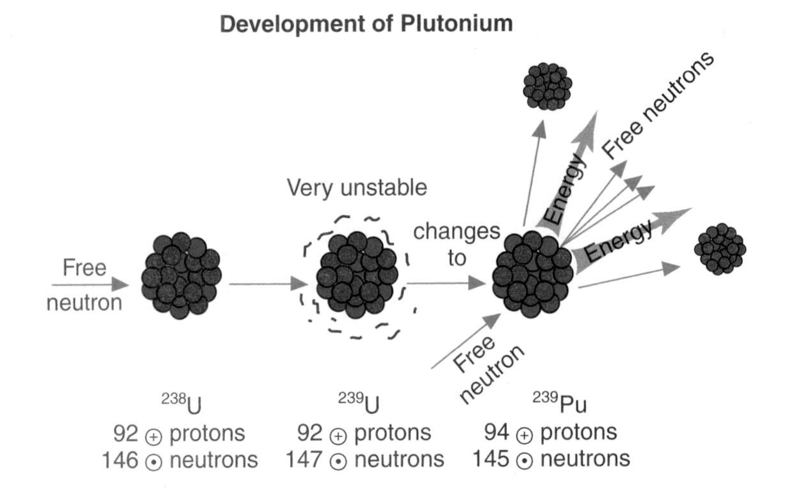

If it sounds too good to be true, it may be. Like all energy sources, there are trade-offs. One of the biggest trade-offs with breeder reacting technology is the fact that, while Pu-239 can be used as a fuel source, it is also the material of which nuclear bombs are made. Additionally, the production and consumption of Pu-239 yields some of the worst by-products known to humanity. The half-life of high-level waste extracted from Pu-239 used as nuclear fuel would be expected to last over 24,000 years. This waste would either have to be reprocessed into something far less harmful or isolated from society for thousands of years. The possibility of proliferation is frequently mentioned when breeder reacting is discussed. ***Proliferation*** is the illegal acquisition of nuclear fuel for potentially harmful purposes. Since plutonium is extremely difficult to produce, those wishing to acquire plutonium for illegal purposes would have incentive to try to acquire it through any means possible. The more plutonium produced, the more likely it would be that someone with harmful intentions could acquire plutonium to fashion into a crude nuclear bomb.

Proliferation: The use of by-products of nuclear power for the production of nuclear weapons.

This has not hindered other countries, such as Japan, Russia, and particularly pronuclear France, from pursuing and developing breeder reactor technology. Japan has been converting several of its more than 50 operating nuclear reactors to run on a mixture of U235 and Pu-239. The mixture is presently coming from reprocessing plants in France and the United Kingdom, thus creating some environmental controversy regarding shipping of the fuel. Japan has plenty of reason to pursue this technology. It has virtually no domestic energy resources. Japan's electrical demand continues to increase, however, at about 4% per year.

Nuclear Fusion

Nuclear fusion is considered by some knowledgeable scientists and technologists to be the ultimate energy source for electrical power generation. The initial concept of fusion, or the combining of light atoms, has been around as long as the nuclear fission process. As was described earlier, the fission process involves the splitting of an atom with a large nucleus, such as U235. The mass of the two smaller nuclei weigh less than the original nuclei. The loss of mass is converted into energy. Fusion is the combining of two nuclei into a larger nucleus. The large nucleus weighs less than the two smaller nuclei that formed it. The result of this process yields a large energy release. See **Figure 4-15.** One reason so many people are optimistic about the potential for nuclear fusion is that it represents a long-term solution to power production that cannot be matched by any other energy source, with the exception of solar energy. This is due in large part to the fact that the fuel source for nuclear fusion is found in water.

An ***isotope*** is a variation of an element. U235 and U238 are both isotopes of uranium. They have similar chemical properties and atomic structures, but their atomic weights vary slightly. Deuterium is an isotope of hydrogen. When found in water, its chemical configuration is H_3O instead of H_2O. Deuterium is found in about 1 out of every 6000–7000 hydrogen atoms. The fuel potential contained in deuterium found in just

Isotope: One of two or more atoms with the same number of protons but with different numbers of neutrons.

Figure 4-15. A fusion reaction involving deuterium (D) and tritium (T) results in the release of large amounts of energy. (Joint European Torus/European Fusion Development Agreement)

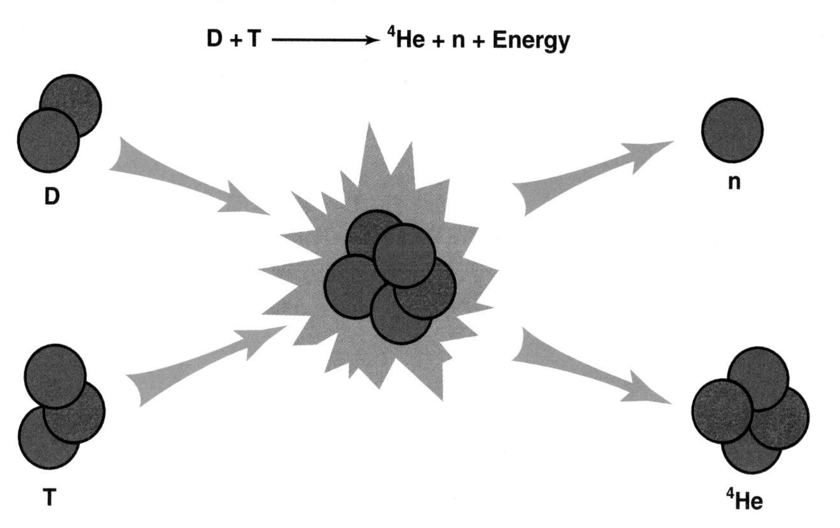

8 gallons of water is equivalent to the consumption of 2400 gallons of gasoline. Additionally, deuterium is easily and inexpensively extracted from water. Water is so plentiful, it is considered almost limitless.

A fusion reactor would be inherently safer than a fission reactor. Unlike the chain reaction of a fission reactor, the fusion reactor could be continuously fed with fuel for fusion. If a problem were to occur, simply closing the fuel feed could stop the fusion, similar to the way a candle goes out when starved for oxygen. In theory, this would make a fusion reactor much safer than a fission reactor loaded with a dangerous fuel.

Again, if it sounds to good to be true, it probably is. Fusion research has been ongoing since the 1940s. A fusion reaction that has yielded more energy than it has consumed, however, has yet to take place. Billions of dollars have been spent on research over six decades, without a reaction capable of the kind of sustained energy suitable for producing electricity. The technical requirements to sustain a fusion reaction include extremely high temperatures and confinement of materials. The temperature required for fusion is considered to be at least 50 million degrees Celsius. At these temperatures, the electrical bond between neutrons and ***electrons***, the negatively charged particles, is broken. The gas decomposes into free electrons and nuclei. This ionized gas with an equal number of positive and negative charges is known as ***plasma***. The plasma must be contained so the nuclei can fuse. Confinement at such high temperatures has proven difficult, however, since a heated gas expands to create great pressures. Confining the plasma using magnetism may yield the answer to a technological problem that has plagued fusion power development. The Tokamak Fusion Test Reactor, located at the Princeton Plasma Physics Laboratory, had some moderate success with fusion testing in the 1990s. A new type of magnetic confinement reactor began operation at Princeton in 1999. The world's largest tokamak is located in the United Kingdom. See **Figure 4-16.**

Electron: A negatively charged atomic particle.

Plasma: Ionized gas with an equal number of positive and negative charges.

Figure 4-16. The tokamak vessel for fusion reactions, shown in this cutaway view, must safely contain the great pressure resulting from plasma with a temperature of more than 100 million degrees Celsius. This tokamak, located in the United Kingdom, is called the Joint European Torus (JET) and is a cooperative venture involving scientists from 15 nations that are part of the European Fusion Development Agreement (EFDA). (JET/EFDA)

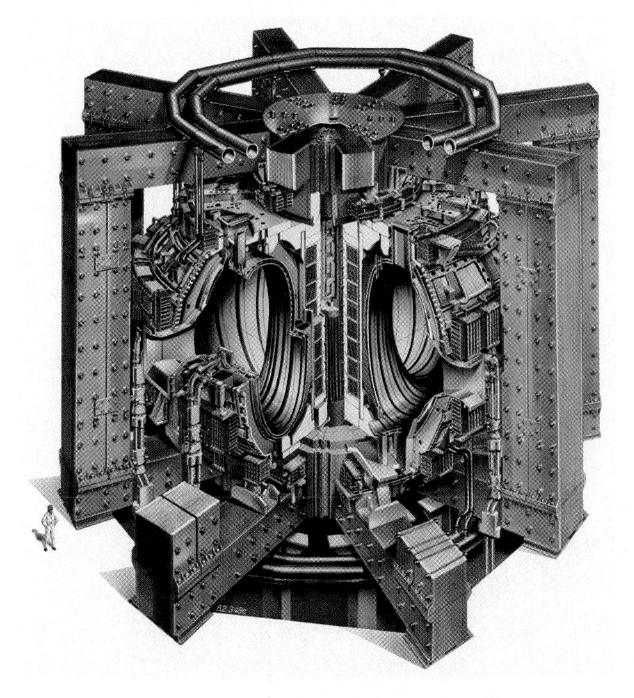

Advantages of fusion power over fission power include that it is a virtually limitless supply of energy, with little radioactive waste and no possibility of a runaway reactor. The advantages over fossil fuels include the fact that fusion power would not contribute to global warming and that fusion power plants would operate at much higher efficiencies than conventional power plants. For all these reasons, fusion power research may be worth pursuing and may be considered essential to our continued existence on this planet. It may just be a matter of time before fusion power becomes a reality.

Summary

A nonrenewable element known as uranium 235 (U235) is the primary source of nuclear fuel for power plants. The only nuclear process used to produce nuclear power in use in the United States today is known as nuclear fission. Fission refers to the splitting of atoms in order to convert matter into energy. This process produces tremendous energy for the amount of fuel consumed. It also produces radioactive by-products, however, that must be isolated from society for thousands of years until they reach half-lives that are much safer for the environment. There are about 100 operating nuclear power plants located throughout the United States. Most are concentrated in the Northeast region of the country. These nuclear power plants produce about 20% of all the electricity produced in the United States. Even though nuclear power plants have had a relatively safe track record in the United States, no new plants have been ordered since a major accident at the Three Mile Island (TMI) nuclear power plant in Pennsylvania in 1979. In addition to reactor safety, economic issues and concerns about technical solutions for the long-term storage or reprocessing of nuclear waste have hindered the further development of nuclear power. Even so, the potential for breeder reacting, which can make usable fuel out of elements that cannot be used as nuclear fuel in their natural states, and the long-term potential of nuclear fusion, which uses elements found in ordinary water, make the need to continue nuclear research almost inevitable.

Key Words

All the following words have been used in this chapter. Do you know their meanings?

- boiling water reactor (BWR)
- breeder reacting
- control rod
- electron
- half-life
- isotope
- Kyoto Protocol
- neutron
- nuclear fission
- nuclear fusion
- Nuclear Waste Fund
- Nuclear Waste Policy Act
- nucleus
- plasma
- plutonium 239 (Pu-239)
- pressurized water reactor (PWR)
- primary loop
- proliferation
- proton
- radioactivity
- secondary loop
- shipping cask
- Three Mile Island (TMI) accident
- uranium 235 (U235)
- uranium 238 (U238)
- Yucca Mountain storage facility

Test Your Knowledge

Write your answers on a separate sheet of paper. Do not write in this book.

1. Discuss in two or three sentences how nuclear fuel is produced.
2. Nuclear power is responsible for approximately _____% of all electricity produced in the United States.
3. *True or False?* The roots of the nuclear power industry can be traced to nuclear weapons.
4. *True or False?* Orders for new construction of nuclear power plants have been on the rise in recent years.
5. Recall in two or three sentences how nuclear fission is used to produce electrical power.
6. The term most often used to describe the rate at which a material loses half of its radioactivity is known as its _____.
7. *True or False?* Uranium 235 (U235) is the fuel most often used in nuclear power plants.
8. Rods used to regulate the rate of nuclear fission in a nuclear reactor are known as _____.
9. Summarize the major differences between BWRs and PWRs.
10. *True or False?* Most nuclear power plants in commercial operation in the United States are located in the western United States.
11. *True or False?* All high-level waste is presently stored in a federal repository.
12. Most spent nuclear fuel is stored in _____ at the _____.
13. *True or False?* Recycling nuclear waste is one way to reduce the possibility of proliferation of nuclear waste.
14. Write two or three sentences describing the present status of and future plans for storage of low-level and high-level nuclear waste in the United States.
15. *True or False?* The shipping casks used to transport nuclear waste are highly problematic and frequently leak.
16. Since the Three Mile Island (TMI) accident in 1979, reactors have been redesigned to rely primarily on _____ safety systems, which do not require human intervention.
17. Much has been learned about nuclear reactor safety since the accident at TMI. Describe advancements in nuclear power plant safety and design.
18. *True or False?* Breeder reactors can create more useful fuel than they consume.

19. *True or False?* Nuclear fusion is a process currently used to produce electrical power.
20. Identify the pros and cons of generating more electricity with nuclear power.
21. Compare the risks associated with generating electricity using nuclear power with risks of generating electricity from fossil-fueled power plants.

Activities

1. With an adult, tour a nuclear power plant, if one is located near your geographic area.
2. Participate in a classroom debate to discuss the pros and cons of nuclear power.
3. Research the disposal of nuclear waste within your state.

An Osprey vertical or short takeoff and landing (V/STOL) aircraft is readied for flight aboard a U.S. Navy vessel. The rotors atop the two engines are folded to minimize space needed for storage between flights. The V/STOL technology is currently used for military purposes, but it is expected to be used for civilian aircraft in coming years. (U.S. Navy)

5

Renewable and Inexhaustible Energy Sources

Basic Concepts

- Identify the basic sources of renewable energy and inexhaustible energy.

Intermediate Concepts

- Describe at least three different methods of producing power from renewable energy resources.
- Provide at least six examples of different methods of producing power from inexhaustible energy resources.

Advanced Concepts

- Summarize several factors that influence the development of an energy resource.
- Explain advantages and disadvantages of various conventional and alternative energy resources.
- Discuss the environmental consequences most closely associated with the use of various conventional and alternative energy resources.

It is important to study energy in order to make the best use of our energy resources and to preserve them for future generations. Alternative energy resources are in various stages of utilization. Some, such as hydroelectricity, have been in use for years. In this sense, hydroelectricity is not really an alternative. Other energy resources, such as wind energy, are only now beginning to make a small contribution to the world energy mix. A few of the resources discussed are not used extensively at this time, but they may play a much greater role in the future.

Most energy has been stored in nature and can be traced back to the sun. For instance, the sun stores energy in plants and trees. This stored energy is released when plants or trees are burned. Plants and animals that have died millions of years ago also have stored energy. Today, we use this stored energy in the forms of coal, oil, and natural gas. These are just a few examples. Energy comes from many sources. These sources, as you learned in Chapter 2, have been divided into three categories:

- Nonrenewable energy sources.
- Renewable energy sources.
- Inexhaustible energy sources.

We will study energy resources in this chapter by classifying them as either renewable or inexhaustible. Nonrenewable energy sources were explained in Chapter 3.

Renewable Energy Sources

Renewable energy resources are those energy sources that can be replaced as needed on a relatively short-term basis. See **Figure 5-1.** This energy comes from plants and animals. Management of renewable energy sources is critical to their effective use. Once a tree is cut down, it is important to replant one to sustain this form of energy. The basic renewable energy sources include animals, food, wood, and alcohol, including methanol. See **Figure 5-2.** Energy can also be created from waste products, in a process called bioconversion.

Figure 5-1. This forest is an example of a renewable energy resource. Trees can be harvested, and new trees can be planted for future harvesting.

Animals

Animals were used as a main source of energy for work in this country into the 1900s. They are still used to produce power for farming in many third world nations and in certain cultures within the United States. Animals are capable of pulling heavy loads on sleds or wheeled vehicles that humans are not able to pull. See **Figure 5-3.** Animals served as a major source of mechanical power until the development of mechanical machines. In the United States and other developed nations, the development and refinement of the external and internal combustion engines slowly led to the replacement of using animals for work. Animals are also used for food.

Food

Food is another source of energy. Physical labor, exercise, and recreation all require energy. Energy from food also keeps our bodies warm. Our bodies use the energy from the food to produce heat, which keeps our bodies at the correct temperature. Food is a renewable source of energy because it can be regrown.

Figure 5-2. Basic renewable energy resources. A—Animals. (U.S. Department of Agriculture) B—Food. (U.S. Department of Agriculture) C—Wood. (U.S. Department of Agriculture) D—Alcohol. (DaimlerChrysler)

Figure 5-3. Less than 100 years ago, animals were the energy source for most farmwork in the United States. In many countries, animal power is still widely used. This photo shows mules being used for plowing in the 1920s on a farm in the American Midwest. (U.S. Department of Agriculture)

The food supply continues to sustain us. Farmers plant and harvest their crops, including corn, soybeans, wheat, or other plant life. See **Figure 5-4.** Some foods not only supply humans with energy, but are also used to supply vehicles with energy. For example, ***gasohol*** is made from grains to fuel automobiles. It is described more extensively in the section on alcohol fuels. Food product waste can also be burned or converted into fuel. Successful experiments in producing fuels from gardening and agricultural wastes have been performed. They are described in the bioconversion section later in this chapter.

Gasohol: An automobile fuel made from grains.

Wood

Wood is a very old source of energy. It has been used as an energy source since the prehistoric times, and it continues to be used as an energy source today. For many years, it has been used to heat homes and for cooking. By the early 1900s, the use of wood as a fuel declined. Other sources of energy that offered more energy per volume, such as oil and natural gas, were beginning to be used. Today, however, you can find wood-burning stoves and fireplaces in modern homes. See **Figure 5-5.** Wood pellet stoves are also a popular source of supplemental heat for some people. Once again, wood has become a useful source of energy.

When harvesting wood, the wood must be split; transported; dried for optimal efficiency; stacked; and manually loaded into a fireplace, stove, or furnace. Burning wood requires periodic stoking of the fire, and temperature control within the structure can be difficult to regulate. Depending on availability, wood can be an inexpensive energy source in comparison to other fuels. Denser woods, like oak and hickory, tend to yield a greater amount of energy per volume than many other species of wood.

Figure 5-4. The yearly cycle of planting, growing, and harvesting crops provides an abundant supply of food to serve as an energy source for people and animals. Some crops, such as these soybeans, can be used to produce other types of energy sources, such as biodiesel fuel for vehicles.

One major disadvantage of burning wood is that it produces great amounts of air pollution. Wood does not burn as cleanly or efficiently as other energy sources. The fact that it is burned by individual homeowners and not burned at a central location the way coal is burned at a power plant means it is less subject to environmental regulation. It is also less subject to pollution control devices. Woodstoves require periodic cleaning and disposal of ashes and are regarded as a more dangerous method of heating than other methods. Many residential fires occur each year as the result of woodstoves and fireplaces. ***Creosote*** is a tarlike substance that can build up on the walls of a chimney when wood is burned. Wood-burning stoves are particularly susceptible to creosote buildup. See **Figure 5-6.** On the positive side, wood is a renewable energy source that can be harvested regularly, if properly managed.

Figure 5-5. Modern wood-burning stoves have an efficient design that provides a pleasant, comfortable heating source. In certain situations, wood can be a less expensive heating source than other fuels. (Hearth and Home Technologies—Quadra-Fire)

Wood can also be converted to another hydrocarbon fuel. It can be changed into a liquid and used to fuel vehicles. One type of wood-based fuel that is being used is methanol fuel, which is further explained later in this chapter. If wood were to be used as a large-scale energy source, it is possible that the competition between using wood as an energy source and as a construction medium could increase. Large forests would need to be planted and maintained in order to have enough wood for all the uses.

Creosote: A tarlike substance that can build up on the walls of a chimney when wood is burned.

Alcohol

Alcohol is a liquid hydrocarbon that may be used as fuel. It is made from different crops, such as corn, sugar beets, and sugar cane. It can be used to power internal combustion engines, including those in

Figure 5-6. Since flammable creosote builds up in the chimneys of wood-burning stoves and fireplaces, periodic inspection and cleaning are important. Chimney sweeps do this work. The history of the chimney sweep traces back hundreds of years. This instructor at the Chimney Safety Institute of America (CSIA) annual Chimney Sweep School is explaining proper fireplace cleaning procedures. (CSIA)

automobiles. In some foreign countries, such as Brazil, automobiles operate with 100% alcohol fuel. In the United States, we use a mixture of alcohol and gasoline, known as gasohol, to fuel some vehicles. Gasohol is a mixture of unleaded gasoline and ethyl alcohol, or ***ethanol***. Ethanol is sometimes referred to as grain alcohol. Gasohol does not, however, provide as much energy per volume as gasoline.

Ethanol

Ethanol is produced from biomass by the conversion process known as ***fermentation***, similar to the way alcohol made for consumption is produced. Fermentation involves the decomposition of carbohydrates found in plants with the production of carbon dioxide (CO_2) and acids. An ethanol plant is a plant where the crops are distilled and processed into alcohol for use as fuel. The alcohol is then often mixed with gasoline. Gasohol is used to power many automobiles today. It is a mixture of approximately 10% ethyl alcohol and 90% gasoline. Ethanol can be refined from cornstalks and other domestically grown plants.

Methanol

Methanol is a clean-burning liquid. It is also known as ***methyl alcohol***. Methanol can be made from nonrenewable sources of energy, such as coal, but it can also be made from renewable sources of energy known as biomass, such as wood, plants, and manure. It is used as a fuel to power vehicles. Methanol produces more energy per volume than ethanol. Therefore, it does not need to be mixed with gasoline to produce good power. Methanol also burns more slowly than gasoline. The redesign of engines specifically engineered to run on methanol has given methanol the ability to produce as much power as gasoline. Because methanol burns more slowly, it has smoother engine performance than gasoline. Methanol is being substituted for gasoline in some transportation systems, but a large infrastructure for the refinement and distribution of methanol is not yet in place in the United States. Furthermore, the amount of land that would have to be occupied to harvest plant products specifically for conversion to methanol would be tremendous, if the United States or any other country attempted to replace gasoline with methanol on a large-scale basis.

Bioconversion

The process that produces energy from the waste products of our society is known as ***bioconversion***. The waste is known as ***biomass***. Biomass is organic material, such as trees, plants, grains, and algae. Wastes, such as manure, garbage, sewage, and paper, are also sources of biomass. All these sources of biomass can go through a bioconversion. They can be burned or converted into alcohols, such as ethanol and methanol. Biomass conversion can also yield petroleum substitutes and methane gas.

A ***methane digester*** converts shredded organic materials into methane gas that can be used for heating, used for power generation, or purified and stored for distribution. See **Figure 5-7.** The type of bioconversion used by a methane digester is also known as ***anaerobic digestion***, or decay, which refers to decay without the use of oxygen. A less complex way of using biomass as an energy source is simply to burn the biomass.

Ethanol: Ethyl alcohol, sometimes referred to as grain alcohol.

Fermentation: The decomposition of carbohydrates found in plants with the production of carbon dioxide (CO_2) and acids.

Methanol: A clean-burning liquid used as fuel to power vehicles. It can be made from nonrenewable sources of energy, such as coal, or from renewable sources of energy, such as wood, plants, and manure. Methanol produces more energy than ethanol, per volume, and burns more slowly than gasoline.

Methyl alcohol: See *Methanol*.

Bioconversion: The process that produces energy from the waste products of our society.

Biomass: Waste products that can be used in bioconversion. Examples are organic material (such as trees, plants, grains, and algae) and wastes such as manure, garbage, sewage, and paper.

Figure 5-7. A methane digester produces usable methane gas from decaying organic matter.

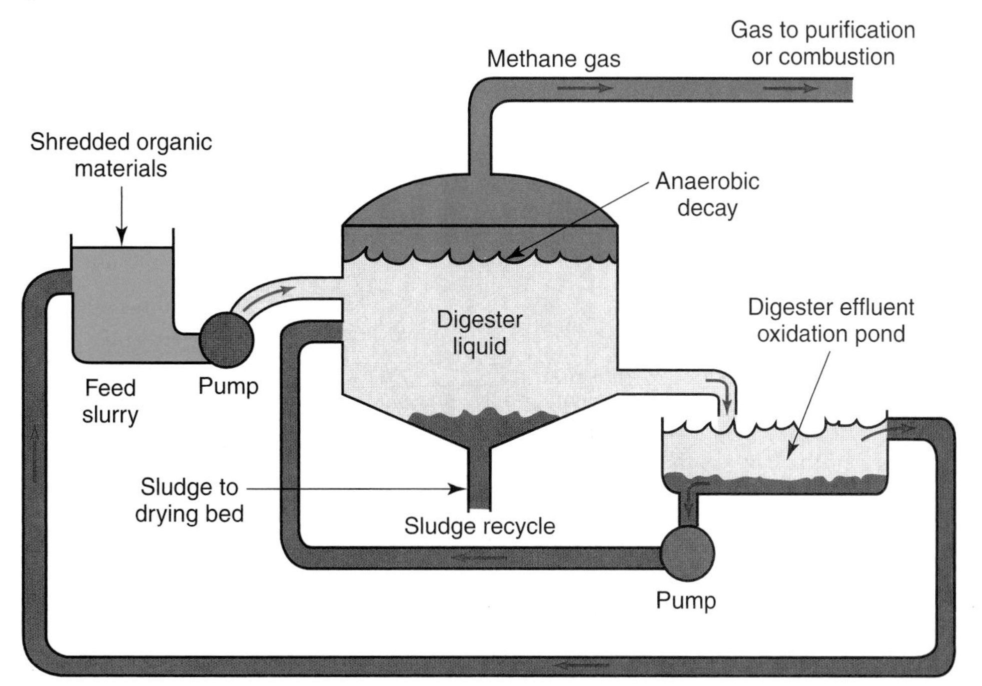

Methane digester: A vessel that converts shredded organic materials into methane gas, which can be used for heating, used for power generation, or purified and stored for distribution.

Anaerobic digestion: Decay without the use of oxygen.

Several municipalities throughout the United States now shred and burn waste. See **Figure 5-8.** The heat energy is then used to produce electricity or for industrial processes. This type of bioconversion also helps solve a growing societal problem. As landfills reach capacity, there is a growing concern of how to get rid of unwanted waste generated by society. These ***waste-to-energy plants*** have gained in popularity in recent years because they are economically viable.

Waste-to-energy plant: A plant that shreds and burns waste. The heat energy is then used to produce electricity or for industrial processes.

Inexhaustible Energy Sources

Inexhaustible energy sources are those sources of energy that will never run out. We are fortunate to have such energy sources, and it is in the best interest of our nation and the world to learn how to develop them for long-term use. The roots of most inexhaustible energy sources can be traced back to solar energy. It is solar energy that produces the changes in temperature that ultimately create tides and winds. The most frequently used of the inexhaustible energy sources is hydroelectricity, but others include geothermal energy, the wind, tides, Ocean Thermal Energy Conversion (OTEC), and hydrogen. Since forces beyond our control renew these sources, these sources are considered to be inexhaustible.

Hydroelectric Energy

Years ago, flowing water was used as a source of energy. See **Figure 5-9.** Waterwheels have few applications in contemporary America, but they were a major source of mechanical power for factories of colonial America. Water tapped from a nearby river is diverted into a sluiceway. A sluiceway is a channel that carries the water to a waterwheel. The force

Figure 5-8. Municipal waste can be converted directly to energy by burning it. The heat energy can then be used for industrial processes or electrical generation. This waste-to-energy plant also recovers metals from the shredded waste before the burnable waste is sent to the furnace. (Union Electric Company)

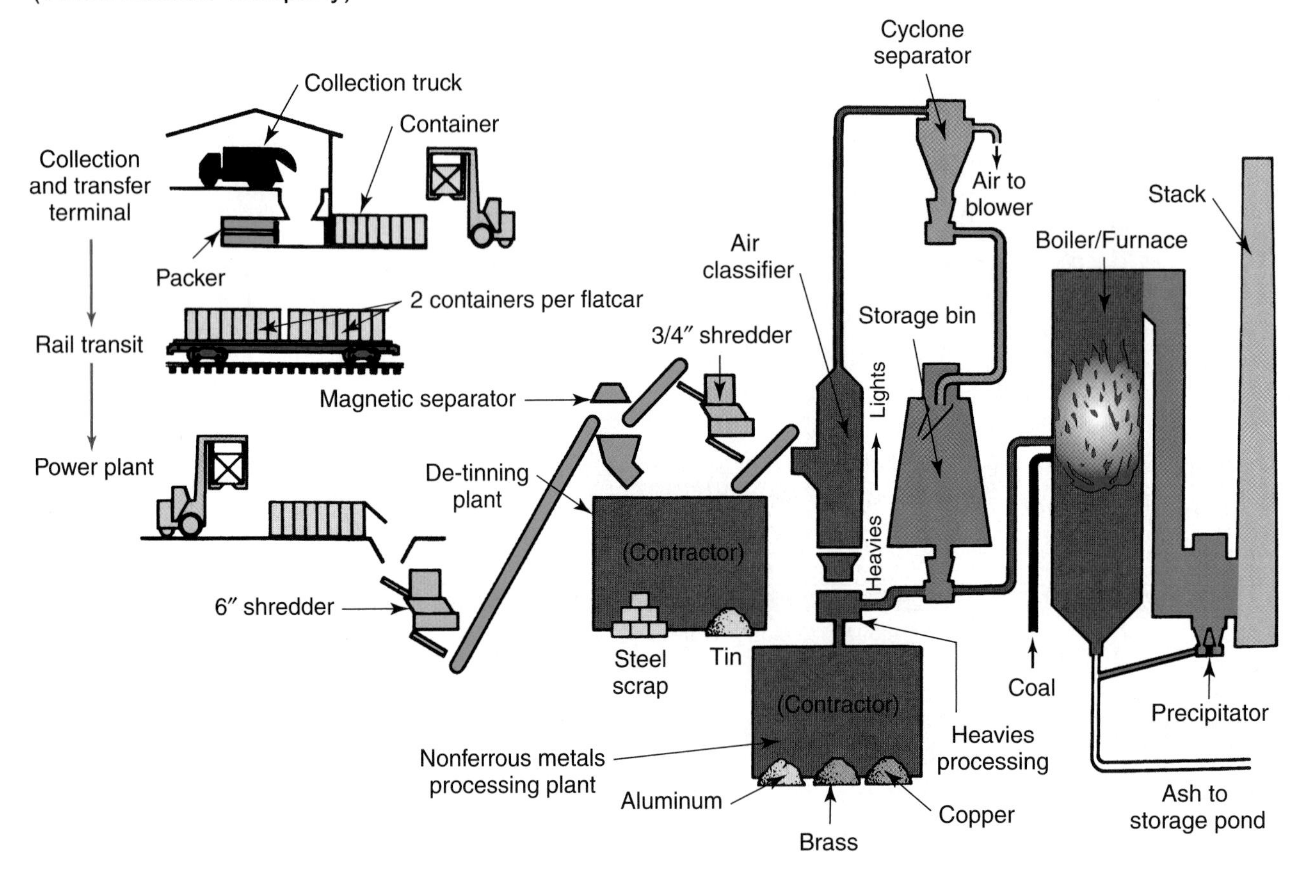

Figure 5-9. For thousands of years, waterwheels have been used to harness the energy of flowing water. This large wheel is part of a restored gristmill located alongside a creek. The wheel is more than 10′ in diameter.

Technology Link

Construction: Preventing Infiltration

Insulating products help seal a structure to prevent infiltration of cold air in the winter and hot air in the summer. Infiltration is discussed in terms of *air exchanges per hour*. One air exchange per hour is approximately equivalent to all the air within the heated or cooled structure escaping and being replaced by fresh air at the outside temperature. Most buildings experience one to three entire air exchanges per hour. By reducing the number of air exchanges per hour, infiltration products conserve energy and reduce the associated costs to the building owner. The following energy conservation products help to protect structures from heat loss or cool air loss by infiltration:

- *Caulk* is one of the most common and least expensive infiltration prevention products available. Silicone-based caulks tend to have exceptional adhesive properties to a variety of building materials, including wood, metal, and glass.
- *Expandable foam* is designed for gaps that are too large to seal effectively with caulk. Typical opportunities to use expandable foam often occur around the outsides of windows and door frames, as well as where the sill plate meets the foundation in older buildings. Manufacturers of expandable foam recommend practicing before using foam on a live application, as it continues to expand for a period of time after it has been applied to a gap. It is easier to trim dry foam with a utility knife than to try to manipulate the foam when it is wet.
- *Weather-tight doorstop molding* is similar to similar doorstop molding, but it has an additional foam rubber bead attached. Attach weather-tight molding so the foam bead is wedged snugly between the door and frame, creating a tight seal.
- *Weather-tight threshold* is used to replace older thresholds to prevent a gap underneath an exterior door.
- *Switch and receptacle sealers* are inexpensive and effective products for preventing infiltration. In a wall, junction boxes take up the space where insulation would normally be placed, so switch and receptacle sealers compensate for the lack of insulation and block the path for infiltration through the drywall opening.
- *Infiltration wraps* are a means of protecting exterior walls from infiltration during new construction. They are installed starting at one corner of the building and then literally wrapping the structure like a present. While doors and windows are cut out of the wrap, the rest of the walls have an extremely effective infiltration barrier.
- *Sill seal gaskets* are used during new construction, when they are applied to the top of the foundation, prior to erecting the exterior walls. When the walls are bolted to the foundation, the sill seal gasket is compressed between the concrete and the wall to tightly seal this notoriously leaky area.

of the water is used to spin the wheel, creating mechanical power. This mechanical power was often transferred into a factory by a shaft attached to the waterwheel. From there, it was connected to machinery, often by a leather belt. This allowed many machines, such as saws, mills, and lathes, to be powered at a time that predated the electric motors often used to power such machinery today. There were two primary types of waterwheels. See **Figure 5-10.**

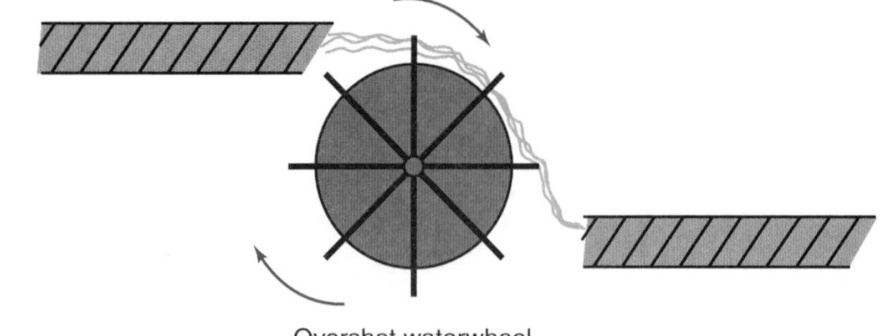

A Overshot waterwheel

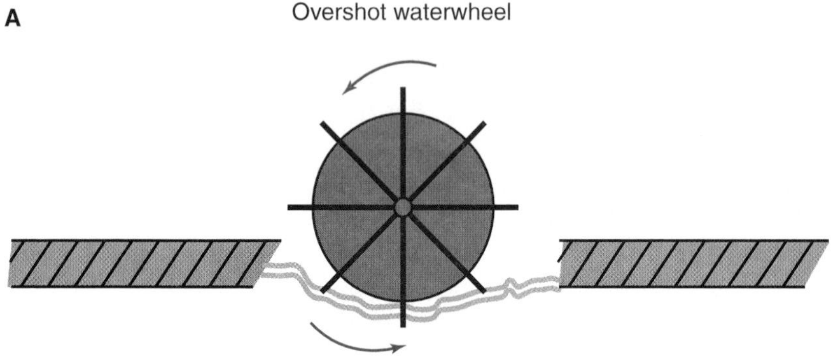

B Undershot waterwheel

Figure 5-10. The two types of waterwheels. A—Overshot waterwheels are driven by water falling from above. B—Undershot wheels are driven from below.

Overshot waterwheel: A waterwheel that relies on an elevation change and makes use of the weight of the water, in addition to the water's force.

The ***overshot waterwheel*** relies on an elevation change and makes use of the weight of the water, in addition to the water's force. The ***undershot waterwheel*** does not require a significant elevation change and primarily makes use of the force of the flowing water. Although flowing waters do not power manufacturing plants anymore, flowing water is still a valuable source of energy. ***Hydroelectric energy*** is the use of flowing waters from waterfalls and dams to produce electricity. See **Figure 5-11.**

Figure 5-11. Hydroelectric power can be even generated by small rivers and low dams. This plant, operated by a small midwestern city, generates enough electricity to operate a sewage treatment facility serving several adjoining communities.

How a hydroelectric plant works

Hydroelectricity cannot really be considered an alternative form of energy because it produces a significant amount of U.S. energy. Dams are used to trap and store water. See **Figure 5-12.** A gate lets the water in. When the gate is open, the water rushes through a tunnel known as a ***penstock***. At the end of the penstock is a water turbine. The turbine turns as the water comes rushing in, causing the generator to spin. The mechanical energy that spins the generator is converted into electricity. The electricity is then sent off to power companies through electric lines and distributed to customers. This process is described in much greater detail in Chapter 8.

Undershot waterwheel: A waterwheel that does not require a significant elevation change and primarily makes use of the force of flowing water.

Hydroelectric energy: The use of flowing water from waterfalls and dams to produce electricity.

Penstock: A dam tunnel through which stored water rushes to a water turbine.

Figure 5-12. Hydroelectric power generation. A—Very large dams, such as the Hoover Dam in Nevada, impound huge amounts of water for use in electrical generation. (U.S. Department of Agriculture) B—This schematic shows how water stored behind a dam is used to drive turbines and electrical generators.

A

B

To customer
Generator
Gate
Dam
Turbine
Water in
Penstock

Curricular Connection

Social Studies: Building the Hoover Dam

The Hoover Dam and Hoover Powerplant produce 4 billion kilowatt-hours (kWh) of hydroelectric power annually. The power generated provides electricity for Nevada, Arizona, and California. Upon its completion, the Hoover Dam was the largest dam ever constructed and remains one of the world's largest dams.

Construction of the Hoover Dam began in 1931—two years into the Great Depression. After the stock market crashed in 1929, an economic depression began that affected the entire United States. Before the Great Depression, the unemployment rate was about 3% of the population. Once the economy was affected by the Great Depression, it is estimated that about 25% of the U.S. population was unemployed. With so many Americans out of work, assembling construction crews for the Hoover Dam was not a difficult task. Men came from all over the country with their families and all their possessions, hoping to find employment with Six Companies, Inc., to help build the dam.

Thousands of people and families moved to the area around the construction site for jobs. Initially, everyone lived in tents within camps. Safety and sanitation quickly became a serious concern with so many people making due with so little. In response, officials expedited plans to build a town for the workers. The town included housing, churches, a school, and other structures. Though this town was specifically built to accommodate the laborers and their families during the Hoover Dam construction, the town still exists today and is known as Boulder City, Nevada.

During the dire economic times of the Great Depression, the construction of the Hoover Dam offered people hope. Even though the working conditions were not always ideal and the pay may not have been considered "fair," consistent work meant steady income. This kind of financial stability was uncommon at the time. Families stayed together and developed a community that continues to grow to this day.

Hydroelectricity provides a vital percentage of electrical generating capacity in the United States. To date, almost all good opportunities for large-scale hydroelectric production have been developed, although Canada still has some excellent potential sites for future development. There is also considerable potential for the development of small-scale hydroelectric development in the United States. The generating capacity of most sites would, however, be quite limited.

The fact that hydroelectric plants can be taken off-line and brought on-line quickly is a great advantage. If the flow of the river is not strong, utilities will often let the water build up behind their hydroelectric dams all night long, to run generators only during the working day, when they are needed to meet maximum load capacity. This technique has proven so effective at meeting the peak load demand of some utilities that they have invested in hydroelectric power plants designed only to meet the peak demand. These peaking plants can actually consume more electricity than they generate, but they are still economical. Peaking plants pump water uphill all night long, when there is excess power in the electrical grid produced by fossil fuel– and nuclear-powered generating facilities. The water is then allowed to flow back down through a penstock to spin a turbine generator. The reservoir may only be large enough to provide generating capacity for several hours per day. The plant may be brought on-line quickly, however, when power is needed to meet peak demands. Even though this type of plant operates at a net energy loss, it is relatively inexpensive to construct, compared to a base-load plant that produces power all the time. The fact that this type of plant allows for supplemental power when the utility needs it most makes it economical.

Hydroelectricity and the environment

Hydroelectricity produces no fossil fuel pollutants, and the facilities are fairly quiet and perhaps less obtrusive than other generating plants. Hydroelectric power plants, however, are polluting. They just produce a different form of pollution by having the potential to alter the ecology of rivers.

If a river is dammed, migratory fish may not be able to return upstream to their spawning grounds. To counteract this problem, a series of ***fish lifts*** were installed in the Susquehanna River in Pennsylvania in the 1990s to help American shad return upstream. The fish lifts are in operation for only a two- to three-week period out of the year. They are like giant elevators that lift the fish and water up onto the top of the dam, where they dump into a sluiceway. The fish can then swim again until they reach the next dam, several miles upstream, where another fish lift has been installed. They instinctively swim toward a downstream current, which is artificially created during the times the fish lift is in operation. The fish lift is an example of the extent to which utilities may be required to accommodate nature in exchange for damming the river. The utilities may also be responsible for maintaining the river level at a safe depth for recreational boating and water sports. Draining the river too quickly could also create fish kills, as fish and other aquatic life become trapped in areas that could stagnate and evaporate if not replenished with flowing water.

Fish lift: A giant elevator-like device installed on dammed rivers to help fish return upstream during spawning season.

Tidal Energy

Tidal energy is another form of hydroelectric energy that offers tremendous generating potential if it can be harnessed. The gravitational pull of the moon and the rotation of the earth cause the changing tides. Significant tide changes occur about twice daily. There are only a few places in the United States where the tide change is so significant that it can be used to generate power.

How a tidal power plant works

The concept of a tidal power plant is simple. Make use of the flowing tide to spin a generator in one direction during the rush of high tide, and then spin the generator in the opposite direction with the exiting water during low tide. Of course, a natural bay that could be sealed by a permanent or floating dam known as a ***barrage*** is most ideal for using tidal power. See **Figure 5-13.** The extreme capital cost of tidal energy and the lack of ideal locations are the primary limitations to more extensive use of

Barrage: A permanent or floating dam that seals a natural bay in order to use tidal power.

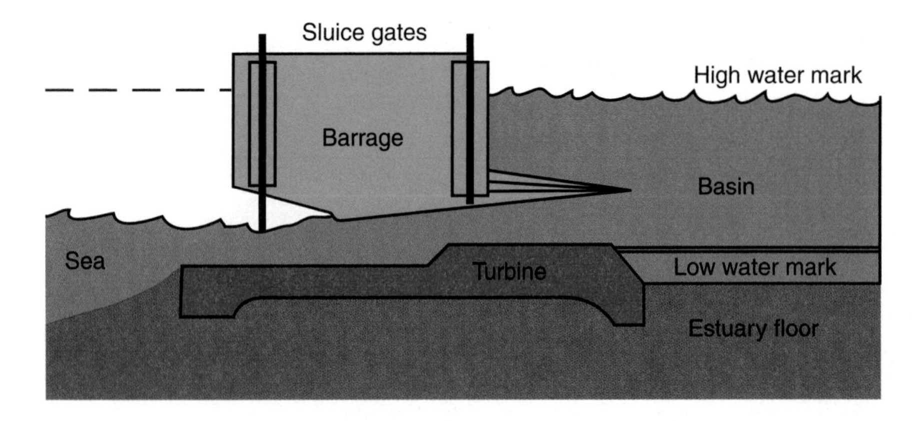

Figure 5-13. Tidal power can be generated by allowing the incoming tide to be impounded behind a dam (called a barrage). As the tide recedes on the seaward side of the barrage, the trapped water is allowed to flow out, rotating a turbine to generate electricity.

tidal energy. Factors in favor of using more tidal energy include that it is environmentally friendly, in comparison to other forms of power production, and the cost of maintenance and operation is low. The tidal energy is essentially free for the taking if it can be captured. Underwater ***tidal fences*** would have less impact on the environment, as they do not require flooding a basin to generate power.

Tidal fence: A barrier intended to prevent the power of tides from escaping back into the ocean.

Tidal energy and the environment

There is only one major environmental concern associated with tidal energy. It has to do with a possible ecological disturbance. The use of dams or barrages may create disturbances to the surrounding ecological environment.

Geothermal Energy

Geothermal energy: Heat from the earth.

Mantle: Rock that conducts heat coming from the earth's core.

Magma: Molten rock located miles beneath the earth's surface.

Geothermal energy is heat from the earth. *Geo* means "earth," and *thermal* relates to heat. The heat from the earth's core continuously flows outward. It is conducted by the surrounding rock known as ***mantle***. The mantle becomes molten and is then known as ***magma***. Magma is located miles beneath the earth's surface. Heat from magma is trapped underground. Sometimes, magma erupts from the earth in a volcano. Underground water turns to steam when it comes near this molten rock. Great amounts of steam are produced from the earth. The high pressure from the steam is used to turn turbines connected to generators. The generators then produce electricity. The production of electricity is the greatest use of geothermal energy. The United States leads the world in installed capacity of geothermal plants. Further development of geothermal energy for electrical generation is limited to locations where the heated steam is close enough to the earth's surface in order to make it economically recoverable.

Heat pump: An application of the use of geothermal energy for residential heating and cooling. A system of these pumps consists of pipes buried in the shallow ground and located near a home. Water flows through the pipes and conducts heat from the ground for transfer into the home.

How a geothermal heat pump works

Heat pumps are another application of the use of geothermal energy. They are used extensively for residential heating and cooling. The geothermal heat pump is based on the fact that ground temperature remains an approximate 55°F. A geothermal heat pump system includes pipes buried in the shallow ground and located near a home. Water flows through the pipes and conducts heat from the ground for transfer into the home. If the water enters the home at 55°F and that heat can be captured, less conventional energy is needed to heat the home. See **Figure 5-14.** The energy from the earth reduces heating costs significantly, but the system has another big advantage over more conventional heating systems. It can be reversed and used to provide central air conditioning in the warm summer months.

Hydrogen sulfide gas: A gas that smells like rotten eggs, found within the steam of geothermal energy plants.

Geothermal energy and the environment

Geothermal energy plants have raised the ire of a number of environmental groups that regard them as unsightly, smelly, and extremely noisy. The ***hydrogen sulfide gases*** within the steam smell like rotten eggs, and the minerals commonly found in steam with little moisture have the capability of poisoning lakes and streams if they become airborne. The moisture from wet steam plants tends to be heavily laden with salt, which can create maintenance problems, due to buildup, and waste disposal problems, once the steam is brought aboveground. On a positive note, geothermal plants emit far less CO_2 into the environment than burning fossil fuels, and they make use of a heat source that is present for the taking.

Wind

For years, people built windmills in places where the wind blew much of the time. See **Figure 5-15.** The windmills pump water, grind grain, or do other useful tasks. Unfortunately, the wind has never been a totally

Figure 5-14. A geothermal heat pump has its evaporator coil buried in the ground. Refrigerant in the evaporator coil picks up heat from the ground (even in cold weather) and is carried inside, where it passes through a compressor. In the condenser coil, the heat is given up to the flow of indoor air, warming the indoor spaces.

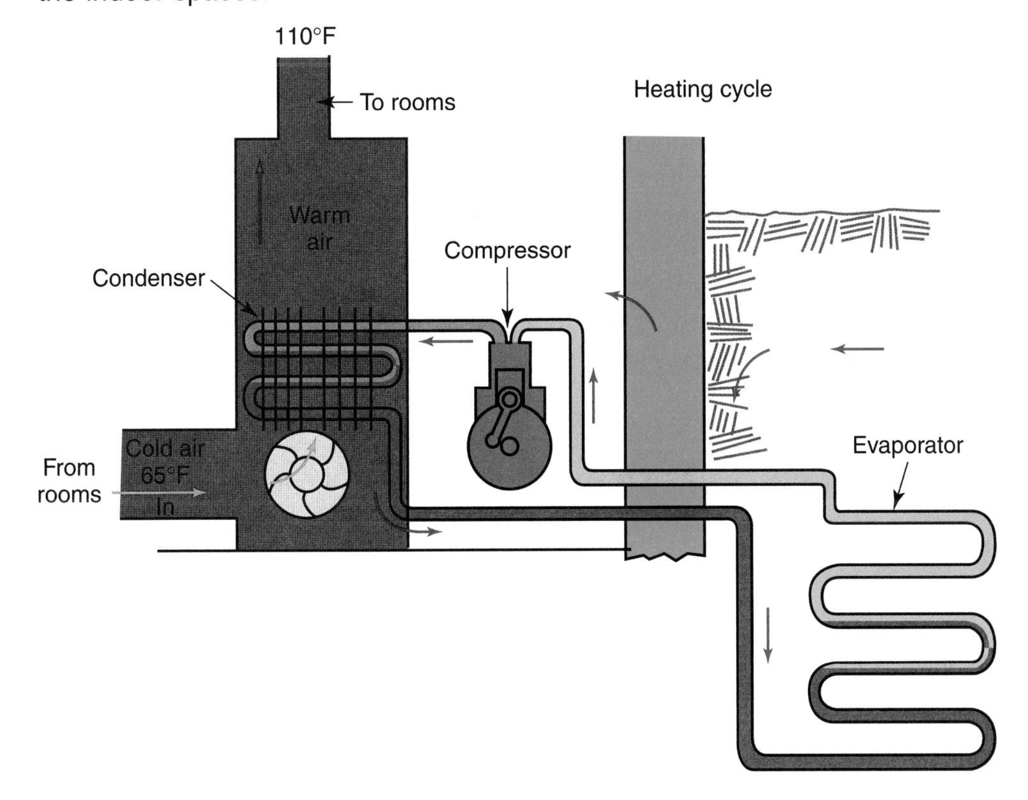

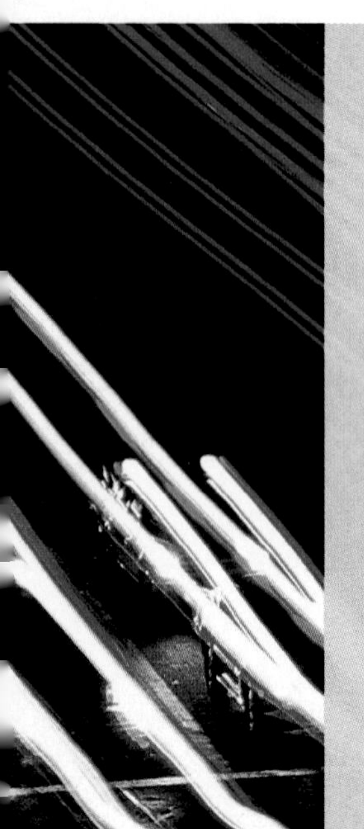

Career Connection

Meteorologists

The function of meteorologists is to give accurate predictions of the weather. The aviation industry depends extensively on weather reporting. Air traffic controllers and pilots need to be kept aware of changing weather patterns for the safety of all passengers. Meteorologists' predictions are typically broadcast on television or radio stations for public use and interest. In order to provide up-to-date information, meteorologists analyze reports from specialized equipment. Using the technology of radio detecting and ranging (radar) and satellites, meteorologists can forecast both short- and long-term climate activity.

Knowledge of physics, geography, and math is essential for a meteorologist. Also important are communication skills, particularly the ability to speak clearly. Meteorologists frequently collect data useful in energy, power, and transportation. For instance, average wind speed charts can indicate which locations would be most favorable for a power company to consider using for wind power generation. Rain patterns can be used to help predict generating capacity for hydro plants, and accurate weather information is essential to most of the transportation industry. A bachelor's degree is required to be a worker in this field. The yearly salary may range from $30,000 to $92,000.

Figure 5-15. Windmills were fixtures on farms for many years, primarily providing energy to pump water. Much less efficient than today's wind generators, they were almost totally replaced by fossil fuel power sources during the second half of the twentieth century.

reliable energy source. Windmill usage decreased when fossil fuels that offered more convenience became available.

Today, the windmill is back in use, but it is mainly used to generate electricity. A way to harness this free energy produced by the wind is by a ***wind turbine***. Today, most windmills are referred to as wind turbines. A wind turbine has two different designs. See **Figure 5-16.** In both designs, the wind drives a propeller, or turbine, connected to a generator. The wind makes the turbine turn the generator, which produces electricity.

Proponents of wind as an energy source view it as an environmentally friendly alternative to fossil fuels. Generally, wind speeds in excess of 11 mph are necessary for any type of wind turbine connected to the power grid. Large-scale wind generators, such as those found on wind farms, require wind speeds of at least 13 mph, as determined by a ***wind velocity profile***, which characterizes the number of expected hours of a given wind speed for a particular location. This limits the potential for wind farming in some areas of the country, but wind energy is contributing to overall generating capacity of more than half the states in America. Other countries are tapping the potential of the wind at a faster growth rate than here in the United States.

Figure 5-16. Wind farms make use of many high-capacity wind turbines to generate significant amounts of electrical energy. (Shell Wind Energy)

Wind turbine: A propeller driven by the wind and connected to a generator. The wind makes it turn the generator, which produces electricity.

Wind velocity profile: Data characterizing the number of expected hours of a given wind speed for a particular location.

How a wind-turbine generator works

Regardless of design, all wind generators work essentially the same way. They all have some type of blades that capture the force of the wind. Unlike the windmills of previous eras, most modern wind generators are of a two- or three-blade design. See **Figure 5-17.** This is due to research that indicates multiple blades actually capture less wind than two- or three-bladed designs. The reason is that, every time a blade cuts through the air, it creates some turbulence or swirling air that opposes the penetration of the next blade. Wind turbines that have fewer blades actually allow the turbulence to stabilize before the next blade approaches.

Once the wind energy is converted to mechanical power by the blades, it is transferred to a gearbox for conditioning. The gearbox steps up the revolutions per minute (rpm) of the mechanical power and then transfers the power to spin the generator. The ***nacelle*** houses the gearbox, the generator, and a variety of equipment necessary to keep the wind turbine properly positioned into the wind and spinning at a safe speed. See **Figure 5-18.**

Nacelle: An enclosure that houses the gearbox, the generator, and a variety of equipment necessary to keep a wind turbine properly positioned into the wind and spinning at a safe speed.

When it comes to wind generation, increased wind speed is generally a good thing, but only to a point. A wind generator spinning at excessive speeds creates undue stress on the machine. Therefore, all wind generators must be created with a means of slowing themselves to prevent overspin. This can be done by feathering the pitch of the blades so they do not catch as much wind or by banking the wind turbine out of the wind to

Figure 5-17. Today's wind generators are usually designed with two or three blades for maximum efficiency. This small three-blade generator serves the needs of a single home, supplementing other energy sources.

create the same result. Frictional braking systems and systems that use a combination of means may also be used.

Wind and the environment

One obvious concern about increased wind generating capacity would be the amount of land that large wind farms would occupy. One recent study suggested that wind farms would need to occupy about 16,000 square miles to produce 20% of the electrical demand for the United States. Only a small percentage of this land would be physically occupied, however. About 95% of the land could serve dual use for farming or ranching, in addition to wind generation. Large wind turbines are also known to produce some noise pollution, which could be a detriment if located close to populated areas.

Hydrogen

To some, ***hydrogen*** is considered the ultimate energy source. This is because hydrogen is found in water, which covers about two-thirds of the earth's surface. It is also one of the most common elements in the galaxy. Therefore, hydrogen is considered inexhaustible. A problem is that hydrogen rarely

Hydrogen: The first and simplest element on the periodic table. It is one of the most common elements in the galaxy.

Figure 5-18. The major parts of a horizontal wind turbine.

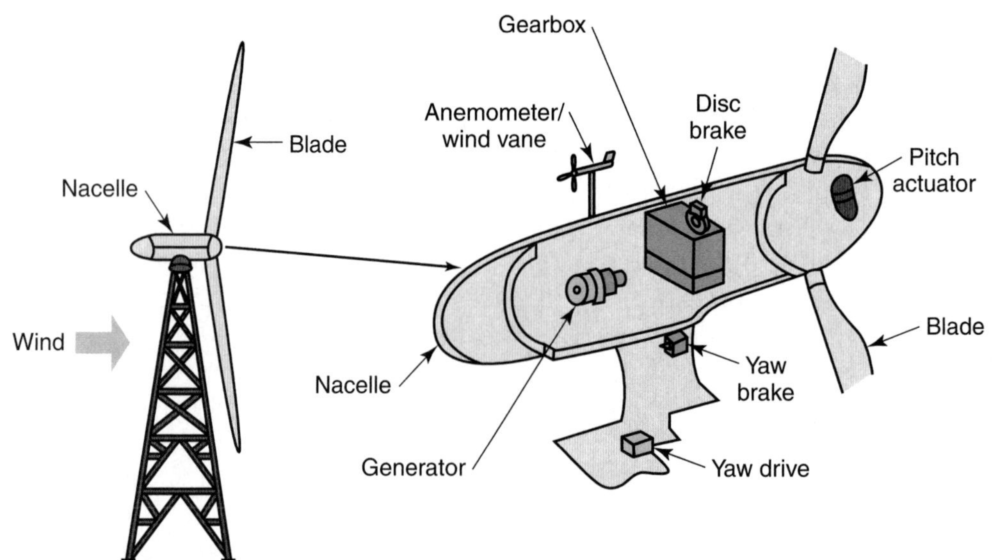

exists in its pure form. Usually, it is locked in a compound. In the case of water, the two hydrogen atoms are bonded tightly with one atom of oxygen. See **Figure 5-19.**

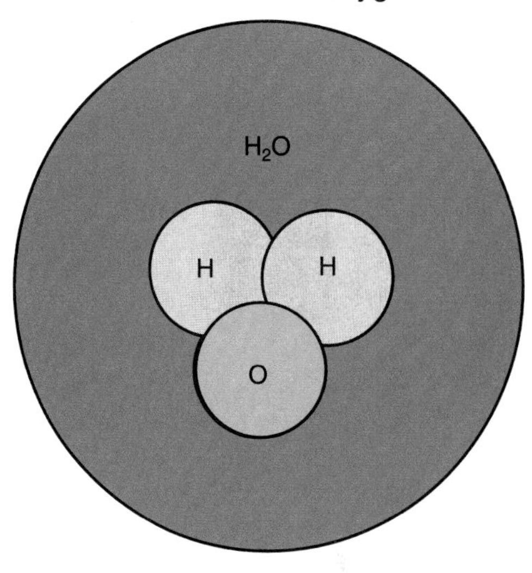

Figure 5-19. The water molecule consists of two atoms of hydrogen bonded to one atom of oxygen.

How hydrogen is collected

Breaking this bond and collecting the hydrogen can be done in several ways, all of which require energy in order to obtain another form of energy. So the question becomes, how much energy does it take to separate the hydrogen-oxygen bond? The answer has proven frustrating because the bond that holds water molecules together is very strong. Heating can break the bond, in a process called ***pyrolysis***. The use of an electrical current can also break the bond, in a process called ***electrolysis***. The bonds need to be broken in order for the hydrogen to be released and collected for use as an energy source. The pursuit of hydrogen as an energy source is worthy for several reasons. Hydrogen is a clean-burning combustible, so it could take the place of fossil fuels to power generators and automobiles. It is very plentiful, and techniques for safely storing hydrogen are improving. See **Figure 5-20.**

Pyrolysis: The process of separating the hydrogen-oxygen bond in water using heat.

Electrolysis: The process of separating the hydrogen-oxygen bond in water using an electrical current.

Hydrogen and the environment

The by-product of inefficient hydrogen combustion is water, about as environmentally friendly as it gets. Breaking the hydrogen-oxygen bond has proven so difficult, however, that more recent efforts to capture

Figure 5-20. Cars operating on hydrogen fuel cells have moved beyond the experimental phase and are being operated in small numbers in Germany, Japan, Singapore, and the United States. This is a demonstration vehicle based on a Mercedes-Benz automobile model. (DaimlerChrysler)

Tech Extension

Ocean Thermal Energy Conversion (OTEC)

Any scuba diver knows that there are often vast differences between the temperature of surface waters and the temperature of waters some depth below. Imagine having the ability to take advantage of that temperature change and produce electricity! This is what Ocean Thermal Energy Conversion (OTEC) really does. See **Figure 5-A.**

The OTEC system of power generation relies on a refrigerant, such as ammonia, that vaporizes at a temperature less than the surface water temperature. The refrigerant also must recondense back into a liquid at a certain temperature, in accordance with the depth waters. The gas is locked in a closed loop. Warm surface waters are used to vaporize the refrigerant. The force of the expanding gas is used to spin a turbine coupled to a generator. Once the energy from the expanding gas has been transferred to the turbine, the gas is cooled back into a liquid using the temperature of the depth waters. The entire process requires about a 40°F temperature differential and goes on continuously to provide steady power generation. Obviously, this type of temperature differential is not available in all waters. OTEC is being tested domestically in Hawaii.

OTEC does not provide any fossil fuel pollutants. Small generating plants can run quietly and efficiently. They must, however, be located near land, in order to feed electricity onshore. Also, they must be tethered to the ocean floor, thereby creating a form of environmental and visual pollution. The changes in temperature created by OTEC generators can kill nearby marine life. Lastly, these generators appear to work best for small-scale power generation, leaving their ability to someday replace major power-producing sources in question.

hydrogen have used a process called steam reforming. This process involves vaporizing fossil fuels and mixing them with high-pressure, high-temperature steam. A nickel-based catalyst is introduced, and the process yields hydrogen. It also, however, yields carbon monoxide (CO) and CO_2, the primary greenhouse gas. Additionally, the source of the hydrogen yielded from this process is a nonrenewable form of energy. Scientists working with this process argue that it is worthy for two primary reasons. First, what has been learned about fossil fuels could soon be applied to other renewable hydrocarbon fuels, such as biomass. Second, the waste by-products, such as CO_2, are more easily controlled in a single large-scale production facility than they are coming out of automobile tailpipes. Therefore, converting car engines to run on only

Figure 5-A. A cutaway view of an Ocean Thermal Energy Conversion (OTEC) plant. The floating plant uses volatile fluids to generate electricity by taking advantage of the temperature difference between water at and below the surface. Such differences are most pronounced in tropical areas. (Lockheed Missiles and Space Company, Inc.)

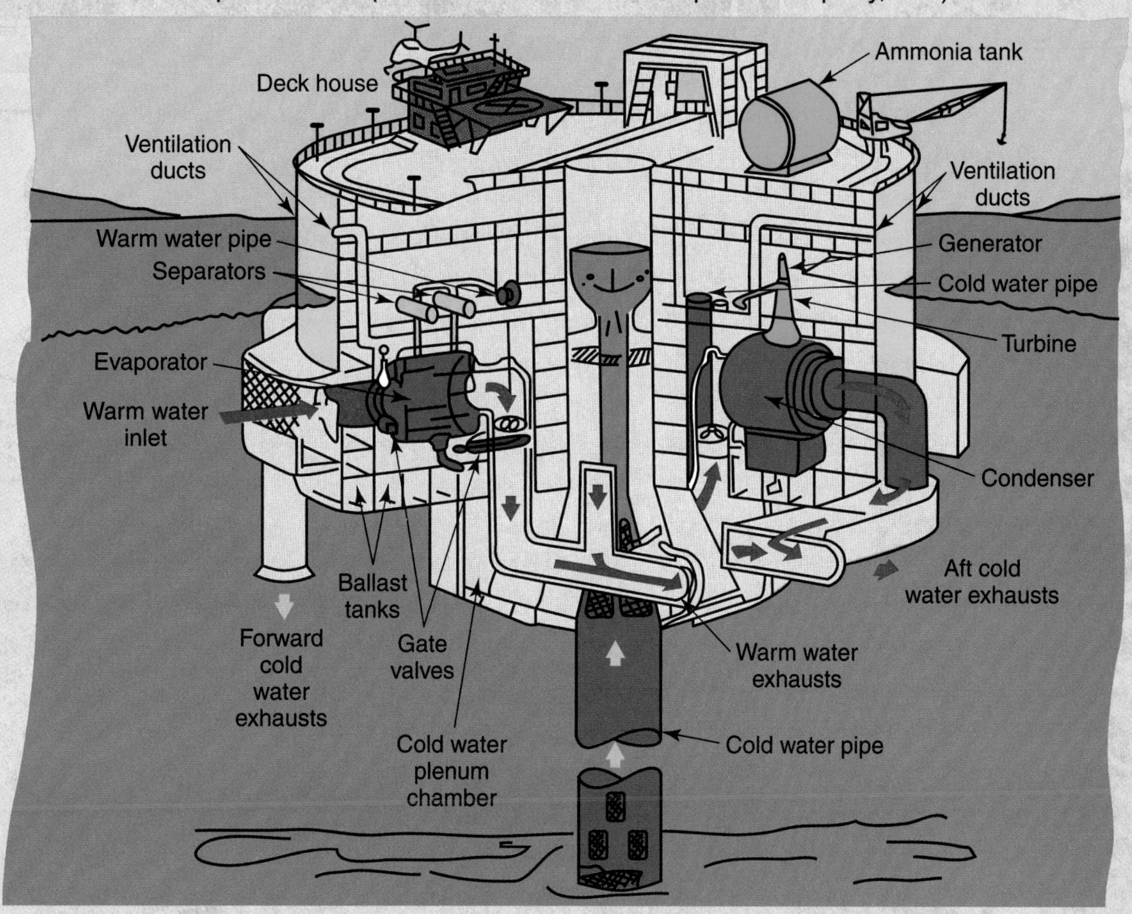

hydrogen from fossil fuels could be much more environmentally friendly than burning the fossil fuels as they are currently burned. So, one plan could be to begin slowly converting over to a hydrogen-based economy. The first step would be to use fossil fuels for the production of hydrogen. We would then use a renewable source of energy, like biomass, to yield hydrogen. Ultimately, we would develop ways to cost-effectively yield hydrogen from water. Do not bet on this happening any time soon. Fossil fuels are still plentiful, and their associated environmental concerns have not yet gained enough prominence to influence society on a large-scale basis. Other nontechnical issues, such as tax credits, money allocated for research, and advantageous legislation, could help to spur the development of hydrogen as a major energy source for the future.

Summary

The origin of most forms of energy can be traced to the sun. Energy resources can come in many forms. Renewable energy sources are those sources that can be replaced. They mainly come from plants and animals. Inexhaustible energy sources are those sources of energy that will not run out. Some examples of these are the sun, the wind, geothermal energy, hydroelectric energy, and hydrogen. Virtually every energy source has some advantages and some disadvantages. Factors affecting the development of energy resources include technological factors, environmental factors, economic factors, and sometimes even political factors.

Key Words

All the following words have been used in this chapter. Do you know their meanings?

anaerobic digestion
barrage
bioconversion
biomass
creosote
electrolysis
ethanol
fermentation
fish lift
gasohol
geothermal energy
heat pump
hydroelectric energy
hydrogen
hydrogen sulfide gas
magma
mantle
methane digester
methanol
methyl alcohol
nacelle
overshot waterwheel
penstock
pyrolysis
tidal fence
undershot waterwheel
waste-to-energy plant
wind turbine
wind velocity profile

Test Your Knowledge

Write your answers on a separate sheet of paper. Do not write in this book.

1. The origins of most forms of energy can be traced to:
   A. the sun.
   B. gravity that creates pressure.
   C. heat from within the earth.
   D. water.
2. How are fossil fuels formed?
3. *True or False?* Food can be considered an energy source.
4. Describe several factors that can influence the development of a certain energy resource.

5. This fuel is a hydrocarbon fuel, but it is not a fossil fuel.
   A. Natural gas.
   B. Oil.
   C. Alcohol.
   D. Coal.
6. What is an advantage of using more alcohol in a gasoline mixture?
7. _____ and _____ are two types of alcohol fuels that could serve as substitutes for gasoline.
8. *True or False?* Biomass fuels can be made from industrial, societal, and agricultural waste.
9. _____ refers to the decomposition of material without the use of oxygen.
10. *True or False?* As a means of power generation and trash reduction, some municipalities have constructed waste-to-energy plants to burn garbage.
11. State the difference between renewable and inexhaustible energy sources.

For Questions 12 through 21, label each source of energy as nonrenewable (N), renewable (R), or inexhaustible (I).

12. Alcohol. _____
13. Animals. _____
14. Coal. _____
15. Hydrogen. _____
16. Natural gas. _____
17. Oil. _____
18. Sun. _____
19. Water. _____
20. Wind. _____
21. Wood. _____
22. *True or False?* Waterwheels once provided significant mechanical power for some industrial processes.
23. What is hydroelectric energy?
24. *True or False?* Large-scale hydroelectric power generation growth is predicted to increase in the next 25 years.
25. Hydroelectric dams do not produce fossil fuel pollutants. They are most closely associated with:
    A. noise pollution.
    B. low-level radioactive waste.
    C. water pollution.
    D. ecological damage.

26. The _____ provides one common use for residential heating with the use of geothermal energy.
27. *True or False?* Geothermal is an ideal energy source because it poses no environmental affects.
28. *True or False?* Hydrogen could be considered the ultimate energy source because it can be found in seawater.
29. _____ and _____ are two current methods of breaking the bond between hydrogen and oxygen in ordinary seawater.
30. _____ is a form of inexhaustible energy that makes use of the temperature difference between surface water temperatures and depth water temperatures.

Activities

1. Design an experimental wind generator. Prepare drawings for building the experimental device and develop a bill of materials for construction. Construct the demonstration project. Test and refine the project.
2. Convert a small gas engine to run on an alternative fuel, like methanol or ethanol. This project will require research, testing, and modification.
3. Construct, test, and demonstrate a device that will change organic wastes into a usable energy form.
4. Tour a hydropower, waste-to-energy, or wind turbine generating facility.

6

Solar Energy

Basic Concepts

- Identify the reason solar energy is considered to be an inexhaustible energy resource.
- State why solar energy is one of the only long-term options for energy independence.

Intermediate Concepts

- Explain the difference between open loop solar collection and closed loop solar collection.
- Summarize how solar energy creates heat.
- Describe three basic types of active solar collectors.
- Give examples of three passive solar collection schemes.
- Discuss how a photovoltaic cell works to convert sunlight to electricity.

Advanced Concepts

- Differentiate between active solar energy and passive solar energy techniques.
- Discuss advantages and disadvantages of various solar collection schemes.
- Perform calculations to determine the payback period associated with the cost of installing solar collection equipment and techniques.

The sun is considered an inexhaustible source of energy. Life on earth would not exist without the energy from the sun. The sun supplies the earth with an extremely large amount of energy. Its diameter is 110 times that of the earth. If the sun was 18″ in diameter, the earth would be 1/16″ across. The sun is so huge that it is twice the weight of the earth, even though it is gaseous. It is 93 million miles away from the earth. Considering the distance, it is hard to imagine that we receive so much energy from the sun. Only one of every 2 billion portions of the sun's rays reaches the earth. Yet, it is estimated that 2.5 times the amount of energy necessary to sustain a home for a year strikes that home in the form of sunlight every year.

Collecting Solar Energy

The sun provides us with several forms of energy, including light and heat, often referred to as solar energy. If we knew a way to collect a bigger portion of the sun's energy, we would have all the energy we would need. Solar energy is not that easy to collect because it is spread out all over the surface of the earth. The current methods used to collect the sun's energy are insufficient and rather expensive, compared to other energy sources. It is also hard to store solar energy.

Solar energy does provide a great amount of heat for some homes and industries. Solar panels are used to trap solar energy. See **Figure 6-1.** The energy is used to heat water or air. Hot water or air can then be used directly. Heated water can be stored in a tank for later use. Heat can also be stored in a bin full of rocks or in storage media, such as ceramic tile, concrete, or water tubes, for heating the surrounding environment.

Figure 6-1. Solar energy can be captured to provide heating or electrical power for a structure. Two different capture methods are being used in this home, which was designed to make extensive use of solar energy. The flat-plate collector on the left captures solar energy and uses it to heat water for household uses. The photovoltaic panels on the right convert the sun's energy to electricity, which is then stored in batteries for later use.

The development of solar energy collection schemes has been around since ancient times. Many civilizations, including the ancient Greeks and Native American Indians from the West, incorporated passive solar concepts into their architecture. These techniques often included the orientation of their living structures to capture and store the radiant energy, which is energy traveling as electromagnetic waves, from the sun. They also included the use of overhangs to block the sun from entering during summer months when it is high in the sky, while allowing the sun to enter during winter months when it is low in the sky.

More recent technological developments in solar energy were brought about as a result of the oil embargos of the 1970s. Solar research boomed in the 1970s, but it began to fade in the 1980s, as tax credits for the installation of solar energy devices expired and were not renewed. The stable fossil fuel prices of the 1990s further diminished interest in solar energy. Yet some significant research in solar energy continues, and many people feel it is in America's best interest to continue to pursue solar energy as a key resource for the future, as fossil fuels dwindle.

Measuring Solar Energy

There are three terms that are used most frequently when discussing the measurement of solar energy. The *solar constant* represents the amount of energy in all forms of radiation reaching the earth's outer atmosphere. See **Figure 6-2.** The solar constant can be measured in British thermal units (Btu) per square foot (ft^2)

per hour, often represented as Btu/ft²/hr. It is not often used, as it is primarily a theoretical term representing a value that exists at the outer edge of our atmosphere.

A more useful term is the *insolation value*, sometimes referred to as the value of incident solar radiation. Insolation can also be measured in Btu/ft²/hr, but it represents the amount of energy available on a specific square foot of earth in a given location. See **Figure 6-3.** The insolation value is more useful than the solar constant because it is a measurement of energy that can actually be collected. This is generally somewhere between 0–360 Btu/ft²/hr, depending on cloud cover, location, altitude, air temperature, time of day, and angle at which the sunlight is measured.

A third term for measuring solar energy, known as the *Langley*, is primarily used by weather agencies such as the National Oceanic and Atmospheric Administration (NOAA). One Langley is approximately equal to 221 Btu/ft²/hr. Charts often refer to solar intensity over periods of time, such as months or years, in Langleys. In countries using the metric system, the values are expressed in megajoules per square meter per hour (MJ/m²/hr). See **Figure 6-4.** The higher the number is, the greater the exposure to solar radiation is.

Figure 6-2. The solar constant is a measurement of the amount of solar energy reaching the earth's outer atmosphere.

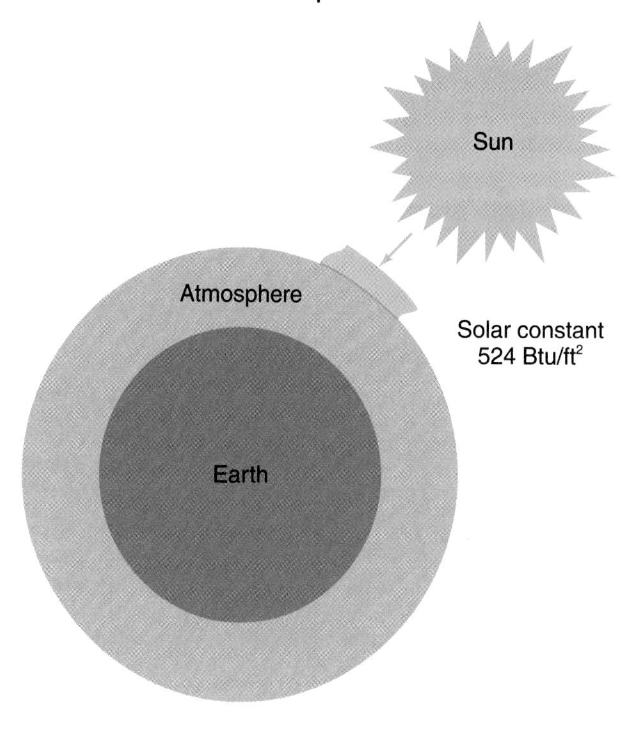

Solar Energy Collection Concepts

There are two types of solar energy collection: active solar energy collection and passive solar energy collection. ***Active solar energy collection*** systems use circulating pumps and fans to collect and distribute heat. Some active solar heating systems are capable of

Active solar energy collection: Systems that use circulating pumps and fans to collect and distribute heat.

Variations in Insolation for Selected Cities

		December		June	
City	**Latitude**	**I_H***	**I_T****	**I_H**	**I_T**
Miami	26°N	1292	1770	1992	1753
Los Angeles	34°N	912	1496	2259	1920
Washington, D.C.	38°N	632	1068	2081	1790
Dodge City	38°N	874	1652	2400	2040
East Lansing	42°N	380	638	1914	1646
Seattle	47°N	218	403	1724	1465

*Insolation on a horizontal surface (in Btu/ft²/d)
**Insolation on a surface tilted at an angle equal to the latitude (in Btu/ft²/d)

Figure 6-3. The amount of energy reaching a square foot of the earth at a given location is the insolation value for that location. This table shows insolation values for cities across the United States during selected winter and summer months. Values are shown in Btu/ft²/day.

Figure 6-4. Differences in solar radiation reaching the earth's surface vary dramatically with latitude and season. These maps of Canada show the average daily solar radiation across the country during the months of January and July. Since Canada uses metric measurements, the values on these maps are shown in megajoules per square meter per hour (MJ/m²/hr). One Langley is equivalent to 23.88 MJ/m²/hr. (Canadian Solar Industries Association)

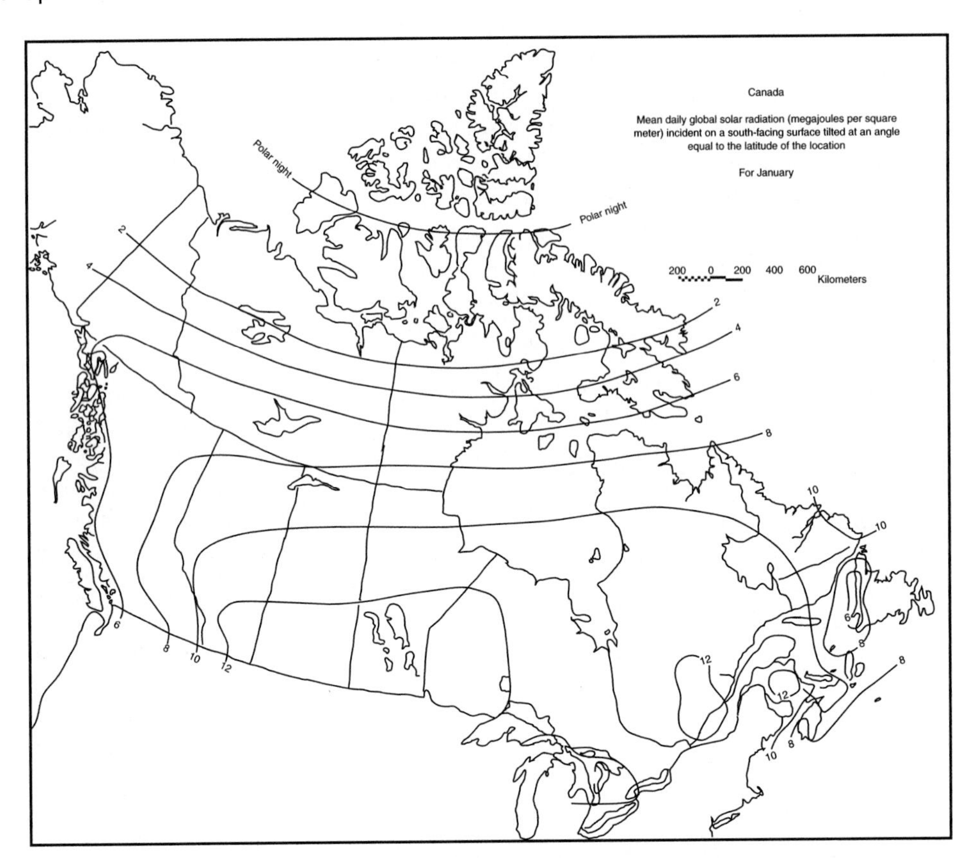

January

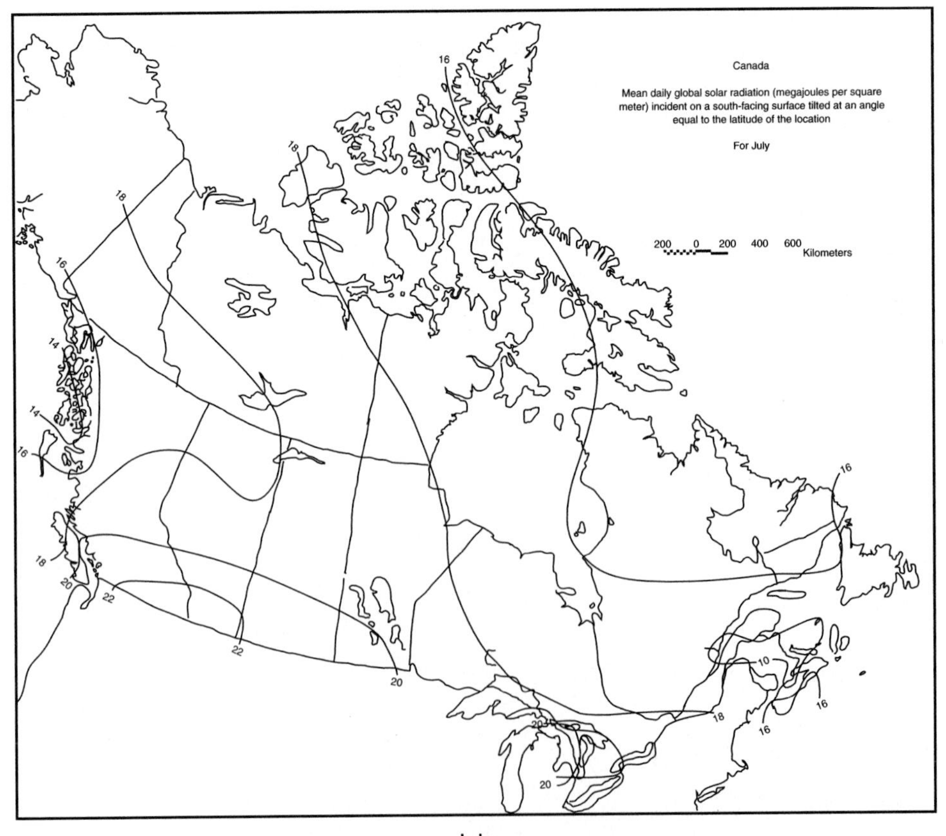

July

concentrating solar energy. Thus, they can reach higher temperatures than passive solar heating systems. ***Passive solar energy collection*** techniques do not make use of any externally powered, moving parts, such as circulation pumps, to move heated water or air. A passive solar system typically makes use of gravity; natural principles of heat movement, such as convection (the fact that hot air rises); evaporation; and architectural design to store and move heat. Passive solar architecture frequently incorporates concepts for capturing heat into the design of a structure. See **Figure 6-5.** Both active and passive solar collection systems can also be utilized as either open loop or closed loop systems, which are described in the following sections.

Passive solar energy collection: Systems that do not make use of any externally powered, moving parts, such as circulation pumps, to move heated water or air.

Open loop solar collection

In an ***open loop solar collection*** system, the heated water or air is directly distributed for use. See **Figure 6-6.** In the example, the water that flows through the collector is actually what will come out of your water faucet. The solar collector is not required to provide all the heat for the water. On a very sunny day, in the right climate, the collector might actually be able to send water down to the tank at 130°F. If this is the case and you call for hot water by turning on a sink, the water from the tank will

Open loop solar collection: Systems in which the heated water or air is directly distributed for use.

A

B

Figure 6-5. This new home is designed for efficient use of solar energy. A—In addition to active solar and photovoltaic panels on the roof, this home is designed to make use of passive solar principles. The south-facing side of the house has large windows to gather maximum sunlight during the winter months, when the sun is lower in the sky. Wide overhangs shade the windows during peak sunlight hours in the summer. B—Inside the home, ceramic tile floors and furniture absorb solar energy during daytime hours and then release it slowly during the evening. This photo was taken on a late summer afternoon, when the roof overhangs limit the amount of sunlight falling on the floor tile.

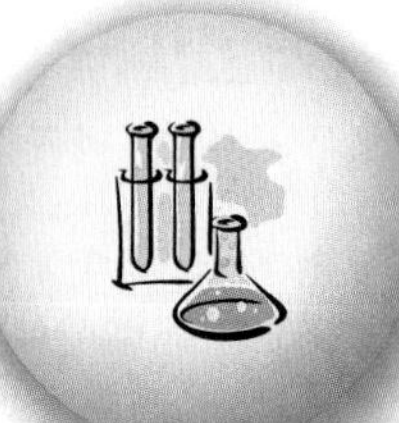

Curricular Connection

Science: Principles of Heat Movement

There are three basic principles of heat movement worthy of explanation. These three principles are known as *conduction, convection,* and *radiation.* Conduction refers to the transfer of heat from molecule to molecule, straight through a material or group of materials, such as those that comprise a wall. If the temperature outside of a structure is colder than the temperature inside the structure, heat will conduct through the wall toward the colder temperature. In warmer climates, heat often conducts inward from the exterior of a structure toward the climate-controlled interior of a structure.

Convection refers to the fact that a heated gas, such as air, tends to move in an upward direction. As the gas is heated, it expands and is pushed up by denser and cooler air. From a structural standpoint, this is why the upper floor of a home tends to be warmer than the lower floor of the home.

Radiation refers to the fact that heat travels with a frequency much the same as visible light. Heat transferred via radiation does not require a medium to travel through. Many types of structural heating devices, such as baseboard heaters, transmit heat to the surrounding environment via radiation.

be replenished with hot enough water that the thermostat will never turn on the conventional heating source (such as electricity or natural gas). Of course, this is not always possible. On days when it is cloudy, it might not be possible to heat the water flowing through the collector to above 100°F.

Figure 6-6. An open loop solar collection system. In this system, water heated by the sun flows directly from the collector to a storage tank for direct use by the household. On cloudy days or in periods of high demand, water in the tank can be heated by electricity or a gas burner.

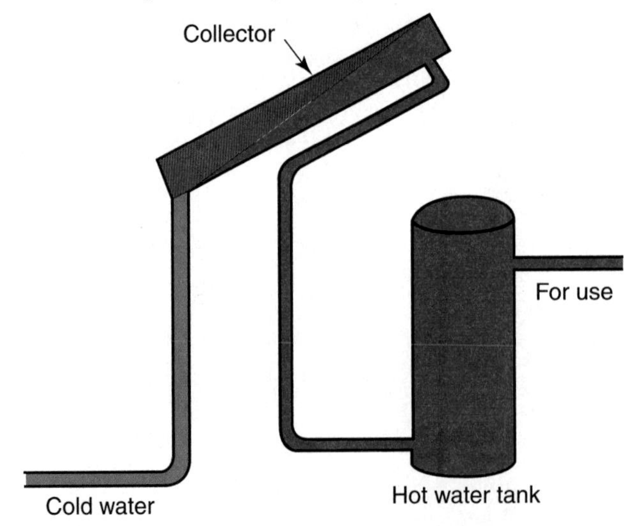

Even so, any heat collected will result in less consumption of conventional energy to heat the water to optimum temperature. Both energy and money will be saved.

Closed loop solar collection

In a ***closed loop solar collection*** system, a collection medium is used to collect the heat and transfer it to water or air for end use. The liquid within the closed loop simply circulates from the collector down to a heat exchange tank and then back up to the collector again to collect more heat. It never comes in direct contact with the water being heated for end use in your home. See **Figure 6-7.** A closed loop solar collection scheme offers several advantages over an open loop system. It can avoid freezing by using a water-alcohol antifreeze solution in the closed loop. The water-alcohol solution serves as a collection medium to transfer the heat from the collector down to the storage tank. If the temperature on the roof is not suitable for collecting energy, the circulation pump attached to the closed loop does not have to run.

Closed loop solar collection: Systems in which a collection medium is used to collect the heat and transfer it to water or air for end use.

Types of Active Solar Energy Collection

There are essentially three types of active solar energy collectors. Each type may have some variation, but active solar collectors can be classified as flat-plate collectors, linear-concentrating parabolic collectors, or parabolic dish (point-focusing) collectors. Each type has certain advantages and disadvantages.

Figure 6-7. A closed loop solar collection system uses a heat exchanger to transfer energy from the solution used in the collection system to the water that will be stored for household use. Such a system is more practical than the open system in cold climates, since antifreeze can be added to the solution flowing through the collector.

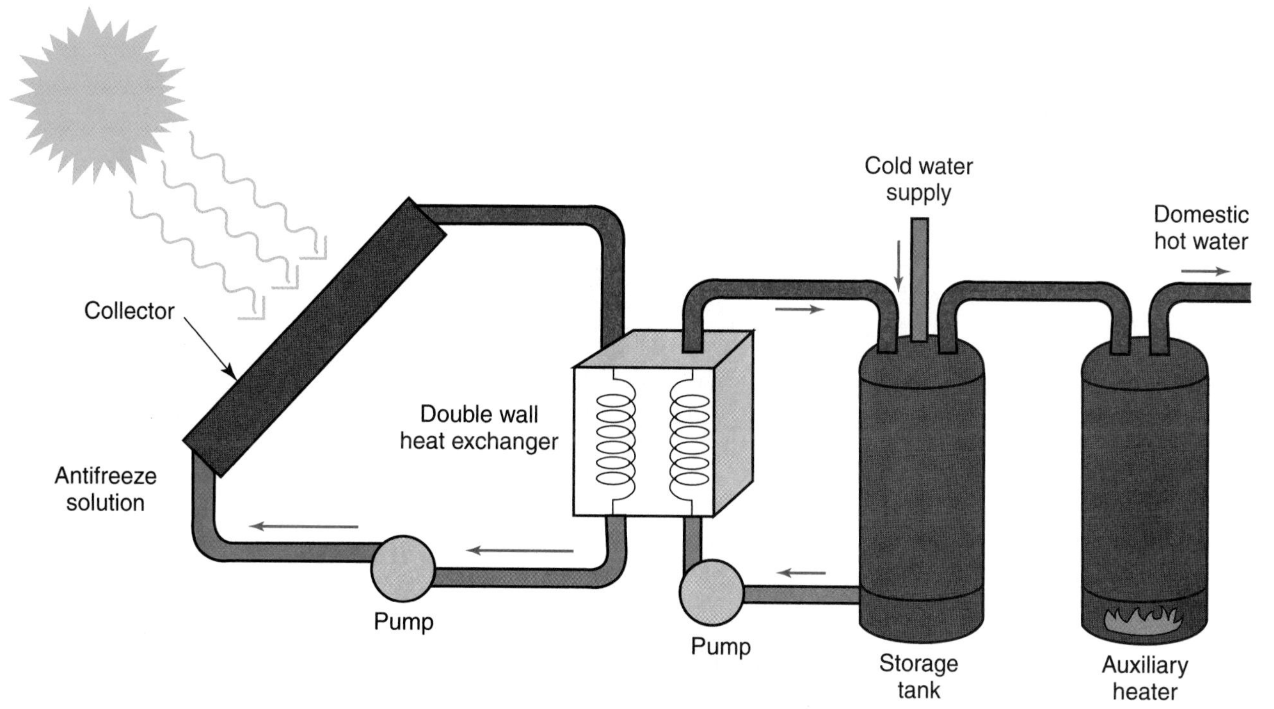

Flat-plate collectors

Flat-plate collector: The most common type of active solar collector. It is typically stationary, mounted on a rooftop, facing south in the northern hemisphere. It has the ability to collect heat from diffuse sunlight, even on cloudy days.

The most common type of solar collector is the ***flat-plate collector***. See **Figure 6-8.** The flat plate offers several advantages, including an ability to collect heat from diffuse sunlight, which is sunlight that bounces off of clouds, even on cloudy days. Additionally, the flat-plate collector is typically stationary, mounted on a rooftop facing south in the northern hemisphere. It is not required to track the sun. Simplicity and reliability are distinct advantages of the flat-plate collector over other types of collectors. For these reasons, the flat-plate collector is the most popular collector, even though it has a relatively low collection ratio for the space it occupies. A collection ratio is the amount of energy collected per square foot compared to the amount of energy available for collection per square foot. It allows for the comparison of one collector to another.

Conduction: The transfer of heat from molecule to molecule, straight through a material or group of materials.

As shown in **Figure 6-9,** visible light is radiated at a frequency that readily passes through glass. Once the light passes through the glass, it strikes the absorber plate, which is typically painted flat black. The color is important, since lighter colors tend to reflect light, whereas darker colors tend to absorb light and reradiate it in the form of heat. Even glossy black does not convert light to heat as well as flat black. Once light is converted to heat, it is a fundamental law of physics that the heat from a mass at a higher temperature will move to a mass at a lower temperature. This phenomenon is known as ***conduction***, the transfer of heat by molecule to molecule from a body at a higher temperature to that of a lower temperature. Some of the heat is not absorbed by the absorber plate, but rather, it is radiated outward. The infrared wavelengths are longer than those of visible light, however, and they cannot pass through glass as readily. Thus, a partial trap has been created, and the heat is held within the collector. The same principle of converting light to heat is applicable to other types of collectors and passive solar architecture techniques. Note that infrared energy has a lower frequency than visible light radiation. This means it has a longer wavelength.

Linear-concentrating parabolic collectors

The next type of active solar collector is the ***linear-concentrating parabolic collector***. Parabolic collectors must track the sun from east to west in what is known as the *azimuth path*. The ***azimuth path*** is the path

Figure 6-8. The flat plate is the most common type of active solar collector and is usually mounted in large panels on the south-facing roof. Flat-plate collectors may heat either water or air, depending on the system design.

Figure 6-9. Cross section of a flat-plate collector, showing how the sun's heat energy is transferred to the water inside the tubes.

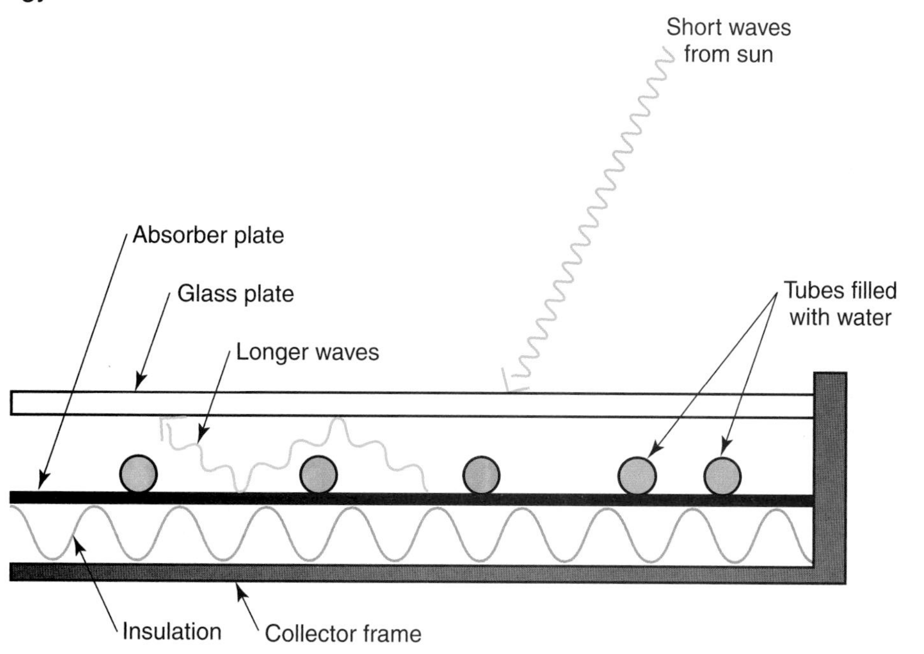

the sun travels daily. In a country in the northern hemisphere, such as the United States, the sun rises in the east and sets in the west, traveling across the sky throughout the day. See **Figure 6-10.** The collector can be mounted to a roof facing south, like a flat-plate collector. It must be capable, however, of moving with the sun. See **Figure 6-11.** If the collector does not move a little bit every few minutes, the rays of the sun will not hit the focal line of the collector. Mounted along the focal line is often a single collection tube through which water flows. When linear-concentrating collectors are on track, they can collect much more energy than a flat-plate collector, in much shorter periods of time, while

Linear-concentrating parabolic collector: An active solar collector that tracks the sun's movement, generating temperatures of hundreds of degrees Fahrenheit on a clear day. Mounted along the focal line is often a single collection tube through which water flows. This type collects much more energy than a flat-plate collector, in much shorter periods of time, while occupying much less space.

Azimuth path: The sun's movement from east to west.

Figure 6-10. The azimuth path is the sun's movement from east to west, which a linear-concentrating parabolic collector must follow to operate most efficiently. The zenith angle is the vertical angle of the sun from the horizon. In the winter, the zenith angle is low; in the summer, it is high.

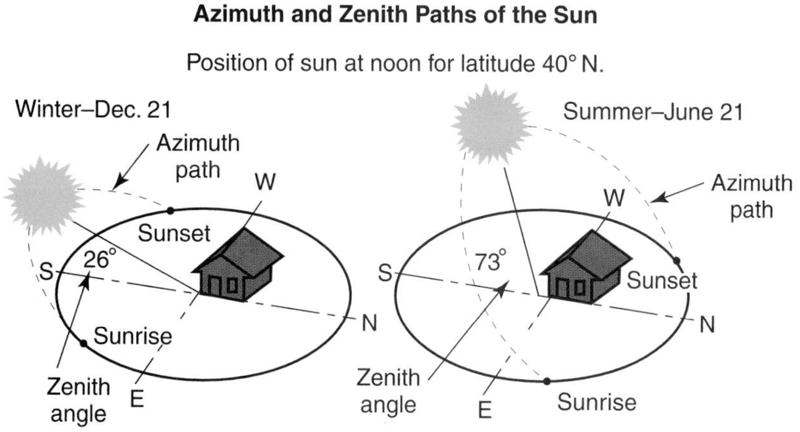

Azimuth and Zenith Paths of the Sun

Position of sun at noon for latitude 40° N.

Winter–Dec. 21
Azimuth path
W
Sunset
26°
S
N
Sunrise
Zenith angle
E

Summer–June 21
Azimuth path
W
73°
S
Sunset
N
Zenith angle
E
Sunrise

Figure 6-11. The linear-concentrating parabolic collector mounted on this home tracks the sun's azimuth path. The water heated by the sun is stored for household needs, such as dish-washing, showering, and washing clothes. (Howard Bud Smith)

occupying much less space. Linear-concentrating collectors can generate temperatures of hundreds of degrees Fahrenheit, when tracking on a clear day. Can you think of disadvantages of such a collector design? Disadvantages include the fact that this type of collector requires direct sunlight. Therefore, it is not suitable for all climates. Also, the tracking mechanism may need to be calibrated periodically. If the tracking mechanism drifts off, the collector will miss the focal line, and heat will not be collected.

A ***compound parabolic collector*** is a combination of a flat-plate collector and a parabolic collector. This type of collector offers advantages of both types of collectors previously described. It is stationary mounted and does not need to track the sun. This type of collector is also a linear-concentrating collector, so it offers a greater collection ratio than a flat-plate collector. It is best used in direct sunlight, like a parabolic collector, however, and the curvature of the reflectors must be generated for a specific location. For instance, the curve that works in Phoenix, Arizona could not be expected to work on a roof in New York City.

Compound parabolic collector: A combination of a flat-plate collector and a parabolic collector, offering advantages of both types. It is stationary mounted and does not need to track the sun. It is also a linear-concentrating collector, offering a greater collection ratio than a flat-plate collector.

Parabolic dish collectors

The third main type of active solar collector is known as a ***parabolic dish collector***. See **Figure 6-12.** Parabolic means it is point focusing. This type of solar collector has a tremendous collection ratio. Large point-focusing collectors are capable of producing temperatures of thousands of degrees Fahrenheit, but they obviously require a very sophisticated tracking mechanism. Not only must this type of collector track along the azimuth path of the sun, but it must also track the height of the sun in the sky. This is referred to as the ***zenith path*** of the sun. Refer to Figure 6-10. The net result is that the collector is almost constantly making some minor adjustment to track the sun. When in focus, it has a collection ratio that far exceeds other types of collectors. This type of collector is actually used in some countries along the equator to collect heat for the production of steam to generate electricity. It is not, however, a collection scheme easily modified for residential use.

Parabolic dish collector: An active solar collector that is point focusing. It has a tremendous collection ratio, permitting large point-focusing collectors to produce temperatures of thousands of degrees Fahrenheit. This type of collector tracks the height of the sun in the sky, as well as the azimuth path.

Types of Passive Solar Collection

Passive solar collection involves three primary methods. These are the direct gain approach, the indirect gain approach, and the isolated gain approach. Each method has advantages and disadvantages.

The direct gain approach

The most common type of passive solar collection is the ***direct gain approach.*** Direct gain means there are no significant architectural provisions made to collect solar energy, other than facing windows in the proper direction. An example of the direct gain approach is a car with the

Figure 6-12. Researchers at Sandia National Laboratories in New Mexico make final adjustments to a parabolic dish collector being developed for use by Native American farmers in remote areas of the American Southwest. The collector concentrates the sun's energy on a receiver unit. (U.S. Department of Energy photo by Randy Montoya)

Zenith path: The height of the sun in the sky.

Direct gain approach: The most common type of passive solar collection. In this type of system, there are no significant architectural provisions made to collect solar energy, other than facing windows in the proper direction.

windows rolled up. In such a situation, the sun can heat the car up significantly. Solar energy passes through the glass and is converted to heat. Just like solar collectors, the darker the interior of your car is, the more sunlight gets converted to heat. Similar to the way a car collects heat, architects and designers can design a home to collect solar energy by properly orienting the home. See **Figure 6-13.**

Some basic requirements of passive solar architecture include the following:

- Good southern exposure.
- Window areas facing south, east, and west and representing about 10%–15% of the home's square footage.
- Very few windows facing north.

These requirements are necessary because, by the time the sun comes around to the north side of a home in the northern hemisphere, it is probably too low in the sky to provide any significant solar gain. These homes are also generally well insulated and tightly sealed. Note that orientation and window space are the key factors considered for solar gain, when using the direct gain approach. An intentional storage medium is not used with the direct gain approach.

Figure 6-13. Proper orientation is important for capturing solar energy. South-facing walls with plenty of glass area take fullest advantage of the sun's rays, especially during the winter months. A properly insulated home with few window and door openings on the north side will shelter against winter winds and minimize heat loss.

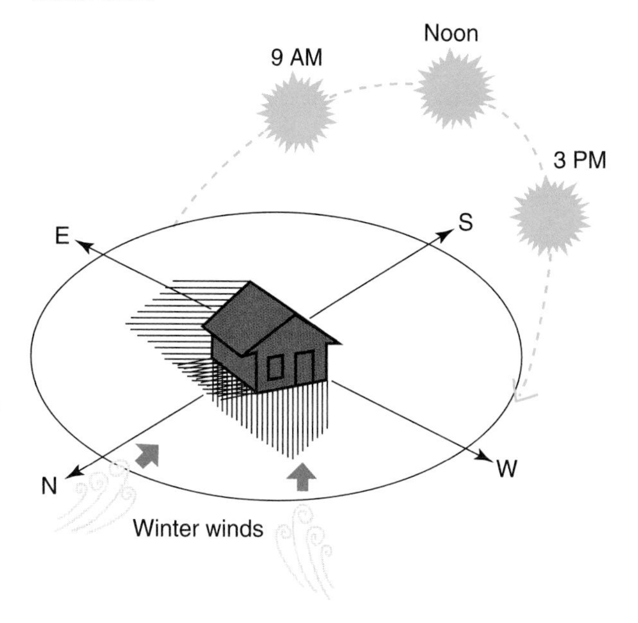

Indirect gain approach: A type of passive solar collection that makes use of a storage medium to store heat for later use. The storage medium may be rocks, concrete, or water.

Trombe wall: A common application of the indirect gain approach. The Trombe wall is heated throughout the day. The storage medium takes a long time to heat up, but it gives off its heat slowly throughout the night, requiring less use of conventional energy.

Phase change: A change in a substance's state (liquid, solid, or gas).

Passive solar architecture also makes significant use of roof overhangs. Using a solar angle reference manual, it is possible to calculate the ideal overhang for a structure in any location to maximize solar gain in winter and minimize solar gain in summer. Another concept frequently used to block excessive summer sun is the planting of deciduous trees to provide shade in summer, but allow sunlight in winter. Deciduous trees are trees that maintain their leaves throughout the warm months and shed their leaves annually each fall or winter.

The indirect gain approach

The ***indirect gain approach*** makes use of a storage medium to store heat for later use. The storage medium may be rocks, concrete, or water. A ***Trombe wall*** is a common application of the indirect gain approach. It is a wall specifically designed to act as a thermal storage medium. A Trombe wall is often made of concrete or masonry material, but it can also be made of water stored in tubes or barrels. See **Figure 6-14.** The Trombe wall is heated throughout the day. The storage medium takes a long time to heat up, but it gives off its heat slowly throughout the night, requiring less use of conventional energy. Less conventional applications of the indirect gain approach include the use of storage pools and specialty products, such as eutectic salts, which melt when heated and give back heat to the environment as they experience a phase change when cooling back into a solid. A ***phase change*** occurs when a substance changes its state (liquid, solid, or gas) to another state. Such a change occurs when liquid water freezes and becomes a solid in the form of ice.

The isolated gain approach

The third passive solar collection scheme is known as the ***isolated gain approach***. See **Figure 6-15.** In the isolated gain approach, the solar collector is isolated from the structure to be heated. The collector is typically located next to or beneath the home, and the system relies on convection to carry heated air up to the structure. Cold air return vents

Career Connection

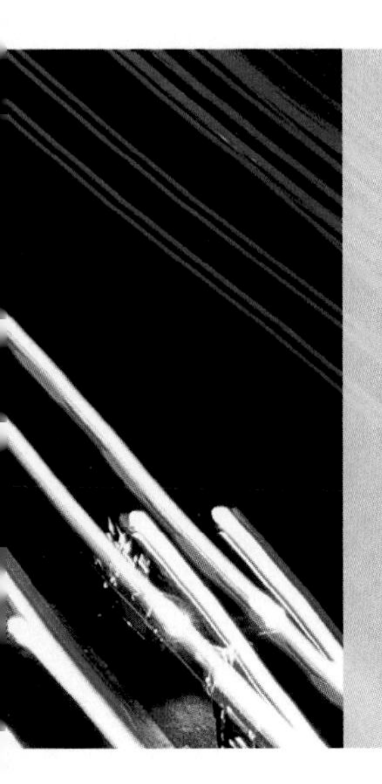

Architects

Architects are needed to design buildings. Their responsibilities also include planning for energy efficiency. When an architect is designing a structure, heat loss must be taken into account to more accurately plan for energy conservation.

The job of an architect consists primarily of designing structures, but architects must also prepare for construction. The proper contracts and permits must be obtained in order for work to begin, and architects must perform frequent checks on the site once construction begins. Apart from these tasks, architects must also work directly with the clients to meet their needs for the projects.

Architects should have good communication skills in order to work with clients. They must also have knowledge of engineering, mathematics, and design. A bachelor's degree is required for architects. The yearly salary may range from $37,000 to $98,000.

Tech Extension

R-Values

An R-value is a convenient measure of the resistivity to heat flow that is used to compare the heat loss allowed by various types of insulation and other construction materials. See **Figure 6-A.** If a building's structural components have high R-values, the building has a high resistivity to heat loss by conduction. By increasing the R-values of building materials or adding insulation with higher R-values, the rate of heat loss in a building is reduced, which helps to conserve energy.

Material	R-Value
.5″ gypsum board	.5
5.5″ fiberglass insulation	19–23
3.5″ fiberglass insulation	11–13
8″ concrete block	1.6
Vinyl siding	.33
.5″ wood siding	.5
.5″ sheathing (oriented strand board)	.5
.5″ sheathing (insulating board)	.5

Figure 6-A. the R-values of some typical products used in construction.

Isolated gain approach: A type of passive solar collection in which the solar collector is isolated from the structure to be heated. The collector is typically located next to or beneath the home, and the system relies on convection to carry heated air up to the structure. Cold air return vents typically allow cooler air to return to the collector.

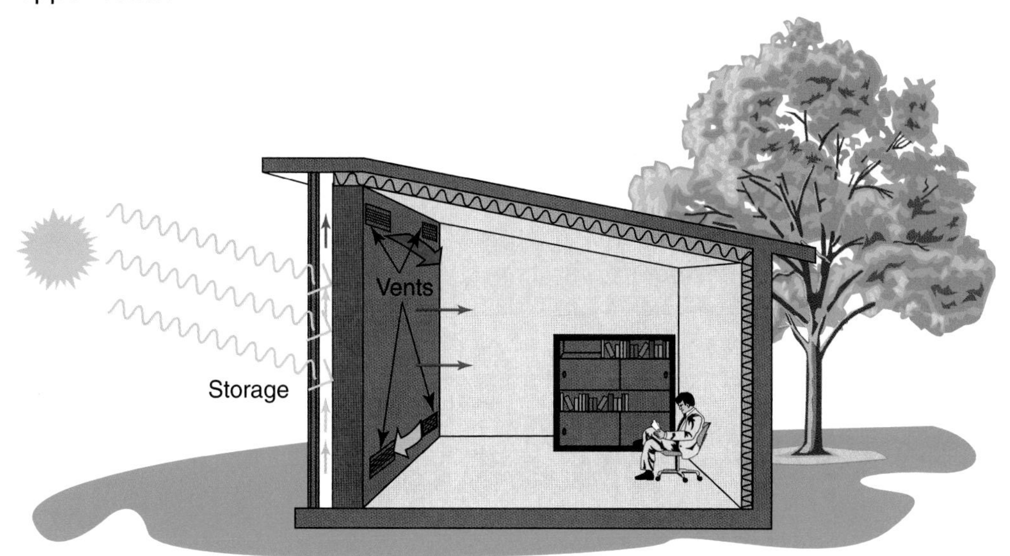

Figure 6-14. A Trombe wall is a thermal mass used for heat storage in an indirect gain system. It is typically a masonry wall located a few inches behind a large expanse of glass. Sunlight slowly heats the mass during daylight hours, and then the mass gradually releases stored heat to the living space at night. Vents allow the passage of cooler air into the space between the wall and the glass. As the air is heated, it rises and exits to the room through the upper vents.

Figure 6-15. The thermal mass in an isolated gain system might be a concrete floor and wall, bins of rocks, or barrels of water (water has greater heat storage capacity than masonry). Heat moves to living areas by convection.

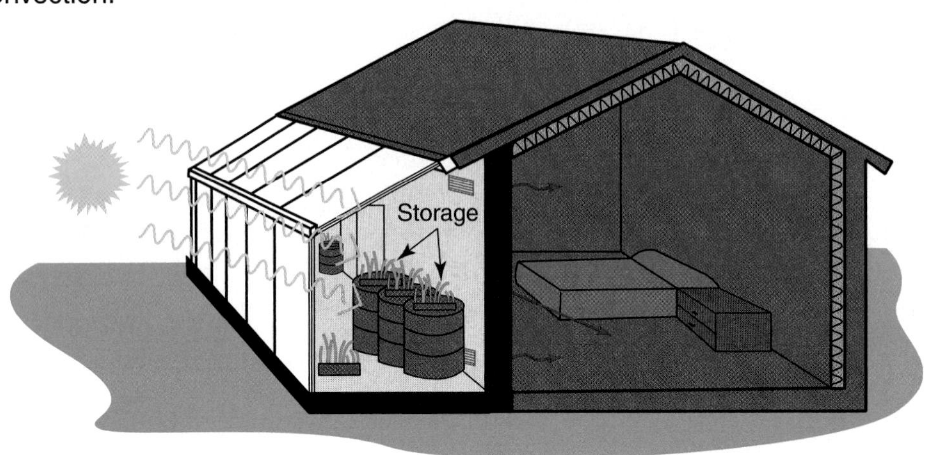

Curricular Connection

Math: Calculating a Payback Period

Even though solar energy is a free energy source, it costs a lot, due to the expense of the equipment necessary to collect and distribute it. These costs may or may not be justifiable, depending on a number of factors, including geography, installation costs, the demand for heated air or hot water, the cost of conventional energy to do the same amount of heating, and the life expectancy of the system. Since active solar systems are often used to heat hot water, we will perform a payback calculation for the installation and use of a solar hot water system. A payback calculation (also sometimes referred to as the *breakeven point*) is used to determine how many years something will take to pay for itself. The following is the information you need to perform a payback calculation for an active solar hot water system being considered for installation by a family of four.

- The cost of a domestic solar hot water system = $2800.
- An average family of four uses 100 gallons of hot water per day.
- Hot water is heated from 55°F (average ground temperature below the frost line) to 125°F for use all 365 days per year.
- The conventional heating source is a natural gas hot water heater.
- Natural gas for this area costs $.010 per cubic foot (ft^3).
- 120 ft^3 of natural gas can produce 100,000 Btu.
- 1 Btu = the energy necessary to raise 1 lb. of water (H_20) 1°F (1ΔT°F).
- The solar collector manufacturer estimates savings of conventional energy in your region at 40%, if a solar collection system is installed.

Step 1. Determine how much energy is needed to heat the water for 1 day.

$$\frac{\text{Btu}\,(125 - 55\ \cancel{\Delta T°F})}{1\ \cancel{\text{lb. } H_20\ 1\Delta T°F}} \times \frac{100\ \cancel{\text{gal}}}{\text{day}} \times \frac{8\ \cancel{\text{lbs. } H_20}}{\cancel{\text{gal}}} = 56{,}000\ \text{Btu/day}$$

typically allow cooler air to return to the collector. In this way, a ***convective loop*** is formed that does not require any circulating pumps or fans. A storage medium, such as masonry, crushed stone, or water, can be used to store some heat for later use. Note that the collector need not be a part of the structure when using the isolated gain approach to passive solar heating.

Convective loop: The loop created when convection carries heated air from an isolated gain solar collector up to a structure, and cold air return vents allow cooler air to return to the collector. It does not require any circulating pumps or fans.

Photovoltaic Cells

The process of converting sunlight to electricity begins with tiny bundles of light known as photons. It has been known since the late 1800s that, when photons strike certain metals, free electrons are emitted. This is referred to as the ***photoelectric effect***. A photon strikes a ***photovoltaic cell***, which is a semiconductive material that emits free electrons when exposed to light. The free electrons result in the flow of electricity. If

Step 2. Determine how much it will cost to heat the water for 1 year using a conventional energy source.

$$\frac{56{,}000\ \text{Btu}}{\text{day}} \times \frac{120\ \text{ft}^3}{100{,}000\ \text{Btu}} \times \frac{\$.010}{\text{ft}^3} \times \frac{365\ \text{days}}{\text{year}} = \$438/\text{year}$$

Step 3. Determine what percentage of domestic hot water will be saved using an active solar collection system. This number must be estimated based on geographic location (to predict the insolation value) and system efficiency. We will assume an anticipated yearly savings of 40% for the Midwest. Actual savings will vary, based on system efficiency and geographic location.

Step 4. Determine dollars saved by installing a solar collection system.

.40 × $438 = $175/yr anticipated savings

Step 5. Determine the payback period for solar collection system installation.

$2,800/$175 = 16-year payback at the current cost of energy

Whether or not this is an acceptable payback period depends on a number of factors, but generally speaking, improvements to conserve energy should pay for themselves in less than 10 years. Otherwise, they are usually not considered worth doing. If this system was being installed in a home in Albuquerque, New Mexico, and the anticipated savings in yearly energy was estimated at 80% because of the number of sun hours Albuquerque receives, the payback period would be reduced by half, to about 8 years. This is why solar collectors are frequently installed on residential homes in the southwestern United States. It is economically viable to install them and expect homeowners and new home buyers to pay for them.

Remember to factor in maintenance costs and life cycle renewal of parts. One of the reasons longer payback periods are ill advised is that, as a system ages and wears, it will be subject to maintenance and repair. These costs could also be factored into a more extensive payback calculation, although a conventional hot water heater heating system might also need maintenance, repair, and perhaps even replacement within 10 years.

Photoelectric effect: The emission of free electrons when photons strike certain metals.

Photovoltaic cell: A cell that converts sunlight directly into electricity. This occurs when positively charged photons strike the cell and displace electrons from the valence shell of the material making up the cell. Sometimes referred to as a solar cell.

enough cells are combined, the flow of electricity can be substantial enough to perform useful work. Photovoltaic cells, sometimes referred to as solar cells, have the ability to convert sunlight directly into electricity without any moving parts. This occurs when positively charged photons strike the cell and displace electrons from the material making up the cell. The free electrons begin to collect along thin copper strips that feed a heavier copper conductor. The electrons are drawn away when placed in a circuit and connected to an electrical load, such as a lightbulb. **Figure 6-16** shows the operation of a photovoltaic cell. These cells are used most frequently in remote locations, where it is not feasible to power something with conventionally generated electricity. See **Figure 6-17.** One major purchaser of photovoltaic cells is the U.S. Coast Guard, which uses them to power navigational devices, such as buoys located along waterways. Since the electricity produced by photovoltaic cells is direct current (DC) electricity, it may be used immediately or stored in batteries for later use. It may also be inverted to alternating current (AC) electricity, such as the type used in a residence or commercial building.

Other Applications for Solar Energy

Solar energy has some other applications that may surprise you. Of course, one application of solar energy used much more extensively than solar heating is the use of solar energy for natural lighting. This is common in most structures. Did you know that solar energy can be used for cooling applications as well?

Figure 6-16. Solar cell construction and operation. The cell converts heat energy from sunlight to electrical energy. (Solar Power Corporation)

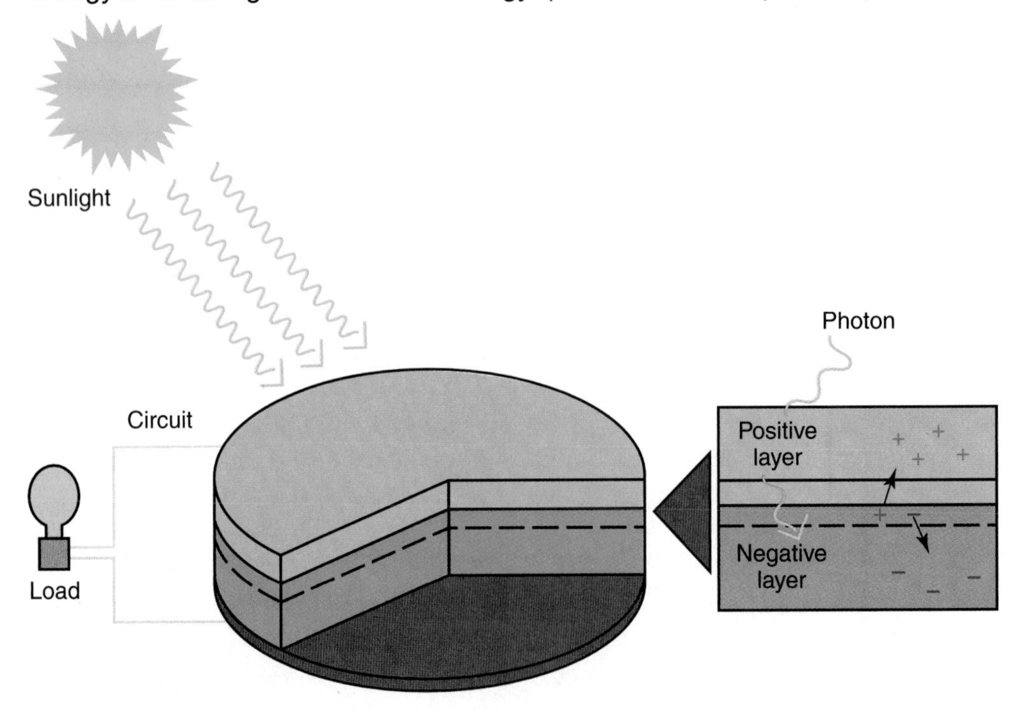

Figure 6-17. Photovoltaic panels are used to provide power in locations where electrical power is not readily available or where a separate power supply is needed that will not be affected by outages of the electrical distribution. This panel provides electricity to power a small radio transmitter that sends data from monitors at a large natural gas storage site. It permits the radio to operate independently and not depend on power from the electrical grid.

Solar Cooling

In cooling operations, the heat from the sun is used to boil a refrigerant, turning it into a gas. The gas is condensed into a liquid and transferred to the ***evaporator*** (freezing unit). The pure liquid refrigerant absorbs the heat from the surrounding area and turns back into a gas in the evaporator. It is then transferred to the ***condensing unit***, where it is condensed back into a liquid and gives up its heat. The evaporator is located within a large room or building insulated from exterior heat for refrigeration. The condenser is located outside the refrigeration area and transfers heat out of the area to another environment.

Evaporator: A freezing unit used in solar cooling operations. Pure liquid refrigerant absorbs the heat from the surrounding area and turns back into a gas in the evaporator. The evaporator is located within a large room or building insulated from exterior heat for refrigeration.

Condensing unit: In solar cooling operations, the unit where refrigerant is condensed from a gas back into a liquid and gives up its heat. The condenser is located outside the refrigeration area and transfers heat out of the area to another environment.

Large-Scale Power Generation

Solar thermal energy conversion (STEC) is an experimental process used to generate large-scale electricity, such as that which could be used to power a community. See **Figure 6-18.** A STEC collector uses many mirrors, all aimed at one common focal point. Each mirror is controlled by a computer to position the mirror for maximum solar gain throughout the day. When the STEC mirror array is properly focused in direct sunlight, it is capable of generating thousands of degrees of heat at the focal point. This much heat can be used to create high-pressure steam to spin a turbine and generate electricity. The steam cycle process would be used to generate electricity the same way electricity is generated when steam is made by burning fossil fuels. Only the heat source would change.

Technology Link

Manufacturing: Solar Cell Production

Energy technology is important to our daily lives, as it provides us with light, heat, and other necessities and conveniences. In order to have the energy we use every day, we depend on many applications of manufacturing technology. Energy supplies, such as solar cells, need to be developed, produced, and distributed for use.

There are several types of solar cells. The most common are made of silicon. The process of manufacturing solar cells has four main parts, which occur in the following sequence:

- **Casting and wafering.** This typically involves metallurgical processes, such as crystal growing and casting. Molten silicon is processed at extremely high temperatures. The silicon is then shaped into wafers.
- **Solar cell manufacturing.** At a solar cell plant, the wafers are taken through a semiconductor processing sequence, where they become working solar cells. They must go through etching, diffusion, and screen-printing steps before being tested and graded.
- **Module assembly.** This usually involves soldering cells together to produce a string of cells, which is then laminated between glass plates and framed so it can be easily mounted.
- **Solar energy system assembly and installation.** The solar module has to be integrated into the structure for which it was designed. This may involve mounting it onto a roof or other structure. The module also has to be integrated into the other parts of the solar energy system, which involves connecting the inverters, batteries, wires, and regulators. Often, a computer software program is used to calculate the electrical load required by the customer.

Solar energy can significantly impact our society, economically and environmentally. Because it is so dependent on manufacturing technology, it has the potential to improve employment rates by creating long-term jobs. Lower prices and higher efficiency suggest that solar energy will be used more and more in the coming years.

Figure 6-18. Solar Thermal Energy Conversion (STEC) is an experimental program being evaluated at the Sandia National Laboratory of the Department of Energy (DOE) in New Mexico. A—Nearly 2000 large mirrors are arranged in an arc across 72 acres and are kept oriented to the sun. B—Sunlight reflected from the mirrors is focused on a receiver near the top of a tall tower. The resulting heat energy is used to generate steam and drive a turbine. Electrical energy to power 10,000 homes is generated by this system. (U.S. DOE)

A

B

Summary

The sun is considered an inexhaustible source of energy. Life on earth would not exist without the energy from the sun. Plenty of solar energy reaches the earth every day, but it is difficult to collect and store. Present solar collection techniques may be divided into the categories of active and passive solar energy collection. Passive solar collection techniques do not require external power sources to help in the collection or distribution of heat. These collection techniques include the direct gain approach, the indirect gain approach, and the isolated gain approach. Active solar collection systems use pumps for circulation. These collection systems include the use of flat-plate, linear-concentrating, and point-focusing collectors. Active solar collection systems may be configured as open loop systems or closed loop systems. Some active solar collectors must be capable of tracking the sun on its azimuth path. Photovoltaic cells have the ability to convert sunlight into direct current (DC) electricity. This electricity may be used, stored, or converted to alternating current (AC) electricity. Photovoltaic arrays are typically used to provide power in remote locations. All solar collection techniques offer some advantages and some disadvantages. While solar energy is not presently used extensively, the fact that it is an environmentally friendly inexhaustible resource makes it worthy of continued exploration and research. Solar energy will remain a viable source of energy as long as humans inhabit the planet.

Key Words

All the following words have been used in this chapter. Do you know their meanings?

active solar energy collection
azimuth path
closed loop solar collection
compound parabolic collector
condensing unit
conduction
convective loop
direct gain approach
evaporator
flat-plate collector
indirect gain approach
isolated gain approach
linear-concentrating parabolic collector
open loop solar collection
parabolic dish collector
passive solar energy collection
phase change
photoelectric effect
photovoltaic cell
Trombe wall
zenith path

Test Your Knowledge

Write your answers on a separate sheet of paper. Do not write in this book.

1. Write one or two sentences explaining why solar energy is considered to be an inexhaustible energy resource.
2. *True or False?* Ceramic tile, concrete, and water tubes can all be used as storage media for solar energy.
3. Why is solar energy one of the only long-term options for energy independence?
4. *True or False?* Passive solar techniques rely on tracking mechanisms and circulating pumps.
5. A passive solar home typically makes use of which method of heat movement to circulate air throughout the home?
   A. Conduction.
   B. Convection.
   C. Radiation.
   D. Infiltration.
6. What are the major differences between active and passive solar techniques?
7. *True or False?* An open loop solar collection system works best in a southern climate that is not subject to extensive cold weather and freezing.
8. Describe the purpose of the absorber plate in a flat-plate collector.
9. Flat-plate collectors work on the principle that:
   A. shorter light waves get longer when converted to heat.
   B. longer light waves get shorter when converted to heat.
   C. heat can be converted to light.
   D. wavelengths stay the same when light is converted to heat.
10. *True or False?* Visible light passes through glass easier than infrared energy.
11. *True or False?* A compound parabolic collector needs to track the sun.
12. *True or False?* The zenith path refers to the altitude of the sun in the sky.
13. *True or False?* Parabolic dish collectors are typically used to collect energy for residential use.
14. Passive solar collection techniques include:
    A. indirect gain.
    B. isolated gain.
    C. direct gain.
    D. All of the above.
15. *True or False?* A Trombe wall is often associated with the indirect gain approach to passive solar energy.

16. Eutectic salts may be used as:
    A. tracking devices.
    B. measuring devices.
    C. photovoltaic cells.
    D. storage media.
17. Explain three concepts associated with passive solar architecture.
18. Identify the advantages and disadvantages of the different types of solar collection techniques.
19. You install a small passive solar greenhouse on the rear of your home. Through a series of temperature-controlled vents, the warm air from the greenhouse is transferred to the living space within the home. You estimate the heat gained from the greenhouse will save about $300 in structural heating per year, and the greenhouse costs $2800 to purchase and install. What is the payback period for the installation of the greenhouse?
20. _____ have the ability to displace _____, causing the flow of electricity in a photovoltaic cell.
21. List three current applications for photovoltaic cells.

Activities

1. Construct, test, and evaluate scale models of various types of collectors.
2. Construct a solar hot dog cooker.
3. Calculate the payback of a professionally installed solar hot water system.
4. Incorporate passive solar design techniques into architectural design.

Workers install solar panels that will power the lighting system used to illuminate a bicycle parking area in Chicago's lakefront Millennium Park. (Photo by Bill Lyons, Spire Solar Chicago)

7

An Introduction to Power

Basic Concepts

- Identify the difference between work and power.
- Define power.
- Identify the basic power systems.
- List the basic elements of all power systems.
- Define horsepower (hp).

Intermediate Concepts

- Recognize the various power components in electrical circuits and fluid circuits.
- Summarize the advantages and disadvantages of various forms of power.

Advanced Concepts

- Describe various forms of power for specific applications.
- Diagram the basic power components in an electrical circuit or a fluid circuit.
- Calculate the efficiency of power systems and conversion devices.
- Compute power and hp for various forms of power.

Power is needed in our technological society. It lights our cities, cooks our food, and washes our clothes. Power is also needed in our transportation systems. Without power in transportation, cars could not run, airplanes could not fly, and subways could not operate. Power is something that is easily taken for granted. We do not often think about where the power needed to perform specific tasks comes from. As you ride your bicycle, think about what makes it move:

- You supply the energy to turn the pedals.
- The energy you supply is then changed into a form of power.
- As the pedals turn, the chain and gears power the bicycle. See **Figure 7-1.**

Figure 7-1. Pedaling a bicycle converts human energy into mechanical power to drive the chain and gears, moving you and the bicycle forward.

Types of Power Systems

Work is the application of force that moves an object a certain distance. Power is the rate at which work is done. Sources of energy, such as wind, solar, and heat, are harnessed to perform useful work. When energy is harnessed, converted, transmitted, and controlled to perform useful work, we call this a ***power system***. In the harnessing of energy sources, machines are used to convert energy into movement.

Electrical Systems

Electrical systems are power systems that use electrical energy to do work. The most common electrical power components include switches for controlling the flow of electricity, fuses or circuit breakers for protecting electrical circuitry, wires for transmitting electricity, and loads (such as lights, heaters, motors, or appliances) for utilizing electricity. Consider what your world would be like if electricity was not available. You would not have light to allow you to easily function when it is dark outside. To get hot water for a shower, you would need to build a fire to heat the water. Using a toaster to make toast for breakfast would not be an option. Electrical systems affect each of us daily!

Consider, also, how electricity is important to our technological world. Manufacturing plants need electricity to run machines. Without electricity, products like stereos and bicycles could not be produced. The communication systems around us could not operate without electricity. The television, telephone, and radio would be useless pieces of equipment without electricity. Our system of transportation would not be as efficient and safe as it is today without an electrical system. An electrical system is needed for traffic signals and for the air bag safety system in automobiles. Electricity propels several modes of public transportation, as well. It is needed in most everything we use today. Electrical systems are discussed further in Chapter 8.

Work: The application of force that moves an object a certain distance.

Power system: A system in which energy is harnessed, converted, transmitted, and controlled to perform useful work.

Mechanical Systems

Mechanical systems are power systems that use mechanical energy to do work. This is energy created by motion. ***Machines*** are devices used to manage mechanical power. Six simple machines are used to control and change mechanical power: the lever, the pulley, the wheel and axle, the

Technology Link

Communication: Radios and Satellites

Several kinds of communication technology rely on energy and power to work properly. Radios and satellites transmit signals by altering electromagnetic waves with frequencies below those of light. These waves travel through the atmosphere and space.

In the early 1900s, it was discovered that messages could be combined with waves of electromagnetic energy that radiate through space at the speed of light. The discovery of radio waves allowed people to instantaneously communicate over great distances. Within a few years, millions of people all over the world were tuning in to radio broadcasts for news and entertainment.

Satellites use super high frequency (SHF) electromagnetic waves, or microwaves, to transmit information across great distances. These waves travel directly from the satellite to its primary coverage area. Satellites can provide the signals for radio detecting and ranging (radar) systems, global positioning systems (GPSs), cellular telephones, and television.

The communication devices we use on a daily basis depend on the technology of electromagnetic energy. Without energy technology, we would not be able to send and receive information nearly as easily or quickly. Communication technologies are continuously evolving and improving, and in the coming years, we may see new technologies that allow us to communicate even better.

inclined plane, the wedge, and the screw. See **Figure 7-2.** These simple machines are explained in more detail in Chapter 9.

Machines used to produce work create mechanical energy. A machine or combination of machines can change the size, direction, and speed of force. Machines can also change the type of motion produced.

These systems are often used to harness energy from the wind, the force of water behind a dam, and high-pressure steam. When potential energy is harnessed and converted into mechanical power, it can be used to do work. Sometimes, mechanical power is put directly to work without any changes. For example, the blade on a lawn mower is connected directly to the power source—the crankshaft of a small gas engine. Most often, however, a change in the mechanical power is necessary before putting it to work. Mechanical power always has a direction of motion. This may be linear (straight and in one direction), reciprocating (back and forth), or rotational (spinning).

Fluid Systems

Fluid systems perform work using the energy created by liquids and gases. Fluid power can accomplish the movement of very heavy objects. Entire buildings and houses have been moved by the use of fluid power. This is why fluid power is referred to as the "muscles of industry." Examples of fluid power components include valves, hoses, air compressors or hydraulic pumps, cylinders, and motors.

Electrical system: A power system that uses electrical energy to do work.

Mechanical system: A power system that uses mechanical energy to do work.

Machine: A device used to manage mechanical power.

Fluid system: A power system that uses the energy created by liquids and gases to do work.

Pneumatic: A type of fluid power system that uses a gas, such as air, to transmit and control power.

Hydraulic: A type of fluid power system that uses a liquid, such as oil, to transmit and control power.

Figure 7-2. These six simple machines control all mechanical energy. Complex machines are combinations of two or more simple machines.

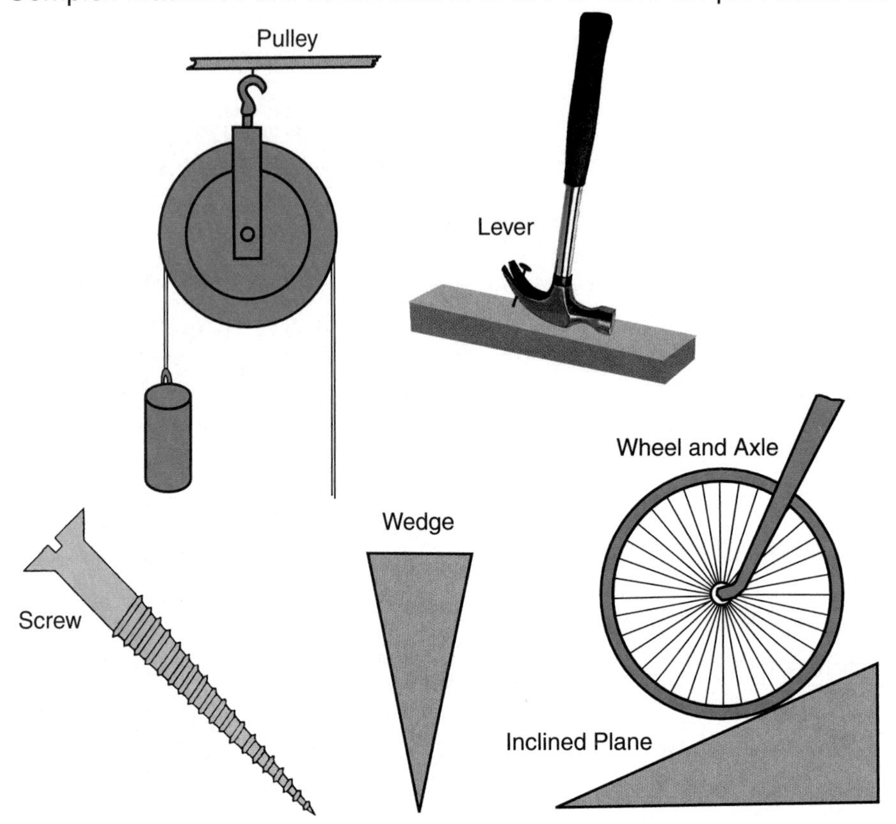

Figure 7-3. Many machines used in construction, such as this end loader, use hydraulics to transmit power and control machine functions.

There are two types of fluid power systems: pneumatic and hydraulic. ***Pneumatic*** systems use a gas, such as air, to transmit and control power. ***Hydraulic*** systems use a liquid, such as oil, to transmit and control power. See **Figure 7-3.** A hydraulic system controls and operates the landing gear on airplanes. The process of forging parts in industry is accomplished by the use of pneumatics or hydraulics.

Fluid power has many advantages over the other forms of power. Mechanical power is often slow and awkward. Electrical power is often expensive and complex and is often converted back to mechanical power to do useful work. Fluid power systems are easily operated and controlled, durable, and accurate in their control. Chapter 10 explains the uses and principles of fluid power in more detail.

Characteristics of Power Systems

Power systems come in various sizes and perform a wide variety of tasks. See **Figure 7-4.** An automobile is a power system because the fuel is converted into power. A motor is a power system because electricity is

Figure 7-4. Examples of power systems. A—An animal's muscle energy can be used to power a vehicle or operate a machine. (Howard Bud Smith) B—Chemical energy from petroleum is converted to mechanical energy to power this snowmobile. (Bombardier, Inc.) C—Electrical energy is changed to mechanical power to operate this large shipboard winch. (Howard Bud Smith) D—Heat energy (steam) is converted to mechanical energy in this turbine to generate electric power. (Siemens) E—Chemical energy, in the form of natural gas, is converted to mechanical power to fuel this city bus. F—A welding generator converts chemical energy into electric power. (Hobart Brothers Company)

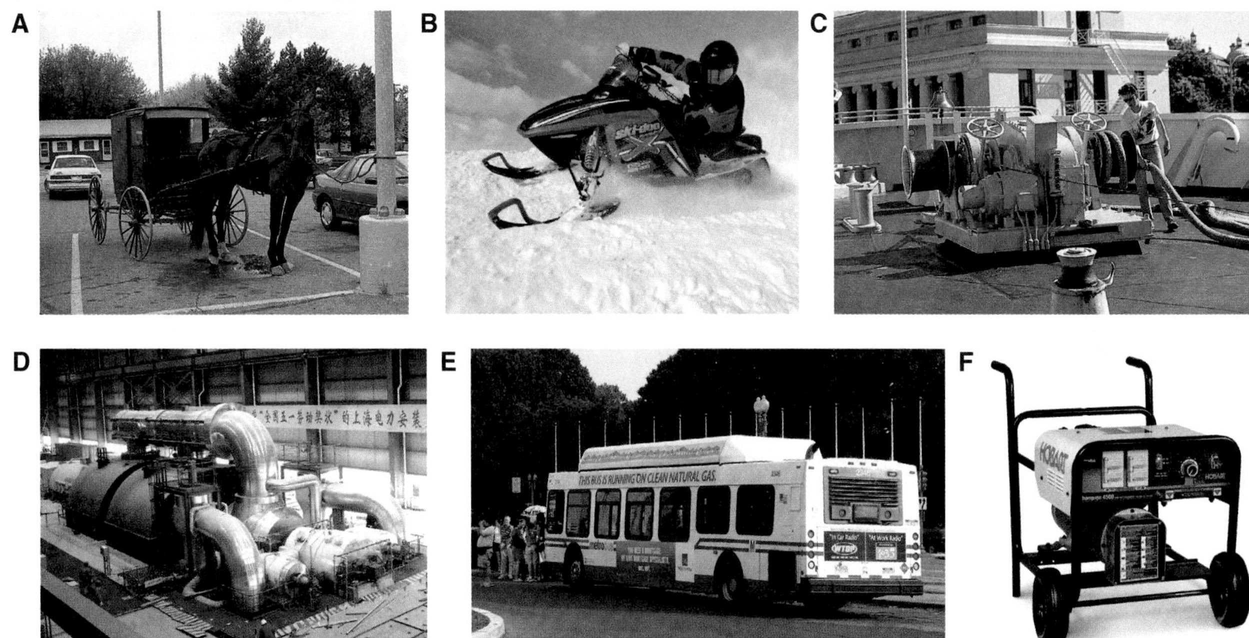

converted into power. An electric power plant is another good example of a large power system. See **Figure 7-5.** Power can be produced in three forms: electrical, mechanical, or fluid. In any of these three forms, power is comprised of two basic, measurable characteristics: effort and rate.

Figure 7-5. A large electrical generating plant in Spain. Coal is pulverized before being burned, providing maximum energy recovery, while decreasing pollution. (Siemens)

Effort

Effort: The force behind movement in a power system.

Force: Effort, in mechanical power.

Torque: Effort, in rotary mechanical power.

Foot-pound (ft.-lb.): The amount of force necessary to move a 1-lb. load a distance of 1′.

Pressure: Effort, in fluid power.

Pounds per square inch (psi): A unit used to measure pressure.

Effort is the force behind movement in a power system. See **Figure 7-6.** In linear mechanical power, this effort is usually known as ***force*** and is usually measured in pounds. In rotary mechanical power, the term for effort is ***torque***. Torque is a twisting or turning force. It is typically measured in foot-pounds (ft.-lbs.). A ***foot-pound (ft.-lb.)*** is the amount of force necessary to move a 1-lb. load a distance of 1′. You can see and feel the effects of force in mechanical power. When lifting an object that weighs 50 lbs., you need to generate more than 50 lbs. of pulling force to overcome the weight of the object.

In fluid power, effort is referred to as ***pressure***. It is usually measured in ***pounds per square inch (psi)***. Imagine turning on a garden hose. Without pressure, water would not flow from the hose. Water from a drinking fountain that has too little pressure is difficult to drink because

Figure 7-6. Some examples of the units of measure for effort in power systems.

Electrical	*Potential difference (voltage)* — causes the displacement of electrons	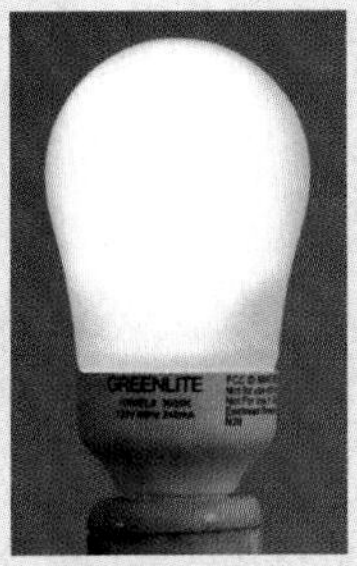
Fluid	*Pressure difference* — causes the displacement of a fluid (liquid or gas)	
Mechanical (Linear)	*Force* — causes the linear displacement (vertical or horizontal) of a mass	
Mechanical (Rotary)	*Torque* — causes rotary displacement of a mass	

the water barely flows above the nozzle. On the other hand, water from a drinking fountain that has too much pressure may spray or overshoot the water basin.

In electricity, the effort behind the movement of electrons is called ***voltage***. Voltage within a wire cannot be seen. Imagine a garden hose with marbles flowing out of it. The marbles represent the flow of electrons, and the pressure behind the marbles, pushing them through the hose, is the voltage. Just like a water fountain with too little pressure is not easy to drink from, an electrical circuit without adequate voltage does not work properly. A flashlight with old batteries that provides only dim light is a good example of inadequate voltage.

Voltage: In electricity, the effort behind the movement of electrons.

Rate: The characteristic of power that expresses a certain quantity per unit of time.

Ampere: In electrical power, the measurement for rate of flow.

Rate

Rate is the characteristic of power that expresses a certain quantity per unit of time. See **Figure 7-7.** Regardless of the unit of measure, all rate characteristics include both a quantity (gallons, electrons, revolutions, or distance) and a time element (seconds, minutes, or hours).

In electrical power, the measurement for rate of flow is the ***ampere***. As ampere flow increases, a greater amount of work can be accomplished. For instance, a lightbulb that draws 2 amperes produces more light than the same type of bulb that draws just 1 ampere when the same voltage is applied to both bulbs.

Figure 7-7. Some examples of the units of measure for rate in power systems.

Electrical	*Ampere rate* = $\frac{\text{Electrons displaced}}{\text{Time}}$	
Fluid	*Fluid rate* = $\frac{\text{Flow}}{\text{Time}}$	
Mechanical (Linear)	*Linear mechanical rate* = $\frac{\text{Mass displaced}}{\text{Time}}$	
Mechanical (Rotary)	*Rotary mechanical rate* = $\frac{\text{Revolutions}}{\text{Time}}$	

Revolutions per minute (rpm): A measurement of movement in a rotational mechanical system.

Gallons per minute (GPM): One of the most common measurements of rate of flow in a fluid system.

Energy source: A force that has the capacity to do work.

In mechanical power, the rate characteristic involves a type of movement per time. If the mechanical system is a rotational system, the movement is usually measured in ***revolutions per minute (rpm)***. The rate is often determined in a measure of distance per time, such as feet per second (fps) or feet per minute (fpm), if the mechanical system is linear (like a conveyor belt). Since cars can travel great distances, it is necessary to use a larger scale for measuring rate. Most people are familiar with the speedometer in a vehicle that measures in miles per hour (mph).

Water flowing from a garden hose is an example of flow in a fluid system. The water flows from the hose as a result of pressure. The rate at which the water is flowing can be measured in terms of volume per time. One of the most common measurements of rate of flow in a fluid system is ***gallons per minute (GPM)***.

Basic Elements of All Power Systems

The three basic power systems are electrical, mechanical, and fluid. No matter how big or small these power systems are, they each have basic elements, or functions, in common. From a lawn mower to a power plant, every power system incorporates these basic functions:

- An ***energy source*** is required for a power system to function. Fuel is often used as an energy source, but other sources, such as water or wind power, may also be used.
- A ***conversion method*** is necessary to convert energy so some type of work is produced. For example, falling water spins a turbine that operates an electrical generator to produce electricity.
- A ***transmission path*** is needed to move energy to the point where it is supposed to produce work. An example of a transmission path is the electrical power lines often seen strung across the landscape. See **Figure 7-8**. These paths provide a means of transporting energy from the point where it was generated to where it will produce work.
- A ***storage medium*** is necessary when power must be stored for use at a later point in time. A battery is a common type of power storage device. A spring is a mechanical power storage device.
- ***Protection devices*** shield components in power circuitry from excessive effort or rate of flow. For example, fuses protect an electrical circuit from too many amperes flowing through the circuit.
- ***Advantage-gaining devices*** modify the effort and rate characteristics of power in order to

Figure 7-8. Electrical power lines are the transmission paths for electrical energy from the generating station to the end users.

achieve a goal. A lever, for example, can multiply force to create additional leverage. Similarly, a step-up transformer can create a high voltage suitable for transmission over great distances. A step-down transformer reduces the voltage, while providing higher amperage suitable for household use. Advantage-gaining devices cannot produce more power than they utilize. They can only modify the effort and rate characteristics of the total amount of power provided.

- ***Control systems*** are needed to control the power within a system. The simplest type of control is an on-and-off motion control. More complex control systems include one-way motion control and variable motion control. The throttle on a lawn mower is a control system. See **Figure 7-9.**
- ***Measuring devices*** are required in power systems and provide a source of feedback to monitor how well the system is functioning. They include meters, indicators, and gauges.
- A ***load,*** or output, is the final goal of the power system. It is the work done by the system. For example, a lawn mower's load is the cutting blade. Electrical loads include lighting, motors, heating elements, and appliances. Fluid power loads include cylinders for linear motion and motors for rotary motion.

Calculations of Power Systems

The power available in a system can be measured or calculated for each form of power—electricity, fluidics, or mechanical. The ability to measure power is important since it can tell us many things about how a power system is performing. For instance, measuring mechanical power can tell us if an engine is producing power to specification. Calculating the electrical power used in a circuit can help us estimate how much using a particular machine or appliance for a number of hours will cost.

Figure 7-9. Control systems may be one-way or variable. The throttle of this lawn mower (inset) is used to vary the amount of power from the source (engine) to the load (cutting blade).

Conversion method: A necessary process to convert energy so some type of work is produced.

Transmission path: A means of transporting energy from the point where it was generated to the point where it will produce work.

Storage medium: A device that is necessary when power must be stored for use at a later point in time.

Protection device: A device that shields components in power circuitry from excessive effort or excessive rate of flow.

Advantage-gaining device: A device that modifies the effort and rate characteristics of power in order to achieve a goal.

Control system: A system necessary to control the power within a system.

Measuring device: A device required in power systems that provides a source of feedback to monitor how well the system is functioning.

Career Connection

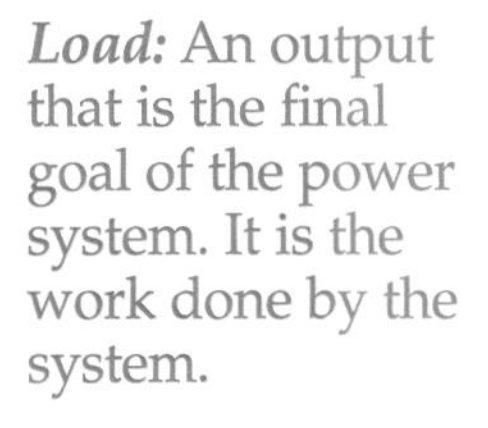

Power Line Repairers

A power line is an example of a transmission path. Energy is transported through power lines over great distances before it can be made useful as work. Power lines distribute electricity across the country, from generators to buildings. Power line repairers are necessary to ensure the public has continued use of electricity.

The job of a power line repairer involves a great deal of physical work. Power line repairers must replace poles and wires. They must repair and then test the power lines in order to ensure safety and functionality.

Power line repairers must have knowledge of mechanical and electrical principles. They must be able to work the machines necessary to aid in their repair work. Power line repairers have to work within safety limits so as to not hurt themselves or anyone else in the area. To be a worker in this field, long-term on-the-job training is required. The yearly salary may range from $27,000 to $66,000.

Load: An output that is the final goal of the power system. It is the work done by the system.

Work

Work creates movement by using a form of energy. Calculate work using the following formula:

- work = distance × force (W = D × F)

Force times the distance through which the force acts equals work. Work is measured in ft.-lbs. for conventional U.S. measure.

Suppose a rider in a canoe weighs 120 lbs. How much work is being done, if the canoe is paddled 600′?

- W = D × F
- W = 600′ × 120 lbs.
- W = 72,000 ft.-lbs.

If 2000 lbs. have to be moved 30′, how much work has to be done?

- 2000 lbs. × 30′ = 60,000 ft.-lbs. of work

Power

Power is the amount of work performed over a period of time. It can be calculated by using the following generic formula:

- power = work/time (P = w/t)

When referring to mechanical power, the equation may be written as: P = F × D/t

In this equation, *F* stands for force and is usually measured in pounds. *D* stands for distance and is usually measured in feet. The following is an example of using the mechanical power formula:

If a crane operator is going to move a 1000-lb. barrel of nails up 40′ to a fourth story window in 30 seconds, how much power is developed?

- P = 40′ × 1000 lbs./30 seconds
- P = 40,000 ft.-lbs./30 seconds
- P = 1333 ft.-lbs. per second

In electrical power systems, power is measured in ***watts***. The quantity of effort in determining watts is known as voltage. Voltage is the force pushing the electrical current through the conductor. The rate factor is typically referred to as amperage. ***Amperage*** refers to the rate at which electrons or coulombs move through a conductor. The ***coulomb*** is a unit of electrical charge equal to the amount of electricity transported by 1 ampere in 1 second. As an example, assume that an electric heater operates at 220 volts and draws 10 amperes. How much electrical power is being utilized?

- effort × rate = power
- 220 volts × 10 amps = 2200 watts

This formula is known as ***Watt's law*** and will be discussed in greater detail in Chapter 8.

Watt: A measurement of power in electrical power systems.

Amperage: The rate at which electrons or coulombs move through a conductor.

Coulomb: A unit of electrical charge equal to the amount of electricity transported by 1 ampere in 1 second.

Watt's law: Power equals effort multiplied by rate.

Efficiency

Efficiency is the relationship between input energy, or power, and output energy, or power. Measuring and calculating the efficiency of a power system is important to the process of improving efficiency. Efficiency is important to all forms of power devices, as well as energy and transportation devices. It is determined by using the following generic formula:

- output/input × 100 = percentage of efficiency

Use the efficiency formula to determine the efficiency of an electrical transformer designed to double the voltage fed into the transformer:

- input voltage = 120 V
- output voltage = 240 V
- input amperage = 20 A
- output amperage = 8.7 A

Both voltage and amperage are needed to measure wattage. Watt's law provides an equation (wattage = amperage × voltage) to measure electrical power in watts.

1. Use Watt's law to determine power in and power out.
   W = V × A
   120 V × 20 A = 2400 W (in)
   240 V × 8.7 A = 2088 W (out)
2. Use the efficiency formula to determine the efficiency of the transformer.
   output/input × 100 = percentage of efficiency
   2088 W/2400 W × 100 = 87% efficient

Calculate the efficiency of a gear set with the following characteristics:

- Drive gear: 20 ft.-lbs. torque at 100 rpm
- Driven gear: 9.7 ft.-lbs. torque at 200 rpm
  20 ft.-lbs. × 100 rpm = 2000 (in)
  9.7 ft.-lbs. × 200 rpm = 1940 (out)
  1940/2000 × 100 = 97% efficient

Calculating the efficiency of a power system is slightly more complicated than calculating the efficiency of an individual component, such as the transformer or gear, previously presented.

Determine the total system efficiency of a wind generator. See **Figure 7-10.** Each component of the wind generator system is described below, with a corresponding efficiency.

- The blades of a wind generator convert about 55% of all air flowing through them into useful mechanical power.
- The generator converts about 90% of all the mechanical power into electricity.
- The inverter changes the electricity into a suitable voltage for household use. It is about 85% efficient.

To find the efficiency of the entire system once the efficiency for each component is known, multiply the efficiency of each component with the next component, until all have been factored into the equation. Note that the efficiencies are represented in decimal form.

- .55 × .90 × .85 × 100 = 42% total system efficiency

Horsepower (hp)

Horsepower (hp): The standard measuring unit of power equal to the energy needed to lift 33,000 lbs. 1′ in 1 minute.

Horsepower (hp) is one standard measuring unit of power. The energy needed to lift 33,000 lbs. 1′ in 1 minute equals 1 hp. Calculate hp by using the following formula:

- hp = work/(time in minutes × 33,000)

If 200 lbs. are lifted 165′ in 1 minute, how much hp is developed? To find the answer, follow these steps:

1. Determine how much work was done.
   W = D × F
   W = 165′ × 200 lbs. = 33,000 ft.-lbs.
2. Find the hp generated.
   hp = work/(time in minutes × 33,000)
   hp = 33,000/(1 × 33,000)
3. The answer is 1 hp.

Once the hp values are known, the efficiency can be determined using the following formula:

- output hp/input hp × 100 = percentage of efficiency

Figure 7-10. The total efficiency of a wind generator is calculated from the combined efficiencies of its components. A—The blades and the generator. B—The inverter.

A

B

A motor consumes 2 hp and produces 1.75 hp of mechanical power. How efficient is the motor?

- 1.75 hp/2.00 hp × 100 = 87.5% efficient

Using hp as a unit of measurement allows one form of power to be compared with another form of power. To determine the efficiency of an electric motor, for example, measure the wattage used by the motor to determine the input power. With the proper instruments, measure the output rotary mechanical power produced by the motor. Because these units of power measurement are different from each other, the efficiency of the motor cannot be determined. In this scenario, using the hp unit of measurement is particularly useful. By converting both the input wattage of the motor and the output mechanical power into hp, the motor's efficiency can be calculated.

Curricular Connection

Math: Canceling Units

Canceling units, or unit analysis, is based on the principles that multiplying anything by 1 does not change its value and that anything divided by itself equals 1. You know that 1 minute equals 60 seconds. Therefore, the following is true:

$$\frac{1\text{ min}}{60\text{ sec}} = \frac{60\text{ sec}}{1\text{ min}} = 1$$

The fact that the conversion ratio equals 1 no matter which units are on top is essential to the process of canceling units. If you are given a measurement in feet per second (fps), and you want to know the speed in miles per hour (mph), you will need to convert the units. For example, suppose a vehicle is moving at 80 fps, and you need to know how many mph it is going. Knowing that there are 12 inches in 1 foot, 60 seconds in 1 minute, 60 minutes in 1 hour, and 5280 feet in 1 mile, you can convert the units in the following way:

$$\frac{80\text{ }\cancel{\text{feet}}}{1\text{ }\cancel{\text{sec}}} \times \frac{60\text{ }\cancel{\text{sec}}}{1\text{ }\cancel{\text{min}}} \times \frac{60\text{ }\cancel{\text{min}}}{1\text{ hour}} \times \frac{1\text{ mile}}{5280\text{ }\cancel{\text{feet}}}$$

You can cancel units just like you cancel factors when multiplying fractions. Make sure the units cancel correctly. You are left with the units you wanted to find: miles/hour, or mph.

$$= \frac{80 \times 60 \times 60 \times 1\text{ miles}}{1 \times 1 \times 1\text{ hour} \times 5280} = 54.5\ \frac{\text{miles}}{\text{hour}}$$

When you encounter an energy or power calculation that requires you to convert units, just set up the problems so the units you do not need cancel out. This may take some rearranging, and you may need to find out additional conversion ratios. Once you have the problem set up correctly, you will be left with the units you need for your answer. Remember that you should be left with only one unit above the line and one unit below the line. All other units must be canceled.

While there are many formulas to convert units of power to hp, the following are the most commonly used:

- **Electrical power.**
  watts/746 = hp
- **Fluid power.**
  psi × GPM × .000583 = hp
- **Rotary mechanical power.**
  torque (ft.-lbs.) × rpm/5252 = hp
- **Linear mechanical power.**
  $\frac{550 \text{ ft.-lbs.}}{\text{sec}} = \text{hp}$ or $\frac{33{,}000 \text{ ft.-lbs.}}{\text{min}} = \text{hp}$

In using any of these formulas, the units must be exact. For example, if a formula specifies that time be measured in minutes, it must be recorded in minutes, not seconds or hours. This applies to any unit of measure included in a formula. If force is to be measured in pounds, other units, such as tons or ounces, will not result in an accurate calculation.

Measurement Conversion

It is important to know how to use measurements in power and other areas. The two measuring systems we use in the United States are U.S. customary and SI metric. It is important to know how to convert these measurements from one system to another. The chart in **Figure 7-11** provides a guide for converting many of the most common measurements used in energy, power, and transportation.

Figure 7-11. A conversion chart for U.S. customary and SI metric values.

	U.S. Customary to SI Metric					SI Metric to U.S. Customary				
	Customary Units	×	Conversion Factor	=	Metric Units	SI Metric Units	×	Conversion Factor	=	Customary Units
Length	inches	×	2.54	=	centimeters	centimeters	×	.4	=	inches
	yards	×	.9144	=	meters	meters	×	1.1	=	yards
	miles	×	1.609	=	kilometers	kilometers	×	.6	=	miles
Weight	pounds	×	.4536	=	kilograms	kilograms	×	2.2	=	pounds
Force	pounds	×	4.448	=	newtons	newtons	×	.2248	=	pounds
Torque	pound-feet	×	1.356	=	newton-meters	newton-meters	×	.7376	=	pound-feet
Pressure	pounds per sq. inch	×	6895	=	pascals	pascals	×	.000145	=	pounds per sq. inch
Work	foot-pounds	×	1.36	=	joules	joules	×	.7376	=	foot-pounds
Heat	British thermal units	×	252	=	calories	calories	×	.003968	=	British thermal units
Mechanical Power	horsepower	×	746	=	watts	watts	×	.001341	=	horsepower
Electrical Power	watts	×	1	=	watts	watts	×	1	=	watts

Summary

We often take power for granted and do not always realize it affects our everyday lives. Power is energy that has been converted to produce useful work. When energy is converted, transmitted, and controlled to do useful work, it is referred to as a power system. There are three basic types of power systems in our society: electrical systems, mechanical systems, and fluid systems. Electrical power systems range in size from toasters to manufacturing plants. Mechanical power systems convert mechanical energy into work—the energy of motion. Machines are devices that change and control mechanical power. Fluid power systems are referred to as the "muscles of industry" and may be pneumatic or hydraulic. Each type of power system contains the same basic elements, including an energy source, a conversion method, and a means of transmission, control, and storage. Electrical power is measured in watts. All forms of power can be measured in or converted to a standard unit known as horsepower (hp).

Key Words

All the following words have been used in this chapter. Do you know their meanings?

advantage-gaining device
amperage
ampere
control system
conversion method
coulomb
effort
electrical system
energy source
fluid system
foot-pound (ft.-lb.)
force
gallons per minute (GPM)
horsepower (hp)
hydraulic
load
machine
measuring device
mechanical system
pneumatic
pounds per square inch (psi)
power system
pressure
protection device
rate
revolutions per minute (rpm)
storage medium
torque
transmission path
voltage
watt
Watt's law
work

Test Your Knowledge

Write your answers on a separate sheet of paper. Do not write in this book.

1. What is the difference between work and power?
2. Define power.
3. Name the three basic types of power systems.
4. Summarize how different forms of power are used for specific applications.

5. Give examples of three components of an electrical circuit and three components of a fluid circuit.
6. Name the two types of fluid power.
7. Discuss briefly the advantages and disadvantages of different forms of power.
8. The term for the rate characteristic of electricity is known as _____.

Matching questions: For Questions 9 through 17, match the examples on the left with the correct term on the right.

9. _____ Fuel, water, and wind.
10. _____ Falling water that spins a turbine to produce electricity.
11. _____ Electrical power lines.
12. _____ Batteries and springs.
13. _____ Fuses.
14. _____ Levers, transformers, and transmissions.
15. _____ Throttles.
16. _____ Meters, indicators, and gauges.
17. _____ Cutting blades, motors, and televisions.

A. Advantage-gaining devices.
B. Control systems.
C. Conversion methods.
D. Energy sources.
E. Loads.
F. Measuring devices.
G. Protection devices.
H. Storage media.
I. Transmission paths.

18. Electrical power is measured in _____.
19. Explain the concept of efficiency.
20. The efficiency of a component is calculated by dividing the power output by the _____.
21. A machine has a motor that draws 9.7 amps at 120 volts. The machine can produce the mechanical equivalent of 900 watts of power. How efficient is the machine?
22. Calculate the efficiency of a hydraulic motor, given the following characteristics:
    - input effort: 100 psi
    - input rate: 75 GPM
    - output effort: 12.5 ft.-lbs.
    - output rate: 1200 rpm
23. A power system is comprised of four components. The efficiency for each component is as follows: 90%, 93%, 67%, and 85%. What is the total efficiency for the power system?
24. What is the formula for determining work?
25. A person moves 350 lbs. of roofing shingles 100′. How much work has been performed?
26. A standard unit of measurement for all forms of power is known as _____.

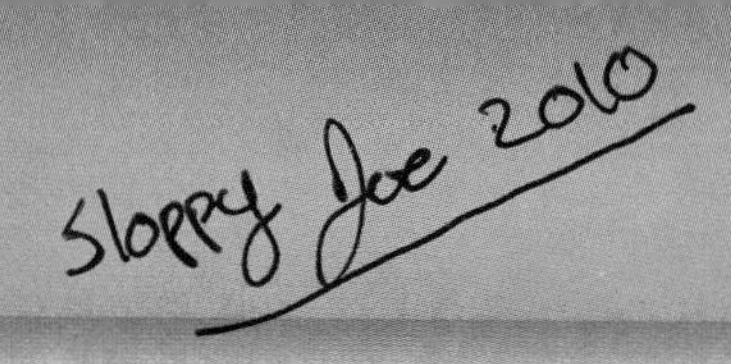

27. If the work in the Problem 20 is accomplished in 25 minutes, what is the average hp utilized?

28. Motors are rated in terms of their horsepower (hp) output. In order to calculate the efficiency, the power input and output of a motor must be in the same unit of power measurement. An electric motor on a lift is rated at 8 hp. The electrical current required by the motor is 28 amps at 240 volts. What is the efficiency of the motor?

29. A compressor motor provides 10 hp to a pump. If the motor is 85% efficient, what is the power input to the motor?

30. A lift can move 3000 lbs. a distance of 8′ in 10 seconds. How much hp is delivered to the lift, if the lift is 90% efficient?

31. An alternator in a car produces 40 amps at 12 volts. How much power is being produced? How much hp is being produced?

32. A hydraulic pump produces a pressure of 300 pounds per square inch (psi) and a flow rate of 75 gallons per minute (GPM). What hp is the pump producing?

Activities

1. Design a machine that uses multiple forms of power. The machine should serve a practical purpose and include each of the basic elements of power systems discussed in the chapter. Include a sketch of your machine and a sample workflow.

2. Calculate the kilowatt-hour (kWh) consumption of your home for one month. Note the number of people living in your home and the square footage of the home. Be prepared to share your findings in class.

Coast Guard helicopters perform vital search and rescue missions in natural disasters, such as hurricanes and floods. In the aftermath of the 2005 hurricane and flood in New Orleans, Louisiana, this crewman is searching rooftops for victims waiting for rescue and evacuation. (U.S. Coast Guard)

8

Electrical Power Systems

Basic Concepts

- List the types of current and explain how they are produced.
- State how electrical power is measured.
- Name different types of electrical circuits and give examples of their uses.
- Identify how electricity and magnetism are related.
- Identify the types of electrical circuits used in transportation vehicles.

Intermediate Concepts

- Describe how atoms act to produce electrical current.
- Perform electrical calculations using Ohm's law, Watt's law, and Kirchoff's law.
- Summarize applications of various switching schemes.
- Explain the purpose of common electrical components, such as fuses, breakers, wires, and batteries.
- Select wire size and other electrical components for specific applications.
- Discuss safety factors associated with live electricity.

Advanced Concepts

- Use electrical instrumentation safely and properly.
- Troubleshoot basic electrical circuits to identify and remedy problems.
- Design simple electrical circuits to meet specified criteria.

An electrical system uses electrical energy to perform work. Electricity is the most widely used and versatile type of energy. Both simple and very complex electrical systems can be found in all aspects of modern technology. Electrical systems are used to power home appliances that make our lives easier and safer. To save and maintain lives, hospitals use electrical systems to run special machines. Manufacturing facilities rely on electrical systems to run the machines that produce needed products. See **Figure 8-1.**

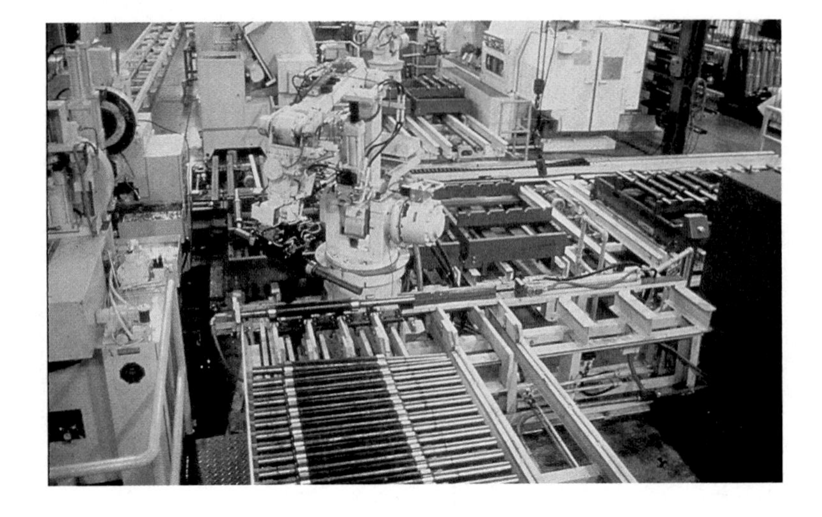

Figure 8-1. Electricity powers most modern machinery.

Electrical systems are also an essential part of most energy, power, and transportation systems. An electrical system in an automobile lets us start the engine and keep it running. Separate electrical systems operate the lights, dashboard displays, radio, and other accessories. Electrical systems can provide the source of propulsion for transportation systems such as elevators, moving sidewalks, and escalators. See **Figure 8-2.** Often, vehicles are propelled by electric motors. Auto manufacturers are beginning to develop electric automobiles as an alternative to the gasoline or diesel engine.

Atom: The "building block" of everything we know of on earth. Atoms are made up of protons, neutrons, and electrons.

These systems rely on the use of atoms. An ***atom*** is the "building block" of everything we know of on earth. Rocks, trees, people, air, metal, and plastic objects are just a few examples of things made up of atoms. Since atoms are at the root of all electrical systems, this is where we will begin our study.

Figure 8-2. Electricity powers some forms of transportation. Electrical energy powers the people mover and the elevator system shown here. (United Airlines and Murphy/Jahn, Architects)

Career Connection

Power Plant Operators

Power plants use generators to produce electricity. This electricity is then distributed across power lines to substations and ultimately to homes and buildings throughout the community. Power plant operators help monitor and maintain machines to ensure the machines are functioning correctly.

By careful monitoring of the machines, operators are able to predict and often avoid malfunctions. The workers in a power plant not only monitor the machines, but they also assist in mechanical and electrical repairs. In performing these tasks, they are involved in working with both electrical and mechanical equipment. Their jobs also include testing the machines in order to maintain safe operation.

Power plant operators must have a mechanical aptitude, and they must understand the chemical and physical properties of substances. Necessary skills for this job include engineering and mathematics. A knowledge of computer and electronic equipment is also essential. To be a worker in this field, long-term on-the-job training is required. The yearly salary may range from $33,000 to $70,000.

Atomic Structure

Polarity: The type of charge an atomic particle has.

Ring: The path followed by electrons in an atom.

Valence ring: The outermost ring of electrons in an atom.

Atoms are extremely small, yet they are made up of several particles: protons, neutrons, and electrons. A proton has a positive (+) electric charge. An electron has a negative (–) electric charge. A neutron has no charge at all. Neutrons are said to be neutral. In an atom, unlike charges attract each other, and like charges repel one another. The type of charge an atomic particle has is called its ***polarity***. See **Figure 8-3.**

At the center is the nucleus, which contains the protons and neutrons. Electrons travel around the nucleus in elliptical paths. There are exactly as many electrons as there are protons in the nucleus. The number of separate paths followed by the electrons depends on their number. Each path is called a ***ring***. As the rings fill with electrons, new rings form to allow room for more electrons. Compare the two atoms illustrated in **Figure 8-4.** Notice in each atom how the number of protons equals the number of electrons. Note, too, that the atom with more electrons has more rings.

Atoms remain stable (neutral) by keeping the numbers of protons and electrons equal. Sometimes atoms need to lose electrons or gain electrons to maintain their stability. Electrons in the outermost ring, called the ***valence ring***, are the ones gained or lost. If an atom has more electrons than protons, it will lose some electrons from its valence ring. When the number of protons is greater than electrons, the valence ring will pick up electrons from nearby atoms

Figure 8-3. The parts of an atom. Atoms contain a nucleus of protons and neutrons, while electrons orbit around the nucleus.

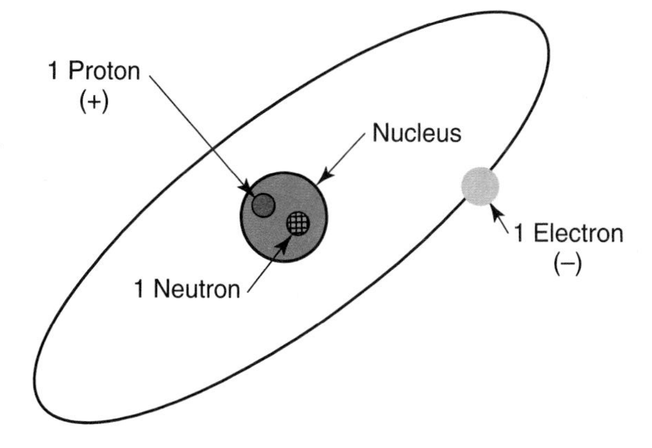

Figure 8-4. Atoms always try to remain stable. This means that the number of electrons will balance the number of protons.

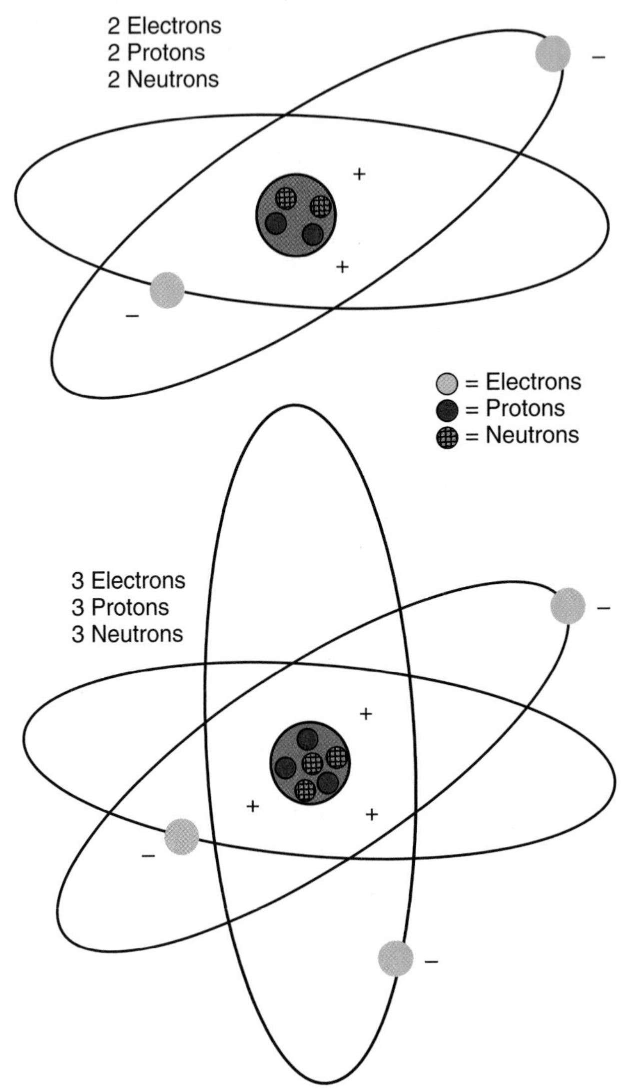

Conductor: A material made of atoms that transfer electrons easily.

Insulator: A material made of atoms that do not transfer electrons easily.

Semiconductor: A material that is both a conductor and an insulator.

so the atom is stable. It is the electrons in the valence ring that lie behind the theory of electricity.

Some atoms are able to lose and gain electrons easily, while others have a difficult time. A material made of atoms that transfers electrons easily is called a ***conductor***. Wires used to carry electricity are good conductors. They are often made of copper. A material made of atoms that does not transfer electrons easily is called an ***insulator***. Insulators resist the flow of electricity. Some materials are both conductors and insulators. This type of material is called a ***semiconductor***.

Electrical systems depend on the action of electrons in relation to insulators and conductors. A semiconductor is a good example of how materials can be adapted to control movement of electrons from one atom to another. **Figure 8-5** shows one type of semiconductor (called a diode) and how it operates.

Electron Theory and Current

Normally, an atom is neutral (has no charge). This tells us that the number of protons equals the number of electrons. When we are able to force electrons from their valence rings, electricity is produced. **Figure 8-6** diagrams the movement of electrons through a conductor. As a negatively charged electron is forced from its valence ring, the atom becomes positively charged (there is one more proton than there are electrons). The electron forced from the valence ring is attracted to the positively charged atom to its right. Next, the atom that just lost an electron is positively charged and picks up an electron from an atom on its left.

According to the ***electron theory***, electrons flow from a negative point to a positive point. The negative point has an abundance of electrons, and the positive point has a shortage. The flow of electrons in a conductor is called ***current***.

There are two types of current: direct current (DC) and alternating current (AC). Each is based on how current moves through a conductor (wire). In ***direct current (DC)***, electrons move only in one direction. ***Batteries*** are common devices that produce DC. This type of current is used in modern automobiles, as well as in many other devices that require portable power, such as radios and flashlights. See **Figure 8-7.**

Alternating current (AC) involves electrons flowing first in one direction and then reversing and flowing in the other direction. See **Figure 8-8.** AC is easier than DC to send long distances through wires. This type of current is used in houses and businesses to provide energy for lights, appliances, and machinery.

Electron theory: Electrons flow from a negative point to a positive point.

Current: The flow of electrons in a conductor.

Figure 8-5. A diode is a type of semiconductor. It allows electrons to jump from one atom to another in one direction, but not in the opposite direction. The diode controls electron flow the same way a check valve controls the flow of water in a plumbing system.

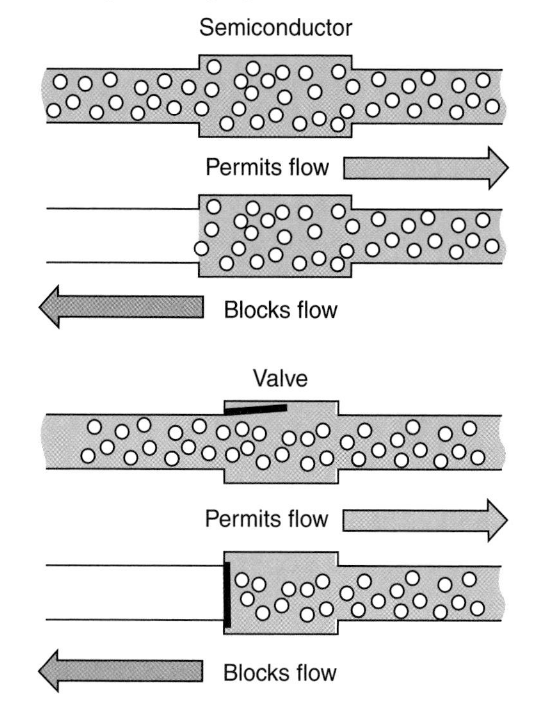

Figure 8-6. When electrons migrate from one atom to another in a conductor, an electric current is produced.

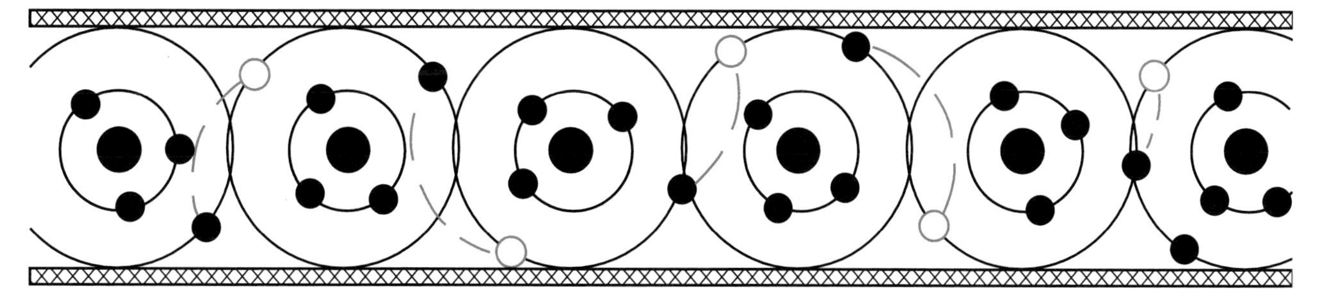

Figure 8-7. Batteries chemically store electricity and release it as direct current (DC). (Exide)

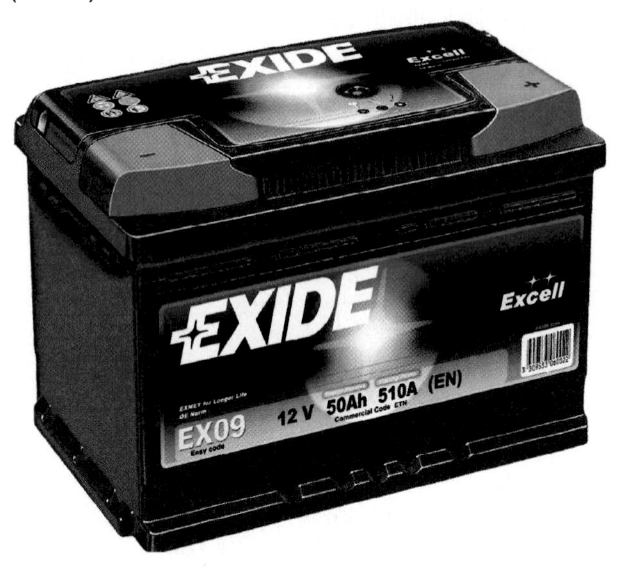

Direct current (DC): A type of current in which electrons move only in one direction.

Battery: A common device that produces direct current (DC).

Alternating current (AC): A type of current in which electrons flow first in one direction and then reverse and flow in the other direction.

Current flow: The rate at which electrons move, or amperage.

Resistance: Opposition to the flow of current.

Effort, Rate, and Opposition in Electrical Systems

Like other forms of power, such as mechanical or fluid power, electricity is composed of an effort characteristic and a rate characteristic. The effort characteristic in electricity is referred to as *voltage*. Other terms used to describe electrical effort include *electromotive force, potential,* and even *force*. Voltage is the force behind the movement of electrons. The rate at which the electrons move is referred to as *amperage*, or ***current flow***.

To try and understand the relationship between voltage and amperage, picture a garden hose with marbles flowing out of it. The pressure pushing the marbles can be thought of as the voltage. The actual marbles themselves represent the ampere flow. Now, to push the marbles through the hose, some opposition will have to be overcome. This opposition is referred to as ***resistance*** in electrical terminology, and it is often represented by the omega (Ω) symbol.

Electrical Circuits

An ***electrical circuit*** is the heart of any electrical system. A simple electrical circuit is made of several components: a power source, a load, and conductors. The components are connected so electrical current flows in a complete path. See **Figure 8-9.**

This type of circuit must have a power source that has both negative and positive terminals (connection points). Electricity flows from the negative terminal, through the circuit, to the positive terminal. Remember, according to electron theory, electricity flows from negative to positive.

We can manipulate the electricity in the circuit by adding components, such as lightbulbs, motors, and switches, to name a few. By placing these objects in the circuit, we allow electricity to flow through them so they can

Figure 8-8. The forward and reverse flow of alternating current (AC) is represented by a sine wave, like the one shown. Household current goes through 60 full cycles every second.

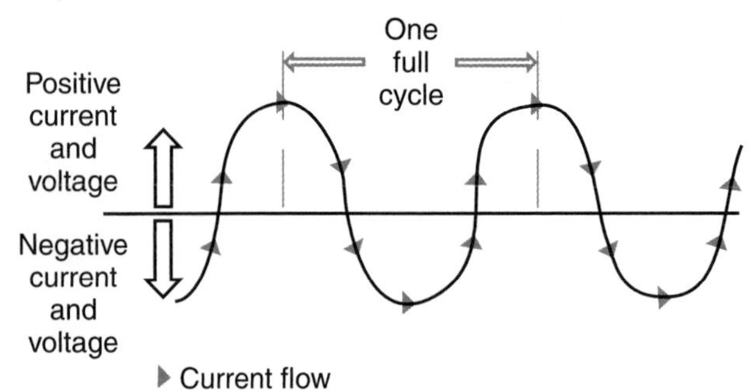

operate. A component that uses electricity in a circuit is called a *load*. Conductors connect the load to the power source.

Figure 8-9. A simple circuit consists of a power source, a load, and conductors.

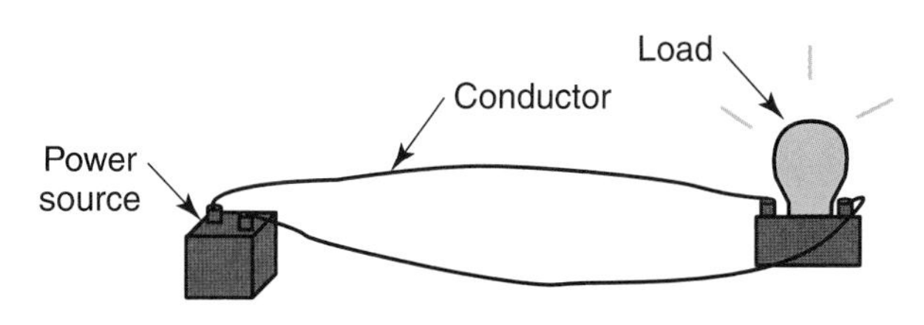

Schematics

When planning and describing electrical circuits, it is easier to do it graphically than in words. ***Schematic drawings*** are used to represent an electrical circuit graphically. See **Figure 8-10.**

Schematic drawings are like road maps. Road maps show the paths taken by travelers, while schematics trace the path electron flow will take in an electrical or electronic circuit. Schematics include symbols that represent the components in the circuit. **Figure 8-11** explains some of these symbols.

Electrical circuit: A power source, a load, and conductors connected together so electrical current flows in a complete path.

Open Circuits, Closed Circuits, and Short Circuits

Some terms that are helpful in describing the way a circuit is functioning are *closed circuit*, *open circuit*, and *short circuit*. A ***closed circuit*** is a properly functioning circuit in which all loads are energized. An ***open circuit*** is a circuit or part of a circuit that is not energized. Open circuits may be created intentionally, as in the case of a switch that is positioned to turn off a light. They may also occur accidentally when connections are not made properly. Such open circuits result in a loss of continuity through the circuit. ***Continuity*** is the continuous flow through a component or through an entire circuit. Accidental open circuits cause problems and can sometimes be difficult to troubleshoot.

Schematic drawing: A drawing that traces the path electron flow will take in an electrical or electronic circuit. Symbols are included that represent the components in the circuit.

Figure 8-10. Schematic drawings are the "road maps" of electrical and electronic circuits.

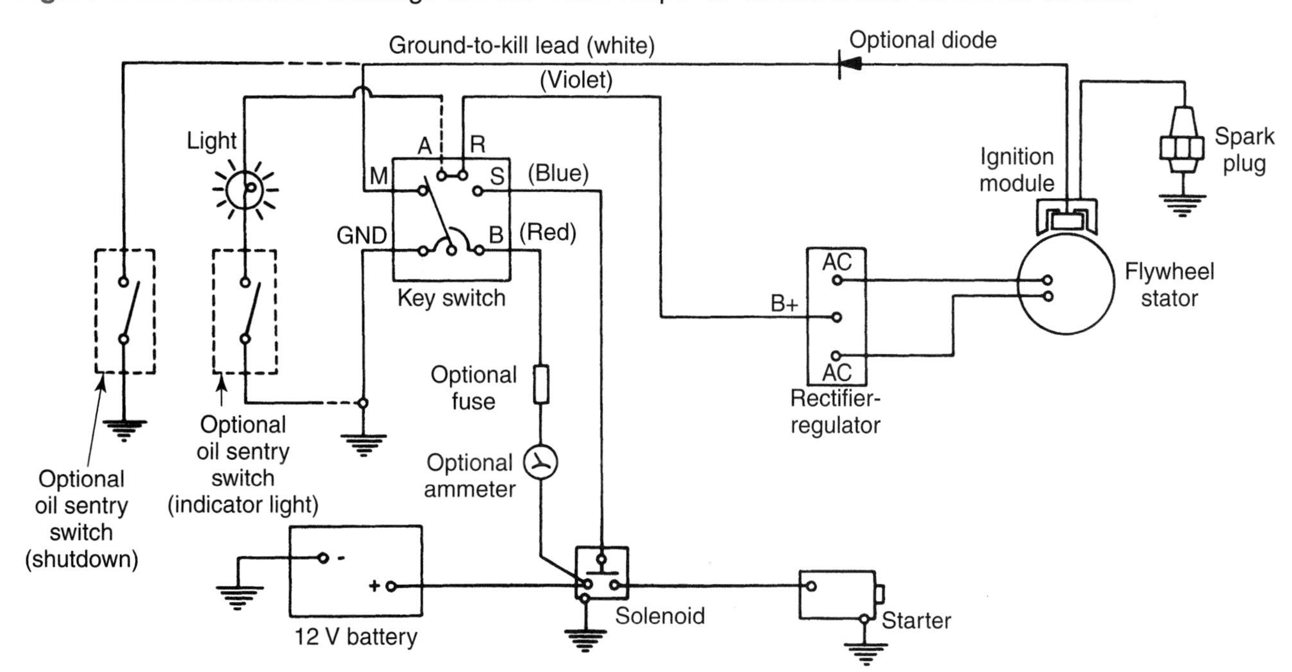

Figure 8-11. Some electrical and electronic symbols used in circuit schematics.

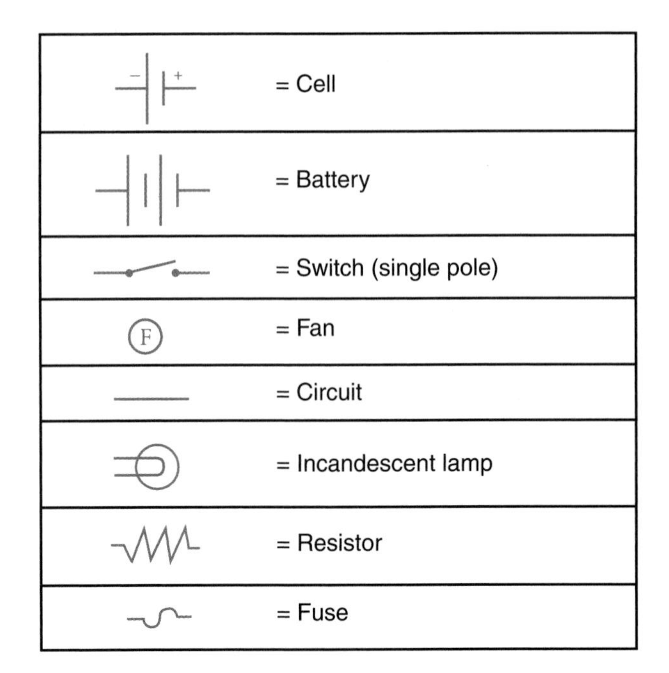

Closed circuit: A properly functioning circuit in which all loads are energized.

Open circuit: A circuit or part of a circuit that is not energized.

Continuity: The continuous flow through a component or an entire circuit.

Lastly, you may hear of a short circuit. A ***short circuit*** occurs when the load is bypassed and the hot wire comes directly into contact with the return leg or with something grounded. When short circuits happen, sparks may fly, and many amps (current) flow, since there is virtually no opposition to current flow. **Figure 8-12** shows all these conditions.

Laws that Describe Electricity

There are several laws that describe the way the characteristics of electricity, such as voltage and amperage, behave within a circuit. The first law we will describe is Ohm's law. ***Ohm's law*** states that voltage (E) can be determined by multiplying current (I) by resistance (R):

voltage = current × resistance ($E = I \times R$)

The equation can also be rewritten to calculate either current or resistance. The Ohm's law circle helps in identifying the correct equation to use. See **Figure 8-13.** To use this tool, follow these steps:

1. Cover the variable you want to find. For example, to find current, cover the *I*.
2. Review the values on the circle. In this example, you will see E/R.
3. Note the applicable equation. For example, current equals voltage divided by resistance ($I = E/R$).

With a simple formula like Ohm's law, it is important to recognize that two out of the three quantities must be known to determine the unknown quantity. The unknown quantity can be identified in the formula through simple

Figure 8-12. Circuit conditions.

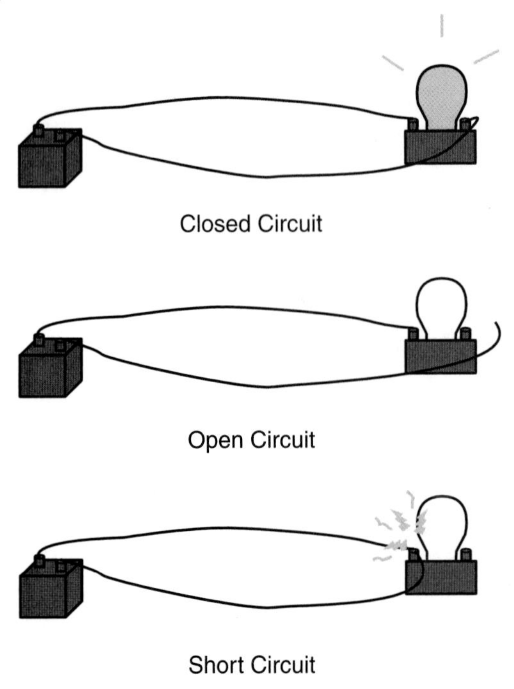

multiplication or division. Examples of all potential calculations are provided next to the Ohm's law circle in Figure 8-13.

Figure 8-13. The Ohm's law circle can be used to calculate voltage (E), current (I), or resistance (R).

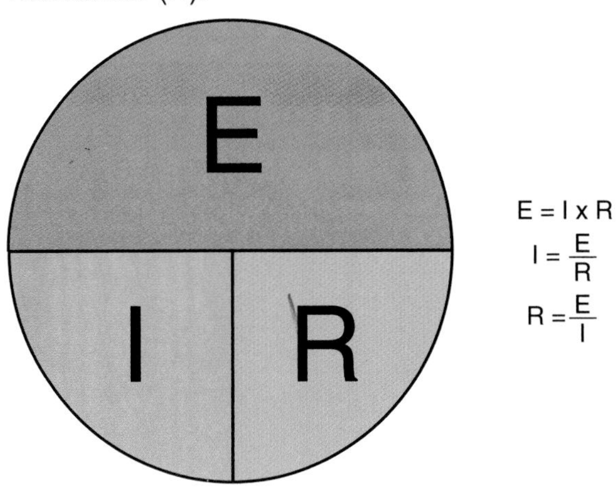

Ohm's law expresses the relationships between the characteristics of electricity. In the Ohm's law formula, $E = I \times R$, you can see that current and resistance are directly related to voltage. This means that an increase or decrease in either current or resistance will affect voltage in the same way. In other words, an increase in amperage or resistance will result in an increase in voltage. A decrease in current or resistance will result in a decrease in voltage. For example, if current flow through a load increases, the amount of voltage dropped across the load will also increase.

Looking at the formula $I = E/R$, you can see that resistance and current are indirectly related to each other. This means that a change in resistance will cause an opposite change in current. For example, if the resistance of a load increases, the amount of current flowing through the load decreases. If the resistance of the load decreases, the amount of current flowing through the load increases. Understanding the relationships found in Ohm's law could help a technician predict the outcome of a change within a circuit and help a technician troubleshoot a circuit.

Short circuit: A circuit in which the load is bypassed and the hot wire comes directly into contact with the return leg or with something grounded.

Ohm's law: Voltage (E) can be determined by multiplying current (I) by resistance (R).

Let us say a 15-amp circuit breaker that feeds your toaster and microwave keeps tripping. The microwave is rated at 120 vac (volts alternating current), 10 A. The toaster is unlabeled, but you know it plugs into a standard 120-vac receptacle and the heating element has a heat resistance of 20 Ω. You can now use Ohm's law to determine if there is something wrong with one of the appliances or the breaker, or if the circuit is simply overloaded. To do this, you need to determine the total draw on amperage. You already know that the microwave draws 10 A. The amperage for the toaster, however, is unknown. To find the amperage for the toaster, you can use the following Ohm's law equation and fill in the known values for voltage and resistance:

$$I = E/R$$
$$I = 120 \text{ vac}/20\ \Omega$$
$$I = 6 \text{ A}$$

Now, to find the total draw on amperage, add the two amperage values:

$$6 \text{ A} + 10 \text{ A} = 16 \text{ A}$$

The total amperage is 16 A. Since 16 A is greater than the amperage rating of the circuit breaker (15 A), you can conclude that the circuit is simply overloaded when both appliances are turned on.

The next law that we will investigate is Watt's law. Watt's law provides an equation to measure electrical power. It is often referred to as the electrical power formula:

$$P = I \times E$$

Figure 8-14. The Watt's law triangle can be used to calculate wattage (P), voltage (E), or current (I).

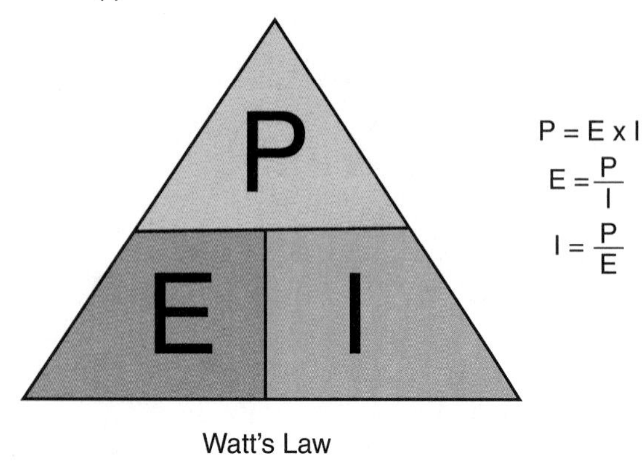

In this formula, power (P) can be calculated by multiplying current (I) by voltage (E). The *P* represents power. Electrical power is measured in watts (W). A flow of 1 ampere of electrical current at a pressure of 1 volt produces 1 watt of power.

The equation can also be rewritten to calculate either current or voltage. The Watt's law triangle helps in identifying the correct equation to use. See **Figure 8-14.**

Let us say current of 20 A flows through a portable heater that is plugged into a standard 120-vac receptacle. To find the amount of power used by the heater, multiply the amount of current flowing through the heater by the voltage provided by the receptacle:

$$P = I \times E$$
$$P = 20\text{ A} \times 120\text{ vac}$$
$$P = 2400\text{ W}$$

An all-inclusive formula for electricity combines Ohm's law and Watt's law. It provides many different methods of calculating voltage, current, resistance, and wattage. See **Figure 8-15.**

Kilowatt-hour (kWh): Wattage multiplied by the number of hours the wattage is used and then divided by 1000.

Wattage: Power being used.

Calculating Kilowatt-Hours (kWh)

When electrical consumption is calculated for billing purposes, it is measured in ***kilowatt-hours (kWh)***. The kWh formula includes both the ***wattage***, or power being used, and the amount of time the wattage has been used. The prefix *kilo-* means "thousand." Therefore, 1 kilowatt (kW) represents 1000 watts. For the time portion of the calculation (hours), remember that 1 hour comprises 60 minutes. The following are examples of 1 kWh of electricity:

Figure 8-15. A helpful reference that contains both Ohm's law and Watt's law formulas.

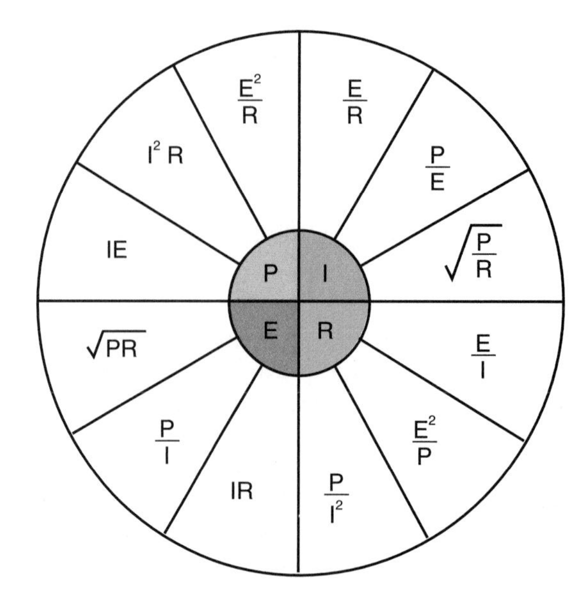

Curricular Connection

Math: Ohm's Law and Watt's Law

Through the wonders of algebraic manipulation, it is possible to combine Ohm's law and Watt's law so one or more unknowns (resistance, voltage, amperage, or wattage) may be found if any two of the other variables are known. Let us say you want to determine the wattage utilized by a baseboard heater. You know the voltage for the heater is 208 V, and the information on the heater says to replace the heater coil with a 15 Ω coil if it goes bad. Using the power formula, you can find the answer through any of the following methods:

$$P = \frac{E^2}{R} = \frac{208 \times 208\text{ V}}{15\ \Omega} = \frac{43{,}264\text{ V}}{15\ \Omega} = 2885\text{ W}$$

Solve for amperage using Ohm's law, and then solve for power using Watt's law.

$$I = \frac{E}{R} = \frac{208\text{ V}}{15\ \Omega} = 13.87\text{ A} \qquad P = I \times E = 208\text{ V} \times 13.87\text{ A} = 2885\text{ W}$$

Solve for power using a combination of Ohm's law and Watt's law.

$$P = I^2 \times R = 13.87\text{ A} \times 13.87\text{ A} \times 15\ \Omega = 2885\text{ W}$$

- One 1000-watt heater running continuously for 1 hour.
- Ten 100-watt lightbulbs left on for 1 hour.
- One 2000-watt heater running continuously for 30 minutes.
- Five 200-watt lightbulbs left on for 1 hour.
- One 4000-watt central air conditioner running continuously for 15 minutes.

To determine the costs associated with using electricity, kWh must be calculated, and the cost per kWh must be known. The kWh is determined by multiplying wattage by the number of hours the wattage is used and then dividing by 1000. Next, the cost of electricity is calculated by multiplying the kWh by the cost per kWh:

kWh = (watts × hours)/1000
cost of electricity = kWh × (cost/kWh)

Let us say twelve 100-watt lightbulbs are left on in a work area for 7 1/2 hours per day, and the cost of electricity is $.09/kWh. How much would the total cost be to light the area for a full workweek, Monday through Friday?

1200 watts × 7.5 hours × 5 days = 45,000 watt-hours
45,000 watt-hours/1000 = 45 kWh
45 kWh × $.09/kWh = $4.05 for the week

A dial meter, known as a *watt-hour meter*, is one type of meter typically used by electric utility companies to measure the power or wattage used within a home over a period of time. The meter readings are usually recorded by a meter reader once per month. Some meters can now simply

Technology Link

Manufacturing: Peak Load Demand Billing

Imagine a day when most of the household chores get done. The laundry gets washed, the dishwasher is run, and plenty of electricity is utilized. Now imagine receiving your electric bill. Inside the bill, it states your maximum electrical power for the month was used on that particular day. This is no surprise to you. The bill goes on to say you will be billed for having used that much power every day of the month, even though both you and the power company know that you did not use that much power every day. It does not seem fair. Fortunately, this is not the way most residential customers are billed by their electric utility providers, but it is the way most industrial consumers of electricity are billed. This method of billing for electricity is known as *peak load demand billing*.

Compare two hypothetical manufacturing firms. Company A consumes roughly the same amount of electricity over an entire 24-hour period. Company B consumes the same total amount of electricity, but almost all of it is consumed over an 8-hour period.

Even though both companies may consume the same amount of electricity to manufacture their products, the power-generating utility has much more money invested in generation, transmission ,and distribution equipment to meet the needs of Company B than Company A. The reason is simple. Since equipment used to supply Company B is only used about one-third of the day, it yields a relatively poor return on investment. Conversely, the equipment used to supply Company A is used to full potential throughout the majority of the day and, therefore, yields a much better return on investment. The cost of producing electricity to meet the peak demand brought about by companies like Company B is also considerably more substantial than the cost of generating base-load electricity that a generating utility must produce to meet its demand all the time.

The power companies could average all costs among all consumers and still yield a positive return on their investment, but this would not be fair to those consumers that use equipment most efficiently, like Company A. Therefore, most companies bill their

be scanned with a reader that automatically records a measurement. Reading a dial meter is not difficult. See **Figure 8-16.** To read a dial meter, follow these rules:

- Always read the dials on a meter from left to right.
- Interpret the first dial on the left-hand side as thousands, the second dial as hundreds, the third dial as tens, the fourth dial as ones, and the fifth dial as tenths.
- If the pointer on a dial is not positioned directly over a number, use the lower number on the dial you are reading. For example, if the pointer is positioned somewhere between three and four, use the lower number, three.

industrial customers, such as manufacturing firms, according to each customer's maximum demand for a given period. That period is usually one month, but sometimes it can be a quarter of a year. A demand meter monitors electric utility usage and records the usage. A measurement is usually taken every 15 minutes throughout the day. At the conclusion of the billing cycle, the maximum demand for any 15-minute interval is used to compute the utility bill.

It may not seem fair, but the power-generating utility's rationale is that if X amount of power was used during one period in the billing cycle, in theory, that amount of power must be made available for the entire billing cycle. After all, if a company consumes a very large amount of power for a part of one day, it certainly has the ability to consume that much power every day. The bottom line is that power companies would prefer to see a stable demand profile over a 24-hour period. This way, they would be using their generating, transmission, and distribution equipment most efficiently. Too many companies like Company B in a territory can cause a utility to have to add generating capacity, even though the utility can only sell the electricity for one-third of the day.

Energy management is a term that describes how utility costs are managed in a manufacturing environment. Large companies may make use of computers and software to manage the demand. For instance, if a company has four large machines and three of them are running, the energy management system might not allow the fourth machine to be started until one of the first three is shut off. This will help to keep the peak demand down to an acceptable level. In turn, this may save thousands of dollars on the monthly or quarterly utility bill. In some cases, it is even more cost effective to have a company manufacturing 24 hours per day. The power-generating utility can offer electricity at a discounted cost, as long as the electricity is not being consumed during the peak hours of 8 AM–5 PM, when it is in the greatest demand.

- If you cannot tell if the pointer is positioned over a number, check the dial to the immediate right. If the pointer on that dial is positioned between the nine and zero, use the lower number on the dial you are reading. If the pointer on the dial is positioned on or past the zero, use the higher number on the dial you are reading.

With the dial readings from the previous month, the total kWh consumption for the month and average daily kWh consumption can be calculated. The following is an example of this calculation:

current month's reading – previous month's reading = kWh consumed
kWh consumed /30-day billing period = kWh per day consumed
6453 kWh (read on May 5) – 5691 kWh (read on April 5)
= 762 kWh consumed
762 kWh / 30-day billing period
= 25.4 kWh per day (on average) consumed.

Figure 8-16. An electric meter records usage in kilowatt-hours (kWh). A—The center disk rotates, driving the dials that record the kWh of electricity used. B—Dials are read from left to right. This meter shows a reading of 3405.5 kWh.

Figure 8-17. A series circuit has only one path for electric current. When the switch is closed, the electrons flow through the circuit, and the lamps are on.

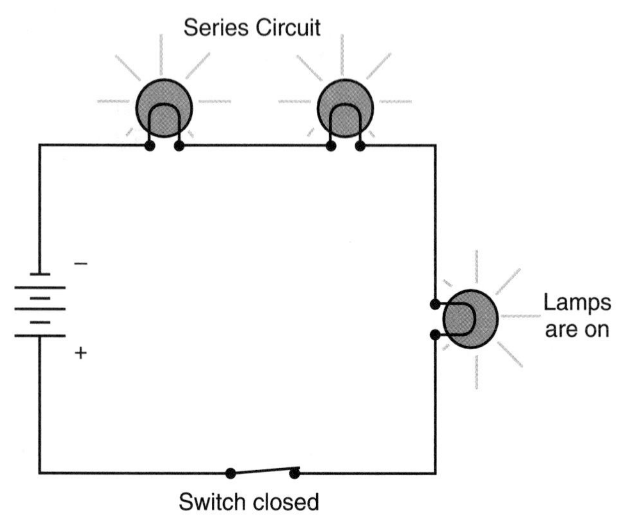

Series, Parallel, and Series-Parallel Circuits

There are three basic types of electrical circuits: series, parallel, and series-parallel circuits. These circuits have different characteristics and can be used for many applications. Circuits can be compared by the ways they allow current to pass through them.

Series Circuits

Series circuits have one continuous path for electrical current. If there is a break in the conductor or a bad connection, the whole circuit becomes useless. **Figure 8-17** illustrates a series circuit. Notice the single path for current between the components of the circuit. Electrical current flows from the negative terminal on the power source, through a single switch and the lightbulbs, and then to the positive terminal. One switch operates all three lightbulbs in the circuit at the same time. When it is opened, no current can pass through the circuit to power the lights. Refer to the series circuit in **Figure 8-18.**

Now let us investigate the characteristics of a series circuit. Voltage in a series circuit "drops" across each load. If you add up all the voltage drops across each load, they should equal the source voltage, as seen in **Figure 8-19.** The amount of each voltage drop varies with the resistance of each load. Opposition to electron flow, known as resistance, is easily calculated for a series circuit using a simple formula:

$$R_T = R_1 + R_2 + R_{3\ldots}$$

This means that the total resistance in a series circuit can be determined by adding the resistive value of each load in the circuit to determine the total resistance. If the voltage and total resistance in a circuit are known, it is easy to calculate the current or ampere flow in a series circuit using Ohm's law. Measuring amperage in a series circuit is also easy. Since there is only one path for current flow in a series circuit, amperage can be measured anywhere in the circuit. To say it another way, amperage is the same at all points within a series circuit, regardless of where it is measured.

Parallel Circuits

Unlike series circuits, parallel circuits allow more than one path for electrical current. Because there is more than one path, a break in the conductor or a bad connection might only shut off part of the circuit. Notice in **Figure 8-20** how the components of this circuit are arranged in branches. By adding a switch for every light, we can turn each one on or off individually, without affecting the other lights. See **Figure 8-21.**

Examining the characteristics of a parallel circuit is slightly more complicated than examining a series circuit, since there are multiple paths for current flow. The easiest characteristic to understand in a parallel circuit is voltage, since the potential difference between the hot, or feed, leg and the return leg remains constant. This concept is explained by Kirchoff's voltage law for parallel circuits, which states that the voltages across each branch of a parallel circuit are equal. See **Figure 8-22.** The concept is important because it allows for standardization. For instance, most appliances in the United States run on standard 120 vac, which is available at convenience receptacles located throughout residential and commercial structures. If these receptacles were wired in series, the voltage available at each receptacle would vary based on the loads in the circuit. Also, if any load in the circuit were switched off, all the loads in the circuit would switch off.

Neither of these scenarios is very practical. Both scenarios are, however, avoided in a parallel circuit. This is because the voltage or potential difference remains constant between the feed and the return leg and because various branches of the parallel circuit can be energized and de-energized without affecting one another.

Kirchoff's law states that current in a parallel circuit is the sum of current flowing in each of the circuit's branch circuits. A load such as a lightbulb can be turned off on one branch of the circuit without affecting bulbs that are turned on across other branches of the parallel circuit. Look again at Figure 8-22. It shows the amperage that would be expected to be measured at various points in a parallel circuit with several loads, based on the calculations provided.

Figure 8-18. This is the same circuit as in Figure 8-17. When the switch is opened, the lamps will not light.

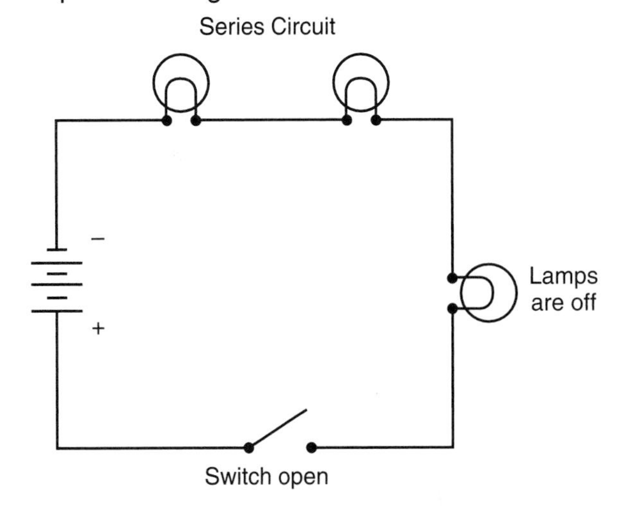

Figure 8-19. Voltage, amperage, and resistance in a series circuit.

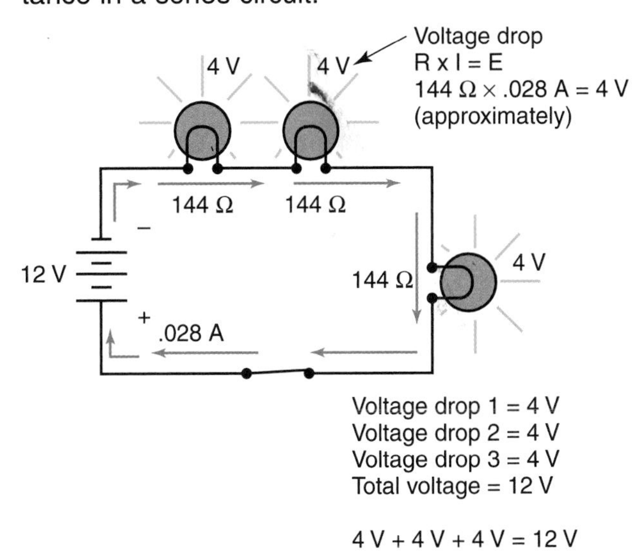

Figure 8-20. An example of a parallel circuit. Even if one of the switches is opened, the other lamps will remain lit.

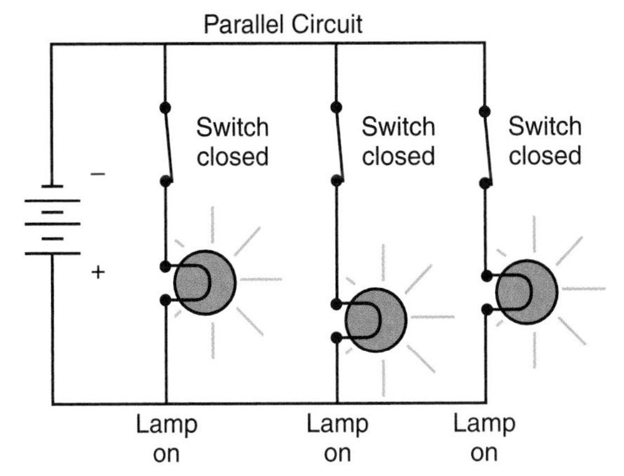

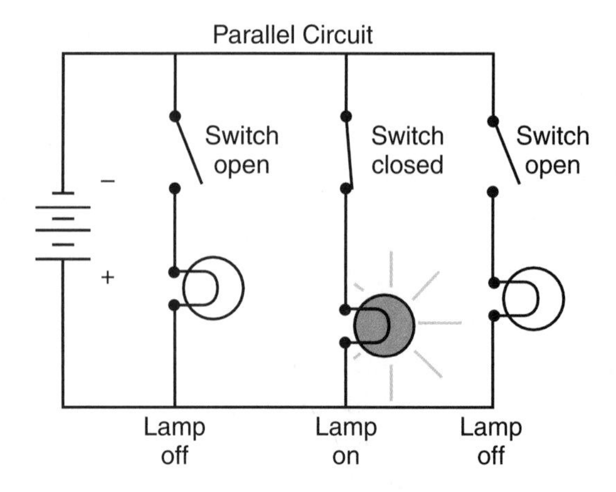

Figure 8-21. This is the same circuit shown in Figure 8-20 with two switches open. Can you trace the current in this illustration?

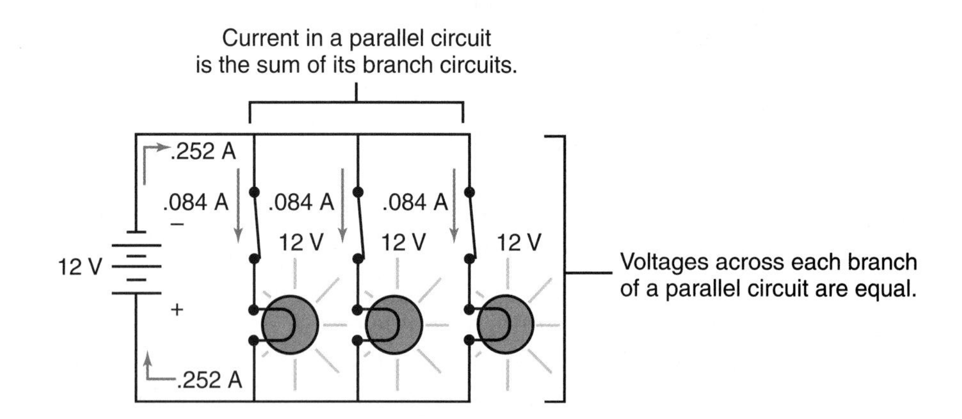

Figure 8-22. This circuit illustrates Kirchoff's voltage and current laws for parallel circuits. Notice that the voltages across each branch are equal and that the total current is equal to the sum of current flowing through each branch.

Resistance in a parallel circuit is perhaps the least intuitive characteristic to understand. This is because, as the load in a parallel circuit increases, resistance actually decreases! Every time a load is added to a parallel circuit, another path is created that allows the current to flow from the positive to the return leg. Thus, more opportunity for electrons to flow between the positive and negative legs is created, decreasing total resistance. The formula for calculating resistance in a parallel circuit is as follows:

$$\frac{1}{\frac{1}{R_1} + \frac{1}{R_2} + \frac{1}{R_3} \ldots}$$

The reciprocals within the formula ensure that the value of the total resistance in a parallel circuit is always less than the value of any individual resistance within the circuit. This includes the resistive value of the smallest load. See **Figure 8-23.**

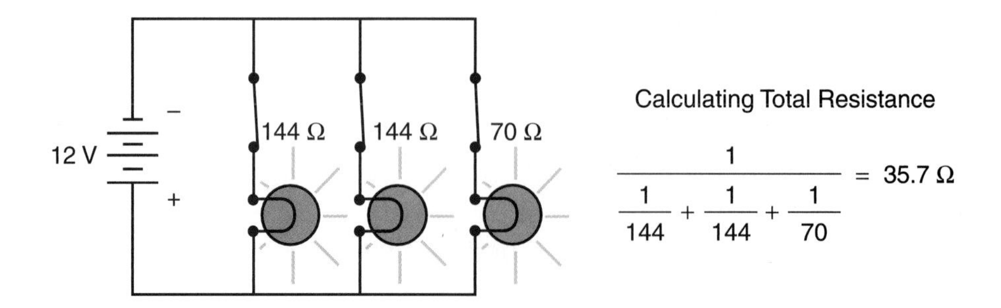

Figure 8-23. Calculating resistance in a parallel circuit. Notice that total circuit resistance is smaller than the resistance of any single component.

Series-Parallel Circuits

See **Figure 8-24** for an example of a series-parallel circuit. You should be able to recognize that it is a combination of the two circuits discussed earlier. In the circuit illustrated, the first lightbulb after the negative terminal is in series with everything else in the circuit. The other two lightbulbs in the circuit are arranged in parallel with each other. This means that a break at the first lightbulb would shut off the whole circuit. A break at any other lightbulb would still allow current to flow through some parts of the circuit. Follow the path of current to see how this can work.

The means by which resistance is calculated in a series-parallel circuit is as follows. First, calculate the parallel portions of the circuit, and then add that value to the value of the resistance in series within the circuit. If the resistance is calculated and the source voltage is known, the current can also be calculated using Ohm's or Kirchoff's law. An example is provided in **Figure 8-25.**

Figure 8-24. In a series-parallel circuit, part of the circuit is in series, and part is in parallel. A—All parts of the circuit are working. B—No current is found in the circuit. Do you know why? C—By looking at this circuit, can you tell which parts of it are parallel?

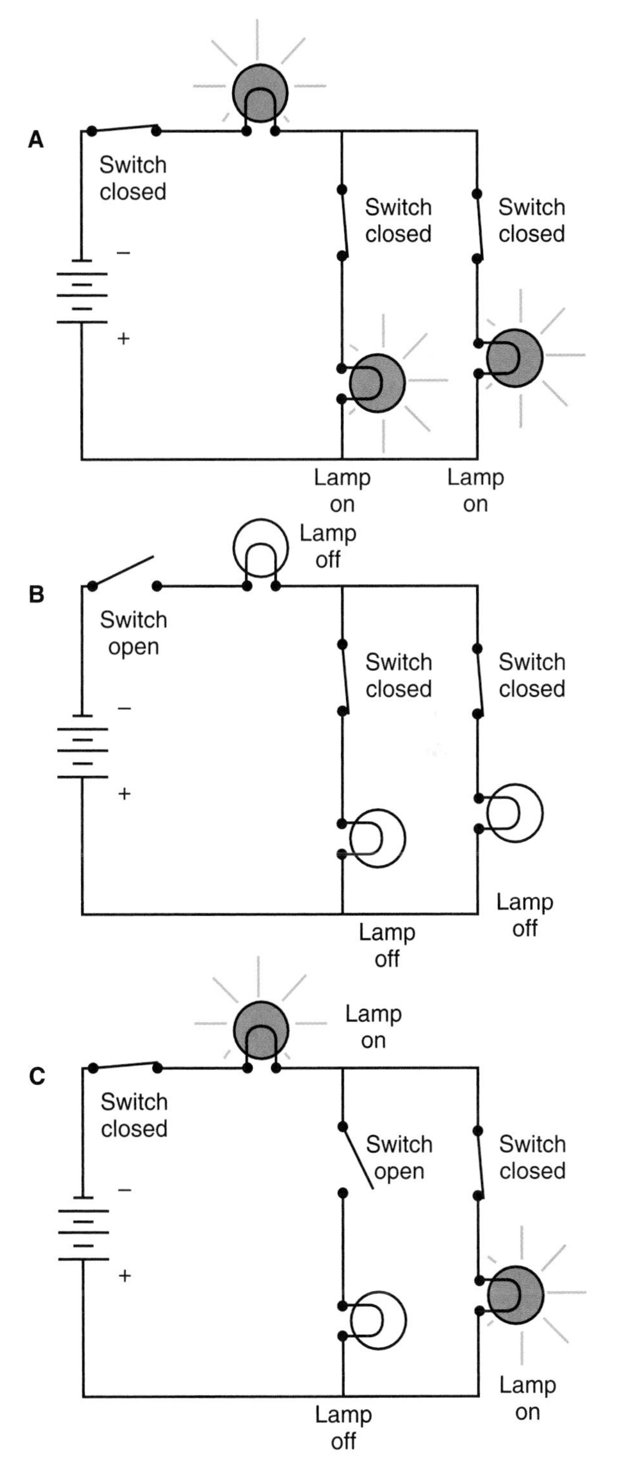

Magnetism

Magnetism, or the property, quality, or condition of being magnetic, is important in the study of electricity because the two can affect each other. Electrical current produces magnetism. Magnets can induce, or cause, electrical current in conductors.

A ***magnet*** is a material attracted to any metal containing iron. It has an invisible force field that makes it stick to these metals. Magnets usually contain two poles: north and south. These poles are on opposite ends of the magnet. ***Lines of force*** run between the poles on the outside of the magnet. These lines are also called ***flux***. You cannot actually see the flux, unless you place the magnet on a piece of paper and then sprinkle iron filings on the paper. See **Figure 8-26.** The iron filings form arcs between the north poles and south poles of the magnet and represent the lines of force.

When two magnets are brought together, their magnetic poles influence each other. Like poles repel each other, and unlike poles attract each other. In other words, a north pole and a

Magnet: A material attracted to any metal containing iron.

Line of force: A theoretical line running between the poles on the outside of a magnet.

Flux: The lines of force on a magnet.

Electromagnet: A conductor wrapped around an iron core. The two ends of the conductor are attached to a power source. When current passes through the conductor, the iron core becomes magnetized.

Electromagnetic induction: The production of electricity in conductors with the use of magnets.

Figure 8-25. Calculating unknown values in a series-parallel circuit.

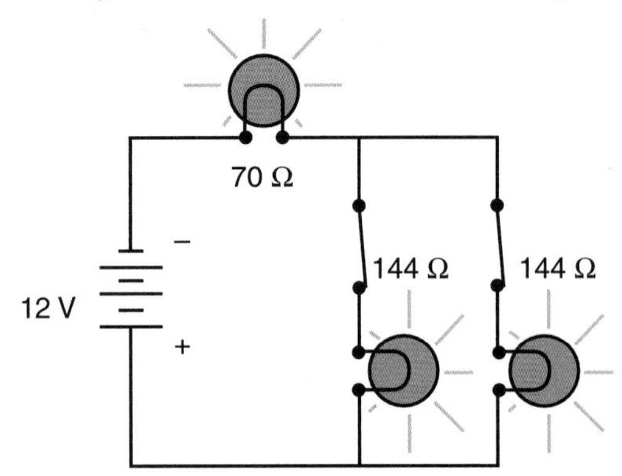

Calculating Total Resistance

First calculate total resistance of the parallel portion of the circuit:

$$\frac{1}{\frac{1}{144} + \frac{1}{144}} = 72\ \Omega$$

Then add the total resistance of the parallel circuit to the resistance in series:

$72\ \Omega + 70\ \Omega = 142\ \Omega$

Calculating Total Current

Use Ohm's Law to calculate total current:

$$\frac{E}{R} = I$$

$$\frac{12\ V}{142\ \Omega} = .085\ A$$

south pole are attracted to each other. Two north poles or two south poles force themselves away from each other.

Electromagnets

More than 150 years ago, Hans Oersted, a Dutch scientist, found that electricity produces a magnetic field. This discovery brought about the electromagnet. An ***electromagnet*** consists of a conductor wrapped around an iron core. The two ends of the conductor are attached to a power source. When current passes through the conductor, the iron core becomes magnetized. When this happens, the iron core is attracted to anything that has iron in it. **Figure 8-27** shows one application of an electromagnet.

Figure 8-26. Lines of flux, indicated by the dotted lines, run between the poles of a magnet.

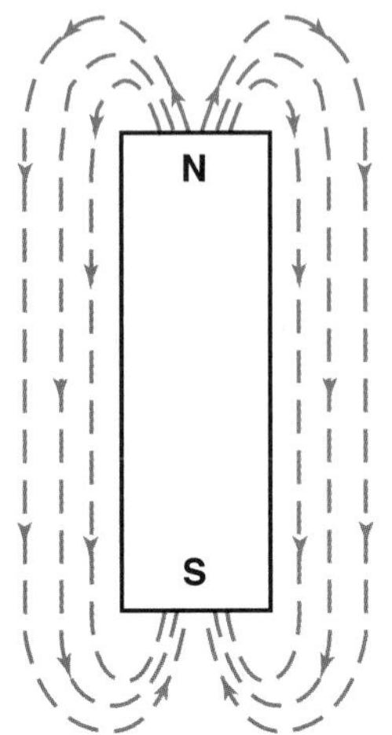

Electromagnetic Induction

As mentioned earlier, we can produce electricity in conductors with the use of magnets. This process is called ***electromagnetic induction***. Most of our electricity is produced this way. Remember that electricity is made when we are able to force electrons from their valence rings. When we pass a magnetic field through a conductor or a conductor through a magnetic field, the flux causes electrons to be forced from their atoms, producing electricity. See **Figure 8-28** for a description of electromagnetic induction.

Electrical Power Sources

We have seen that an important part of an electrical system is a power source. In fact, we could not have a working system without one. The electrical power source provides a method for producing electrical current in a circuit. Some examples of electrical power sources are cells, batteries, AC generators, and DC generators.

Figure 8-27. Scrap iron and steel can be picked up by an electromagnet.

Cells and Batteries

You have probably had to put cells inside many devices, like watches, calculators, smoke detectors, and radios. A ***cell***, often mistakenly called a *battery*, is a common device for storing electrical power. There are many different types of cells and batteries. You may already recognize the cells used in flashlights. The same cell may be used in some clocks and smoke detectors. Batteries are used in transportation vehicles. These are tougher and more powerful than flashlight cells. All cells and batteries produce DC.

A battery is made of one or more cells. Cells and batteries are devices that change chemical energy to electrical energy. They consist of two different materials called electrodes and electrolytes. The terminals on each end of a battery are actually the ***electrodes***. Remember that one electrode has a positive charge, and the other has a negative charge. The ***electrolyte*** is usually a liquid or paste that surrounds and touches both electrodes. The chemical reaction between the electrodes and electrolyte produces electrical current. See **Figure 8-29.**

Cell: A common device for storing electrical power. A cell changes chemical energy to electrical energy.

Electrode: A terminal on a cell or battery.

Electrolyte: A liquid or paste that surrounds and touches electrodes, causing a chemical reaction between the electrodes and electrolyte, which produces an electrical current.

Figure 8-28. A simple electrical generator demonstrates the principle of electromagnetic induction. The armature loop (conductor) of the generator rotates through lines of force from a magnet to produce an electric current.

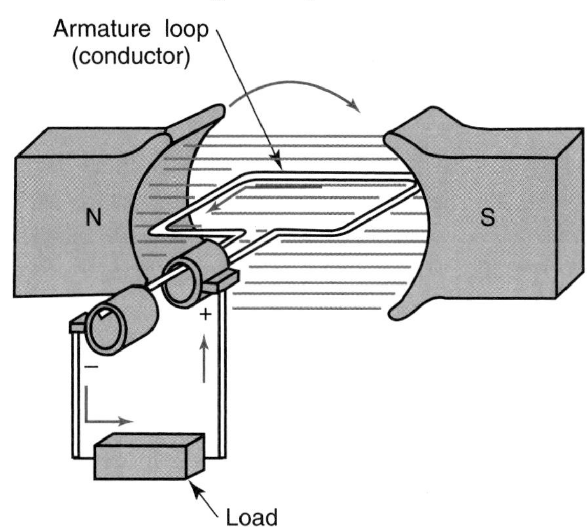

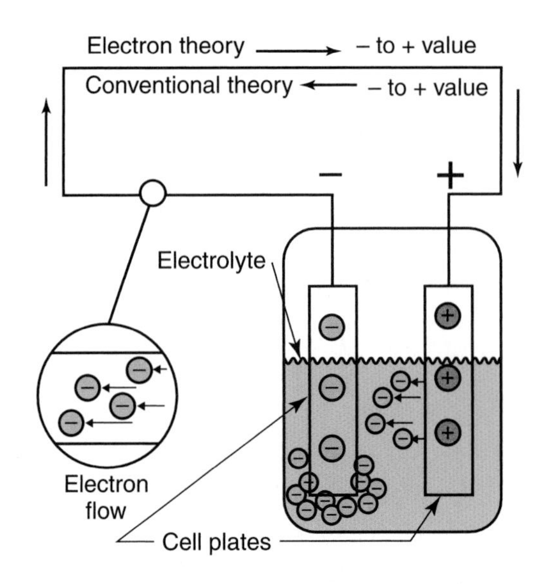

Figure 8-29. A simple battery illustrates how a difference in electrical potential can cause electrons to flow in a circuit. In a chemical reaction between battery terminals and the electrolyte, electrons migrate from the positive to the negative terminal. If the other ends of the electrodes are connected, the electron flow causes a current through the circuit. (Kohler Company)

Primary cell: A type of cell that cannot be recharged.

There are two kinds of cells: primary and secondary. A ***primary cell*** cannot be recharged. These cells can produce electrical current only while the chemicals in the electrolyte are reacting with each other. When the reaction stops, the cells are discharged and must be replaced. It is this type of cell that powers toys and flashlights.

Carbon-zinc battery: A type of primary cell in which the carbon is the positive electrode and the zinc is the negative electrode.

A ***carbon-zinc battery*** is the most common primary cell used in these applications. In this type of cell, the carbon is the positive electrode, and the zinc is the negative electrode. Other types of primary cells are alkaline, lithium, silver-oxide, zinc-air, and zinc-chloride cells. Alkaline batteries are the typical batteries found in almost any convenience store. They are cheap and dependable, but they do not offer longevity or recharging capability. The very small batteries used in watches and some calculators are lithium cells. Silver-oxide batteries are typically used for watches and calculators, although some bigger ones can have other applications. Zinc-air batteries are tiny batteries typically used for applications such as hearing aids. Zinc-chloride batteries come in many varieties, including a 1.5 V AA-sized battery that is ideal for portable games, hand-held radios, and calculators.

Secondary cell: A type of cell that can be discharged and recharged many times.

A ***secondary cell*** can be discharged (used up) and recharged many times. To recharge a secondary cell, electrical current is sent through the cell in a reverse direction from normal electron flow. The number of times a secondary cell can be recharged depends on its size, type, and the conditions under which it operates.

Lead-acid battery: A common secondary cell used in automobiles. It is a combination of several cells. This type of battery includes a series of positive and negative metal plates in a weak sulfuric acid electrolyte.

Figure 8-30 shows a common secondary cell used in automobiles. This cell is known as a ***lead-acid battery***. It is a combination of several cells. This type of battery includes a series of positive and negative metal (lead) plates in a weak sulfuric acid electrolyte. Other types of secondary cells are lithium-ion (Li-ion), nickel-cadmium (NiCd), nickel-iron (NiFe), nickel-metal hydride (NiMH), and rechargeable alkaline cells. Li-ion, NiCd, and NiMH batteries are among the most popular of the rechargeable batteries commonly found in cell phones and cordless power tools. They are even used as reusable replacements for conventional alkaline batteries of the common A, AA, AAA, C, and D sizes in countless

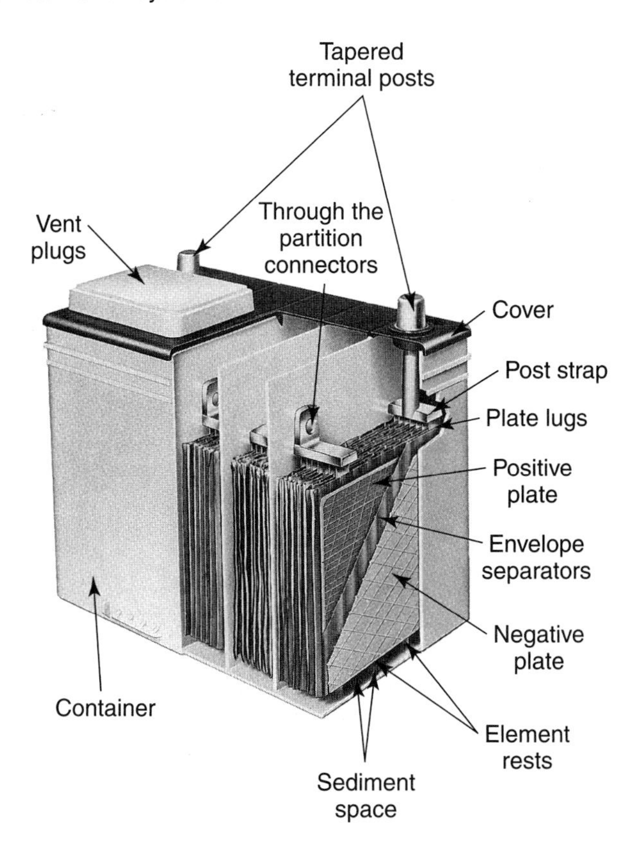

Figure 8-30. A lead-acid battery is made up of a number of cells. Each cell consists of positive and negative plates, or electrodes. They are held apart by a separator and surrounded by conductive fluid called electrolyte. This cutaway shows three of the six cells used in a 12-volt automotive battery. (Battery Council International)

applications. These batteries offer the advantage of having the ability to be charged and discharged many times over the life of the battery, when compared to a typical battery that is used until it is drained and then disposed.

AC Generators

An ***AC generator*** is a device that converts mechanical energy into electrical energy. This device is also called an ***alternator***. It converts mechanical energy into electrical energy by using electromagnetic induction. In a simple AC generator, an armature loop (conductor) passes through a magnetic field. See **Figure 8-31.** As the conductor rotates through the magnetic field, current flows through the conductor to the

AC generator: A device that converts mechanical energy into electrical energy by using electromagnetic induction.

Alternator: See *AC generator*.

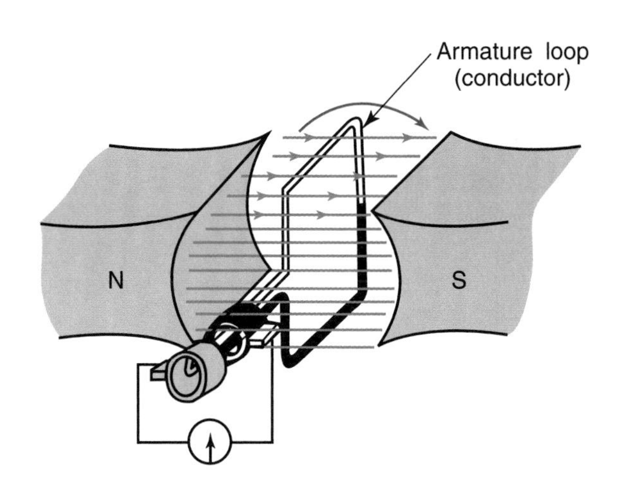

Figure 8-31. A diagram of a simple alternator. As the armature loop (conductor) cuts through the magnetic field, magnetism causes electrons to flow in the armature loop and through the external circuit.

circuit. Current flows through the circuit in one direction for one half of a revolution and in the opposite direction for the second half of a revolution. This action produces AC.

Power plants produce electricity with the aid of AC generators. See **Figure 8-32.** Mechanical energy is supplied to the generator through water or steam channeled to a turbine. See **Figure 8-33.** The turbine is connected by a shaft to the generator. As the turbine spins, it rotates the generator. The blades of the turbine spin the magnets inside the generator. The rotating magnetic field has the same effect as does a rotating armature loop in a fixed magnetic field.

Figure 8-32. The generators at the Hoover Dam hydroelectric plant. The huge generators are powered by turbines, which are situated beneath the generators and are driven by the force of moving water. (U.S. Bureau of Reclamation)

Figure 8-33. The power generation and transmission process.

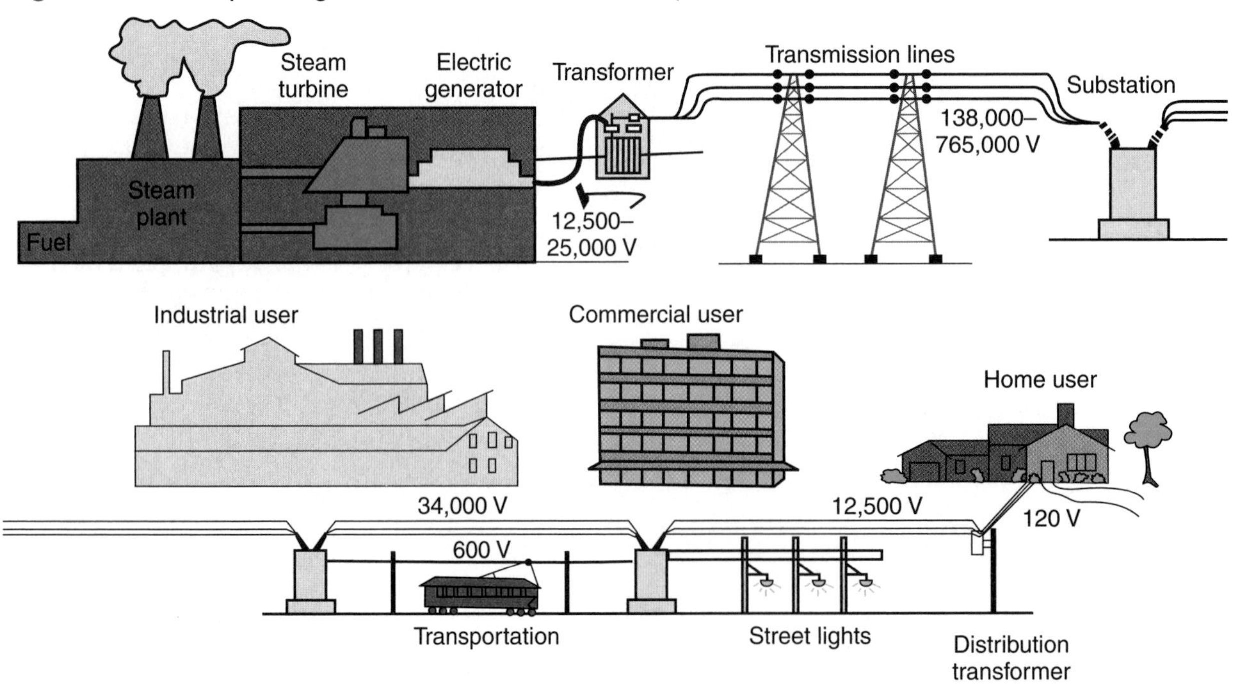

DC Generators

Like an AC generator, a ***DC generator*** relies on the principle of electromagnetic induction to create DC. See **Figure 8-34**. A DC generator consists of an armature loop (conductor) mounted on a shaft that rotates. On two sides of the armature loop are magnets. A north pole and a south pole are directed toward the armature loop. As the armature loop rotates between the magnetic fields, a current is produced.

DC generator: A device that relies on the principle of electromagnetic induction to create direct current (DC). It consists of an armature loop mounted on a shaft that rotates. On two sides of the armature loop are magnets. A north pole and a south pole are directed toward the armature loop. As the armature loop rotates between the magnetic fields, a current is produced.

Figure 8-34. A direct current (DC) generator. A—A simplified drawing of a DC generator, with parts labeled. B—A cutaway view of a DC generator. Can you recognize the different parts? (Baldor Electric)

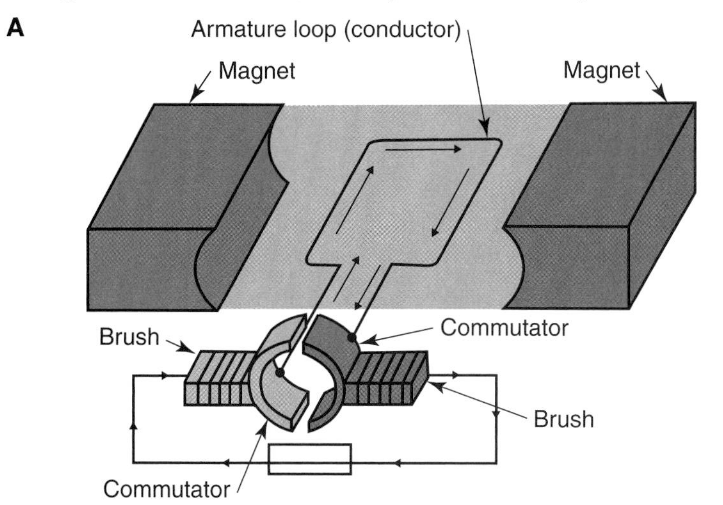
A
Armature loop (conductor)
Magnet
Magnet
Brush
Commutator
Brush
Commutator

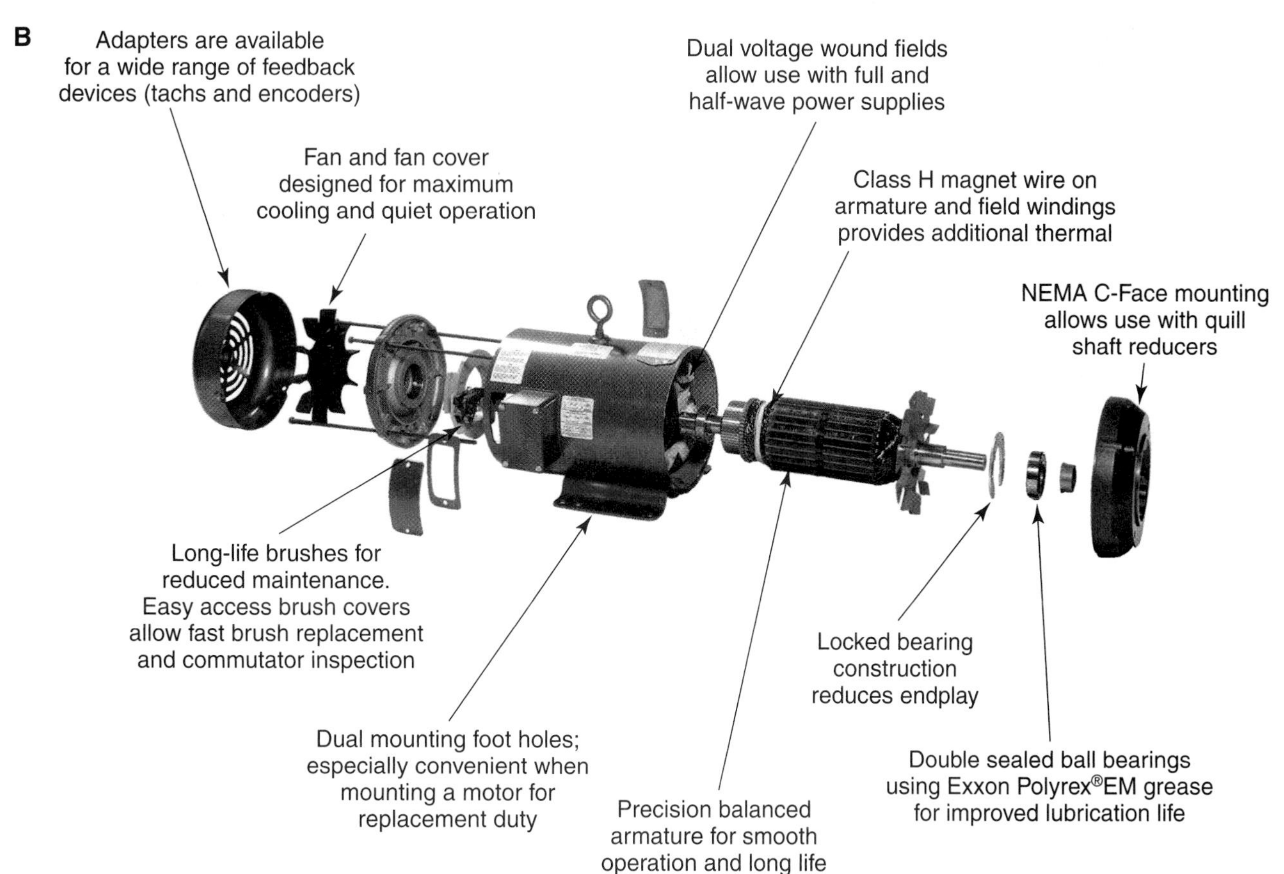
B
Adapters are available for a wide range of feedback devices (tachs and encoders)
Dual voltage wound fields allow use with full and half-wave power supplies
Fan and fan cover designed for maximum cooling and quiet operation
Class H magnet wire on armature and field windings provides additional thermal
NEMA C-Face mounting allows use with quill shaft reducers
Long-life brushes for reduced maintenance. Easy access brush covers allow fast brush replacement and commutator inspection
Locked bearing construction reduces endplay
Dual mounting foot holes; especially convenient when mounting a motor for replacement duty
Precision balanced armature for smooth operation and long life
Double sealed ball bearings using Exxon Polyrex®EM grease for improved lubrication life

Remember that different magnetic poles produce different directions of current flow. To solve this problem and produce DC, DC generators make use of a commutator and brushes. Look again at Figure 8-34.

As the armature loop rotates through the magnetic field, it turns the commutator. Brushes that touch on each side of the commutator carry the electric current away from the armature loop. Notice that the commutator is split so the armature loop is not a complete circuit. The two parts of the commutator touch the brushes at different times. This way, the current produced by the north magnetic polarity will always be carried by one brush. The same is true for the opposite magnetic field. Its induced current will always be carried by the opposite brush. Can you see how this makes one brush negative and the other positive, thus producing a DC output?

Single-pole, single-throw (SPST) switch: A switch that makes or breaks one set of contacts to turn a load on and off.

Single-pole, double-throw (SPDT) switch: A switch used to control a load from two different locations.

Double-pole, double-throw (DPDT) switch: A switch used in conjunction with two single-pole, double-throw (SPDT) switches, allowing one or more loads to be controlled from three or more locations.

Momentary contact switch: A switch that makes or breaks a circuit based on the input or touch of the switch. When the switch is not depressed, the switch returns to its default status.

Push-button make/break (PBMB) switch: A switch that retains its status until pressed again.

Basic Control Elements for Electricity

There are many basic control elements that allow us to effectively use electricity to do useful work. Perhaps the most fundamental control elements are switches, since they are routinely used to turn loads on and off. Other types of control elements are diodes and transformers.

Switches

The simplest of all switches is the ***single-pole, single-throw (SPST) switch***. This switch makes (closes) or breaks (opens) one set of contacts to turn a load on and off. A slightly more complex switching scheme makes use of a pair of ***single-pole, double-throw (SPDT) switches*** to control a load from two different locations. This type of switching is typically used to turn lights on and off from either end of a staircase. It is commonly referred to as a *three-way switch.*

Now let us suppose that the lights needed to be controlled from three or more locations, such as in a large room or garage. This is an application for a ***double-pole, double-throw (DPDT) switch***, often referred to as a *four-way switch*. When a DPDT switch is used in conjunction with two SPDT switches, one or more loads can be controlled from three or more locations. All these switching schemes are diagrammed in **Figure 8-35.**

There are many other types of common switches used in everyday life. Some are known as ***momentary contact switches***. These switches make or break a circuit based on the input or touch of the switch. When the switch is not depressed, the switch returns to its default status. The two most common momentary contact switches are known as the push-button normally opened (PBNO) and the push-button normally closed (PBNC). See **Figure 8-36.**

A further extension of this type of switch is the ***push-button make/break (PBMB) switch***. This switch retains its status until pressed again. In this sense, it is said to have memory. Some examples may be the button on a lamp or on the front of your stereo. If a PBMB switch is pushed in once, the appliance stays on. If it is pushed again, the appliance goes off and stays off.

Sometimes, more complex on/off control capability is required. For example, let us say you have a table saw that has a start and stop button on the front of the saw. Did you ever wonder why there are two separate

Figure 8-35. Basic switching schemes. Look at the various switching schemes and imagine what would happen if any one of the switch positions changed. Would the light turn on or off?

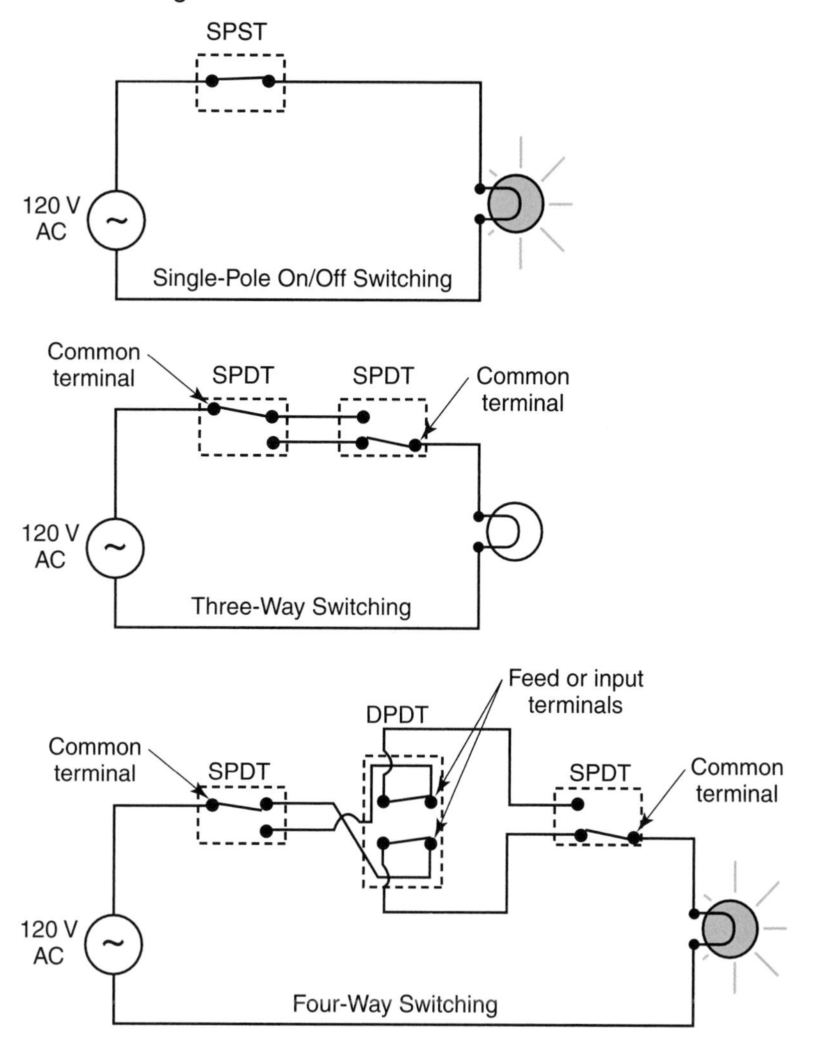

buttons, instead of a simple SPST switch to turn the machine on and off? The answer has to do with safety. If an SPST were used and the power went off, the saw could inadvertently come back on when power is restored to the machine. Using momentary contact switches and a magnetic motor starter to start the saw ensures that, if the power to the saw goes off during operation, it cannot turn on at an inopportune time. The only way to start the saw is to push the start button again. The start button triggers a magnetic contactor that closes and remains latched until the stop button is depressed or the power is interrupted. The entire process is known as a ***latching relay***, and it forms the heart of a magnetic motor starter.

Additionally, most motor starters provide overload protection for the motors they feed. If the motor draws excessive amperage, the motor starter can automatically open the magnetic contactor that allows power to flow to the saw. The result is that the saw stops and cannot be started again until the start button is pushed. See **Figure 8-37.**

Latching relay: The process of using momentary contact switches and a magnetic motor starter to start a device. This ensures that, if the power goes off during operation, the device cannot turn back on at an inopportune time. The only way to start the device is to push the start button again. The start button triggers a magnetic contactor that closes and remains latched until the stop button is depressed or the power is interrupted.

Figure 8-36. A momentary contact switch operation. The normal state is the state the switch is in if it is left alone and nothing is pushing on it. A—A push-button normally opened (PBNO) switch is used to feed a motor that quickly mixes two parts of a plastic resin together so the liquid plastic will harden. When the button is depressed, the motor spins, and the indicator light goes on. When the operator takes his finger off the button, the motor stops spinning, and the light goes off. B—A push-button normally closed (PBNC) switch is used to keep a light on. This is the type of switch that can be found inside your refrigerator door. When the door is opened, the switch pops out to its normal, or closed, state, and electricity flows to the bulb. When the door is closed, the PBNC switch is held in its alternate state (opened), and the bulb goes off.

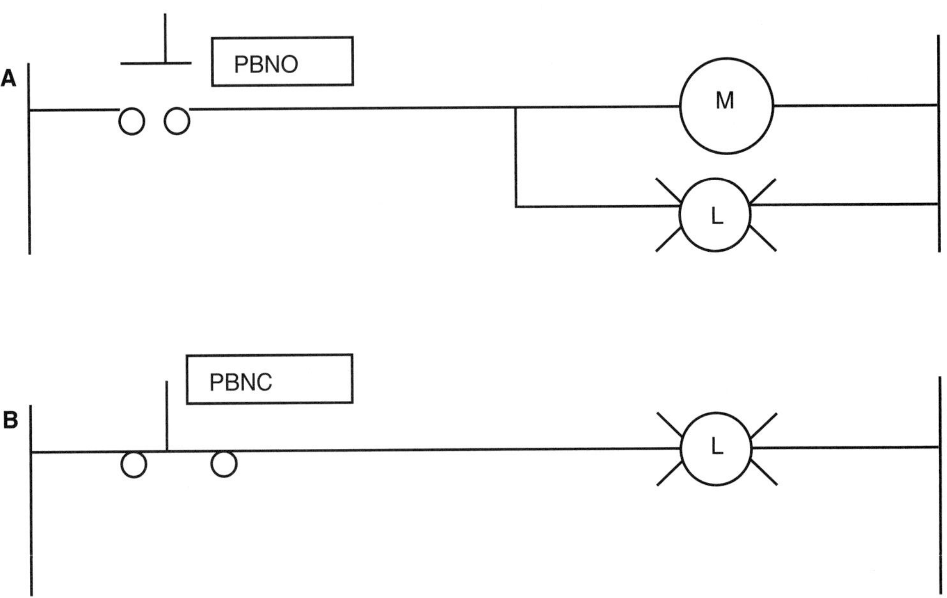

So far, all the switches described are examples of on/off control devices, in the sense that they only have the ability to turn loads on or off at full power. An example of a switch that can provide variable motion control is a dimmer switch, more appropriately referred to as a rheostat, or ***potentiometer***. See **Figure 8-38.** The potentiometer can vary the resistance within the switch, which in turn affects both the current and voltage flowing out of the switch. Potentiometers can also be used to control the speed of certain types of motors.

Potentiometer: A switch that can provide variable motion control. It can vary the resistance within the switch, which affects both the current and voltage flowing out of the switch.

Diodes

A control element that only allows electricity to flow in one direction is known as a diode. Diodes have a multitude of applications in electricity and electronics. Let us say you have to charge a battery, so you plug the charger into the wall and attach it to the battery. Several hours later, the battery is fully charged. The battery did not discharge through the charger and into the power source because a diode ensures that the electricity used to charge the battery can only flow through it in one direction. Diodes are also used in power supplies to convert AC to DC.

Figure 8-37. A start/stop motor control operation. A push of the push-button normally opened (PBNO) button closes the motor starter and causes the motor to turn on. Once the motor starter closes, the sealing contacts labeled "M1" remain held magnetically in place to keep the motor starter closed and the motor starter running. The machine is stopped by pushing the push-button normally closed (PBNC) button at the beginning of the circuit. This stops feeding electricity to the magnetic contacts. Without being fed by the control circuit, the contacts spring open, stopping the flow of power to the motor. The only way to start the motor again is to press the PBNO (start) button and reenergize the coil to make the sealing contacts latch again. Note that if the motor is overloaded, the overload protection can also cause the control circuit to shut the motor down.

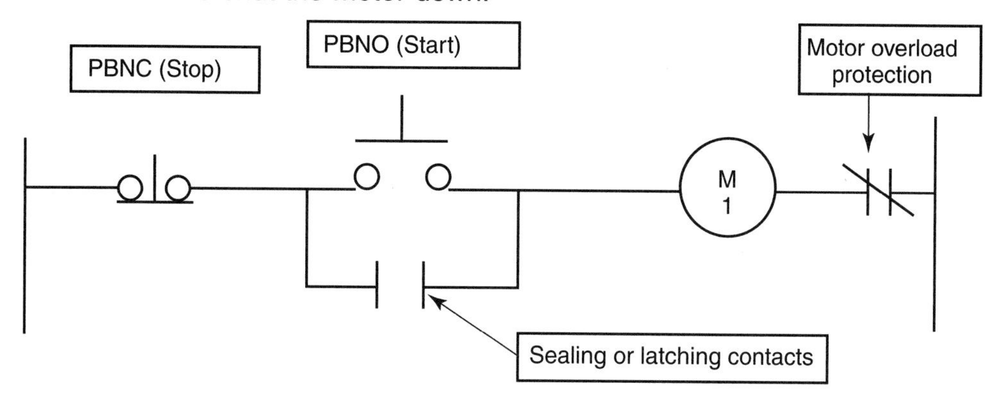

Transformer: A device used to increase or decrease voltage supplied to a circuit.

Step-down transformer: A transformer used to decrease voltage supplied to a circuit.

Step-up transformer: A transformer used to increase voltage supplied to a circuit.

Transformers

Many times, the electrical voltage supplied to a circuit is either too much or too little. If there is a need to increase or decrease voltage or current, we use a ***transformer***. When we need to reduce the amount of voltage, we use a ***step-down transformer***. One common household application for a step-down transformer is to power a doorbell system. When we have too little voltage and need more, we can use a ***step-up transformer***, which boosts the amount of voltage available. The utility company uses step-up transformers when it is necessary to transport electricity over long distances to reach communities. This is because transporting electricity at a high amperage would require wires massive in diameter.

Transformers work on the principle of electromagnetic induction. See **Figure 8-39**. Both transformers are made of an iron core with separate coils wrapped around each side of the core. The coil attached to the input side is called the ***primary coil***. The coil attached to the output side is called the ***secondary coil***.

Notice that the primary coil is smaller than the secondary coil on the step-up transformer.

Figure 8-38. A dimmer switch.

Figure 8-39. A transformer is made of an iron core with wrappings of wire to increase the voltage. The two coils of wire are not electrically connected. Electromagnetic induction caused by a current in one of the coils causes a current in the other coil.

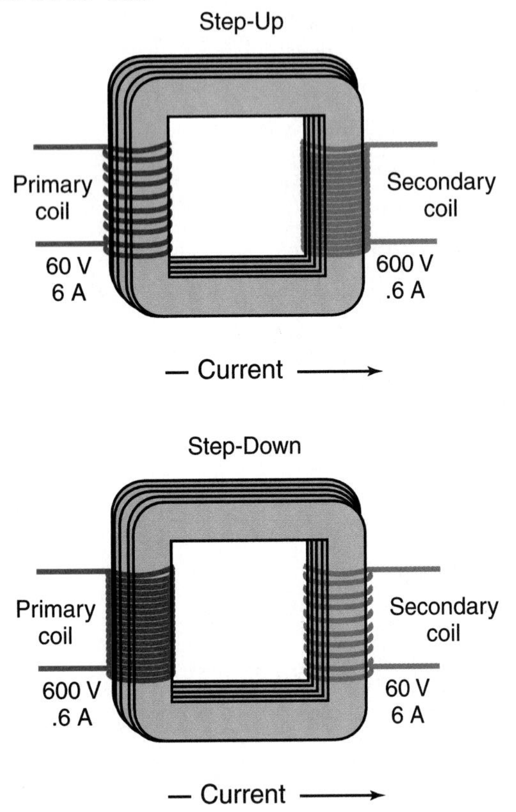

As the current in the primary coil induces a magnetic field, the magnetic field is carried to the secondary coil. The extra wraps on the secondary coil allow the magnetic field to induce more voltage. This causes the transformer to exude more electrical voltage than it was given.

Now, notice that the primary coil is larger than the secondary coil on the step-down transformer. Again, the induced magnetic field travels from the primary coil to the secondary coil. This time, the smaller secondary coil is not able to induce as much voltage as the larger primary coil supplied. It is important to recognize that transformers cannot produce any more power than what is put into them. They can only modify the characteristics of the power to achieve a goal, such as stepping up the voltage for long-distance transmission.

Wattage into a transformer must equal wattage out. This is assuming 100% efficiency. In the examples provided in Figure 8-39, notice how voltage may be increased or decreased on the output side of the transformer, but amperage is affected accordingly. In this sense, the electrical transformer is an advantage-gaining device, similar to a gear set that has the ability to increase effort, while decreasing rate. In this case, the effort is voltage, and the rate is amperage.

Primary coil: The coil attached to the input side of the iron core of a transformer.

Secondary coil: The coil attached to the output side of the iron core of a transformer.

Fuse: A filament that breaks the circuit if too much electrical current passes through it. It prevents damage to the rest of the circuit in the event of an overload.

In reality, transformers are not 100% efficient. To calculate the efficiency of a transformer, you must be able to determine its theoretical maximum potential and compare that to its actual measured output. For instance, if a transformer receives 660 volts at 200 amps on the input side, the maximum power it could possibly produce on the output side is 660 V × 220 A, which equals 132,000 W, or 1.23 kW. A step-up transformer can produce more voltage, but at a lower amperage. A step-down transformer can produce more amperage, but at a lower voltage. There is no transformer that can produce more than 1.32 kW. This is the maximum potential output of the transformer. Note that the maximum output of the transformer can only be determined if the transformer is under load. The efficiency of a transformer can be calculated with the following formula:

$$\text{efficiency} = \frac{\text{output power}}{\text{input power}} \times 100$$

Protecting Electrical Circuitry

Fuses or circuit breakers are an essential part of any electrical circuit, no matter what the circuit's purpose is. A *fuse* is made of a filament that breaks the circuit if too much electrical current passes through it. It prevents damage to the rest of the circuit, should there be an overload. Placing the fuse between the energy source and the rest of the circuit is important. Fuses are made with different ratings. This means that some fuses allow more current to pass than others before they break the circuit. **Figure 8-40** shows various types of fuses used in vehicles, homes, industries, and many other electrical applications.

Figure 8-40. A variety of fuses used to protect electrical circuitry in machines, automobiles, and homes.

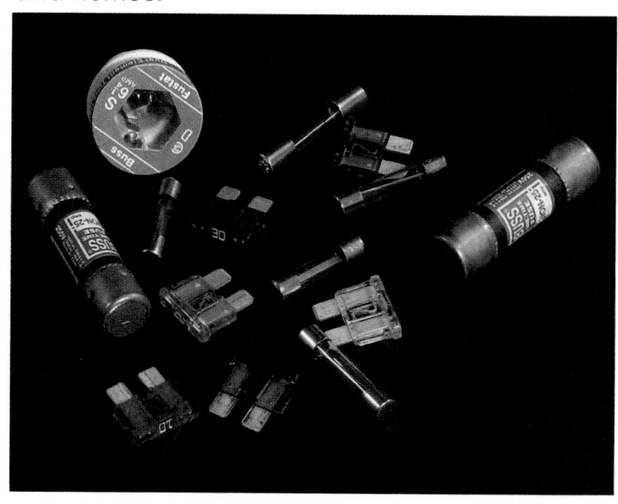

A ***circuit breaker*** performs the same function as a fuse, except a circuit breaker is restorable. Circuit breakers are also designed to withstand the initial surge of current associated with starting a motor when it is at rest. Only a special fuse known as a ***slow-blow fuse*** can be used for this function. Unlike fuses, circuit breakers do not need to be replaced when they break the circuit. They simply need to be reset. Circuit breakers can, however, go bad.

Circuit breaker: A restorable device that breaks the circuit if too much electrical current passes through it.

Slow-blow fuse: A fuse designed to withstand the initial surge of current associated with starting a motor when it is at rest.

One easy way to check if a circuit breaker or fuse is in working order is with a ***continuity checker***. This checker is powered from a battery within the meter, so power should not be applied to a fuse or breaker being checked. The continuity checker simply checks for continuous flow through a device or circuit and indicates continuous flow with a steady beep from the meter. Since fuse elements are destroyed when they fail, a beep will not occur when the meter probes are placed at opposite ends of a fuse.

Sometimes the fuse element in a fuse is visible, but with some fuses, the element is not visible. Either way, the continuity check will provide positive proof as to whether the fuse is good or bad. If the meter is used to check a circuit breaker, be sure the circuit breaker is turned to the on position. Since circuit breakers are designed to be fail-safe, you will not receive a beep if the circuit breaker has failed internally, even when its position is indicated as on. The continuity checker can be used to determine the status of many other components, including switches and diodes. See **Figure 8-41.**

A ***Ground Fault Circuit Interrupter (GFCI)*** can trip to open a circuit in much the same way as a circuit breaker or fuse can open a circuit in the case of an overload. This is, however, where the similarities end. Unlike circuit breakers or fuses designed to protect equipment, the GFCI is designed to protect people. To do so, it must be more sensitive than a circuit breaker or fuse and provide a quick reaction time.

Figure 8-41. Using a continuity checker to see if a fuse is good or bad. A light or a beep indicates continuity (a good fuse).

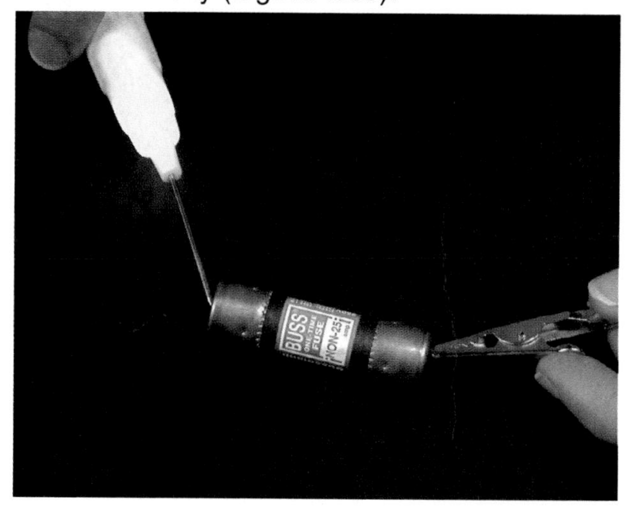

Continuity checker: A device used to check if a circuit breaker or fuse is in working order.

Ground Fault Circuit Interrupter (GFCI): A device that can trip to open a circuit in case of an overload. It is designed to protect people and so is more sensitive than a circuit breaker or fuse and provides a quick reaction time.

To accomplish this goal, the GFCI uses a sensing coil, which monitors the ampere flow going to the load on the hot leg and returning to the load on the return or neutral leg. Since amperes flow in a complete loop, incoming and outgoing amperage should be identical. See **Figure 8-42.**

In the event that these values differ, it is evident that a leak or fault has occurred. Such a situation could be extremely dangerous. A person could be in contact with the electricity and providing a return path or causing the fault current to flow to the ground. Therefore, the sensor on the GFCI is designed to sense ampere flow imbalances between the hot and neutral legs. If an imbalance of greater than 6 milliamperes (mA) occurs, the GFCI will trip, thereby disconnecting power from the source to the load. Since a shock of 6 mA or less is not regarded as life threatening, the GFCI is an ideal personnel protective device for many applications.

Figure 8-42. Ground Fault Circuit Interrupters (GFCI) are designed to protect people from electrical shock. A—A typical GFCI used in a bathroom or other location within 6′ of a sink or other water source. B—The GFCI uses a sensing circuit to detect an imbalance between current flowing in the hot and neutral sides of the circuit.

A

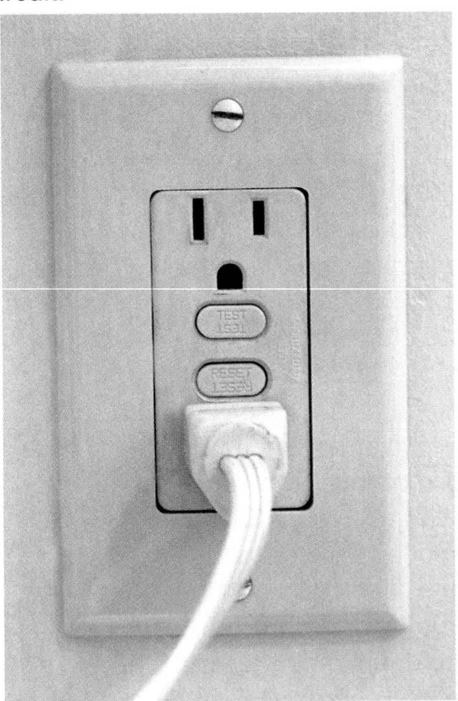

B

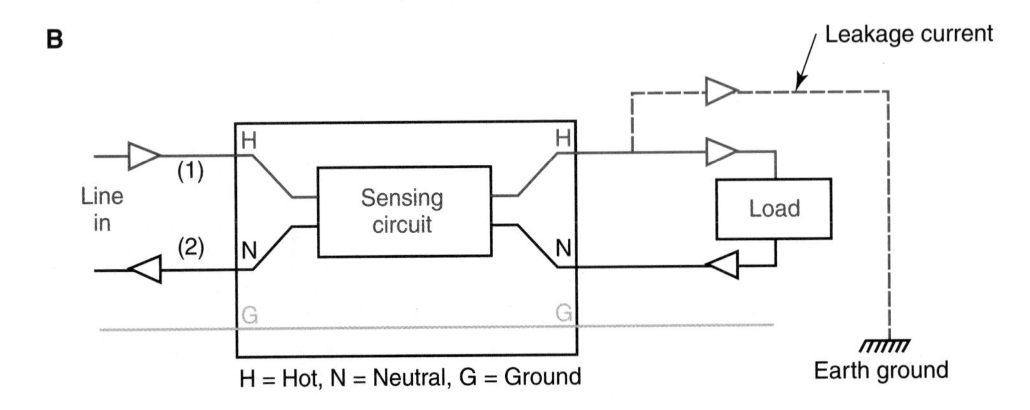

Sizing Wire and Components According to Load

Wire and other permanently installed electrical components must be sized in accordance with the load they are intended to power. If these components are undersized, they will generate excessive heat, possibly tripping a circuit breaker or blowing a fuse. On the other hand, if the components are oversized, they may not be able to appropriately connect to the components they are intended to feed. Oversizing also results in the inefficient use of materials and may create unsafe conditions.

For instance, using the table in **Figure 8-43,** let us say a heater is designed to draw 22 amps at full operation. If it is wired with wire that can feed 40 amps and the wire is protected with a 40-amp circuit breaker, the heater could overheat and cause a fire without the circuit breaker ever tripping. These are some of the reasons that correctly sizing wire and other components in accordance with the anticipated load is so vitally important. **Figure 8-44** provides the maximum amperage for common wire gauges used for residential, commercial, and industrial applications.

Electrical Safety

Every year, thousands of people are injured, and many are even killed, working with voltages as low as 120 vac. The National Safety Council reports that approximately 20% of electricity-related deaths each year are the result of faulty household wiring. Many of the remaining deaths occur from exposure to voltages as low as 120 vac. These figures do not include deaths caused as a result of fires started by improper electrical installations. When studying electricity, no topic is more important than that of electrical safety. Here is some practical advice on how to work with and use electricity safely and efficiently:

- Use the appropriate size fuse or circuit breaker for the given application.
- Do *not* assume that a circuit breaker or fuse will protect your body from excessive amperage flow.

Figure 8-43. Wire gauges and typical high-voltage applications.

Wire Gauge	Typical Application
24–20 Cu	Phones and communications
18 Cu	Doorbells and thermostats
16 Cu	Light extension cord wire
14 Cu	Convenience circuits throughout most areas of the home
12 Cu	Heavier branch circuits for the kitchen, the garage, and dedicated circuits for appliances such as refrigerators and room air conditioners
4–10 Cu	Dedicated circuits that power a specific appliance, such as a hot water heater, electric range, or air conditioner
2–0000 Al*	Service entrance cable for feeding electricity into the home
*Note that aluminum (Al) service entrance cable is typically used to feed homes in most parts of the country. All other wire is typically copper (Cu).	

Figure 8-44. Amperage ratings of common wire gauges.

Wire Type	Wire Gauge	Maximum Ampacity
THHN,UF	14	15
THHN,UF	12	20
THHN,UF	10	30

- Troubleshoot circuits in a de-energized state.
- When troubleshooting a de-energized circuit, be sure to identify and properly drain all capacitors.
- Do *not* ground yourself while working on an electrical system.
- Ensure that GFCI protection is installed in outdoor electrical receptacles and receptacles near water sources.

Always use the appropriate size fuse or circuit breaker for the given application. Fuses and circuit breakers are designed to perform one specific function: they will break the energized (hot) leg of a circuit when an overload or short circuit occurs. They perform their task extremely well if properly sized in accordance with the wire and other components they are designed to protect. If a fuse or breaker is oversized in relation to the components it is supposed to protect, the protective function of the fuse or breaker is significantly diminished.

Do *not* assume that a circuit breaker or fuse will protect your body from excessive amperage flow. Circuit breakers and fuses are designed to protect permanently installed wire and components, such as switches and receptacles, against heat buildup from excessive ampere flow. They are not designed to protect people. Circuit breakers and fuses are essentially single-purpose devices and cannot sense what load is drawing the electricity that flows through them. In other words, a 20-amp circuit breaker has no idea if a person is being electrocuted. The circuit breaker or fuse only knows to shut the circuit off if more than 20 amps flow through it for a period of time. This is important, since many applications require amperage well in excess of that which is required to cause physical harm to humans. See **Figure 8-45** for the effects of electrical current on the human body.

Figure 8-45. Physical effects of different levels of electrical current on the human body.

Effects of Electrical Current on the Human Body	
Current Value	**Physical Effects**
1 milliampere (mA)	Threshold of feeling. Only mild sensation.
1mA –5mA	Sensation of shock and surprise. Not painful. Person can let go of contact point.
5mA–10mA	Painful shock and reflex action. Person can still let go of contact point.
20mA–50mA	Severe, painful shock and muscular contraction. Breathing is difficult.
50mA–100mA	Possible ventricular fibrillation and breathing difficulty.
100mA–200mA	Almost certain heart stoppage and death.
Above 200mA	Severe muscular contractions and burns. Heart stops as long as current flows through body. Breathing stops.

Remember that circuit breakers are designed to protect only permanently installed components and not all components. This is why many plug-in appliances may include their own replaceable fuses or restorable circuit breakers and why most motors come with their own form of protective devices. In these instances, severe damage could occur without ever tripping a circuit breaker or blowing a fuse.

You should always troubleshoot circuits in a de-energized state. Although troubleshooting live circuitry has certain analytical advantages, troubleshooting a de-energized circuit is much safer. Troubleshooting an energized electronic circuit that operates at low voltages and amperages is often desirable. Most electrical circuitry and components can, however, be analyzed in a de-energized state. With an ohmmeter and a continuity tester, it is often just as easy to troubleshoot a de-energized circuit as it is to troubleshoot an energized circuit with a voltmeter. Remember, only well-qualified individuals should troubleshoot energized, high-voltage circuits and only when there is no other alternative.

Be aware that a de-energized circuit can still cause harm. Some electrical or electronic circuitry includes capacitors. Since capacitors are designed to store current even after a circuit is de-energized, they have the potential to deliver a severe shock. From a safety standpoint, learning how to identify and drain capacitors is extremely important. Capacitors are frequently used in electronic circuitry, but their use in high-voltage electricity is often limited to machines that use capacitors to help start motors spinning.

Do *not* ground yourself while working on an electrical system. When you connect an electrical conductor to the earth, it is said to be grounded. Since the earth offers a potential of zero volts, connecting the metal frame or housing of any device that uses electricity to the ground is often done for safety. You do not have to run your own wire to the ground to ground your load. Any device that has a three-terminal plug should make the connection to the ground for you. The third, rounded connector on a three-terminal plug does this. Grounding loads is important because if a hot wire came off of a terminal and touched the grounded part of the device, the electricity would immediately be passed on to the ground. To avoid the possibility of an electrical shock, it is important to ground your loads. When devices are not properly grounded, the electricity can flow through a person to the ground, causing possible injury or even death. Electricians that have to work with live electricity, such as utility workers, go to great lengths through the clothing they wear and the equipment they use to ensure they are *not* grounded while working around live electricity. If an accident were to occur, the severity of the shock would be based in large part on how much a person was grounded.

The National Electrical Code (NEC) requires GFCI protection for many locations, including outdoor electrical receptacles and those near water sources, such as the kitchen and bathrooms. You should therefore ensure that GFCI protection is installed in outdoor electrical receptacles and receptacles near water sources. GFCI protection should also be used in the laboratory when working with electricity.

Figure 8-46. Meter placement. A—Amperage is measured in series. B—Voltage is measured in parallel. C—Resistance is measured in parallel with the circuit de-energized.

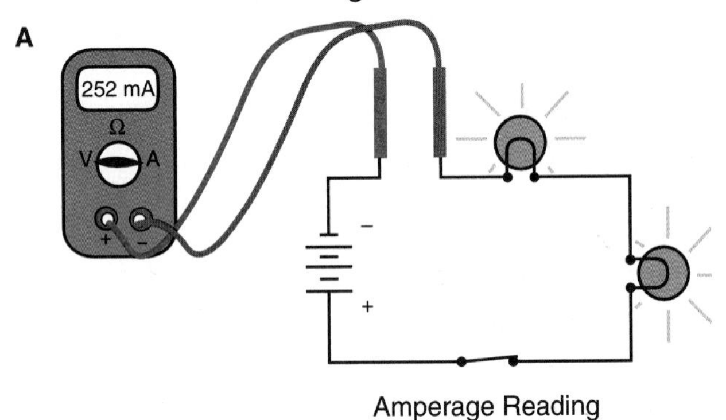

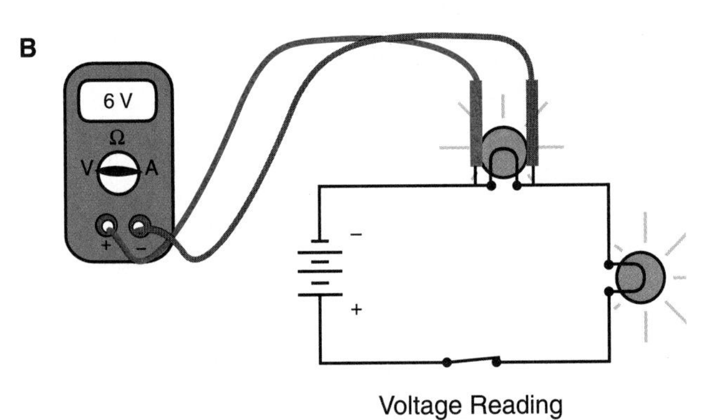

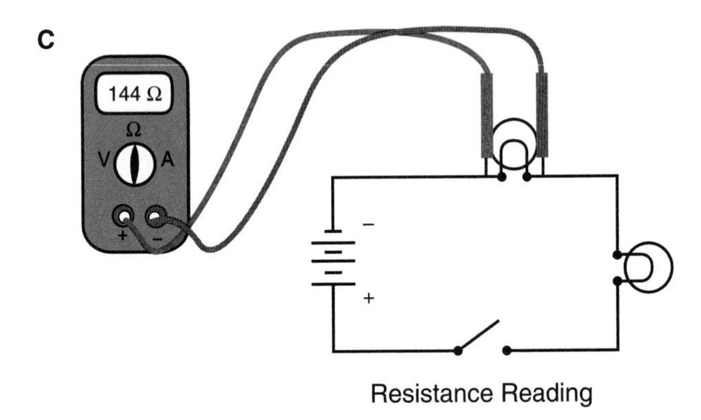

Using Instrumentation Safely and Properly

Several types of instruments can be used when measuring electricity. The primary instrument is the multimeter, which is a meter that can measure voltage, resistance, and sometimes amperage. Other meters sometimes used include voltmeters, ammeters, amp clamps, ohmmeters, wattmeters, and watt-hour meters.

A few general rules apply to the use of all instrumentation. Observing these rules will help ensure accurate measurements and protection of the instrumentation and of the person doing the measuring. Failure to observe these rules could damage an expensive piece of equipment and possibly cause bodily injury.

- Determine the proper instrument for measuring the desired quantity. It is important to be sure the meter you choose has the capability to measure the quantity (voltage or amperage) you desire to measure.
- Determine whether the instrument should be inserted into the circuit in series or in parallel. Amperage is always measured in series, and voltage is always measured in parallel. Both measurements require the circuit to be energized. Observe the placement of the meters in the diagram in **Figure 8-46.**

Safety Note

Working with live electricity can be very dangerous. Be certain you are not grounded when working around live electricity. Always wear rubber-soled shoes when working with live electricity known to be above 24 volts, and always use the proper instrumentation and equipment.

- Resistance is always measured with the circuit de-energized and the meter in parallel with the load. See Figure 8-46.
- Always start on the highest range when measuring unknown quantities with a multirange instrument. Many of the contemporary digital meters automatically switch ranges. Even with these types of meters, however, you may need to switch ranges for an accurate reading.

- If using an analog meter, determine whether the scale you will be reading is divided into even increments or whether it is skewed. It is common for one scale to be used to represent several ranges with a multirange instrument. You may need to add or subtract decimal places to obtain the correct reading on such instruments. See **Figure 8-47.**
- Use surge suppressors to protect delicate equipment, such as computers and stereo systems, from excessive voltages. Remember, circuit breakers protect electrical circuits only from excessive amperage. They are not voltage protection devices. It is possible to protect a circuit from voltage fluctuations and excessive voltage, using devices other than circuit breakers or fuses. For instance, when a constant voltage is critical, a constant voltage transformer may be used. This device has the ability to take in voltages higher or lower than the desired voltage and produce a consistently desired voltage on the output side of the transformer. One application for constant voltage transformers would be to protect delicate hospital equipment. These transformers can receive anywhere from 90 vac to 140 vac and produce a consistent 120 vac.

Figure 8-47. Reading an analog meter scale.

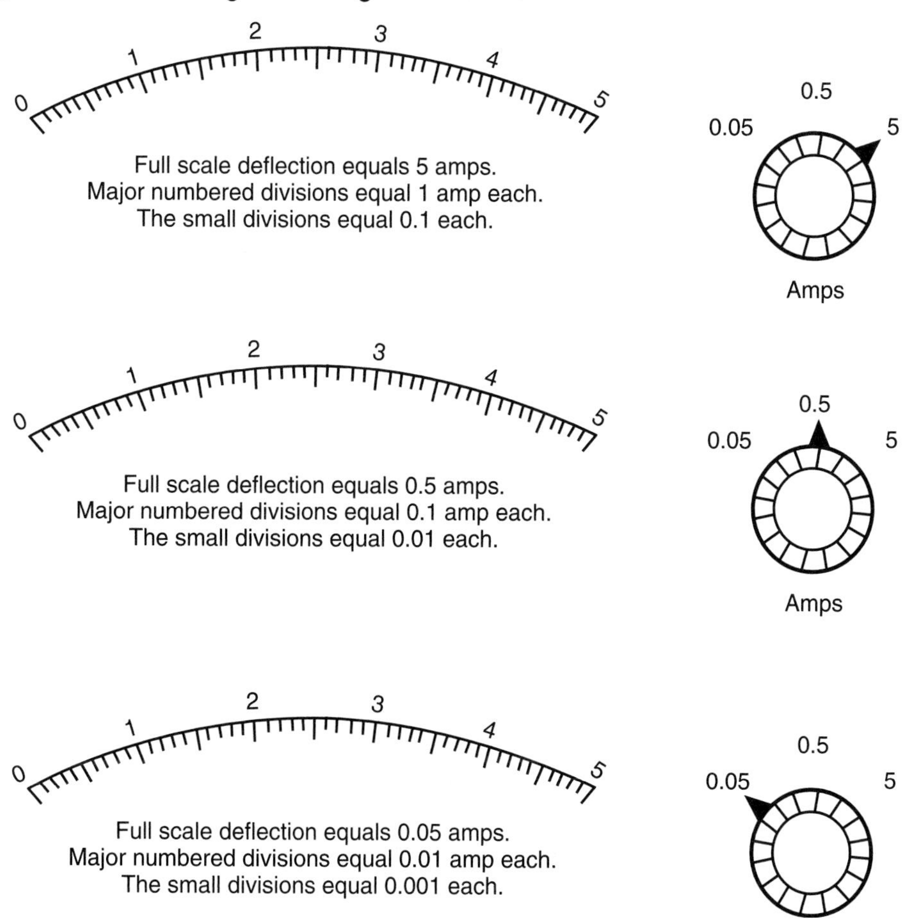
0
1
2
3
4
5
Full scale deflection equals 5 amps.
Major numbered divisions equal 1 amp each.
The small divisions equal 0.1 each.
0.05
0.5
5
Amps
Full scale deflection equals 0.5 amps.
Major numbered divisions equal 0.1 amp each.
The small divisions equal 0.01 each.
Full scale deflection equals 0.05 amps.
Major numbered divisions equal 0.01 amp each.
The small divisions equal 0.001 each.

Summary

Most communication, transportation, and manufacturing systems rely heavily on one or more electrical systems for proper operation. Electricity provides lights for night-time operation. It powers electric motors that operate various components in a vehicle. Electricity is necessary for the operation of most heat engines. Onboard computers need a reliable source of electricity to perform their many functions. Likewise, radios and other accessories depend on a reliable source of electricity.

Electricity is produced when atoms trade electrons between their valence rings. Conductors are materials that carry electrical current easily. Devices like bells, lights, motors, switches, and fuses are connected with a conductor and then attached to a power source to form a circuit. These circuits are the heart of an electrical power system when they do work, create light, or create heat.

Magnetism is important in the study of electricity. Magnets are used to create electrical current by a process called electromagnetic induction. Iron can be magnetized by passing an electrical current around it. When iron is magnetized in this way, it is called an electromagnet. Electricity-producing generators, alternators, and electric motors, use the principles of electricity and magnetism to operate.

Batteries convert chemical energy to electrical energy. Both types of energy may be used for transportation vehicles. Some batteries are also able to store electricity for later use. Transformers are used to increase or decrease the amount of electrical current or voltage supplied to a circuit.

Key Words

All the following words have been used in this chapter. Do you know their meanings?

AC generator
alternating current (AC)
alternator
atom
battery
carbon-zinc battery
cell
circuit breaker
closed circuit
conductor
continuity
continuity checker
current
current flow
DC generator
direct current (DC)
double-pole, double-throw (DPDT) switch
electrical circuit
electrode
electrolyte
electromagnet
electromagnetic induction
electron theory
flux
fuse
Ground Fault Circuit Interrupter (GFCI)
insulator
kilowatt-hour (kWh)
latching relay
lead-acid battery
line of force
magnet
momentary contact switch
Ohm's law
open circuit
polarity
potentiometer
primary cell
primary coil
push-button make/break (PBMB) switch
resistance
ring
schematic drawing
secondary cell
secondary coil
semiconductor
short circuit
single-pole, double-throw (SPDT) switch
single-pole, single-throw (SPST) switch
slow-blow fuse
step-down transformer
step-up transformer
transformer
valence ring
wattage

Test Your Knowledge

Write your answers on a separate sheet of paper. Do not write in this book.

1. Recall three types of electrical circuits found in vehicles.
2. What are three parts of an atom?
3. _____ are materials that transfer electrons easily and are used to connect the parts of an electrical circuit.
4. According to electron theory, electricity moves from a(n) _____ point to a(n)_____ point.
5. DC stands for _____.
6. AC stands for _____.
7. _____ is a measurement of electrical power.
8. What type of drawing is used to represent electrical circuits on paper?
9. A switch can create which type of circuit when electricity does not flow through it?
10. What law would you use to calculate voltage, if current and resistance are known?
11. A current of 47 A flows through an electric welder with a resistance of 10 Ω. What power is dissipated in the circuit?
12. A heating element requires 220 volts. Its heat resistance is 6 Ω. What power does the heater require?
13. _____ is the unit by which we are billed for electricity.
14. How much will it cost to run 14 75-watt lightbulbs for 5 days at 10 hours per day, assuming the cost of electricity is $.09 per kilowatt-hour (kWh)?
15. Calculate the total resistance of a series circuit that contains the following three resistors: 10 Ω, 100 Ω, and 1000 Ω.
16. What is the difference between series and parallel circuits?
17. Calculate the total resistance of a parallel circuit that has three lightbulbs that offer the following resistances: 17 Ω, 42 Ω, and 90 Ω.
18. _____ use an electrical current to produce a magnetic field in a piece of iron.
19. The process of using magnetic fields to create electrical current in a wire is called _____.
20. Make a sketch showing how moving water causes a turbine to create electricity.
21. What electrical component will only allow electricity to flow through it in one direction?
22. Sketch a simple DC generator and describe how it works.

23. Identify the type of switch used to turn a load on or off from two different locations.
24. A(n) _____ switch used in conjunction with two single-pole, double-throw (SPDT) switches can control a load from three or more locations.
25. Diagram a pair of three-way switches that control two lights in parallel.
26. If you want to make an indicator light come on only when a button is depressed and go off after the button is released, use a(n) _____ switch.

Matching questions: For Questions 27 through 32, match the phrases on the left with the correct term on the right.

27. _____ Converts chemical energy to electrical energy.
28. _____ Surrounds electrodes in a battery to help produce current.
29. _____ Can be discharged (used up) and recharged many times.
30. _____ Produces alternating current (AC) using mechanical energy.
31. _____ Produces direct current.
32. _____ Reduces or increases electrical current in a circuit.

A. Electrolyte.
B. Battery.
C. Transformer.
D. Alternator.
E. DC generator.
F. Secondary cell.

33. Calculate the efficiency of a transformer, given the following characteristics: input voltage = 120 vac, input amperage = 2 A, output voltage = 24 vac, and output amperage = 6.5 A.
34. Which devices will protect an electrical circuit from excessive ampere flow?
35. What type of tester can be used to check if a switch, diode, fuse, or breaker is functioning properly?
36. Identify the primary factor to consider when sizing wire and other electrical components.
37. Is 120 vac a safe voltage with which to work?
38. Almost certain death or heart stoppage can occur from a current as little as _____ mA.
39. _____ is always measured with the meter in series with the load.
40. _____ is always measured with the meter in parallel with the load and with the circuit energized.
41. _____ is always measured with the circuit de-energized.

Activities

1. Connect two flashlight cells (batteries) in series (positive poles to negative poles). Ask your instructor to help you use a voltmeter or volt-ohmmeter (VOM) to measure voltage. Be sure the function switch on the voltmeter is turned to the direct current (DC) voltage position. Use rubber bands to hold meter leads to the cell's terminals. After measuring the voltage of the two cells, measure each cell individually. Report the voltage readings to the class. Research circuit theory in an electricity reference book and explain why the measured voltage is different from the individual voltage of either battery.
2. Study and set up a simple parallel circuit using a suitable power source (battery or cell), switches, conductors, and lights.

Safety Note

Wear eye protection for Activity 3. Automobile batteries contain acid that could cause blindness. Even a film of battery acid on the case could cause serious injury if it comes in contact with the skin or an eye.

3. Check an automotive battery for leakage across its top. Set a voltmeter at its lowest setting. Attach the probe with the black conductor to the negative terminal of the battery. Touch the positive probe to the battery case on the opposite side near the positive battery terminal. If the voltmeter registers voltage, there is an electrical leak because of a dirty top. Using a brush, clean the top of the battery case with a solution of warm water and baking soda. Retest as before. Is there any leakage now?

This large power plant in Kuwait generates electricity with turbines driven by the burning of natural gas, a plentiful source of fuel in the Middle East. (Siemens)

9

Mechanical Power Systems

Basic Concepts

- List the six simple machines and give an example of each.
- List three types of gears.
- Name the two primary characteristics of power.
- Identify two mechanical transmission devices and describe how each operates.
- Define mechanical advantage and give an example.
- Recognize the difference between ideal mechanical advantage (IMA) and actual mechanical advantage (AMA).

Intermediate Concepts

- Discuss force and rate in a mechanical system.
- Describe the difference between scalar and vector quantities.

Advanced Concepts

- Design a mechanical system for a specific application.
- Predict the result of a mechanical system based on knowledge of balanced and unbalanced loads.
- Calculate the mechanical advantage of a simple machine.
- Compute the mechanical advantage of compound machines.
- Solve for the percentage of frictional loss in a mechanical system.

Mechanical systems produce work using one or more machines. Machines can change the size, direction, and speed of forces and can also change the type of motion produced. Sometimes, machines are needed to perform important functions, but very little force is necessary to produce work.

A block and tackle is a machine that enables a person to lift extremely heavy loads with little effort. This machine multiplies the force acting on it. A seesaw is a machine that changes direction of force. Pushing down on one end of the seesaw lifts the other end. An eggbeater changes the speed of an applied force. The blades of the beater move fast, even when you turn the crank slowly.

Machines and vehicles make use of three types of motion: reciprocating, rotary, and linear. See **Figure 9-1.** Reciprocating motion is back and forth along the same line. Rotary motion moves in circles. Linear motion is straight and in one direction.

Simple Machines

When thinking of machines, washing machines, table saws, and drill presses may come to mind. These are examples of machines, but they are complex machines. ***Complex machines*** use more than one simple machine to accomplish their tasks and have multiple subsystems. We can study complex machines by identifying the smaller parts of each that actually perform different types of work using mechanical energy. These smaller parts are the ***simple machines***.

Complex machine: A machine that uses more than one simple machine to accomplish its tasks.

Simple machine: A lever, a pulley, a wheel and axle, an inclined plane, a screw, or a wedge.

There are six simple machines used to control mechanical energy: the lever, the pulley, the wheel and axle, the inclined plane, the screw, and the wedge. See **Figure 9-2.** All six types of simple machines rely on the principles of only two: the lever and the inclined plane. The sections that follow explore the different simple machines you use every day.

Figure 9-1. Machines have three types of motion. A—Reciprocating. B—Rotary. C—Linear.

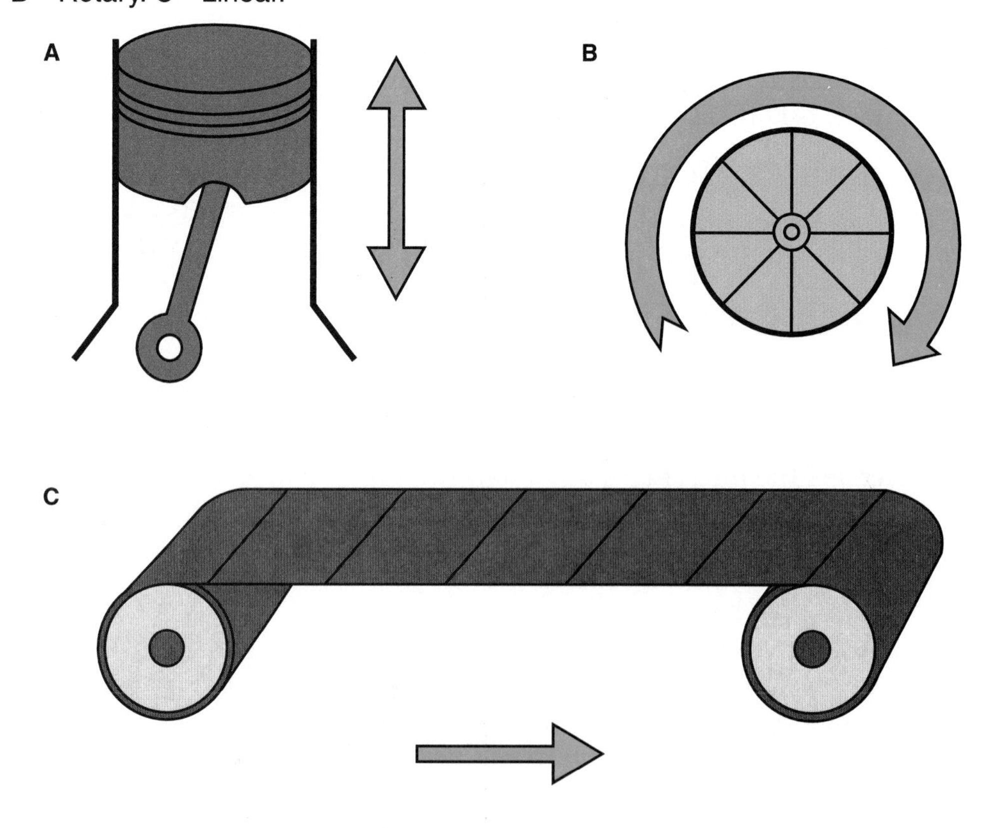

Figure 9-2. Six simple machines control all our mechanical energy. Compound machines are combinations of any of these six. A—A lever. B—A pulley. C—A wheel and axle. D—An inclined plane. E—A screw. F—A wedge.

Levers

A ***lever*** is a rigid bar that rotates (turns) around one fixed point. The fixed point is called the ***fulcrum***. We use levers to apply force on loads. There are three classes of levers: first, second, and third. The position of the fulcrum, load, and input force determines the lever class. See **Figure 9-3.**

Lever: A rigid bar that rotates around one fixed point.

Fulcrum: The fixed point around which a lever rotates.

Figure 9-3. The relationship between the load, fulcrum, and force determines the type of lever.

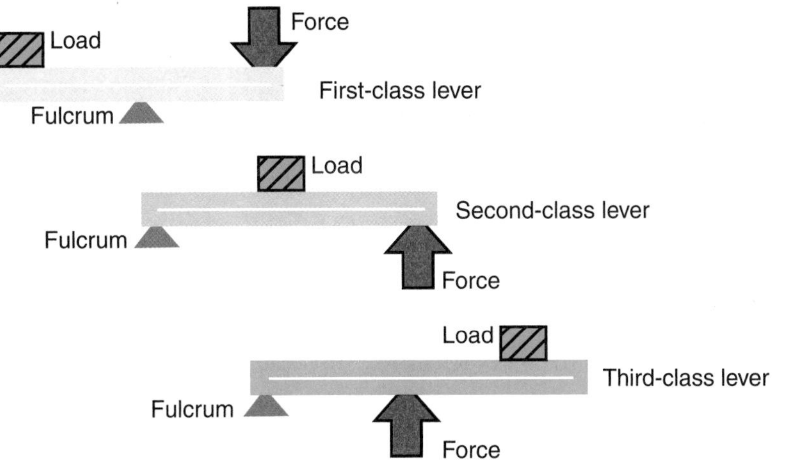

Technology Link

Construction: Cranes

Cranes of various types and sizes can be found at almost any construction site. Some cranes are attached to delivery trucks so they can lift roof trusses into place or unload construction materials from the truck bed to the ground. Other cranes that are sometimes present at construction sites can rise hundreds of feet into the air to lift steel beams, concrete, and large tools onto the upper levels of a building under construction. These tower cranes must be assembled on the construction site, since they are too massive to transport. Tower cranes are especially impressive because they can often lift up to 20 tons of weight. They can also move that weight in or out, sometimes by more than 200′, and up or down, sometimes by hundreds of feet. To move this massive load requires tremendous mechanical advantage.

Before the crane ever arrives on-site, a pad for the crane is constructed. This pad can be as large as a 30′ square that is at least 4′ thick with reinforced concrete. It can weigh as much as 200 tons, forming a sturdy anchor for the crane base, which is bolted to the pad with the use of massive anchor bolts.

Next, the crane arrives in sections of steel that are constructed in triangulated fashion for increased strength. A mobile, truck-mounted crane is used to assemble the jib. The jib is the long horizontal working arm of the crane. It also contains the motor and machinery that allow the crane to operate and the counterweights necessary to stabilize the load when the crane is in use. The jib is then placed on a section or two of mast. At the top of the mast is a slewing unit, a gear-driven device that allows the crane to rotate. The mast is the upright section of the crane that allows it to stand tall. Each section of mast is about 20′ long. The assembly is placed onto the pad in the upright position and bolted down.

From this point, the crane must "grow itself" in height if it is required to be taller. Growing the crane is a three step process. First, a weight is hung on the jib to balance the counterweight. Next, the slewing unit is detached and lifted by large hydraulic cylinders to the top of the mast. These hydraulic cylinders are capable of lifting the slewing unit more than 20′ above the top of the crane. The crane then lifts another section of mast into place between the top of the mast and the bottom of the slewing unit. When the jib and slewing unit are lowered, the crane is bolted back together with the mast, but this time it is 20′ taller. Using this technique, a free-standing tower crane can grow to more than 25 stories tall! It can also grow much taller if the mast is secured to the structural steel of the building it is helping to construct. This process can be repeated in reverse order when the crane is ready to be disassembled.

As for how the crane lifts a load, it is a balancing act between the working side of the jib and the counterweight or machinery side. A tower crane can safely lift more weight closer to the mast than it can as the load is transferred toward the far end of the jib. Can a tower crane become unbalanced enough to tip? It is unlikely because there are safety sensors to ensure that the crane does not become overloaded and unstable.

First-class lever: A lever that has the fulcrum positioned between the input force and the load.

First-class levers have the fulcrum positioned between the input force and the load. A seesaw is an example of a first-class lever. A screwdriver might be used as a first-class lever to open a can of paint. Pliers are an arrangement of two first-class levers using one fulcrum.

With ***second-class levers***, the load is placed between the fulcrum and the input force. A wheelbarrow is an example of a second-class lever. See **Figure 9-4.** Unlike first-class levers, second-class levers always provide an increase in force. They cannot, however, increase the distance the force moves.

Second-class lever: A lever that has the load placed between the fulcrum and the input force.

With ***third-class levers***, the input force is positioned between the fulcrum and the load. The distance the load is moved is greater than the movement of the input force. Many applications require increasing the distance, rather than the strength. Shovels, rakes, and other gardening tools use third-class levers. See **Figure 9-5.** Other examples include baseball bats, golf clubs, and hammers.

Figure 9-4. A wheelbarrow is a good example of a second-class lever. The fulcrum is the wheel, and the load is positioned between the wheel and the lifting force.

Pulleys

Pulleys consist of solid discs that rotate around a center axis. The discs usually have a groove around the outside edge that allows ropes or belts to easily ride around them. Pulleys are used to change forces in two ways:

- A single fixed pulley changes the direction of force. See **Figure 9-6.** Flags are raised on poles with the help of fixed pulleys.
- Single moveable pulleys change the size of a force. Moveable pulleys do not change the direction of the force. See **Figure 9-7.**

These simple machines operate on the principle of levers. This means they have an input force, a fulcrum, and a load. See **Figure 9-8.** Several pulleys used together make up a block and tackle. With several pulleys in the same system, the input force is greatly multiplied. Piano movers and construction workers use this type of system to lift heavy loads to great heights.

Figure 9-5. A third-class lever, such as this rake, is always a distance multiplier.

Wheels and Axles

The ***wheel and axle*** system is also based on the principle of levers. The large-diameter wheel and its small-diameter axle are attached to each other to move as one unit. Wheel and axle systems can be used to change the size or distance of a force. If the input force is applied to the wheel, it multiplies the turning force as it turns the axle. See **Figure 9-9.** The steering wheel of an automobile is an example of a wheel and axle system that resembles a second-class lever. See **Figure 9-10.**

Figure 9-6. A fixed pulley changes the direction of force to lift a load.

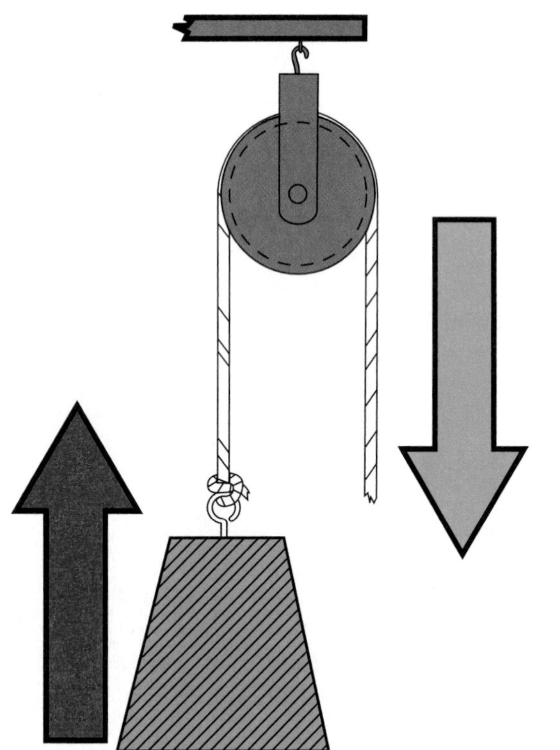

Figure 9-7. A moveable pulley does not change the direction of force.

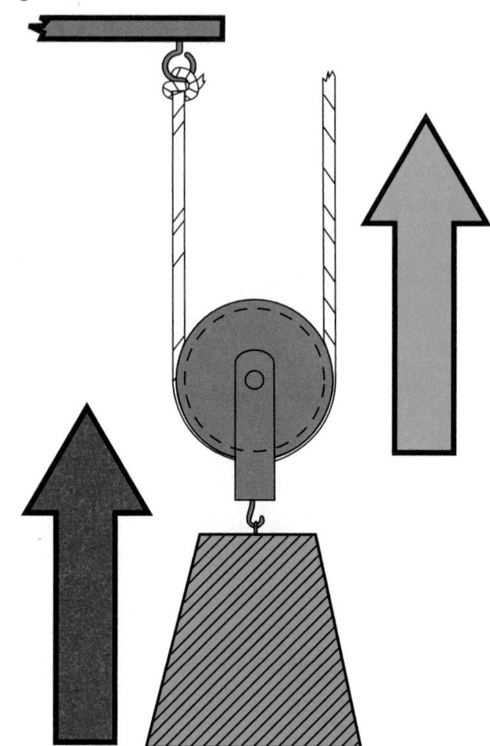

Third-class lever: A lever that has the input force positioned between the fulcrum and the load.

Pulley: A solid disc that rotates around a center axis. It usually has a groove around the outside edge that allows ropes or belts to easily ride around them.

Wheel and axle: A large-diameter wheel and its small-diameter axle are attached to each other to move as one unit.

Figure 9-8. Pulleys operate like levers.

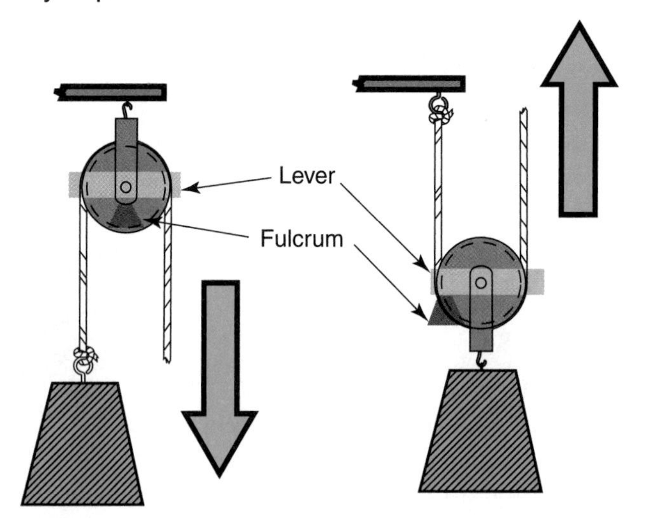

When the input force is applied to the axle, however, the applied force is transferred to the outside diameter of the wheel, and the distance it travels grows. See **Figure 9-11.** This action resembles a third-class lever. Doorknobs, screwdrivers, winches, and drills are a few of the common examples of wheel and axle systems.

Figure 9-9. The steering wheel of an automobile is a type of lever. It multiplies the force used to turn the front wheels.

Figure 9-10. Without the leverage of the steering wheel, it would be impossible to keep an automobile on a winding road.

Inclined Planes

An ***inclined plane*** is a simple machine that makes use of sloping surfaces. Loading ramps are an example of this simple machine. It is difficult for a person to lift a 100-lb. object up to the tailgate of a truck. See **Figure 9-12.** By rolling the object up a gently sloped ramp, much less force is exerted to achieve the same result. The one drawback in this system is that the amount of energy used is spread over a greater distance.

Inclined plane: A simple machine that makes use of sloping surfaces.

Figure 9-11. A wheel and axle is a distance multiplier when the force is applied to the smaller-diameter axle.

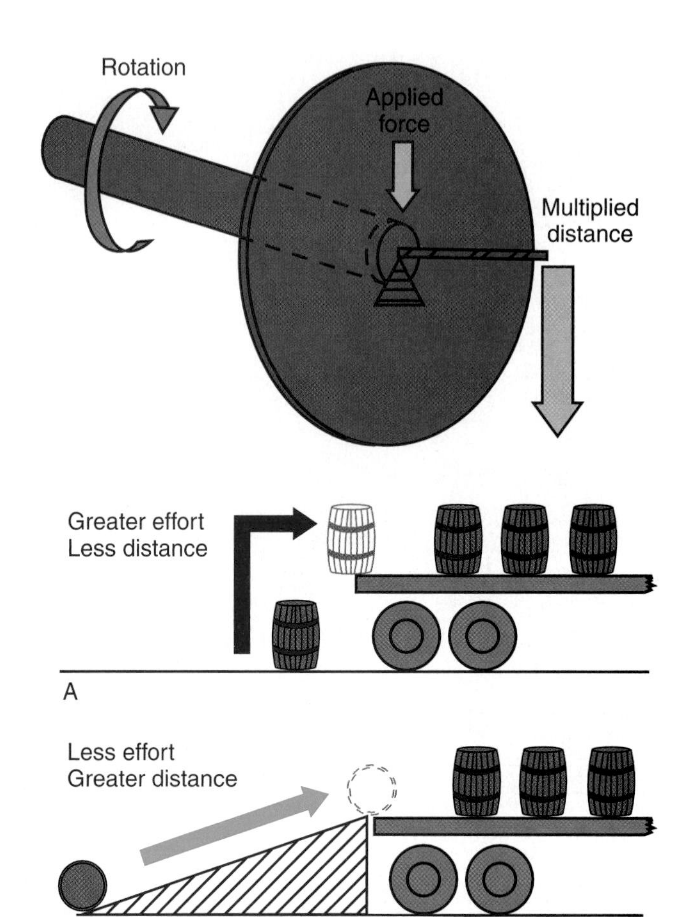

Figure 9-12. The inclined plane increases mechanical advantage. A—Greater force is needed to lift this barrel than to roll it up an incline. B—Rolling the barrel up a ramp (inclined plane) requires moving it a longer distance. Less force, however, is used. The inclined plane is a force multiplier.

Inclined planes are often found in the roads up steep hills or mountainsides. The roads switch back and forth, making a more gently angled slope. This allows vehicles going up to use less effort. See **Figure 9-13.** Wheelchair ramps use the same principle. Patients who use wheelchairs often have the necessary energy to negotiate a ramp, but would need assistance to mount steps.

Figure 9-13. The ramp that makes this community art center accessible to people with handicaps is a practical example of the inclined plane. To prevent a steep slope, the builder used a switchback design.

Screws

A *screw* is a simple machine that operates on the principle of inclined planes. It is a very long inclined plane wrapped around a shaft. See **Figure 9-14.** The longer the slope in the inclined plane on a screw, the more turns are required to advance 1″ up its shaft. Screws that have 16 threads per inch apply greater force than screws that have 2 threads per inch. They also create more surface area to produce friction. Screws are commonly used as mechanical fasteners for wood, metal, and plastic. In this type of application, friction is desirable to grip and hold parts together better. When used in vises and car jacks, screws function as force multipliers.

Screw: A simple machine consisting of a very long inclined plane wrapped around a shaft.

Wedge: A simple machine based on the principle of the inclined plane.

Gear: A metal wheel with small notches cut into its rim.

Figure 9-14. Looking at the threads on this screw, it is easy to see it is an inclined plane wrapped around a shaft.

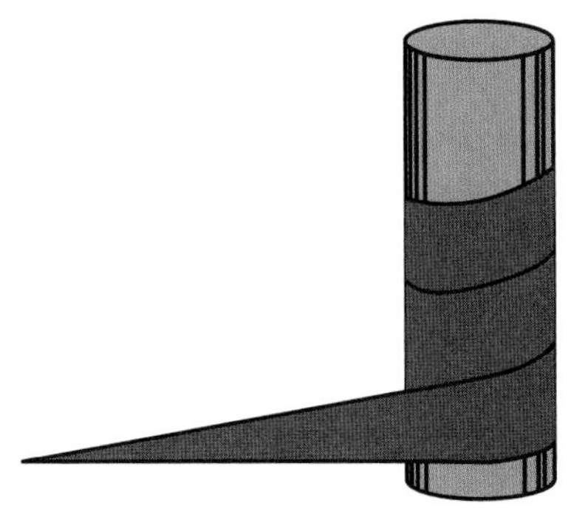

Wedges

A *wedge* is a simple machine consisting of two inclined planes placed back to back. Wedges are often used to split materials. See **Figure 9-15.** If you have ever driven a nail or chopped wood with a hatchet, you have used a wedge. As the hatchet enters the wood, it forces the wood fibers apart until they separate, or split.

Figure 9-15. A hatchet uses the principle of a wedge. Its weight and fast movement combine to enable it to split the piece of wood.

Gears

A *gear* is a metal wheel with small notches, or teeth, cut into its rim. **Figure 9-16** shows some typical gears. Gear sets are made so the gear teeth interlock and drive each other. A gear powered by the engine or motor is referred to as the *drive gear.* The gear to which power is transferred is called the *driven gear.* As one tooth from the driver gear meshes with a tooth from the driven gear, power is transferred from the driver to the driven gear. A pair of gears is all that is needed to modify the characteristics of power coming from a hydraulic motor, an electric motor, or an internal combustion engine.

Figure 9-16. Several of the many types of gears available. A—A worm wheel. B—A spur gear. C—A bevel gear. (Boston Gear)

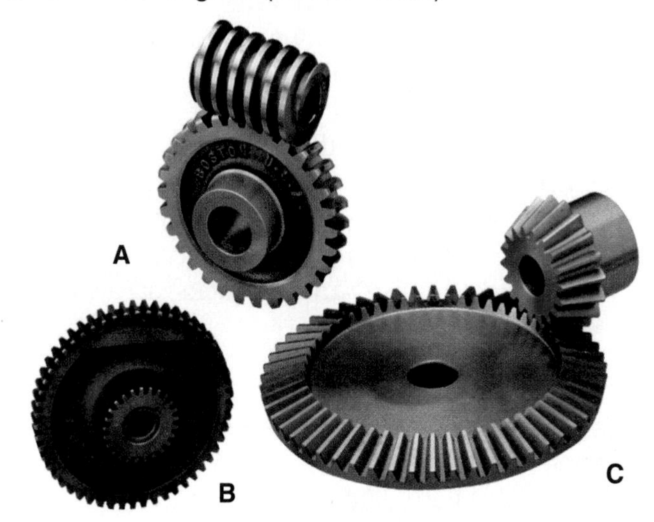

Although there are many different types of gears, the most common are the spur, helical, rack and pinion, worm, and bevel gears. Gears can control mechanical power in the same way as belts and pulleys—by changing the direction of power, speed, and torque. Like chains and sprockets, gears do not slip, and therefore they produce positive power transfer.

Characteristics of Mechanical Power

All forms of power, including mechanical power, have two primary characteristics: effort and rate. In a linear mechanical system, the effort is referred to as force and is usually measured in pounds, in the U.S. Customary system. In a rotary mechanical system, the effort characteristic is referred to as *torque* and is usually measured in foot-pounds (ft.-lbs.) or inch-pounds (in.-lbs.) for some low-torque applications.

The measurement of rate in a mechanical system varies, based on the application. Feet per second (fps) is a commonly used rate measurement for linear systems, such as conveyors and other material-processing and handling systems. Miles per hour (mph) is a measurement of linear mechanical rate associated with the automobile and is measured by the speedometer. Rate in a rotary mechanical system is usually measured in revolutions per minute (rpm) and is measured with a tachometer. Rotary mechanical rate may also be measured in revolutions per second (rps) or revolutions per hour (rph) in other, less common applications.

Quantities of Measurement

There are two categories of measurement for quantities: scalar quantity and vector quantity. A ***scalar quantity*** represents a physical quantity specified by the magnitude of the quantity and expressed by a number or unit. Examples include 75°F, 12.5 amperes, 65 mph, and 45 lbs. These types of expressions are referred to as scalar statements. Quantities that have both magnitude and direction are referred to as ***vector quantities***. There are three common vector quantities:

- ***Displacement*** includes both distance and direction. An example is 100 miles east.
- ***Velocity*** includes speed and direction. Some examples are 100 mph south and 25 fps east.
- Force includes the magnitude of the force (usually expressed in pounds) and the direction of the force. An example is 250 lbs. upward.

Scalar quantity: A physical quantity specified by the magnitude of the quantity and expressed by a number or unit.

Vector quantity: A quantity that has both magnitude and direction.

Displacement: A vector quantity that includes both distance and direction.

Velocity: A vector quantity that includes speed and direction.

Torque

Force that produces a twisting or turning effect or rotation is referred to as *torque* (torsional) force, or *moment force*. A torque measurement has two distinct components:

- A measurement of how much force is applied to the lever arm.
- A measurement of the radius of the lever arm itself. See **Figure 9-17.**

For example, picture a person trying to remove the lug nut that holds a wheel onto a car. The person generates 50 lbs. of force at a distance of

Figure 9-17. Torque can be increased by applying more effort or by extending the length of the lever. A—Applying 50 lbs. of effort to a 1.5′ lever does not generate enough torque to loosen the lug nut. B—Increasing the length of the lever to 3′ and applying the same amount of force doubles the torque and loosens the nut.

A

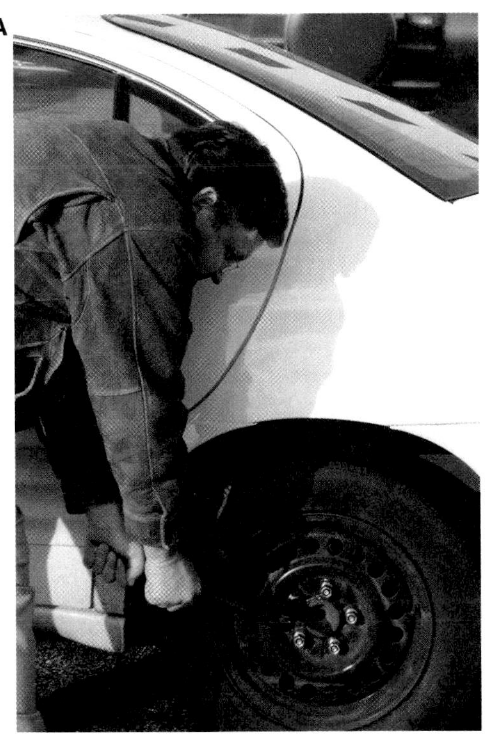

50 lbs. × 1.5′ = 75 ft.-lbs.

B

50 lbs. × 3′ = 150 ft.-lbs.

Prony brake: A device used to measure the effort produced by a twisting or turning force. It is based on the principle that if an opposite force equals the effort produced by a spinning object, movement will cease.

Indicated horsepower (ihp): The maximum potential hp produced by an engine under ideal conditions.

Brake horsepower (bhp): The amount of power available at the rear of the engine under normal conditions.

Frictional horsepower (fhp): The amount of hp necessary to overcome the internal friction of an engine and other forms of frictional loss.

1.5′ from the fulcrum of the wrench. If the lug nut does not move, there are two options to increase the torque applied to the bolt:

- Apply more force.
- Extend the effective lever arm of the wrench.

By increasing the length of the lever arm, more torque is generated when the same amount of force is applied.

It may be difficult to measure the effort produced by a twisting or turning force. An instrument more specific for the task than a simple spring scale must be used. For example, consider a flywheel attached to a pneumatic motor. The radius of the flywheel, one component of the torque measurement, is easily measured with a ruler. The second component in the measurement of torque is the force with which the flywheel is spinning. This can be obtained with a measuring device called a ***Prony brake***. See **Figure 9-18.** The Prony brake is not a precise measuring instrument, but it will give a reasonable approximation of the force produced by a rotating shaft. It is based on a simple principle. If an opposite force equals the effort produced by a spinning object, movement will cease.

The Prony brake is adjustable and provides increasing tension on the flywheel until the flywheel stalls. At the moment the flywheel comes to a complete stop, the opposing force applied to it is approximately equal to the force with which the flywheel was spinning. A reading is taken from the scale and multiplied by the radius to determine the approximate torque produced by the flywheel.

Horsepower (hp)

Horsepower (hp) is the rate at which output work is performed. This unit of measurement was initially associated with mechanical power, but it is now commonly used to represent all forms of power, including fluid and electrical power. Today, devices that produce mechanical power, such as electric motors and internal combustion engines, are all rated in terms of their hp output. There are several different types of hp ratings:

- ***Indicated horsepower (ihp)*** is the maximum potential hp produced by an engine under ideal conditions.
- ***Brake horsepower (bhp)*** is the amount of power available at the rear of the engine under normal conditions.
- ***Frictional horsepower (fhp)*** is the amount of hp necessary to overcome the internal friction of an engine and other forms of frictional loss, such as tire resistance or wind drag. It represents the percentage of power inevitably lost within the internal workings of the engine and by other processes.

Figure 9-18. Using a Prony brake to measure torque.

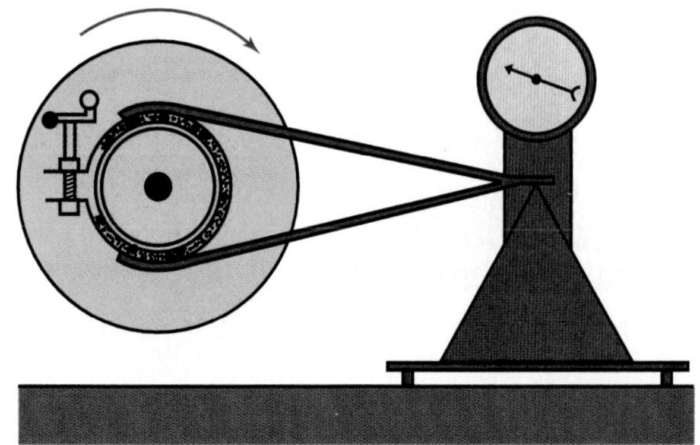

Net Forces of Balanced and Unbalanced Loads

When forces are balanced, they are said to be in a state of equilibrium. This means that all movement ceases. Observe the two examples of how balanced forces create equilibrium shown in **Figure 9-19.**

Curricular Connection

Social Studies: The History of Horsepower (hp)

James Watt, a Scottish engineer and inventor, coined the term *horsepower (hp)*. He developed this unit of power as a way of comparing machines to the existing form of power generation for work purposes, the horse. In the early days of the Industrial Revolution in England, horses were used to raise coal from the mines. Watt was working on improving the steam engine, and he tried to sell it to the coal companies to replace their horses. He wanted to know how powerful his new engines were, so he measured how long it took a horse to lift a weight a certain distance. Watt then compared this with how long it took his steam engine to lift the same weight the same distance.

After many tests with horses, he determined that a horse, on average, can haul coal at the rate of 22,000 ft.-lbs. per minute. He decided to raise this number by 50%, to err on the conservative side and, at the same time, make his engines seem more powerful than a horse so they would sell. Therefore, 33,000 ft.-lbs. per minute became the standard unit of measurement for linear mechanical hp. If an engine can push or pull 33,000 lbs. of load 1′ in 1 minute, it is a 1 hp engine. The energy outputs of cars, lawnmowers, vacuums, and nuclear power plants are all measured in hp.

Figure 9-19. Here are two balanced loads. The diagram on the right shows a fixed pulley. One fixed pulley creates only a change in direction, but no mechanical advantage. Pulling on the handle with 50 pounds of force will stabilize the load to provide equilibrium. Additional force will be necessary to create an imbalance and move the load.

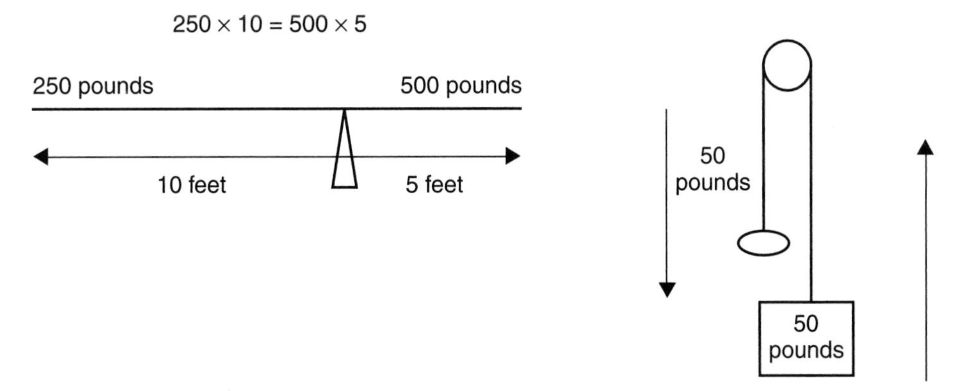

When the forces of effort and opposition are totally balanced, no movement will occur. A change in effort creates an unbalanced load, which results in movement. When this type of imbalance results along a straight line, calculating the results is easily performed with simple addition and subtraction. Forces in parallel are simply added to create what is known as a resultant force. If you were pulling on the load in Figure 9-19 with 50 pounds of force and your friend was pulling with another 20 pounds of force, the resultant force would be 70 pounds of effort.

Transmission of Mechanical Energy

Compound machines make use of two or more simple machines. To transmit or pass power from one part of a vehicle to another, variations on these simple machines are used. The devices described in the following sections are specialized components in industrial machines and transportation vehicles. The operation of these devices relies on the principles of one or more simple machines.

Clutches

Clutch: A mechanical device that connects the power source to the rest of the machine.

A ***clutch*** is a mechanical device that connects the power source (a motor or engine) to the rest of the machine. In many transportation vehicles, the clutch connects the power system to the drive system. When the operator of the vehicle wants to disconnect these two systems, he simply engages the clutch. This device is needed so vehicles can remain at rest with the engine running, start slowly without stalling, and shift gears while moving.

Clutches operate on the principle of friction. Usually, two surfaces rub against each other and then lock against each other. This enables the two parts to move at the same speed, powered by one source. The two types of clutches widely used in transportation vehicles are the diaphragm clutch and the centrifugal clutch.

Pulleys and Belts

If you have ever looked under the hood of an automobile, you have probably seen many different belts that move around pulleys. See **Figure 9-20.** The belt and pulley systems under the hood of a car transmit

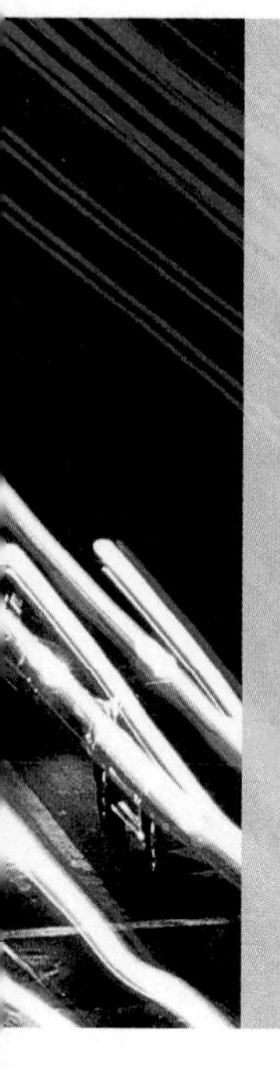

Career Connection

Mechanical Technicians

The power used in machines of mechanical systems produces work. Since the work machines do play such a large role in our daily lives, the equipment must function properly. Mechanical technicians are necessary to maintain and service machines.

Maintenance of equipment can include monitoring, repair, and design of mechanical systems. Technicians may be responsible for preparing proposals and for sketching and designing equipment. Once the machines have been assembled, technicians subject the equipment to several tests. Testing may be conducted on machines already in use. Technicians determine from these tests whether improvement is needed. Machines also sometimes require repairs. Technicians must stay within safety limits when working with this equipment.

Technicians who design mechanical systems must have a thorough knowledge of engineering and design techniques. The ability to communicate their plans and ideas clearly is also important. To repair machines, technicians must understand mechanics and possess mathematical skills. An associate's degree is required to be a worker in this field. The yearly salary may range from $25,000 to $61,000.

power from the engine to drive engine components, such as the water pump and fan. Other belt and pulley systems are used to power accessories, such as air conditioning.

Belts and pulleys control mechanical energy through any of five different arrangements. Through their use, we can connect and disconnect power like a clutch, change direction, reverse rotation, change speed, and change torque. **Figure 9-21** illustrates the different arrangements to accomplish these tasks.

Figure 9-20. Automobile engines make use of belt and pulley systems. Older engines often had several belts driving different devices, but most modern engines, such as this one, use a single long belt called a *serpentine belt*.

Chains and Sprockets

The chain and sprocket setup shown in **Figure 9-22** is a very familiar sight. This mechanical transmission system is found on bicycles, mopeds, and motorcycles. Chains and sprockets are usually used as the drive system to bring power to the driving wheel of the vehicle.

Like belts and pulleys, chain and sprocket systems can change speed and torque. By shifting the chain to different sprockets on a bicycle, the input force can be controlled. Making this adjustment can make it easier to climb hills or deliver torque to the wheels when riding downhill. Chain and sprocket systems have an advantage over belt and pulley systems when used for drive systems. They provide positive power transfer, which means that the chain cannot slip like a belt on a pulley.

Figure 9-21. Pulley and belt arrangements transfer power from one point to another. They also can change torque and speed through pulley sizes. Sometimes, belts are used to change direction of a force.

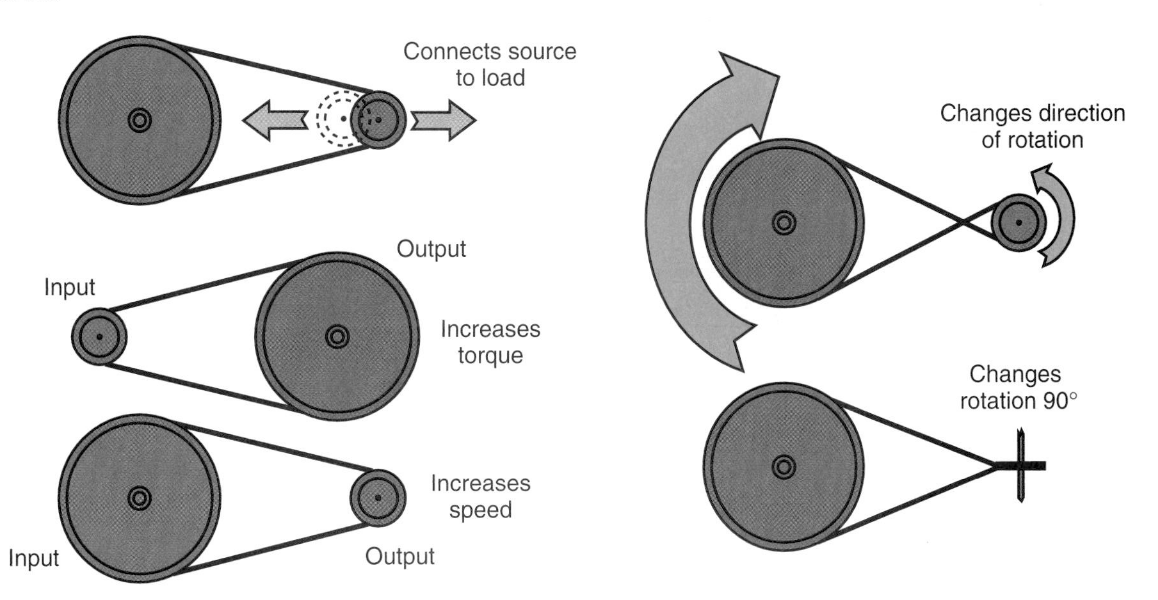

Figure 9-22. A chain and sprocket transfer power from the bicycle pedals to the rear wheel.

Shafts and Bearings

Shafts are long, cylindrical pieces of metal used to transfer mechanical energy in many types of machines. They are vital parts of automobile engines and drive systems. See **Figure 9-23.** Because of their solid construction, shafts permit positive power transfer.

Bearings are specially shaped pieces of metal used to support shafts and reduce friction between metal parts as they move past or revolve around each other. They are made to be strong, but they also allow the shaft to turn inside them. Although there are many types of bearings, the most common types in vehicles are smooth circles of metal shaped to exactly fit the outside of the shaft.

These are very important in the operation of any shaft-driven machine. They must conduct heat away from the shaft, as well as resist softening from heat. Bearings must also be soft enough to glide over uneven shaft surfaces and tough enough to resist the corrosive properties of some lubricants. **Figure 9-24** illustrates a typical shaft and bearing relationship.

Because the shaft is a solid piece of metal, it is not easily bent. If the shaft is bent, it cannot accurately transfer mechanical energy. In cases where flexibility is needed, ***universal joints*** are placed on the ends of the shaft. **Figure 9-25** is an illustration of a common universal joint. Universal joints allow connected shafts to spin freely, while permitting a change in direction.

Shaft: A long, cylindrical piece of metal used to transfer mechanical energy in many types of machines.

Bearing: A specially shaped piece of metal used to support shafts and reduce friction between metal parts as they move past or revolve around each other.

Universal joint: A joint that allows connected shafts to spin freely, while permitting a change in direction.

Figure 9-23. The drive shaft of a rear-wheel drive car shows the normal relationship of a shaft and bearing.

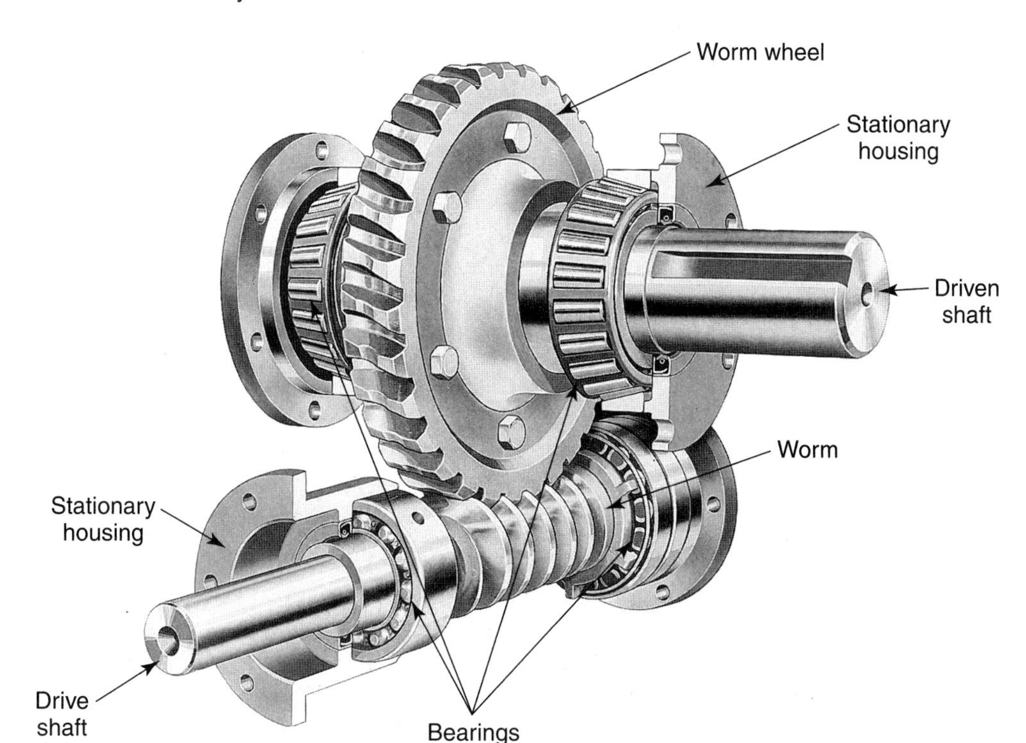

Figure 9-24. The shafts in this assembly are supported in their housings by bearings. The bearings reduce friction between the rotating and stationary parts. The worm gear, comprised of a worm and a worm wheel, transmits power between drive and driven shafts. (Cleveland Gear)

Mechanical Advantage

Mechanical advantage is increased force, speed, or distance and the benefits of that increase created by using a machine to transmit force. The force you apply to a machine is actually multiplied. For example, imagine trying to move a large rock that is partially embedded in the ground with only your arms, legs, and body weight. The task seems extremely difficult, if not impossible. If you use a long metal bar as a lever in combination with your physical effort, you increase the amount of force produced. The increase in force is a gain in mechanical advantage.

Figure 9-25. A universal joint is often placed between two shafts not on the same plane. It allows the two shafts to turn together without binding or breaking. (Boston Gear)

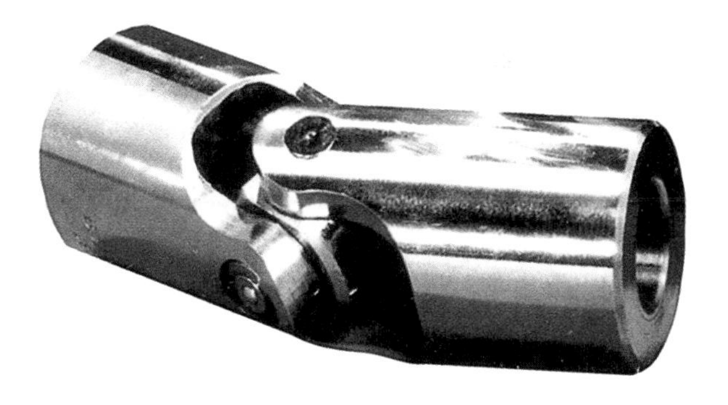

This advantage can be calculated by dividing the input distance by the output distance. For example, suppose you are using the metal bar wedged between two rocks. See **Figure 9-26.** You exert force 4′ away from the smaller rock. The smaller rock is the fulcrum, or pivot point. The load is 2′ away from the fulcrum. This means the machine provides twice the force you put into it. Using mechanical advantage usually results in less effort required to perform a task than if a mechanical device was not used.

Mechanical advantage: An increased force and the benefits of that increase created by using a machine to transmit force.

Using Simple Machines to Gain Mechanical Advantage

Simple machines can make lifting and other strenuous activities easier by multiplying force. They cannot, however, produce more actual power than the power put into them. These machines simply modify the

Figure 9-26. Levers give their users a mechanical advantage. Less energy is required than if trying to move the load by lifting it.

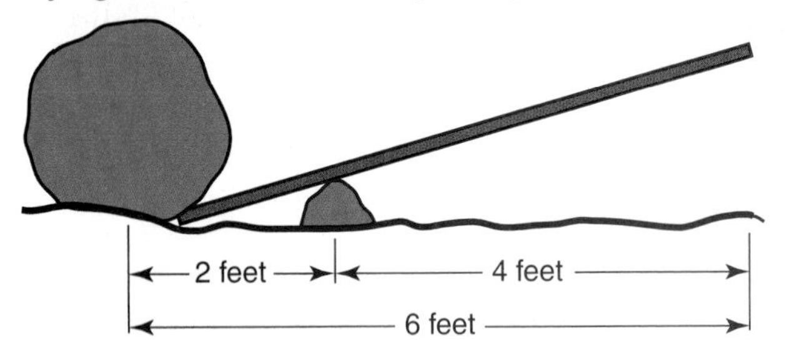

characteristics of the power provided in order to achieve a specific goal, such as making lifting easier. There is a trade-off when using simple machines. The trade-off for gaining force is a loss of speed or distance.

This trade-off is true for all advantage-gaining devices, including those associated with other forms of power, like fluidics and electricity. If effort is made easier, rate will be sacrificed. An example is the transmission in an automobile. If the car is stopped at a traffic light and the light turns green, torque is required to get the vehicle moving, not speed. A low gear produces that torque, but it will not get the car moving very fast. Once the car is moving, the transmission is shifted into mid-range gears that provide less torque and allow for more speed. If driving on a highway, the car may be shifted into a high gear, such as fourth or fifth gear. These higher gears provide much higher speeds, but much less torque. To climb a steep hill, it may be necessary to shift into a lower gear to provide more torque. Today, many modern automobiles use an automatic transmission. This type of transmission has the ability to sense strain on the engine and shift the transmission into the appropriate gear range automatically to meet the power characteristics required for the terrain (more speed or more torque).

Levers

We can change the force produced by first-class levers by changing the fulcrum's position between the input force and the load. If the fulcrum is closer to the load, the force is increased. When the fulcrum is exactly centered between the two, the force on both the load and the input force are equal. As the fulcrum moves toward the input force, the machine loses strength, but the distance we can move the load increases. Rowing teams use this positioning to increase the distance their oars move through the water. See **Figure 9-27.**

Figure 9-27. Levers in use. A—Oars are based on the principle of first-class levers. They are designed to be distance multipliers. (Harvard University) B—Are these paddles examples of a first-class lever or a third-class lever?

A

B

Pulleys

Pulley systems provide a mechanical advantage. For example, in a pulley system that uses four pulleys, a 4:1 mechanical advantage is produced. This means that every 25 lbs. of input force results in about 100 lbs. of lifting, or output force. The trade-off, however, is that 4′ of rope must be reeled in to create just 1′ of lift on the output side of the system. The distance between the fulcrum and input force is larger on a moveable pulley system. This is why it has greater mechanical advantage in force than a fixed pulley system.

Wheels and axles

The wheel and axle is a simple machine consisting of two parts. The larger circular part is called the wheel, and the smaller circular part is called the axle. Practical examples of the wheel and axle at work include large crank handles. **Figure 9-28** shows a crank handle that can help to generate a large mechanical advantage. Such crank handles are often used on retractable boat lifts. Through a cable system, a small boat can actually be lifted right out of the water.

Inclined planes

Using a ramp makes it easier to load packages onto a loading dock. The distance that must be covered is much greater, however, than if the load could simply be lifted directly onto the loading dock. Using a ramp makes loading easier, but more time-consuming, because of the increased distance that must be covered to reach the desired height.

Wedges

If you calculate the mechanical advantages of different-shaped wedges, you find that the longer the wedge (compared to thickness), the greater the advantage. See **Figure 9-29.** This is the reason you can split logs easier with thinner wedges.

Gears

Gears are often used to modify the characteristics of mechanical power and not always used to gain torque. **Figure 9-30** presents a diagram of an automobile engine coupled to a manual transmission. A belt or chain may be

Figure 9-28. Calculating mechanical advantage with a wheel and axle is actually easy. If the diameter of the axle is 3″ and the diameter of the wheel is 4′, the mechanical advantage can be calculated as follows: 4′/.25′ = 16:1 mechanical advantage. This means that if the wheel is driving the axle, it will convert every pound of effort on the handle into 16 lbs. of force at the axle. On the other hand, if the axle is driving the wheel, every inch the axle moves, the wheel will move 16″.

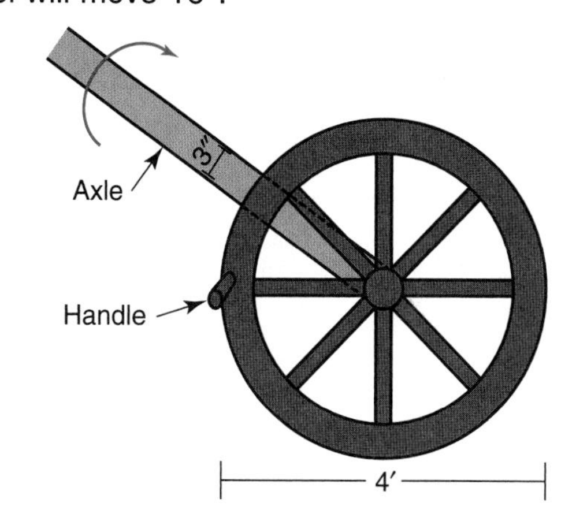

Figure 9-29. A long, thin wedge has more mechanical advantage than a short, fat one.

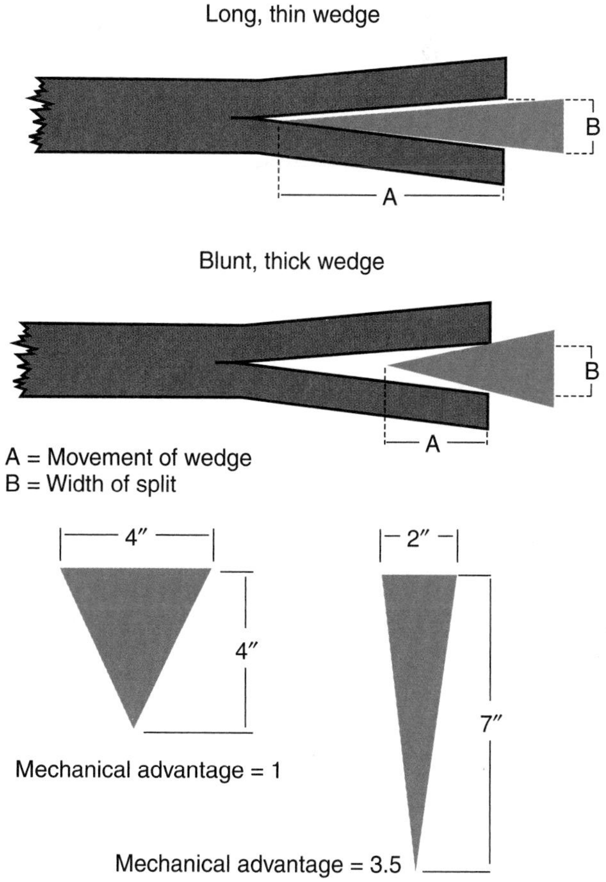

Figure 9-30. A transmission modifies the characteristics of the automobile's engine power, changing speed and torque.

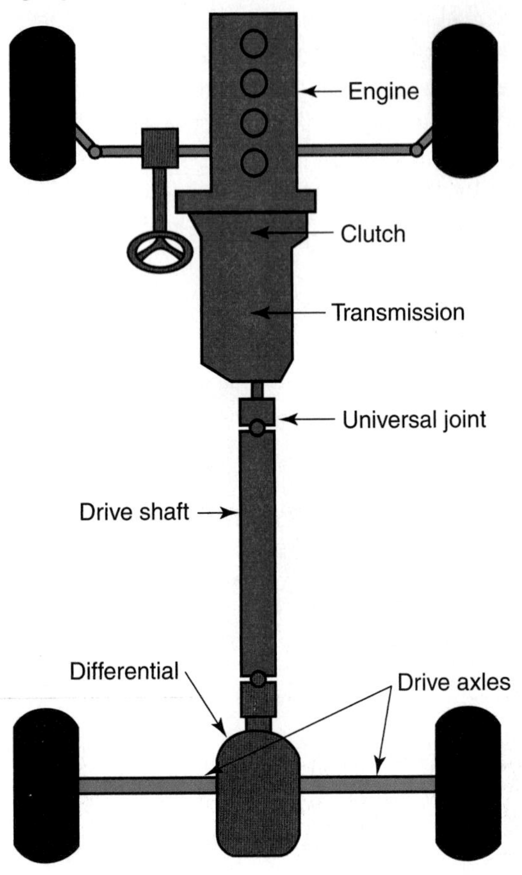

used to span the distance between two gears, but the number of teeth on the two gears determines the mechanical advantage produced by the gear set. See **Figure 9-31.** For example, a machine is chain driven by a motor that rotates at 1750 rpm. Attached to the motor is a 50-tooth drive gear. The machine is designed to rotate at a speed of 2500 rpm. How many teeth must the driven gear have? See **Figure 9-32.**

A simple gear train may be created using three or more gears. When three or more gears are used, the gears between the driver gear and the driven gear are referred to as ***intermediate gears***. They are also sometimes called ***idler gears***. See **Figure 9-33.** The idler gear is often used in the middle of a gear set to change the rotation of the driven gear to spin in the same direction as the driver gear. Note that, no matter how many idler gears are used between the driver gear and the driven gears, none of them has any effect on the mechanical advantage of the gear train. The impact is based on the relationship of the driver gear to the driven gear.

Gears arranged in a series gear train provide only limited mechanical advantage. When gears are arranged in a compound gear

Intermediate gear: A gear between a driver gear and a driven gear.

Idler gear: See *Intermediate gear*.

Figure 9-31. Different combinations of gears modify the characteristics of power from a motor or other power source.

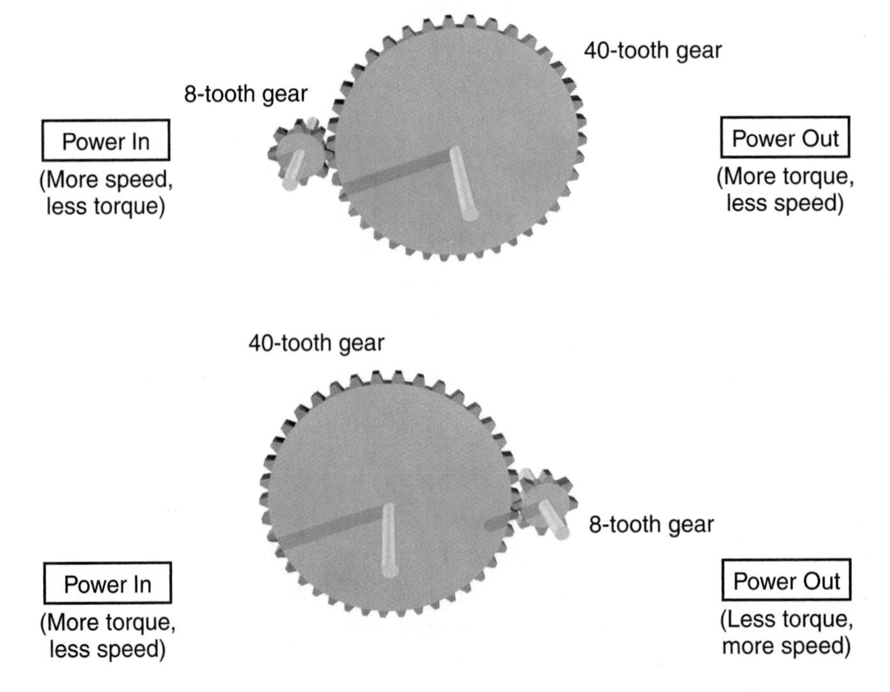

Figure 9-32. Calculating the mechanical advantage of a gear set. In this example, torque is actually sacrificed in order to make the machine spin the proper revolutions per minute (rpm). The mechanical advantage is in terms of speed, in this instance, instead of torque, as is usually the case when mechanical advantage is discussed.

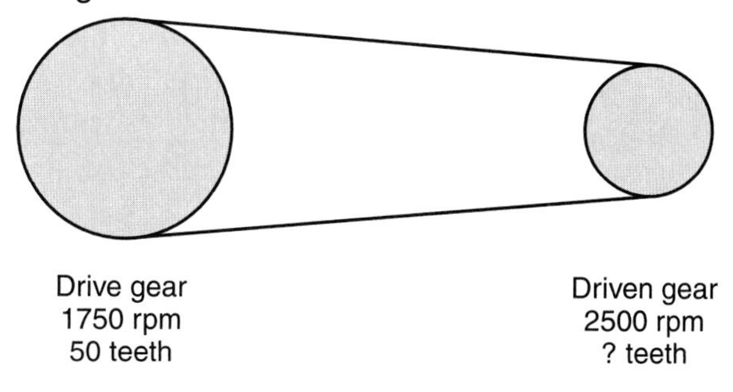

1. $1750 \text{ rpm} \times 50 \text{ teeth} = 87{,}500$
2. $\dfrac{87{,}500}{2{,}500 \text{ rpm}} = 35 \text{ teeth}$
3. $\dfrac{50}{35} = 1.43$ mechanical advantage of speed

Figure 9-33. A gear train consists of the drive gear, one or more idler gears, and the driven gear. Idler gears do not alter mechanical advantage.

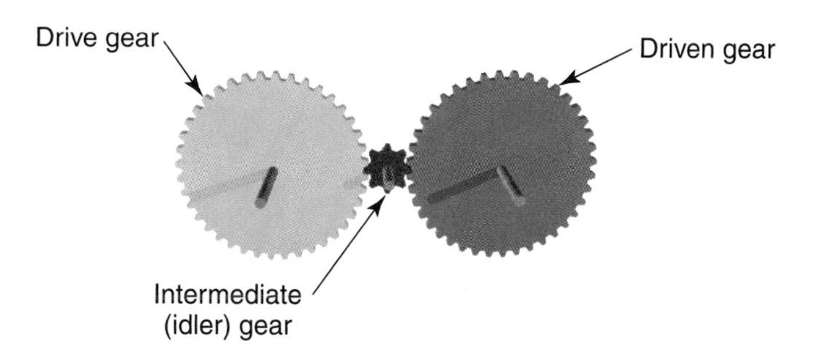

cluster along parallel shafts, however, force is greatly multiplied and creates much greater mechanical advantage. A ***compound gear cluster*** consists of more than two gears arranged in such a way as to gain significant mechanical advantage of force or speed. See **Figure 9-34.** Gear boxes often use compound gear clusters in order to generate much greater mechanical advantages than could be created by a two-gear set. When the technique of compound gearing is utilized in conjunction with specialty gears, such as a worm gear, tremendous mechanical advantages can be created.

Compound gear cluster: More than two gears arranged in such a way as to gain significant mechanical advantage of force or speed.

Calculating the Mechanical Advantage of Compound Machines

A compound machine employs two or more simple machines. We can determine the mechanical advantage of a compound machine. To do so, complete the following steps:

Figure 9-34. Compound gear clusters are used to increase mechanical advantage. This is the reduction gearbox used in the power system of a helicopter. (Bell Helicopter Textron)

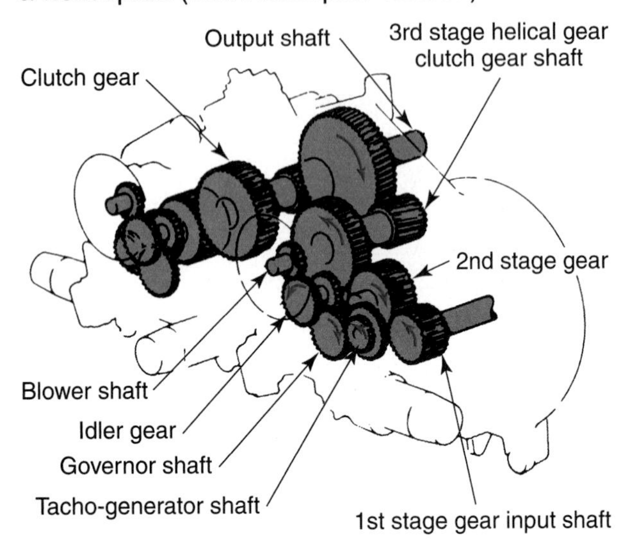

1. Identify each of the simple machines contained within the compound machine.
2. Calculate the mechanical advantage of each simple machine.
3. Multiply the mechanical advantages of each simple machine together. The product of this calculation is the total mechanical advantage for the compound machine, as demonstrated in **Figure 9-35.**

Ideal Mechanical Advantage (IMA) vs. Actual Mechanical Advantage (AMA)

Ideal mechanical advantage (IMA) is a ratio of the forces or the distances involved in a mechanism. In the real world, however, we must account for the amount of energy lost through friction. Friction is the heat energy that is a common by-product of mechanical energy. ***Actual mechanical advantage (AMA)*** is the ratio of the increase of force or effort by a machine, including energy lost through friction. AMA is always less than IMA because IMA assumes 100% efficiency, and AMA accounts for frictional losses.

Ideal mechanical advantage (IMA): A ratio of the forces or the distances involved in a mechanism. It assumes 100% efficiency.

Actual mechanical advantage (AMA): The ratio of the increase of force or effort by a machine, including energy lost through friction.

Figure 9-35. Calculating the mechanical advantage of compound machines.

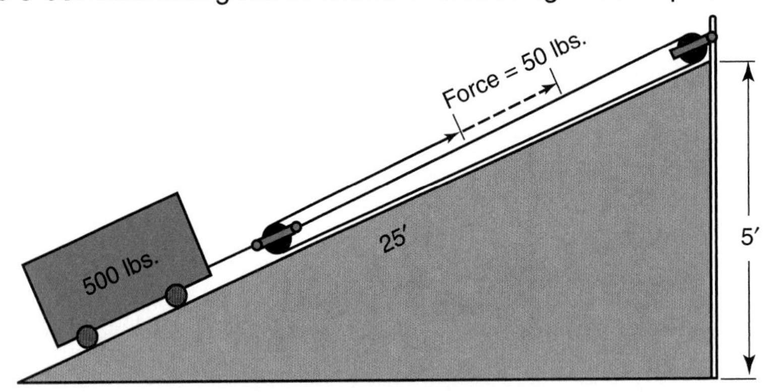

Inclined Plane and Pulleys

Mechanical advantage of inclined plane = 25′:5′
= 5:1

Mechanical advantage of pulley system = 2:1

Mechanical advantage of compound machine = 5:1 × 2:1
= 10:1

Force on rope needed to move load = Load ÷ Mechanical advantage
$= 500 \text{ lbs.} \div \frac{10}{1}$
= 50 lbs.

To calculate AMA, a formula that involves the forces in the system must be used. The force is often measured in pounds of weight. In order to determine the total effort in the system, force is multiplied by the distance of the lever on the input side of the fulcrum. Divide the output effort by the input effort to find AMA.

All the mechanical advantage calculations we have done in this chapter are really IMA calculations. We have not taken any frictional loss into account. When we do live measurements to determine mechanical advantage, sometimes the numbers do not work exactly because of frictional loss in the real world. For instance, let us say a pulley system is supposed to provide a 4:1 mechanical advantage. In theory, only a little more than 25 pounds should lift 100 pounds. In reality, you may find that 28 pounds are required.

The efficiency of machines is a comparison of the input effort to the amount of output effort. One way to calculate efficiency (as a percentage) is to divide AMA by IMA and multiply the answer by 100.

$$\frac{100 \text{ lbs.}}{28 \text{ lbs.}} = 3.57 \text{ AMA} \quad \frac{100 \text{ lbs.}}{25 \text{ lbs.}} = 4.0 \text{ IMA} \quad \frac{3.57 \text{ AMA}}{4.0 \text{ IMA}} \times 100 = 89\% \text{ efficient}$$

This simple calculation can determine both system efficiency and percentage of effort lost as a result of friction. See **Figure 9-36.**

Figure 9-36. Calculating the efficiency of a mechanical system.

$$\%E = \frac{\text{AMA}}{\text{IMA}} \times 100$$

Summary

Mechanical power systems control mechanical energy to do work. They consist of machines that can change size, direction, and speed of forces, as well as change the type of motion. The three types of motion studied in transportation technology are reciprocating, rotary, and linear.

There are six simple machines used to control mechanical energy: the lever, the pulley, the wheel and axle, the inclined plane, the screw, and the wedge. All six rely on the principle of either the lever or the inclined plane. Mechanical power systems move mechanical energy through special transmission devices. These devices include clutches, pulleys and belts, chains and sprockets, and shafts and bearings.

Mechanical advantage is the increase in force a machine provides. Ideal mechanical advantage (IMA) is the ratio of forces assuming the mechanism is 100% efficient. Actual mechanical advantage (AMA) considers energy loss due to friction.

Key Words

All the following words have been used in this chapter. Do you know their meanings?

actual mechanical advantage (AMA)
bearing
brake horsepower (bhp)
clutch
complex machine
compound gear cluster
displacement
first-class lever
frictional horsepower (fhp)
fulcrum
gear
ideal mechanical advantage (IMA)
idler gear
inclined plane
indicated horsepower (ihp)
intermediate gear
lever
mechanical advantage
Prony brake
pulley
scalar quantity
screw
second-class lever
shaft
simple machine
third-class lever
universal joint
vector quantity
velocity
wedge
wheel and axle

Test Your Knowledge

Write your answers on a separate sheet of paper. Do not write in this book.

1. Machines can change the _____, _____, and _____ of forces.
2. The six simple machines are _____, _____, _____, _____, _____, and _____.
3. A wheelbarrow is an example of a(n) _____ lever.
4. Explain the difference between first-, second-, and third-class levers.
5. Third-class levers always increase the _____ of the load.
6. Shovels and baseball bats are examples of _____ levers.
7. In what ways do single fixed and single moveable pulleys differ?
8. Screws and wedges operate on the principle of the _____.
9. Name three types of gears and state a common application for each.
10. Identify the two primary characteristics of power.
11. How much torque is required to drive a pump rated at 60 horsepower (hp) that must operate at a speed of 350 revolutions per minute (rpm)?
12. Summarize the difference between scalar and vector quantities.
13. How much force would it take to stabilize the following load?

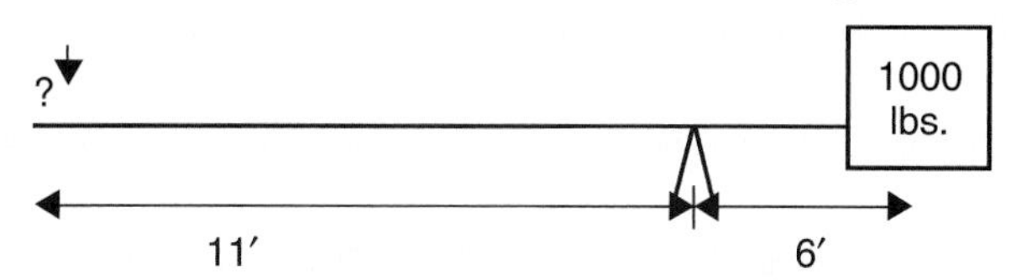

14. A(n) _____ is the device that links the power system to the drive system in many vehicles.
15. List the five ways that pulley and belt systems control mechanical energy.
16. A mechanical system that provides _____ means that the power-transmitting device will not slip when used.
17. _____ are used on the ends of shafts at points where flexibility is needed.
18. A motor spins at 3550 rpm and is connected to a 4″ diameter pulley. This pulley feeds a 20″ diameter pulley that drives an auger blade for digging holes in the ground. What is the mechanical advantage that the pulleys create?
19. Do gears significantly change the amount of power produced by a motor or engine? Explain your answer.
20. When using a pulley system, you determine that you must reel in about 6′ of rope in order for the load to be elevated by only 1′. Based on this information, what is the approximate mechanical advantage of the pulley system?
21. The drive gear on a motor has 36 teeth and revolves at a speed of 1150 rpm. How many teeth must the driven gear have if it must rotate at 575 rpm?

22. The rotating portion of a conveyor drive system spins at 80 rpm and is driven by a gear containing 100 teeth. If the drive gear on the motor has 20 teeth, what is the speed of the motor?

23. Calculate the mechanical advantage of the following compound machine.

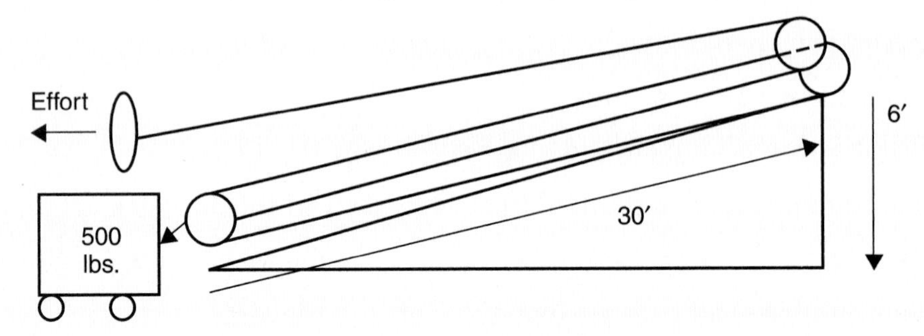

For questions 24–28, match the term in the left-hand column with the correct definition in the right-hand column.

24. _____ Complex machine.
25. _____ Fulcrum.
26. _____ Velocity.
27. _____ Mechanical advantage.
28. _____ Actual mechanical advantage (AMA).

A. An increase in force provided by a machine.
B. A ratio that includes energy loss in a mechanical system.
C. A measurement that includes speed and direction.
D. Made up of many simple machines.
E. The point around which a lever rotates.

29. A pulley system develops a 6:1 ideal mechanical advantage (IMA), but you measure the AMA at only 5.75:1. What percentage of loss has occurred due to friction?

Activities

1. Survey the lab area in your classroom and list the various types of equipment contained. Compile one master class list of equipment and determine the simple machines found in each piece of equipment.
2. Design and explain the operation of a useful, compound machine that combines two or more simple machines to perform the following tasks:
   A. Lifting a boat out of the water.
   B. Steering a go-cart.
   C. Raising a ladder.
3. Use LEGO® toys or a similar mechanism kit to design and build the following:
   A. A slow car. Design a car that goes as slow as possible from the starting point. You must use only parts within your kit, and the car must make forward progress.
   B. A "lift-a-lot." Design a lifting device that can lift the most weight.
   C. A clean sweeper. Design a device to pick up small balls of paper from a tabletop and deposit them in a bin that can be emptied.

10

Fluid Power Systems

Basic Concepts

- Recognize the differences between hydraulic and pneumatic power systems.
- Identify the advantages of using fluid power systems.
- List the uses of valves in fluid power systems.
- State the physical characteristics of liquids used in fluid power systems.
- Recognize why hydraulic cylinders can increase mechanical advantage.

Intermediate Concepts

- Give examples of various applications of fluid power systems.
- Explain the difference between gauge pressure and absolute pressure.
- Describe the operation of directional-control valves.
- Summarize the differences between how single-acting and double-acting cylinders operate.
- Discuss the safety concerns when working with liquids and gases under pressure.
- Interpret a basic fluid power circuit.

Advanced Concepts

- Calculate the mechanical advantage created by using liquids under pressure.
- Compute the size of cylinders necessary to perform a specific application.
- Design simple fluid power circuits that are controlled by electricity.

Ancient societies found many uses for the movement of water. Common uses included simple plumbing systems to circulate fresh water and to move crude transportation vehicles, such as rafts and hollowed-out logs. Early societies found that by using the fluid properties of water, their work could be made easier. See **Figure 10-1.** Fluid power systems use the energy found in liquids and gases to perform work.

Figure 10-1. The waterwheel is a fluid power system, since the water (fluid) operates the machines. A—This Illinois gristmill was built in the 1850s and is powered by an external waterwheel. B—An unusual mill with an interior waterwheel is on George Washington's Mount Vernon plantation in Virginia. Opening the sluice gate allowed water from the millrace to turn the wheel that powered grinding stones.

A

B

This chapter focuses on the technology involved in using fluid power systems to perform work. The principles behind hydraulic and pneumatic systems will be explained. The devices used to control the fluid power will also be described.

What Are Fluid Power Systems?

Hydraulic system: A system that controls and transmits energy through the use of liquids.

The study of fluid power systems includes both hydraulics and pneumatics. ***Hydraulic systems*** control and transmit energy through the use of liquids, such as oil and similar liquids. Various components have been developed to control liquids under pressure so they can do work for us. These components include cylinders, pumps, and transmission lines. See **Figure 10-2.** In pneumatic systems, gases are used in place of liquids. Air from our atmosphere is the gas most commonly used. The properties of gases and liquids are the same in many ways, and both can be described as fluids.

Career Connection

Distributors

Power converters are used to change one form of power into another form of power. An example of power conversion is the changing of mechanical power into electricity. A generator is the machine power distributors use to convert power in this way.

Power distributors operate and monitor the equipment used in electrical distribution. It is their job to maintain and repair the machines. In order to keep equipment functioning properly, it is imperative to monitor and sustain quality. In order to perform these tasks, distributors must be proficient in the use of hand tools.

To properly manage electrical distribution, distributors must be familiar with mathematical functions. They should also have an understanding of mechanics and engineering. To be a worker in this field, long-term on-the-job training is required. The yearly salary may range from $34,000 to $75,000.

Why Use Fluid Power?

Fluid power systems involve the transfer of mechanical energy. The primary advantage of fluid power is the ability to multiply force in order to generate strength. Also, the components used in fluid power systems experience less wear than purely mechanical systems. Other advantages of fluid power systems include the following:

- An almost unlimited amount of power can be produced and maintained in a fluid power system.
- Easy and complete control can be maintained over a wide range of power in a fluid system. This allows for smooth, quick control of energy transfer.
- Gases used under pressure have a natural springiness, which produces a cushioning effect. This reduces shock in the system.
- The components in fluid power systems can be located far apart from each other, but still quickly transfer power.

Fluid power systems offer a wide range of mechanical functions. They are used to produce linear and rotary motion, while remaining mechanically efficient. Many industries that use modern technology employ fluid power systems, such as manufacturing plants, construction sites, farms, and manufacturing parts of vehicles that transport people and cargo worldwide. See **Figure 10-3.**

Figure 10-2. Hydraulic cylinders are able to lift heavy loads, such as the bed of this large dump truck.

Figure 10-3. Fluid power is widely used on construction projects. A—Several hydraulic cylinders are used to operate this end loader. B—One of the uses of pneumatics is to power tools, such as this jackhammer.

A

B

Figure 10-4. The type of water tanks used by communities to store water and provide good flow into water taps must be built to withstand tremendous pressure due to the weight of the water stored in them.

The Physics of Fluid Power Systems

The liquids and gases used in fluid systems have some similar characteristics. They exert pressure, take the shape of their container, and can flow freely from one container to another. These characteristics have been studied and used for centuries. Learning these characteristics is important to understand fluid power systems.

Fluids Exert Pressure

Long ago, scientists found that liquids exert pressure in all directions, not only on the bottoms of their containers. Because pressure is caused by the amount (weight) of liquid in a container, more pressure is exerted on its lower sides. The more liquid a container holds, the more weight pushes down and out. For this reason, large tanks that hold water and oil are made so the bottom is stronger than the top. See **Figure 10-4.**

How Fluids Act

Water and other fluids will flow from a higher level to a lower level, until both levels are the same. Fluids will also flow from an area of high pressure to an area of low pressure. The fluid will continue flowing until the pressure in both areas are equal. In **Figure 10-5,** notice that the containers are connected by a pipe at the bottom. The water begins at different levels, but water flows between the containers until the water in each container reaches the same level. If one container constantly held a higher level of water, it would exert more pressure than the container holding a lower level of water. The system is balanced when each container is filled with water to the same height, regardless of the shape or size of the container.

Measuring Effort and Rate in Fluid Systems

As with all forms of power, specialized instruments allow power to be measured. See **Figure 10-6.** A ***pressure gauge*** measures the difference between the pressure within a fluid circuit and the pressure in the surrounding atmosphere. This is typically measured in pounds per square inch (psi). The rate of flow in fluid power systems is typically measured with a ***flowmeter***. Standard units of measurement for rate of flow include gallons per minute (GPM) for liquids and cubic feet per minute (CFM) for gases. Other units of measure that indicate some type of flow per amount of time may be used as well.

When measuring pressure, there are two classifications of pressure gauge measurements that can be used: pounds per square inch gauge (psig) pressure and pounds per square inch absolute (psia) pressure. Most measurements are taken with a psig instrument. ***Pounds per square inch gauge (psig)*** pressure instruments do not

Figure 10-5. Imagine two containers connected by a tube. A—If you were to start filling one of the containers, it would also start filling the other one. B—Once the flow of water stops, the water flow continues into the other container until the water levels are equal.

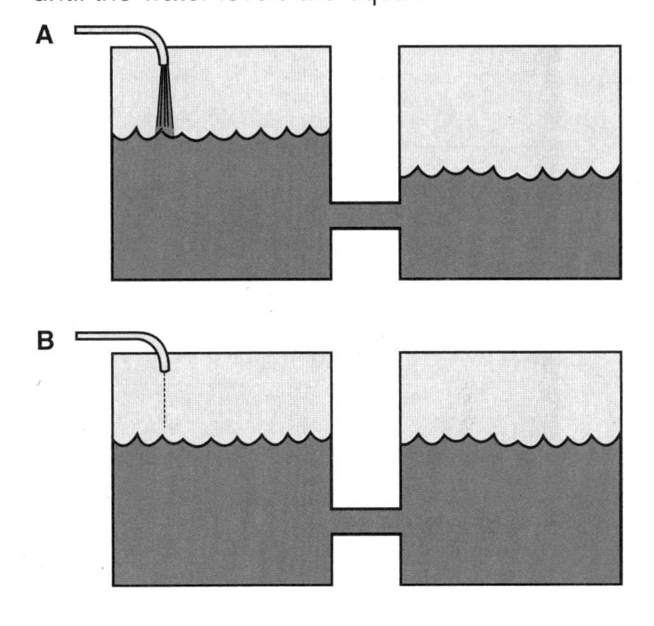

Pressure gauge: A device that measures the difference between the pressure within a fluid circuit and the pressure in the surrounding atmosphere.

Flowmeter: A device that measures the rate of flow in fluid power systems.

Pounds per square inch gauge (psig): A pressure instrument that determines the approximate pressure developed in a fluid circuit but does not take atmospheric pressure into account.

Figure 10-6. Fluid power systems make use of flowmeters and pressure gauges. A—This flowmeter measures the rate of flow of carbon dioxide (CO_2) gas in cubic feet per minute (CFM). (Miller Electric Manufacturing Company) B—A pressure gauge used with welding gas cylinders. One gauge measures the higher pressure of oxygen gas, and the other measures acetylene, which must be delivered at a lower pressure. Both are measured in pounds per square inch (psi). (Uniweld)

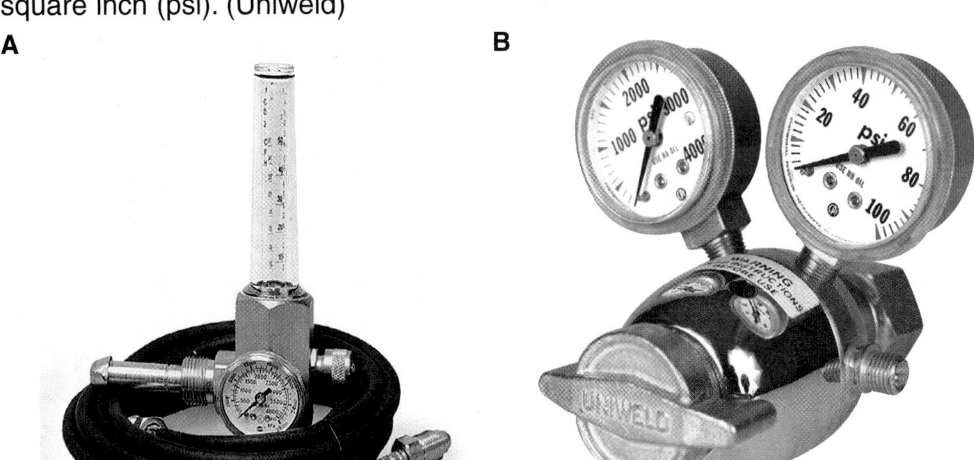

take atmospheric pressure into account, but they are adequate for determining the approximate pressure developed in a fluid circuit. A ***pounds per square inch absolute (psia)*** pressure gauge accounts for atmospheric pressure in its measurement. While a psig reading indicates 200 psi, for example, a psia gauge reading of the same pressure would indicate 214.7 psi. The psia gauge reading accounts for 14.7 psi at sea level. The psia gauge reading would decrease slightly if taken above sea level and increase slightly if taken below sea level.

Pounds per square inch absolute (psia): A type of pressure gauge that accounts for atmospheric pressure in its measurement.

Viscosity of Liquids

Viscosity: A measurement of internal friction, or the resistance of a fluid to flow.

Viscosity is a measurement of internal friction, or the resistance of a fluid to flow. For example, pancake syrup does not flow as freely out of a bottle as water does. This is because water has a lower viscosity rating than pancake syrup. Likewise, different hydraulic oils have different viscosity ratings. The viscosity rating is based on the length of time it takes a given quantity of fluid, such as a quart, to flow through a fixed opening at a specified temperature. This is exactly how engine oil gets its winter-summer rating, such as 10W-40. The viscosity requirement for different applications depends on temperature, pressure, and the specifications of the hydraulic components within the fluid circuit.

Opposition to Flow in Fluid Systems

There are two types of opposition to flow that occur in fluid power systems:

- Friction occurs when fluid is forced under pressure through pipes, hoses, and fittings.
- Opposition to flow can occur within the liquid itself and is known as ***turbulence***.

Turbulence: The opposition to flow that occurs within a liquid itself.

You may consider fluids, particularly hydraulic oils, to be relatively friction free. Plenty of effort or pressure can be lost due to friction, however, particularly if hydraulic lines or fittings are undersized. When fluids flow smoothly, they flow in stratified layers that may not be visible to the eye. This type of smooth flow is referred to as ***laminar flow***. Turbulence typically occurs when the fluid reaches a juncture point, such as a tee in the fluid circuit. The laminar lines of flow become interwoven. Internal losses are much greater in turbulent flow than in laminar flow. Other factors that affect opposition to flow in fluid circuits include the following:

Laminar flow: The smooth flow of liquids in stratified layers that may not be visible to the eye.

- Length of the fluid circuit.
- Friction factor of the conductors used to carry fluid.
- Interior diameter of the conductors.
- Head loss (losses due to pumping fluid upward).

Components of Fluid Power Systems

Like other power systems, fluid power systems require various components to operate together to perform work. When planning and designing these circuits, the engineers or drafters make schematic drawings that describe fluid circuits on paper. See **Figure 10-7.** Some of the components are used to produce the pressure, some components control

Figure 10-7. Drafters and engineers use symbols to represent different parts of pneumatic or hydraulic systems when drawing or designing systems.

Typical Fluid Power Symbols			
Connector	•	Flowmeter	
Line, flexible		Pressure gauge	
Motor, oscillating		Temperature gauge	
Line, joining		Pump, single fixed displacement	
Line, passing		Motor, rotary fixed displacement	
Flow direction hydraulic pneumatic		Motor, rotary variable displacement	
Manual shut-off valve		Electric motor	M
Cylinder, single-acting		Lever	
Cylinder, double-acting		Spring	

the pressure, and others make the pressure do the necessary work. While most devices are designed specifically for use in either hydraulic or pneumatic power systems, some are used in both types of systems.

Hydraulic Pumps

Hydraulic pumps supply and transmit the pressure needed to operate a hydraulic power system. These pumps convert mechanical energy into fluid power, creating the necessary flow in the system. Pumps have many applications in modern industry and technology:

- Moving water and coolant around boilers and nuclear reactors.
- Powering hydraulic cylinders for industrial and earthmoving equipment.
- Moving water out of ship bilges.
- Supplying the power to assist automobiles in braking and steering.
- Moving bulk liquids from the holds of ships.
- Powering pneumatic clamping and drilling equipment.
- Moving clean water in the plumbing systems of skyscrapers.

Hydraulic pump: A pump that supplies and transmits the pressure needed to operate a hydraulic power system. It converts mechanical energy into fluid power, creating the necessary flow in a system.

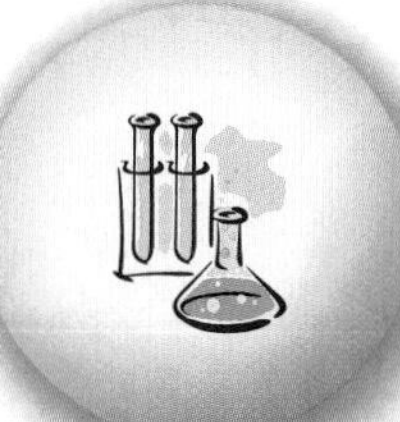

Curricular Connection

Science: The Human Heart

The human heart is the ultimate fluid pump. A healthy heart is only slightly larger than your fist, yet it will beat about 100,000 times per day, resulting in the flow of approximately 2000 gallons of blood. Over the course of a 70-year lifespan, it is estimated that the heart will beat more than 2.5 billion times.

The heart consists of four chambers and four valves. See **Figure 10-A.** The upper two chambers are known as the left and right atria. The lower two are the left and right ventricles. The four valves are as follows:

- The tricuspid valve is located between the right atrium and the right ventricle.
- The pulmonary valve is located between the right ventricle and the pulmonary artery.
- The mitral valve is located between the left atrium and the left ventricle.
- The aortic valve is located between the left ventricle and the aorta, which is the main artery of the body carrying refreshed oxygen-rich blood to all major organs of the body.

Each valve has a set of "flaps" called *cusps*. The cusps are one-way flow control devices, similar to check valves in fluid power systems. Dark blood that is low in oxygen (and appears blue in your veins), flows back to the heart after circulating through the body. The veins return the blood from your body to the right atrium of the heart. This chamber empties blood through the tricuspid valve into the right ventricle when the heart beats. The right ventricle then pumps the blood under low pressure through the pulmonary valve into the pulmonary artery. The pulmonary artery is a main artery that takes the blood to the lungs, where the blood gets fresh oxygen.

After the blood is refreshed with oxygen, it is bright red. It returns by the pulmonary veins to the left atrium. From there, it passes through the mitral valve and enters the left ventricle of the heart. This is the last stop for the oxygen-rich blood in the heart. The left ventricle pumps the refreshed blood out through the aortic valve into the aorta, which routes the blood to the rest of the body. Since the left ventricle is the last stop for blood within the heart, the blood pressure in the left ventricle is the same as the pressure measured in your arm.

The heart itself could be thought of as the first two-stage pump ever created. The right ventricle pumps the blood under lower pressure through the pulmonary arteries and veins to the left ventricle, where it is pumped at a higher pressure for circulation throughout the body. When the heart muscle contracts, or beats, it pumps blood out of the heart. This is

All pumps can be classified as either positive-displacement or nonpositive-displacement pumps. Positive-displacement pumps deliver an identical amount of fluid for every stroke of a linear pump or every rotation of a rotary pump. Nonpositive-displacement pumps deliver varying amounts of fluid to the load. If a valve is closed, preventing fluid from flowing to the load, a nonpositive-displacement pump can simply circulate fluid within the pump chamber. The fluid does not have to go anywhere, but this could cause it to heat up over time.

referred to as the *systole*, or *systolic action*. The heart contracts in two stages. First, the right and left atria contract at the same time, pumping blood to the right and left ventricles. The two ventricles then contract together to propel blood out of the heart. Lastly, the heart muscle relaxes before the next heartbeat. This allows blood to fill up the heart again. This relaxation period is referred to as the *diastole*. You may have heard of the systolic and diastolic readings when your blood pressure is taken. These readings are measurements of your heart at work and heart at rest. The heart is a very efficient fluid pump in a small package.

Figure 10-A. Each chamber of the heart has a one-way valve at its exit, which prevents blood from flowing backward. The valve at each exit opens when the chamber contracts. It closes when the chamber is finished contracting, so blood does not flow backward.

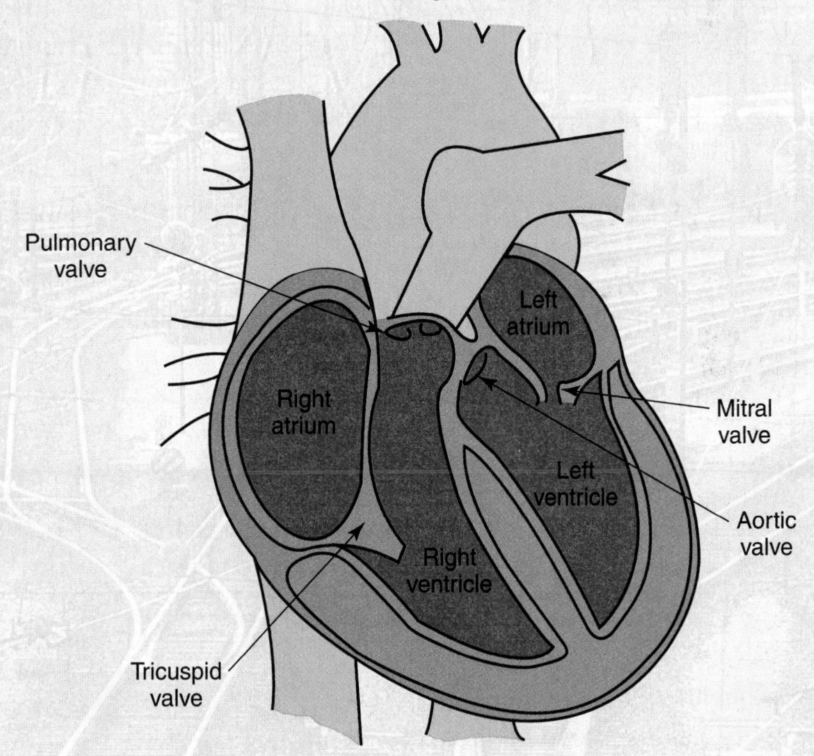

Gear pumps

Gear pumps create the pressure needed to operate many hydraulic systems. In this type of pump, two gears are positioned so they mesh with each other inside a housing. See **Figure 10-8.** One gear is turned by a power source and turns the other gear in the opposite direction. Oil enters the pump through the low-pressure port on one side of the housing. It is drawn through the housing by the two spinning gears. This action forces the oil around the gears and to the other side of the housing, where it exits through the high-pressure port. The oil under pressure is then used to do work.

Gear pump: A pump in which two gears are positioned so they mesh with each other inside a housing.

Figure 10-8. The inside of a gear pump. Two gears, one driven by a power source, pull the fluid into the pump housing and place it under pressure at the outlet port.

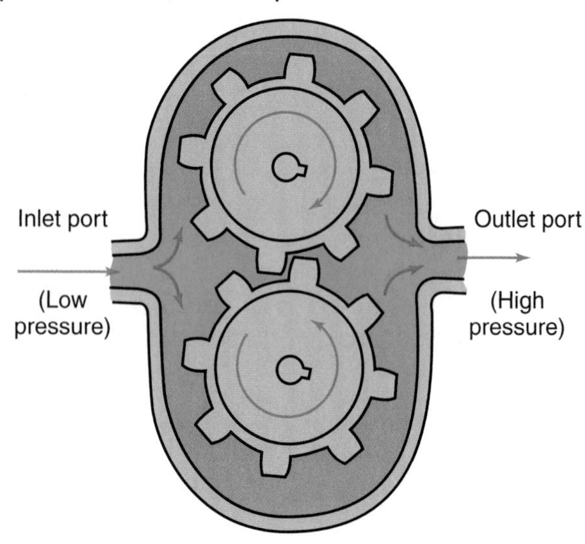

Centrifugal pumps

Centrifugal pumps are also used to produce pressure in a hydraulic system. This type of pump uses centrifugal force to move fluids in the system. ***Centrifugal force*** is the energy that makes objects fly outward when spinning around. It is this force that keeps water in a bucket when you swing it over your head very fast. Centrifugal pumps use a device commonly found in many propulsion systems and other pumps called an *impeller*. An ***impeller*** is a device that has many small blades mounted on a shaft. As the impeller spins inside its housing, it draws liquid through the inlet port from a ***reservoir***, or container. Because the impeller is driven from an outside power source, it can be made to spin very fast. The movement of the impeller forces the liquid outward against the housing, through the outlet port, and to the rest of the system. See **Figure 10-9.**

Centrifugal pump: A pump that uses centrifugal force to move fluids in a system.

Centrifugal force: The energy that makes objects fly outward when spinning around.

Reciprocating pumps

Reciprocating pumps use a piston that moves back and forth in a cylinder to move hydraulic fluid. Each stroke of the piston moves a certain volume of liquid through the system. See **Figure 10-10.** Each time the piston moves in the cylinder, it forces oil out of the cylinder. ***Check valves*** are needed when using reciprocating pumps to keep fluid from moving backward in the system. A hand water pump is a simple type of reciprocating pump you may have used to move water. See **Figure 10-11.**

Air Compressors

Pneumatic systems use ***air compressors*** in the same way hydraulic systems use hydraulic pumps. Air compressors convert the mechanical energy put into pneumatic power systems. This conversion creates the necessary flow to make the system work. Compressors are often teamed up with reservoirs or pressure tanks to store and transmit compressed air to the power system whenever it is needed.

The most common air compressor is also the reciprocating type. See **Figure 10-12.** A piston moves up and down (reciprocates) inside a cylinder. The piston is powered by an external power source, such as an electric motor. There are two ***valves*** at the top of the cylinder that let air pass in only one direction. As the piston moves down, it creates suction in the cylinder that opens the intake valve and lets air into the cylinder. As the piston moves

Figure 10-9. The inside of a centrifugal pump. As the impeller spins, fluid is forced off the vanes and becomes pressurized as it moves out of the pump housing.

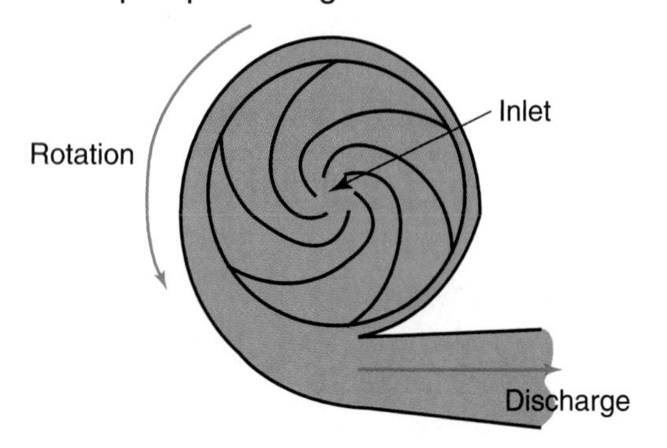

Figure 10-10. A reciprocating pump has a piston driven by a shaft, located off center, on a revolving wheel or crankshaft. Note the action of the intake and delivery valves as the direction of the piston changes.

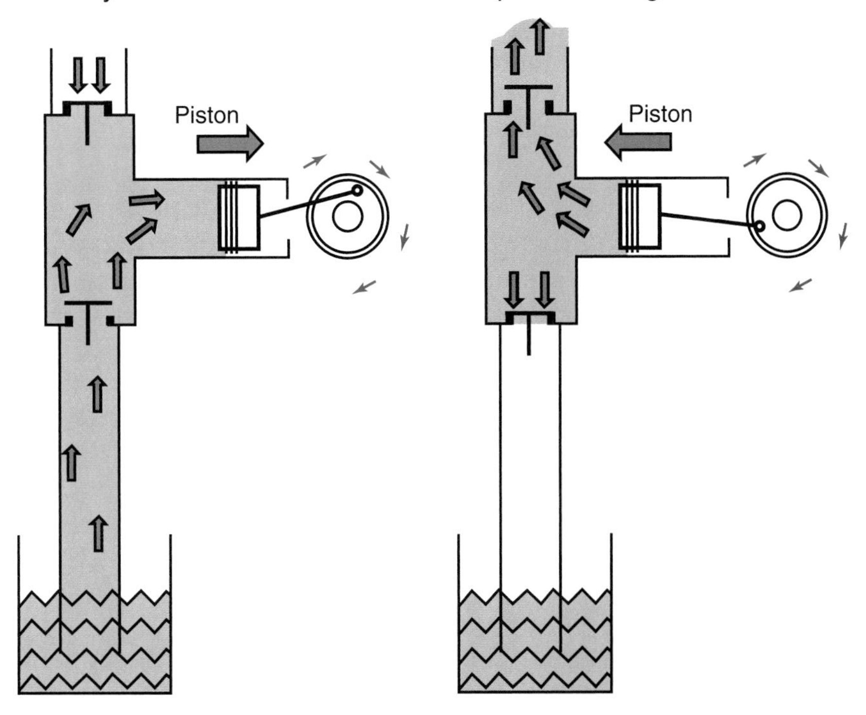

Figure 10-11. A simple hand-operated water pump. You may have used one like it.

Figure 10-12. Air compressors that use a reciprocating-type pump are common in automotive shops.

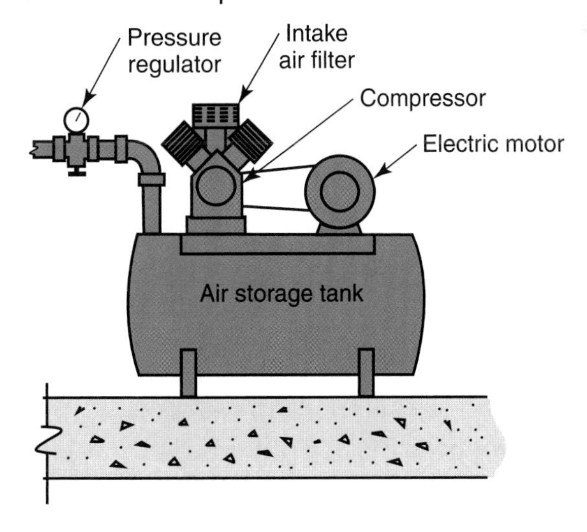

Impeller: A device that has many small blades mounted on a shaft. It spins inside its housing, drawing liquid through the inlet port from a reservoir.

Reservoir: A container to hold liquids to be used again in a system.

Reciprocating pump: A type of pump with a piston that moves back and forth in a cylinder to move hydraulic fluid.

Check valve: A valve needed when using reciprocating pumps to keep fluid from moving backward in the system.

Air compressor: A device that converts the mechanical energy put into pneumatic power systems creating the necessary flow to make the system work.

Valve: A part of an air compressor at the top of the cylinder that lets air pass in only one direction.

up, the intake valve shuts, and air is compressed in the cylinder until the outlet valve opens. The air then enters the receiver, where it is kept at a constant pressure.

Some compressors are multistage types. This means air passes through more than one piston or cylinder arrangement before it enters the receiver. As air passes between the stages, its pressure is gradually increased. In multistage compressors, air reaches the desired pressure in the last compression stage.

It is important that moisture does not enter a pneumatic power system. This may cause the system's component to rust and corrode. Filters and separators are placed in the system to remove any moisture from the air as it is compressed. These devices are located in the compressor, between the outlet valve and the receiver.

Controlling Fluid Power

In a fluid power system, pressurized fluid is transported by way of transmission lines. Piping or tubing capable of withstanding high pressure is used for this purpose. Transmission lines carry the high-pressure fluids to where they will be used to do work. Components are placed in the system to control pressure, flow, and direction.

All fluid power systems must use control devices to be functional. Valves are devices that control fluid power. They are used for the following:

- Flow control.
- Pressure control.
- Directional control.

Liquids and gases used in fluid systems have many characteristics in common. Therefore, the valves for both hydraulic and pneumatic systems are similar in design. The important difference is that valves for hydraulic systems require different seals to prevent loss of liquid.

Flow-Control Valves

Flow-control valve: A valve used to start or stop the flow of fluid in a system.

Variable-flow restrictor: See *Flow-control valve.*

To start or stop the flow of fluid in a system, a ***flow-control valve*** is used. This type of valve is also known as a ***variable-flow restrictor***. When you turn on the water at a kitchen sink, for example, you are using a flow-control valve. Many times, there are two flow-control valves at a sink—one for hot water and one for cold. Some flow-control valves are made to control both hot and cold water simultaneously. Flow-control valves also control how much fluid passes through a system. If the valve is only partially opened or closed, it does not allow full flow. Flow-control valves can be helpful in controlling the output speed of a fluid motor or in controlling how fast a cylinder is allowed to extend or retract.

Pressure-Control Valves

Pressure-reducing valve: A valve that reduces the pressure within a fluid circuit to levels suitable for use.

Pressure-reducing valves reduce the pressure within a fluid circuit to levels suitable for use. For instance, the pressure coming off a particular pump is 300 psi. Several components within the circuit require high pressure, but a control circuit is comprised of components that operate between

0–30 psi. This is an ideal application for a pressure-reducing valve. The pressure-reducing valve could provide lower pressure for the control circuit, while maintaining the high pressure needed for the rest of the circuit.

A ***pressure-regulating valve*** controls pressure coming from the compressor. This device can vary pressure and is often used in conjunction with lubrication and filtration devices. Such a device is typically referred to as a ***filter, regulator, lubricator (FRL) device***. In addition to regulating pressure, the FRL device also filters harmful moisture from the air and adds lubrication to the air so the wear on pneumatic equipment is minimized. Pressure-regulating valves are also used in hydraulics and are particularly critical, since liquids cannot be compressed within the circuit. Excess pressure could cause component failure. See **Figure 10-13.**

Pressure-regulating valve: A valve that controls pressure coming from the compressor.

Filter, regulator, lubricator (FRL) device: A device that controls pressure coming from the compressor, filters harmful moisture from the air, and adds lubrication to the air so the wear on pneumatic equipment is minimized.

Pressure-Relief Valves

Pressure-relief valves are placed in a system to make sure the pressure does not get too high. If pressure increases to dangerous levels, relief valves automatically open to reduce it. In fluid power circuits, pressure is the primary characteristic of the power that requires protection. Too much effort or pressure built-up may cause damage or entire system failure. In hydraulic systems, the relief valve directs extra liquid to the holding tank, or reservoir. In pneumatic systems, extra air can be released into the atmosphere. Some steam-activated pressure-relief valves, such as those on car radiators and hot water tanks, work this way as well. See **Figure 10-14.**

Pressure-relief valve: A valve placed in a fluid power system to make sure the pressure does not get too high.

Directional-Control Valves

Fluid power systems are often designed with more than one path for fluid to travel. A ***directional-control valve*** is used to control which path fluid takes in a circuit. These valves can be operated manually, mechanically, or electrically. Directional-control valves are often referred to as ***spool valves*** because of the way they are constructed. On early models of directional-control valves, the interiors resembled spools that hold thread. Grooves are cut into the interior to guide the fluid to the proper outlet

Directional-control valve: A valve used to control which path fluid takes in a circuit in a fluid power system.

Spool valve: Another name for a directional-control valve. On early models of directional-control valves, the interiors resembled spools that hold thread.

Figure 10-13. A filter, regulator, lubricator (FRL) device is often installed in pneumatic systems. (Monnier, Inc.)

Figure 10-14. A pressure-relief valve protects the fluid circuit from excessive pressure buildup. This valve is part of a small air compressor.

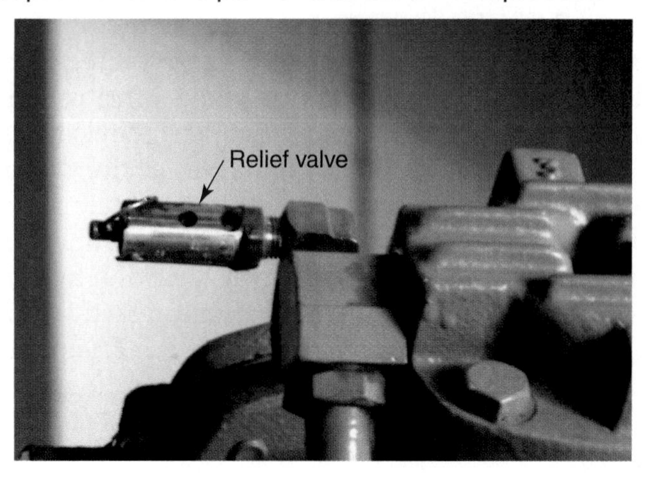

Four-way valve: A common type of spool valve that allows both the pressure and return lines to reverse themselves when the valve is triggered.

Cam-activated valve: An automatically operated valve. When the cam is triggered, the valve shifts its position, allowing another action to occur.

ports. The most common spool valves are a group of valves known as ***four-way valves***. These valves allow both the pressure and return lines to reverse themselves when the valve is triggered. This provides directional-control for fluid cylinders and motors. Many four-way valves also include a neutral position. Neutral allows the pump to run without moving the cylinders or motors that the pump powers. When a neutral position is included in the construction of a valve, neutral is almost always the default position for the valve. See **Figure 10-15.**

Note that, when reading the schematic, the number of boxes that comprise the valve equate to the number of positions the valve can offer. The number of ports indicate the number of hoses that attach to the valve. This is how the proper name of the valves indicated are derived.

Figure 10-15. A four-way spool valve is a common type used for directional control. The labels show where the hoses go: Hose P goes to the pressure, or feed; hose T goes to the tank, or reservoir; and hoses A and B go to the load. The fact that only two boxes are present makes this a two-position valve. It is actually called a *two-position, four-port valve.*

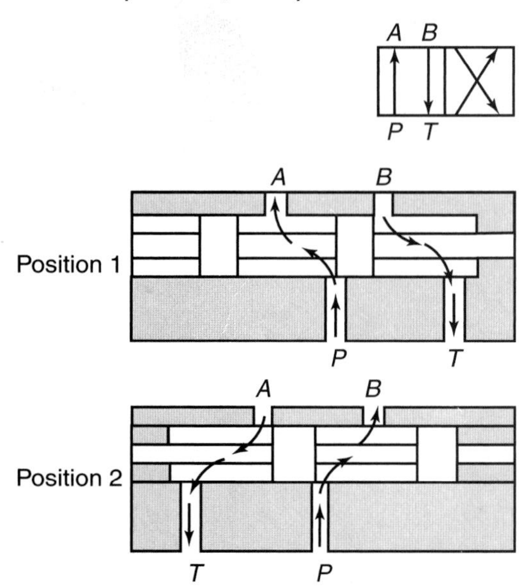

Other Flow-Control Valves

Manually operated valves are the most common valves used. These valves are operated by manually moving a shift lever. Typically, the valve returns to its default state by spring action when the manual lever is released. This type of valve is ideal for use when the control valve is in close proximity to the operation being performed, such as with a hydraulic press or log splitter.

Automatically operated valves can be triggered in a number of ways. A simple, automatically operated valve is the ***cam-activated valve***. Something pushing on the cam causes the valve to shift. Sometimes, a part slides into place and pushes down on the cam. When the cam is triggered, the valve shifts its position. This allows another action to occur. This action may change the direction or pressure of the flow, or it may stop the flow all together. ***Solenoid-operated valves*** are shifted by an electrical signal. The electrical signal may be sent from a switch located remotely on an assembly line or from a programmable control device, such as a programmable logic controller (PLC) or a computer. One great advantage of solenoid-operated valves is that the trigger mechanism may be located a considerable distance from the valve and hydraulic power unit. For instance, a hydraulic bailing unit is attached to a farm tractor. The unit is powered by a small gas engine located on the bailing rig. The rig must be triggered, however, by the driver of the tractor located a considerable distance from the bailing rig. A solenoid-controlled valve allows virtually all the hydraulics to be located in close proximity to the power unit. See **Figure 10-16.** Two light gauge wires can be strung to the cab of the tractor and attached to a push button. When the tractor driver pushes the button, the valve will shift, and the bailing rig will be activated.

Solenoid-operated valve: A valve shifted by an electrical signal.

Check valves are another type of directional-control valve that allow fluid to flow in just one direction. If fluid begins to move backward in a system, check valves close and stop the backward movement of fluid. See **Figure 10-17.**

Transmitting Fluid Power

Fluid power may be transmitted through a variety of different devices known as *conductors*. See **Figure 10-18.** Hoses are used to transmit fluid because they allow for both high pressure and tremendous movement and

Figure 10-16. Solenoid-controlled valves. A—A three-position, four-way valve. Solenoids a and b move the spool either in or out from its neutral position to open or close different combinations of the four ports (A, B, P, and T). B—A double-solenoid, proportional valve. (Bosch Automation Technology)

A

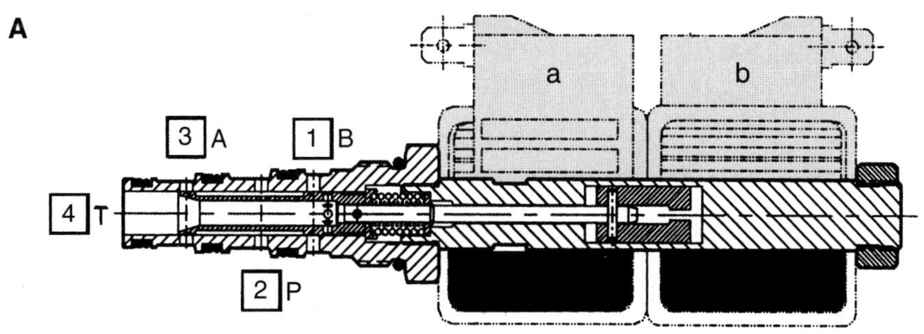

B

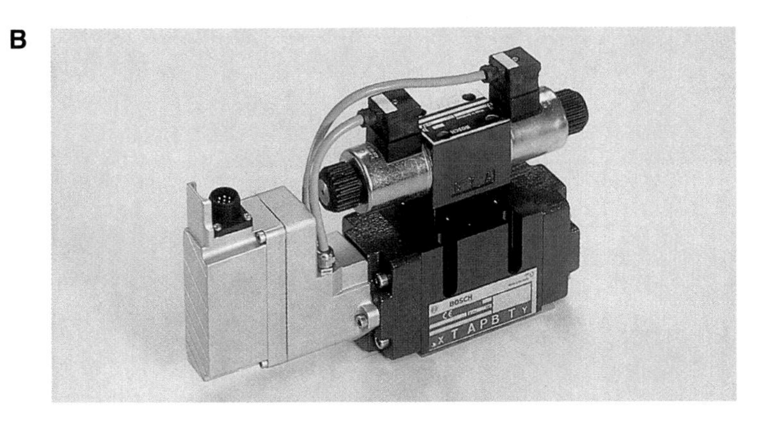

Figure 10-17. The basic construction and action of a check valve in a fluid power system.

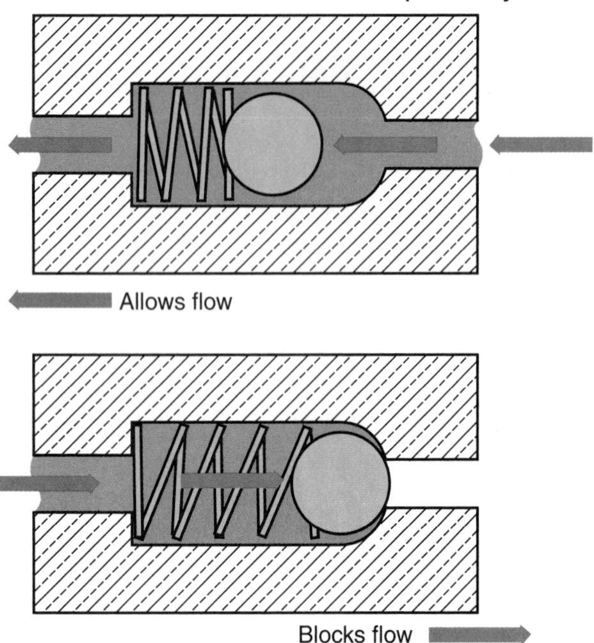

flexibility. Movement and flexibility allow hoses to absorb shock very effectively. Rigid pipes are also used to transmit fluid power. Pipes can handle high pressures and are typically used in permanent installations where flexibility is not required. When using pipe to transmit fluid power, it may be necessary to use an accumulator as a shock absorber in the system. The rigidity of pipe cannot absorb shock as well as hose or tubing. Tubing is semirigid and provides some of the strength of pipe, as well as allowing for some flexibility. When coiled, tubing can absorb and dissipate shock much better than rigid pipe. Duct work is used to transmit high volumes of gases under relatively low pressures. Ducting is commonly used for climate control applications, such as structural heating, but it is not appropriate for transmitting liquids under pressure.

Sizing Fluid Power Conductors

There are a number of important factors involved in sizing fluid power conductors. The most important factors include the following:

- The volume the conductor must carry.
- The velocity at which the fluid must travel.
- The pressure the conductor is designed to withstand.

Flow Considerations

The flow rate capacity for a particular conductor is primarily determined by the inside diameter of the conductor. The rate increases significantly as the inside diameter is increased even slightly. Consider the flow

Figure 10-18. Examples of fluid power conductors. A—Flexible hoses allow movement of parts in machines operated by fluid power. B—Rigid piping is used in fixed installations, such as this compressed air system in an auto repair shop. C—Ducting is used to transmit large volumes of air or other fluids at low pressure, as in this forced-air heating installation.

A

B

C

capacity a garden hose with a 1″ interior diameter has, compared to a garden hose with only a 1/2″ interior diameter. You may assume that the increased capacity would be twice as much, but that is not correct. The increase in flow capacity is actually four times as much. As the diameter is doubled, the cross-sectional area increases fourfold. See **Figure 10-19.**

Velocity Considerations

Velocity is a measurement of the rate of motion in a particular direction. The velocity of fluid flow is typically measured in feet per second (fps). A measurement of velocity varies directly with the rate of flow. If the rate of flow increases, the velocity also increases. The surface area of a conductor has the opposite effect, however, on velocity. A conductor with a larger diameter allows greater flow capacity, but the velocity decreases because the conductor can now hold more fluid in the larger area available.

Nomograph: A chart helpful in determining the inside diameter of conductors or for estimating flow or velocity, if two of the three variables are known.

A ***nomograph*** is a chart helpful in determining the inside diameter of conductors or for estimating flow or velocity, if two of the three variables are known. This type of chart could be used to determine the required interior diameter of a conductor, if an application such as a pressure washer calls for fluid flow at a rate of 15 GPM and a velocity of 10 fps in order to be effective. Since the desired flow and velocity are known, the diameter of the hose to use can be determined using the nomograph. Similarly, let us say you have a 2″ diameter pipe flowing water at a constant velocity of 5 fps and you would like to calculate the volume of flow. Since the interior diameter of the pipe and the velocity are known, the flow rate can be determined using a nomograph.

Pressure Considerations

Working pressure: The normal operating pressure for which a conductor is designed.

Test pressure: The maximum pressure that a conductor is designed to withstand.

Burst pressure: The pressure at which a conductor will fail by rupturing.

Knowing the amount of pressure a conductor can withstand is critical from an operating safety perspective. Pressure ratings vary, based on the composition of the construction material and the wall thickness of each conductor. Conductors often have three pressure ratings:

- The ***working pressure*** indicates the normal operating pressure for which the conductor is designed.
- The ***test pressure*** indicates the maximum pressure that the conductor is designed to withstand.
- The ***burst pressure*** indicates the pressure at which the conductor will fail by rupturing.

When selecting conductors, it is important to remember that intermittent pressures may significantly exceed normal working pressure, and hoses should be designed to withstand not only working pressure, but also excess pressure. The conductors should also have a large enough diameter to provide adequate flow. Properly sized conductors will prevent pressure drops greater than 10%, which result from excessive friction and turbulence generated inside undersized conductors.

Figure 10-19. As the diameter of a conductor is doubled, the cross-sectional area available for flow increases by four.

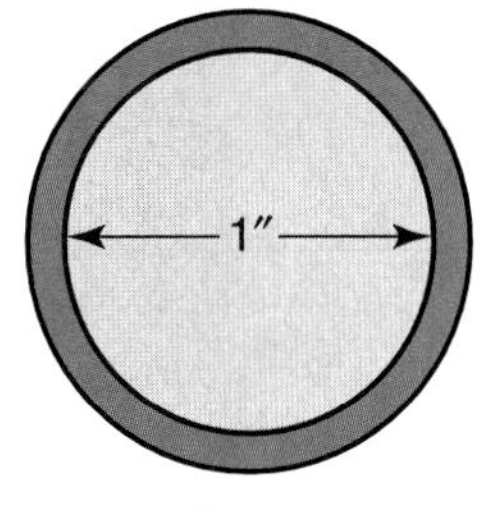

Figure 10-20. Actuators convert fluid power to mechanical power. A—A butterfly valve with a hydraulic piston actuator. The piston moves to close or open the valve. B—This valve has a pneumatically operated diaphragm actuator. (Fisher Controls International, Inc.)

A

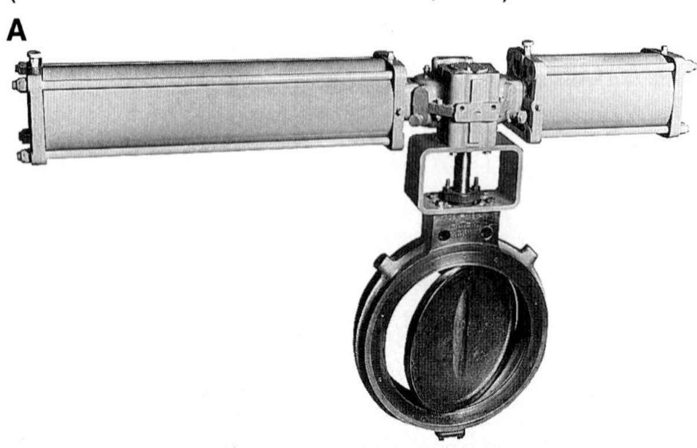

B

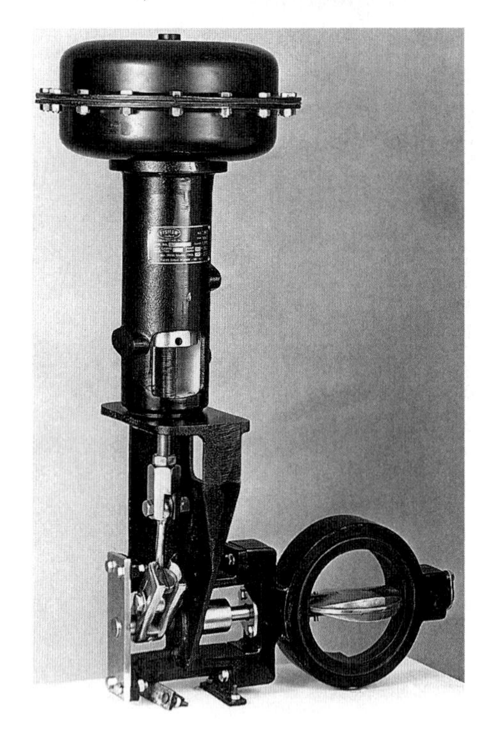

Making Fluid Power Work

Remember that the purpose of any power system is to perform work. In the previous sections of this chapter, the parts of fluid power systems that control fluid flow and pressure were discussed. This section presents the system components that convert fluid energy to mechanical energy of solid parts.

Actuators

Actuators are devices that convert fluid power to mechanical power in both hydraulic and pneumatic systems. These devices help to make fluid power systems easy to design and use. Fluid power is capable of creating almost any type of mechanical motion. Actuators produce the reciprocating and rotary motion that allows fluid power systems to do work. See **Figure 10-20.**

Cylinders are one type of actuator. They are classified as either single-acting cylinders or double-acting cylinders. Both types contain pistons that reciprocate inside them. ***Single-acting cylinders*** use the force of fluid to move the piston in one direction. The weight of the load then forces the piston to its original position when pressure is not applied to the piston within the cylinder. Other types of single-acting cylinders use an internal spring to return the cylinder to its original position when pressure is not being applied to the cylinder. ***Double-acting cylinders*** use the force of the fluid to move the piston in both directions. Fluid pressure can be exerted on either side of the piston. These cylinders are used where there is a need for complete control.

Hydraulic and pneumatic components are very similar in design. The characteristics of available fluids, liquid or gas, determine which type of system is used. Remember that liquid is not compressible, and gas is. Because of this, hydraulic cylinders are used in systems where quick and precise control of power is needed. The liquids used in these systems act as a solid link between the fluid and the solid parts. Pneumatic cylinders are used in systems in which a certain amount of cushioning is needed, such as with clamping operations where you would not want to crush the material being held.

Actuator: A device that converts fluid power to mechanical power in both hydraulic and pneumatic systems.

Single-acting cylinder: A cylinder that uses the force of a fluid to move the piston in one direction.

Fluid Motors

Fluid motors are devices that convert fluid power into rotary mechanical motion. Two basic types of fluid motors are the gear motor and the vane motor. Gear motors operate like a gear pump, but the process is reversed. Fluid is forced into a housing that contains gears. As the fluid flows around the outside of the gears, against the housing, it forces the gears to spin. One of the gears is connected to an output shaft. The speed of the output shaft depends on the pressure of the fluid in the system.

In a vane motor, a rotor is offset inside a round housing. See **Figure 10-21.** The rotor has spring-loaded vanes, so they always touch the inside of the housing as the rotor spins. Fluid is forced through the housing. The fluid pushes against the vanes as it flows past the rotor, causing the rotor to spin. The speed of the output shaft attached to the rotor depends on the pressure of the fluid in the system.

Double-acting cylinder: A cylinder that uses the force of a fluid to move the piston in both directions.

Fluid motor: A device that converts fluid power into rotary mechanical motion.

Pressure tank: A tank that stores air under pressure in pneumatic systems.

Accumulator: A device that stores hydraulic liquid under pressure, even when the hydraulic pump is not running.

Storing Fluid Power

Pneumatic systems store air under pressure in tanks known as ***pressure tanks***. These tanks typically include a pressure-sensitive switch that automatically turns the compressor motor on when the pressure in the tank drops below a preset level. The compressor produces more compressed air and pumps it into the pressure tank until the pressure-sensitive control valve reaches its high-pressure limit and shuts off the compressor motor. This way, a constant amount of pressurized air is maintained within the system. Pressure tanks usually come equipped with a pressure-relief valve, to ensure that excess pressure does not build up in the tank. A bleeder valve at the base of the tank allows any water that condenses inside the tank to drain.

Normally, hydraulic systems do not store liquid under pressure. The liquid is kept in a storage tank known as the *reservoir*. The reservoir is vented into the atmosphere so pressure does not build up within the reservoir and so a vacuum cannot be created. The reservoir is also equipped with baffles. The baffles deflect fluid entering under high pressure and prevent it from creating turbulence in the reservoir. Turbulence in a reservoir is created similarly to the way a garden hose flowing openly in a bucket creates lots of air bubbles. If the air bubbles get into the hydraulic lines, the entire system will become spongy, or compressible, and the results could be disastrous, given the high-strength applications that typically employ hydraulics. Air can be compressed, but hydraulic fluid cannot be compressed. Imagine air in the brake lines of your car. When you step on the brake, the air could compress, creating a delayed reaction.

An ***accumulator*** is a device that stores hydraulic liquid under pressure, even when the hydraulic pump is not running. Accumulators serve two purposes in hydraulics:

- To store pressurized fluid, providing it to the circuit on demand.
- To reduce pressure shocks in a hydraulic system.

This device works by allowing hydraulic oil to force a piston to pressurize a gas. Since liquids cannot be compressed, pressure is provided by the force of the expanding gas

Figure 10-21. A vane motor powered by air pressure from a pneumatic power source.

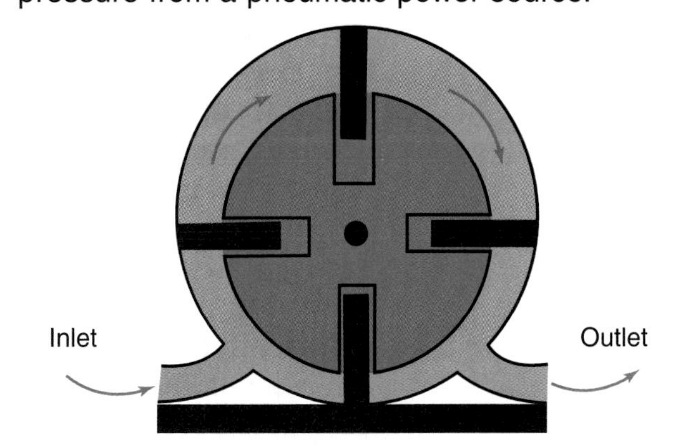

when the control valve is opened. Accumulators allow pressure to be stored so the pump does not run the entire time the hydraulic system is in use. Accumulators are also helpful in providing backup pressure to a system in which pump failure could result in serious injury. An accumulator can store varying amounts of fluid. This allows small actuators to be used as shock absorbers in hydraulic systems.

Working Safely with Fluid Power

There are a number of safety considerations to observe when working with liquids and gases. This is particularly true if the fluids are under pressure. The following safety cautions should be followed when working with fluid power circuits:

- Always adjust pressure-relief valves to provide a safe operating pressure. These adjustments should be made in accordance with the ratings on the components used in the fluid circuit. Many times, pumps and compressors can generate greater pressures than the components in the circuit can actually handle.
- Always wear safety glasses when working with fluid power to protect your eyes from unexpected hazards. Pressurized pneumatic hoses can burst from their fittings, causing hoses to whip violently in the air until the pressure is released or shut down by closing a valve. Hydraulic hoses can burst under high pressure, sending pressurized streams of oil into the work environment.
- Make necessary changes and adjustments to fluid circuits when they are not under pressure.
- Always respect the pressure that fluid power circuits can exert and the speed with which fluid cylinders can extend and retract.

Comparing Liquids and Gases

Liquids and gases act differently because the molecules in liquid are closer together than the molecules in gases. This can be observed in boiling water. The high temperature causes the water molecules to speed up and spread further apart, turning water (a liquid) into steam (a gas). See **Figure 10-22.**

Because pressure is involved in doing work with fluid systems, the compressibility of fluids is a factor. ***Compressibility*** is the extent to which any substance can be packed down into a smaller size, or volume. Liquids cannot be compressed much at all, even under enormous pressure. For all practical purposes, liquids are not compressible. Gases, however, may be compressed because of the greater amount of space between the molecules.

Compressibility: The extent to which any substance can be packed down into a smaller size or volume.

Temperature also affects the compressibility of gases. The hotter a gas becomes, the more active its molecules become. Therefore, the molecules have more space between them, which makes the gas more compressible. An English scientist named Robert Boyle studied this relationship between pressure, volume, and temperature. He concluded that if the temperature remains constant, increasing the pressure on a gas reduces its volume. In fact, if the pressure is doubled, the gas is compressed to half its original volume. This concept is an important consideration when designing pneumatic systems.

Figure 10-22. Heating water will cause it to boil and change its state from a liquid to a gas (steam).

Another important difference between hydraulic and pneumatic power systems is that pneumatic power can use an open system, while hydraulic power must be used in a closed system. Pneumatic systems, such as the pneumatic cylinder on a door, can be open because the air put into the system is drawn from the atmosphere and, therefore, can be returned to the atmosphere. See **Figure 10-23.** There is no need for a reservoir to contain the air. A hydraulic system must be closed so the liquids used can be returned to a reservoir to be used again in the system.

Pascal's law: When there is nothing but liquid in a container, compression applied to any part of the container is distributed equally in all directions. The initial pressure multiplies as it is distributed through a container with a larger diameter.

Blaise Pascal and Mechanical Advantage

In the seventeenth century, a French scientist named Blaise Pascal studied how liquids act while in closed containers. From his findings, Blaise Pascal developed ***Pascal's law***. He found that when there is nothing but liquid in a container, compression applied to any part of the

Figure 10-23. Door closers and air tools are examples of open pneumatic systems, since they use atmospheric air in their systems.

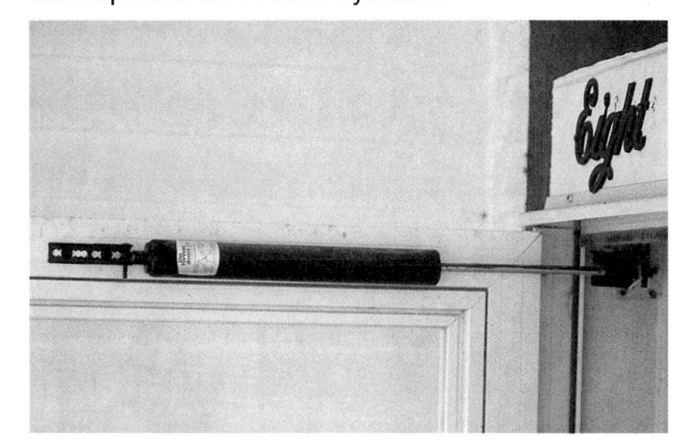

Figure 10-24. Containers representing different sizes of hydraulic cylinders. The force applied to the smaller cylinder is increased 100 times on the large cylinder.

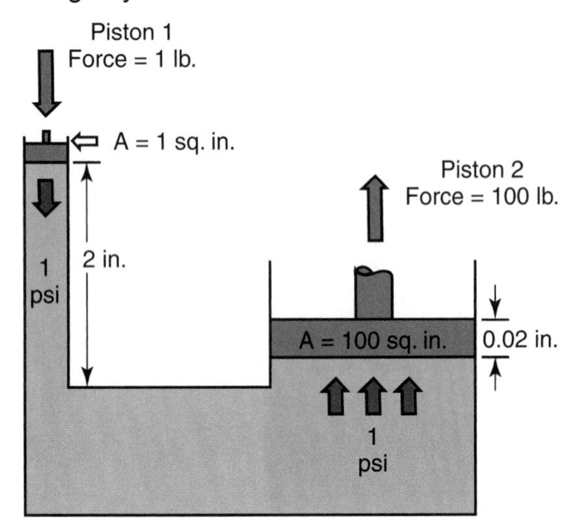

container is distributed equally in all directions. Pascal also noted that this initial pressure multiplies as it is distributed through a container with a larger diameter.

The properties of confined fluids make hydraulic devices capable of increasing mechanical advantage. In **Figure 10-24,** two closed containers are displayed that represent hydraulic cylinders. Assume that one is 100 times smaller than the other. If a force of 1 lb. is exerted on the smaller container, it will transfer the pressure to the larger cylinder. Because the surface area of the other cylinder is 100 times larger and pressure is exerted in all directions, there is 100 times more pressure in the larger cylinder. The force on each square inch of both cylinders is the same. Because the large cylinder spans more square inches, it exerts a larger total force. See **Figure 10-25.**

Using Pascal's Law to Calculate Mechanical Advantage

The formula to apply Pascal's law states that force is equal to pressure multiplied by area. To apply Pascal's law to a real problem, assume we need to know how many pounds of force need to be applied to lift the

Figure 10-25. A backhoe uses hydraulics to control the digging claw and bucket, the lift, and the extension of the bucket.

Technology Link

Medicine: Jaws of Life

Thousands of vehicle accidents occur on a daily basis. Fortunately, only a few of those accidents involve entrapment, which is when the occupants cannot escape the vehicle. These instances are when the jaws of life are dispatched to assist in the rescue. This tool uses mechanical advantage and fluids under pressure to save lives.

The jaws of life is a hydraulically powered, multipurpose tool that can cut or spread metal. It is frequently used to assist in cutting victims out of a mangled vehicle. This tool can also be used as a lift or ram to free people caught under collapsed debris. The jaws of life is one of the most practical applications of the tremendous mechanical advantage that can be created with the use of fluids under pressure.

Liquids under pressure are virtually incompressible. This gives them the strength of solids, yet allows for the flexibility that a liquid can provide. The liquid used in hydraulic circuitry is almost always oil, but the jaws of life use a special liquid known as phosphate-ester. This fluid is both fire resistant and nonconductive to electricity. Both of these characteristics make it ideal for use with rescue equipment, in situations where the operators could be working in close proximity to fire or live electricity.

The jaws of life is powered by a small gas engine coupled to a small, but powerful pump, which pressurizes the fluid. The control valve is a typical three-position, four-port valve attached to a double-acting cylinder. The valve can drive the cylinder in or out. The cylinder itself is often coupled to a cutter and spreader device that can spread with more than 15,000 lbs. of force or cut through metal as easily as scissors cut through paper. Rescue personnel operate the valve to drive the cylinder in or out, using it as a cutter or spreader, depending on what the situation warrants. The cutter on the jaws of life can easily snap doorposts on cars right in half to help extricate a person.

Sometimes the rescue situation calls for another type of end effector to do the work. If a dashboard has the legs of a victim trapped, a ram device might be employed to push the dashboard back up, closer to its normal position, so the victims can be extricated. The ram is another double-acting cylinder, but it is only coupled to a metal plate and not to a cutter or spreader. This plate can be used to lift heavy loads or to bend crushed metal. There are many practical applications for using liquids under pressure, but perhaps none are as vital as the jaws of life.

side of a 3000-lb. car. Since the jack must only lift half the car, we can estimate the load to be 1500 lbs. The input, or drive, cylinder is .25″ in diameter, and the output, or driven, cylinder is 1.25″ in diameter. Pascal's law can be applied to either the input or the output side of the jack, but two of the three pieces of information are required to work this type of formula. In this example, we will apply Pascal's law to the output (right) side of the jack, where more information is provided:

- Pascal's law says that force (lbs.) / area (in^2) = pressure (psi)
- area = πr^2
- 1500 lbs. / [(3.14 × .625″ × .625″) = 1.23 in^2] = 1220 psi

Pressure is the same throughout a hydraulic circuit because liquids under pressure exert pressure evenly in all directions when they are not flowing. Therefore, if we solve for pressure on the output (right) side of the circuit, we have also solved for pressure on the input (left) side of the circuit. Once pressure is determined, Pascal's law can be applied to the other side of the jack because two out of three pieces of information are now available to work the formula.

- Pascal's law says that area × pressure = force
- .049 in^2 × 1220 psi = 59.8 lbs. of force is required to lift the car

This represents an approximate 25:1 mechanical advantage, which can be determined by dividing the output force (1500 lbs.) by the input force (59.8 lbs.). If a jack handle is placed on the input side of the jack, it creates an additional 3:1 mechanical advantage. This is because the jack handle is actually a first class lever that multiplies force by 3:1. The input force required is only about 20 lbs., which results in a total mechanical advantage of 75:1.

Sizing Actuators

Pascal's law can also be helpful in sizing cylinders and determining the stroke length of a cylinder for specific applications. In **Figure 10-26,** a lightbulb will be picked up from a tester by a clamping device and then rotated away for packaging if it is good. Generating a big mechanical advantage for this application is important because it requires very little force to lift the bulb. The distance the driven piston moves, however, is obviously very critical. Overextending the piston could easily crush the lightbulb. If the driven cylinder is underextended, the gripper will not pick up the lightbulb. A cam will provide intermittent motion to power the drive cylinder in and out.

Figure 10-26. The driven cylinder has to move out only 1/4″ to properly grip the bulb. If the drive cylinder has a 1/2″ diameter, what diameter must the driven cylinder be in order to extend only 1/4″ when the drive cylinder is inserted 1″? Remember that if the diameter of a cylinder is doubled, volume will be quadrupled. If the diameter of the driven piston was exactly double that of the drive piston, the driven piston would extend only 1/4″ for every inch the drive piston is pushed in. Therefore, the driven piston should be 1″ in diameter to properly grip the bulb.

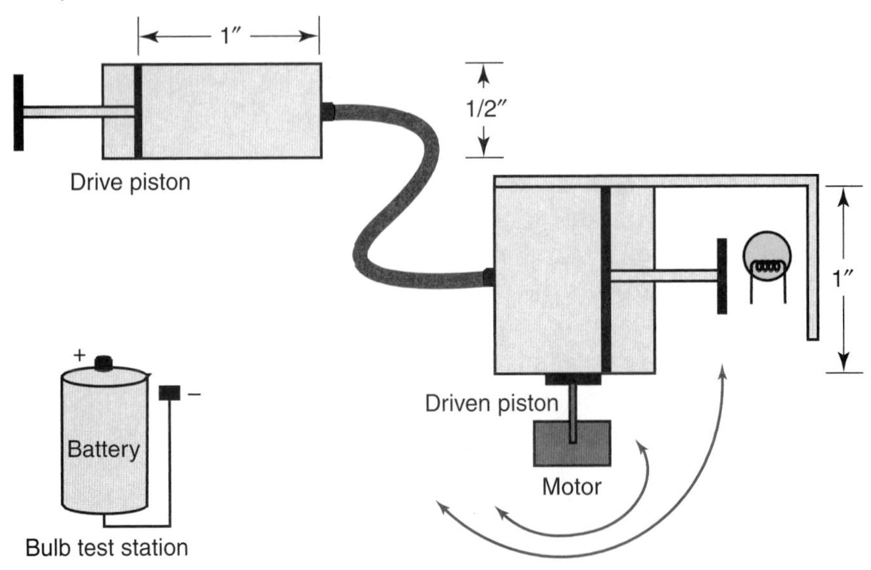

Summary

Fluid power systems use the power of pressurized liquids and gases to do work. Systems that use liquid are called hydraulic, and those that use gas are called pneumatic. Both types of fluids have similar characteristics that let the systems work with very little wear on mechanical parts. The basic difference, which must be considered when designing the systems, is the compressibility of the fluids. Liquids cannot be compressed, while gases may be compressed into as much as half their original volume.

Like any power system, fluid power systems require several components to perform work. Pumps and compressors are devices that convert mechanical energy to fluid power and supply the pressure and flow needed to operate the system. Fluid pressure is sent through conductors, such as pipes and hoses, to other parts of the system. Valves are devices within the fluid circuit that control flow, pressure, and direction of fluid power. Actuators and fluid motors are components of a fluid power system that convert fluid power back to mechanical power for end use. These are the parts of the system that usually do the actual work.

Key Words

All the following words have been used in this chapter. Do you know their meanings?

accumulator
actuator
air compressor
burst pressure
cam-activated valve
centrifugal force
centrifugal pump
check valve
compressibility
directional-control valve
double-acting cylinder
filter, regulator, lubricator (FRL) device
flow-control valve
flowmeter
fluid motor
four-way valve
gear pump
hydraulic pump
hydraulic system
impeller
laminar flow
nomograph
Pascal's law
pounds per square inch absolute (psia)
pounds per square inch gauge (psig)
pressure gauge
pressure-reducing valve
pressure-regulating valve
pressure-relief valve
pressure tank
reciprocating pump
reservoir
single-acting cylinder
solenoid-operated valve
spool valve
test pressure
turbulence
valve
variable-flow restrictor
viscosity
working pressure

Test Your Knowledge

Write your answers on a separate sheet of paper. Do not write in this book.

1. Fluid power systems involve the transfer of _____ energy.
2. What are the advantages of using fluid power systems?
3. Fluids exert pressure in all _____.
4. Fluids try to reach balanced pressure by flowing from an area of _____ pressure to an area of _____ pressure.
5. Fluid flow rate is measured with a(n) _____.
6. Summarize the difference between gauge pressure and absolute pressure.
7. The term _____ is used to describe smooth fluid flow.
8. Two types of hydraulic pumps are _____ and _____.
9. What is an impeller? How is it used in hydraulic pumps?
10. Two types of air compressors are _____ and _____.
11. _____ are devices that control fluid power within transmission lines.
12. A(n) _____ valve protects a fluid circuit from excessive pressure buildup.
13. If asked to create a fluid circuit that can provide both forward and reverse motion, use a(n) _____ valve.
14. How is the operation of a solenoid-operated valve different from the operation of a cam-activated valve?
15. _____ are devices that transmit power in both hydraulic and pneumatic systems.
16. A(n) _____ is a chart helpful in determining the interior diameter of fluid conductors when a flow rate is specified.
17. Describe the differences between how single-acting and double-acting cylinders operate.
18. A storage device used in pneumatic systems is a(n) _____.
19. Identify three safety concerns that should be considered when dealing with fluids under pressure.
20. Of the two types of fluids used in power systems, _____ can be compressed, and _____ cannot be compressed.
21. Explain the difference between an open fluid power system and a closed fluid power system.
22. Identify three common applications each for both hydraulic power circuits and pneumatic power circuits.

Matching questions: For Questions 23 through 29, match the phrases on the right with the correct term on the left.

23. _____ Oil.
24. _____ Atmospheric air.
25. _____ Robert Boyle.
26. _____ Viscosity.
27. _____ Working pressure.
28. _____ Accumulator.
29. _____ Blaise Pascal.

A. A type of fluid used in hydraulic power systems.
B. Studied the relationship between temperature, volume, and pressure.
C. A device that can store liquid under pressure.
D. A gas used in pneumatic power systems.
E. Described how hydraulic cylinders are able to increase mechanical advantage.
F. The term used to describe the ease with which liquid flows.
G. A term that describes the normal operating pressure for a conductor.

30. Write two or three sentences explaining how hydraulic cylinders increase mechanical advantage.
31. Calculate the amount of force that can be generated on the driven cylinder in the circuit below.

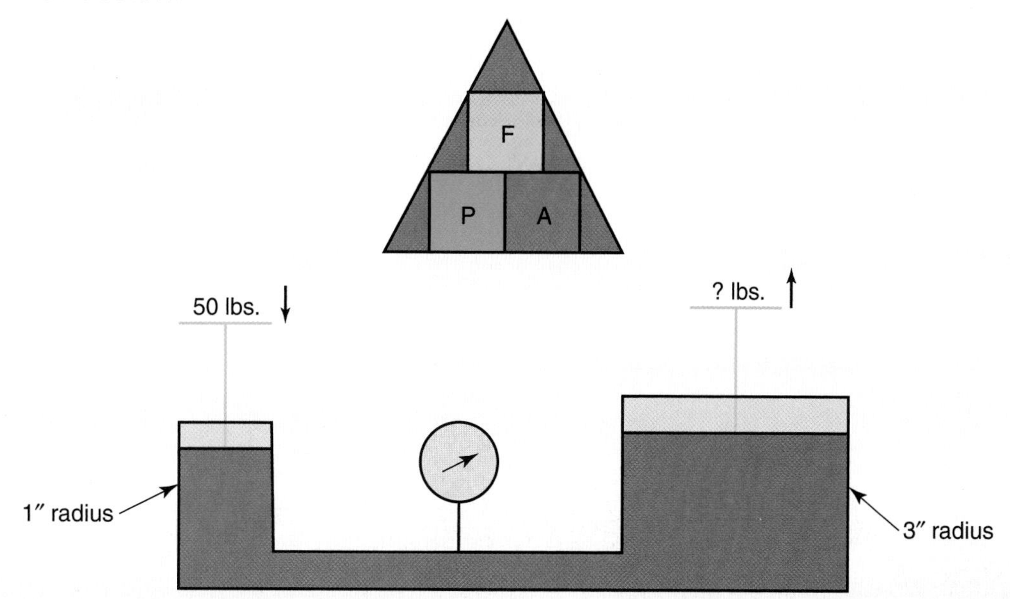

32. Calculate the mechanical advantage created in the circuit in Question 31.
33. Calculate the distance that the drive cylinder must be inserted in order to achieve 1/2″ of lift on the driven cylinder in the circuit in Question 31.
34. You are asked to create a hydraulic circuit that will lower the landing gear on an aircraft. The landing gear is located a considerable distance from the cockpit of the airplane. What type of valve will you use?

Activities

1. Connect two clear, 2-liter bottles with a length of clear plastic hose. Clamp the hose that connects the two bottles and fill one of the bottles with water. Release the clamp and observe the water levels in both bottles reach a balance.
2. Research the fluid clutch system used in many vehicles. Write a short essay on how a fluid clutch system operates and provide a sketch of a common fluid clutch system.
3. Construct a model door opener out of some large syringes connected by a fish tank tube. Glue the tube between the two syringes. Inserting the drive syringe should make the driven syringe extend. Retracting the drive syringe should pull the driven syringe back in. Use popsicle sticks or scrap wood to represent the door and attach the door to the end of the driven syringe to create the model.

11 Control Technology and Automation

Basic Concepts

- List some impacts of control technology and automation on society.
- Explain open loop and closed loop feedback in control circuitry.
- Describe different levels of control technology.
- Discuss the history of control technology.
- Identify applications of control logic.
- Define OR, AND, and NOT logic.

Intermediate Concepts

- Describe how sensors provide feedback in closed loop control circuits.
- Explain the advantages of computer or programmable control over other methods of control technology.
- Read values expressed in the binary-coded decimal (BCD) system.
- Use line diagrams and control logic to design basic control circuitry.
- Identify the functions of NOR and NAND logic in control circuits.

Advanced Concepts

- Demonstrate the use of five forms of control logic in control circuitry.
- Design and construct control circuitry for specific applications.

Control technology and automation are everywhere in our daily lives. Not only are control technology and automation at the heart of modern industry, but they have also made their way into most other areas of society. Automation is now a common feature in the home. Think about the many sensors that control things in your house. A thermostat controls the air temperature automatically, operating the furnace or air conditioner without any human intervention. Another thermostat automatically keeps the temperature inside your refrigerator cold enough to prevent food from spoiling. Floodlights outside your home may turn on as your car pulls into the driveway and activates a motion sensor. Timers, electronic eyes that sense darkness, or even sensors that detect heat may control lights.

Other devices around the home use more complex control systems. For instance, a washing machine has a control system that takes it through a number of steps. The system must start the water flowing, adjust the water temperature, sense when the machine's tub is full of water, run the agitator in different directions and for varying lengths of time, drain the water from the machine, and spin excess water out of the clothing. To do these tasks, the control system uses a series of automated valves, a timing mechanism, and various sensors.

All control systems have input and output devices. Additionally, all control systems make decisions. With simple control circuits, the decisions are often made by the way the input devices are arranged in the circuit. For example, assume you have a heater that is to be controlled automatically by a thermostat. The simple control circuit would look like the one in **Figure 11-1.** What if the heater is located, however, in your basement, which you do not use very often? You want the heater to operate only when you are using the basement. You can add a master switch to the circuit, so the thermostat can function only if the master switch is "on." The revised circuit is shown in **Figure 11-2.**

What Is a Control System?

A control system is a group of components working together to produce desired results. It does this by monitoring inputs and regulating outputs. **Figure 11-3** shows the most basic control system elements.

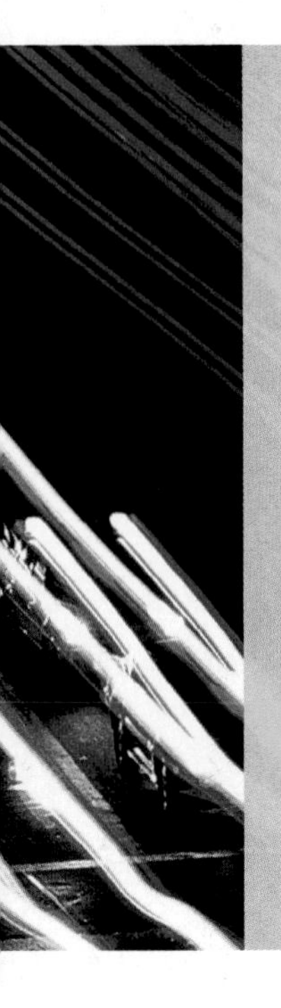

Career Connection

Electrical Engineering Technicians

The use of control systems is dependent on technology. People rely on their thermostats and motion detectors. It is important for these technological systems to function properly for continued use. Part of the job of an electrical engineering technician is to install and repair control systems.

Electrical engineering technicians maintain the electrical circuitry used in control systems. They study the work of engineers and make changes to designs where necessary. These technicians assist engineers throughout the design, assembly, and testing processes. They evaluate data received through testing, in order to fix any developmental problems. Electrical engineering technicians have the authority to recommend approval for electrical engineering projects. They are also responsible for ensuring safety in designs.

These technicians must have a thorough background in engineering. Their knowledge must include design and electronics principles. Also, since their duties may include language-oriented tasks, such as reviewing proposals, they must have well-developed written communication skills. An Associate Degree is required for this position. The yearly salary may range from $26,000 to $64,000.

Figure 11-1. A simple heater circuit with a temperature-controlled switch (thermostat).

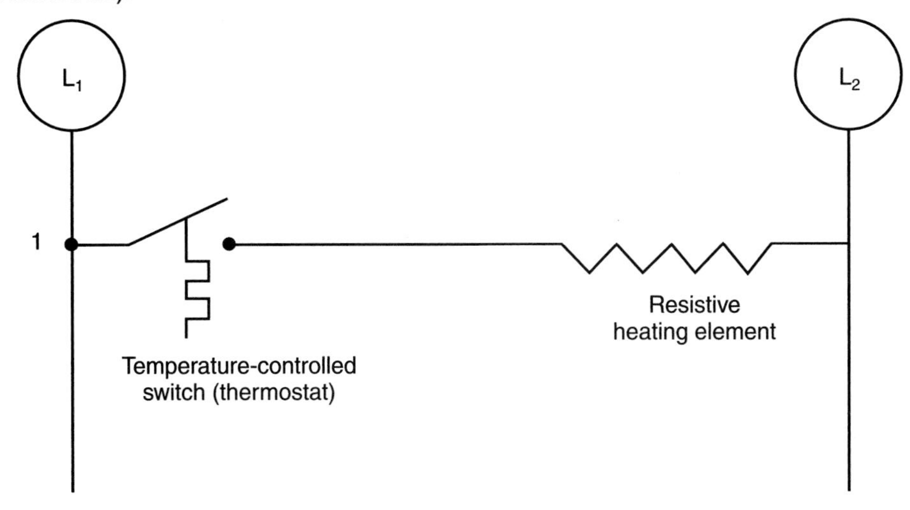

Figure 11-2. A simple heater circuit with thermostatic control and a master switch. The thermostat can control the heater only if the master switch is closed.

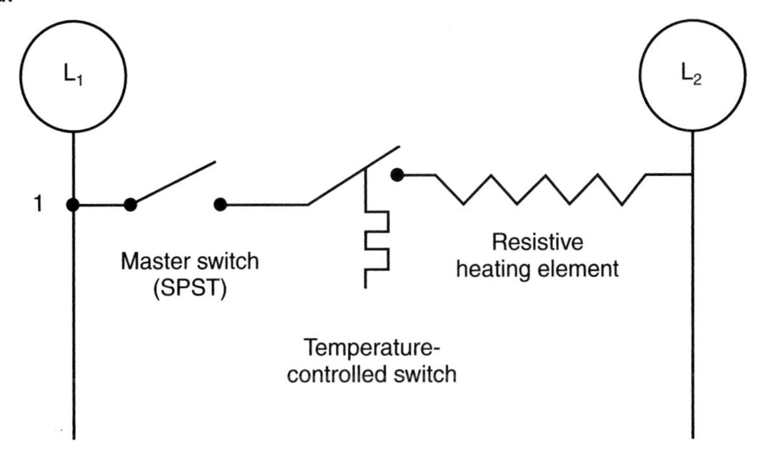

Figure 11-3. The main elements of a control system.

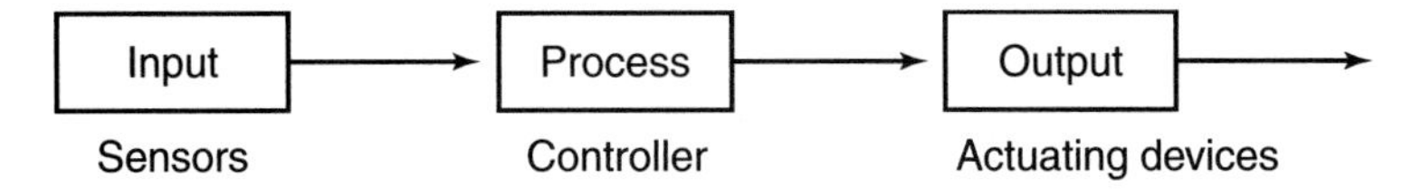

They are input from sensors, processing through a controller, and output of the results using some form of actuating device (such as a lamp, horn, or motor).

Understanding Line Diagrams

Line diagram: A pictorial representation of a control circuit.

The control circuits in Figure 11-1 and Figure 11-2 are shown in the form of ***line diagrams***. Such diagrams are often referred to as *ladder diagrams* because of their appearance. Line diagrams are pictorial representations of control circuits. They are the key to understanding and creating control circuitry. When you read or create line diagrams, keep these simple rules in mind:

- Read all line diagrams from top to bottom and left to right.
- The rails of the ladder (L_1 and L_2) represent the power for the control circuit (usually 12 V, 24 V, 110 V, or 220 V). If the control circuit has a hot leg and a neutral or cold leg, the hot leg is assumed to be the left hand rail of the ladder (L_1). The hot leg may be labeled + for positive, and the neutral leg (L_2) may be labeled – for negative.
- All inputs (switches and sensors) are shown to the left of the output to be controlled.
- There can be any number of inputs on a line, but only one output per line is allowed. That output should be connected to L_2.
- Lines are typically numbered so they can be referenced.

Figure 11-4 is a simple line diagram for a heating element control circuit. Note that the heating element can be controlled manually with the use of a toggle switch (SW1) in line 1, or it can be controlled automatically by the temperature switch (T1) in line 2. Switch T1 closes on falling temperature and will energize the heating element (H1) when the temperature drops to 60°F. If the heater has already been energized manually with switch SW1, switch T1 is bypassed. Note the switch in Line 1 labeled T2. This is the high-temperature cutoff switch, which will open when the room temperature reaches 80°F, turning off the heater.

Figure 11-4. This line diagram shows a heating element that can be controlled manually by closing SW1 or automatically by the normally open thermostat T1, which closes when the room temperature drops to 60°F. Thermostatic switch T2 is normally closed, but it will open if the temperature reaches 80°F, turning off the heating element.

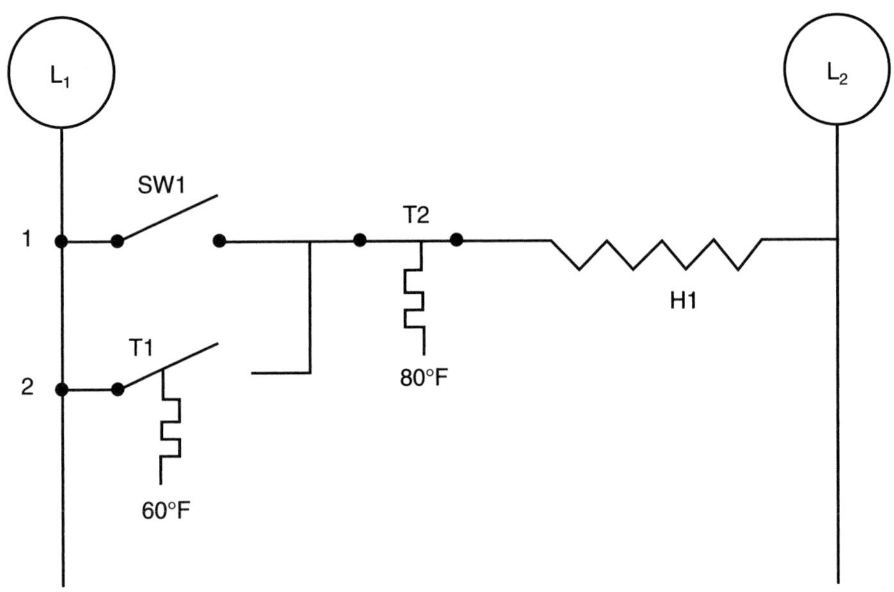

Open Loop Control Circuitry

There are two basic types of control circuits, open loop and closed loop. In an ***open loop system***, the system output has no effect on the control (input). For example, when you operate a light switch, the room light may or may not come on. See **Figure 11-5.** The electrical circuitry has no way of checking to determine if the lightbulb is burned out or is operating properly. There is no feedback mechanism built into the system to identify the state of the bulb.

Closed Loop Control Circuitry

Some control systems can consider the output of a system and make adjustments based on that output. They use information provided by sensors, called *feedback*, to predict the best way to control a process. This type of control is referred to as a ***closed loop system***. See **Figure 11-6.** For example, the thermostat in a water heater turns on the heating element when the water temperature in the tank drops to a predetermined level (the low ***set point***). The heating element will stay on until the temperature reaches the *high* set point, when the thermostat will turn it off.

Basic Control Elements

To understand control technology and automation, you must be familiar with basic electricity, fluid power, and mechanics, along with their corresponding control elements. These control elements include transmission devices, storage devices, protection devices, motion control devices, and advantage-gaining devices. Without the use of these devices, control and automation circuitry simply could not function. See **Figure 11-7.**

Figure 11-5. A light switch is an example of an open loop control system. There is no feedback to indicate whether the lamp did or did not turn on.

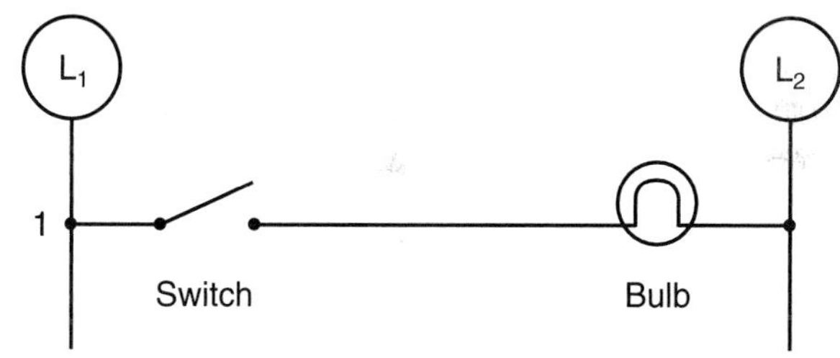

Figure 11-6. Closed loop control systems make use of sensor information (called feedback) to make any necessary adjustments in processing.

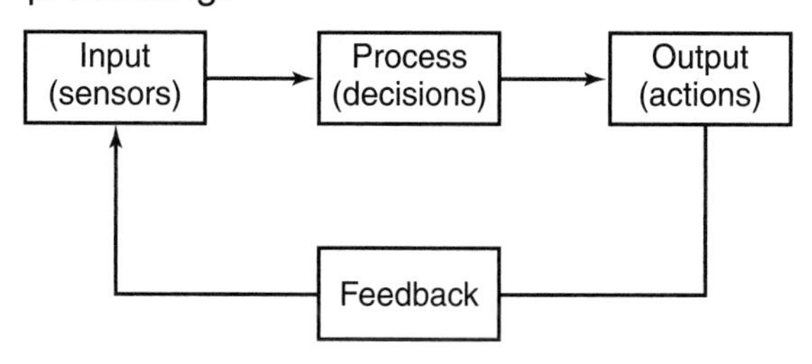

Levels of Control Technology

There are four different levels of control technology. See **Figure 11-8.** Each level builds on the previous one.

The first and simplest type of control technology is referred to as ***manual control***. Manual control requires human input in order to function. Pressing the start button of your microwave, opening a faucet, or turning the ignition key of your car are all examples of this type of control.

Figure 11-7. Basic power control devices for the various forms of power.

	Forms of Power		
	Electrical	**Fluid**	**Mechanical**
Transmission Devices	Wires Cables Buss bar	Hoses Pipe Tubes	Shafts Rods Belts
Storage Devices	Batteries Capacitors	Pressure tanks Accumulators	Springs Flywheels
Protection Devices			
Effort	Surge suppressors Breakers	Pressure relief valves	Shear pins, keys Torque limiting clutches
Rate	Fuses, circuit breakers	Velocity fuses	Governing mechanisms
Advantage-Gaining Devices	Transformers	Flow amplifiers	Gears, levers, pulleys Other simple machines
Motion Control Devices			
On/Off	Switches	Spool valves	Sprague clutches
One-way	Diodes	Check valves	Overrunning clutches
Variable	Potentiometers	Variable flow restrictors	Frictional clutches

Open loop system: A control circuit in which the system output has no effect on the control.

Closed loop system: A control system that considers the output of a system and makes adjustments based on that output.

Set point: A predetermined output level at which a closed loop system makes an adjustment.

Manual control: The simplest type of control technology, requiring human input in order to function.

The second level of control technology is ***automatic control***. As the name implies, automatic control is achieved by using a sensor or another automatically functioning device, such as a timer, to turn a machine on or off. This eliminates or reduces the need for input from humans. An electric eye that turns on lights when it becomes dark outside and a sensor that prevents a car from being placed in gear unless the driver's foot is on the brake pedal are examples of automatic control.

The third level of control technology is ***programmable control***. This level of control typically uses a dedicated microprocessor or computer as the brains of the system. ***Programmable logic controllers (PLCs)*** are microcomputers designed exclusively for control purposes. As such, they offer several distinct advantages for control technology over the desktop microcomputers with which we are all familiar. A desktop microcomputer is versatile and can perform many different kinds of tasks. A PLC, however, is designed to do *one thing* well: receive signals (inputs), process those inputs based on the program, and provide outputs. Since PLCs were developed to operate in industrial environments, they are less subject than microcomputers to interference from noise, vibration, or humidity. A very important feature of the PLC is the programming method. Programming is done in a line-by-line format, similar to the line or ladder diagrams familiar to electricians and control technicians. For this reason, the PLC is regarded as easier to program than a microcomputer, which may require knowledge of a complicated programming language.

Regardless of whether a microcomputer or a PLC is used, one characteristic makes programmable control stand out from lower levels of control technology: the control functions or instructions can be easily modified. For example, you can program the operating sequence of a

Figure 11-8. The four levels of control technology.

Levels of Control		
	Type	**Description**
	Manual	Human as controller.
	Automatic	Machine is self-acting or self-regulating.
	Programmable	Control instructions easily changed by humans.
	Intelligent	Machine emulates human abilites to solve problems and assign meaning.

traffic light using a PLC or microcomputer. The program would cycle through the red light/yellow light/green light sequence and then begin the process again. Once the traffic light is installed, it would be easy to modify the program to make the light behave differently. You might want a longer yellow light or decide that only a blinker was necessary late at night. Using a programmable device, you could easily make these things happen without ever touching the hardware used to construct the traffic light. You would do this by simply changing the program in the computer. This powerful concept of easily modified instruction makes modern automation possible.

The highest level of control technology is known as ***intelligent control***. This form of control uses machines and programming techniques that are capable of solving complex problems without human intervention. The technology emulates human thought processes using sophisticated software that makes use of artificial intelligence principles.

Automatic control: The level of control technology achieved by using a sensor and another automatically functioning device, such as a timer, to turn a machine on or off.

Programmable control: The level of control technology that typically uses a dedicated microprocessor or computer as the brains of the system.

Programmable logic controller (PLC): A microcomputer designed exclusively for control purposes.

Intelligent control: The highest level of control technology. It uses machines and programming techniques capable of solving complex problems without human intervention. The technology emulates human thought processes using sophisticated software that makes use of artificial intelligence principles.

Medicine: Intoxilyzers

Complicated sensors are most commonly associated with industrial applications. There are many other applications for automated sensors, however, that have nothing to do with industrial control. These sensors are used for other types of important applications. One such application is to perform a sobriety test.

An intoxilyzer is one type of detection device that can determine a person's blood-alcohol level. Generically, these devices are usually referred to as *breathalyzers*, but a true breathalyzer uses a chemical reaction to determine blood-alcohol level. An intoxilyzer measures blood-alcohol level with infrared (IR) spectroscopy. This process identifies the level of alcohol in the blood by passing an IR beam of light through a sample chamber that a person breathes into. On the opposite side of the sample chamber is a filter wheel that measures certain variations in the wavelength of the IR beam. The variations are created by the effect that alcohol has on the IR beam of light. More alcohol creates greater variations. The wavelengths are then converted to electrical impulses that are sent to a microcomputer for interpretation. Ultimately, a blood-alcohol level reading is digitally displayed.

A Brief History of Control Technology

Manual control is the oldest and most widely used form of control technology. It was in use long before the development of sensors and programmable control devices. It would be a mistake, however, to assume that all manual control circuits are simple—they can become quite complex. Modern control circuits that use sensors and programmable control devices are often based on knowledge of manual control circuitry. Thoroughly understanding manual control circuitry is essential to working with the more advanced levels of control technology.

Tech Extension

Intelligent Control

Researchers are making progress in developing machines that are not only reprogrammable, but that can learn while executing instructions—in some cases, correcting their own programs. This important new form of control technology is referred to as *intelligent control.* All control systems emulate the human thought process in some way. A control system that can solve problems and assign meaning to complex inputs is an intelligent control system. The point at which true artificial intelligence starts and ends, however, is not easily identified. After all, even an expert system capable of making complex decisions was created by a human being.

It is also important to note that not all forms of automatic control technology rely on electronic sensors or the microchip. For instance, James Watt's fly ball governor is an early example of automatic control that represented a real technological breakthrough. The governor functions automatically, but it is totally mechanical. Before Watt developed the fly ball governor in 1788, steam engines had to be controlled manually, a particularly difficult and dangerous task. The governor allowed the steam engine to feed itself with more steam automatically, while producing a relatively constant amount of power from the engine. The main shaft of the engine turns the governor. See **Figure 11-9.** As the engine speeds up, centrifugal force causes the fly balls (weights) to move away from the axis of rotation. This outward movement closes the steam valve, slowing down the engine. Decreased centrifugal force, coupled with gravity, then causes the weights to pull back toward the axis of rotation, opening the steam valve and starting the closed loop cycle all over again.

Figure 11-9. Watt's fly ball governor is an example of a totally mechanical automatic control. Although this mechanism was invented more than 200 years ago, it is still in use today for various types of engine control applications.

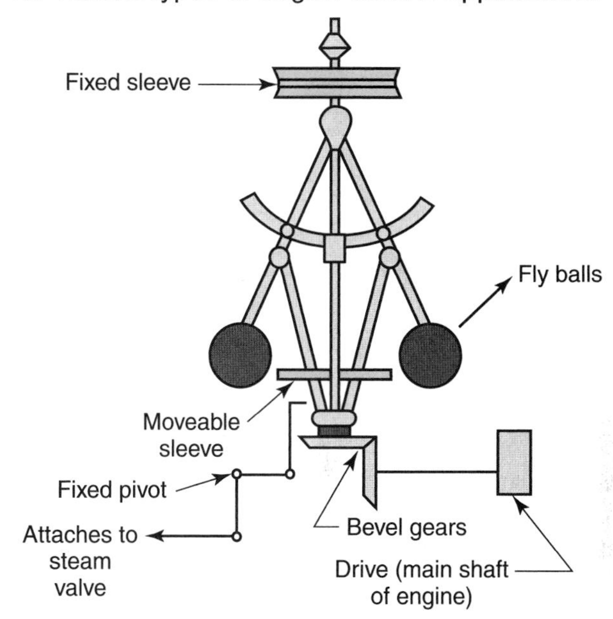

Programmable control is most closely associated with the development of the microchip and the microcomputer. Early forms of programmable control, however, used mechanical methods. One of the most important systems of programmable control was the punch card developed in the early 1800s by Joseph-Marie Jacquard, a French weaver. He automated the weaving of complex patterns by using different patterns of holes, punched in a series of cards, to control operation of the weaving loom. The cards were strung together to form a loop that automatically produced the same pattern over and over. See **Figure 11-10.** To produce a different pattern, another set of punched cards was mounted on the loom. The idea of punch card control was used in early computers and in industrial machinery control through the first half of the twentieth century.

From the early days of the computer, it was understood that one of the best applications for the device would be for industrial control purposes. Prior to computer control, heavy manufacturing industries, such as the automotive industry, had to totally reconstruct assembly lines whenever their products changed. The computer offered the power to simply modify programs and have the machines along the assembly line react differently. See **Figure 11-11.** General Motors pioneered the use of PLCs in the automotive industry, installing its first unit on an assembly line in 1968. Early PLCs were very large and very costly, but they could replace up to 100 relays and eliminate the need for scores of timers and counters along the assembly line. Modern PLCs are smaller and less expensive. Today, if a PLC can take the place of several timers or counters, it is less costly to use the PLC than the stand-alone components.

Figure 11-10. An early form of programmable control that predates the computer by hundreds of years is the Jacquard punch card system. It was used to control a loom weaving textiles with complex patterns. A—The punched cards were strung together to run through the loom, with the hole patterns governing which threads were woven into the design. B—Textiles with very complicated patterns could be consistently reproduced with the Jacquard method. To change patterns, a different set of punched cards was mounted on the loom.

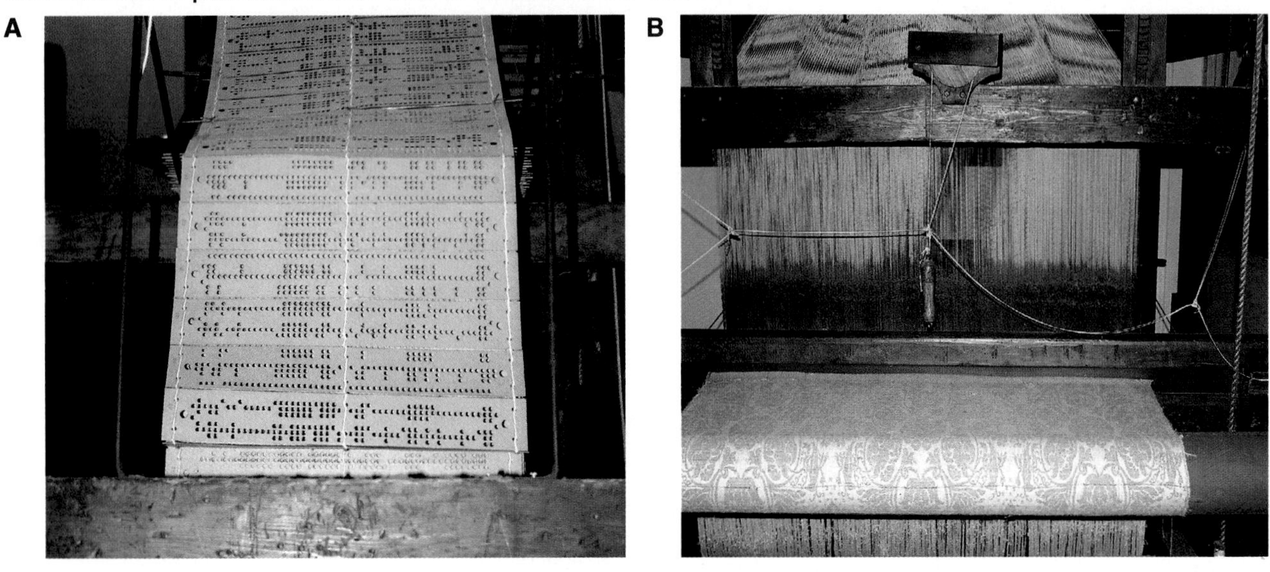

Figure 11-11. Modern programmable control is most closely associated with the microcomputer and the microchip. A computer is controlling these welding robots on an automotive assembly line. The computer can be easily reprogrammed to adapt the robots to produce a different size or type of vehicle. (Siemens)

Inputs, Processes, and Outputs

Programmable control systems are based on a computer or PLC that receives signals from external sensors (input), performs processing functions on those signals, and then sends a signal (output) to some form of actuating device. **Figure 11-12** is a schematic representation of such a system, showing examples of inputs and outputs, as well as listing typical processing functions.

Inputs

Input devices allow the computer or PLC to receive signals from external sensors, such as touch, temperature, light, or rotation sensors. There are two primary types of input signals. The first and easiest type to understand is a ***digital signal***, which has only two possible states—the sensor either sends a signal or does not send a signal. Examples are a switch (on or off) and a thermostat designed to turn something on or off

Digital signal: An input signal that has only two possible states—the sensor either sends a signal or does not send a signal.

Figure 11-12. This diagram of a programmable logic controller (PLC) shows typical processing functions and examples of input and output devices that can be attached to the PLC to form control systems.

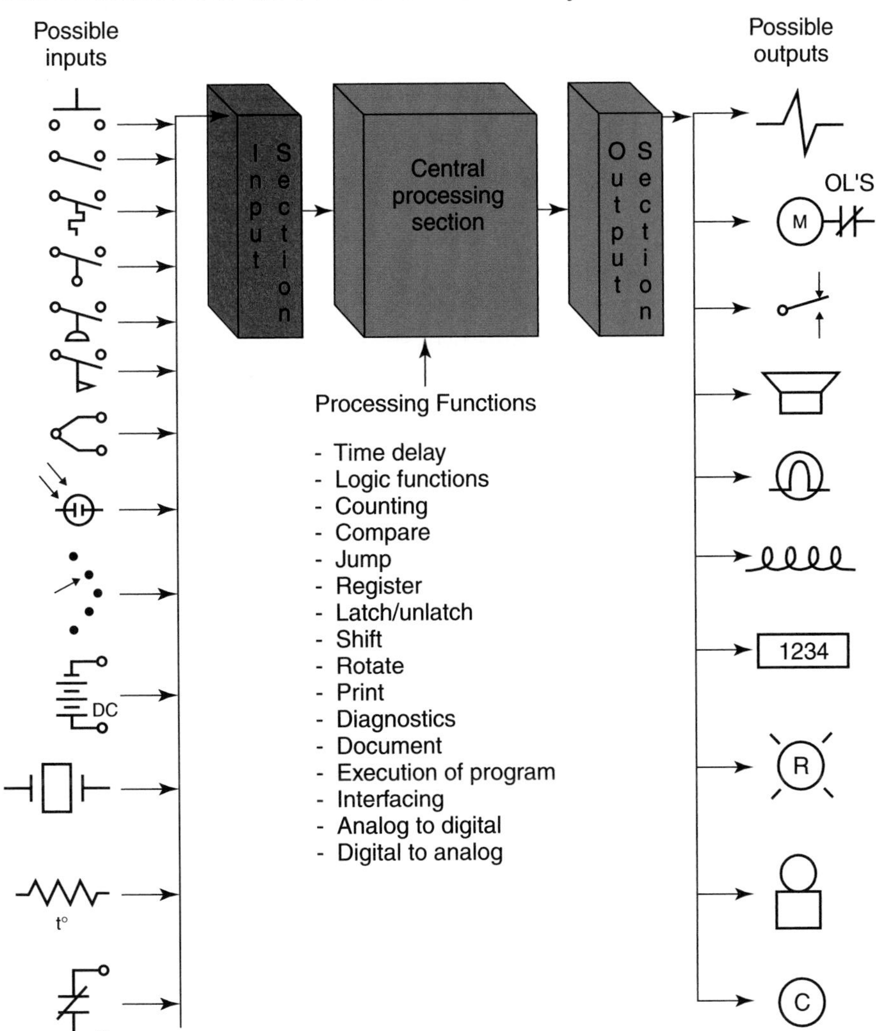

at a specific temperature. When a switch is moved to "on" or a sensor reaches its set point, a positive signal is sent to the processor. This signal is often referred to as a ***true signal***, or *high signal*. If the switch remains in the "off" position or the thermostat does not reach its set point, no signal is being sent. This is often referred to as a ***false signal***, or *low signal*.

The second type of input signal that can be sent is an ***analog signal***. Analog signals are used to transmit variable data, such as percentage of light, loudness of a sound, or weight of an object. In such cases, the data being transmitted can vary greatly within a given range. A simple true/false signal is not useful, since it does not supply enough information to the controller. Since computers are devices that understand only digital signals, the analog data must first be converted into digital form. This conversion can be accomplished in various ways. One way is to vary the voltage or current returned to the controller from the sensor. The computer or PLC can then interpret the voltage or current variations as representing particular values. Industrial control devices work this way, using ***analog-to-digital converters (A/D converters)*** for signal processing. A/D converters are electronic circuits that convert the analog information sent from sensors to digital representations the computer can understand.

A more cumbersome conversion method is to use multiple input ports to represent a digital value in the ***binary-coded decimal (BCD)*** numbering system. This system represents values using only the digits 0 and 1 (*false* and *true*). Assume a process was to occur six times. Signals at the four input ports would read *0110*, or *false/true/true/false*. In the BCD system, *0110* represents the number *6*. In this way, a variable (analog) value can be represented digitally, without the use of an A/D converter. The tradeoff is that BCD inputs require multiple input ports. Using four input ports, the range of numbers that can be represented is only from *0* to *15*.

Sensors are also important because they provide feedback for control purposes. Feedback often represents the difference between something that is accurately controlled and something that is controlled more crudely. For example, imagine that you have a greenhouse you want to keep cool in the summer. You could create a simple circuit to turn a fan on all day and switch it off at night, but this could prove wasteful and inefficient. Similarly, you could use slightly more complex circuitry to turn the fan on and off at set times throughout the day, but this too could prove less than ideal. What if it was a cool day? The fan would continue to cycle on and off anyway. Both of the previous examples represent open loop control technology. No feedback has been employed to control the fan. It simply turns on and off, without regard to the environment. A more accurate and efficient control method would be based on the temperature within the greenhouse, and this is where sensory technology and feedback come into play. The feedback from a temperature sensor represents closed loop control technology. The actual greenhouse temperature determines when the fan turns on and off. With the use of feedback and closed loop control technology, it will be much easier to keep the greenhouse within the desired temperature range.

True signal: A positive signal sent to the processor when a switch is moved to "on" or a sensor reaches its set point. Also referred to as a high signal.

False signal: No signal being sent, such as when a switch remains in the "off" position or a thermostat does not reach its set point. Also referred to as a low signal.

Analog signal: An input signal used to transmit variable data, such as percentage of light, loudness of a sound, or weight of an object.

Analog-to-digital converter (A/D converter): An electronic circuit that converts the analog information sent from sensors to digital representations a computer can understand.

Binary-coded decimal (BCD): A numbering system that represents values using only the digits 0 and 1 (*false* and *true*).

Curricular Connection

Math: Counting in Binary-Coded Decimal (BCD)

Beginning in kindergarten, you learned our conventional (Base 10) system of counting. You also have been exposed to other numbering systems, such as the Roman numeral system. What does IX stand for in the Roman numeral system? If you said it means *9*, you are correct. Like the Roman numeral system, the binary-coded decimal (BCD) system is simply another way of counting. The BCD system allows analog information—such as temperature variations, rotation speeds, humidity levels, or pressures—to be converted into digital information. To do so, the BCD system allows for the conversion of any number into a series of *0*s and *1*s. BCD values are read from right to left. The place values in BCD, from right to left, are 1, 2, 4, 8, 16, 32, and so on. See **Figure 11-A.** The actual digit shown in each place, however, is either *0* or *1*—a *0* literally means 0, but a *1* indicates a value equal to its place in the sequence.

Try counting in BCD. Read the following number from right to left: 01001. What number does this sequence represent? If you said the number *9*, you are correct. How was this determined? The first number on the right indicates a *1*. The second and third numbers would represent *2* and *4*, if *1*s were present. Since they are both *0*s, however, they represent *0*. The fourth digit has a *1* present, which represents an *8*. The fifth digit (that could represent *16)* is a *0*. So adding 1 + 8, you get the number *9*.

Can you see the value of BCD when it comes to inputting information into a computer or programmable logic controller (PLC) that relies on digital information? It would take only five digits to represent a numerical value up to *31*, six digits to represent a numerical value up to *63*, and only seven digits to represent a value up to *131*. The PLC can be programmed to interpret digital inputs in BCD. If it could not, can you guess how many inputs would be required to represent the number *131*? You probably guessed right: it would require 131 inputs. Because of the BCD numbering system, however, we can represent any number between *0* and *131* with only seven input ports.

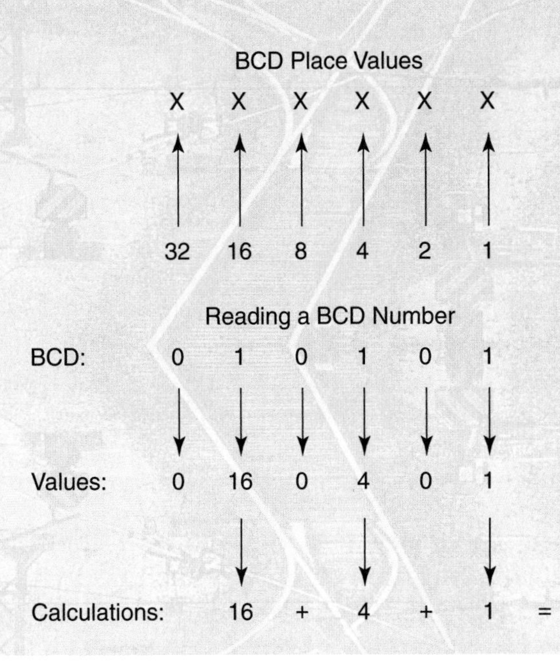

Figure 11-A. Binary-coded decimal (BCD) place values determine how a number is read.

Tech Extension

Smart Sensors

Most switches and sensors operate in a way that is fairly easy to understand. Push a switch, and the contacts on the switch open or close to create a desired effect. The functioning of some sensors, such as electric eyes, motion detectors, and proximity sensors, however, is more difficult to comprehend. Here is a quick look at how one of those smart sensors works.

Proximity sensors are becoming common switching devices because they can detect the presence or absence of almost anything, without requiring physical contact. They have a reputation for accuracy, reliability, and longevity. Proximity sensors are often used to replace less accurate and more cumbersome mechanical switches. There are several types of proximity sensors, but the capacitive proximity switch has the greatest variety of applications. The proximity switch is often located along an assembly line or conveyor belt. See **Figure 11-B.** The capacitor emits and receives magnetic lines of flux, similar to the way a magnet emits magnetic lines of flux beyond the presence of the magnet itself. Unlike a magnet, these lines of flux are not designed to attract metal objects, but simply to detect the presence of an object. When an object is within the flux lines, the lines become distorted. A solid state switching device interprets the distortion and sends a signal indicating the presence of the object.

Figure 11-B. A proximity sensor detects the presence of a part and sends a signal that can be used to perform a function, such as counting or timing. (Balluff)

Processes

Processing functions are at the heart of programmable control. They generally include timing, counting, and recursive functions that can be easily programmed and quickly modified. This ability to easily program functions and modify the program without having to change hardware provides programmable control with a huge advantage over other forms of control. Assume you have a certain process along an assembly line that requires a heat lamp to come on for 30 seconds to dry wet paint every time a part passes a ***proximity sensor*** located alongside the conveyor belt. This is a typical ***timing function***. Such a task could be accomplished with a cycle timer, but it is easily accomplished via programming with a microcomputer or a PLC. Now, assume an indicator bell is supposed to ring as every twelfth part passes down the same conveyor belt. This ***counting function*** could be accomplished with a simple counter, but a microcomputer or a PLC could also perform this task. Finally, assume we want the whole process of paint drying and parts counting to start over again after the twelfth piece has been counted. This ability to make something happen over and over again is known in programming terms as a ***recursive function***, or recursion. All programming languages used for control and automation provide for a means of creating a recursive loop.

Many languages that use lines of numeric code allow a GOTO statement. The GOTO statement sends the program back to a particular line of code, where a sequence can then begin again. Using a microcomputer or PLC can eliminate a great deal of expensive hardware. Both can perform many timing, counting, and recursive functions, eliminating the need for independent timers and counters. Factor in the power, ease, and convenience of programming, and you will really get a feel for the benefits of programmable control over other forms of control. Processes requiring multiple timing and counting functions can all be controlled from the same location, and the processes can be easily modified, simply by changing the program. For example, one common timing function is known as an ***on-delay function***. An on-delay is typically created to provide a margin of safety from the time the start button on a machine is pushed until the machine actually begins running. As the name implies, something will be turned on after the controller receives an input and a specified period of time has elapsed. **Figure 11-13** displays the code for

Proximity sensor: A device that responds to physical closeness and transmits a resulting impulse.

Timing function: A computer subroutine that observes and records the elapsed time of a process.

Counting function: A computer subroutine that indicates by units so as to find the total number of units involved.

Recursive function: The ability to make something happen over and over again.

On-delay function: A common timing function typically created to provide a margin of safety from the time the start button on a machine is pushed until the machine actually begins running.

Figure 11-13. A program written in the Logo programming language to perform an on-delay function in controlling a motor.

Description	Syntax
• name of program	to demo.on-delay
• waits until touch sensor 1 is pressed and sends a true signal to interface box	waituntil [touch1]
• creates a 10 second delay	wait 100
• orients interface to motor port C	tto ``motord
• turns motor C on for 45 seconds	onfor 450
• ends program	end

performing an on-delay function, using the Logo programming language. Notice that altering one line of the program could easily change the delay. Using a new value in the appropriate line could also change the amount of time that the motor operates.

Cycle timer: A device that turns a load on and off on a continuous cycle.

Latching: A function used to start a machine at the touch of a button and have it remain running indefinitely.

Unlatching: A function used to turn a machine off.

Memory: The ability of a control circuit to remember the last command it received.

Typical processing functions include counting up, counting down, cycle timing, on-delay timing, off-delay timing, recursive loops, latching, and unlatching. A ***cycle timer*** turns a load on and off on a continuous cycle. A traffic light is a good example of a cycle timer. Other examples are a timer that turns a hot water heater on and off and a timer that turns lights on during certain hours. Latching and unlatching functions are best described by relating them to the latch, or locking device, on a door. A ***latching*** function would be used when you want a machine to start at the touch of a button and remain running indefinitely, similar to the way a door would remain latched if someone locked it. An ***unlatching*** function will be used to turn the machine off. The start button would be programmed to latch, thereby providing power and keeping it flowing to the machine, even when the button has been released. In terms of control logic, this is referred to as ***memory***, or the ability of a control circuit to remember the last command it received. The stop button would be responsible for unlatching the control circuit, thereby stopping the machine.

Outputs

Output device: A small load that makes things happen, such as a relay, an indicator light, a horn, a bell, a solenoid-controlled valve, a heating element, and a small motor.

Electrical output from a microcomputer or PLC does not allow for the kind of high-amperage current flow that can directly power a heavy load, such as an electric motor. The small amounts of current that can flow through a microcomputer or PLC can, however, power smaller loads. Such loads are the ***output devices*** that make things happen. Typical output devices include relays, indicator lights, horns, bells, solenoid-controlled valves, heating elements, and small motors. The schematic (graphic) symbols for some common output devices are shown in **Figure 11-14.**

Figure 11-14. Schematic symbols for various output devices.

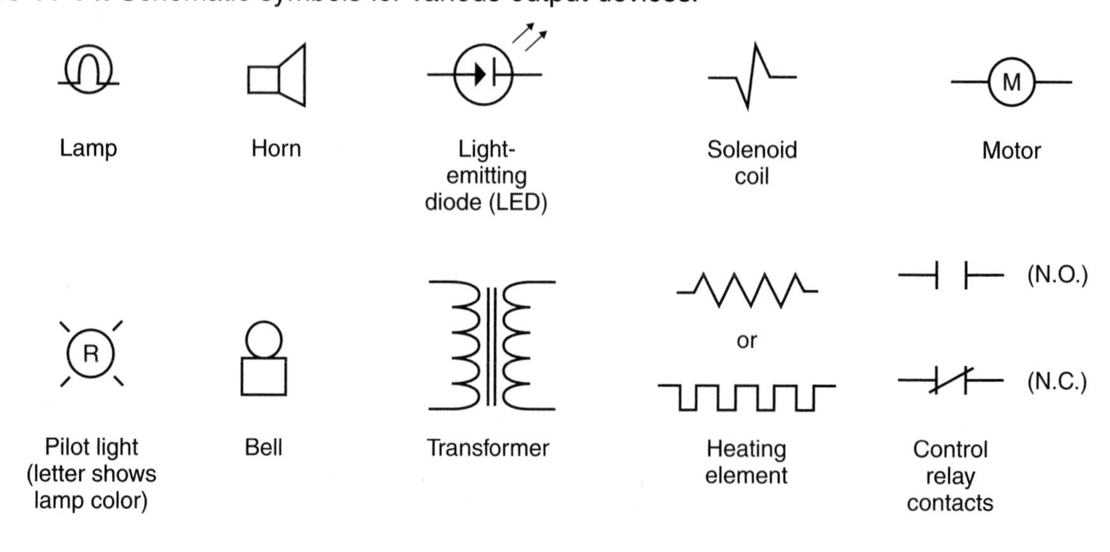

If a PLC or microcomputer must control a heavy load, the electricity from the control circuit is used to trigger an output device called a *relay*. Relays are switching devices that can be used to control high-amperage current flow to a heavy load, such as a motor. When the current from the processor energizes the coil on the relay, it closes the heavy relay contacts. This permits current flow to the motor. See **Figure 11-15.**

Figure 11-15. Using a relay in a low-voltage control circuit to control a motor or similar high-amperage load. The relay coil on the 24-volt side of the circuit is energized by a signal from a programmable logic controller (PLC). This closes the relay contacts on the 120-volt side of the circuit, sending current to the motor.

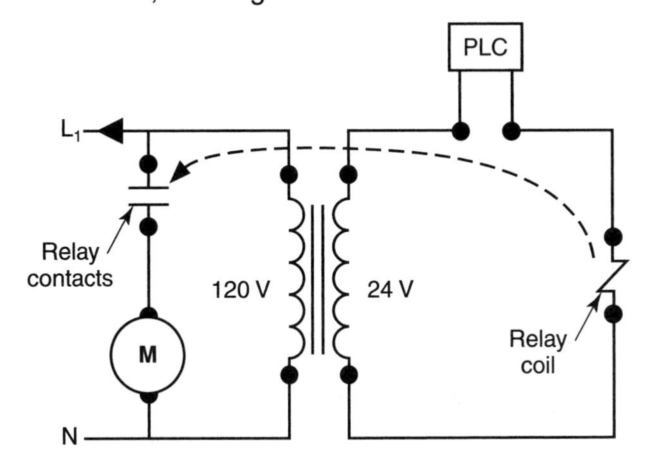

Control Logic

To fully understand, interpret, and design control circuitry, you must be familiar with the logic of such circuitry. The logic of control circuitry involves the way the input devices (switches and sensors) are arranged, in order to achieve the desired control function. For instance, in the example used at the beginning of this chapter, a heater was to function automatically, based on the set point of the temperature switch or thermostat. The circuit also called for a manual override switch. This gave the option of turning the heater on, regardless of whether the thermostat was calling for heat or not. The logic employed in this circuit is known as *OR logic* and is described in the next section.

OR Logic

One of the most common forms of control logic is ***OR logic***. Input must be received from either push button 1 OR push button 2 before the remainder of the program will execute. See **Figure 11-16.**

Figure 11-16. An OR logic circuit and truth table.

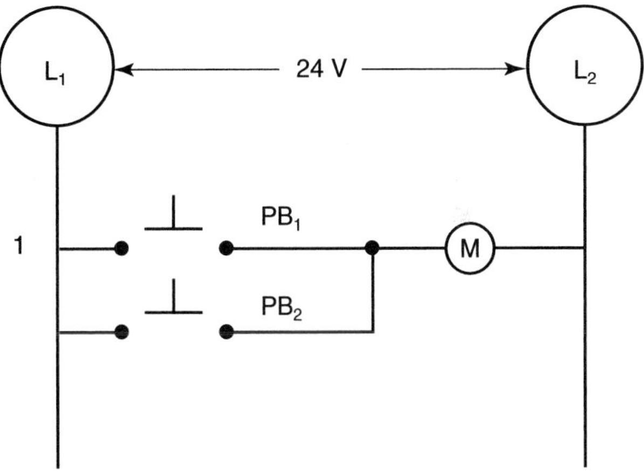

Truth Table

Input A	Input B	Output
0	0	0
0	1	1
1	0	1
1	1	1

Below the circuit diagram is a ***truth table***, a graphic method of representing the possible results from inputs. When reading truth tables, remember that *1* represents a true, or positive, signal, and *0* represents no signal. In the first row of the table in Figure 11-16, input A and input B both show as *0*. Since there are no inputs, there is no output—the output column also shows a *0*. In each of the other rows, input A or input B (or both) shows a *1*, so each results in an output.

OR logic: A form of control logic in which input must be received from either 1 OR more devices before output will occur.

AND Logic

When input from two or more devices is required before an action can take place, ***AND logic*** is used. In the example provided, both push button 1 AND push button 2 must be pressed before any ouput will occur. See **Figure 11-17.**

Truth table: A graphic method of representing the possible results from inputs.

AND logic: A form of control logic in which input from two or more devices is required before an action can take place.

Figure 11-17. An AND logic circuit and truth table.

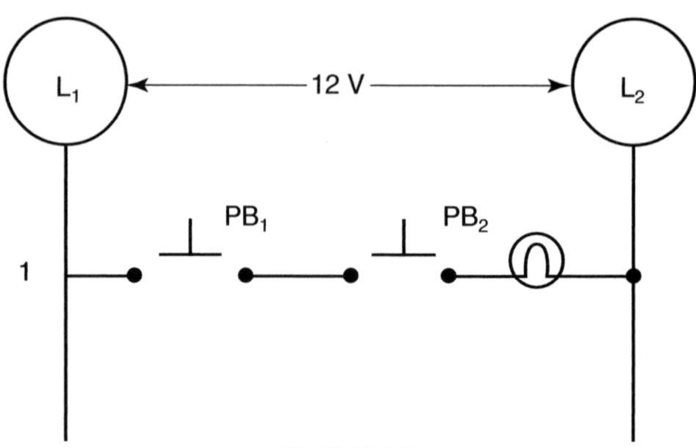

Truth Table

Input A	Input B	Output
0	0	0
0	1	0
1	0	0
1	1	1

NOT Logic

NOT logic: A form of control logic in which an output signal is provided *unless* there is an input signal.

NOT logic is employed when output is required *unless* there is a signal from an input device. This type of logic is often employed with interlocks and safety switches. So long as everything is in a normal state, the machine will function as intended. If a sensor signals that a guard is not in place or a door is open, however, NOT logic will shut down the machine. In the example shown in **Figure 11-18**, output will cease to occur if there is input to limit switch 1, which would cause the switch to open.

Figure 11-18. A NOT logic circuit and truth table.

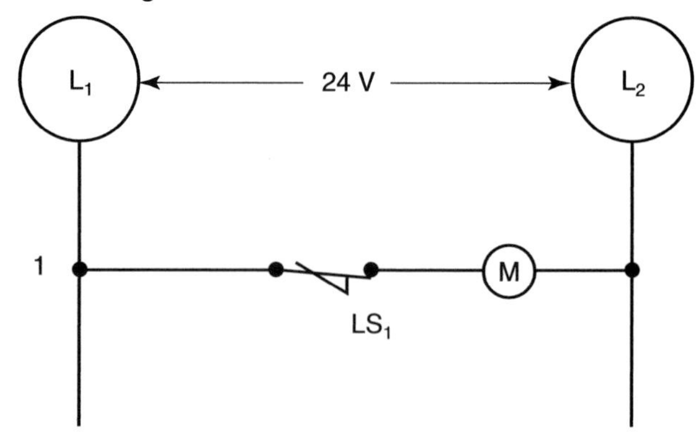

Truth Table

Input	Output
0	1
1	0

NOR Logic

NOR logic is simply a combination of other forms of logic. It is a combination of NOT logic and OR logic. Review the truth table and line diagram in **Figure 11-19** to determine how NOR logic works in control circuits.

NOR logic: A combination of NOT logic and OR logic.

NAND Logic

NAND logic is a combination of two other forms of logic. It is a combination of NOT logic and AND logic. Review the truth table and line diagram in **Figure 11-20** to determine how NAND logic works in control circuits.

NAND logic: A combination of NOT logic and AND logic.

Creating Memory in a Control Circuit

Many control circuits require memory. The switch on your stereo or computer works this way. Push it once, and it turns on. Push it again, and it turns off. In such switches, the memory is created mechanically—the switch contacts are simply held in place by the design of the switch. Industrial control circuits often create memory with the use of a control relay or motor starter. These devices are designed to magnetically create memory. **Figure 11-21** shows a standard magnetic motor control circuit with start and stop buttons.

Pushing the start button sends power to the coil, causing the contacts to close. The contacts provide power to the load (in this case, a motor). The motor starter also includes a special set of contacts (known as ***sealing contacts***, or auxiliary contacts) that are part of the control circuit. When these contacts close, they maintain memory, bypassing the start button and feeding power to the coil of the motor starter. Pushing the stop button interrupts the flow of power to the sealing contacts, opening all contacts on the motor starter. Once the sealing contacts are open, they have no way to close themselves unless the start button is pressed again.

Sealing contact: A contact in a motor starter that is part of the control circuit. When these contacts close, they maintain memory, bypassing the start button and feeding power to the coil of the motor starter.

Figure 11-19. A NOR logic circuit and truth table.

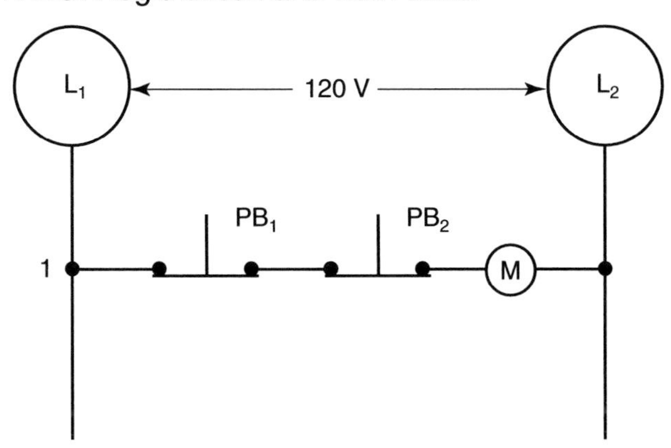

Truth Table

Input A	Input B	Output
0	0	1
0	1	0
1	0	0
1	1	0

Figure 11-20. A NAND logic circuit and truth table.

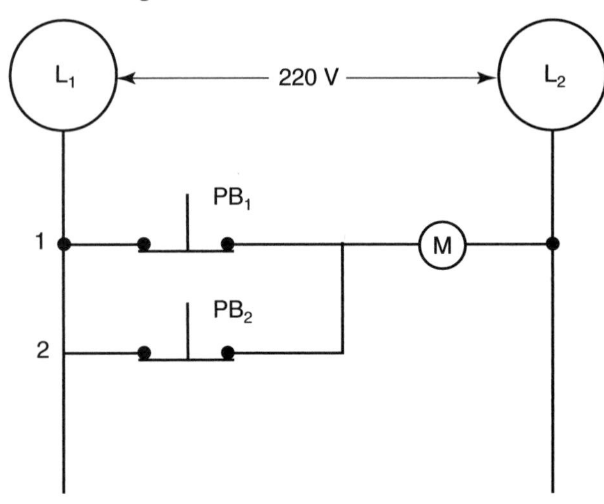

Truth Table

Input A	Input B	Output
0	0	1
0	1	1
1	0	1
1	1	0

Magnetic contacts, such as the one in this example, have several advantages over mechanical switches. First, if the power goes out and you are operating something with a mechanical switch, what happens when the power comes back on? The machine would start running again without warning! With a magnetic contactor, the control circuit would open when the power goes out, and the machine would not start again until the start button is pushed. Now, what if you are operating a large machine like a printing press? You might want the ability to stop the press from many locations, but to start it from only one. This is easily accomplished with the use of a magnetic motor starter, such as the circuit shown in Figure 11-21.

Figure 11-21. A schematic for a start/stop station with a magnetic motor starter.

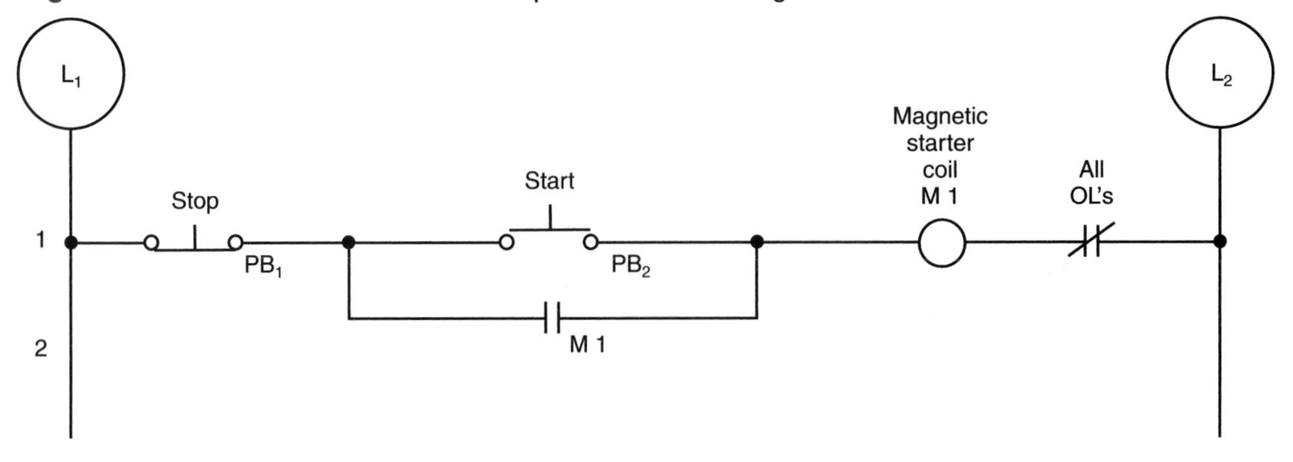

Summary

Control technology and automation play a significant role in contemporary society. Control systems that function without regard for the system outputs are known as open loop systems. Those that use sensors to provide feedback are known as closed loop systems. There are four levels of control technology. They are manual, automatic, programmable, and intelligent control. Each level builds on the previous level, while providing for increased capability. All control systems include inputs, processes, and outputs. Inputs can be arranged within control circuitry in many ways to create various forms of control logic, including AND, OR, NOT, NOR, and NAND logic. Control circuits are often represented by line diagrams, which are sometimes referred to as ladder diagrams. These diagrams indicate all components in a control system in a visual way that allows for the creation and troubleshooting of the circuit.

Key Words

All the following words have been used in this chapter. Do you know their meanings?

analog signal
analog-to-digital converter (A/D converter)
AND logic
automatic control
binary-coded decimal (BCD)
closed loop system
counting function
cycle timer
digital signal
false signal
intelligent control
latching
line diagram
manual control
memory
NAND logic
NOR logic
NOT logic
on-delay function
open loop system
OR logic
output device
programmable control
programmable logic controller (PLC)
proximity sensor
recursive function
sealing contact
set point
timing function
true signal
truth table
unlatching

Test Your Knowledge

Write your answers on a separate sheet of paper. Do not write in this book.

1. List three rules for creating line diagrams.
2. Information sent back to a programmable logic controller (PLC) or computer from sensors is referred to as _____.
3. Describe why PLCs are so prevalent for control and automation applications in modern industry.
4. Explain the difference between the four levels of control technology.

5. Sophisticated software that emulates the human thought process is most commonly associated with _____.
6. Variable information, such as the percentage of light available in a room, would be considered _____ data.
7. Timing and counting are common _____ performed by a microcomputer or a PLC.
8. A process that cycles on again and off again every 30 seconds is an example of a(n) _____.
9. A program that will perform the same function over and over again is in a(n) _____.
10. Provide an example of an application for on-delay logic.
11. On-delay and off-delay are two common types of _____ functions.
12. Latching and unlatching functions can be used to create _____ in a control circuit.
13. Define inputs, processes, and outputs.
14. A horn may be sounded by pushing either one of two switches. This control circuit was designed to use _____ logic.

Matching questions: For Questions 15 through 21, match the phrases on the left with the correct term on the right.

15. _____ Visual representation of a control circuit.
16. _____ Level of control that makes use of sensors to replace manual inputs.
17. _____ Computer used exclusively for control purposes.
18. _____ Primary application for a PLC.
19. _____ Positive signal from a sensor.
20. _____ Numerical means of converting analog data to digital data.
21. _____ Pilot lights, relays, and motors.

A. PLC.
B. Automatic control.
C. Binary-coded decimal (BCD).
D. Counting function.
E. Line diagram.
F. True signal.
G. Programmable control.
H. Output devices.
I. Manual control.
J. AND logic.

Activities

1. Determine the types of control circuitry necessary for an appliance, such as a washing machine or dishwasher, to work.
2. Look at a piece of industrial equipment, such as a saw, a mill, or a printing press. Draw a schematic indicating how you think the control circuitry operates.
3. Arrange for a tour of a factory or some type of production facility. Speak with the person who is responsible for keeping the machinery functioning properly about the control circuitry of the machines and the kinds of skills necessary to troubleshoot these machines.

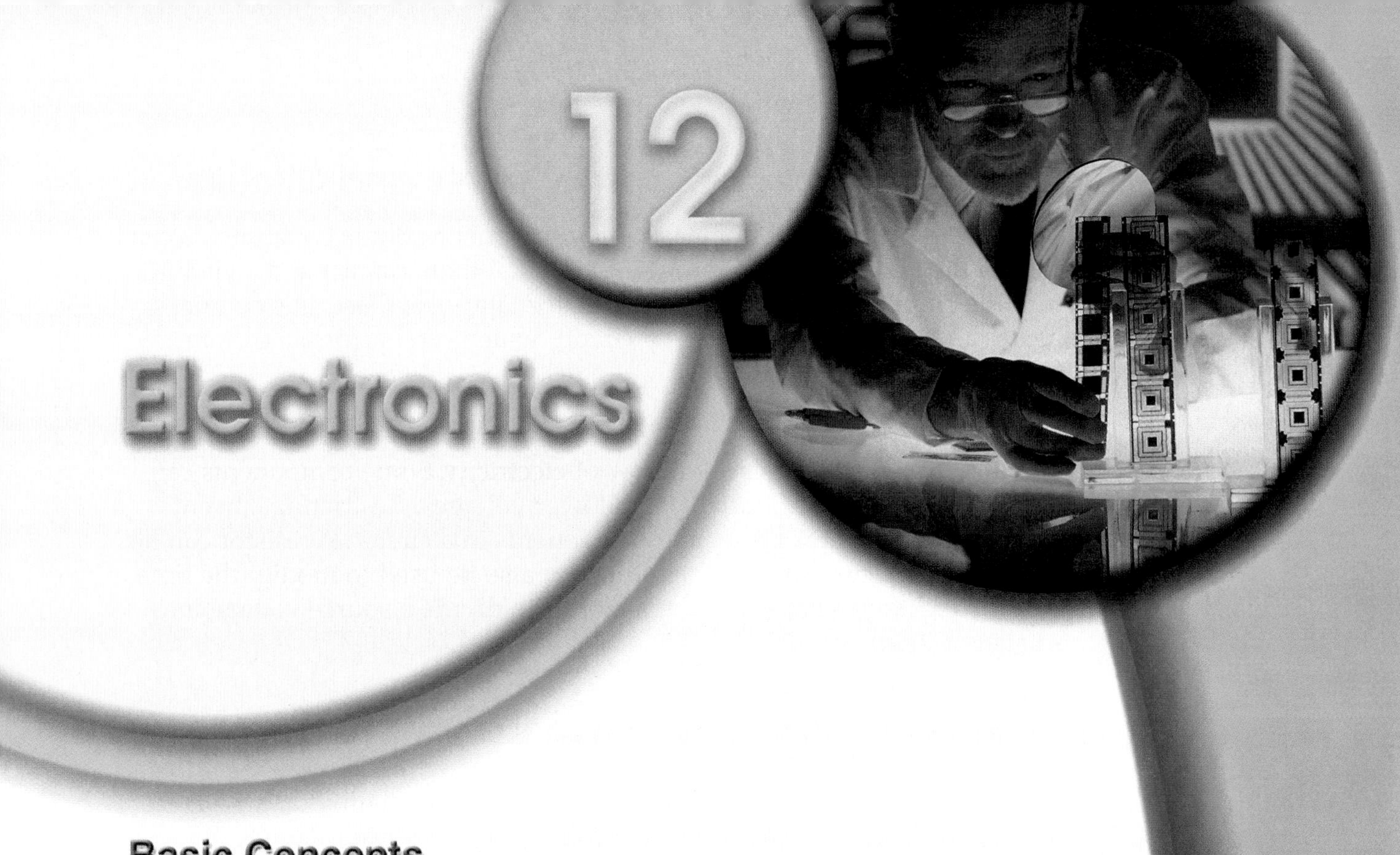

Basic Concepts

- Identify the values for resistors based on the resistor color code.
- State the purpose of using capacitors in conjunction with resistors for timing functions.
- Recognize the purposes of common electronic components, including resistors, capacitors, transistors, and diodes.

Intermediate Concepts

- Explain the purpose of specific integrated circuits (ICs).
- Convert basic schematics into electronic circuits.

Advanced Concepts

- Construct electronic circuits that include ICs.
- Troubleshoot electronic circuitry in a systematic way.

Electronics is the science that deals with electron flow. The science becomes a technology when it is applied to serve a useful purpose, in devices such as microwaves, cameras, stereos, and personal organizers. Electronics helped to usher in the age of miniaturization and the age of information.

An ***electronic device*** is made of one or more electronic circuits. An ***electronic circuit*** is a group of electronic components, such as resistors, capacitors, and diodes, connected together in such a way that they work together to perform a specific function. For example, a television set may have one circuit to receive a TV transmission and another circuit to produce the transmitted images on the TV screen. Another circuit may control the audio portion of the transmission. Each circuit may consist of similar electronic components. Because of each circuit's unique characteristics and arrangement, however, they are each able to perform a specific function.

Electronic device: A device made of one or more electronic circuits.

Electronic circuit: A group of electronic components arranged on a circuit board so that they work together to perform a specific function.

Resistor: A device that resists the flow of electricity.

Common Electronic Components and Circuits

There are many types of electronic components. Some components can control the flow of current through a circuit. Others can act as a switch to turn the circuit on or can store a charge for later use. Electronic components can be combined to create a circuit, which can perform a certain function.

Resistors

A ***resistor*** simply resists the flow of electricity. Some components can only work properly and safely from lower voltages. Restricting current flow can be useful for protecting components in a circuit. A resistor can be used to reduce current flow. Resistors can also be used to modify the time it takes to charge a battery or capacitor or to divide a source voltage into smaller voltages.

Resistor values

The unit of resistance is known as the ohm ( ). The amount of resistance offered by an individual resistor can be determined by reading the four or five color bands on the resistor and then calculating the resistance. See **Figure 12-1.** Using the chart in Figure 12-1B, you can calculate the resistance of any resistor labeled with color bands. Look at the example provided with the chart. Notice that the first two bands represent a numeric value. In the example, the first band, which is red, represents the numerical value 2. The second band, which is violet, represents the numerical value 7. These values are put together to create a single numerical value, 27. The third band is a multiplier. In this example, the third band is green and represents a multiplier value of 100,000. This value is multiplied by the value derived from the first two bands, which gives the resistor a total resistive value of 2,700,000 Ω, or 2.7 MΩ. The last band is known as the tolerance band. It indicates the tolerance or variance of the resistor. The tolerance depends on the specifications that the resistor must adhere to during manufacturing to be within acceptable limits of resistance for proper operation.

On a five-band resistor, the first three bands represent digits, the fourth band is the multiplier, and the fifth band indicates the tolerance. Any resistor that does not have a fourth or fifth band has a tolerance of +/– 20%. This level of tolerance may be acceptable for many applications. Any resistor that ends with a silver band means the resistor should measure within +/– 10% of its indicated value. If greater accuracy is required, a resistor with a gold band must be used to maintain tolerance to within +/– 5% of its indicated value. Lastly, if a resistor ends in a color band other than silver or gold, the tolerance would be equal to the number on the color code. Look again at the example in Figure 12-1. Notice that the fourth band is silver, which represents a +/– 5% tolerance. This means that the resistor's value will measure anywhere from 2.43 MΩ to 2.97 MΩ. The letter *k* (meaning "kilo-") can be used to substitute for thousands and the letter *M* can be used to substitute for millions.

Of course, resistance can also be measured with an ohmmeter. In order to accurately measure resistors in circuits, it is necessary for a technician to disconnect one lead from the circuit before measuring. When checking

Figure 12-1. A resistor and a resistor color-code chart. A—Color-code bands encircle a resistor and represent the resistive value of the resistor. B—A color-code chart for resistors with four or five bands and an example of how to interpret the color code.

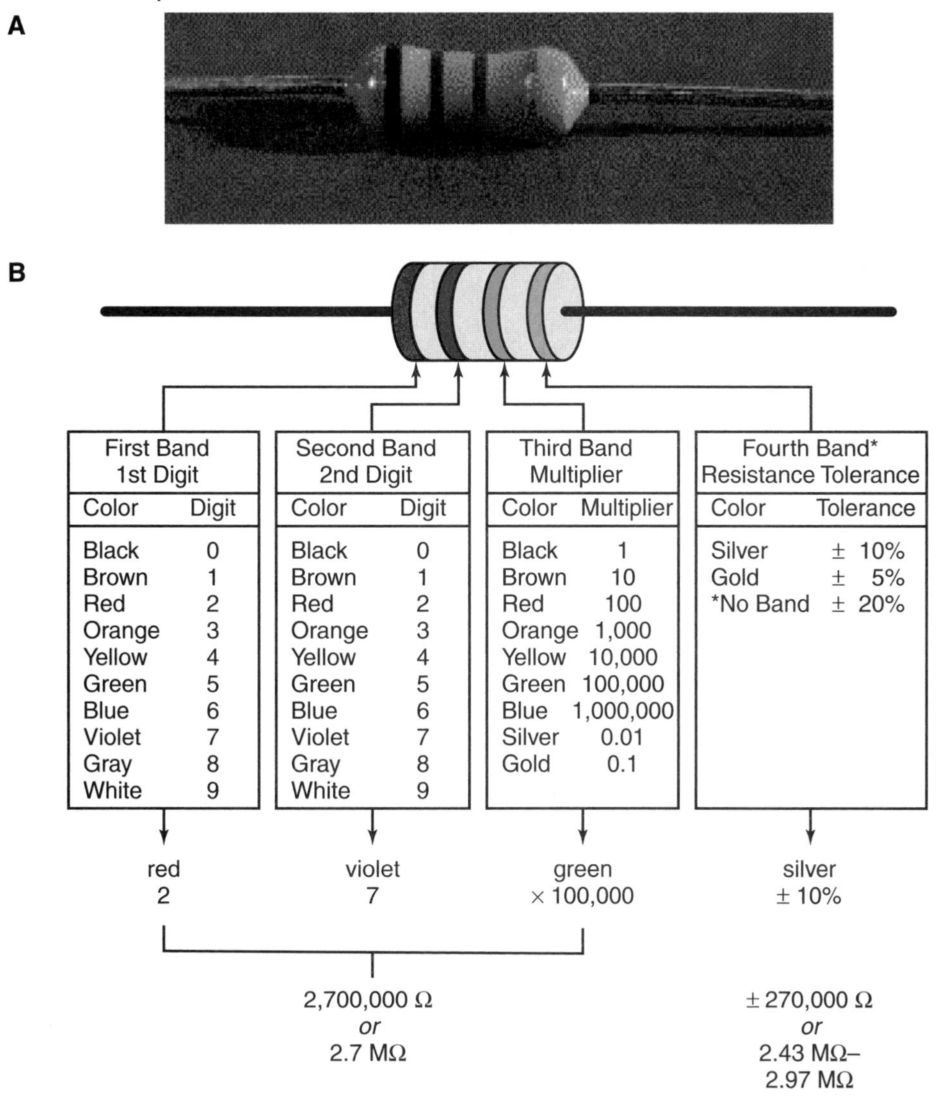

at the component level, it is common to isolate each individual resistor and check its value independently. See **Figure 12-2.** Troubleshooting resistors can also be performed with an ohmmeter. First, calculate the rated resistance from the color code on the resistor. Compare the calculated value of the resistor to the measured value, using the ohmmeter. Remember to take the tolerance of the resistor into consideration when comparing the calculated and measured values.

Safety Note

Before measuring resistance in a circuit, be sure the circuit is de-energized. Measuring resistance in an energized circuit could cause personal injury, damage to the ohmmeter, or both.

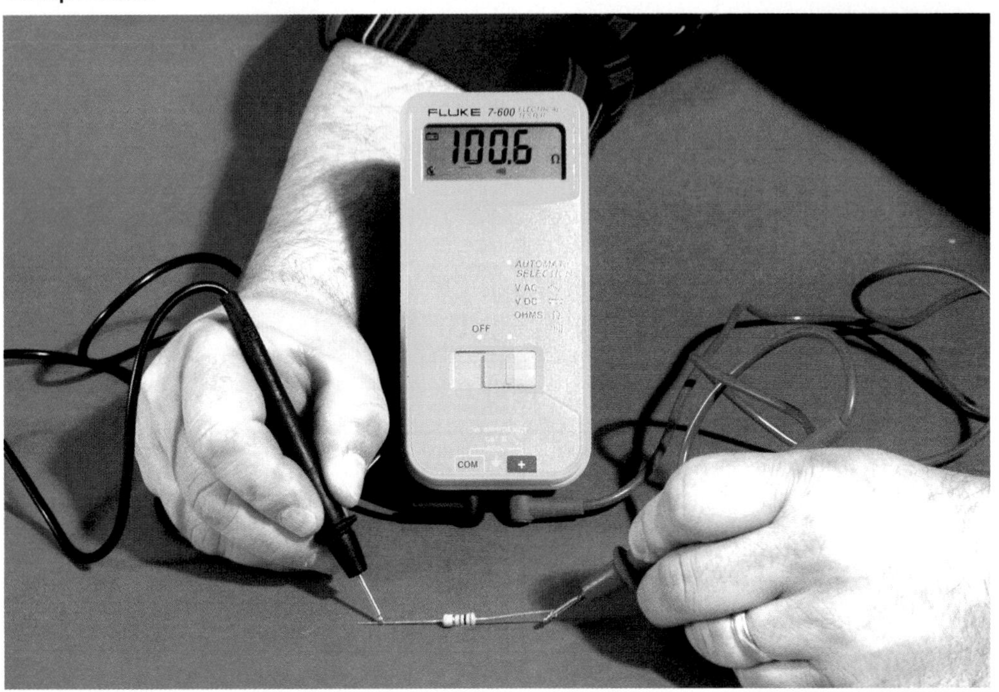

Figure 12-2. Measuring resistance with an ohmmeter. Note that the component to be tested is isolated from the circuit and tested when it is not energized. The power from the battery in the meter is enough to test the component.

Variable resistors

Variable resistors perform the same function as fixed resistors, except the resistive value can be varied. This is typically performed by rotating a knob or moving a sliding switch, such as a dimmer switch for a light. See **Figure 12-3.** The variable resistor is made of a piece of resistive material, often carbon, with an electrical connection (terminal) at each end of the resistive material, similar to a fixed resistor. A variable resistor, however, includes another electrical connection (terminal) known as a ***wiper***. The wiper can change position along the resistive material, based on the position of the knob or slider. This, in turn, changes the amount of resistive material between the wiper and each end terminal of the variable resistor. The result is a resistor that can offer a varying resistive value as the position of the variable resistor is changed.

Wiper: The electrical connection on a potentiometer that can change position along the resistive material, based on the position of the knob or slider.

Capacitors

A ***capacitor*** has the ability to store an electrical charge. Unlike batteries, capacitors can store and discharge electricity very quickly. The electrical charge can last for very long periods of time. Capacitors in a circuit can also be used to smooth out variations in power pulses and to block continuous, direct current (DC) flow, while allowing for current pulses to flow.

Capacitor: A device that has the ability to store an electrical charge. Unlike batteries, capacitors can store and discharge electricity very quickly.

There are several types of capacitors, but two of the most common are ceramic disk and electrolytic. The ceramic disk capacitor is made of ceramic and silver, and the electrolytic capacitor is made of an electrolyte

Figure 12-3. Variable resistors are often used in dimmer switches. A—A dimmer switch with a slide control. B—A dimmer switch with a knob (dial) control. C—The inside of a variable resistor. D—Each of the schematic symbols identified may be used to represent a variable resistor.

A

B

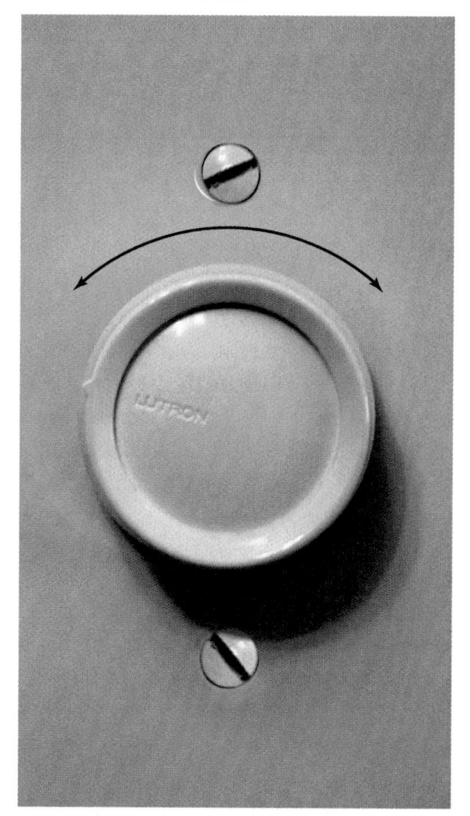

C

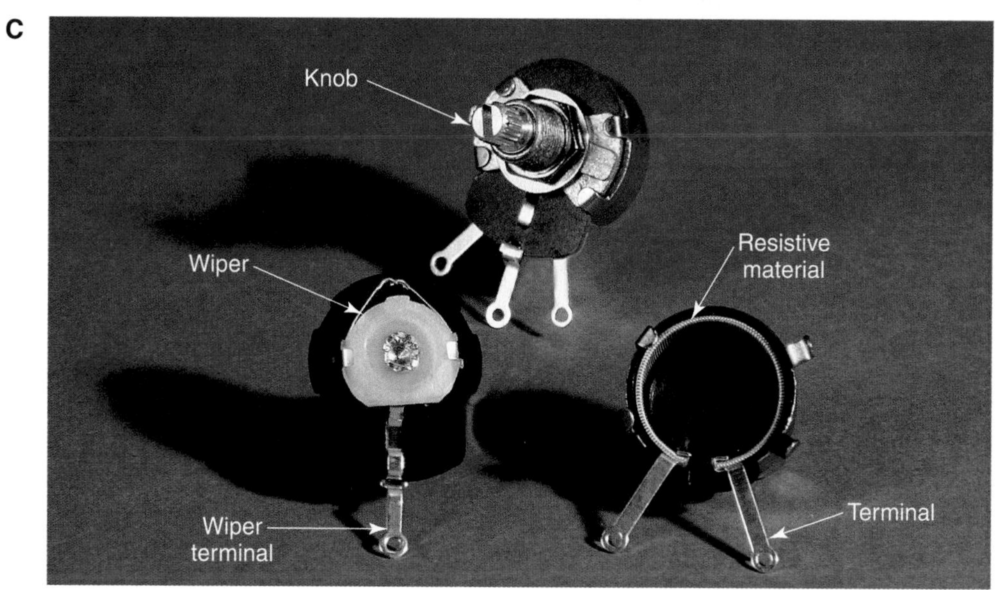

D

All these symbols represent variable resistors

Technology Link

Medicine: Electrical Shocks

Working with electronics can be dangerous. Several precautions can be taken to prevent accidents from occurring, but medical technology is very important when accidents do occur. Power technology depends on medicine and treatment for the safety and well-being of anyone working with live electricity.

Electricity can cause vital organs, such as the heart and brain, to malfunction. It can also lead to fatal burns. Electric shocks are responsible for about 1000 deaths in the United States every year. The severity of an injury from an electric shock depends on several factors: the voltage and amperage, the body's resistance to the current, the current's path through the body, and how long the body remains in contact with the current.

Various types of medical technology are important for treating victims of electric shocks. Neurological problems (injuries to the brain, spinal cord, or nerves) are the most common forms of nonlethal damage caused by electric shocks. Extensive damage can also be done to the respiratory and cardiovascular systems. Electric shocks can even cause cataracts, kidney failure, destruction of muscle tissue, and violent muscle spasms that can break and dislocate bones.

After an electrical accident has occurred, emergency medical help should be called immediately. Bystanders trained in cardiopulmonary resuscitation (CPR) may be required to perform first aid until help arrives. First, check to see if the victim is breathing and if there is a heartbeat. Feel for a pulse and watch the victim's chest to see if it is rising and falling. If the victim has no heartbeat or breath, CPR must be performed. If the heart is beating but the victim is not breathing, rescue breathing should be started immediately.

Once medical help has arrived and the victim is in stable condition, his or her cardiovascular system and kidneys must be monitored. Neurological activity also must be observed closely for changes. Medical technology, such as a computerized tomography (CT) scan or magnetic resonance imaging (MRI), may be necessary to check for brain injuries. Treatment for kidney failure may be required.

The victim also may suffer from third degree burns from an electric shock. Burn victims typically require treatment at a burn center. To restore lost fluids and electrolytes, fluid replacement therapy is necessary. Tissue that has been severely damaged can be repaired surgically and may include skin grafting or amputation. To prevent infections, antibiotics and antibacterial creams are used.

Depending on the extent of the injuries, a victim of an electric shock may require physical therapy in order to recover. He or she may also need psychological counseling to cope with the tragedy and, in some cases, the disfigurement or other long-term effects. Prevention is the best medicine for electrical accidents, so every precaution should be taken when dealing with this dangerous form of power.

material and aluminum. See **Figure 12-4.** Both types can store electricity, but the electrolytic capacitor has more storage capacity. Other types of capacitors are made of other materials, such as Mylar® film, metalized film, and polyester film.

The unit of capacitance is known as the farad (F). The majority of capacitors have a capacitance of only a few millionths of a farad. The range of capacitance varies so widely that capacitors are typically rated in

Figure 12-4. Ceramic disk and electrolytic capacitor construction.

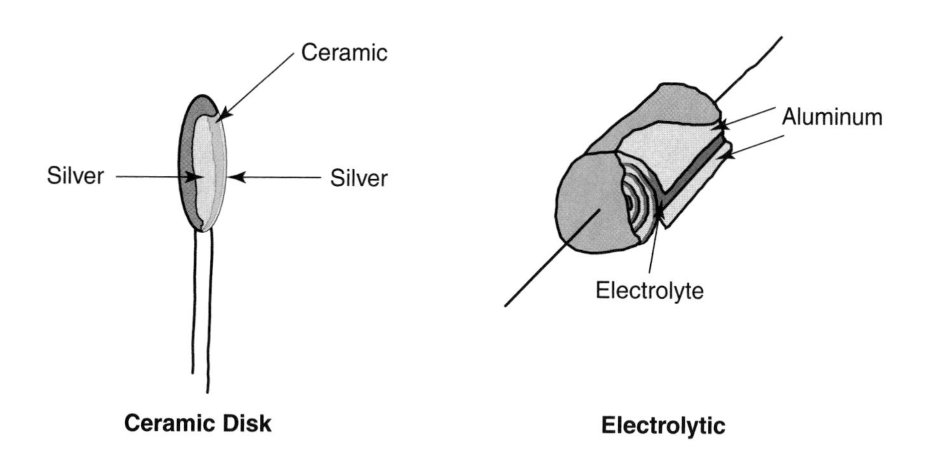

terms of millionths of a farad (microfarads, or μF) or trillionths of a farad (picofarads, or pF). See **Figure 12-5.**

Electrolytic capacitors are typically large enough to have their value in μF marked on their casing. Ceramic disk capacitors may be marked with a value in μF, or they may be marked with a three-digit code, indicating their value in pF. The first two digits of the code represent numerical values, and the last digit represents the number of zeros. For instance, a capacitor stamped *224* would calculate as 22 + 0000, translating to 220,000 pF, or .22 μF. Some capacitors use a color-coded system similar to resistors.

Diodes

A standard ***diode*** allows for the one-way flow of electricity. It is similar to a check valve in a fluid power circuit. Diodes can also perform a switching function, in that electricity cannot flow through them until it reaches a certain voltage. For small diodes, this may be about .6 V, but larger diodes may require higher voltages before allowing electricity to

Diode: A device that allows for the one-way flow of electricity. It can also perform a switching function, in that electricity cannot flow through it until it reaches a certain voltage.

Figure 12-5. This table shows the relationship between farads (F), microfarads (μF), and picofarads (pF).

Farads (F)	Microfarads (μF)	Picofarads (pF)
.001	1000	1,000,000,000
.0001	100	100,000,000
.00001	10	10,000,000
.000001	1	1,000,000
.0000001	.1	100,000
.00000001	.01	10,000
.000000001	.001	1,000

Figure 12-6. Typical rectifier diodes. Notice that the band on the diode indicates the cathode end. In the schematic symbol of a diode, the arrow represents the anode, and the line represents the cathode.

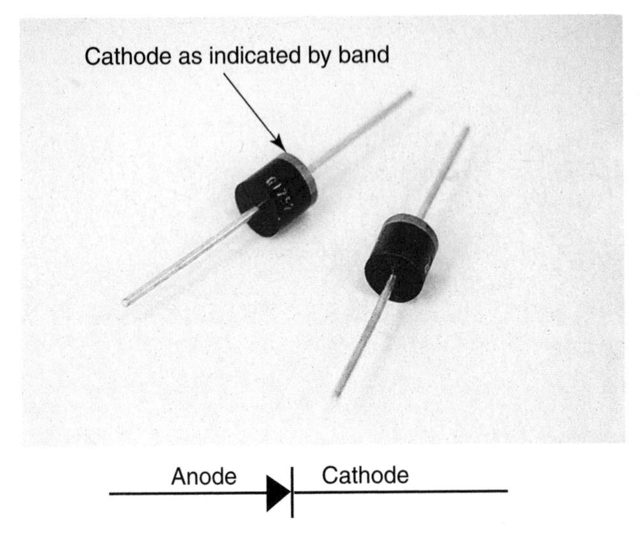

flow. The physical size of a diode is usually indicative of its amperage rating. A band on the diode indicates the cathode of the diode. See **Figure 12-6.** The cathode is the negative terminal into which electrons will flow. The anode is the positive terminal from which electrons will flow. The line in the schematic symbol of a diode represents the cathode, while the triangle represents the anode.

The flow of electricity occurs through a standard diode only when electricity of the proper polarity is applied. This is called a ***forward bias***. *Polarity* is defined as the condition of being electrically positive or negative, with respect to ground. DC flows through the diode in circuits in which the diode is forward-biased, but it does not flow through the diode in circuits in which the diode is reverse-biased. Alternating current (AC) has a charge that constantly changes between positive and negative. When AC flows in a direction in which the diode is forward biased, current flows through the diode. When AC flows in a direction in which the diode is reverse biased, the current cannot flow through the diode.

Forward bias: The flow of electricity occurring through a standard diode when electricity of the proper polarity is applied.

Testing diodes

Diodes can be tested in a de-energized state using an ohmmeter or continuity tester. Some continuity testers are labeled as both "continuity check" and "diode check." The diode should indicate a low level of resistance when the meter probes are placed on the diode in the forward-biased position and a high level or infinite level (overload) of resistance when the meter probes are placed on the diode in the reverse-biased position. See **Figure 12-7.**

Figure 12-7. Testing a diode. The diode should only allow for continuity in one direction.

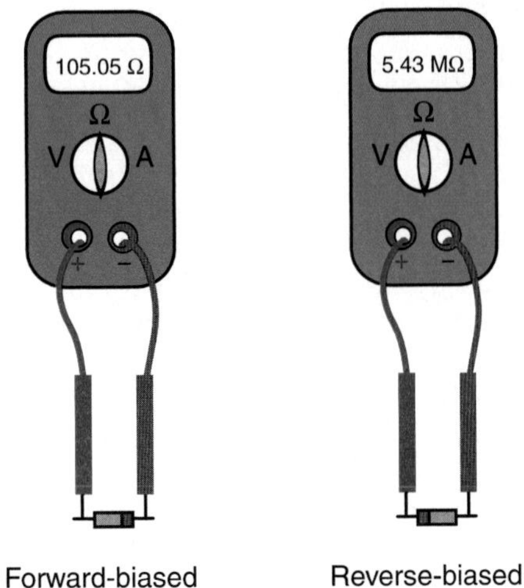

Rectifier circuits

Many appliances use DC power instead of AC power. A diode is a useful tool for converting AC current into DC current. A half-wave rectifier is the simplest form of AC-to-DC converter. The half-wave rectifier allows only half the AC cycle to progress past the rectifier. The result produces a pulsating, DC current. See **Figure 12-8.** In Figure 12-8A, the positive alternation of an AC waveform is applied to the circuit. The diode allows current to flow because it is forward biased. In Figure 12-8B,

Figure 12-8. Using a diode to create a half-wave rectifier.

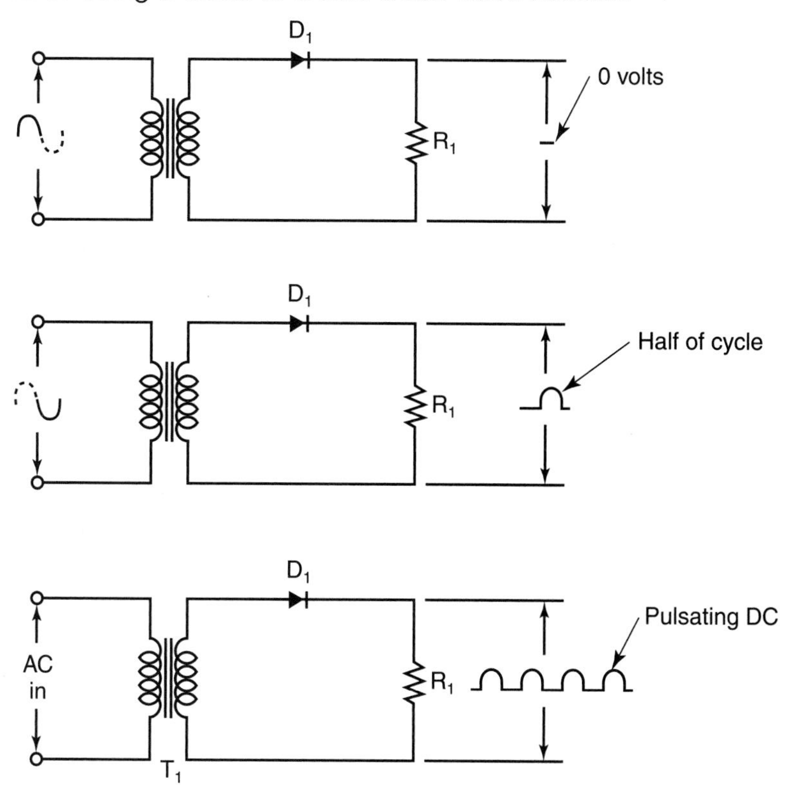

the negative alternation of an AC waveform is applied to the circuit. The diode prevents current flow. Figure 12-8C shows the continual output of this circuit, which is a pulsating DC.

Zener diode: A diode that conducts current in the reverse-biased direction. It blocks current until the voltage exceeds a certain level, and then it allows the current to flow. Once the zener diode allows current to flow, it is able to maintain a steady voltage.

Zener diodes

Another type of diode is referred to as a zener diode. Like a standard diode, a ***zener diode*** only allows for electricity to flow in one direction. The zener diode, however, conducts current in the reverse-biased direction. Therefore, zener diodes are connected in the reverse-biased direction. They block current until the voltage exceeds a certain level, and then they allow the current to flow. Once the zener diode allows current to flow, it is able to maintain a steady voltage. Zener diodes are often used in voltage regulation circuits because of this characteristic. See **Figure 12-9.**

Light-emitting diodes (LEDs)

The third type of diode is extremely popular for visual displays. It is known as a light-emitting diode (LED). LEDs are often used as indicator lights and can be produced in a variety of colors. Some can emit more than one color. The cathode lead from an LED is often shorter than the anode lead and is typically indicated by a flat area on the base of the LED.

Figure 12-9. A zener diode is properly inserted into a circuit in the reverse-bias direction.

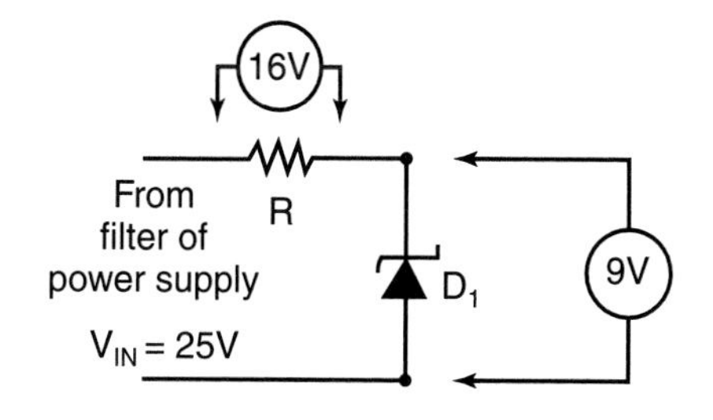

Curricular Connection

Science: Light-Emitting Diodes (LEDs)

Light is emitted in waves of energy. Higher energy lights have higher frequencies, resulting in light beyond the visible spectrum. The frequency of a light determines its color. In the electromagnetic spectrum, the lowest visible frequency is red, while violet is the highest. Light waves with a frequency lower than those of red light are infrared. Higher frequency light waves than those of violet light are ultraviolet (UV). The wavelength, or size of a wave, is determined by measuring the distance between peaks of waves.

One of the most common methods of producing light is by an incandescent bulb. This type of bulb contains a filament of tungsten, which glows white when heated. Producing light in this way requires that a great deal of energy supplied to the filament be converted to heat rather than light.

A more efficient means of lighting is the use of light-emitting diodes (LEDs). LEDs typically produce light in a single color and are used in several practical applications, including remote controls and traffic lights. In the 1960s, LEDs were created to produce infrared light. Shortly afterward, visible red, yellow, and green light could be produced by LEDs. More recently, violet and UV lights have been achieved by using LEDs.

LEDs operate more efficiently than incandescent bulbs to provide light because they do not produce heat. The semiconductor material is placed within a casing that is typically transparent. The current that flows through the diode must be forward biased in order to produce a wavelength of a single color. The casing is usually formed so the light reflects off the sides and so it is focused outward.

One popular form of alphanumeric display is a seven-segment LED. See **Figure 12-10.** The seven segments of the visual display are each powered by an individual LED. Lighting different combinations of the seven segments can create numbers *0* through *9* on the visual display. The seven-segment LED can also be used to create many letters and symbols.

Transistors

Solid-state: A type of device that can perform a switching function without any physical moving parts.

Bipolar transistor: A transistor with three junction points: an emitter, a base, and a collector.

Transistors are ***solid-state*** switching devices. This means they can perform a switching function without any physical moving parts. There are two types of transistors: bipolar and power field-effect. A ***bipolar transistor*** usually has three junction points. These points are known as the emitter, base, and collector. See **Figure 12-11.** Current typically flows between the emitter and the collector. This current flow can be switched on or off by a current delivered to the base. Bipolar transistors can sometimes be used for amplification purposes. When used for amplification, a small amount of current applied to the base of the transistor controls a larger amount of current across the collector and emitter. In other words, a gain in base current will produce a proportional gain in collector/emitter current.

Figure 12-10. Light-emitting diodes (LEDs) are often used to display numbers, letters, and symbols. A—A device that uses several, seven-segment LEDs. B—A seven-segment LED. Notice how the seven segments can be used in any combination to portray numbers *0* through *9*.

A

B

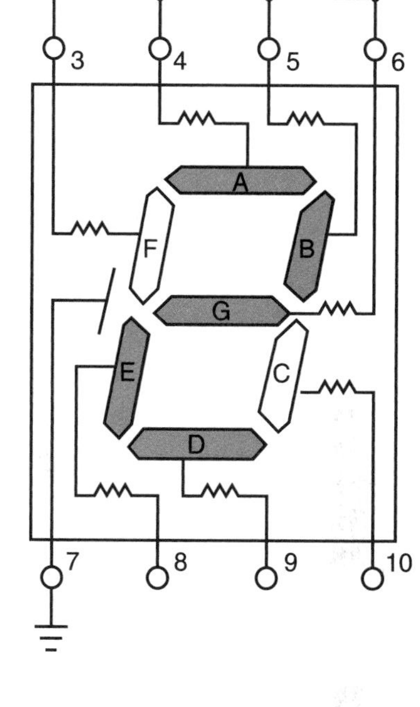

A

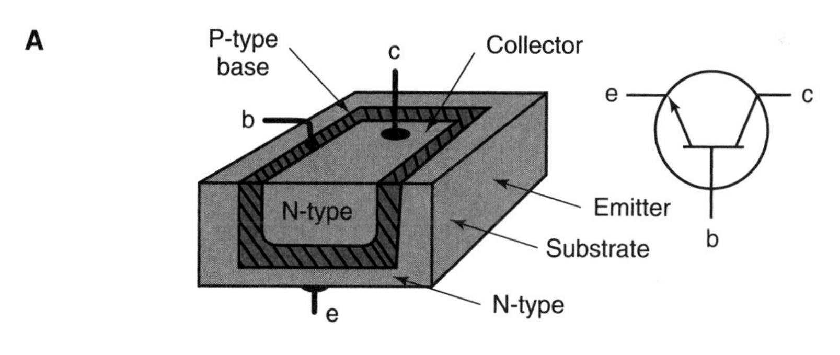

B

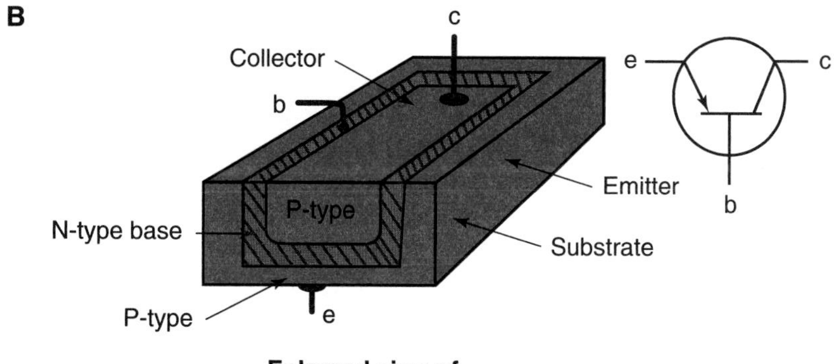

Figure 12-11. Bipolar transistor construction and schematic symbols. The three junction points are the emitter (e), base (b), and collector (c). A—An NPN transistor. B—A PNP transistor.

Field-effect transistor: A switching device often used because it can carry much more current than a bipolar transistor. It has three terminals: a gate, a drain, and a source.

Integrated circuit (IC): A collection of electronic circuits, undistinguishable to the naked eye, etched into a thin layer of silicon and installed into a plastic or ceramic housing.

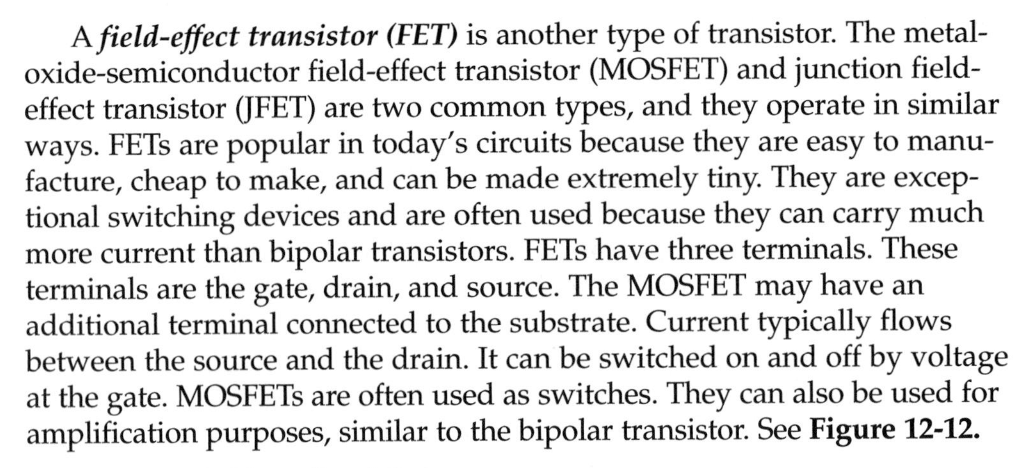

A ***field-effect transistor (FET)*** is another type of transistor. The metal-oxide-semiconductor field-effect transistor (MOSFET) and junction field-effect transistor (JFET) are two common types, and they operate in similar ways. FETs are popular in today's circuits because they are easy to manufacture, cheap to make, and can be made extremely tiny. They are exceptional switching devices and are often used because they can carry much more current than bipolar transistors. FETs have three terminals. These terminals are the gate, drain, and source. The MOSFET may have an additional terminal connected to the substrate. Current typically flows between the source and the drain. It can be switched on and off by voltage at the gate. MOSFETs are often used as switches. They can also be used for amplification purposes, similar to the bipolar transistor. See **Figure 12-12.**

Integrated Circuits (ICs)

An ***integrated circuit (IC)*** is a collection of electronic circuits, undistinguishable to the naked eye, etched into a thin layer of silicon, and installed into a plastic or ceramic housing. See **Figure 12-13.** The ceramic housing contains a series of protruding leads. These leads are known as pins. The pins allow the IC to be installed into a base, called a ***socket***, so the IC can be connected with other components in a circuit.

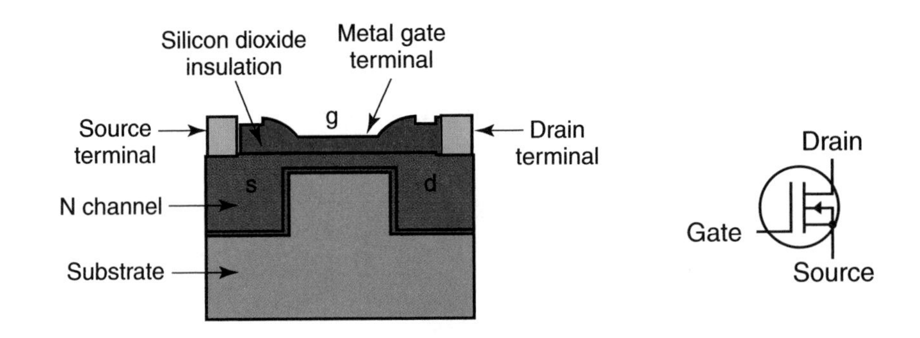

Figure 12-12. N-channel metal-oxide-semiconductor field-effect transistor (MOSFET) construction and its schematic symbol.

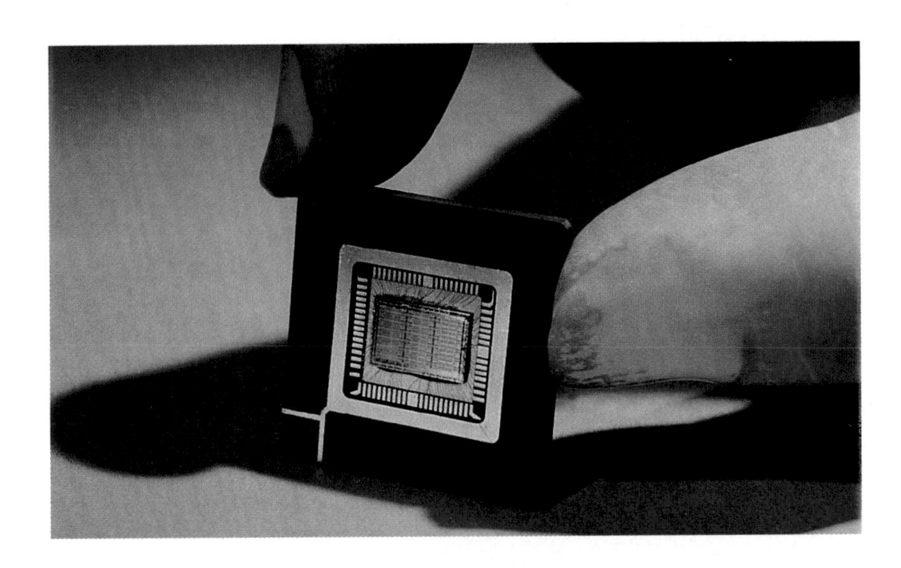

Figure 12-13. An integrated circuit (IC). (Miller Electric Manufacturing Company)

Always be careful not to bend or force the pins of an IC when installing them into a socket. IC pins can easily be damaged. A dot or notch usually identifies Pin 1 on an IC. Pin 1 is always the bottom pin on the left-hand side of the index marker. A ***pin-out*** shows all the pins on a given IC and may indicate their purpose. See **Figure 12-14.**

Common applications for ICs include counting functions, timing functions, and logic functions. Logic functions include AND, OR, and NOT logic, identical to the circuit logic studied in Chapter 11. Other ICs convert analog signals into digital signals.

ICs come in two basic types. The first type of IC is known as transistor-transistor logic (TTL), and the second type is known as complementary metal oxide semiconductor (CMOS). The ***transistor-transistor logic (TTL)*** IC works on low voltage, typically 5 V or less. An advantage of the TTL IC is it is relatively easy to work with because it is not subject to damage by static electricity. A downside to using a TTL IC is that it has a higher current draw than a CMOS IC.

Complementary metal oxide semiconductor (CMOS) ICs typically work with voltages up to 18 V and draw very little current. This provides an advantage when powering the circuit from a remote source, such as a battery. CMOS ICs are easily damaged by static electricity, however, and must be handled and stored with care. IC labels that begin with the number *4*, like the 4046, are typically part of the CMOS family. TTL ICs and CMOS ICs are generally incompatible with one another. Selection of the appropriate IC family is made based on the particular application.

ICs labeled beginning with the number *74*, such as the 7448, are usually part of the TTL family. The 7448 IC is a decoder-driver for a seven-segment LED. It interprets signals received and illuminates certain segments of an LED, thereby creating numbers, symbols, and letters based on the interpreted signals. A seven-segment LED would be of little use without a decoder-driver IC to sort out the signals and illuminate the correct segments on the LED.

The 741 series of ICs are operational amplifiers (op-amps). An ***operational amplifier (op-amp)*** has the ability to take in an AC or DC signal and amplify the output by as much as 100,000 times the input. This makes op-amps ideal for use in radios, TVs, and other sound-producing media devices. The 741 IC has two inputs: inverting and noninverting. A signal applied to an inverting input results in a signal of the opposite polarity at the output. One applied to a noninverting input results in a signal of the same polarity at the output.

The 555 IC chip is based on CMOS technology and is among the most popular ICs. It can be used for all sorts of timing operations, including monostable and astable applications. A ***monostable application*** turns something on or off for a specific period of time, such as the indicator light for seat belts in a car. An ***astable application*** provides continuous pulses at timed intervals. A 555 IC used to make an indicator light blink continually is an example of an astable application.

Socket: A base into which an integrated circuit (IC) is installed, allowing the IC to be connected with other components in a circuit.

Pin-out: A device that shows all the pins on an integrated circuit (IC) and may indicate their purpose.

Transistor-transistor logic (TTL): A type of integrated circuit (IC) that works on low voltage, typically 5 V or less.

Complementary metal oxide semiconductor (CMOS): A type of integrated circuit (IC) that typically works with voltages up to 18 V and draws very little current.

Figure 12-14. A pin-out for a typical integrated circuit (IC) chip.

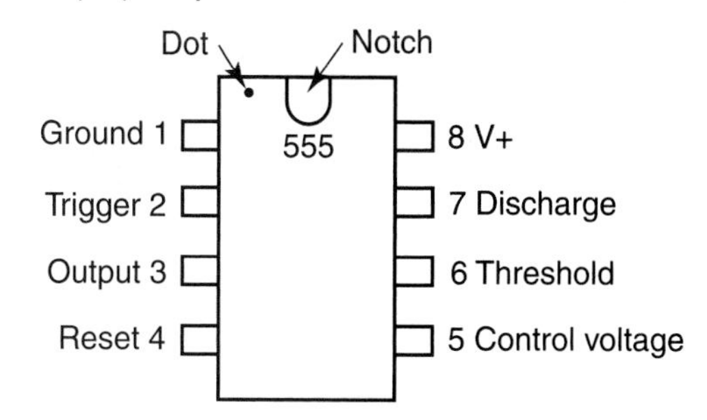

Operational amplifier (op-amp): An integrated circuit (IC) that has the ability to take in an alternating current (AC) or direct current (DC) signal and amplify the output by as much as 100,000 times the input.

Monostable application: An integrated circuit (IC) application that turns something on or off for a specific period of time.

Astable application: An integrated circuit (IC) application that provides continuous pulses at timed intervals.

The 556 IC is actually a pair of 555 ICs combined into one 14-pin package. This IC is extremely useful when two timing functions are required. For instance, if you wanted an indicator light to blink, it would require the use of one 555 IC in astable mode. If you wanted the blink to occur for a specific duration of time, it would require a second 555 IC operating in monostable mode.

Fuses and Circuit Breakers

Fuses and circuit breakers protect electronic circuits from excessive current flow (amperage). Most often, fuses are used in electronic circuits, but occasionally restorable circuit breakers are used to protect the circuit. Regardless of which protection method is used, it is important that the protection device be properly sized to protect all components of the circuit from excessive current flow that could damage components in the circuit.

Circuit Boards and Solderless Breadboards

Circuit boards provide a platform for electronic circuitry. They serve as a base for mounting components and serve as part of the circuitry. Tiny copper paths on the circuit board make many of the connections between electronic components. See **Figure 12-15.**

A solderless breadboard is ideal for testing electronic circuits for educational purposes and for testing circuits prior to permanent installation. See **Figure 12-16.** Since components are not soldered in place, they can be easily removed and replaced. The solderless breadboard saves a lot of time and allows components to be easily installed and removed. Solderless breadboards sometimes use a coordinate system made of letters and numbers to indicate junction points. This can be helpful for creating circuits from specific plans that state, for instance, "Insert Pin 1 in Slot F-12."

Figure 12-15. A circuit board provides a platform for electronic circuitry. A—Components are mounted to the circuit board. B—Copper paths on the circuit board connect the electronic components and form a circuit.

A

B

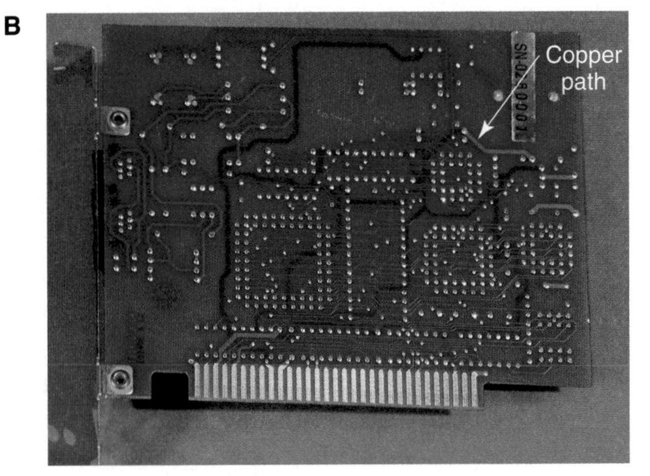

Figure 12-16. A solderless breadboard.

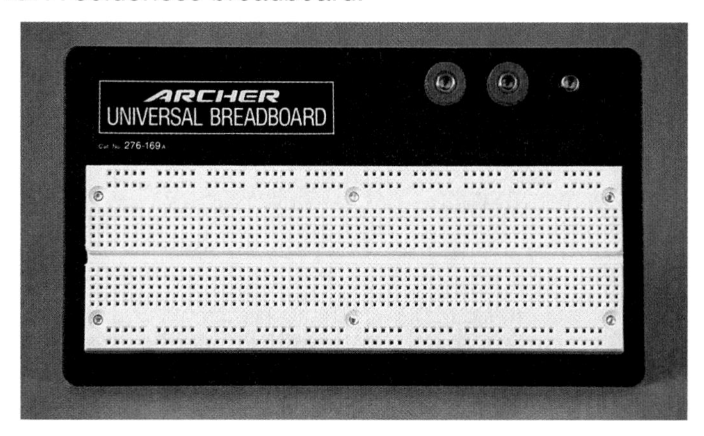

Wire

Wire leads are often used to feed an electronic circuit and sometimes used to make connections within an electronic circuit. The solderless breadboard uses a series of wire leads to make connections to the various components of a circuit. Copper paths often serve this purpose on a conventional electronic breadboard, but even a conventional breadboard may require wires to power the breadboard or to power components like speakers, which are located remotely from the breadboard. Wire leads, like all electrical components, must be sized appropriately to handle the anticipated ampere flow without overheating.

Schematic Diagrams

Schematic diagrams allow for the construction of electronic circuitry. Reading and interpreting schematic diagrams—from the simplest circuits involving only a few components to more complex circuits involving dozens of components—is essential. See **Figure 12-17.** Buzzers and relays are examples of loads.

Even complex schematics that appear intimidating at first glance can be broken down into simple elements. For instance, if asked to construct the circuit in **Figure 12-18,** you might begin with the first component, or you might begin with the simple components, such as switches you are absolutely sure you know about. You may also seek some additional assistance. For instance, if the circuit includes an IC, locating the pin-out for the chip may be necessary to ensure it is installed properly. In time, your knowledge of schematic symbols and proper component installation will grow, and your confidence in constructing circuits from schematics will grow as well.

Troubleshooting Electronic Circuits

Troubleshooting electronic circuits requires more advanced skills than simply constructing electronic circuits. To effectively troubleshoot, it is important to know how to operate a multimeter properly. It is also important to have a thorough understanding of how the components are intended to work in their proper state so faults can be correctly identified.

Figure 12-17. Schematic symbols for common electronic components.

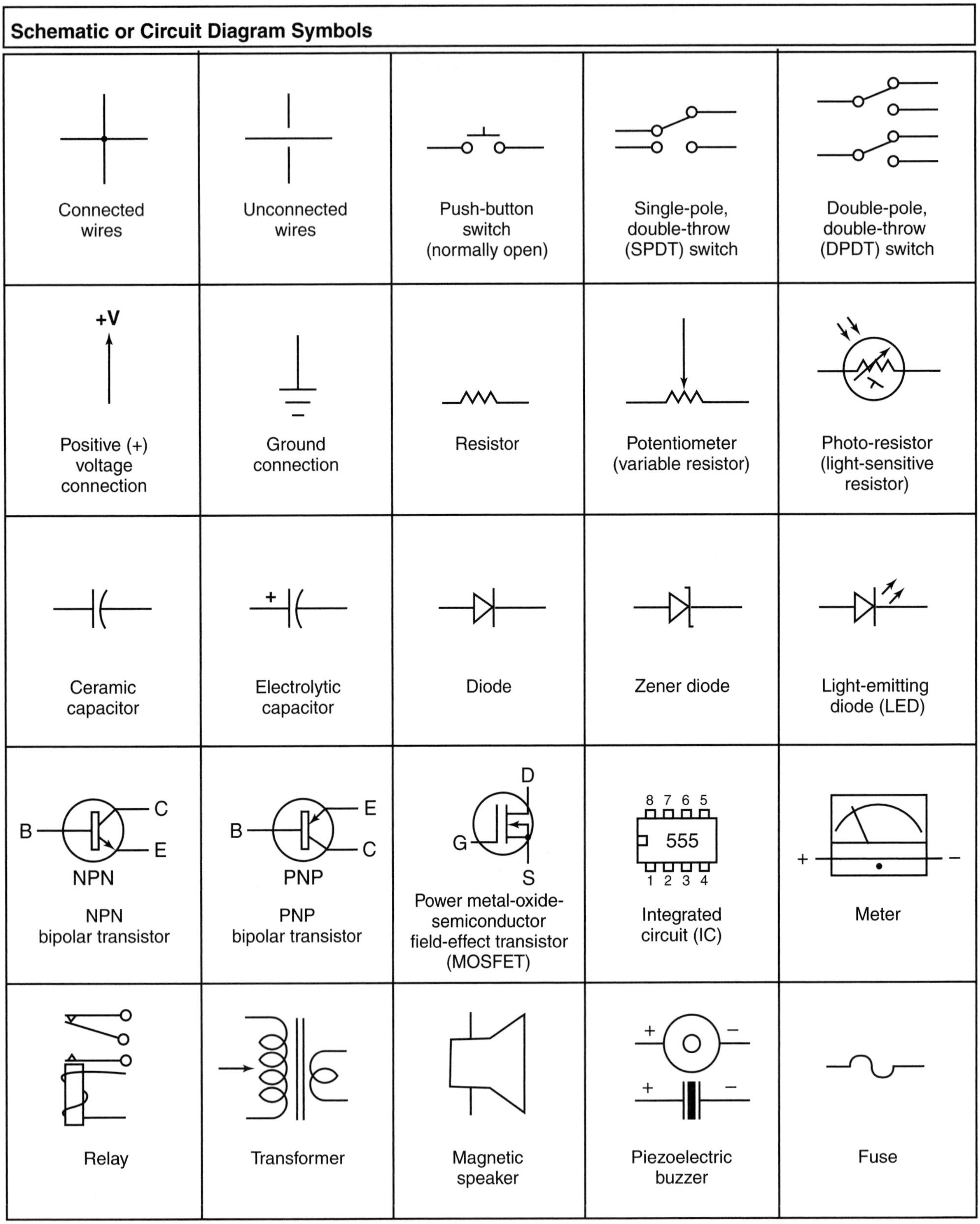

Schematic or Circuit Diagram Symbols				
Connected wires	Unconnected wires	Push-button switch (normally open)	Single-pole, double-throw (SPDT) switch	Double-pole, double-throw (DPDT) switch
+V Positive (+) voltage connection	Ground connection	Resistor	Potentiometer (variable resistor)	Photo-resistor (light-sensitive resistor)
Ceramic capacitor	+ Electrolytic capacitor	Diode	Zener diode	Light-emitting diode (LED)
B C E NPN NPN bipolar transistor	B E C PNP PNP bipolar transistor	D G S Power metal-oxide-semiconductor field-effect transistor (MOSFET)	8 7 6 5 555 1 2 3 4 Integrated circuit (IC)	+ − Meter
Relay	Transformer	Magnetic speaker	+ − + − Piezoelectric buzzer	Fuse

Figure 12-18. A schematic of a buzzer activated by a push button. Notice that a relay separates the alternating current (AC) control circuit from the direct current (DC) buzzer circuit. When the push button is pressed, the coil in the relay energizes, and the relay contacts close. This energizes the buzzer.

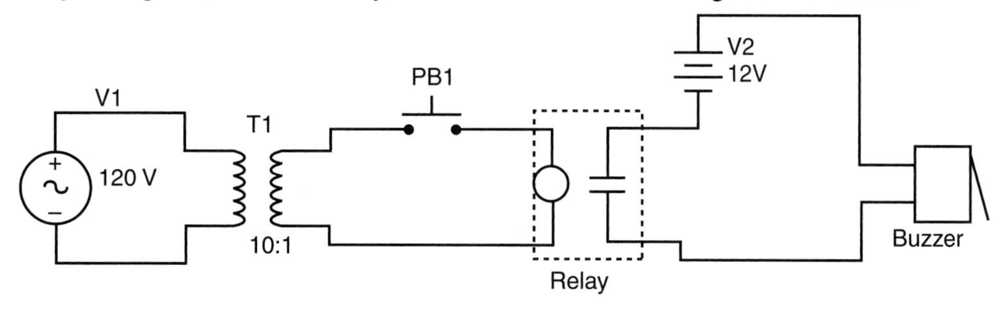

Lastly, it is important to perform troubleshooting in a systematic manner. This means checking only one component at a time, verifying it is functioning properly, and then moving on to the next component. See **Figure 12-19.**

Figure 12-19. An electronic circuit for troubleshooting.

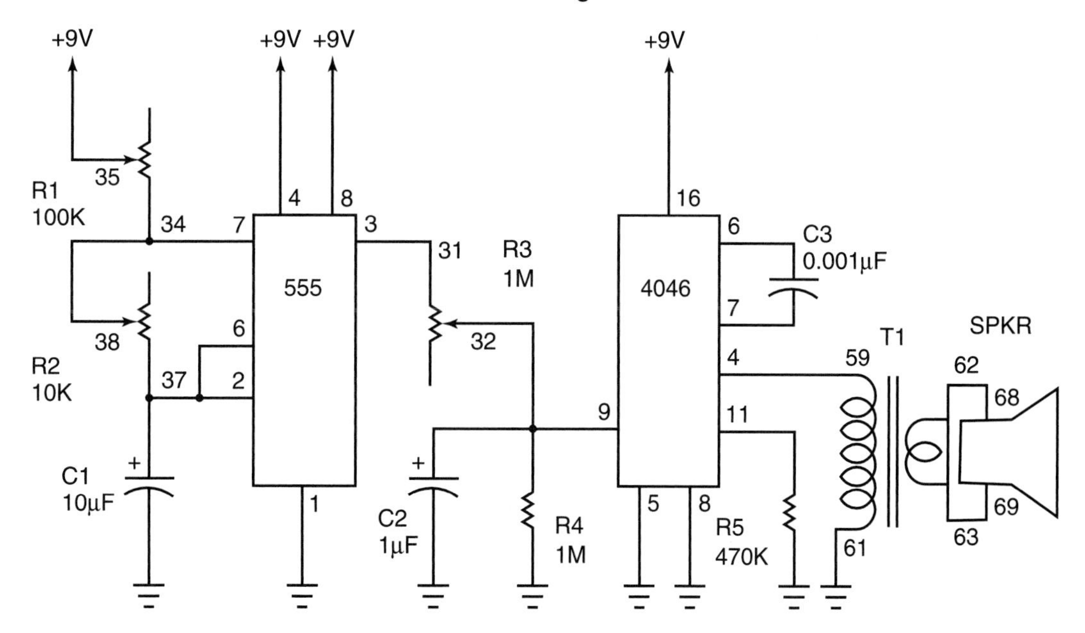

Career Connection

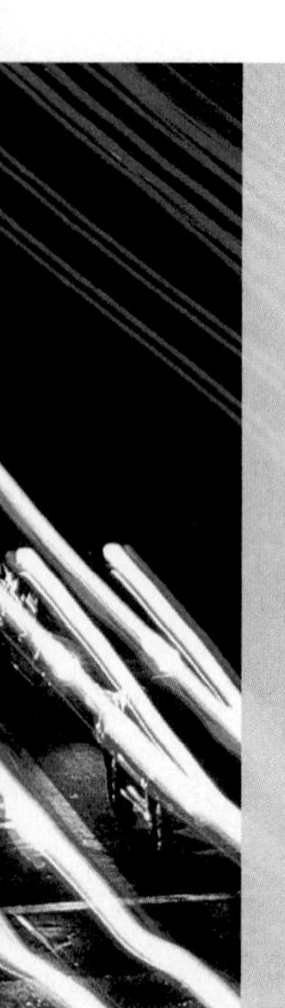

Power Plant Maintenance Electricians

Power plant maintenance electricians are responsible for the installation, inspection, repair, and testing of all major equipment associated with an electric power generating station. The equipment typically includes high-wattage generators and high-voltage transformers. This work also includes training and supervising coworkers and outside contractors during major power outages. Being an electrician requires a knowledge of electrical and electronic theory and practices, as well as a high degree of safety.

One of the most desirable aspects of this job is the opportunity to work with all types of equipment, both very new and very old. Electricians use their training to meet the challenges of the job. They also often work with a multitalented workforce.

This job, like any other, also has some disadvantages. One of the least desirable parts of this job is having to deal with the weather. On any given day, a maintenance electrician may be outside working in the freezing cold or blazing heat. Working inside a power plant is also very hot because of the generating equipment that gives off heat.

Entry-level technicians need to have at least an Associate Degree in Electronics or Electrical Systems. From there, a Journeyman Certification in Powerhouse Electricity would typically be pursued. This leads to a state licensure in many states. It is also beneficial to have a Bachelor of Science Degree in a field such as Industrial Technology. The estimated entry-level salary is about $40,000.

To troubleshoot this circuit, you might begin by setting your multimeter to DC volts and verifying that proper voltage is supplied to the circuit. Next, you might remove the voltage source and use the continuity function of the meter to verify that the fuse is not blown and that the switch works properly. From this point, you could use the ohms function of the meter to compare resistor readings to their calculated values. You could also check the diode with the continuity checker to ensure that it works in the proper bias and is installed correctly. It may also be most effective to check some ICs simply by switching them with another IC that is known to be functioning properly. This is an easy, but potentially expensive, method of troubleshooting.

Safety Note

Be sure to use antistatic precautions before touching or handling a CMOS IC. CMOS ICs are easily damaged by static electricity.

Summary

Electronic components drive our technological society. Some knowledge of how electronic components work is essential to technological literacy. Components such as resistors, capacitors, diodes, transistors, and integrated circuits (ICs) are used in combination to produce thousands of electronic devices. Schematic diagrams are used to visually represent how these components are connected to produce electronic circuits for practical use. Interpreting these diagrams requires knowledge of the various symbols used to represent the electronic components. Troubleshooting electronic circuitry requires knowledge of how electronic components are supposed to function, the ability to properly interpret schematics, the skill to test electronic components, and the ability to analyze a circuit in a systematic way.

Key Words

All the following words have been used in this chapter. Do you know their meanings?

astable application
bipolar transistor
capacitor
complementary metal oxide semiconductor (CMOS)
diode
electronic circuit
electronic device
field-effect transistor (FET)
forward bias
integrated circuit (IC)
monostable application
operational amplifier (op-amp)
pin-out
resistor
socket
solid-state
transistor-transistor logic (TTL)
wiper
zener diode

Test Your Knowledge

Write your answers on a separate sheet of paper. Do not write in this book.

1. On a resistor with four color bands, the third band represents:
   A. the tolerance.
   B. the multiplier.
   C. the last digit of the total resistance.
   D. the anode end of the resistor.
2. On a resistor with four color bands, the fourth band represents:
   A. the tolerance.
   B. the multiplier.
   C. the last digit of the total resistance.
   D. the anode end of the resistor.

3. If a resistor has a color code of "red-orange-brown-gold," it offers _____ ohms of resistance.
4. Devices that change resistance in relation to the position of a knob or slider are known as _____.
5. A(n) _____ can store electricity.
6. The unit of storage for capacitance is known as the:
   A. volt.
   B. ampere.
   C. farad.
   D. ohm.
7. A(n) _____ is a device that permits current to flow in only one direction.
8. The cathode end of a component is considered to be the _____ end.
9. Describe how to test a diode.
10. Describe the purpose of a rectifier circuit.
11. The cathode lead on a light-emitting diode (LED) can be identified by:
    A. a color band.
    B. a flat spot at the base of the LED.
    C. the shorter lead from the LED.
    D. Both *B* and *C*.
12. List the letters and numbers that can be displayed using one seven-segment LED.
13. Transistors can perform a(n) _____ function without any physical moving parts.
14. A MOSFET is a type of:
    A. integrated circuit (IC).
    B. diode.
    C. transistor.
    D. capacitor.
15. *True or False?* An IC can be used in a circuit to replace hundreds or thousands of individual components, such as resistors and transistors.
16. Summarize the two types of ICs and provide the advantages and disadvantages of each.
17. One disadvantage of complementary metal oxide semiconductor (CMOS) ICs is they can be easily damaged by _____.

18. This IC is among the most widely used of all ICs. It is used for single timing functions:
    A. 4046.
    B. 7447.
    C. 555.
    D. 556.

19. Devices that protect electronic circuitry from excessive amperage include _____ and _____.

Matching questions: For Questions 20 through 27, match the schematic symbols on the right with the correct term on the left.

20. _____ Resistor.
21. _____ Potentiometer.
22. _____ Capacitor.
23. _____ Diode.
24. _____ LED.
25. _____ Transistor.
26. _____ IC chip.
27. _____ Fuse.

A.

B. B C E NPN

C. 8 7 6 5 555 1 2 3 4

D.

E.

F.

G.

H.

28. If you wanted to build the following circuit, what parts would you need?

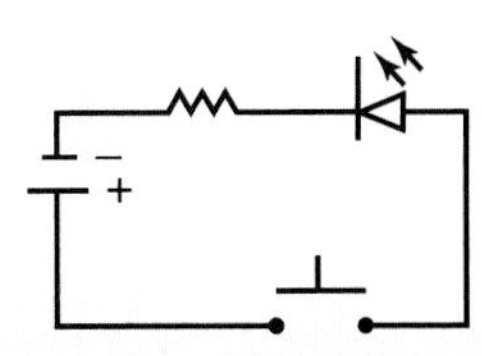

29. Devices such as buzzers and relays, which actually use electricity, are known as _____.

30. Discuss the process for systematically troubleshooting electronic circuitry.

Activities

1. Interpret the color codes on various resistors, and then calculate their resistive values. Next, measure their resistance, using an ohmmeter. Lastly, compare the calculated values to the measured values to determine if all the resistors are within tolerance.
2. Using the continuity function on a multimeter, test a diode to determine the forward bias and the reverse bias.
3. Design a simple electronic circuit to perform a practical use. For instance, you might design a photosensor to turn on a light when it is dark in the room. Next, construct the circuit using a solderless breadboard and various electronic components to see if it works. If it does not, the solderless breadboard will allow you to quickly change components. Each component can be tested to ensure it is in working order, and new components can be substituted to try to get the circuit to work as you had intended.

Basic Concepts

- List the types of energy and power conversions that can occur.
- Identify devices used to convert various forms of energy or power into other forms of energy or power.

Intermediate Concepts

- Describe the operation of devices used to convert various forms of energy or power into other forms of energy or power.

Advanced Concepts

- Explain new conversion devices not yet routinely used.
- Calculate the efficiency of various energy and power conversion devices.

A converter is simply a device that allows for the changing of one form of energy or power into another form of energy or power. The lightbulbs in your classroom convert electrical power into a form of ***radiant energy***, or energy in the form of electromagnetic waves, known as visible light. The engine in your car converts the chemical energy stored in gasoline into fluid power and then into mechanical power. Some devices perform a conversion in one step and are known as ***direct converters***. Other devices are considered ***indirect converters*** because it takes several intermediate steps for them to perform conversions. This chapter will examine some of the most commonly used converters and explain how they work.

Types of Conversions

There are four basic types of conversions involving power and energy:

- Power conversion (power to power).
- Energy conversion (energy to power).
- Frequency conversion (energy to energy).
- Energy inversion (power to energy).

Radiant energy: Energy transferred by radiation, including infrared rays, visible light rays, ultraviolet (UV) rays, X rays, radio waves, and gamma rays.

Direct converter: A device that changes one form of energy or power into another form of energy or power in one step.

Indirect converter: A device that changes one form of energy or power into another form of energy or power using several intermediate steps.

Power converter: A conversion device used to change one form of power to another.

Energy converter: A device that changes a form of energy into a useful form of power.

Power converters are used to change one form of power to another and are the most common conversion devices. Because it is so widely used, an electric motor is probably the best example of a power converter. The motor converts one form of power—electricity—into another form of power—mechanical power.

Another type of conversion involves changing a form of energy into a useful form of power. Devices that do this are known as ***energy converters***. A good example of an energy converter is the photovoltaic cell. It converts visible light energy directly into electricity.

Frequency converters change one frequency (wavelength) of radiant energy directly into another frequency of radiant energy. One example is a solar collector. It converts visible light energy into infrared energy.

Sometimes, it is useful to invert a form of power back into a form of energy. One example of an ***energy inverter*** is the electric space heater. It converts electric power into infrared energy.

Power Conversions

The three forms of power are electrical, fluid, and mechanical power. Power conversions involve changing one of these forms of power into another. See **Figure 13-1.**

Electrical to mechanical

Devices that convert electricity into mechanical movement have many practical applications. One of the most widely used electrical-to-mechanical power converters is the ***solenoid***. See **Figure 13-2.**

The solenoid consists of a coil of wire with a moveable steel core in the center. When the coil is energized with electricity, it creates a magnetic field and causes linear movement of the steel core. This linear movement can perform many useful purposes. For instance, the moving core could strike a chime or bell and create sound. This is the way many doorbells work. The solenoid's core movement could also be used to trigger a release mechanism. When you push a button at an apartment building for someone to open the entry door and let you in, the button he or she pushes energizes a coil that pulls on a latch, allowing the door to open.

The most common application of this type of electrical energy to linear mechanical movement is to control a relay. A ***relay*** is an electromechanical switching device. When the coil of a relay is energized, the steel core is pushed in one direction. This pushing force is used to open and close switch contacts. Why not just use a manually operated switch to open and close the contacts? Consider this situation: you want to be able to start a conveyor system from one location, but be able to stop it from several locations. Assuming the conveyor is driven by a powerful electric motor, the use of a relay would be more desirable than using manual switches. This is because a relay can be controlled with very little electricity. If you wished to control a powerful motor from many locations by using switches, you would have to

Figure 13-1. The types of power conversions that can occur.

Types of Power Conversions
Electrical ⇨ Mechanical
Mechanical ⇨ Electrical
Mechanical ⇨ Fluid
Fluid ⇨ Mechanical
Fluid ⇨ Electrical
Electrical ⇨ Fluid

Figure 13-2. The operation of a solenoid. A—When no electric current is applied to the solenoid coil, the plunger is at rest, and the switch remains open. B—When the solenoid is energized (current is applied to the coil), the plunger is pulled upward by magnetic force, closing the switch.

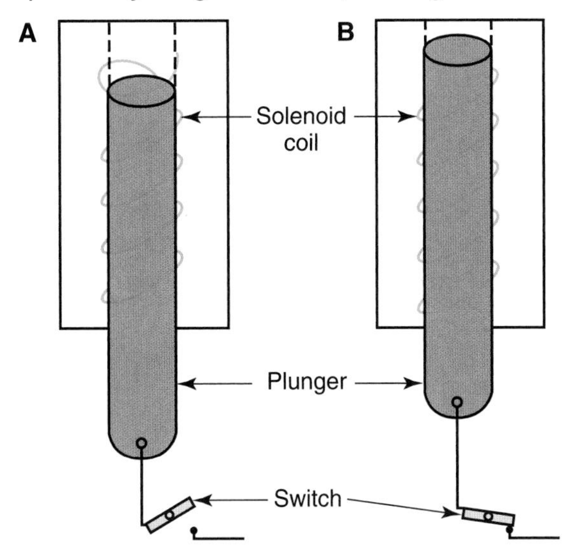

Frequency converter: A device that changes one frequency of radiant energy directly into another frequency of radiant energy.

Energy inverter: A device that inverts a form of power back into a form of energy.

Solenoid: A device that converts electricity into mechanical movement.

Relay: An electromechanical switching device.

run the 120-volt electric current that powers the motor through all those switches before it goes to the motor.

Relays are typically identified by the number of contacts they provide, the number of positions they can offer, and the current ratings for their contacts. For instance, the relay in **Figure 13-3** provides two sets of contacts, so it is known as a two-pole relay. Additionally, this relay offers a normally open (NO) and a normally closed (NC) contact for each pole. The relay can permit electrical flow through the NO contacts when the coil is energized and closes the contacts. In the same way, it can stop electrical flow through the NC contacts when the coil is energized and cause the contacts to open.

Figure 13-3. A two-pole relay that offers both normally open (NO) and normally closed (NC) contacts.

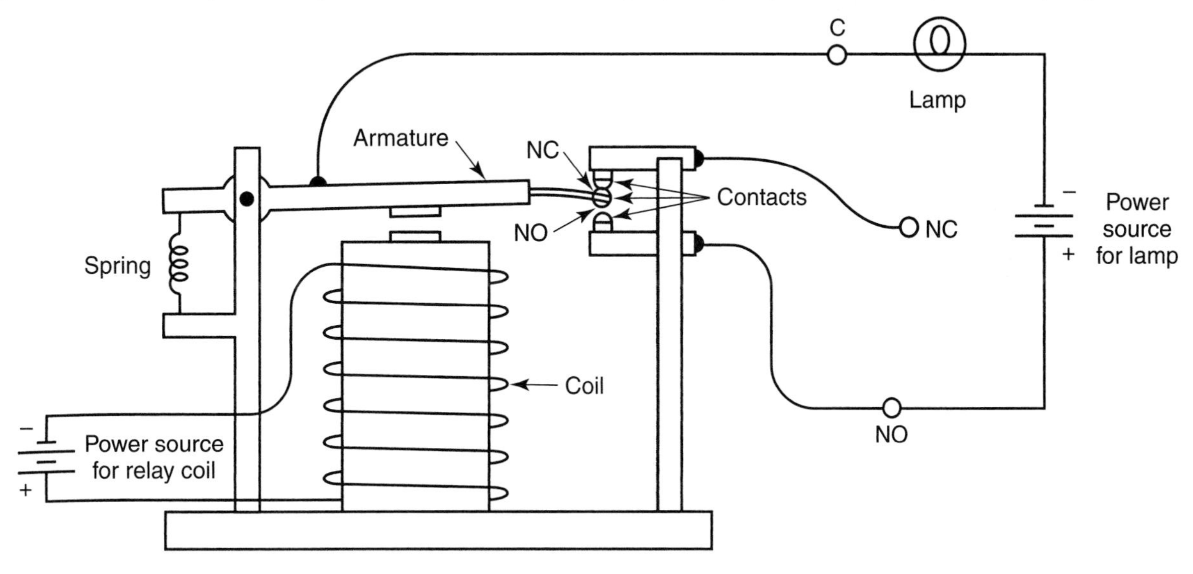

Induced: Made to flow.

Armature: A series of wires around a metal core in which electricity is induced, which is free to rotate.

Brush contactor: The part of a direct current (DC) motor that supplies electricity to the armature.

Commutator: A circular conductive strip connected to the end of an armature loop.

Slip ring: See *Commutator*.

One common relay used for a very different purpose is the solenoid on the starter motor of a car. When you turn the key to start the car, a simple 12-volt control circuit operates the relay, closing heavy contacts that allow electricity to flow from the battery to the powerful starter motor. When you release the key, the coil of the solenoid is no longer energized. A spring attached to the solenoid's steel core pulls the core backward. This opens the contacts that feed the starter motor. Imagine how large and heavy-duty the ignition switch would have to be if all the electricity that flows to the starter motor had to pass through that switch.

Solenoids are also frequently used to open and close valves. This allows for remote control of valves, since the operator does not need to be near the valve to push a lever. Many valves can be controlled from a central location.

When we think of electric motors, we usually do not think of solenoids and relays, but rather of motors that can provide rotary power. See **Figure 13-4.** More than 70% of all electricity generated nationwide is used to operate electric motors. There are many types of motors, but they work on the same basic principles. Simple motor construction begins with the use of magnets that provide a north and south magnetic pole. If a wire is placed between the north and south poles, a current will be ***induced*** (made to flow) in the wire. When current flows in the wire, another magnetic field is generated that spirals around the wire. In some places, the two magnetic fields aid each other, and in other places, they oppose each other. This constant aiding and opposing of magnetic fields creates movement of the wire. If the electricity flowing through the wire is reversed or if the poles of the stationary magnets are quickly reversed, the wire in the center will be continuously pulled and pushed so as to sustain movement.

Figure 13-4. Electrical motors are widely used to provide rotary power. This powerful motor turns the blades of a large industrial fan.

A simple direct current (DC) motor is depicted in **Figure 13-5.** DC motors and generators are almost identical in construction. In fact, if mechanical power is supplied to the rotating shaft of many DC motors, they will act as generators and produce electricity. The most basic motor consists of permanent magnets, mounted with attracting poles facing one another. In between the poles is a single loop ***armature***, a series of wires around a metal core in which electricity is induced, that is free to rotate. An actual working motor consists of many armature loops.

This motor uses ***brush contactors*** to supply electricity to the armature. The brush contactors slide along the ***commutator***, a circular conductive strip connected to the end of the armature loop. A commutator is sometimes known as a ***slip ring*** because of its design and the function it serves. In Figure 13-5, you can see how the commutator reverses the flow of electricity in the armature every half-turn of the coil (in larger motors, this can be done more frequently). When the armature loop is totally vertical, it is being neither attracted nor repelled by magnetic forces. The inertia of the motor in motion, however, carries the armature past the neutral point, where it is again influenced by the magnetic forces.

More powerful motors, like alternating current (AC) induction motors, replace the permanent magnets with electromagnets. The electromagnets are capable of producing much stronger magnetic fields than permanent magnets, and stronger fields result in a more powerful motor. The two primary parts of an AC motor are the rotor and the stator.

Stator windings are so named because they remain stationary and do not rotate. The ***stator*** is typically connected to the housing of the motor and the motor frame. The ***rotor*** is the spinning coil of the motor that is connected to the motor shaft. An induction motor does not require that electricity be provided to the rotor. The motor is built so that, as the stator winding is energized, it induces electricity in the rotor. This causes the rotor to spin. Single-phase induction motors require a special set of ***start windings***. The start windings draw more current than the ***run windings***, which power the motor once it is up to speed. They are needed because a ***single-phase induction motor*** only has one set of run windings. If the rotor is not spinning, the run windings do not create enough magnetic attraction and repulsion to move the rotor past the neutral point of rotation. Once the motor builds up speed, a centrifugal switch opens, taking the start windings out of the circuit. At full operating speed, the inertia of the motor easily carries the rotor past the neutral point, and it can maintain speed. The start windings and their heavy current draw are no longer needed.

Three-phase induction motors, sometimes called *polyphase motors*, are used for many industrial applications. These motors are constructed using three sets of stator (run) windings, instead of one set. See **Figure 13-6.** The windings are 120° out of phase with one another. The result is that this type of motor does not require separate start windings, since (unlike a single-phase motor) the neutral points are almost nonexistent. This is because if one stator winding is in the neutral position, there are two other windings engaged. The three magnetic fields working in attraction and

Figure 13-5. Components of a simple direct current (DC) motor.

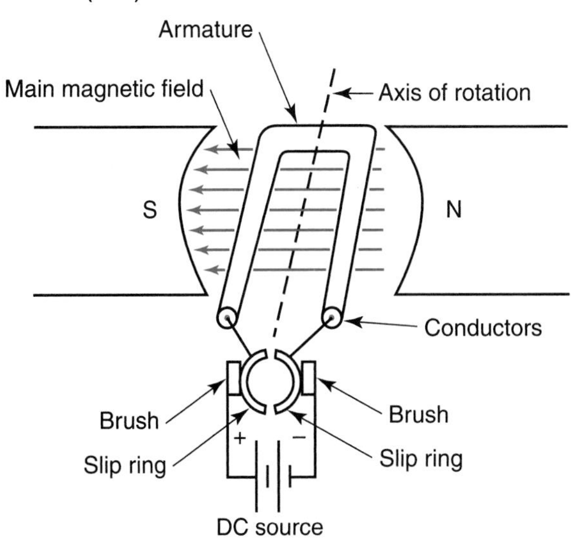

Stator: The part of an alternating current (AC) motor that remains stationary and does not rotate. It is typically connected to the housing of the motor and the motor frame.

Figure 13-6. The stator windings in a three-phase motor are in pairs and are arranged 120° out of phase with each other. No separate start winding is needed.

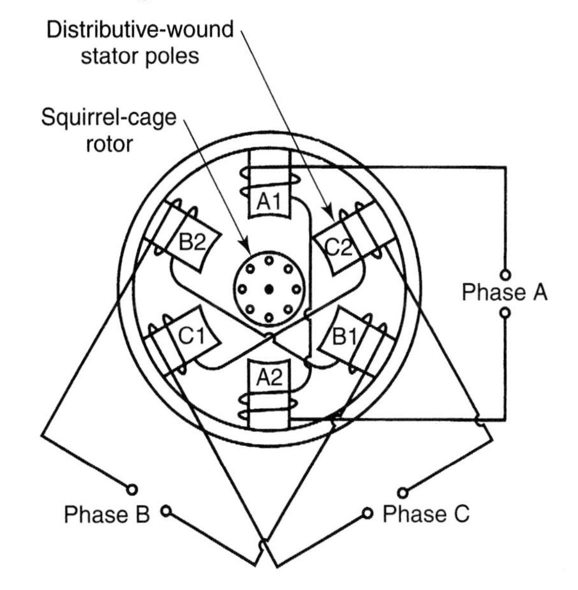

Rotor: The spinning coil of an alternating current (AC) motor that is connected to the motor shaft.

Start winding: A special winding required by a single-phase induction motor because its single set of run windings do not create enough magnetic attraction and repulsion to move the rotor past the neutral point of rotation.

Run winding: A winding that powers a motor once it is up to speed.

Single-phase induction motor: A type of alternating current (AC) motor that only has one set of run windings.

Three-phase induction motor: An alternating current (AC) motor constructed using three sets of stator windings.

opposition to one another also produce much more power than a single-phase motor of comparable physical size.

Mechanical to electrical

A generator is the principal means of producing electricity from rotary mechanical power. The basic theory behind a generator involves moving a wire through a magnetic field. When this happens, a voltage is induced in the wire. See **Figure 13-7.**

If the wire is moved parallel to the magnetic lines of flux, no current or voltage will be generated. A current can be measured if the wire cuts through the magnetic lines of flux. Depending on the direction in which the lines are cut, the current in the wire will flow one way or the other. When magnetism is converted to electricity by this process, the electricity is known as *induced voltage*. Increasing the speed of the wire movement or the magnetic field will produce a stronger induced voltage. In an actual generator, powerful electromagnets are used to produce a tremendous magnetic field, and the armature spins at several thousand revolutions per minute (rpm). The armature may be powered by a steam turbine, a hydro-turbine, or even (in the case of a portable generator) a small gas engine. See **Figure 13-8.**

Mechanical to fluid

A pump is used to convert mechanical power to fluid power. Gear and centrifugal pumps are the most common. Reciprocating piston pumps are also sometimes used to pressurize fluids. **Figure 13-9** shows how mechanical power is converted to fluid power in a rotary vane pump. This pump has a circular chamber with a rotor positioned slightly off center. The vanes, or pump paddles, are able to slide in and out of slots in the rotor. As the pump spins, centrifugal force pulls the vanes outward to press against the walls of the pump housing. Since the rotor is slightly off center, the volume of space between each pair of adjoining vanes varies. Fluid enters the pump at a point where a compartment (space between vanes) is at its greatest capacity. As the pump rotates, the compartment becomes progressively smaller, increasing pressure on the fluid until it is forced out of the pump outlet port.

Figure 13-7. When a simple wire armature cuts through the magnetic lines of flux, an electric current is induced in the wire.

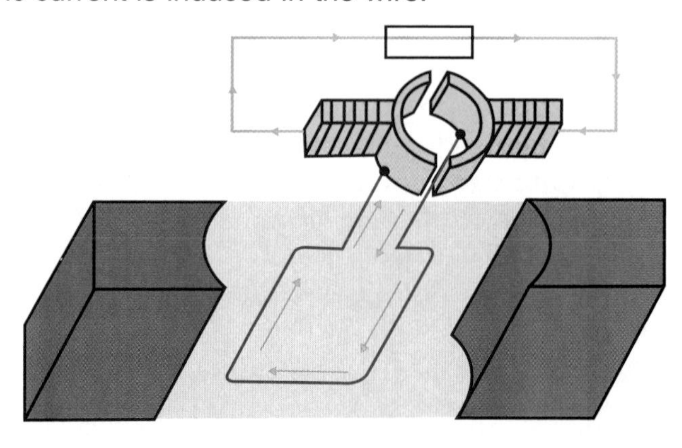

Figure 13-8. A small gasoline engine can power a portable generator. This portable generator can be used on construction projects where electricity is not yet available or for equipment repair in areas far from electrical lines. (Miller Electric Manufacturing Company)

Figure 13-9. A rotary vane pump converts mechanical power into fluid power. As the pump rotates, the compartments between vanes become smaller, pressurizing the liquid, until it is expelled through the outlet port.

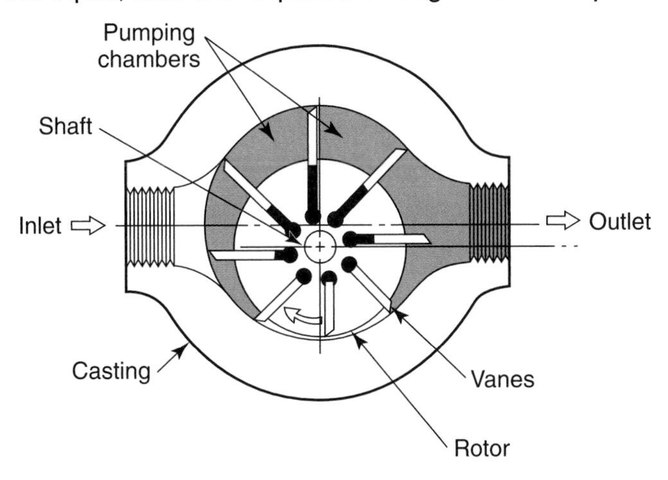

Fluid to mechanical

Fluid power is converted back to mechanical power with the use of cylinders, known as *actuators*, that permit linear mechanical movement. See **Figure 13-10.** Fluid motors are also used for converting fluid power into mechanical power. The vane motor in **Figure 13-11** is the reverse of the vane pump described in the preceding paragraph. Fluid enters the inlet port under high pressure and propels the vanes or pump paddles around until it exits the outlet port. The rotor that holds the vanes is coupled to a rotating shaft that provides mechanical power to the load.

Figure 13-10. Linear actuators are a type of hydraulic cylinder. A—Several examples of actuators. (Bimba Manufacturing Company) B—Parts of a typical actuator.

A

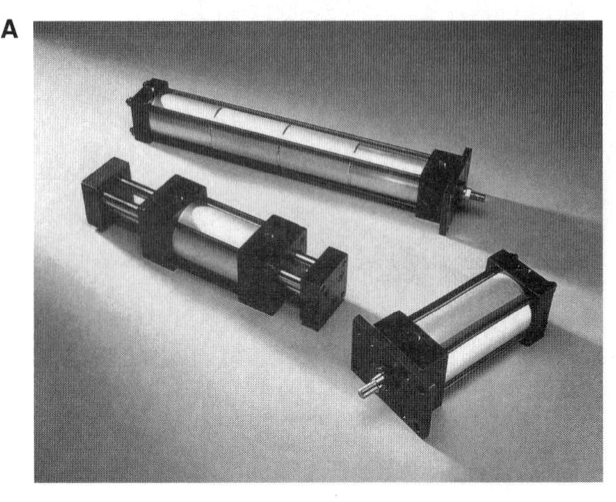

B

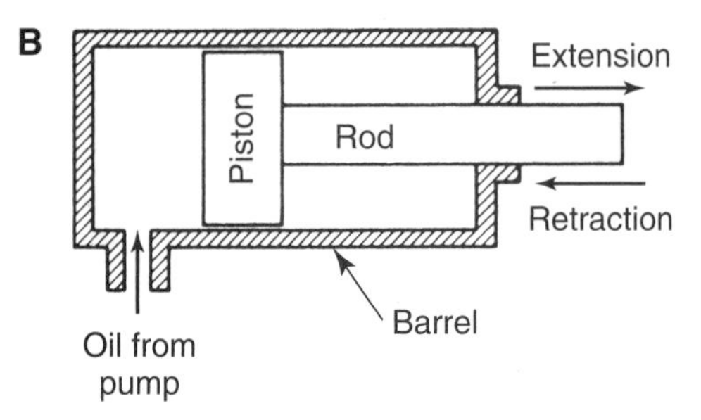
Extension
Piston
Rod
Retraction
Barrel
Oil from pump

Fluid to electrical

Converting electricity directly into fluid motion or fluid motion directly back into electricity can be done, but these conversions have far fewer applications than the other power conversions previously described. A ***magnetohydrodynamic (MHD) generator*** is an advanced system for generating electricity from fossil fuels. See **Figure 13-12.** While it is still in the research stage, the appeal of an MHD generator is that it operates at a much higher efficiency than a conventional generating plant. An MHD generator is considered a direct conversion device because it produces electricity directly from the heat source. It does not require a steam generator to spin a turbine, as is the case in conventional power plants. In the MHD process of power generation, fuels are burned at very high temperatures. This permits greater efficiency and causes less pollution. In the MHD process, a very hot ionized gas takes the place of the copper windings of conventional electric generators. Gases from the high temperature ***combustion***, or burning, of fossil fuels are made electrically conductive by seeding them with conductive chemicals. The hot gases then travel at high speed through a magnetic field to produce a DC. Waste heat can be used to produce steam for a conventional turbine generator. This helps to increase the total efficiency of the generating system to one and one-half times that of a conventional power plant.

Magnetohydrodynamic (MHD) generator: An advanced, highly efficient direct conversion device that generates electricity from fossil fuels.

Combustion: The process of burning.

Figure 13-11. A fluid vane motor is similar in construction to a vane pump, but it operates in the opposite manner. Instead of the rotary motion of the device moving the fluid, the fluid movement generates rotary motion.

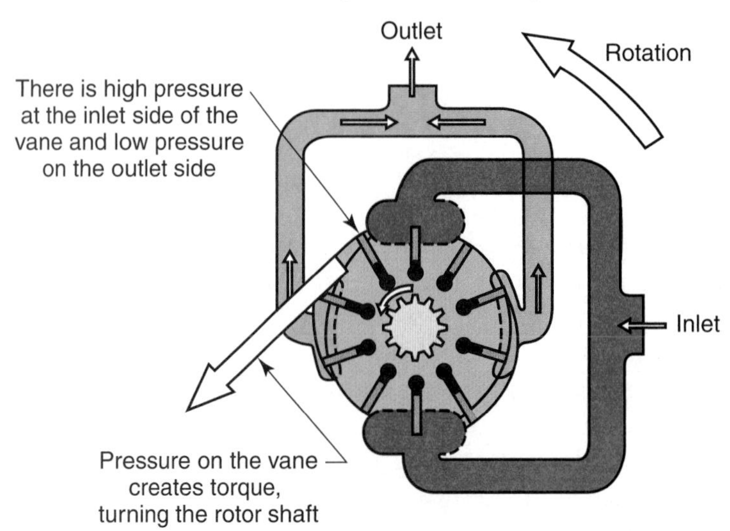
Outlet
Rotation
There is high pressure at the inlet side of the vane and low pressure on the outlet side
Inlet
Pressure on the vane creates torque, turning the rotor shaft

Figure 13-12. In a magnetohydrodynamic (MHD) generator, an electrically conductive heated gas flows inside an insulated duct through an induced magnetic field. As the gas moves through the magnetic field, an electric current is induced and collected by a pair of electrodes on opposite sides of the duct.

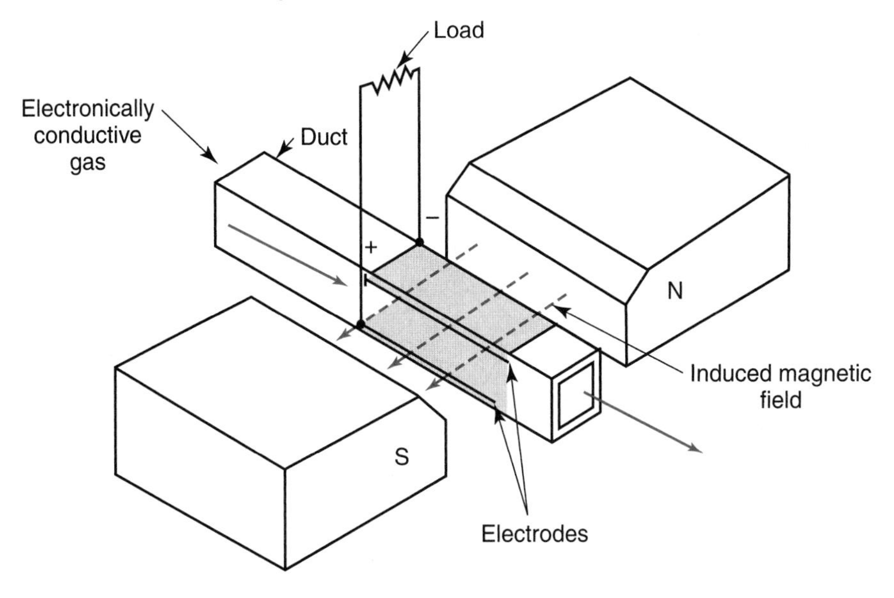

Electrical to fluid

An electromagnetic induction pump without any moving parts can propel a liquid, as long as the liquid can be polarized. Liquid metals, such as mercury, can be easily polarized. The essential principle of the electromagnetic induction pump can be visualized by moving mercury around on a tabletop using a permanent magnet.

Energy Conversions

Energy converters are responsible for changing a form of energy into a form of power. Some practical examples of energy converters include the photovoltaic cell, which converts sunlight directly into electricity, and the internal combustion engine, which converts chemical energy into heat, then into fluid power in the form of expanding gases, and ultimately into mechanical power. Some energy converters are direct converters. Others are indirect converters.

Visible light energy to electrical power

Photovoltaic cells have the ability to convert sunlight directly into electricity. An array of cells can be used to produce enough electricity to power a load. See **Figure 13-13.** Photoelectric "eyes" (sensors) also make use of the photovoltaic effect. In these devices, photons striking the sensor generate a small current. This current triggers a ***transistor*** (solid-state switching device) to open or close a circuit. In the case of a photoelectric eye used for streetlights or security floodlights, the lights will remain off as long as the eye is generating enough current to influence or bias the transistor. When the sun sets or light levels fall to a selected point, the current weakens, and the transistor switches to turn the lights on. The lights will remain on until the morning light becomes strong enough to produce enough current to bias the transistor again.

Transistor: A solid-state switching device.

Figure 13-13. Photovoltaic cells convert light energy from the sun into electricity. This view shows part of a large solar array made up of many thousands of photovoltaic cells.

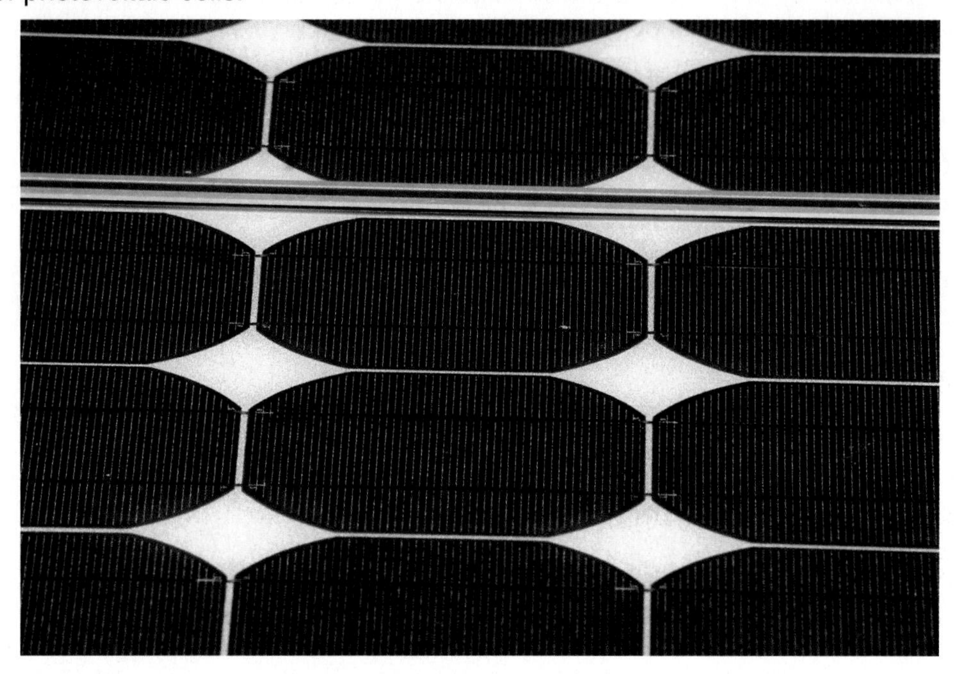

Infrared energy to mechanical power

Bimetallic coil: The part of a thermostat formed from strips of two metals with differing coefficients of expansion.

Coefficient of expansion: The rate at which a metal expands and contracts as temperature rises or falls.

A thermostat is a device that can detect temperature and convert changes in temperature into mechanical movement. This movement is often used to open or close electrical switch contacts to turn a furnace or air conditioner on or off. At the heart of the thermostat is a ***bimetallic coil***. The coil is formed from strips of two metals with differing ***coefficients of expansion***. This means that the metals will expand and contract at differing rates as the temperature rises or falls. This expansion and contraction can be calibrated. If a needle is attached to the bimetallic coil, a thermometer is created. This will indicate temperature, but that is all. To make the bimetallic coil perform switching functions, a little glass bulb of mercury is attached to it. Mercury is a highly conductive liquid metal. The mercury bulb sits between two wires and acts as a switch. If the temperature decreases, the bimetallic coil will contract (coil more tightly). This movement will cause the ball of mercury to move to one end of the glass bulb, closing switch contacts that send a signal to the furnace calling for heat. As the room temperature rises, the coil expands (coils more loosely) and moves the mercury bulb the other way, opening the switch contacts and causing the furnace to shut down. See **Figure 13-14.**

Sound waves to electrical power

Sound: Vibration traveling through matter in the form of a wave motion.

Sound is vibration traveling through matter in the form of a wave motion. The vibrations we hear as sound are movements in a gas, liquid, or solid. Sound can travel through any form of matter, but it cannot travel though a vacuum because it would not have any medium in which to generate a wave motion. Air is perhaps the most important medium through which sound travels, since it brings the sound to our ears. It is

Figure 13-14. Operation of a thermostat in a home heating system. A—While temperature in the room remains above the set point, the bimetallic coil holds the bulb in the position shown. The mercury in the tube remains at the end opposite the electrodes. The switch used to operate the furnace remains open. B—When the room temperature drops below the set point, the bimetallic coil moves the bulb, allowing the mercury to flow to the electrode end. Since mercury is a conductive material, it completes a circuit with the electrodes and closes a switch that operates the furnace.

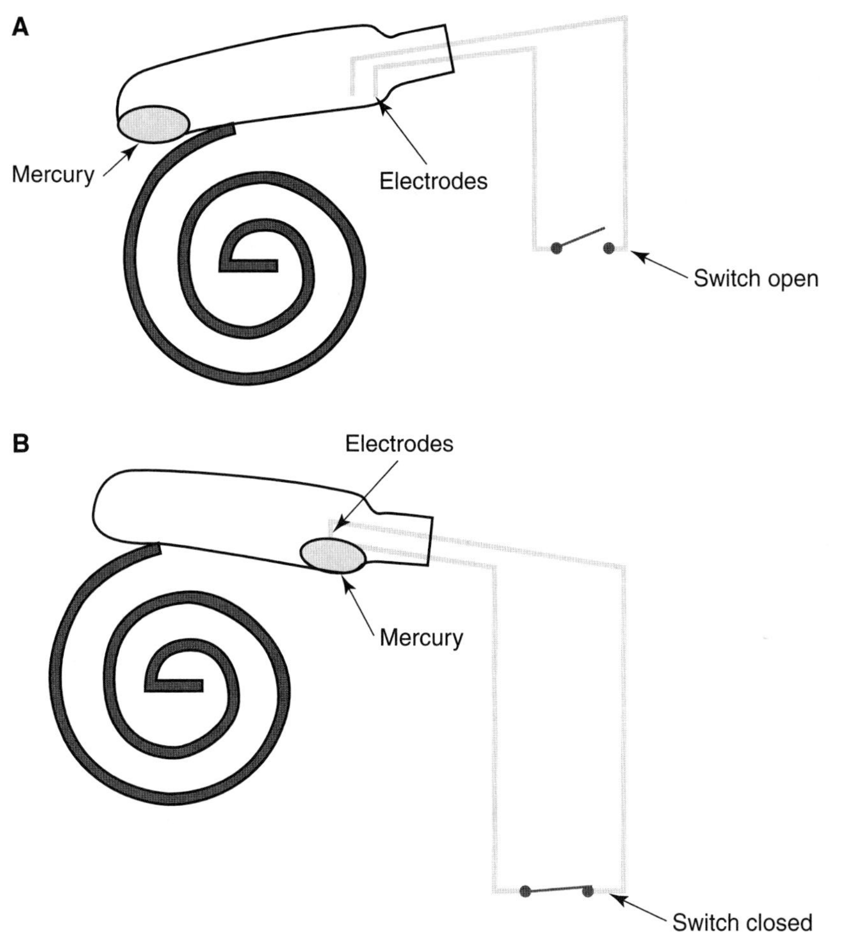

not, however, the fastest medium for sound travel. Sound travels much more quickly through water, for example. See **Figure 13-15.** The speed of sound can be calculated using the velocity formula:

$$\text{velocity} = \frac{\text{distance}}{\text{time}}$$

It is important to remember there are variables that affect the actual speed at which sound is transmitted.

The speed sound can travel through substances such as steel may appear impressive in comparison to sound traveling through air. Steel is, however, not a very useful means of transmitting sound over distance. Assuming 100% efficiency, a conversation transmitted through a solid piece of steel could only travel a little more than 3 miles per second. This means, if grandma lives about 35 miles away, you could expect a 10-second delay when speaking to her by telephone. This delay would

Figure 13-15. The speed of sound through air, water, and steel.

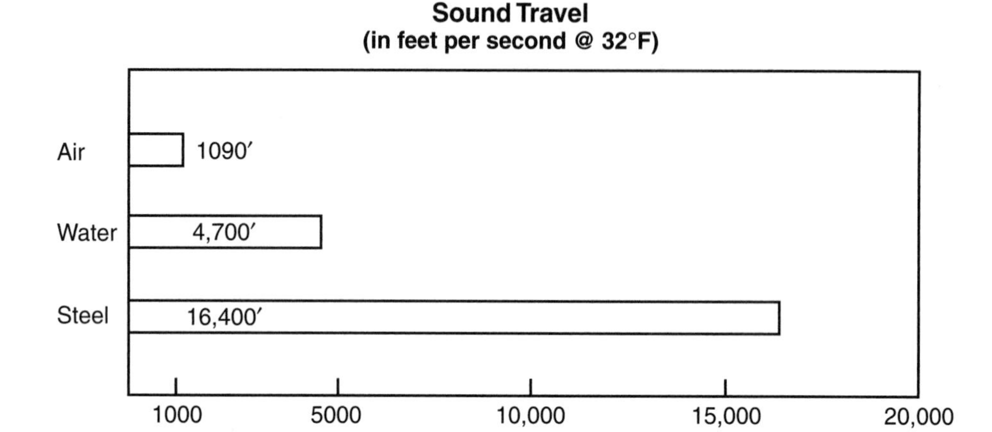

make it quite hard to carry on a conversation. Some means for speeding up sound transmission would be needed. This can be done if the sound waves are converted into a form of energy that travels much faster than sound. Both visible light and electricity can travel at 186,000 miles per second, approximately more than 10 times faster than the speed of sound through steel.

The simplest explanation for how sound is converted to electricity or light is the operation of a microphone. A microphone is a device that converts mechanical movement (created by changes in air pressure caused by sound waves) into electrical impulses of equivalent strength and duration. There are many types of microphones, but all of them convert the mechanical vibrations into equivalent electrical waves that can travel much faster than sound itself.

Curricular Connection

Science: The Speed of Light and Sound

During a thunderstorm, the lightning and sound of thunder actually occur close together. Yet, lightning is always seen before the thunder can be heard. If you are close to the location of the strike, you will hear the thunder before others who are farther away. You can tell how far you are from the actual strike by the amount of time between seeing the flash of lightning and hearing the thunder. This delay is caused by the difference between the speed of light and the speed of sound.

Sound and light both travel in wave form. While air is the fastest medium for light waves, the same is not true of sound waves. Sound travels more quickly through materials with more density, such as water. The speed of sound traveling through air is about 0.2 miles per second, but the speed of light traveling through air is 186,000 miles per second. When lightning strikes, therefore, it can be seen before the thunder can be heard.

One common type of microphone is a ***dynamic coil microphone***. See **Figure 13-16.** Like most microphones, the dynamic microphone has a ***diaphragm***—a thin membrane that will receive the sound waves in the air and vibrate accordingly. If you shout, it will vibrate vigorously. It will hardly vibrate at all, if you whisper. A ***voice coil*** is a wire attached to the diaphragm that moves back and forth as the diaphragm vibrates. It is surrounded by a magnetic field. As vibrations move the voice coil back and forth in the magnetic field, voltages are induced in the voice coil. These voltages are an electrical reproduction of the sound waves that struck the diaphragm. The electrical waves are then transmitted to a speaker that is the reverse of the microphone, converting the electrical waves back into sound with the use of a voice coil and diaphragm. If the electrical waves must travel a long distance or if the sound needs to be significantly increased, they may be amplified. An ***amplifier*** receives the electrical waves and increases the power of the signal, so it has greater strength, before sending it on to the speaker. The telephone system uses a specialized amplifier known as a ***repeater***, which receives electronic communication signals and sends out corresponding amplified signals.

Dynamic coil microphone: A common type of microphone with a voice coil attached to the diaphragm.

Diaphragm: The part of a dynamic microphone that is a thin membrane designed to receive sound waves in the air and vibrate accordingly.

Voice coil: The wire in a dynamic coil microphone that is attached to the diaphragm and moves back and forth as the diaphragm vibrates.

Amplifier: A device that receives electrical waves and increases the power of the signal, so it has greater strength, before sending it on to the speaker.

Repeater: A specialized amplifier used in a telephone system. It receives electronic communication signals and sends out corresponding amplified signals.

Chemical energy to mechanical power

The internal combustion engine is a good example of an energy converter. It could not, however, be considered a direct converter by any means. The engine consumes gasoline, a form of chemical energy with plenty of potential in it, and converts it first to heat and then to fluid power in the form of expanding gases. Finally, the expanding gases are converted into mechanical power through the piston, connecting rod, and crankshaft assembly. This mechanical power can be used to perform useful work. See **Figure 13-17.**

Frequency Conversions

Frequency conversions occur when one ***frequency*** (wavelength) of radiant energy is directly converted to another frequency of radiant energy. The electromagnetic spectrum is made up of radiant energies with varying frequencies, including infrared rays, visible light rays, UV rays, X rays, radio waves, and even gamma rays. The term *frequency* describes the number of cycles in a given time interval, such as one second. The number of ***cycles*** per second can be measured. It is usually measured in a unit known as ***Hertz***. A complete ***wavelength*** from start to finish is one Hertz.

Frequency: The number of cycles in a given time interval.

Cycle: One complete performance of a periodic process.

Figure 13-16. The components of a dynamic microphone.

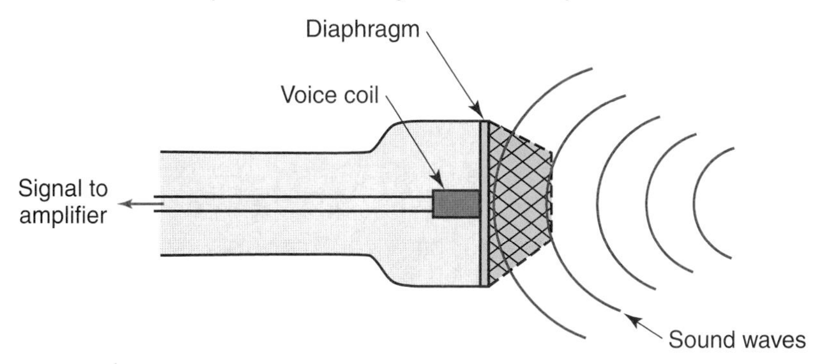

Figure 13-17. An internal combustion engine converts chemical energy from gasoline into mechanical power. Gasoline vapor is drawn into the engine cylinder (1). It is compressed and then changed to heat by combustion (2). Fluid power of the expanding gases drives the piston downward. The piston, connecting rod, and crankshaft assembly transmit mechanical motion (3).

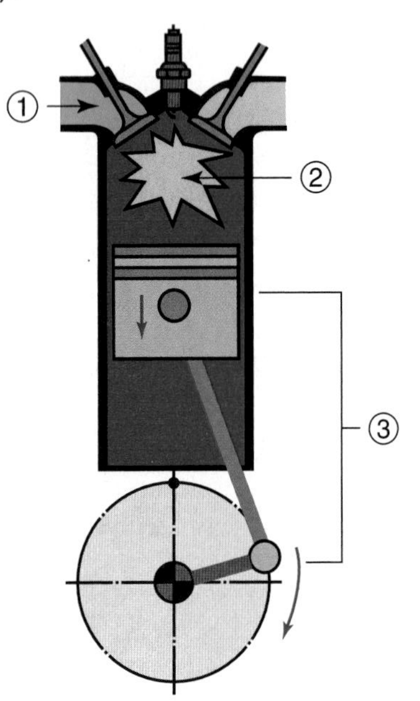

See **Figure 13-18.** For example, AC electricity has a wavelength of 60 cycles per second, which is also referred to as 60 Hertz.

Radiant energy to chemical energy

Radiant energy is converted into chemical energy all the time. The type of change is a purely scientific conversion, not a technological conversion. In other words, this conversion occurs naturally and is not the result of a human-made product. The process known as photosynthesis is essential to life. Without it, all the hydrocarbon fuels we use and all the food we eat would not exist. ***Photosynthesis*** is the process by which carbohydrates are compounded from carbon dioxide (CO_2) and water in the presence of sunlight and chlorophyll. From a practical standpoint, photosynthesis is the primary method for bioconversion of solar energy into forms of energy such as hydrocarbon fuel sources and agricultural crops. Photosynthesis occurs when organisms convert CO_2 to organic material by reducing these gases to carbohydrates. The carbohydrates are compounds of hydrogen, oxygen, and carbon (such as sugars, starches, and celluloses), which are formed by green plants. Energy for this process is provided by light, some of which is absorbed by the pigments of chlorophyll, the green matter that comprises many plants. See **Figure 13-19.**

Hertz: A unit of frequency equal to one cycle per second.

Wavelength: One frequency of radiant energy.

Photosynthesis: The process by which carbohydrates are compounded from carbon dioxide (CO_2) and water in the presence of sunlight and chlorophyll.

Figure 13-18. A waveform—in this case, 2 Hertz, or 2 cycles per second.

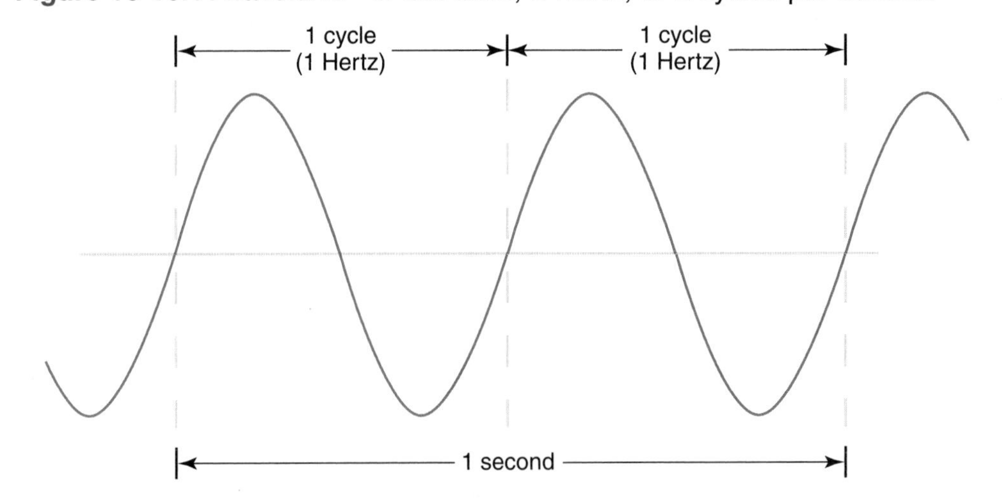

Chemical energy to radiant energy

Chemical energy is converted to radiant energy through the process known as *combustion*. Scientifically, it is described as a rapid chemical reaction in which heat and light are produced. Combustion requires ***oxidation***, a union between fuel and oxygen, in order to occur.

Visible light to infrared energy

The most common frequency conversion occurs when visible light is converted to infrared energy upon striking something. Solar collectors are designed to maximize this conversion and capture the energy from the heat. See **Figure 13-20.**

Ultraviolet (UV) radiation to visible light

When electricity is passed through a fluorescent lightbulb, electrodes are heated by the current flowing through them. These electrodes emit free electrons, which strike atoms of mercury vapor stored in the bulb. The mercury vapor emits radiant energy when it is stimulated by the free electrons. This radiant energy is not in the form of visible light, however, but in the form of UV ***radiation***. The UV energy strikes a phosphor coating on the

Figure 13-19. Plants, such as this cactus, convert sunlight into organic matter through the process known as photosynthesis.

Figure 13-20. An active solar energy collector system performs a frequency conversion, changing visible light into infrared energy.

Oxidation: A union between fuel and oxygen.

Radiation: Energy radiated in the form of waves or particles.

tube, and a frequency conversion to visible light occurs. More specifically, the UV energy stimulates the electrons in the phosphorous atoms, and the atoms emit visible ("white") light. This process of converting one form of light to another is known as ***fluorescence***, providing the name of the fluorescent lightbulb. See **Figure 13-21.**

Fluorescence: The process of converting one form of light to another.

Energy Inverters

Energy inversions involve converting a form of power back into a form of energy. An example of an energy inverter is a lightbulb that changes electricity into visible light. Among the most useful inversion devices are those that provide the ability to convert electricity into heat, light, and sound.

Electrical power to visible light energy

There are two basic methods of producing artificial light with the use of electricity. One is to heat something to the point where it gives off a glow. In a standard incandescent lightbulb, the filament is made of tungsten, a conductive metal element. See **Figure 13-22.** Tungsten can be heated to the point that it becomes white-hot, but does not melt. The second method of producing artificial light is to pass electricity through a gas or vapor, causing the tiny charged particles within the atoms to emit radiation. The color that such lights emit is specific to the gas within them. Halogen lights work in this manner.

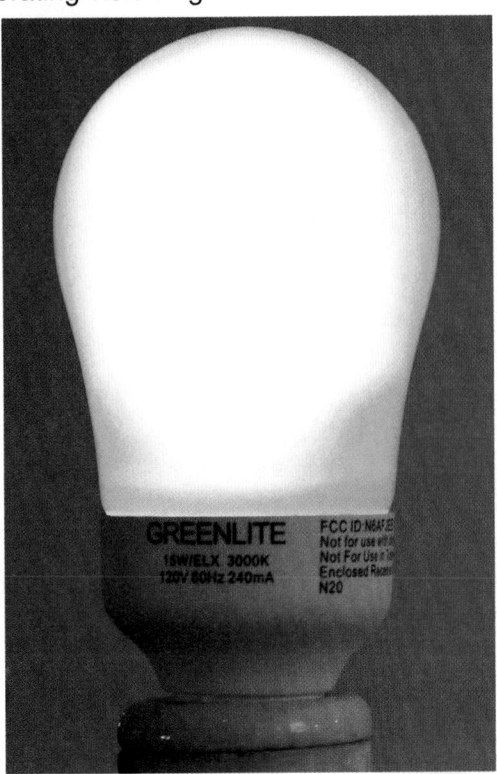

Figure 13-21. In a fluorescent lightbulb, ultraviolet (UV) radiation strikes a phosphor coating, generating visible light.

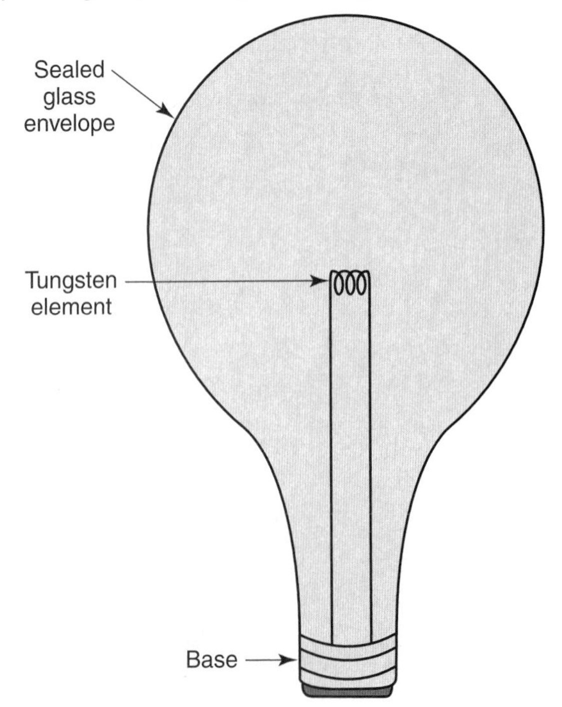

Figure 13-22. The components of an incandescent lightbulb, which is an energy inverter. Electrical energy flowing through the high-resistance tungsten element causes it to heat up and glow, emitting visible light.

Technology Link

Communication: Radio Detecting and Ranging (Radar) Guns

Police typically use radio detecting and ranging (radar) guns to identify speeding vehicles in order to protect the public and help keep our roadways safe. A radar gun consists of both a transmitter and a receiver. The transmitter converts electricity into radio waves and broadcasts or aims them toward a moving object, such as a vehicle. The radio waves strike the vehicle and bounce back to the radar gun. The receiver picks up the radio signal and converts it back into electricity. The analog electrical signal is converted into a digital electrical signal, and ultimately a visual readout in miles per hour is provided.

Radar guns measure so accurately because radio waves travel at a constant 186,000 miles per second, or the speed of light, and with a certain frequency, or number of oscillations per second. If the radio waves strike a stationary object, they will return at the same rate of speed and with the exact same wavelength. The distance it takes for them to return can be used to accurately calculate how far they have traveled.

If the radar gun is aimed at an object that is moving rather than stationary, however, the speed at which the radio waves return will be different than the rate at which the signal was sent. The frequency will also vary based on the speed of the object. This is because the radio wave is oscillating. Even though the signal is traveling very fast, because the car is moving, the signal will not reflect from the vehicle as the exact same waveform. This variation in waveform is known as a *Doppler Shift*. If a car is moving away from the radar, the wavelength is stretched. It has to travel a greater distance to return to the receiver. If the vehicle is moving toward the radar gun, the signal is compressed and returns to the receiver more quickly than the rate at which it was transmitted. These signals are calibrated and can measure speed with great accuracy.

Electrical power to infrared energy

Another very common inversion is that of electricity to heat. The way electricity is used to generate heat is easily explained. When electricity flows through a wire, it excites the molecules within the wire. This increased agitation causes the molecules to move, and this movement generates excessive energy, given off in the form of heat. See **Figure 13-23.**

In some homes, resistance heating is the primary heating system. It is used to supplement other heating methods in other homes. Small electrical resistance ("space") heaters may be used to warm specific rooms or such spaces as garages and workshops.

Electrical power to x-radiation

The ***X-ray tube*** is a wonderful device that can convert electricity into x-radiation. X rays have a higher frequency than visible light or even UV light. The X-ray tube produces a stream of negatively charged electrons that strike a tungsten filament, similar to that within a lightbulb. This bombardment of electrons causes the tungsten to emit X rays. The X rays are focused out through a window and aimed at a piece of film or an

X-ray tube: A device that converts electricity into x-radiation. It produces a stream of negatively charged electrons that strike a tungsten filament, causing the tungsten to emit X rays.

Career Connection

Administrative Assistants

Utility companies have the responsibility of providing electricity to the public. As a result of this, these businesses must also have a reliable record-keeping system. Administrative assistants have the task of organizing the paperwork of utility companies.

The duties of an administrative assistant vary from answering phones and opening mail to attending board meetings and performing research for executives. Assistants may be in charge of other clerical staff, which requires them to train their subordinates. Administrative assistants keep track of all incoming and outgoing paperwork and files. They are qualified to answer many questions from customers and other individuals.

Administrative assistants must be skilled in using multiple word processing and spreadsheet systems. They must understand business and management procedures in order to better aid executives. Administrative assistants must also possess a knowledge of grammar and communication skills. They typically receive on-the-job training. The yearly salary for this profession can range from $22,000 to $50,000.

Figure 13-23. The heating element in this toaster is an example of energy inversion. Electricity flowing through the high-resistance heating element is converted to heat.

array of photodiodes to produce an image. See **Figure 13-24.** The radiation is such that it can pass through lighter atoms, such as those that make up flesh, cloth, or canvas. The rays are absorbed, however, by more dense materials, such as metals. Since the calcium in your teeth and bones is a form of metal, X rays will not pass through these areas. This is why bones, teeth, and fillings will appear as negative images on a piece of developed X-ray film. All other areas of the film are exposed to the radiation and appear black.

An airport baggage scanner works in a way similar to the X-ray machine. Instead of exposing a piece of film, however, the scan displays a continuous "live" image. The X rays radiate through more porous substances, such as the luggage, and are received by photodiodes that convert the x-radiation to a displayed image. Denser items within the luggage do not allow X rays to pass through, and the result is that a strong image of such items appears on the screen.

Electrical power to sound waves

A number of different types of speakers are available for converting electricity back into sound. Electrodynamic speakers work on the

principle of magnetism, which states that like forces repel one another and opposing magnetic forces are attracted to one another. See **Figure 13-25.**

Principle of magnetism: Like forces repel one another, and opposing magnetic forces are attracted to one another.

This type of speaker contains both a permanent magnet and an electromagnet (a device that can become magnetic only when electricity is flowing through it). As the electromagnet receives its signal from the microphone or amplifier, it will change its polarity and intensity continuously, based on the sine wave it is receiving. The electromagnet is sometimes referred to as a *voice coil*, because it is attached to the cone of the speaker. The cone acts much like the diaphragm of a microphone, except in reverse. Based on the electrical signal it receives, the voice coil is continuously attracted and repelled by the permanent magnet mounted behind the cone. The cone moves in and out accordingly, producing vibrations in the air that cause sound to be created. See **Figure 13-26.**

Transducers

A device for converting one form of energy to another or one form of power to another is called a ***transducer***. Transducers can be frequency converters or power converters. They are often used to measure quantities in a system (such as pressure, current, or voltage) and convert them to a proportional unit displayed on a meter or scale. A thermometer is one such device. Other transducers are used as switching mechanisms. An example of a common transducer is a ***thermocouple***. See **Figure 13-27.** A thermocouple consists of two different metals joined end-to-end to produce a loop. When there is a difference in temperature between the two junctions (points in the loop where the metals are joined), a small electric current will flow between the two junctions. This electricity can be used to send a signal to turn something on or off.

Figure 13-24. An X-ray tube emits electrons that are converted to x-radiation. If an object (such as the human knee shown here, is placed between the radiation source and a piece of film, a negative image will be formed. More dense materials (such as the bones and the screw used for a medical repair) block the rays in part or in full. They show up darker on the film than less dense substances, such as muscle. (Siemens)

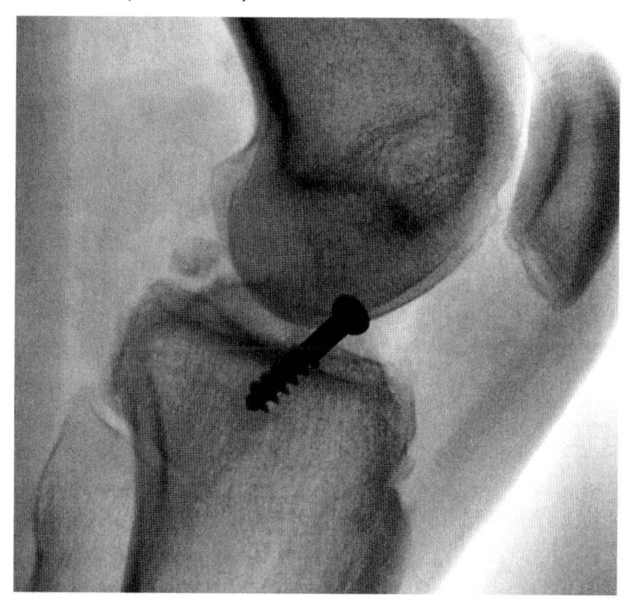

Entropy and Efficiency of Converters

In any conversion, entropy will occur when the conversion takes place. Entropy is a measure of energy lost upon conversion. Simply stated, total system efficiency will decrease every time an energy or power form undergoes a conversion. As described in Chapter 7, efficiency can be calculated using the following formula:

$$\frac{\text{output}}{\text{input}} \times 100 = \text{\% of efficiency}$$

Figure 13-25. The principle of magnetism. A—Unlike magnetic poles are attracted to each other. B—Like magnetic poles repel each other.

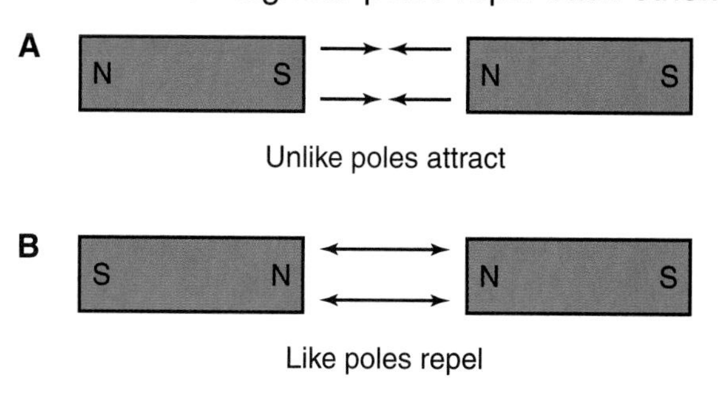

Figure 13-26. The components of an electrodynamic speaker.

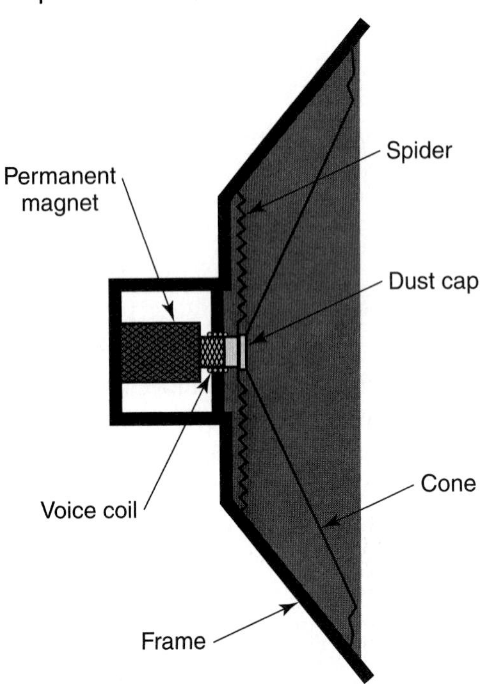

The output and input units must always be the same. For instance, watts can be divided by watts, but watts cannot be divided by horsepower (hp) when using this formula. Units commonly used for calculating efficiency of energy and power devices include hp and British thermal units per minute (Btu/min). When calculating efficiency, sometimes it may be necessary to use a constant to convert some other unit to hp or Btu/min, in order to be able to perform the efficiency calculation. Some constants discussed in Chapter 7 include the following:

$$1 \text{ hp} = 746 \text{ watts} \qquad 1 \text{ hp} = \frac{550 \text{ ft.-lbs.}}{\text{sec}}$$

The following constant is useful because it allows for the comparison of thermal systems to electrical, fluid, or mechanical systems:

$$1 \text{ hp} \times 42.44 = \frac{1 \text{ Btu}}{\text{min}} \quad \text{or} \quad \frac{1 \text{ Btu/min}}{42.44} = 1 \text{ hp}$$

Transducer: A device for converting one form of energy into another. It is often used to measure quantities in a system and convert them to a proportional unit displayed on a meter or scale.

Thermocouple: A transducer consisting of two different metals joined end-to-end to produce a loop.

Figure 13-27. A thermocouple in operation. A temperature difference between the two junctions causes a current to flow. This current can be used to operate a switch.

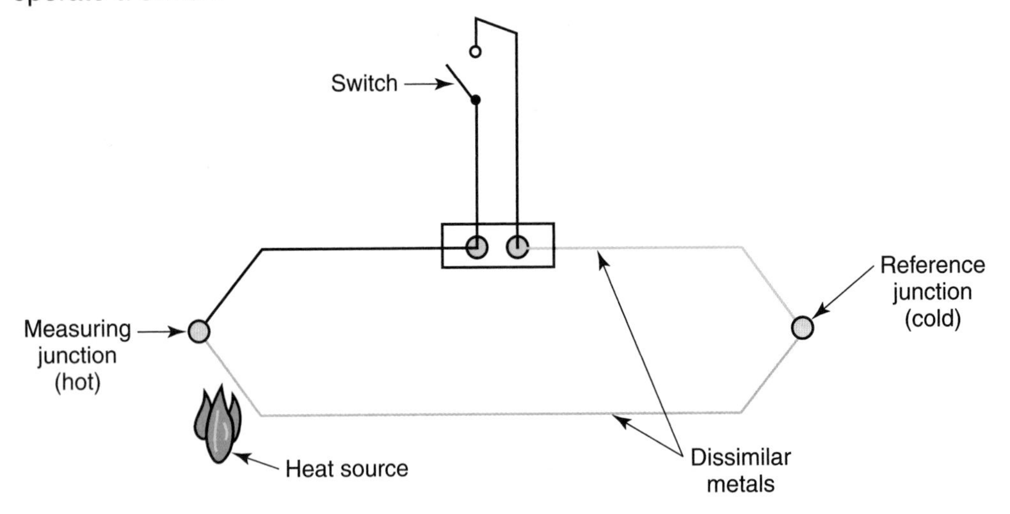

Summary

The ability to convert energy into power and one form of power into another form of power is essential to many of the modern conveniences we take for granted. Some conversions occur naturally, but most conversions are the result of technological innovation. Conversions can be classified as either direct, meaning they take only one step, or indirect, meaning they take two or more steps. Transducers convert one form of power into another. They are typically used for sensing equipment and can also perform switching duties. Some of the most popular conversions are actually termed inversions because a form of power is converted back into a form of energy. Every time a conversion occurs, some energy or power is lost due to entropy.

Key Words

All the following words have been used in this chapter. Do you know their meanings?

amplifier
armature
bimetallic coil
brush contactor
coefficient of expansion
combustion
commutator
cycle
diaphragm
direct converter
dynamic coil microphone
energy converter
energy inverter
fluorescence
frequency
frequency converter
Hertz
indirect converter
induced
magnetohydrodynamic (MHD) generator
oxidation
photosynthesis
power converter
principle of magnetism
radiant energy
radiation
relay
repeater
rotor
run winding
single-phase induction motor
slip ring
solenoid
sound
start winding
stator
thermocouple
three-phase induction motor
transducer
transistor
voice coil
wavelength
X-ray tube

Test Your Knowledge

Write your answers on a separate sheet of paper. Do not write in this book.

1. Explain the difference between a direct converter and an indirect converter.
2. When one form of radiant energy is converted into another form of radiant energy, this is known as a(n) _____ conversion.
3. List the four types of power conversions that can occur, and describe each type of conversion.

4. *True or False?* Power conversion devices are technological, as opposed to natural.
5. *True or False?* Energy converters change one form of energy into another form of energy.
6. Describe at least three uses for a solenoid.
7. Spinning the shaft of a direct current (DC) motor will _____.
8. The windings in a motor that do not move are known as the _____ windings.
9. Induction motors are typically _____-phase or _____-phase.
10. *True or False?* Magnetohydrodynamic (MHD) generators are extensively used to produce power throughout the United States.
11. *True or False?* One advantage of an MHD generator is its increased efficiency over conventional steam turbine power generation.
12. List the sequence of conversions that occur within an internal combustion engine in order to produce rotary mechanical power.
13. *True or False?* A transistor is a solid-state switching device.
14. Sound is converted to electricity using a(n) _____.
15. The signal strength of sound is boosted by a(n) _____.
16. The frequency of a wavelength can be determined by dividing the _____ by the _____.
17. *True or False?* The phosphor coating on a fluorescent lightbulb converts ultraviolet (UV) energy into visible light energy.
18. *True or False?* Electricity generates heat by exciting the molecules in the wire of a heating element.
19. The law of magnetism states that opposite forces attract and like forces _____ each other.
20. In a speaker, the _____ converts electrical signals into movement of the cone to produce sound.
21. A device often used to convert a form of energy or power into another form for measurement or switching purposes is known as a(n) _____.

Activities

1. Construct a simple electric motor from a set of plans.
2. Design and construct a model of simple relay.
3. Disassemble a small electric motor and identify its components.
4. Locate and photograph devices that are examples of each of the four types of energy conversion. Write a paragraph describing each device and how it functions.

14

Small Gas Engines

Basic Concepts

- Identify the differences between internal and external combustion engines.
- Recognize the basic process by which two-stroke engines and four-stroke gasoline engines operate.
- Select the correct tool for a specific application.
- Accurately read a micrometer.

Intermediate Concepts

- Describe the operating procedures of at least five subsystems of the small gas engine.
- Discuss helpful hints for successfully disassembling and reassembling complex machinery, such as small gas engines.

Advanced Concepts

- Troubleshoot and diagnose six common causes of engine malfunctions.
- Perform live adjustments and take measurements from a running engine.
- Use repair manuals to locate specifications, such as torque ratings, clearances, torque patterns, and replacement part codes.

A little more than 100 years ago, internal combustion engines began to replace ***external combustion engines*** (steam engines) as the major source of power for vehicles and many other applications requiring mechanical power production. ***Internal combustion engines*** produce heat inside the cylinder containing the piston. They are more efficient and reliable than external combustion engines, which generate heat in a boiler or other device outside the cylinder. The ***cylinder***, more correctly identified as the *cylinder bore*, is a hole in the block of the engine that directs the piston during movement. See **Figure 14-1.** Internal combustion engines also produce much more power for a comparably sized unit. Today,

External combustion engine: The steam engine used a little over 100 years ago as the major source of power for vehicles and many other applications requiring mechanical power production.

Internal combustion engine: An engine that produces heat inside the cylinder containing the piston.

Cylinder: A hole in the block of an engine that directs the piston during movement.

Figure 14-1. An external combustion engine, such as this old steam tractor, generates heat outside the cylinder. Steam from the boiler (at the front, under the smokestack) is piped to the engine cylinder, where it expands to drive the piston in the power stroke. (Howard Bud Smith)

internal combustion engines power the majority of transportation vehicles in the United States. They also are used extensively for power in the construction and agricultural industries and are even employed to generate limited amounts of electricity.

Technology Link

Agriculture: Farming Equipment

Agricultural technology relies on energy and power to produce crops and raise livestock. Various applications of energy and power technology are essential to the farming industry. Many types of farming equipment require mechanical power, hydraulics, and electricity in order to operate.

Until about the mid-1920s, small gas engines were used in tractors and other farm equipment more than any other types of engines. By the 1930s, however, the shift toward diesel engines started to occur. Diesel fuel is more efficient than gasoline for these applications, so it is the more cost-effective choice. These days, farm equipment uses both types of engines, though diesel engines are still more common.

The main use of engines in the agriculture industry is to power equipment, such as tractors and combines. These machines play a large part in efficiently planting, cultivating, and harvesting crops, as well as tending to livestock. Engines are vital to transporting the products to market once the products have been harvested and prepared for sale. Gas and diesel engines power the trucks, river barges, and railroad cars that ship the goods from the farm to the end users. Engines are also used in farming to pump water for irrigation. The farming industry would not be nearly as efficient or profitable without energy and power technology.

Engine Theory

There are two main types of small gas engines, the four-stroke cycle engine and the two-stroke cycle engine. The two types perform the same function—converting chemical energy to mechanical power—and have many common mechanical elements. They differ considerably, however, in their methods of operation.

The Four-Stroke Cycle Engine

Automobiles use ***four-stroke cycle engines*** as their power source . The automobile engine has 4, 6, 8, or 12 cylinders, which are all coupled to one crankshaft. The ***crankshaft*** is an engine component that converts the reciprocating motion of the piston and rod assembly into rotary motion. It is also the shaft that powers the load. Most small gas engines have only one cylinder powering the crankshaft. See **Figure 14-2.** The ***piston*** is a cylindrical engine component that slides back and forth in the cylinder

Four-stroke cycle engine: An engine that requires four movements of the piston in its cylinder to complete a full cycle.

Crankshaft: An engine component that converts the reciprocating motion of the piston and rod assembly into rotary motion.

Piston: A cylindrical engine component that slides back and forth in the cylinder when propelled by the force of combustion.

Figure 14-2. In most small gas engines, a single piston is connected to the crankshaft.

Stroke: The movement of the piston from the bottom limit to the top limit (or vice versa).

Intake stroke: The downward stroke of the piston that begins the process of producing power.

Atomized: Broken into small droplets.

Fuel-air charge: Small droplets of liquid fuel and air.

Compression stroke: An upward movement of the piston and connecting rod assembly.

Connecting rod: An engine component that connects the piston with the crankshaft.

when propelled by the force of combustion. A four-stroke cycle engine requires four movements (strokes) of the piston in its cylinder to complete a full cycle. A ***stroke*** is the movement of the piston from the bottom limit of its travel to the top limit (or vice versa). See **Figure 14-3.** The four-stroke cycle engine's strokes are the following:

- ***Intake stroke.*** This is the downward stroke of the piston that begins the process of producing power. This movement creates a partial vacuum. The force of this vacuum draws air through the carburetor. Liquid fuel is drawn into the carburetor at the same time and is ***atomized*** (broken into small droplets) to mix with the air. This mixture is called the ***fuel-air charge***. It flows into the cylinder through the intake valve.
- ***Compression stroke.*** This is an upward movement of the piston and connecting rod assembly. The ***connecting rod*** is an engine component that connects the piston with the crankshaft. The fuel-air charge is typically squeezed to about one-ninth of its original volume. When the piston is as low in the cylinder as it can go, it is said to be at ***bottom dead center (BDC)***. When the piston is as high in the cylinder as it can go, it is said to be at ***top dead center (TDC)***. The ***compression ratio*** of an engine is the mathematical relationship between the volume available in the cylinder with the piston at BDC and the volume available in the cylinder with the piston at TDC. See **Figure 14-4.**
- ***Power stroke.*** This is the stroke in which power (mechanical movement) is transferred from the piston to the connecting rod and then to the crankshaft. As the piston approaches TDC on the compression stroke, the ***spark plug*** fires. It takes a fraction of a second for the gases in the ***combustion chamber*** to ignite and expand. This allows the piston to move past TDC, so the expanding gases will push down on the piston with tremendous force.

Figure 14-3. Sequence of events in a four-stroke cycle engine. This type of engine requires two revolutions of the crankshaft and provides one power stroke out of every four strokes.

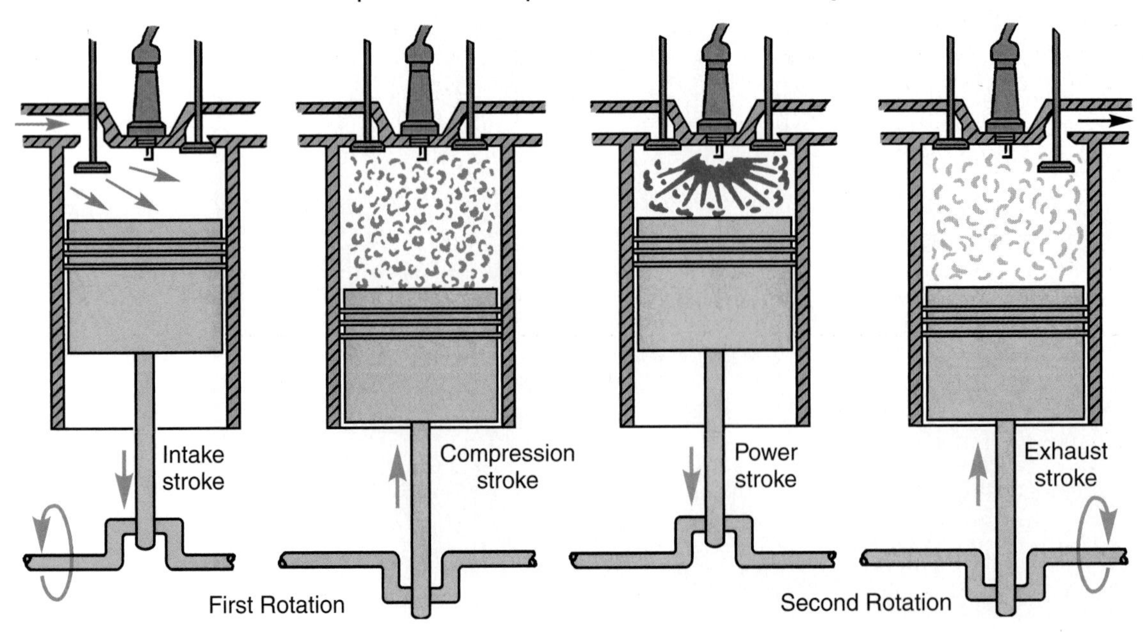

- ***Exhaust stroke.*** This is the final movement in the four-cycle process—an upward stroke of the piston. The ***camshaft*** holds the exhaust valve open. Movement of the piston forces the spent fuel-air mixture out through the exhaust valve. As the piston clears TDC, the camshaft causes the exhaust valve to close and the intake valve to open. A new fuel-air charge is drawn into the cylinder, beginning the four-stroke cycle again.

Figure 14-4. The compression ratio is determined by comparing the volume of the cylinder with the piston at bottom dead center (BDC), which in this example, is 6 in^3, to the volume when the piston is at top dead center (TDC), which in this example, is 1 in^3. This is a compression ratio of 6 to 1, usually written in the form 6:1. (Briggs and Stratton Corporation)

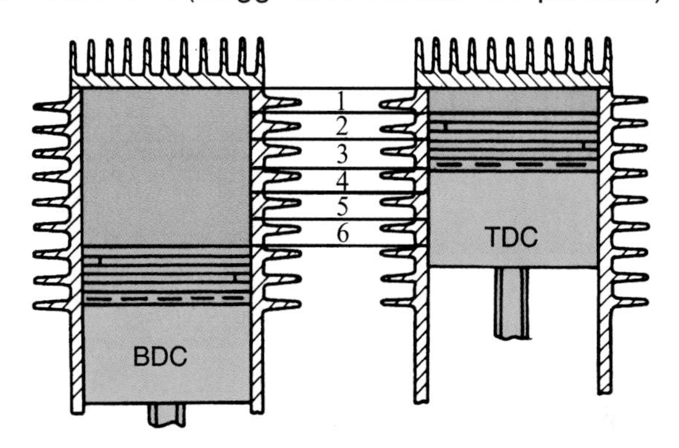

The Two-Stroke Cycle Engine

All the functions that take place in a four-stroke cycle engine—intake, compression, power, and exhaust—also must happen in a two-stroke cycle engine. In a ***two-stroke cycle engine***, however, every upward stroke is a compression stroke, and every downward stroke is a power stroke. Intake and exhaust occur during the compression and power strokes. See **Figure 14-5.** This means it takes only one revolution of the crankshaft to produce a power stroke, instead of the two revolutions required by the four-stroke cycle engine.

The two-stroke design has both advantages and disadvantages when compared to the four-stroke design. See **Figure 14-6.** One advantage of the two-stroke design is that a power stroke occurs on every downward movement of the piston (as opposed to every other). This makes the two-stroke cycle engine very powerful for its size. It is also very good at applications with a high number of revolutions per minute (rpm). The two-stroke cycle engine has a design simpler than that of the four-stroke cycle engine, since it does not require a camshaft and valve train assembly.

Intake and exhaust are accomplished through the placement of ports along the cylinder. Since it has fewer parts, a two-stroke cycle engine is much lighter than a four-stroke cycle engine of comparable power. Because the two-stroke design does not require an oil reservoir, there is a further savings in weight. Oil is mixed with the fuel in this type of engine and burned in the combustion chamber. Since the combustion chamber is sealed tightly and there is no oil reservoir, the two-stroke cycle engine can be operated at any angle. This makes it ideal for string trimmers, chain saws, leaf blowers, and similar applications.

The two-stroke cycle design has some disadvantages as well. Since the lubricant is burned with the fuel, the exhaust of a two-stroke cycle engine is dirtier than that of a four-stroke cycle engine. The fact that every other stroke is a power stroke means that two-stroke cycle engines wear more rapidly than four-stroke cycle engines. It is unrealistic to expect a two-stroke cycle engine to last as long as a four-stroke cycle engine of similar power. Mixing the oil with the fuel is inconvenient, but forgetting to add the oil will result in major engine damage. On some larger two-stroke cycle engines, such as those used on snowmobiles, boats, and dirt bikes,

Bottom dead center (BDC): As low in the cylinder as the piston can go.

Top dead center (TDC): As high in the cylinder as the piston can go.

Compression ratio: The mathematical relationship between the volume available in the cylinder with the piston at bottom dead center (BDC) and the volume available in the cylinder with the piston at top dead center (TDC).

Power stroke: The stroke in which mechanical movement is transferred from the piston to the connecting rod and then to the crankshaft.

Spark plug: A part that fits into the cylinder head of an internal combustion engine and carries two electrodes separated by an air gap across which the current from the ignition system discharges to form the spark for combustion.

Combustion chamber: An enclosed space in which burning takes place.

Exhaust stroke: The final movement in the four-cycle process—an upward stroke of the piston.

Camshaft: A shaft to which a cam is fastened.

Two-stroke cycle engine: An engine in which every upward stroke is a compression stroke and every downward stroke is a power stroke.

Figure 14-5. The sequence of events in a two-stroke cycle engine. Compression and intake occur simultaneously, and then ignition occurs. Exhaust precedes the transfer of fuel during the lower portion of the power stroke. The piston functions as a valve, opening and closing the intake and exhaust ports as it moves up and down in the cylinder. (Rupp Industries, Inc.)

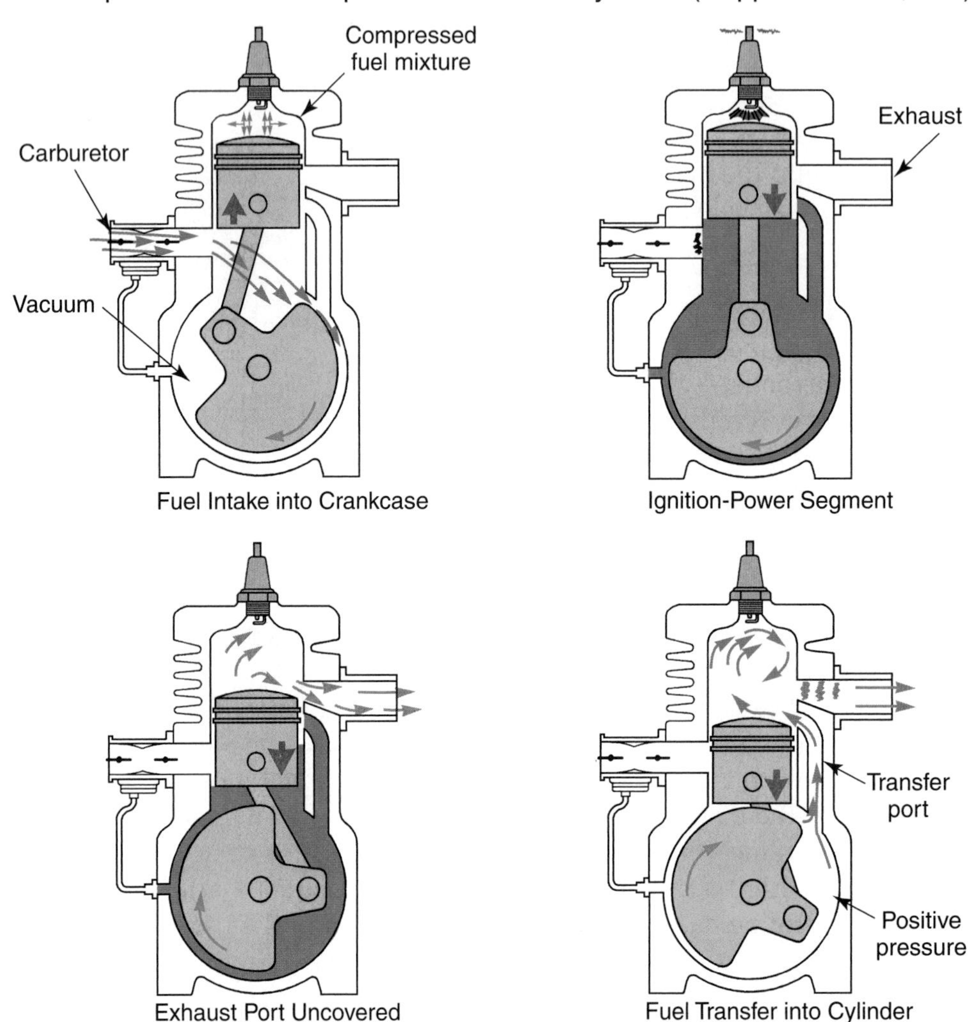

oil injector pumps have been added to mix the lubricant with the gasoline. The pump reservoir can hold enough oil to provide an adequate mix for several tanks of gasoline, making these types of two-stroke cycle engines somewhat more convenient.

Engine Subsystems

An internal combustion engine is a complex machine because it has multiple subsystems, all of which must perform properly for the engine to run at peak performance. These subsystems include the cooling subsystem, the lubrication subsystem, the mechanical subsystem, the electrical subsystem, the governing subsystem, and the fuel subsystem. They are described in the following sections, along with some information critical to diagnosis of problems with each particular subsystem.

Figure 14-6. Differences between two-stroke cycle and four-stroke cycle engines.

Characteristics	Four-Cycle Engine (Equal hp) One Cylinder	Two-Cycle Engine (Equal hp) One Cylinder
Number of major moving parts	Nine	Three
Power strokes	One every two revolutions of crankshaft	One every revolution of crankshaft
Running temperature	Cooler running	Hotter running
Overall engine size	Larger	Smaller
Engine weight	Heavier construction	Lighter in weight
Bore size equal hp	Larger	Smaller
Fuel and oil	No mixture required	Must be premixed
Fuel consumption	Fewer gallons per hour	More gallons per hour
Oil consumption	Oil recirculates and stays in engine	Oil is burned with fuel
Sound	Generally quiet	Louder in operation
Operation	Smoother	More erratic
Acceleration	Slower	Very quick
General maintenance	Greater	Less
Initial cost	Greater	Less
Versatility of operation	Limited slope operation (receives less lubrication when tilted)	Lubrication not affected at any angle of operation
General operating efficiency (hp/wt. ratio)	Less efficient	More efficient
Pull starting	Two crankshaft rotations required to produce one ignition phase	One revolution produces an ignition phase
Flywheel	Requires heavier flywheel to carry engine through three nonpower strokes	Lighter flywheel

The Cooling Subsystem

The ***cooling subsystem*** of the engine is responsible for keeping the engine operating within a comfortable temperature range. Engines are cooled by air or liquid. Most small gas engines are air cooled.

The primary parts of an air-cooled system are the cooling fins on the head and block of the engine, the blades on the flywheel, and various sheet metal parts that enclose the engine. The ***flywheel*** blades create a flow of air that cools the engine. The sheet metal shrouds channel the airflow across the hottest parts of the engine, which are the cooling fins surrounding the combustion chamber. The ***cooling fins*** conduct heat from the combustion chamber and transfer it to the surrounding atmosphere. See **Figure 14-7.** It is important that all sheet metal shrouds are in place, so the airflow produced by the rotating flywheel is channeled to the proper areas of the engine. See **Figure 14-8.**

Some larger engines may use a liquid cooling system that includes a water pump, radiator, and thermostat. Since water is much denser than air, it can absorb and dissipate much more heat from an engine. In this type of system, a water and antifreeze solution is pumped through

Cooling subsystem: The system responsible for keeping the engine operating within a comfortable temperature range.

Flywheel: A heavy wheel for opposing and moderating any fluctuation of speed in the machinery with which it revolves.

Cooling fin: A projecting rib on an engine cylinder that moderates heat.

Figure 14-7. The cooling system on a small gas engine. The vanes on the flywheel generate air movement, which is directed upward by a sheet-metal shroud (removed in this photo to show the flywheel). The air stream moves over the cooling fins on the cylinder to carry away heat. It is important to keep the cooling fins on the engine clean of debris so they can function efficiently.

Figure 14-8. A shroud is fitted over the flywheel to direct air movement over the cooling fins. The air intake screen should be cleaned regularly to provide an unrestricted flow of air. (Tecumseh Products Company)

Water jacket: A space machined into the block of an engine.

Thermostat: A temperature-controlled flow valve.

Radiator: A heat exchanger that transfers the heat from a liquid to the surrounding environment.

water jackets, or spaces machined into the block of the engine, surrounding each cylinder. When the liquid heats up, a ***thermostat*** (a temperature-controlled flow valve) opens, allowing the liquid to flow to the radiator. The ***radiator*** is a heat exchanger that transfers the heat from the liquid to the surrounding environment. The cooled water is pumped back to the engine block to absorb more heat from the engine and transfer it to the radiator.

The Lubrication Subsystem

The ***lubrication subsystem*** of a small gas engine includes the oil distribution mechanism, the oil seals, the piston rings, and the lubricating oil itself. See **Figure 14-9.** There are several ways to distribute oil to the working parts of an engine. Some distribution mechanisms work better for smaller engines, however, and others work better for larger engines. See **Figure 14-10.**

It is important that all moving parts within the engine are lubricated. The splash lubrication method is usually acceptable for a small gas engine. This splash is created by an ***oil dipper*** attached to the bottom of the connecting rod. As the connecting rod moves up and down during the four-cycle process, the dipper drops down into the reservoir in the bottom of the crankcase. As it rises up out of the oil reservoir, it splashes oil onto the moving parts of the engine that must be lubricated whenever the engine is running. If a dipper has been replaced, it is vital to properly torque the connecting rod bolts. Undertorquing the bolts could cause the dipper and connecting rod cap to become loose, leading to catastrophic engine failure.

Lubrication of the cylinder wall is necessary. The ***engine rings*** serve to limit the amount of oil that makes its way into the combustion chamber. Too much oil entering the combustion chamber creates excessive emissions, usually in the form of thick smoke, and it also can foul spark plugs, causing the engine to stop running all together. The oil ring is the bottom ring on the piston. It is designed to allow a small amount of oil to make its way through the holes in the ring, through the matching holes in the piston, and out onto the cylinder wall. The bottom compression ring

Figure 14-9. A thin film of oil provides for almost friction-free movement between closely fitted metal parts, such as the crankshaft journal and connecting rod assembly. This reduces wear on the parts.

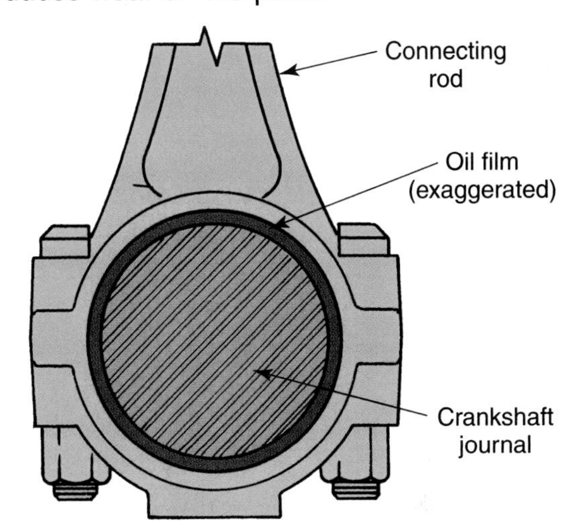

Figure 14-10. Oil distribution mechanisms. A—An oil dipper that attaches to the connecting rod cap is common for many small gas engines. The dipper simply splashes the moving parts of the engine with oil when the engine is running. An oil slinger works in a similar fashion, but it is often driven by the cam gear. The slinger has a series of paddles to move more oil and is often found on higher-horsepower (hp) small gas engines. (Briggs and Stratton Corporation) B—An oil pump can be used to force the lubricant to distant places. Oil pumps are frequently used on larger engines, where the moving parts are not in close proximity to the oil reservoir. Some small gas engines, however, make use of a piston-style pump to provide oil distribution.

A

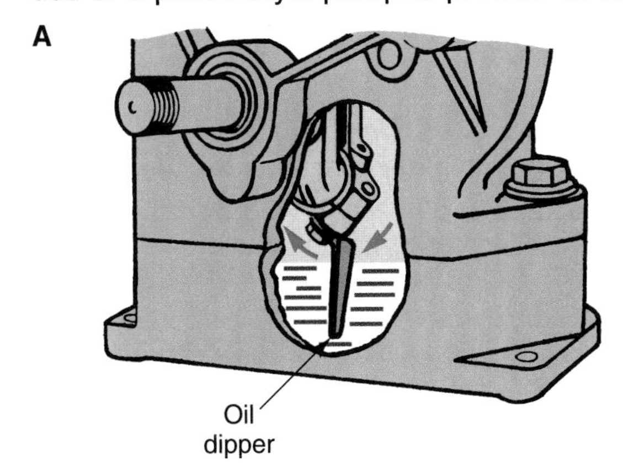

B

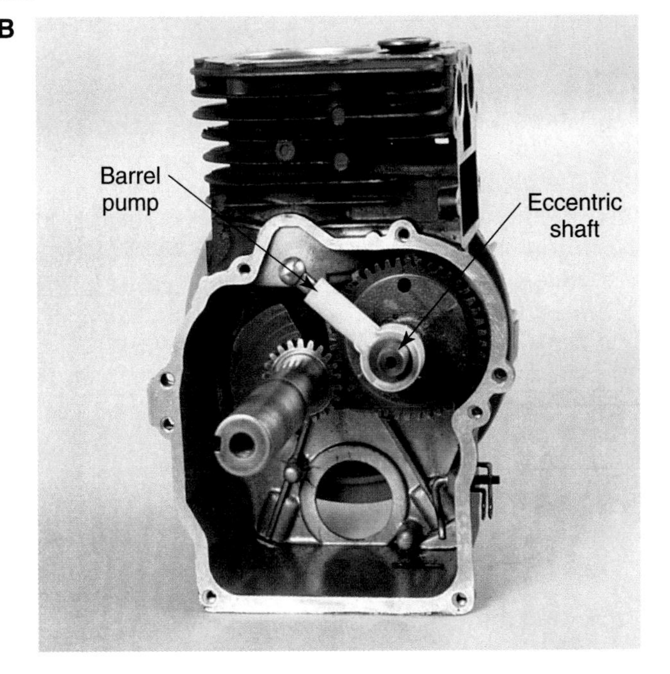

Lubrication subsystem: The oil distribution mechanism, the oil seals, the piston rings, and the lubricating oil itself.

Oil dipper: A device that splashes the moving parts of the engine with oil when the engine is running.

Engine ring: A circular band for sealing and limiting the amount of oil that makes its way into the combustion chamber.

Society of Automotive Engineers (SAE): An organized group of individuals working together because of common training in the branch of engineering relating to self-propelled vehicles.

usually has a groove or a bevel to scrape excess oil off the cylinder wall and send it back down into the block where it is stored. Oil seals serve a similar function on the valve stems.

Lubrication maintenance on small gas engines includes the following:

- Changing the oil at regular intervals.
- Wiping debris and sludge from the bottom of the oil reservoir every few oil changes, since most small gas engines do not contain oil filters.
- Ensuring that rings and seals are installed correctly during engine reassembly.
- Inspecting seals for wear.

It is important to recognize all the functions an engine lubricant performs. In addition to lubricating the engine's working parts, oil also serves to do the following:

- Protect the internal engine parts from corrosion.
- Cleanse the engine of foreign matter by transferring it to the engine block, where the foreign matter settles to the bottom of the oil reservoir and can do little harm.
- Seal the engine by filling the small space between moving parts, such as the piston rings, and the cylinder wall.
- Cushion moving engine parts from the tremendous force of the power stroke on combustion.
- Improve fuel economy.

Oil selection is critical to the performance of an engine. The viscosity of an oil is a measure of its resistance to flow. The viscosity ratings for engine oils were established by the ***Society of Automotive Engineers (SAE),*** but they vary significantly from manufacturer to manufacturer. See **Figure 14-11.** Oils labeled with a winter rating, such as 10W-40, are known as ***multigrade,*** multiweight, or multiviscosity oils.

Figure 14-11. A comparison of viscosity grade recommendations by five major engine manufacturers. These recommendations do not cover all model engines by any particular manufacturer. Note that the *W* stands for a "winter" rating. Oils labeled with a winter rating are blended to operate under the varying temperature conditions that occur during summer and winter in many locations.

<table>
<tr><th colspan="6">Viscosity-Grade Recommendations by Manufacturers for Four-Cycle Crankcase Lubrication</th></tr>
<tr><th>Manufacturer</th><th>Above 40°</th><th>Above 32°F</th><th>Below 5°F</th><th>Below 0°F</th><th>Below –10°F</th></tr>
<tr><td>Briggs and Stratton</td><td>SAE 30 or 10W-30</td><td colspan="4">5W-20 or 10W</td></tr>
<tr><td>Clinton</td><td colspan="2">SAE 30</td><td colspan="2">10W</td><td>5W</td></tr>
<tr><td>Kohler</td><td colspan="2">SAE 30</td><td>SAE 10W</td><td colspan="2">5W or 5W-20</td></tr>
<tr><td>Wisconsin</td><td>SAE 30</td><td>SAE 20 or SAE 20W</td><td colspan="3">10W</td></tr>
<tr><td>Tecumseh</td><td colspan="2">SAE 30</td><td colspan="3">10W-30</td></tr>
</table>

The Mechanical Subsystem

The ***mechanical subsystem*** converts the force of the expanding gases during combustion into mechanical power, delivering the power to the crankshaft. The mechanical subsystem begins with the block of the engine, which is the main housing for the engine components. See **Figure 14-12.**

The piston and connecting rod are within the cylinder of the block. The expanding gases that result from combustion exert force on the piston, which moves downward, transferring power to the connecting rod. The connecting rod attaches to the ***crankpin journal,*** which is an offset on the crankshaft that converts the downward movement of the piston and connecting rod into rotary motion. See **Figure 14-13.**

On four-stroke cycle engines, the crankshaft also powers a camshaft. The purpose of the camshaft is to open the appropriate valve at the correct time during intake and exhaust strokes. See **Figure 14-14.**

Figure 14-12. A combination cylinder block and crankcase. The cylinder head and sealing gasket are bolted to the cylinder block. The cylinder block houses all the mechanical components of the engine and keeps them aligned. The head of the engine contains the combustion chamber. On some small gas engine configurations, it may contain the intake and exhaust valves.

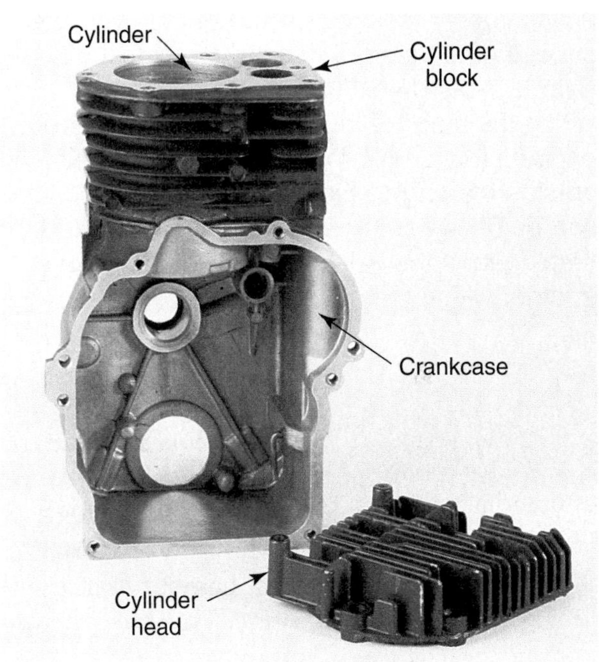

Figure 14-13. Connecting rod and crankshaft. The connecting rod pivots on the crankpin journal. It is subject to severe stress as it transfers power from the piston to the crankshaft during power strokes. The large counterweights on the crankshaft, opposite the crankpin journal, balance the rotational forces of the crankshaft during the power stroke of the piston. When the piston reaches bottom dead center (BDC), the continuing rotary motion will start it traveling upward again in the cylinder.

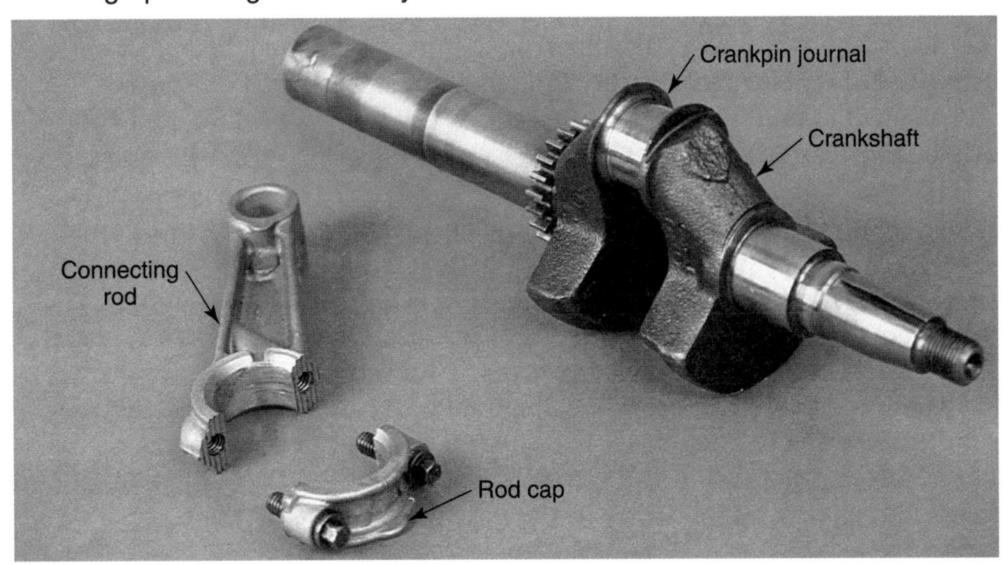

Multigrade: Many positions in a scale of qualities.

Mechanical subsystem: The subsystem that converts the force of the expanding gases during combustion into mechanical power, delivering the power to the crankshaft.

Crankpin journal: An offset on the crankshaft that converts the downward movement of the piston and connecting rod into rotary motion.

Figure 14-14. The camshaft gear is driven from the crankshaft and provides power to the valve train. The camshaft and crankshaft must be properly aligned so the valves are held closed during the compression and power strokes. Otherwise, it will not be possible to achieve compression. The camshaft gear is always twice as big as the crankshaft gear, since valve action is only required on two out of every four strokes of the piston. Usually, the camshaft gear is located next to the crankshaft gear within the block of the engine, but some large engines use an overhead camshaft located outside the engine block. A timing belt or chain can be used to connect the cam gear and crank gear if they are not located next to one another within the block.

Cam lobe: A curved projection of the rotating piece in a mechanical linkage used to open and close the valves by pushing on rods.

Valve lifter: A lifter that transfers power from the cam lobe to the valve.

Inertia: The tendency of a body in motion to remain in motion.

Key: A small metal piece that holds the flywheel in a properly aligned position on the crankshaft for spark to occur.

Micrometer: A basic precision measuring instrument used to check for wear points on engine parts.

Cam lobes on the camshaft open and close the valves by pushing on rods called lifters. ***Valve lifters*** often can be readjusted to maintain optimum clearances as valves and cam lobes wear. See **Figure 14-15.**

The flywheel on a four-stroke cycle small gas engine is proportionally much heavier than the flywheel on a larger engine or a two-stroke cycle engine. This weight is needed because the flywheel is used to store the energy provided by the power stroke in the form of ***inertia*** (the tendency of a body in motion to remain in motion). The inertia stored in the heavy flywheel helps the engine coast through the exhaust, intake, and compression strokes and smoothes out the power produced by the engine so the engine does not appear to constantly slow down and speed up. See **Figure 14-16.**

A small metal piece known as a ***key*** holds the flywheel in place on the crankshaft. The key fits in matching slots on the flywheel and crankshaft, and it serves as both a coupling device and a safety device. It is made of a softer metal than the flywheel and crankshaft. If the engine stops suddenly, the key will shear, allowing the heavy flywheel to freewheel. This protects the internal parts of the engine from serious damage. Another function the key performs is to keep the flywheel aligned properly on the crankshaft.

The mechanical subsystem is subject to the most wear. This wear is often not visible to the eye because it is only a few thousandths of an inch. The ability to measure with micrometers, feeler gauges, hole gauges, and telescoping snap gauges is very important. ***Micrometers*** are the basic precision measuring instruments used to check for wear points on engine parts. ***Feeler gauges***, sometimes referred to as *thickness gauges*, are thin strips of metal machined to a specific thickness, often measured in thousandths of an inch. The metal strips are used to verify a gap between two parts. A feeler gauge that is too thick cannot enter the gap. One that is too small moves around too freely in the gap. Measurements are typically

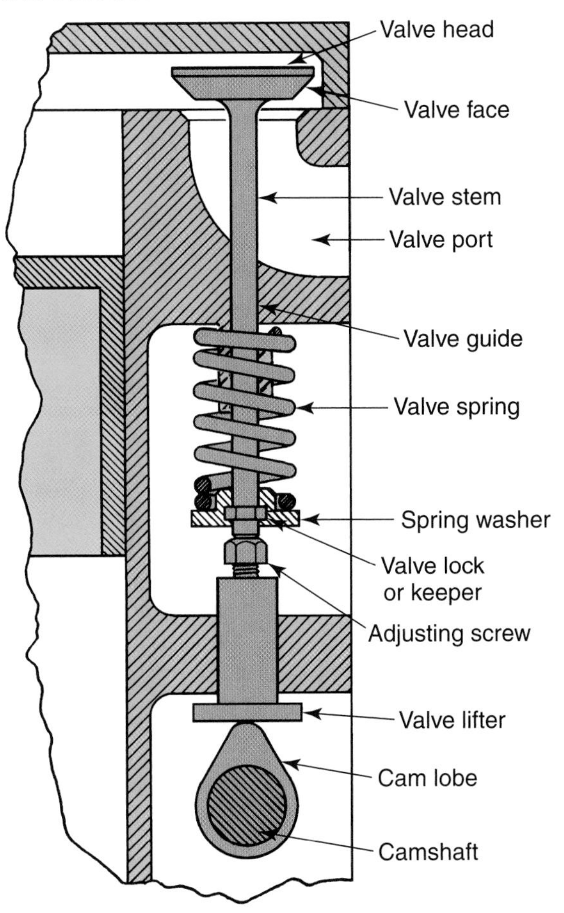

Figure 14-15. A complete valve train with all its components. The lifters transfer power from the cam lobe to the valve. The valve springs hold the valve closed tightly during compression and power strokes.

Figure 14-16. The flywheel on a small gas engine plays a part in most of the engine's subsystems. In addition to the mechanical subsystem, the flywheel is part of the cooling subsystem, the electrical subsystem, and the governing subsystem on some small gas engines.

taken in only the most critical areas, such as valve guides, bearing surfaces, and journals, where wear is most likely to occur. Following is a list of the most critical wear areas in most engines.

- Crankshaft front, rear, and crankpin journals.
- Crankshaft front and rear bearings.
- Camshaft front and rear journals.
- Camshaft front and rear bearings.
- Cam lobes.
- Intake and exhaust valves.
- Intake and exhaust valve guides.
- Cylinder diameter (top).
- Cylinder diameter (bottom).
- Cylinder head (for warpage).

In addition to measuring valve stem wear, check the stems for squareness and concentricity of the valve face to the seat. The ***margin thickness*** on the valve also must be checked because the margin is what wears away as the valve operates. To check the valve stems for squareness, roll them

Feeler gauge: A thin strip of metal machined to a specific thickness.

Margin thickness: The dimension of the degree of difference.

on a surface plate to determine if there is any warping. To check for concentricity between the valve face and the seat, place the valve stem in the valve guide and rotate the valve 360° to determine if it rides up or down. Measure the margin thickness by using a steel rule that measures in sixty-fourths of an inch.

Curricular Connection

Math: Measuring with a Micrometer

The micrometer is a precision instrument used to measure parts and determine whether they are within specifications. In a small gas engine, wear of a few thousandths of an inch on a part could be very critical to engine performance. Before you learn how to read a micrometer, it is important to learn about handling a micrometer. Since it is a precision measuring instrument, it should be held gently and with care. See **Figure 14-A.** Never overtighten a micrometer, and be sure to keep it clean and free of debris, which could cause an inaccurate reading.

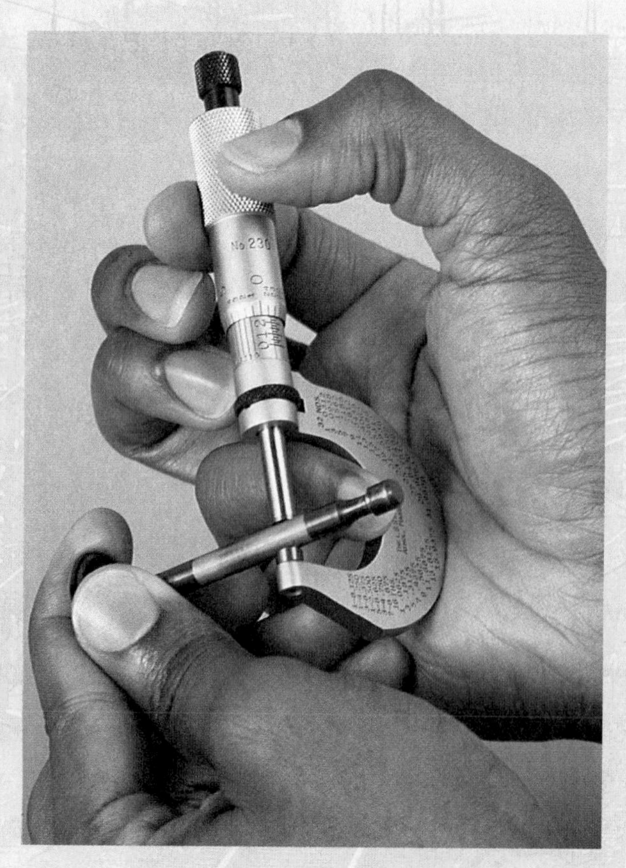

Figure 14-A. The proper way to hold a micrometer is to hold the micrometer in one hand, while holding the piece to be measured in the other hand. Rotating the ratchet at the end of the thimble will tighten the micrometer until it freewheels. The measurement is then taken.

The Electrical Subsystem

The ***electrical subsystem*** produces the current that fires the spark plug. It begins with the permanently mounted magnets within the flywheel. As the flywheel spins, the small amount of magnetism induces a low voltage in the armature each time the magnets pass the armature. This low voltage is then converted to high voltage in the ignition coil when the primary field collapses on the secondary field, causing the spark plug to fire.

Electrical subsystem: The subsystem that produces the current that fires the spark plug.

Reading the micrometer is a three-step additive process. Each small line on the sleeve represents 0.025″, or 1/40″. See **Figure 14-B.** Every fourth line on the sleeve is longer. These lines represent 0.100″, or 1/10″. The rotating thimble on the micrometer is divided into 25 equal parts, each measuring 0.001″ or 1/1000″. The additive process used to obtain a measurement consists of these steps:

1. Count all the long lines you can see on the sleeve. Let us say there are three. They represent 0.300″.
2. Next, add 0.025″ for each small line you can see beyond the last long line. Again, let us say you can see three. These three lines represent 0.075″, collectively (0.025″ × 3).
3. Finally, add the number from the thimble closest to the main measuring line on the sleeve. See **Figure 14-C.** Let us say the number is *18*. This number would represent 0.0018″. The total measurement would equal 0.393″, as shown in the following calculation:

   0.300″
   0.075″
   + 0.018″
   0.393″ total

   Practice by using a micrometer to measure a few spare parts. Ask your instructor to check your readings.

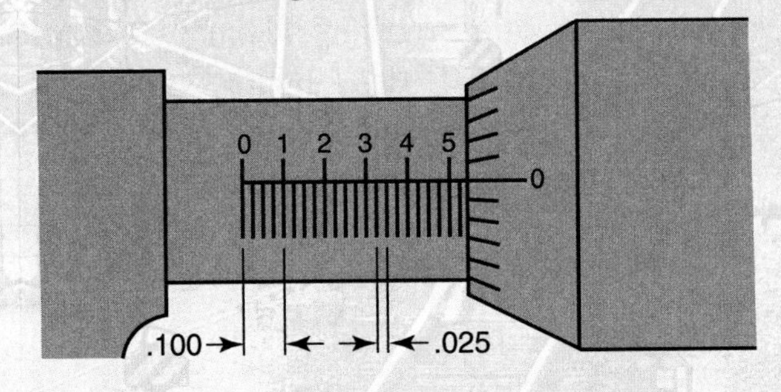

Figure 14-B. Each of the small spaces on the micrometer's sleeve is equal to one-fortieth of an inch. The distance between each large line is .100″, or one-tenth of an inch. The reading on this micrometer is .550″.

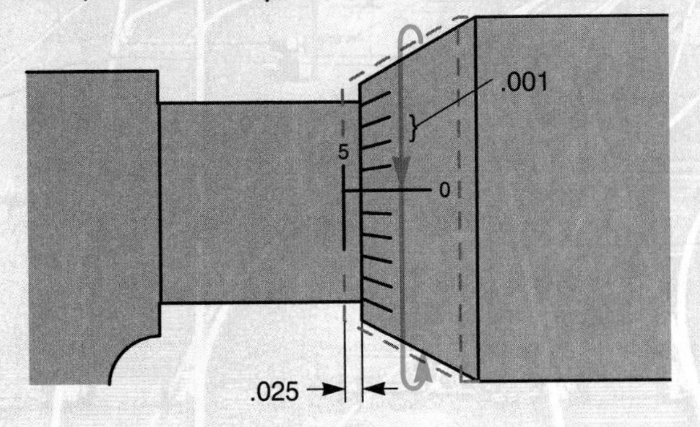

Figure 14-C. Each small line on the thimble is equal to .001″, or one-thousandth of an inch. One complete revolution of the thimble moves the spindle 25 thousandths of an inch, or one complete line on the sleeve.

Figure 14-17. The major parts of a small gas engine ignition system typically include breaker points, a condenser, a coil, flywheel magnets, and a spark plug. On newer engines, a combination coil and a solid-state sensing and switching system eliminate the need for breaker points and a condenser.

Figure 14-18. Spark plug electrode gapping. A—Checking the gap with a spark plug feeler gauge. Do not use a standard flat-bladed feeler gauge. Since the electrodes on a spark plug are rounded, this type of gauge will not provide an accurate reading. B—Bending the electrode to set the proper plug gap.

A

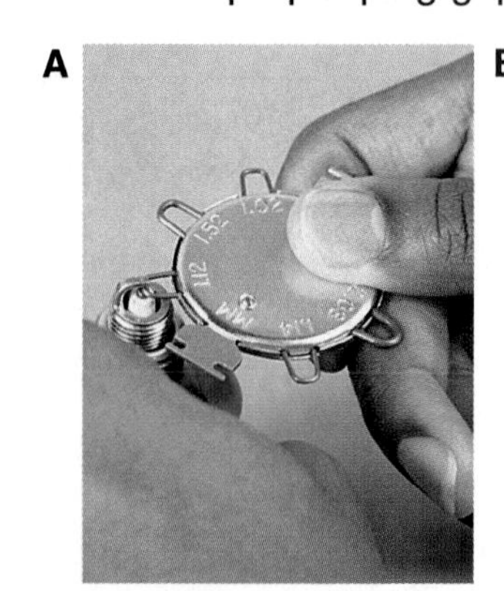

B

The flywheel must be aligned properly on the crankshaft for spark to occur. It is held in place by a small metal piece known as a key. If no key is installed, or if the key shears (breaks), the position of the flywheel on the crankshaft can shift. If the flywheel shifts position, the magnets will not be in the proper location at the time ignition is supposed to occur. See **Figure 14-17.** Also, for spark to occur, the armature must be located the proper distance from the flywheel. If the armature is overgapped (too far from the flywheel), a weak orange spark may result. Undergapping can cause the armature to rub against the flywheel and result in failure to produce a spark. To set the gap on some small gas engines, place an index card of appropriate thickness on the flywheel, and then bring the armature down onto the flywheel and lock it in place. Remove the index card, leaving the proper air gap.

A good spark plug is essential to proper engine performance. If the plug is cracked, the spark will jump through the ceramic insulator and over to the block, rather than jumping between the spark plug electrodes to create ignition. Plugs must be gapped properly. Overgapping or undergapping may result in failure to fire. To properly gap the plug, use a ***spark plug feeler gauge*** and bend the electrode as necessary. See **Figure 14-18.** Here are some generic diagnostic procedures for inspecting and reinstalling the electrical subsystem on a small gas engine:

- Inspect the flywheel key to ensure that the flywheel is seated in the proper location.
- Determine the proper armature air gap. When reinstalling, make sure the armature is properly spaced from the flywheel.
- Remove and inspect the spark plug. Look for cracks around the ceramic insulator ring. Determine the proper spark plug gap, and then use a spark plug feeler gauge and regap the plug as necessary.

- Inspect the thin wire that goes to the kill switch, if the engine is so equipped. Make sure the coating on the wire is not skinned or pinched anywhere along the way to the switch. If this wire is shorted to (in contact with) the block of the engine, spark will not occur.
- Test for spark by disconnecting the plug wire and using a spark plug tester. Be sure that a strong blue spark is being fed to the plug.

Spark plug feeler gauge: A tool used to properly gap a spark plug.

Governing subsystem: The subsystem designed to keep an engine running at a desired speed, regardless of the load applied to the engine.

The Governing and Fuel Subsystems

The governing subsystem and the fuel subsystem are described together because they work in conjunction with each other. The ***governing subsystem*** is designed to keep an engine running at a desired speed, regardless of the load applied to the engine. The ***fuel subsystem*** is responsible for creating the fuel-air mixture used to power the engine and delivering that charge to the combustion chamber, based on how much fuel-air charge the governing system allows. See **Figure 14-19.**

Some engines use a fuel injection system to introduce the fuel to the combustion chamber, instead of a carburetion system. When an engine is fuel injected, air is compressed, and fuel is injected into the cylinder prior to the piston reaching TDC. This type of system is common in many automobile and truck engines, including diesel engines, but carburetion is much more common in small gas engines. Major carburetor components and associated carburetor terms are described below:

- ***Venturi.*** The narrow, restricting section of the carburetor, where air speed increases and drafts the fuel vapor along with it into the combustion chamber.
- ***Choke.*** Usually, a platelike device that varies the amount of air that can enter the carburetor. When the engine is "choked" (the plate is mostly closed), more fuel vapor and less air are entering the combustion chamber. The primary reason for choking an engine is to create a ***rich mixture*** (a mixture with more fuel vapor than normal), which is desirable to get the engine started and warmed up to temperature during a cold start.
- ***Throttle.*** Another platelike device, located in back of the venturi, that regulates the amount of fuel-air mixture entering the carburetor.
- **Load.** The condition under which an engine runs when it is called on to do work. When an engine is running under load, both the choke and the throttle are fully (or almost fully) open.

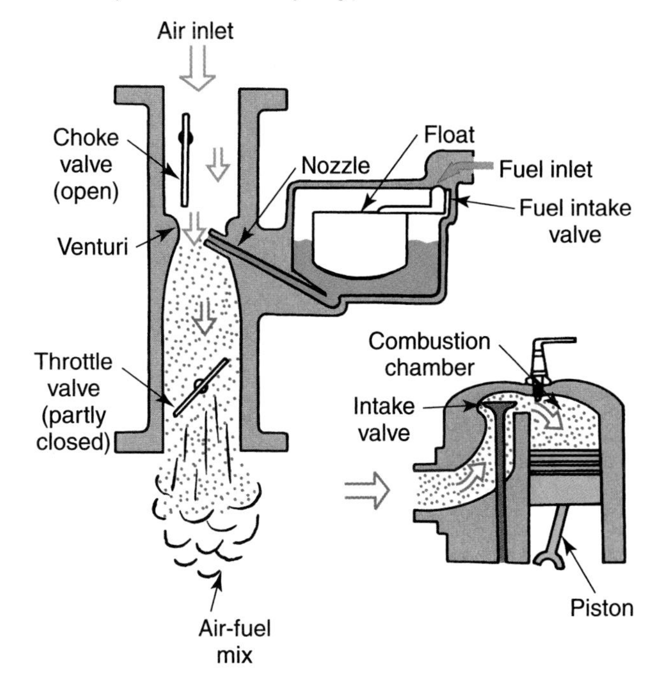

Figure 14-19. The main components and operation of a carburetor. Fuel vapor is drawn through the carburetor by the air that rushes past it as a result of a vacuum created by the intake stroke of the piston. The spark plug ignites the fuel-air mixture, and the expanding gases create a tremendous power surge that is transferred from the piston down to the crankshaft. (Deere & Company)

Fuel subsystem: The subsystem responsible for creating the fuel-air mixture used to power the engine and delivering that charge to the combustion chamber, based on how much fuel-air charge the governing system allows.

Venturi: The narrow, restricting section of the carburetor, where air speed increases and drafts the fuel vapor along with it into the combustion chamber.

- ***Idle.*** The condition an engine will run under when it is warmed up to temperature and not under load. When an engine is at idle, the choke is generally open, and the throttle is generally closed.
- ***Idle bypass circuit.*** A small passageway that allows some fuel-air mixture to escape around the throttle plate and keep the engine running, even when the throttle is closed.

How much fuel-air mixture enters the combustion chamber is the result of the ***governing mechanism***. Most types of governing mechanisms rely on the speed of the engine to determine whether more or less fuel-air mixture is needed. See **Figure 14-20.**

A centrifugal governor is directly linked to the throttle plate. See **Figure 14-21.**The faster the engine turns, the more the governor pulls the throttle toward its closed position, allowing less fuel-air mixture to enter. When the engine begins to slow down, the governing mechanism moves inward, opening the throttle. This allows more fuel-air mixture to enter the combustion chamber.

Measuring, Testing, and Troubleshooting

Like any complex machine, an engine needs maintenance, periodic testing, and troubleshooting. The following sections describe common terminology used when measuring and testing an engine. They also provide some helpful troubleshooting hints for small engine repair.

Figure 14-20. An air vane governing system for small gas engines. Increasing or decreasing airflow from the flywheel vanes is used to adjust the throttle position.

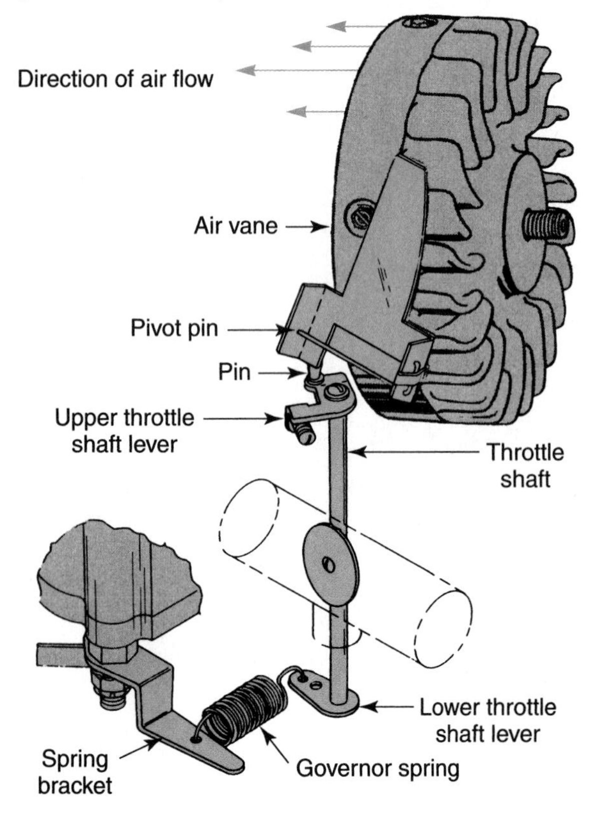

Efficiency

Volumetric efficiency measures how well the engine breathes. Breathing, in this sense, compares the amount of fuel-air mixture actually drawn into the cylinder with the maximum amount of fuel-air that could be drawn into the cylinder if it were completely filled. ***Mechanical efficiency*** is the percentage of power developed in the cylinder compared to the power actually delivered to the crankshaft.

Thermal efficiency, sometimes referred to as *heat efficiency*, is a measurement of how much heat is actually used to drive the piston downward. Internal combustion engines are not particularly efficient conversion devices. You may be surprised to learn that about 75% of all the heat produced during combustion is not transferred to the piston.

Practical efficiency is perhaps the most useful term of all. It is simply a measurement of how efficiently an engine uses its fuel supply. If an engine is used for motive power, this could be expressed in terms of miles per gallon. The practical efficiency measurement of

Figure 14-21. Operation of a centrifugal governor. The governing mechanism automatically regulates the amount of fuel-air mixture required to keep the engine running at a relatively constant speed under load conditions ranging from idle through full load.

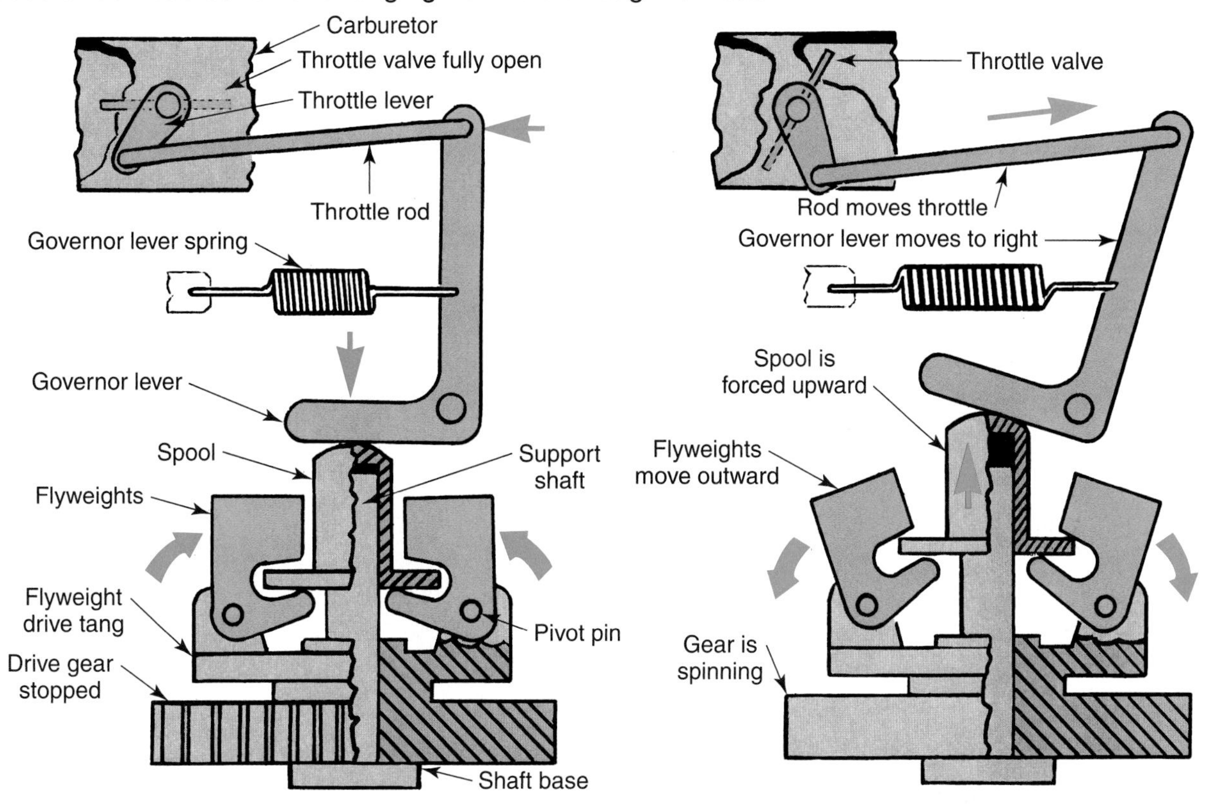

an engine takes into account all losses (such as frictional loss, incomplete combustion, and thermal loss), leaving only a comparison of potential energy in to useful power out.

Horsepower (hp)

There are many different uses of the term *horsepower (hp)*. As described in Chapter 9, the term came about as a means of comparing the power output of James Watt's steam engine to the amount of work a horse could do. Watt's constant for hp was determined to be 550 foot-pounds (ft.-lbs.) per second, meaning that a draft horse could perform about 550 ft.-lbs. of work in 1 second. Engines are typically rated in terms of their hp capability. This capability is affected by many factors, but the two most important are the bore and the stroke of the engine. The ***bore*** refers to the diameter of the cylinder. The stroke refers to the maximum length of piston travel. An engine with a greater bore or stroke (or both) should yield greater hp. Another factor that greatly influences hp output is frictional loss. Antifriction bearings will help to reduce friction, but they are more expensive to use than friction bearings.

The characteristics of power produced by a small gas engine can be measured on a dynamometer and used to calculate hp. See **Figure 14-22.** Today, there are several different terms for hp, all with different meanings. Following are the most common terms for hp, along with a brief explanation of each.

Choke: A platelike device that varies the amount of air that can enter the carburetor.

Rich mixture: A fuel mixture with more fuel vapor than normal.

Throttle: A platelike device, located in back of the venturi, that regulates the amount of fuel-air mixture entering the carburetor.

Idle: The condition an engine will run under when it is warmed up to temperature and not under load.

Idle bypass circuit: A small passageway that allows some fuel-air mixture to escape around the throttle plate and keep the engine running, even when the throttle is closed.

Governing mechanism: A mechanism that determines how much fuel-air mixture needs to enter the combustion chamber, based on the speed of the engine.

Volumetric efficiency: A measurement of how well the engine breathes.

Mechanical efficiency: The percentage of power developed in the cylinder compared to the power actually delivered to the crankshaft.

Thermal efficiency: A measurement of how much heat is actually used to drive the piston downward.

Practical efficiency: A measurement of how efficiently an engine uses its fuel supply.

Bore: The diameter of a cylinder.

Figure 14-22. A dynamometer is used to measure the horsepower (hp) output of a small gas engine. (Go-Power Corporation)

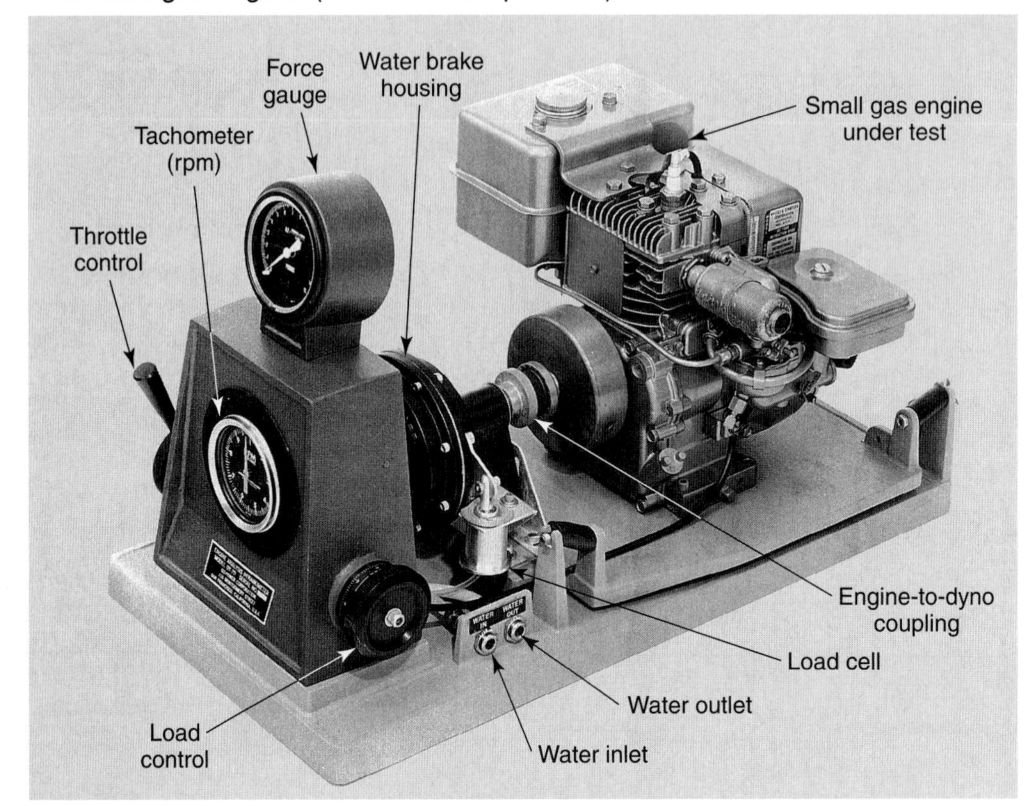

Brake horsepower (bhp) is the hp available for use at the crankshaft. Typically, bhp will increase with engine rpm. It will actually decrease, however, if the engine rpm climbs into a range beyond the engine's normal rating.

Indicated horsepower (ihp) is a more theoretical hp term. It measures the power developed by the fuel-air charge upon ignition in the combustion chamber. This type of hp is actually derived by taking an average of the pressure within the cylinder during all four strokes. Other factors involved when calculating ihp include the volume within the cylinder and the number of cylinders in the engine.

Frictional horsepower (fhp) represents the part of the potential hp or ihp lost due to friction within the engine. It can be calculated by subtracting the bhp available at the crankshaft from the ihp. The ihp is the theoretical hp available from the fuel-air charge in that size cylinder or engine:

ihp – bhp = fhp

Rated horsepower (rhp) is another often-used hp term. It typically represents about 80% of the engine's bhp capability. (Engines should not be run at full load for extensive periods of time. Doing so will cause excessive wear and premature engine failure.) The rhp is what is labeled on the engine. The 80% rule can be helpful in sizing engines for specific applications. For instance, if a portable pump requires a constant 6 hp to operate, it would be desirable to power it with an engine rated at 7.5 hp or greater to provide an adequate safety margin. Using an undersized engine could lead to increased fatigue and premature engine failure.

Engine Testing

Testing hp typically occurs at the factory to ensure that new engines are performing up to specification. A series of more practical tests can be performed in the field to determine the causes of engine malfunctions. Two of the most common field tests are spark tests and compression tests.

A ***spark test*** can be performed to determine if the magneto is producing appropriate spark. It does not indicate whether or not the spark is occurring at the correct time, but it can determine the presence and quality of spark. The tester is connected between the plug wire and the top of the spark plug. See **Figure 14-23.** When the starter cord is pulled to rotate the flywheel, the quality of the spark can be observed through a sight hole (window) in the tester. A strong blue spark is desirable. An orange color indicates a weak spark that may not jump across the plug electrodes. When performing a spark test, be careful to avoid a shock. Do not touch any exposed electrical connections.

Another way to check for spark is to unscrew the plug from the cylinder head. Reconnect the plug wire, and use insulated pliers to hold the bottom electrode of the plug tight against the engine block. Have someone pull the starter cord. Watch as the spark jumps across the electrodes on the plug.

A ***compression test*** can be performed with a pressure gauge covering the spark plug hole. Some compression gauges thread into the spark plug hole. Once the gauge is installed, the engine is rotated as it would be for starting. If the pressure that would build up during compression is significantly less than the manufacturer's specification, a leak is likely occurring. See **Figure 14-24.** A loss of compression could occur from any of the following:

- A cracked compression ring.
- A blown head gasket.
- Worn valves.
- Poor seating of valves.
- A worn cylinder.

Figure 14-23. Using a spark tester to check for spark.

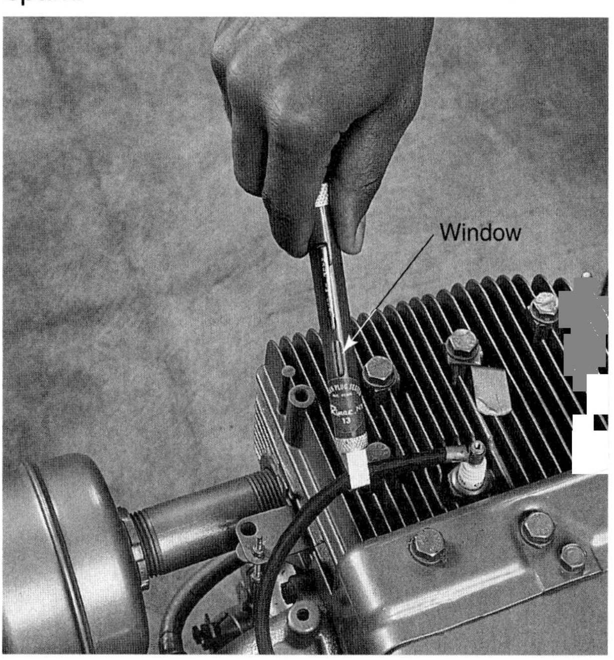

Figure 14-24. Performing a compression test with a handheld compression gauge.

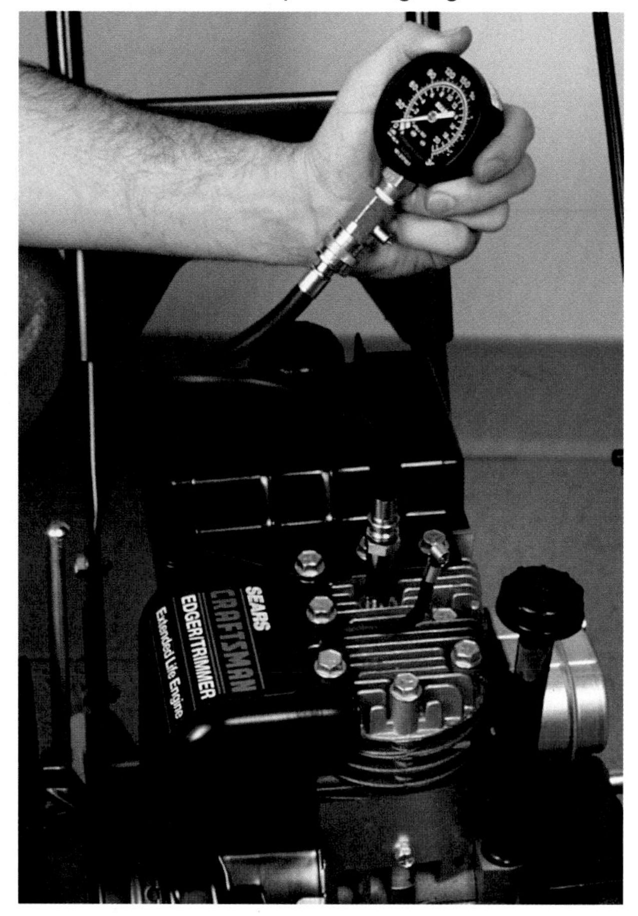

Rated horsepower (rhp): About 80% of the engine's brake horsepower (bhp) capability.

Spark test: A test performed to determine if the magneto is producing appropriate spark.

Compression test: A test performed with a pressure gauge covering the spark plug hole. Once the gauge is installed, the engine is rotated as it would be for starting. If the pressure that would build up during compression is significantly less than the manufacturer's specification, a leak is likely occurring.

Troubleshooting

There are a number of common problems associated with small gas engines. See **Figure 14-25.** Use the list of various possible causes and remedies as a starting point for troubleshooting.

Basic Tools for Small Engine Repair

Prior to disassembling an engine or any other complex piece of machinery, it is always wise to review the tools you will need. This is especially true if the task requires specialty tools. Such is the case with repairing the small gas engine. See **Figure 14-26.** Some of the tools needed are found in most toolboxes, and others are specific to working on engines. Tools for testing and measuring are also needed.

Helpful Hints about Engine Disassembly

When all else fails, you may have to disassemble the engine to perform internal troubleshooting or to replace such parts as gaskets or seals. Engine disassembly is easy (compared to engine reassembly), because you do not need to worry about critical tolerances during disassembly. There is a correct way and an incorrect way, however, to take apart an engine or any other complex piece of machinery with many parts. Seasoned engine technicians use several specific techniques to ensure that parts are reinstalled correctly, in the appropriate sequence, and are not misplaced. Here are a few helpful hints from veteran mechanics about engine disassembly that will prove to be very helpful when it is time to reassemble the engine:

- When possible, replace bolts in the same holes from which they were removed. For example, once the starter housing has been removed, the screws that held the housing to the face of the engine should be screwed back into the engine block. Using this technique, it is virtually impossible to lose a screw or to mistake one screw for another taken from some other part of the engine. This is important because even screws and bolts that are the same size as one another can be made of different alloys for different purposes. For example, some may be intended for high-torque and high-heat applications, while others are not.
- Prior to disassembly, make drawings of complex linkages, such as the governor linkage on the top of the engine. You might think you will remember where all those springs and linkage arms belong, but chances are, you will not. For more complex engines, photographs can be helpful. Pictures probably will not be necessary for small gas engine disassembly, but drawings of complex assemblies might be very useful.
- Use pencil marks and notes to indicate the locations of variable adjustments prior to disassembly. This will help you restore the initial settings following reassembly.
- Pay particular attention to small, easily misplaced parts, such as the key that holds the flywheel onto the crankshaft. You may wish to store such parts in a special place so they are not misplaced.

Figure 14-25. A troubleshooting chart for small gas engines.

Engine Troubleshooting Chart	
Cause	**Remedy**
Engine fails to start or starts with difficulty	
No fuel in tank.	Fill tank with clean, fresh fuel.
Shut-off valve closed.	Open valve.
Obstructed fuel line.	Clean fuel screen and line. In necessary, remove and clean carburetor.
Tank cap vent obstructed.	Open vent in fuel tank cap.
Water in fuel.	Drain tank. Clean carburetor and fuel lines. Dry spark plug and points. Fill tank with clean, fresh fuel.
Engine overchoked.	Close fuel shut-off and pull starter until engine starts. Reopen fuel shut-off for normal fuel flow.
Improper carburetor adjustment.	Adjust carburetor.
Loose or defective magneto wiring.	Check magneto wiring for shorts or grounds; repair if necessary.
Faulty magneto.	Check timing, point gap; if necessary, overhaul magneto.
Spark plug fouled.	Clean and regap spark plug.
Spark plug porcelain cracked.	Replace spark plug.
Poor compression.	Overhaul engine.
No spark at plug.	Disconnect ignition cut-off wire at the engine. Crank engine. If spark at spark plug, ignition switch, or safety switch, interlock switch is inoperative. If no spark, check magneto.
Crankcase seals and/or gaskets leaking (two cycle only).	Replace seals and/or gaskets.
Exhaust ports plugged (two cycle only).	Clean exhaust ports.
Engine knocks	
Carbon in combustion chamber.	Remove cylinder head and clean carbon from head and piston.
Loose or worn connecting rod.	Replace connecting rod.
Loose flywheel.	Check flywheel key and keyway; replace parts if necessary. Tighten flywheel nut to proper torque.
Worn cylinder.	Replace cylinder.
Improper magneto timing.	Time magneto.
Engine misses under load	
Spark plug fouled.	Clean and regap spark plug.
Spark plug porcelain cracked.	Replace spark plug.
Improper spark plug gap.	Regap spark plug.
Pitted magneto breaker points.	Replace pitted breaker points.
Magneto breaker arm sluggish.	Clean and lubricate breaker point arm.
Faulty condenser.	Check condenser on a tester; replace if defective.
Improper carburetor adjustment.	Adjust carburetor.
Improper valve clearance.	Adjust valve clearance to recommended specifications.

(Continued)

Figure 14-25. *Continued.*

Engine Troubleshooting Chart	
Cause	**Remedy**
Engine misses under load	
Weak valve spring.	Replace valve spring.
Reed fouled or sluggish (two cylce only).	Clean or replace reed.
Crankcase seals leak (two cycle only).	Replace worn crankcase seals.
Engine lacks power	
Choke partially closed.	Open choke.
Improper carburetor adjustment.	Adjust carburetor.
Magneto improperly timed.	Time magneto.
Worn rings or piston.	Replace rings or piston.
Air cleaner fouled.	Fill crankcase to proper level.
Lack of lubrication (four cycle only).	Clean air cleaner.
Valves leaking (four cycle only).	Grind valves and set to recommended specifications.
Reed fouled or sluggish (two cycle).	Clean or replace reed.
Improper amount of oil in fuel mixture (two cycle only).	Drain tank; fill with correct mixture.
Crankcase seals leak (two cycle only).	Replace worn crankcase seals.
Engine overheats	
Engine improperly timed.	Time engine.
Carburetor improperly adjusted.	Adjust carburetor.
Air flow obstructed.	Remove any obstructions from air passages in shrouds.
Cooling fins clogged.	Clean cooling fins.
Excessive load on the engine.	Check operation of associated equipment. Reduce excessive load.
Carbon in combustion chamber.	Remove cylinder head and clean carbon from head and piston.
Lack of lubrication (four cycle only).	Fill crankcase to proper level.
Improper amount of oil in fuel mixture (two cycle only).	Drain tank; fill with correct mixture.
Engine surges or runs unevenly	
Fuel tank cap vent hole clogged.	Open vent hole.
Governor parts sticking or binding.	Clean and, if necessary, repair governor parts.
Carburetor throttle linkage or throttle shaft and/or butterfly binding or sticking.	Clean, lubricate, or adjust linkage and deburr throttle shaft or butterfly.
Intermittent spark or spark plug.	Disconnect ignition cut-off wire at the engine. Crank engine. If spark, check ignition switch, safety switch, and interlock switch. If no spark, check magneto. Check wires for poor connections, cuts, or breaks.
Improper carburetor adjustment.	Adjust carburetor.
Dirty carburetor.	Clean carburetor.

(Continued)

Figure 14-25. *Continued.*

Engine Troubleshooting Chart	
Cause	**Remedy**
Engine vibrates excessively	
Engine not securely mounted.	Tighten loose mounting bolts.
Bent crankshaft.	Replace crankshaft.
Associated equipment out of balance.	Check associated equipment.
Engine uses excessive amount of oil (four cycle only)	
Engine speed too fast.	Using tachometer, adjust engine rpm to specifications.
Oil level too high.	To check level, turn dipstick cap tightly into receptacle for accurate level reading.
Oil filler cap loose or gasket damaged, causing spillage out of breather.	Replace ring gasket under cap and tighten cap securely.
Breather mechanism damaged or dirty, causing leakage.	Replace breather assembly.
Drain hole in breather box clogged, causing oil to spill out of breather.	Clean hole with wire to allow oil to return to crankcase.
Gaskets damaged or gasket surfaces nicked, causing oil to leak out.	Clean and smooth gasket surfaces. Always use new gaskets.
Valve guides worn excessively, thus passing oil into combustion chamber.	Ream valve guide oversize and install 1/32″ oversize valve.
Cylinder wall worn or glazed, allowing oil to bypass rings into combustion chamber. Piston rings and grooves worn excessively.	Bore hole or deglaze cylinder as necessary. Reinstall new rings, check land clearance, and correct as necessary.
Piston fit undersized.	Measure and replace as necessary.
Piston oil control ring return holes clogged.	Remove oil control ring and clean return holes.
Oil passages obstructed.	Clean out all oil passages.

- If possible, try to salvage gaskets. An actual engine overhaul will require a replacement set of gaskets and seals to restore the engine to original condition. If the engine is being taken apart simply as a learning experience, however, saving the existing gaskets and seals is much less costly than using replacement items. Sometimes, it is simply not possible to salvage a particular gasket or seal, and using a replacement is the only option.
- Use a repair manual that has been designed for your specific engine, if possible. There are critical specifications and information about replacement parts that will be difficult to track down without the assistance of such a manual.

Figure 14-26. The most essential tools necessary for engine disassembly, inspection, repair, and reassembly. (Monarch Instruments)

Tool		Description
Screwdrivers	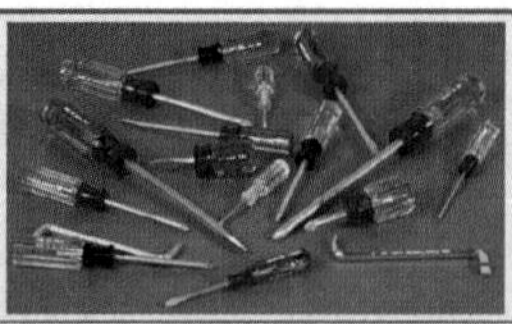	Small hand tools used for turning screws. They may be regular or Phillips head screwdrivers.
Nut drivers	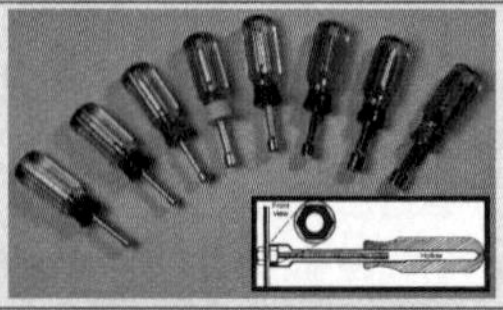	Small hand tools used for turning nuts.
Slip-joint pliers	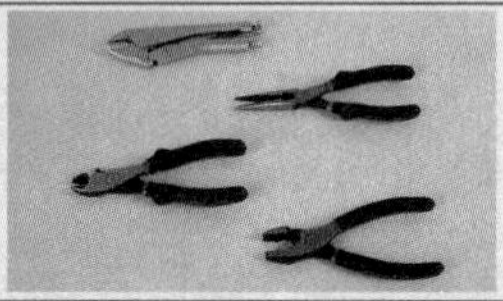	Small hand tools used for gripping and holding objects.
Steel rule	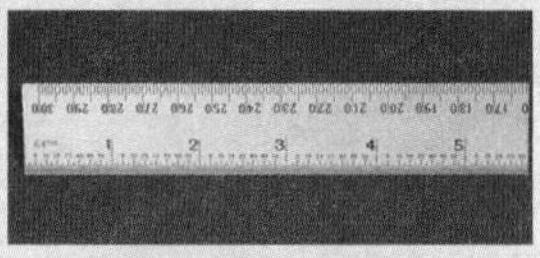	A small hand tool used for measuring in 1/64″ divisions. It can be used to make measurements in both U.S. customary and SI metric units.
Quarter-inch drive socket set	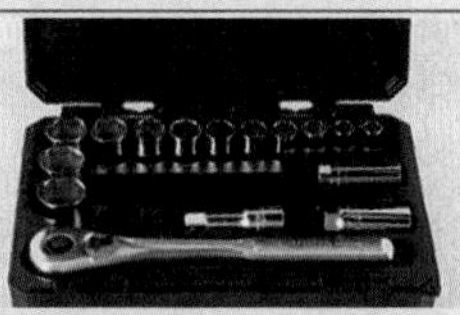	Small hand tools used for installing and removing nuts and bolts in hard-to-reach places.
Combination wrenches	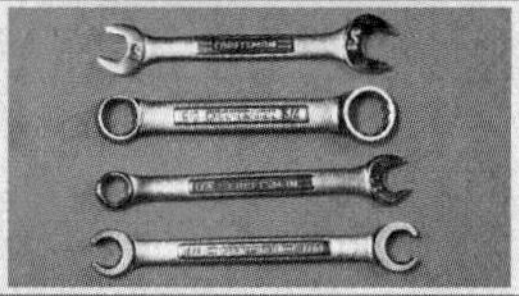	Small hand tools used for loosening and tightening nuts and bolts on the engine that cannot be addressed with sockets.
Adjustable wrench	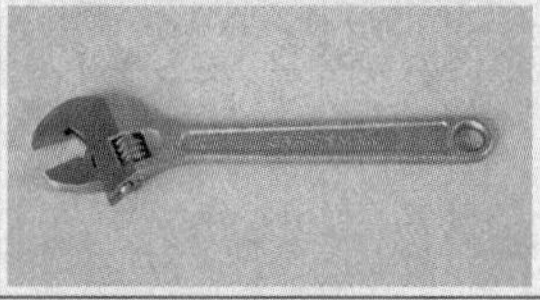	A small hand tool used for loosening and tightening nuts and bolts.
Torque wrench	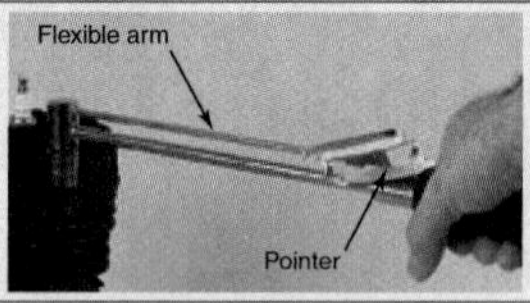 Flexible arm Pointer	An engine tool used for installing fasteners, such as head bolts and connecting rod bolts, which must be torqued to specification. It can measure torque in inch-pounds (in.-lbs.) or foot-pounds (ft.-lbs.).
Ring expander	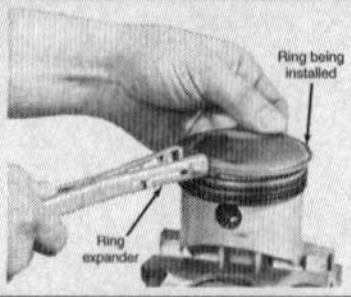 Ring being installed Ring expander	An engine tool used for removing the rings from the piston so they can be examined for wear.
Ring compressor		An engine tool used for holding the rings tightly to the piston. It wraps around the rings so the piston and rod assembly can be reinserted into the cylinder.
Valve spring compressor	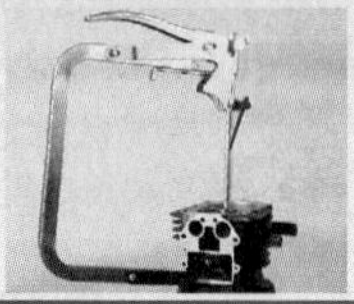	An engine tool used for overcoming tension from the intake and exhaust valves so the valves can be removed for inspection or replacement. It squeezes the valve spring so the keeper can be easily removed.

(Continued)

Figure 14-26. *Continued*

Square and surface plate		A testing tool used for checking parts for warpage. It is a highly polished stone that has been ground perfectly flat. Engine parts, such as the engine head, valves, and valve springs, can be checked for straightness, proper length, and squareness.
Feeler gauge		A testing tool used for checking critical gaps within an engine. The fine blades are used to ensure that gaps are within critical tolerances.
Spark plug feeler gauge		A testing tool used for gapping the electrodes on a spark plug.
Outside micrometer		A precision measuring tool used for checking for critical wear points on engine parts. It is used to measure the external diameter of a part, such as a piston pin.
Inside micrometer		A precision measuring tool used for checking for critical wear points on engine parts. It is used to measure the internal diameter of a part, such as a cylinder.
Telescoping gauge		A measuring tool used for checking the diameter of interior openings, such as valve guides or engine cylinders. It is used in conjunction with a micrometer to yield accurate measurements.
Tachometer		A measuring tool used for measuring the speed of an engine for diagnosis and tune-up.

Career Connection

Diesel Mechanics

Trucks, buses, and other large machines may be powered by diesel engines. When these machines no longer function correctly, mechanics are needed to service and maintain them. Diesel mechanics' tasks are very similar to those of other automotive technicians. These mechanics are required to diagnose and repair problems with such equipment as brakes and suspension systems. Diesel engines, however, are different from small gas engines. To repair an engine, a mechanic in this field must possess working knowledge of the assembly and functions of the diesel engine. This is especially important in cases of rebuilding engines.

Apart from mechanical skills, diesel mechanics must be able to work with customers, which may require both written and verbal communication skills. They must be able to conduct tests within accepted safety limits. A working knowledge of computers and design techniques is also important. A vocational certificate is necessary to be a worker in this field. The yearly salary may range from $22,000 to $57,000.

Small Gas Engines and the Environment

When it comes to protecting the environment, air pollution from small gas engines might not immediately come to mind, but it should. Emissions from lawn mowers, chain saws, and other outdoor equipment powered by small gas engines are significant sources of pollution in the United States. According to one estimate, small gas engines produce almost 7 million tons of air pollution yearly. It is also estimated that mowing a lawn for about half an hour will produce emissions equivalent to driving a car for 172 miles! See **Figure 14-27.** Small gas engines contribute about half of all nonroad exhaust emissions.

The result is that small gas engines have come under increasing scrutiny from environmental agencies. Regulations were first established by the California Air Resources Board in 1995. Because California is such a huge market, the regulations have caused manufacturers to redesign small gas engines to be more environmentally friendly. The Environmental Protection Agency (EPA) proposed tougher federal standards during the 1990s. These standards are being phased in, with all expected to be in place soon. To date, most small gas engine manufacturers claim to be meeting the detailed regulations set in place by California and the EPA through a combination of modifications, including the use of adjustable overhead valve trains, solid-state ignition systems, and better oil control designs. These modifications add cost to a new engine, but they are having some impact. It is estimated that a reduction of 390,000 tons of hydrocarbons and nitrous oxides going into the atmosphere will occur annually by 2027, as a result of these engine modifications. The best part—all this is anticipated to occur, while fuel consumption per engine is expected to decrease!

Figure 14-27. Nonroad exhaust gas emissions by percentage.

Urban Summertime Hydrocarbon Nonroad Sources	
Small spark-ignition engines	50%
Recreational boats	30%
Other nonroad engines	20%

Summary

Engines can be of either internal or external combustion design. Internal combustion engines can be further classified as either two-stroke cycle or four-stroke cycle engines. Regardless of engine type, the processes of intake, compression, power, and exhaust must occur. Two-stroke cycle engines have some inherent advantages over four-stroke engines, but they also have some disadvantages, such as shorter service lives. The type of engine used is often determined by the application. Engines are complex machines because they require several subsystems, all of which must be operational for the engine to function properly. These subsystems include the mechanical, electrical, fuel, governing, lubrication, and cooling subsystems. The power produced by an engine is typically rated in terms of horsepower (hp). There are many different ways to measure hp, but brake horsepower (bhp) and the rated horsepower (rhp) are the two used most often. Frictional horsepower (fhp) refers to hp lost as the result of internal friction within the engine. Working on an engine requires many tools, including specialty items, such as ring compressors, ring expanders, strap wrenches, valve spring compressors, and precision measuring instruments. In the 1990s, several pieces of legislation were passed to protect the environment by regulating emissions produced by small gas engines. These regulations have resulted in the redesign of small gas engines to make them more environmentally friendly.

Key Words

All the following words have been used in this chapter. Do you know their meanings?

atomized
bore
bottom dead center (BDC)
cam lobe
camshaft
choke
combustion chamber
compression ratio
compression stroke
compression test
connecting rod
cooling fin
cooling subsystem
crankpin journal
crankshaft
cylinder
electrical subsystem
engine ring
exhaust stroke
external combustion engine
feeler gauge
flywheel
four-stroke cycle engine
fuel-air charge
fuel subsystem
governing mechanism
governing subsystem
idle
idle bypass circuit
inertia
intake stroke
internal combustion engine
key
lubrication subsystem
margin thickness
mechanical efficiency
mechanical subsystem
micrometer
multigrade
oil dipper
piston
power stroke
practical efficiency
radiator
rated horsepower (rhp)
rich mixture
Society of Automotive Engineers (SAE)
spark plug
spark plug feeler gauge
spark test
stroke
thermal efficiency
thermostat
throttle
top dead center (TDC)
two-stroke cycle engine
valve lifter
venturi
volumetric efficiency
water jacket

Test Your Knowledge

Write your answers on a separate sheet of paper. Do not write in this book.

1. Heat is generated outside the cylinder in a(n) _____ combustion engine.
2. A comparison of the volume in the cylinder with the piston at bottom dead center (BDC) versus TDC is known as the _____.
3. *True or False?* Ignition occurs before the piston reaches top dead center (TDC) on the intake stroke.
4. Describe each step in the four-stroke cycle process.
5. What are the two strokes commonly associated with a two-stroke cycle engine?
   A. Intake and exhaust.
   B. Intake and power.
   C. Compression and power.
   D. Power and exhaust.
6. One complete revolution of the crankshaft will produce a power stroke in a(n) _____ engine.
7. List two advantages of four-stroke engines over two-stroke engines.
8. A(n) _____ attached to the connecting rod provides splash lubrication for many small gas engines.
9. Describe four things that oil does besides provide lubrication.
10. *True or False?* The camshaft is responsible for opening the valves.
11. Identify three things that the flywheel is responsible for in a small gas engine.
12. List at least four critical wear areas within an engine.
13. *True or False?* The choke is responsible for varying the percentage of air in the fuel-air mixture that enters the cylinder.
14. When starting a cold engine, a(n) _____ fuel-air mixture is desirable.
15. Identify six causes of malfunctions in engines.
16. If your small gas engine will not start and you know it has fuel, a logical troubleshooting step would be to:
    A. check for oil.
    B. check the cooling fins.
    C. check the intake valve.
    D. check for spark.
17. *True or False?* An outside micrometer can be used to measure the thickness of a valve stem.

Activities

1. If lab facilities are available, work in a team with several other students to completely disassemble and reassemble a small gas engine. The engine should then be tested and adjusted for optimal performance, if possible.
2. Disassemble a carburetor and identify all the necessary components found within it.
3. Research a topic related to internal combustion engines, such as new engine designs, pollution control devices, or improvements in efficiency. Write a short paper or make a presentation to the class.

Night vision technology makes a "heads-up" display of the road ahead possible in this new automotive system, which is under development. The technology is designed to make night driving safer by helping a driver identify a potential hazard earlier. (Siemens VDO)

15

An Introduction to Transportation Systems

Basic Concepts

- Define a transportation system.
- List the five types of transportation systems.
- Name several transportation system inputs.
- Recognize several transportation system processes.
- State the expected output of a transportation system.
- Identify the types of goals that affect a transportation system.

Intermediate Concepts

- Explain the function of feedback within a transportation system.
- Make a list of devices used to provide feedback in transportation systems.
- Describe the functions of at least one government agency that controls transportation.

Advanced Concepts

- Create a flowchart of the production processes in a specific environment of transportation.
- Give examples of societal and economic goals of transportation systems.

When we think of transportation, we often think of our personal cars and trucks. We may forget there is a "moving" world all around us. The size of the transportation system in the United States alone is enormous. The roads within the system could extend to the moon and back 8 times. The railroad tracks used in just the United States could circle the earth 7 times. Oil and gas pipelines are, quite possibly, the most overlooked type of transportation. The length of these pipelines would wrap the earth over 55 times! Other often overlooked forms of transportation include the following:

- Escalators that move people from one level of a building to another.
- Satellites that circle the earth.
- Conveyors that move products and materials.

Types of Transportation Systems

Transportation system: An organized process of relocating people and cargo using the various modes of transportation.

Transportation is one of our major technological systems. As defined in Chapter 1, a *system* is a combination of parts that work together to accomplish a desired result. A ***transportation system*** is a group of components, including inputs, processes, outputs, and feedback, used together to move people and goods from one location to another. The various transportation systems can be placed into one of the following categories:

- Land transportation.
- Water transportation.
- Air transportation.
- Space transportation.
- Intermodal transportation.

Each of these transportation systems functions as a separate system. See **Figure 15-1.** In addition to the transportation system categories, transportation vehicles fall into one of four transportation system environments—land, water, air, or space. See **Figure 15-2.**

Land Transportation

Land transportation: A transportation system using vehicles on land, including subways, buses, trains, trucks, bicycles, and motorcycles.

Land transportation is the movement of goods and people in a vehicle operating above or under the ground. Think of all the types of transportation vehicles you see on land. Vehicles include subways, buses, trains, trucks, bicycles, and motorcycles. These, and all other types of land transportation vehicles, can be placed into one of five different modes of land transportation:

Figure 15-1. There are five different types of transportation systems. A—This light rail train is a form of land transportation. (Siemens) B—Cargo ships are a type of water transportation vehicle. C—Commercial airliners are the most common type of vehicle in air transportation. D—This is a privately financed space transportation vehicle shown during a test flight. (Scaled Composites) E—Intermodal transportation interchanges people or cargo between different types of vehicles. These containers have been unloaded from a ship and will be reloaded onto train cars or trucks.

A

B

C

D

E

Figure 15-2. There are four different transportation environments. Each has specific vehicles that can travel in that environment.

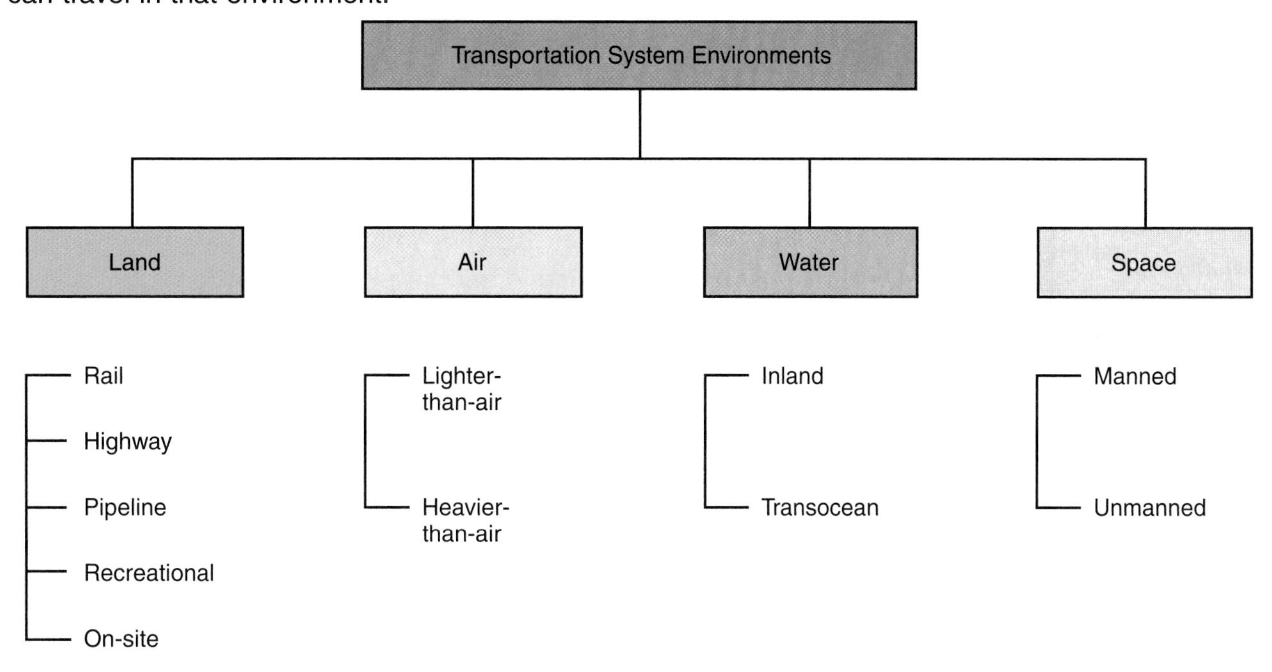

- Highway transportation occurs on local roads and highways.
- Rail transportation involves vehicles that use rails for guidance while traveling.
- Pipeline transportation moves cargo through stationary pipes.
- On-site transportation moves people and cargo short distances, such as between buildings or complexes.
- Recreational transportation is used for fun, sport, and general leisure.

Land transportation is discussed in greater detail in Chapter 17 and Chapter 18.

Water Transportation

Water transportation is the way in which people and cargo are moved on bodies of water. There are two forms of water transportation: inland (rivers, canals, and lakes) and transoceanic (across the ocean). Vessels used in water transportation include ships, sailboats, barges, tugboats, and submarines. Passenger vessels are used for commercial transportation, such as the ferries used by daily commuters, and to transport people for recreation, such as cruise liners and sailboats. The armed forces also use military versions of passenger vessels in many of their branches and operations. Vessels that move cargo typically transport goods within and between countries. Water transportation is discussed in greater detail in Chapter 19 and Chapter 20.

Air Transportation

Air transportation is the movement of people or cargo above the ground and within the earth's atmosphere. When you think about air transportation, one of the first things that comes to mind may be an airplane. Other examples include hot air balloons, airships, hang gliders,

Water transportation: A transportation system in which people and cargo are moved on bodies of water. Vessels used include ships, sailboats, barges, tugboats, and submarines.

Air transportation: A transportation system using either lighter-than-air or heavier-than-air modes of transportation, including airplanes, hot air balloons, airships, hang gliders, military fighter jets, and helicopters.

military fighter jets, and helicopters. All air transportation vehicles may be categorized as either lighter-than-air or heavier-than-air modes of transportation. ***Lighter-than-air*** vehicles rise and float on their own, such as balloons and hang gliders. ***Heavier-than-air*** vehicles require power to create the movement needed to transport people and cargo, such as airplanes, jets, and helicopters. Air transportation is discussed in greater detail in Chapter 21 and Chapter 22.

Lighter-than-air: Of less weight than the air displaced.

Heavier-than-air: Of greater weight than the air displaced.

Space Transportation

Space transportation is the movement of people or cargo within near space and into outer space. Some of the vehicles used in space transportation include missiles, rockets, satellites, space shuttles, and spacecraft. Space transportation vehicles can be categorized as manned or unmanned systems, meaning either a human is inside the vehicle operating the controls or the vehicle does not require onboard human operation. These vehicles serve many purposes. Satellites are commonly used to collect data on the earth's atmospheric and geographical changes for weather prediction and scientific study. Probes are launched into deep space for exploration purposes. Vehicles such as the space shuttle are used to transport people and cargo into space for exploration, scientific studies, and sometimes, for construction and repair of man-made structures in space. Space transportation is discussed in greater detail in Chapter 23 and Chapter 24.

Space transportation: A transportation system in which people and cargo are moved within near space and into outer space.

Intermodal Transportation

While most systems use only one of the environments (land, water, air, or space) for transportation, some systems utilize more than one environment. When more than one environment is used in a system, it is called ***intermodal transportation***. For example, intermodal transportation is used to transport imported products to cities in the midwestern United States. A container of products may cross the ocean on a ship, be transported inland on a railroad car, and be delivered to the final destination on a tractor-trailer. Products transported using intermodal transportation are exposed to several transportation environments and modes. The two types of intermodal transportation are cargo and passenger. Intermodal cargo transportation involves moving cargo from one point to another using various modes of transportation. Intermodal passenger transportation is the process of moving people using various modes of transportation. Think of the various modes of transportation you used on your last out-of-state vacation. During a single trip, you may have been transported by car, taxi, airplane, escalator, moving sidewalk, shuttle bus, and boat. Intermodal transportation is discussed in greater detail in Chapter 25.

Intermodal transportation: A transportation system that uses more than one environment or mode.

Components of Transportation Systems

Transportation systems, like all technological systems, operate based on the needs of society. The development and use of any transportation system consists of inputs, processes, and outputs. See **Figure 15-3.**

Figure 15-3. For every type of technological system, there are goals, inputs, processes, feedback, an output, and a method of control.

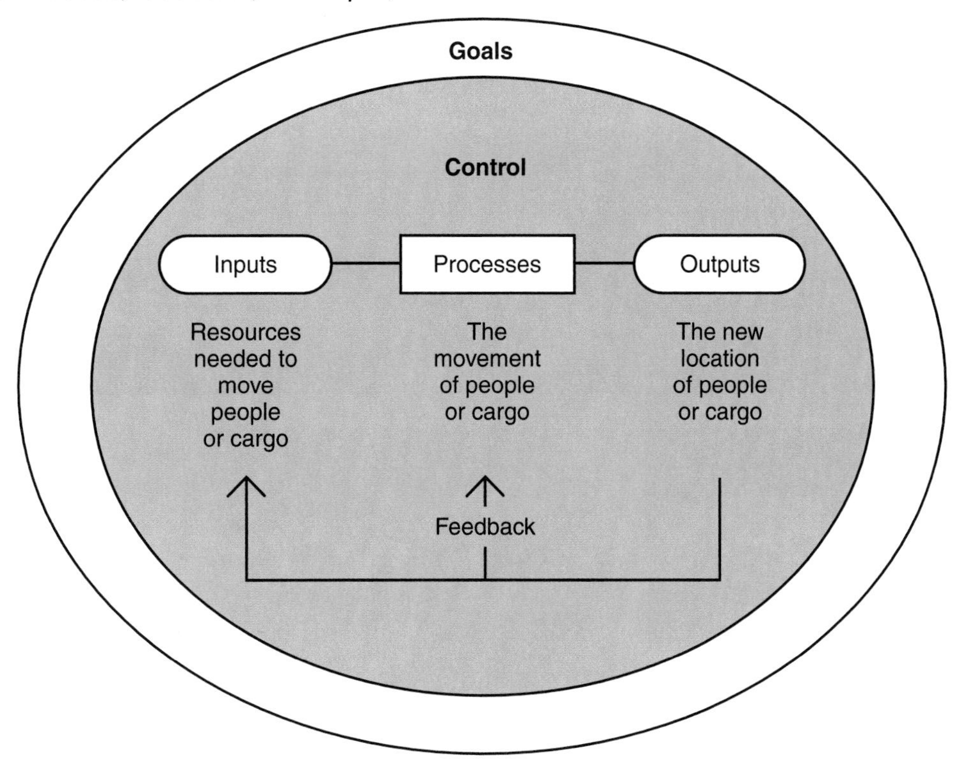

Transportation System Inputs

Transportation system inputs are the various resources needed in order to begin and maintain the use of the system. These resources include the following:

- **People.** They serve as both operators and passengers in transportation systems. Regardless of the mode of transportation, people are needed for repairs, manual operation, ticket sales, and instrument monitoring. See **Figure 15-4.**
- **Capital.** Within a transportation system, this includes the assets used to help operate the system and the possessions of a transportation company. Vehicles, roads, and buildings are all forms of capital in transportation systems.
- **Knowledge.** Within a transportation system, this is information attained through application and experience of the various tasks performed by people. Once information is understood, it becomes knowledge and can be applied to jobs in the system.
- **Materials.** The raw materials used in the construction and functioning of a transportation system include iron, wood, fuel, plastic, and concrete. See **Figure 15-5**. New and innovative materials allow transportation engineers to make improvements to new systems.
- **Energy.** Within a transportation system, this provides power and movement within the system. See **Figure 15-6**. Various forms of energy include heat, mechanical, chemical, nuclear, light, and electrical.

Transportation system input: A resource needed in order to begin and maintain the use of the system, including people, capital, knowledge, material, energy, and finances.

Figure 15-4. What would any transportation system be without people? People repair broken equipment, drive land vehicles, fly airplanes, and assist travelers. More than 7% of the U.S. labor force (9.9 million people) work in a transportation industry. (United Airlines, U.S. Postal Service)

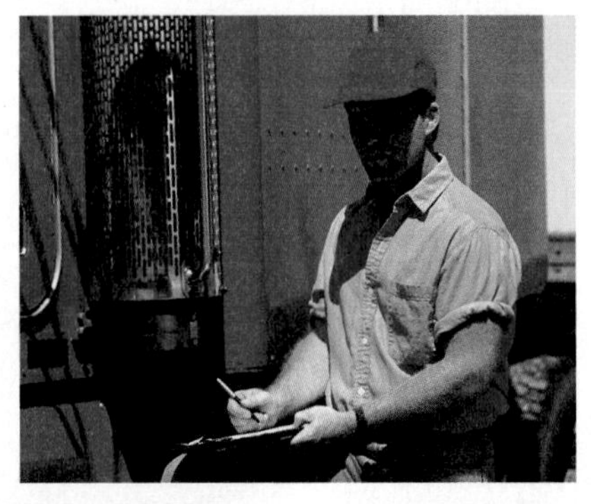

Figure 15-5. Many materials are essential in a transportation system. Transportation means more than powering systems. Pathways, such as roadbeds, must be constructed. This railroad bed requires wood, gravel, and steel.

Transportation system process: An action that converts inputs into desired outputs.

- **Finances.** This includes the money needed to pay for equipment, materials, personnel, and energy sources. As with any system in our society, finances are needed for the system to function as efficiently and effectively as possible.

Whenever a transportation system begins, it must have some input. The input needed may be immediate, such as the need for an energy source before the system can run. It may be long-term, such as the financing of equipment or services. Either way, a transportation system cannot function without all the necessary inputs incorporated into the system.

Transportation System Processes

Transportation system processes are the actions that convert the inputs into the desired outputs. Without processes, the goal of moving people or cargo would not be achieved. There are a number of processes that must take place for people or cargo to reach their destinations. These processes can be divided into two groups: production and management.

Production

The ***production processes*** of a transportation system are the "on the scene" part of the system and are the most recognizable components of a transportation system. Each of the processes involved has a distinct purpose within the transportation system and is needed whether people or cargo are being transported. The processes can be viewed as a cycle of events that include the following:

Production process: The "on the scene" part of a transportation system, including receiving, holding, loading, moving, unloading, storing, and delivering.

- **Receiving.** In the receiving phase, the passenger or cargo is physically placed at the location of departure and enters the destination terminal. Paperwork and tickets are processed to ensure the passengers and cargo will be transported to the correct place.
- **Holding.** After the receiving phase, people and cargo move to a holding area. Holding areas are places for people and cargo to wait for the transportation vehicle. Often, the path to the holding area is determined by the destination. For example, mail at a post office is sorted and placed in different holding areas, depending on the specified destinations. A platform in a subway terminal is an example of a passenger holding area.
- **Loading.** Both people and cargo are moved onto, or are loaded into, the vehicle. Cargo is typically loaded with forklifts and other specialized vehicles. See **Figure 15-7.** With most modes of transportation, passengers board the vehicles themselves. Both people and cargo are secured with safety restraints, such as seat belts or cargo straps.
- **Moving.** This phase involves the actual transporting of people or cargo. It is often the most recognizable part of any transportation system. During this phase of the system, cars, planes, and boats are driven, flown, and sailed.
- **Unloading.** Once the destination is reached, unloading the vehicle begins. Cargo is removed, and passengers exit the transportation vehicle. Unloading the vehicle usually occurs at a terminal within a station or port.
- **Storing and delivering.** These are the final stages within the production process cycle of events. If passengers or cargo have reached their final destination, the transportation process is over. Many times, people and cargo must go through the production process several times before reaching their final destination. If this is the case, the passenger and cargo start over at the receiving stage and enter the transportation production process again.

Figure 15-6. It takes energy to ride a scooter, pedal a bicycle, drive a truck, fly an airplane, and sail a sailboat. This scooter moves only with the addition of human power. Take away the power source, and the scooter remains still.

Figure 15-7. Luggage is loaded onto the vehicle as part of the production process. Conveyors, forklifts, and other specialized machinery may be used.

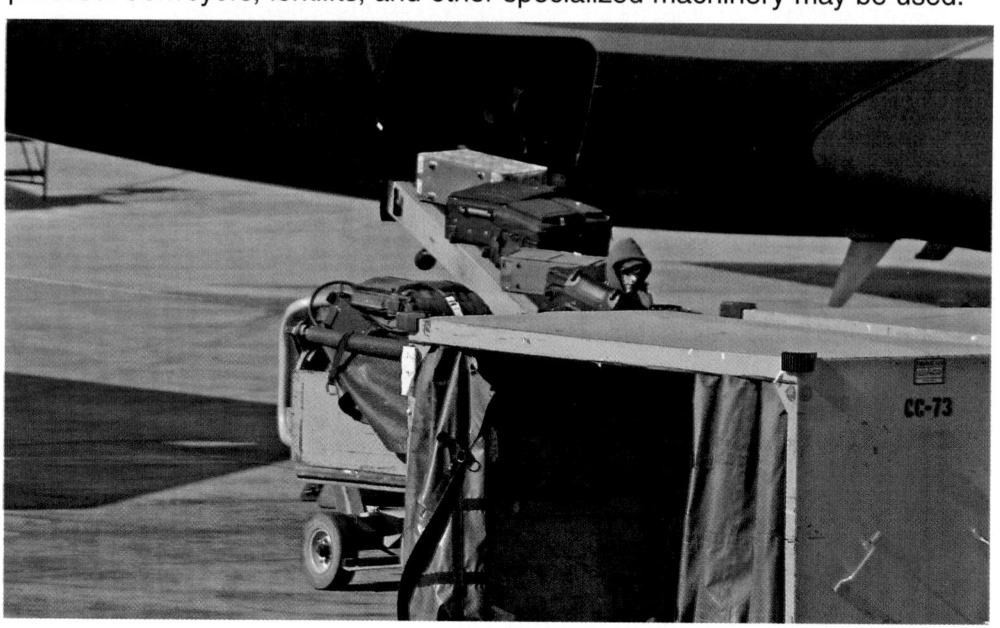

Technology Link

Manufacturing: Product Distribution

Transportation is used in connection with many areas of technology. Possibly the most vital use of transportation by another form of technology is in manufacturing. The purpose of manufacturing is to produce and sell products. In order to move a product throughout the manufacturing process and to the consumer, companies must use transportation systems.

All manufactured products have been moved by at least one form of transportation. Take the book you are reading, for example. This book has been a part of many transportation processes prior to arriving in your classroom. During manufacturing, this book and the materials used to make it were moved throughout the production plant on conveyors, carts, and forklifts. Once the book was complete, it was boxed, crated, loaded onto a truck, and then transported to a warehouse. Inside the warehouse, the crates were moved to the storage location by a forklift.

When your school ordered the book, an automated retrieval system collected all the books in the order and placed them on a conveyor. From the conveyor, the books were again boxed, and shipping labels were attached. The shipping company then picked up the books using a truck or delivery van and took them to a distribution center. Depending on the distance the books needed to travel and the number of books in the order, the books were either loaded on a delivery van, an airplane, or a tractor-trailer. The books may have encountered several additional transportation processes before your school received them. Finally, the books were loaded onto a cart and brought to your classroom. This book you are holding is an example of a manufactured product that would have never reached the consumer without transportation systems and processes.

The production processes are the phases of the transportation process you see most often—for instance, trains speeding down the tracks, trucks and cars on a highway, and speedboats in a local waterway. See **Figure 15-8.** These processes are found in any transportation system, large or small. The production processes can even be identified on a family vacation. Your parents make sure everyone is present and accounted for (receiving) and have everyone wait by the car (holding). You place all the luggage into the car and sit in your seat (loading). The family vehicle then travels to the destination (moving). Once you arrive, you get out of the car and unload the luggage (unloading). The baggage is then moved to the destination, and you relax at the vacation spot (storing and delivering).

Figure 15-8. This boat is in the moving stage of the production process.

Management

The ***management processes*** of transportation systems include the "behind the scenes" activities necessary in keeping people and cargo organized and on schedule. Without the managed portion of the process, our transportation systems could get very chaotic. For example, without a schedule and planned routes for a subway system, you would have no idea which subway train will take you to your destination. See **Figure 15-9.** The subway trains would move down the track, representing the production process, but the system would be too confusing for most people to use. Management processes are needed to plan, organize, and control the system.

- **Plan.** In planning the transportation system, people decide what must be done. People plan the best route and decide how the system will run to be the most efficient. During the planning stage of management, goals are set, and a course of action is determined.
- **Organize.** Organizing a transportation system involves the preparations made for transporting people or cargo. This may be assigning jobs and related personnel or determining a schedule of maintenance for the machinery used in the system.

Figure 15-9. Management is a necessary part of any transportation system. Even with the most efficient running transportation vehicle, someone must schedule stops to ensure that people or cargo arrive at their intended destinations.

Management process: The "behind the scenes" part of a transportation system necessary to plan, organize, and control the system.

- **Control.** In controlling a transportation system, records are kept, computers are used, and systems are monitored. Computers may be used to control the flow of oil through a pipeline. Controlling a transportation system also includes the signs and signals used within the system itself. See **Figure 15-10**. A guard with a stop sign at a school crossing controls the flow of vehicles when children are present.

Management functions also apply to the family vacation example used in the previous section. The travel route needs to be mapped out before leaving home. Stops along the way should be decided (planning). The car may need to be serviced and filled with fuel along the way (organization). During the trip, the performance of the vehicle may be monitored by figuring the miles per gallon and occasionally checking the other gauges (controlling).

Figure 15-10. The flow of traffic is controlled by various signs and signals. This helps to keep transportation systems safe and functioning properly.

Transportation System Outputs

The ***transportation system output*** is the relocation of people or cargo. This output is the result of successful inputs and processes. Upon achieving the output in a transportation system, many events may occur that bring about change. For instance, a log truck transports logs from California to a sawmill in Iowa. Inputs are brought together to begin the system of transporting the logs using a land transportation system:

- Fuel is needed to power the truck.
- A person is needed to operate the truck, read the map, and follow directions.
- Money is needed to operate the truck.

Once the inputs are gathered, the truck enters the processes of the transportation system, or begins the journey of transporting the logs. See **Figure 15-11.** When the truck reaches the sawmill with the logs, the output has been achieved. The logs have been relocated. This is the main purpose of a transportation system—relocating people and cargo. Upon delivering the logs, change will occur with the logs themselves and with the driver of the truck. The driver drops the load of cargo and moves on to a new destination.

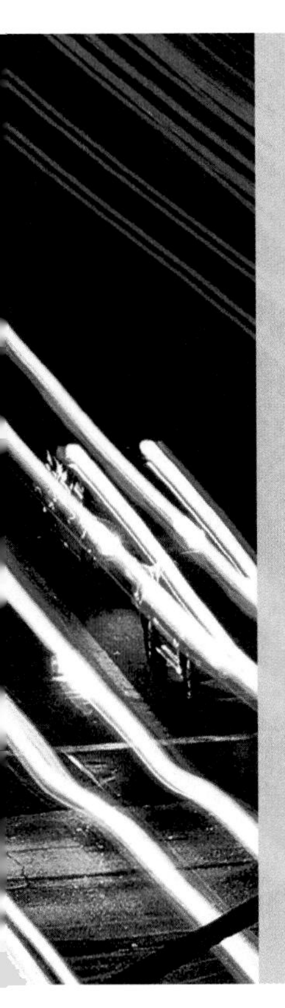

Career Connection

Dispatchers

One of the key roles in most commercial transportation systems is the dispatcher. Dispatchers organize and monitor the movement of their company's vehicles. There are several types of dispatchers, including emergency, trucking, taxi, bus, and railroad dispatchers.

The largest number of dispatchers work in the emergency field. These dispatchers work at communication centers and receive phone calls from people who are in need of emergency services. The dispatcher obtains information from the caller and then sends the appropriate response units to the caller's location. Truck, bus, and railroad dispatchers work for specific transportation companies and plan the routes of all the vehicles. Taxi dispatchers receive calls from prospective customers and then coordinate with their drivers to make sure the customers are picked up and taxied to their desired locations.

All dispatchers use computer mapping and tracking systems. These systems allow the dispatcher to have current location information. Their goal is to ensure efficient movement of the passengers and cargo. The qualifications for most dispatcher positions include good communication skills, good computer and processing skills, a good understanding of maps and charts, and a high school diploma. Some positions, however, especially in the railroad industry, may require an associate or bachelor's degree in logistics or experience in the field. Most positions are paid hourly, with the average yearly salary ranging from around $30,000 to $40,000.

A Functioning Transportation System

Transportation system output: The relocation of people or cargo.

Any type of system is developed based on goals and expected outcomes and is monitored through feedback. This is also true of a transportation system. The goals of a transportation system must be defined before the system can be designed or constructed. Feedback regarding the system's use and performance allow a transportation system to be maintained and improved. Placing controls within the system contribute to its functionality and may help to ensure safety.

Goals

When planning a transportation system, there are certain goals that need to be met:

- **Systemic goal.** The goal of an entire transportation system, regardless of the mode of transportation in use, is to relocate people and cargo to the proper destination at the proper time.
- **Personal goals.** The goals of individuals affect transportation when, for example, selecting a personal automobile. Some vehicles may indicate higher socioeconomic status than others. A personal goal for others may be to purchase some type of recreational vehicle when they retire.

Figure 15-11. These logs are being transported to their final destination, the sawmill. Upon arriving at the sawmill, the truck's participation in the transportation system is complete.

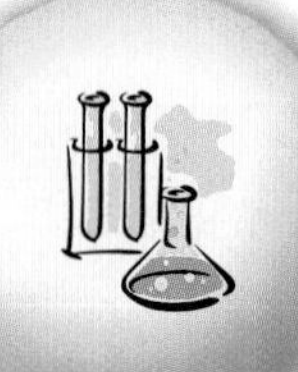

Curricular Connection

Science: Traffic Signals

The first traffic signal was used in London in 1868. It was fashioned from gas lanterns and had to be manually operated. These particular traffic lights did not gain much popularity, but the idea of traffic signals did. By the 1920s, inventors in the United States created working automatic electric traffic lights. The use of red, yellow, and green lights eventually replaced the moving signals of early models.

Incandescent bulbs were used for decades to light traffic signals. These bulbs wasted energy and burned out quickly. Because of the amount of traffic lights per intersection, as well as the number of intersections per city, an efficient alternative was needed. Today, the incandescent bulbs of many traffic lights are being replaced with light-emitting diodes (LEDs). LEDs are more energy efficient than their predecessors and, therefore, will not burn out as quickly.

Traffic signals have also evolved in the sense that they no longer have to rely on timers. Using an inductive loop, traffic lights can be told to change whenever necessary. This technique is typically used to let the traffic signals know if any cars are waiting in a turn lane. The inductive loop uses coils of wire positioned in specific places in the road. When a car is sitting over the wire, a magnetic field is built, which alerts the traffic signal of the car's presence.

- **Economic goals.** These goals consider the potential profits for business and possible income opportunities for the people in the immediate area. Access to highways, railways, airports, and docks can increase property values and allow resources to be imported and exported.
- **Societal goals.** These goals are outcomes from transportation systems that affect the entire society, such as an improved standard of living due to the implementation of a local transportation system. As more communities become accessible by roads and highways, new people and ideas begin to fill the communities. The safe and expedient transport of people and cargo is another important goal of transportation systems. See **Figure 15-12.**

Feedback

Feedback is essential to efficiently operate, maintain, and improve a successful transportation system. Feedback allows the operator or monitor to evaluate how well the system is running, based on the information returned. When driving down the highway, for example, the speedometer provides feedback as you accelerate or decelerate the vehicle. The speedometer helps you determine if you are driving too fast or too slowly, relative to the road conditions and posted speed limit. An instrument panel and various gauges provide feedback needed in the control of a transportation system. See **Figure 15-13.**

Figure 15-12. Any type of transportation system has several different goals.

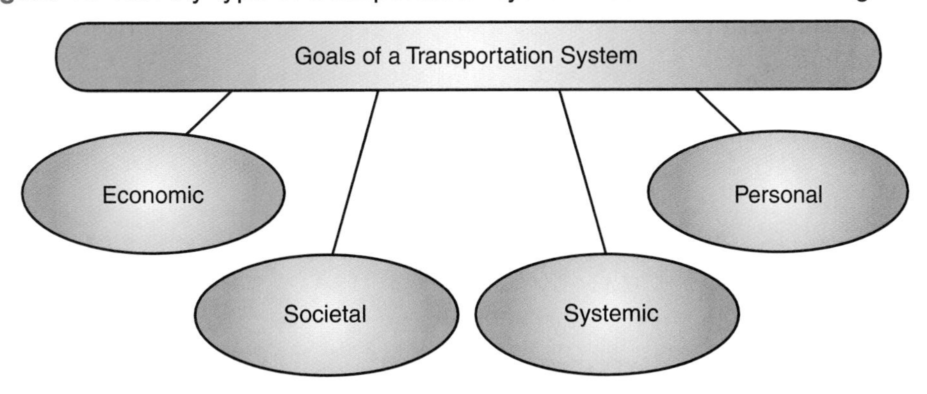

Figure 15-13. The instrument panel on this aircraft provides feedback to the pilot on how the aircraft is functioning.

Control

Like all technological and mechanical systems, a transportation system must have devices in place to control it. Imagine a plumbing system in a house without any controls. Water would flow, but no one would be able to shut the water off. Without controls, transportation systems would not operate efficiently or safely. Most of the control in a nation's transportation system comes from the country's governmental regulations, which help control how the system is used.

In the United States, the Department of Transportation is the organization that sets federal transportation regulations. Within the Department of Transportation, there are several administrations that create and monitor the regulations specific to each mode of transportation. Investigators from some of these agencies study crash tests, vehicle accidents, and normal transportation operations to determine legislative changes for transportation systems. Industry regulations and limitations are set and enforced by these various agencies. See **Figure 15-14.** State governments participate in controlling transportation systems by issuing drivers' licenses and handling the construction of new roads.

Figure 15-14. Some of the agencies that regulate transportation systems within the U.S. Department of Transportation.

U.S. Department of Transportation Agencies
Federal Aviation Administration (FAA) The FAA creates and enforces regulations that apply to the safety, manufacturing, operation, and maintenance of civil (commercial and private) aviation facilities and aircraft.
Federal Highway Administration (FHWA) The FHWA is responsible for the construction, improvement, and preservation of America's system of highways.
Federal Motor Carrier Safety Administration (FMCSA) The FMCSA creates and enforces safety regulations that apply to the motor carrier industries (trucking and buses).
Federal Railroad Administration (FRA) The FRA creates and enforces rail safety regulations and is responsible for the research and development of railroad improvements and rehabilitation.
Federal Transit Administration (FTA) The FTA is responsible for designating federal funds for the development, maintenance, improvement, and operation of public transportation systems.
Maritime Administration (MARAD) MARAD is responsible for developing and maintaining the U.S. maritime transportation system. This agency is also responsible for ensuring that the maritime transportation system can meet the country's future demands.
National Highway Traffic Safety Administration (NHTSA) The NHTSA creates and enforces safety and performance standards for vehicles and related equipment. This agency also assists state and local governments develop highway safety programs and provide consumers with safety information.

Summary

A transportation system is a systematic way of relocating people and cargo. There are four environments of transportation systems: land, water, air, and space. Transportation systems that involve more than one of these environments are considered intermodal transportation systems. Each type of transportation system includes inputs, processes, outputs, feedback, and goals and is controlled by regulations. The determined goals of a transportation system affect its design and use. Various inputs are necessary before beginning the production and management processes of a transportation system. Once the system processes are under way, an output is expected. The expected output of a transportation system is the relocation of people or cargo. Feedback within the system is needed in order to control, maintain, and improve the system.

Key Words

All the following words have been used in this chapter. Do you know their meanings?

air transportation
heavier-than-air
intermodal transportation
land transportation
lighter-than-air
management process
production process
space transportation
transportation system
transportation system input
transportation system output
transportation system process
water transportation

Test Your Knowledge

Write your answers on a separate sheet of paper. Do not write in this book.

1. Define *transportation system.*
2. List the five categories of transportation systems.
3. _____ transportation includes highway, rail, and pipeline transport.
4. Space transportation systems are commonly used for _____and _____.
5. _____ transportation makes use of more than one transportation environment.
6. Name the six transportation system inputs. Provide an example of each.
7. The _____ phase is the portion of the production process cycle in which people and cargo wait for the transportation vehicle.
8. Describe the difference between production and management processes in a transportation system.

9. Management processes are needed to _____, _____, and _____ the transportation system.
10. Write two sentences describing the expected outcome of a transportation system.
11. Identify the four types of goals that affect transportation systems.
12. Give three examples of societal goals for any type of transportation system.
13. What is the main purpose of feedback within a transportation system?
14. Briefly discuss an agency that creates government regulations regarding transportation systems.

Activities

1. Interview someone who works in a transportation industry. Ask him about the knowledge and skills required to properly perform his job. Report your findings to the class and discuss what courses would be most helpful in preparing for such a job.
2. Make a list of the transportation systems used inside your school. Keep in mind the definition of a transportation system when deciding what systems qualify as transportation. Name the inputs and processes of each system. Explain why each system is important.
3. Working alone or in a group, select a transportation vehicle and prepare a visual presentation of the vehicle's systems. Note the inputs, processes, feedback, output, and goals of the system in which the vehicle is used. List any undesirable outputs related to the use of the vehicle.
4. Imagine what our transportation systems would be like without any inputs. Consider the absence of people, for example. If you wanted to fly in an airplane across the country, but there was no pilot to fly the plane, you would not get anywhere. Therefore, that particular air transportation system would not function without the input of a person. List several transportation system inputs and explain how removing them would affect the operation of a transportation system.

16

An Introduction to Vehicular Systems

Basic Concepts

- Identify and define the six separate systems that make up a vehicular system.
- List safety factors in the design and operation of vehicular systems.

Intermediate Concepts

- Give examples of the components used in each of the six vehicular systems.

Advanced Concepts

- Describe how global positioning systems (GPSs) operate.

Vehicles are the "machines" that are the basis of our transportation systems. Everyone recognizes that trains are the vehicles used by the railroad industry to transport passengers and cargo. Airplanes are the vehicles employed to move things by companies involved in the air transportation industry. Ships are used for pleasure cruises, as well as for transportation of large quantities of cargo over waterways, both large and small. See **Figure 16-1.**

All vehicles are comprised of a number of different components that allow them to safely transport people and cargo. These components can be placed into six different systems. The ***vehicular systems*** are a collection of separate systems that allow the machine to move through its environment safely and efficiently. These systems are usually part of the vehicle itself, but they may be external, depending on their purpose. The following is a list of the technical components that make up vehicular systems:

- Propulsion systems.
- Guidance systems.
- Control systems.
- Suspension systems.
- Structural systems.
- Support systems.

Figure 16-1. Many different vehicles are necessary to transport people and goods through the four different types of environments. (Amtrak, Greyhound, British Columbia Transit, Carnival Cruise Lines, Kawasaki, Airbus)

Vehicular system: A collection of separate systems that allow the machine to move through its environment safely and efficiently.

These systems will be defined here, but they will be explained in further detail in later chapters.

Propulsion Systems

Propulsion systems are the components of a vehicular system that produce the power needed to move a vehicle. The main function of the propulsion system is to convert energy into mechanical power that can be used to drive, fly, sail, or move a vehicle in some other way. There are several methods used to convert energy into propulsion, most of which rely on either an engine or a motor. Often the terms *engine* and *motor* are used interchangeably. These devices are, however, different. ***Engines*** are devices that convert heat energy into mechanical energy. ***Motors*** produce mechanical energy by converting electrical energy.

Engines

Because engines use heat to produce mechanical power, they are often referred to as *heat engines*. There are several types of engines used in transportation vehicles. Land and water vehicles most often use gasoline or diesel ***piston engines***. See **Figure 16-2.** These engines produce heat by compressing and igniting a mixture of air and fuel. The heat produced pushes a piston, which then turns a crankshaft. The rotating crankshaft is the mechanical power output of the engine. In land transportation, the power is then sent to the wheels in order to move the vehicle. When piston engines are used in water and air transportation, they are typically connected to a propeller. The mechanical energy produced by the engine is used to spin the propeller. The reaction of the air or water to the spinning propeller produces thrust, which moves the vehicle.

In air transportation, turbines, rather than pistons, are used most often to convert the energy. The engines used are known as ***jet engines***, or *jet turbine engines*. The jet engines use spinning turbines to compress air. The air is then mixed with fuel and ignited. The high-pressure hot air is sent out the back of the engine, which produces forward thrust. The engines most often used in space transportation are ***rocket engines***. Rocket engines produce thrust by expelling hot gases from a rear nozzle. Igniting chemical propellants produces the gases.

Propulsion system: The components of a vehicular system that produce the power needed to move a vehicle.

Engine: A device that converts heat energy into mechanical energy.

Motor: A device that produces mechanical energy by converting electrical energy.

Piston engine: An engine that produces heat by compressing and igniting a mixture of air and fuel.

Jet engine: An engine that uses spinning turbines to compress air.

Rocket engine: An engine that produces thrust by expelling hot gases from a rear nozzle.

Figure 16-2. Propulsion systems use energy to provide force that moves vehicles. The automobile, powered by an internal combustion engine such as this one, is the major means of transportation in North America. (Ford Motor Company)

Motors

Motors are most often used in land transportation vehicles. Vehicles such as golf carts, bumper cars, light-rail trains, and electric cars use motors as their propulsion sources. The energy supplied to the motor is either direct electricity or electricity stored in batteries. The motors are connected to the wheels and convert the electricity into motion by turning the wheels.

Other Types of Propulsion

Propulsion devices other than engines and motors also power vehicles. Some examples include bicycles, sailboats, and canoes. Bicycles are propelled by human power and a system of gears and a chain. Sailboats are powered by harnessing the power of the wind. Humans using paddles and oars propel canoes.

Guidance Systems

Guidance system: A system that provides the information required to make a vehicle follow a particular path or perform a certain task.

Guidance systems provide the information required to make a vehicle follow a particular path or perform a certain task. The vehicle operator reads these systems and then controls the vehicle according to the information given. There are many sources of guidance information. Street signs, road maps, radio detecting and ranging (radar) screens, and airport runway lights all provide guidance information to the vehicle operator. See **Figure 16-3.** These information sources inform the driver about location, road intersections, distance from other vehicles, and where to land. The driver or pilot must be able, however, to understand the information provided. For example, a radar screen is not very helpful if the operator is unsure about how to interpret the dots and blips on the screen. This is why most vehicles require licenses to be able to legally operate them.

Navigation: The act of guiding a vehicle.

Besides providing information, guidance systems also allow vehicle operators to navigate their vehicles. ***Navigation*** is a word used to describe the act of guiding a vehicle. It usually refers to the guidance of a ship or an airplane, but it may also be used for other vehicles. Navigation has its roots in the Latin words *navis*, which means "ship," and *agere*, which means "to drive." Some of the things a vehicle operator needs to know to ensure he is on the right course include direction of travel, speed, and location of destination. These are all forms of navigation information. See **Figure 16-4.** The operator would also need to know where he is at the beginning of the journey.

Global positioning system (GPS): A satellite-based navigation system.

Many devices have been used throughout history to determine navigation information (location, direction, and speed). Compasses were the first devices used to determine direction. There are a number of devices specifically used in each transportation environment. ***Global positioning systems (GPSs)***, however, are used in all types of transportation. A GPS is a satellite-based navigation system. It uses a constellation of 27 satellites that transmit signals to the earth. The signals include a timed sequence. Each GPS receiver on earth gathers the signals from several (usually three or four) of the satellites. The receivers are able to determine the location and distance from each satellite. From this information, the receiver can

Figure 16-3. Guidance systems may be part of a vehicle or separate, such as highway signs, runway lights, maps, control towers, and navigation buoys. (U.S. Navy, U.S. Air Force)

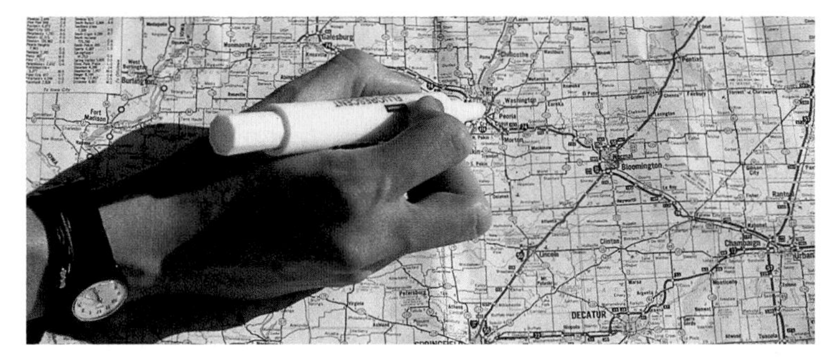

determine exactly where on earth it is located. See **Figure 16-5.** Advanced receivers can also track movements, direction, and speed. The GPS is becoming the standard navigation system for most modes of transportation. Many vehicles can be purchased with built-in GPS receivers and electronic mapping software. This makes navigation much easier.

Figure 16-4. Before operating a transportation vehicle, the operator must have these three kinds of information.

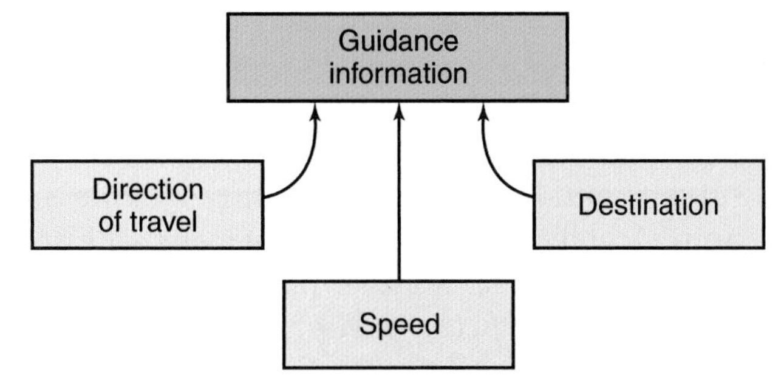

Technology Link

Communication: Telematics

Automobiles were designed to move people and goods from one place to another. As technology has advanced, however, drivers have desired their vehicles to perform an increasing amount of communication functions. The answer to this desire is in the area of telematics. Automobile telematics is a technology that allows the automobile and the driver to be connected to others through the use of wireless communication systems.

Currently the most popular automobile telematics system is OnStar® services, by General Motors (GM). OnStar services is an information and safety system installed on over 14 different foreign and domestic brands of automobiles. The OnStar services system has several components. The "brains" of the system is a communications processor. The processor is linked to the automobile's computer, a global positioning system (GPS) receiver, a microphone, air bag sensors, a three-button keypad, a cellular and digital antenna, and the vehicle stereo. The system in monitored by an OnStar services support system.

The system can either be driver initiated or support center initiated. In driver-initiated uses, the driver presses the OnStar services button located on the rearview mirror or dashboard, which calls the support center. The operator at the support center can aid the driver by calling for roadside assistance, providing directions, or on some GM vehicles, even diagnosing vehicle problems. Because the OnStar services system is linked to an onboard GPS receiver, the support center operator can track the vehicle and provide real-time directions. The GPS receiver is also used to track the vehicle when a driver reports her vehicle as stolen. A driver can also call OnStar services operators if he has locked his keys in the vehicle or cannot find his car in a parking lot. Because the system is integrated with the vehicle computer, the OnStar services operator can remotely unlock the door or sound the horn. In other situations, the OnStar services system automatically notifies an OnStar services operator. In accidents in which the air bag is deployed, the system calls an operator. The operator then checks on the condition of passengers and will notify emergency personnel if necessary.

OnStar services are currently the largest use of automobile telematics. There are, however, many uses that will be further developed in the future. Much of the future development is expected to be based on wireless technology. Possible future uses of the technology include the ability to download music and e-mail to a vehicle, more advanced navigation systems, and the ability to wirelessly play video games in the rear seats of a vehicle. This technology will allow vehicles to be as well linked as many home offices and game rooms.

Control Systems

Obviously, different vehicles have different abilities, regarding the number of directions in which they can be controlled. For example, a train can only move forward and backward on its tracks, while a helicopter has much more freedom. A helicopter can move forward and backward, left and right, and up and down. The number of changes in direction a vehicle is allowed is called its *degree of freedom*. The train has one degree of freedom, and the helicopter has three. See **Figure 16-6.** There is a wide

Figure 16-5. A global positioning system (GPS) receiver and display unit allow a ship's captain, an airplane pilot, or a truck driver to precisely locate the vehicle's current position. (Garmin International)

range of control systems needed to safely operate modern transportation vehicles. Control systems are the parts of vehicles used to change a vehicle's direction and speed.

Changing Direction

Vehicles that are able to move in different directions are normally controlled by the driver or pilot. These vehicles all have devices the operator is able to move in order to steer the vehicle. See **Figure 16-7.** Cars, trucks, and ships have steering wheels. Airplanes have flight sticks and pedals. Space shuttles have hand controllers. All these devices are linked

Figure 16-6. Degrees of freedom. A—Trains must follow tracks. They have only one degree of freedom. (Norfolk Southern Railway) B—A helicopter in flight can move up and down, forward and back, and side-to-side. It has three degrees of freedom.

A

B

Curricular Connection

Social Studies: Longitude and Latitude

To accurately pinpoint a location on a map, there needs to be a definite reference system. For this, we use imaginary lines that run vertically and horizontally over the entire world. These lines are measured in degrees, so each one is identifiable. A small circle (°) is used as the symbol of degree, just like degrees in measuring temperature. The lines form a grid covering the entire planet. The grid of lines around the world has a vertical baseline and a horizontal baseline. The equator is an imaginary baseline running around the middle of the earth, from east to west. It divides the earth into northern and southern hemispheres. The equator and the lines that run parallel to it are called *latitudes*. The reference number given to the equator is 0°. Latitudes to the north of the equator are numbered from 0–90° north. For example, Boulder, Colorado lies on the latitude line 40° north. Latitudes to the south are numbered from 0–90° south.

Lines running from the North Pole to the South Pole are called *longitudes*. The 0° reference longitude is called the *prime meridian*. The prime meridian is drawn from the North Pole to the South Pole and runs through the Greenwich Royal Observatory in Greenwich, England. Because the earth is round, there are 360 longitudinal degrees. Longitudes lying to the east of the prime meridian are numbered from 0–180° east. Those lying to the west of the prime meridian are numbered from 0–180° west.

When locating a fixed position on a chart, the numbers of the closest latitude and longitude are used. These numbers together are called coordinates. If the location is not on an intersection of coordinates, subdivisions of latitudes and longitude are used. The subdivisions are known as minutes. Minutes make up the spaces between longitudes and latitudes. Each space between longitudes or latitudes is divided into 60 minutes. So, if a location was halfway between 175° east and 176° east, it would be 175 degrees, 30 minutes. The foot symbol (′) is used to show minutes. When latitudes and longitudes are used together, as coordinates, you can find any location on the globe or world map.

For example, now that you understand latitude and longitude, you would be able to find that 41°23′ N, 2°9′ E are the coordinates for Barcelona, Spain. If pilots have the coordinates for an airport, they can find it on their charts with no problems. By using this system, we can accurately locate positions on the earth's surface.

to a steering system, control surfaces, or thrusters that are able to change the direction of the vehicle. Steering systems are common in land vehicles. For example, in an automobile, the steering wheel is mechanically linked to a steering system that turns the wheels as the steering wheel is rotated. Air and water vehicles use moveable control surfaces, which change the vehicle's direction. A common moveable control surface in a boat is a rudder. As the rudder is turned, it forces the boat to turn as well. In space transportation, thrusters are used to change direction and attitude, or rotation.

Figure 16-7. Directional control devices. A—A steering wheel for an automobile. (Mazda) B—An aircraft control stick. (Airbus) C—A "spinner" hand control for an industrial lift truck.

A

B

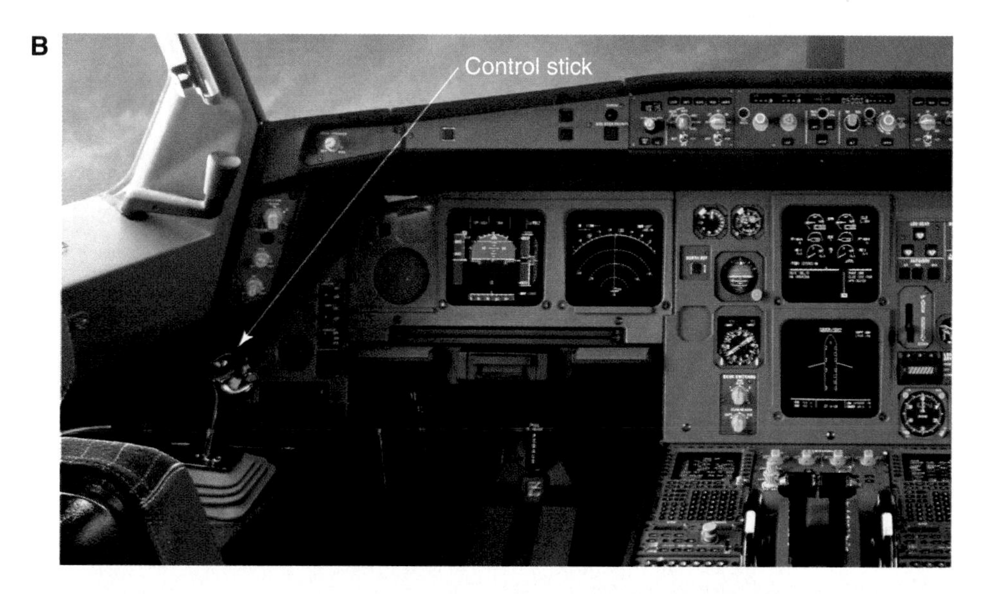

C

Changing Speed

Accelerator: The part of a vehicle that controls speed by changing the amount of power the propulsion system generates.

Driver and pilots must also be able to control the speed at which the vehicles travel. Speed is controlled in two ways. The first method of controlling speed is by changing the amount of power the propulsion system generates. An ***accelerator*** typically controls this power. In automobiles, the accelerator is the gas pedal. As the gas pedal is depressed, it allows more fuel and air to enter into the engine, which creates more power. In other vehicles, the accelerator is known as the throttle. The second method of changing speed is by applying a braking system. In airplanes and space shuttles, flaps are used as brakes. The flaps are raised during landing to increase the amount of surface area in contact with the air. This helps to slow the vehicle down by creating more drag. Most land vehicles use brakes that come in contact with the wheels. The friction created between the wheel and the brake slows the vehicle down.

Suspension Systems

Suspension system: The vehicle system that supports or suspends the vehicle in or on its given environment, providing a method to smooth the ride for passengers and cargo.

Suspension systems on vehicles are designed to support or suspend the vehicle in or on its given environment. The suspension systems for vehicles that fly through the air are completely different from the systems used on cars and trucks. Because of the nature of the environments in which they operate, suspension systems obviously need to be different.

These systems also provide a method to smooth the ride for passengers and cargo. Modern automobiles often have special suspension systems so the rides in them are very comfortable. The suspension systems of automobiles include tires, springs, and shocks. See **Figure 16-8.** The three components work together to provide a comfortable and safe ride for the

Figure 16-8. A car's suspension system consists of tires, springs, and shock absorbers. (Ford Motor Company)

passengers. The tires serve as traction and cushioning for the vehicle. The springs help keep the vehicle traveling forward when the tires hit bumps, potholes, or ruts. The shocks are used to dampen the reaction of the springs. In trains, the wheels and springs are contained on what are known as *trucks*, or *bogies*. The trucks are positioned at the front and rear of the railroad cars and pivot as they follow the track.

The main suspension system for water vehicles is the boat hull. The hull is the underside or lower body of the boat. When sitting in water, boat hulls must be designed to displace their weight in water. For example, if a boat weights 1 ton, it must displace, or push aside, at least 1 ton of water to stay afloat.

Airplane suspension relies on the wings to generate lift. ***Lift*** is the force that keeps aircraft in the air. The wings have an airfoil shape specially designed so, as the wing travels through the air, lift is generated, and the airplane is able to fly. As airplanes land, they require landing gear similar to land vehicle suspension systems (tires and shocks). The landing gear enables the plane to land safely and move from the runway to the gate.

Lift: The force that keeps aircraft in the air.

Possibly the most unique suspension system is that of magnetic levitation (maglev) vehicles. See **Figure 16-9.** Maglev vehicles are suspended from a guideway, or track, by a magnetic field. The magnetic field is generated by electromagnets on both the vehicle and the guideway. This creates a frictionless cushion of air on which the vehicle can travel.

Structural Systems

Structural systems are the parts of vehicles that hold other vehicular systems and the loads they will carry. In most cases, vehicular structures need to be strong and rigid. They provide mounting places for propulsion, control, suspension, and some guidance systems. Structural systems also

Structural system: The parts of vehicles that hold other vehicular systems and the loads they will carry.

Figure 16-9. A magnetic levitation (maglev) train is held suspended above its guideway by magnetic forces. This is an artist's conception of a maglev train (upper level) in the main railroad station in Munich, Germany. The maglev line, now under construction, will connect the railway station and the city's airport. (Transrapid International, Inc., GmbH)

provide the "skin" of the vehicle, which protects the systems, passengers, and cargo from the environment. The environment includes rain, hail, mud, and extremes of heat or cold.

Structures also provide protection from hazards that come about as a result of transportation in society. There is always a danger of accidents. Structural systems are designed to take passenger safety into account. Thorough structural testing is done on most vehicles to improve passenger safety. See **Figure 16-10.**

Most vehicle structures are made of a frame and a body. The frame serves as the connecting place for other vehicle systems and components. For example, the frame of a satellite, known as a bus, serves as a structure for attaching solar panels, thrusters, and antennas. The body of the vehicle is a shell that encloses the vehicle. It usually determines the shape of the vehicle. Car bodies are a good example. See **Figure 16-11.** You are aware of the many styles and shapes automobiles have. Some are plain and basic shapes. Others are flashy, sporty, and even exotic looking. No matter what their styling, one of the functions of cars is to protect their occupants.

The structure of the vehicle often determines the purposes the vehicle serves. In railroad cars for example, the structure could be designed as a hopper car that carries coal, a flatbed that carries large equipment, or a

Figure 16-10. Structural testing. A—Automobile manufacturers conduct crash tests to determine the structural safety of their vehicles. (General Motors) B—Aircraft are crash tested to learn how fabrics and other materials resist burning. (National Aeronautics and Space Administration)

A

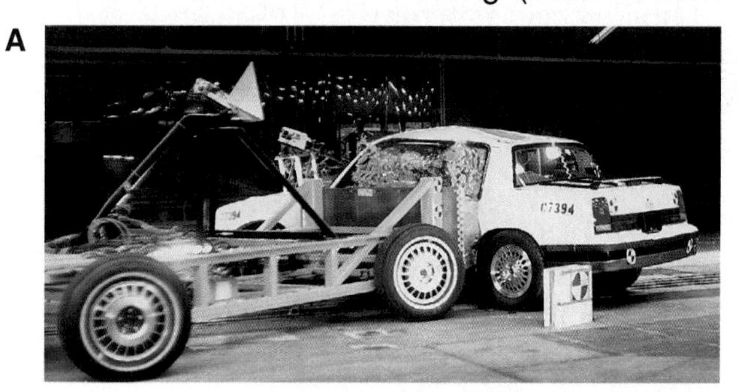

B

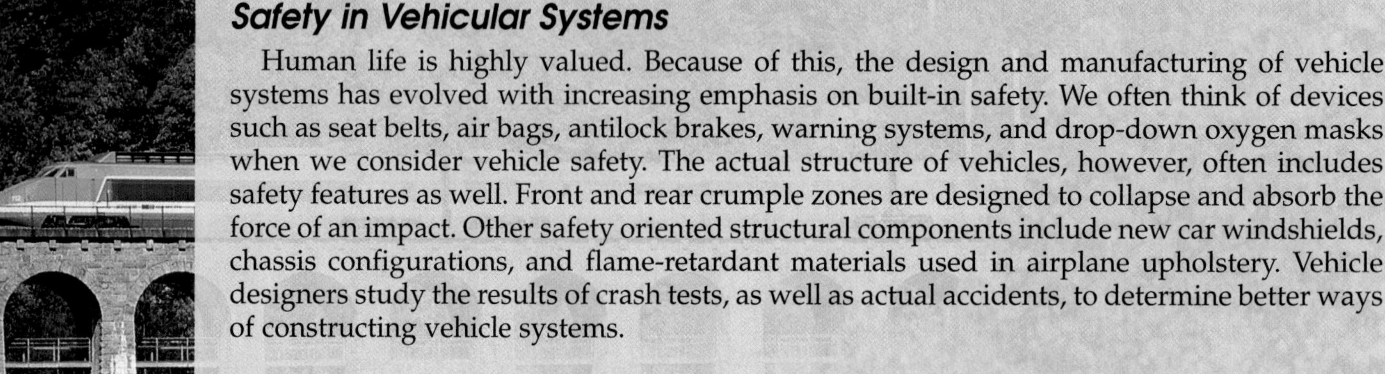

Tech Extension

Safety in Vehicular Systems

Human life is highly valued. Because of this, the design and manufacturing of vehicle systems has evolved with increasing emphasis on built-in safety. We often think of devices such as seat belts, air bags, antilock brakes, warning systems, and drop-down oxygen masks when we consider vehicle safety. The actual structure of vehicles, however, often includes safety features as well. Front and rear crumple zones are designed to collapse and absorb the force of an impact. Other safety oriented structural components include new car windshields, chassis configurations, and flame-retardant materials used in airplane upholstery. Vehicle designers study the results of crash tests, as well as actual accidents, to determine better ways of constructing vehicle systems.

car carrier that transports automobiles. In water transportation, the structure of the hull and the decks of ships and boats can also be designed to serve a number of purposes. The structure of a ship can be designed to carry passengers, in the case of a cruise ship, or cargo, in the case of a containership.

Most vehicle structures are made of steel and other metal alloys, so they are very strong and durable. Plastics and composite materials have also been used by the industry to save weight and enhance the appearance of some vehicles. Advancements in structural designs have made vehicles safer and more efficient.

Figure 16-11. Structural systems provide a body, or support, for all other systems of the vehicle. They also protect the passengers and cargo. (Saab)

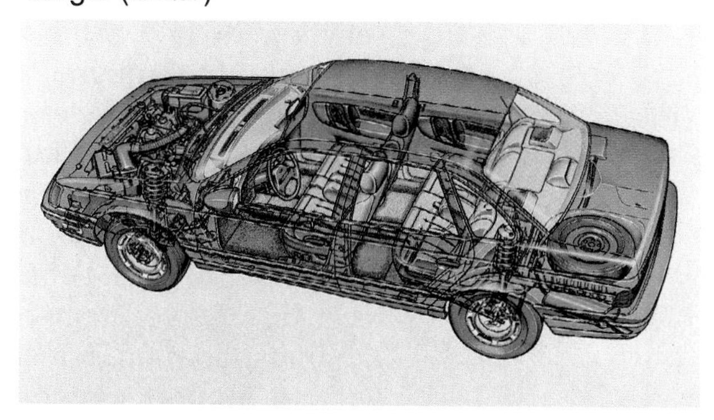

Support Systems

Support systems include all the external operations and facilities that maintain transportation systems. These include passenger and cargo handling, roadway construction, maintenance, life support, economic support, and even legal support. See **Figure 16-12.** Support systems are essential for the operation of any vehicle. Most support systems are not a part of the actual vehicle. Even so, they are an important link in transportation technology.

Support system: The external operations and facilities that maintain transportation systems.

Figure 16-12. Support systems provide protection, services, and repair facilities for vehicular systems. A—A quick-change oil station. B—Railcar repair. (Amtrak) C—Truck stops allow vehicles to refuel and drivers to take meal breaks. D—Rest stops along interstate highways provide drivers with information, restrooms, and sometimes fuel and food services. E—Ports include facilities for loading and unloading ships, as well as for storage of cargo. (Port of Long Beach) F—Repair equipment keeps roadways in good condition.

A

B

C

D

E

F

One of the more obvious support systems for land vehicles is the road and highway network. Without this system of paved, or hard-surfaced, roads, vehicle travel would be uncomfortable and unsafe. Vehicles would not be able to travel at the speeds they do today. For rail vehicles, the same can be said of the system of rail beds and tracks connecting various communities throughout the country. Obviously, this transportation system would be useless without railroads.

Air transportation industries rely on airports. Runways and terminal buildings are important support systems. Overseas and inland shipping companies need harbor and port facilities so they can move their cargo from water to land. People in the shipping industry would have no way to receive or deliver cargo or passengers without access to harbor facilities.

Transportation industries would not be able to survive or compete without support services. People would not be able to own or operate automobiles without support services. Vehicles and their systems are only a part of the whole transportation story. Support systems include all the parts of society devoted to sustaining transportation technology and the ability of people to use it. Although support systems are not a direct part of the vehicles themselves, they are very important.

Career Connection

Automotive Service Technicians

Vehicles are only beneficial if they are in working order. Automotive service technicians accomplish the job of maintaining and repairing cars and light trucks. Automobile dealerships, government organizations, or independent repair shops typically employ service technicians.

The job of an automotive service technician is to diagnose, service, and repair cars and trucks. In order to do so, service technicians must understand diagnostic equipment. Much of the diagnosis is done using electronic and computerized equipment that requires an understanding of electronics. Once the problem has been located, the service technician must then have the ability to use hand and power tools to remove and replace the damaged parts. The hand tools used by service technicians are often the most expensive part of entering into the field. Service technicians are typically expected to provide their own hand tools, while the company they work for supplies the diagnostic equipment and computerized scanners.

The job of a service technician is geared toward people who have a mechanical aptitude and strong analytical skills. Most service technicians begin training in automotive repair classes in high school vocational programs. A great majority then attend a postsecondary vocational or technical school or a community college. Most programs range between 12 and 24 months in length and lead to a certification. Several of the automobile manufacturers provide additional training to the top technical school students. After graduation from an automotive service program, the job market provides good opportunities. The majority of service technicians make between $11 and $20 per hour. Once service technicians have gained several years of experience, they can work at becoming certified by the National Institute for Automotive Service Excellence (ASE). Technicians must have work experience and pass a written exam to become ASE certified in one of eight areas. A service technician that becomes certified in all eight areas is known as an ASE Master Automotive Technician.

Summary

Vehicles can be considered the most important part of transportation in our society. Without them, fast, safe, and efficient movement of people and cargo would not be possible. Vehicular systems are a series of separate, but interrelated systems. The separate components include propulsion, guidance, control, suspension, structure, and support systems. Many of these components are a part of the vehicle itself, although some are not. Support systems include a broad range of facilities and services that are very important to transportation technology in our society.

Key Words

All the following words have been used in this chapter. Do you know their meanings?

accelerator
engine
global positioning system (GPS)
guidance system
jet engine
lift
motor
navigation
piston engine
propulsion system
rocket engine
structural system
support system
suspension system
vehicular system

Test Your Knowledge

Write your answers on a separate sheet of paper. Do not write in this book.

1. _____ are the machines that transportation systems use to move passengers and cargo safely, swiftly, and efficiently.
2. Explain the difference between an engine and a motor.
3. List three types of navigation information.
4. A(n) _____ uses a constellation of satellites to determine the location on earth.
5. Describe the term *degrees of freedom*.

Matching questions: For Questions 6 through 11, match the phrases on the left with the correct term on the right.

6. _____ Vehicle components that provide methods for using energy to propel vehicles.
7. _____ Systems that provide information required by a vehicle to make it follow a certain path.
8. _____ Parts of vehicles that provide methods for changing speed and direction.
9. _____ Systems that provide a smooth ride for passengers and cargo.
10. _____ Parts of vehicles that contain other systems and protect passengers and cargo.
11. _____ All external operations that maintain vehicle and transportation systems.

A. Control systems.
B. Guidance systems.
C. Propulsion systems.
D. Structural systems.
E. Support systems.
F. Suspension systems.

Activities

1. Secure a picture of a lawn mower, and then draw arrows to its different parts. Label each part with its proper name. After the name of each part, write what subsystem of the vehicular system it is.
2. On a sheet of paper, list all the propulsion systems you can. After each system, write down the environment (land, water, air, or space) for which it is suited.
3. Design and construct a working model of a land vehicle that includes all or as many subsystems as you can devise.

17 Land Transportation Systems

Basic Concepts

- Identify the three different types of land transportation routes.
- Cite examples of the three types of pathways.
- List the different modes of land transportation.
- Name the different kinds of land vehicles.

Intermediate Concepts

- Discuss the importance of the wheel.
- Summarize the history of land transportation.
- Explain how diesel-electric locomotives function.

Advanced Concepts

- Create a graph showing the increases in the speed of vehicles over time.

The term *land transportation* includes all methods of transport in which the vehicles travel on or are supported by the earth. Such transport can be long distance, as between cities or coasts. It can also be short-distance travel, such as within a structure or between structures.

The History of Land Transportation

Centuries ago, movement from one place to another was a slow process. The only means of land transportation was by foot. Walking is a very limited mode of transportation. People from long ago could only travel as far as they could walk. They had no vehicles, so all cargo had to be carried in their hands or on their backs.

These early humans soon began to use the knowledge and resources available to improve transportation technology. The development of new transportation vehicles began to emerge. Through the use of the

Sledge: An early example of the modern sled, built using logs.

crude tools available at the time, early travelers created sledges. The ***sledges***, early examples of the modern sled, were built using logs. The early humans dragged sledges across the land. The sledges allowed for the transporting of resources, such as wood, stone, and wild game that was hunted for food.

The Development of Wheeled Vehicles

The invention that most drastically changed the history of transportation was the wheel. The wheel is the basis of almost all land transportation vehicles. It can also be found on vehicles used in other transportation environments, such as the airplane and space shuttle. The invention of the wheel took place in the Middle East, over 5000 years ago. With this advancement came wheeled vehicles drawn by horses and oxen. The first wheeled vehicles were simple carts that had either two or four wheels. The carts were used to move cargo. They were essential to the development of trade among civilizations. The next wheeled vehicle to be developed, the chariot, was also used between civilizations. The chariot was used for war, however, rather than trade. Carts and chariots continued to be innovated into wagons, coaches, and carriages. These wheeled vehicles were common modes of transport through most of the nineteenth century. See **Figure 17-1.**

The Development of Vehicle Power

As the vehicle itself was being innovated, so was the way it was being powered. Early humans pulled the first sledges and carts themselves. As people began to domesticate animals, such as horses, oxen, and donkeys,

Figure 17-1. The development of the wheel brought many advancements in land transportation. Horse-drawn vehicles, arising from the wheel's invention, were in common usage until the early 1900s. (Deere & Company)

the animals were given the task of pulling the vehicles. Animals were used until new technologies, such as the steam engine and the internal combustion engine, were invented. By the mid-1800s, major land transportation advancements were beginning to be developed. In 1830, the first scheduled passenger train began the American railroad era. By 1840, the first electric car had been built. In the 1890s, the first internal combustion engine was utilized. By the early 1900s, Henry Ford devised a method for mass-producing gasoline engine automobiles. See **Figure 17-2.** His method revolutionized the transportation industry by producing automobiles that were affordable to the average family.

As we look at the history of land transportation, we must appreciate it. Inventions like the wheel and cart may seem simple and unsophisticated today. We must remember that these advances were as revolutionary at the time as flying cars would be today. It is important to appreciate each transportation contribution and how it has expanded and matured our society.

The Development of Roadways

Roads are not a new concept. They have been built for thousands of years. The ancient Chinese, Egyptians, and Incas all built roads to connect their cities together. The roads began as paths used by wild animals, migrating people, and nomads. The Romans were the most sophisticated road builders. They built a large system of roads that connected various parts of their territory to Rome, their capital. The Romans, unlike most civilizations before them, built wide roads and paved them with layers of

Figure 17-2. Henry Ford was the first manufacturer to mass-produce automobiles using an assembly line. This was what his assembly line looked like in 1914, in Highland Park, Michigan. (Ford Motor Company)

Curricular Connection

Social Studies: Electric Automobiles

Today, when we see electric automobiles at auto shows and on television, we assume they are a breakthrough idea. While the technology used in the vehicles is state-of-the-art, the concept of electric automobiles has been around for over 100 years. Electric automobiles were actually produced before gasoline vehicles and had their height of glory in the late 1800s and early 1900s. Several people are credited to have built the first electric automobiles in the 1880s and 1890s, including William Morrison of Des Moines, Philip Pratt of Boston, and John Barrett of Philadelphia. Electric cars were even present at the World's Colombian Exhibition at Chicago in 1893. Over the next 10 to 15 years, electric automobiles became quite popular and competed with both steam and gasoline automobiles. By the end of the nineteenth century, electric automobiles had reached top speeds of over 60 miles per hour (mph), which was unheard of at the time.

Electric automobiles were used mainly in cities because rural areas had little, if any, access to electricity to recharge the batteries. One of the more popular uses was for taxicabs. By 1900, 90% of all taxis in New York City were electric. The popularity of electric automobiles ended, however, by the mid-1910s. The decline in electric automobiles is credited to several factors. The main reasons are that internal combustion engines were being greatly improved and the cost of gasoline was decreasing. Gasoline was also becoming more readily available than electricity. While gasoline engines eventually won the popularity race, the existence and performance of early electric automobiles should not be forgotten.

stones. Parts of some of the roads still exist today and are over 2000 years old. The methods of road construction varied little until the nineteenth century. Several engineers in the early 1800s developed new methods of constructing roads and began to use new materials. These methods are the framework that led to many of the advances in roadways today.

The development of roads was critical to the expansion of the United States. It is amazing how large of a role transportation has played in the settling of North America. Without roadways, we would not be as developed as we are today. It was the roads that enabled our founding fathers to continue on their journeys westward. Good roadways encouraged migration, which led to people settling throughout the country.

Today, the United States, as well as many other countries, has a complex system of local roads and highways. The local roads are used to travel within a town or city. Highways are used to travel between cities. The United States has a system of roads that connect most major cities, known as the interstate system. The interstate system, originally constructed to aid in national defense, is comprised of limited access highways that connect to each other. They are termed *limited access* because vehicles can only get on and off the highway at the on- and off-ramps. See **Figure 17-3.**

Technology Link

Construction: Road Construction

Construction technology is often viewed simply as the building of houses. There are, however, several other types of construction. One major area of construction technology is the building of roads and highways. Roads have been built for thousands of years, but in the last century, they have become technologically advanced. The construction of roads today relies on the expertise of engineers and the muscle of road-building construction equipment.

Asphalt roads and highways generally consist of several layers. The bottom layer is known as the subgrade. The leveling and compaction of the subgrade is the first process to take place in the construction of a roadway. The next layer of the roadway is the foundation layer. The foundation is designed and constructed so the center of the roadway is at a higher elevation than the sides of the road. Compactors are then used to compress the foundation. The next layer of the roadway is the surface course, which is typically an asphalt mixture. The asphalt is laid using pavers. Once the paver has laid the asphalt, a series of rollers is used to compress and smooth the final road.

The Development of Railroads

The first railway was developed in England in the sixteenth century. A railway is a road or pathway on which rails are placed. A train's wheels then roll on the rails. The first railways were used to carry heavy loads of cargo on small cars. They were operated by hand and went back and forth

Figure 17-3. Limited access highways, such as those making up the interstate system, permit entry and exit only at certain points. In some areas of the country, exits may be many miles apart.

Figure 17-4. The laying of the first cast-iron rails across a prairie was recorded in this historic photograph. (Burlington Northern Santa Fe)

on short runs in mines. The first rails were no more than narrow wooden strips. Later, rails of wrought iron were placed on the wooden base of the track. In 1767, the first cast-iron rails began to be used. See **Figure 17-4.**

In 1825, Colonel John Stevens of New Jersey built the first small locomotive, or self-propelled railcar, which was powered by steam. George Stephenson built the forerunner of what would become the standard steam engine in 1829. At that time, the locomotive proved itself as a means of motive power. It was then that railways became railroads. ***Railroads*** are permanent roads made of a line of tracks fixed to wooden or concrete ties. In 1869, a railroad track stretching from Omaha, Nebraska to Sacramento, California was constructed. See **Figure 17-5.** It was not long until other railroad companies linked onto this track. This rapid growth soon connected the eastern states with the West Coast. Business began to spread, and towns were established along the railroads. A country still in its early development stages began to grow into a great nation with the help of the railroads and the people who worked for the railroads. A modern railroad provides a track for heavy equipment, such as locomotives and rolling stock. Railroad cars pulled by the locomotive are called ***rolling stock***. Today, railroads are used for transporting large shipments of cargo over long distances.

Railroad: A permanent road made of a line of tracks fixed to wooden or concrete ties.

By the mid-1900s, most steam locomotives had been replaced by diesel-electric locomotives, named after the engine's inventor, Dr. Rudolph Diesel. The diesel-electric locomotives could travel faster and were more powerful and efficient. These locomotives use diesel engines to move large pistons in a generator. The generator converts the mechanical power into electrical power. The electrical power then turns electric motors, which turn the train wheels. Diesel locomotives are used around the world and produce much less air pollution than steam engines. With the use of diesel-electric power, the locomotives can pull upwards of 50–60 train cars. See **Figure 17-6.** By placing more than one locomotive at the front, the train could be even longer. This allows more freight to be transported than previously thought possible.

Figure 17-5. This photograph commemorates the laying of the last section of track that linked the East and the West. (Burlington Northern Santa Fe)

Figure 17-6. Several diesel-electric locomotives are often joined together to pull long strings of freight cars.

Rolling stock: A railroad car pulled by a locomotive.

Pathways

All land transportation vehicles travel on a pathway. These pathways are physical devices arranged in a system. The purpose of the pathways is to restrict the freedom of movement of the vehicles. There are three types of pathways: nonfixed, fixed, and stationary.

Pathways are developed to support vehicles. For instance, roadways and railways are used to support automobiles and trains. Through the use of pathways, vehicles are able to reach their destinations in a safe and efficient manner. Pathways allow natural barriers to be crossed. Mountains are tunneled, and rivers are bridged. See **Figure 17-7.** Pathways allow vehicles to move about on the land without hurting people or bringing damage to property. Our cities, towns, and rural areas would be very dangerous if we did not set aside space to be used as pathways for vehicles. We would not be safe walking or driving!

Nonfixed Pathways

Automobile drivers are free to move their vehicles to the left and right. They are also free to go forward and backward. Automobiles have a lot of freedom of movement. The driver must stay on a road, but she usually has a

Figure 17-7. The problems of natural barriers are solved with the construction of tunnels and bridges.

Nonfixed pathway: A pathway in which a vehicle has freedom to move in various directions.

choice of which road to take. When a vehicle has such freedom to move within a pathway, the pathway is identified as a ***nonfixed pathway***. The main advantages of nonfixed paths are that they best serve human needs and wants.

A disadvantage of nonfixed pathways is that they use excessive amounts of land. The building of roads often destroys a portion of our environment. Also, modes of transportation that use nonfixed pathways consume a lot of energy and cause air and noise pollution. Roads, highways, trails, and sidewalks are all examples of nonfixed pathways.

Fixed Pathways

Fixed pathway: A pathway that has a fixed route.

Stationary pathway: A transporting path in which the structure does not move.

A ***fixed pathway*** does not allow the driver of the vehicle the same freedoms as a nonfixed pathway does. A railway and a subway line are two examples of fixed pathways. See **Figure 17-8.** A train does not have the freedom to go wherever the operator wants. The driver must follow the track. Fixed pathways are not as destructive to our environment. Railroads, for example, do not require the same amount of land as a typical four-lane highway. The vehicles used in fixed pathways are often more efficient than nonfixed pathway vehicles, when you compare the amount of people and cargo they can haul. A disadvantage of a fixed pathway, however, is they are less responsive to meeting human desires. People and cargo are loaded and unloaded at terminals on the fixed pathway. These terminals may not be the final destination, and the cargo or people may then need to take different vehicles to their destinations.

Figure 17-8. A fixed pathway, such as this railway line, allows only linear movement.

Stationary Pathways

There is another type of transporting path called a ***stationary pathway***. The system may have moving internal elements, such as belts or chains, but the basic supporting structure is stationary. Stationary systems usually move goods.

Pipeline structures do not move. The pipe extends from the point where material enters the system to the point where material is discharged. The materials are moved through the system by the use of pressure, gravity, or vacuum.

Conveyors are another example of the stationary system of transporting. They can be designed to move either people or goods. See **Figure 17-9.**

Modes of Land Transportation

Land transportation can be divided into several different modes of transportation. See **Figure 17-10.** Each mode of land

Figure 17-9. This complex conveyor system was built to move passenger baggage at DeGaulle International Airport in Paris, France. A conveyor system is a stationary pathway. (Siemens)

Figure 17-10. All land transportation can be placed in one of these five categories, or modes.

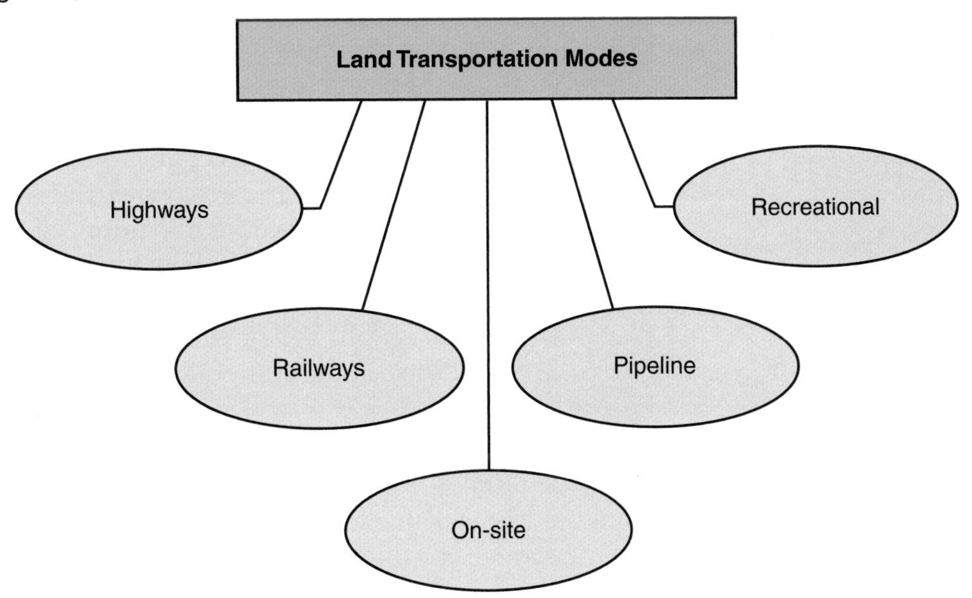

transportation has a specific purpose and includes a number of different vehicles. These modes of transportation will be discussed in more detail within this chapter:

- **Highway land transportation.** This is transportation that occurs on roadways and highways.
- **Railway land transportation.** This is the movement of people and cargo in vehicles that use rails as guidance.
- **Pipeline land transportation.** This is the movement of cargo through stationary pipes.
- **On-site land transportation.** This is the transportation of people and cargo short distances within buildings and complexes.
- **Recreational land transportation.** This is transportation for fun, sport, and recreation.

Highway Land Transportation

Most of the land transportation we commonly use every day falls into the mode of highway land transportation. Whether you ride to school in a car or bus, you are using highway land transportation. Highway land transportation is the movement of people or cargo on roadways and highways. A roadway may be as simple as a dirt or gravel country road or as complex as an eight-lane superhighway. Each road serves a specific need in the land transportation system. Roads and highways make up an enormous part of our world. You need only to look at a road map to get an idea of the miles of highways that crisscross the landscape. Much of our travel would be impossible without these highways.

Figure 17-11. A two-cylinder, four-cycle engine powered Henry Ford's first car, which was invented in 1896. The car had an electric bell attached to the front to warn pedestrians.

Automobiles

In the early twentieth century, the automobile was in developmental stages. See **Figure 17-11.** Ford made the automobile a common sight on early twentieth-century American roads. He made the automobile cheap, easy to operate, and easy to maintain.

The design of the automobile has changed immensely over the past 100 years. From a carriage-like, steam-powered buggy that did a maximum of 10 miles per hour (mph), the automobile has evolved into an aerodynamic turboengine sports car that can reach speeds up to 200 mph. Today, there are also a number of different styles of automobiles available. See **Figure 17-12.**

- ***Coupes*** are small, usually two-door, cars. Some coupes are often termed *sports cars*.
- Sedans are larger than coupes and have more interior room. They are typically found as four-door cars.
- Sport-utility vehicles (SUVs) are somewhat of a combination of sedans and pickups. They are raised, like pickups, and include many of the features and comforts you would find in a sedan.

Figure 17-12. Basic vehicle styles. A—A coupe. (Aston-Martin) B—A sedan. (Lincoln-Mercury) C—A sport-utility vehicle (SUV). (Land Rover) D—A van. (DaimlerChrysler) E—A light truck. (Nissan)

A

B

C

D

E

- Vans are built to hold more cargo and passengers than other automobiles. Minivans and conversion vans are configured to hold passengers and designed for family use. Cargo, or panel, vans are built to carry cargo.
- Light (or pickup) trucks sit taller than most other automobiles and have an open bed to transport cargo.

Coupe: A small, usually two-door, car.

While we have shaped the automobile, it has also shaped our communities. Certainly our neighborhoods would be much different than they are today if there were no automobiles. See **Figure 17-13.**

Figure 17-13. These pictures show how much the automobile has changed our society.

Trucks

Trucks can be divided into three main types: light, medium-duty, and heavy-duty. Light trucks, or pickups, were discussed in the automobile section because they are manufactured, marketed, and sold like automobiles. They are most often used for passenger transportation. Medium- and heavy-duty trucks are used for the transportation of cargo. Medium-duty trucks are configured in a straight truck design. ***Straight trucks*** have one frame that connects both the front and rear axles. The beds of medium-duty trucks are built to fit the needs of the user. Medium-duty trucks are made into delivery trucks, garbage trucks, dump trucks, tank trucks, tow trucks, and recreational vehicles, just by changing the type of bed. See **Figure 17-14.** Heavy-duty trucks are also known as tractor-trailers. These are usually seen on the highways and are called *18-wheelers*, or *semis*. Tractor-trailers are used to carry freight between cities. The tractor is the front part of the truck that houses the engine. It is powered by a diesel engine, rather than a gasoline engine. The reason for this is

Straight truck: A truck that has one frame that connects both the front and rear axles.

Figure 17-14. Medium-duty trucks are offered in many different configurations. A—A television station remote broadcast vehicle. B—A refuse hauler. (Mercedes-Benz) C—A concrete mixer. (Mercedes-Benz) D—A dump truck. (Mack Truck Company)

A

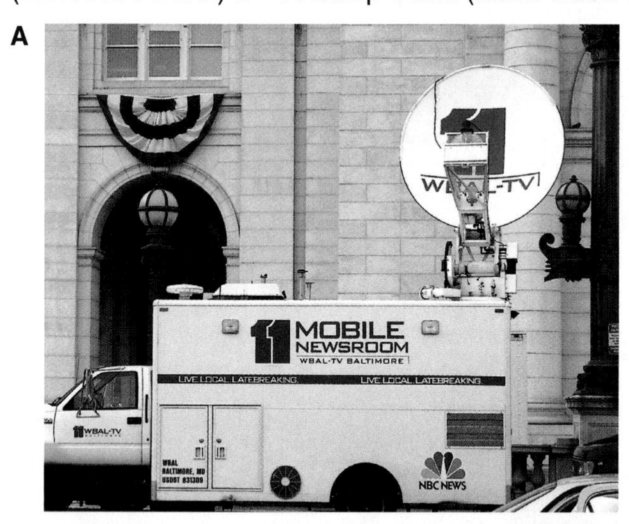

B

C

D

that more power is available for the hauling of heavy freight. The trailer is the back end of the truck. Some trailers are tanks that haul liquid or gas. See **Figure 17-15.** Other trailers are boxlike. They are used for hauling grain, livestock, or packaged goods from manufacturing companies. There are also refrigerator trucks used to transport food products. The trailers are made of lighter metals to reduce the overall weight of the truck. This allows them to carry more cargo. Federal and state governments have regulations that limit the length and weight of the trucks and their cargo.

Figure 17-15. Semitractors haul different types of freight. This one is hauling a tanker loaded with a liquid cargo.

Career Connection

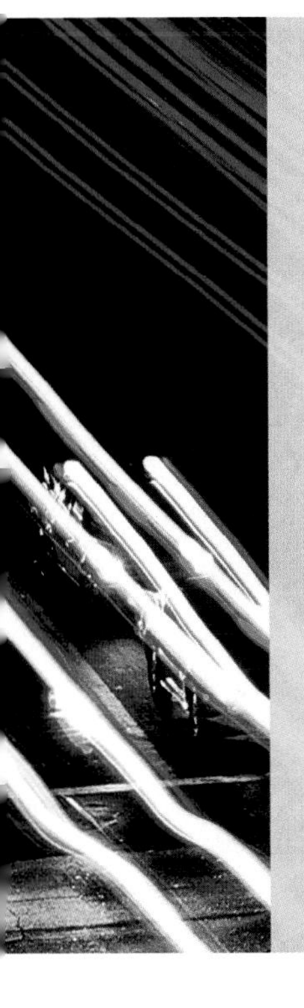

Truck Drivers

Our nation relies on the transportation of goods. Most of the movement of goods within the nation uses highway transportation, and a large amount of goods are moved by tractor-trailer. Without drivers, the goods would never get to their destinations. There are two main types of truck drivers: tractor-trailer drivers and delivery service truck drivers.

Tractor-trailer drivers drive goods from city to city and often across the nation. They often have routes that keep them on the road for days at a time. The average earnings for a tractor-trailer driver equal about $16 per hour. Delivery truck drivers often have daily routes that allow them to be home every night. They deliver goods within cities or to nearby cities in straight trucks or vans. Many of these drivers are stationed at a base and deliver goods to and from the base. The average earnings for a delivery truck driver equal about $11.50 per hour.

The main qualification for all truck drivers is that they must have a commercial driver's license (CDL). The CDL is obtained from the state in which the driver is based. CDLs require drivers to be 18 years old and to take a written test and a driving test. To drive across state lines, the Federal Motor Carrier Safety Administration (FMCSA) regulations require a number of health, vision, and hearing levels, and drivers must be 21 years of age. Most trucking companies require drivers to have a high school diploma, and some look for graduation from a commercial driving school.

An advantage of moving cargo by truck is a reduction in damage to the cargo. Cargo travels more safely in a truck. Packaging is not considered a necessity. Therefore, this saves on the cost of containers. It also saves on the packing material. Trucks are loaded with freight at trucking terminals or warehouses. See **Figure 17-16.** A supervisor checks the loaded freight. Once all information about the load is gathered, the truck is ready to roll on to its destination. In the United States, operation and safety standards are set and regulated by the Federal Motor Carrier Safety Administration (FMCSA), which is a division of the Department of Transportation. A great advantage of the truck, as opposed to any other form of carrier, is its flexibility. It is convenient for a truck to transport products from door to door quickly.

Buses

The first use of the bus was to extend railroad lines. When the tracks ended, a bus would continue transporting people into the city. Bus lines were soon established for crosstown transportation where there were no tracks. Buses also were needed for the transportation of school children. See **Figure 17-17.** These vehicles are a necessity today in school systems. Buses bring rural and suburban children into the cities.

One of the major advantages of a bus is that it can haul many passengers at once. Another advantage is that it has frequent pickup and delivery points for the passengers. Many people feel, however, that the advantages of the automobile are not worth the trade-offs involved in riding a bus. A bus rider would have to trade some convenience and comfort for decreased air pollution and traffic congestion. Bus companies

Figure 17-16. Pallet loads of freight are loaded aboard trucks at warehouses or trucking terminals. Sometimes, a truck will pick up a full load at one location. At other times, smaller quantities of cargo will be loaded at several different locations.

need to consider the scheduling, routing, comfort, and frequency of the bus service. See **Figure 17-18.** Buses must meet the needs of the people, or people will not ride them.

Motorcycles

Motorcycles are another kind of land transportation vehicle. These vehicles are often seen on the highways and streets. Police departments, delivery services, and pleasure riders use them. See **Figure 17-19.** The first successful motorcycle had an internal combustion engine mounted on a three-wheeled bicycle. Today, three-wheeled motorcycles, or trikes, are not as common as two-wheeled motorcycles. Motorcycles can be categorized into two different groups: street and off-road motorcycles. Street motorcycle styles include cruiser, sport, touring, and standard. Many people today travel on motorcycles across the country. The design of the motorcycle has changed to meet the personal needs and preferences of the operator and passenger.

Figure 17-17. Many children travel to school on buses.

Heavy equipment

Many land transportation vehicles are designed for specific purposes other than just moving people or cargo. ***Heavy equipment*** includes large and powerful vehicles used for reasons such as moving earth, farming fields, and conducting warfare. It can be organized into construction, farming, and military equipment. Construction equipment, including bulldozers, paving machines, and wheeled cranes, is used in the building of everything from roads and bridges to homes. See **Figure 17-20.** Farm equipment, such as tractors, is used in the planting, growing, and harvesting of crops. See **Figure 17-21.** Military equipment, including personnel carriers, surveillance vehicles, jeeps, tanks, and other armored fighting vehicles, is the last type of heavy equipment land vehicle. See **Figure 17-22.**

Heavy equipment: A large and powerful vehicle used for reasons such as moving earth, farming fields, and conducting warfare.

Figure 17-18. The bus industry competes with automobiles in the transporting of people. This is a local bus operating in the Netherlands, picking up passengers for the town of Haarlem. (Van Hool)

Rail Land Transportation

Railway lines form a network of tracks across the country. Railroads have been a factor in moving people and cargo for more than 300 years. A railroad system consists of the following elements:

- Miles of roadbed and rails strong enough to carry the heavy weight of the trains and their payloads.
- A system of signal devices so movement of trains on the same track can be safely coordinated.
- A variety of engines to pull the trains.

Figure 17-19. Orange County Choppers, a specialty firm that builds custom motorcycles, created this commemorative motorcycle for the Miller Electric Manufacturing Company. This motorcycle was commissioned to celebrate the 75th anniversary of the company, a major manufacturer of welding equipment. (Miller Electric Manufacturing Company)

Figure 17-20. Paving machines lay down a lanewide ribbon of asphalt in a single pass. Several layers are normally used for roads or parking areas. (Terex/Cedarapids)

- Cars designed for carrying passengers or a variety of different products and materials.
- Stations for loading and unloading passengers and handling cargo (freight).

Rail transportation systems offer services for the transporting of both freight and passengers.

Freight trains

A ***freight train*** is several freight cars joined together and pulled by an engine or locomotive. Several different types of freight cars are used to transport different types of cargo. See **Figure 17-23.** ***Boxcars*** are boxlike cars with doors on both sides. They may be refrigerated to carry frozen foods or any other product that must be kept cool. ***Flatcars*** are sturdy platforms on wheels. They carry such material as steel, lumber, truck trailers, containers, and even very heavy equipment. ***Gondolas*** have high or low sides with no tops. They transport loose material, such as stone, scrap metal, and iron. ***Hopper cars*** have hoppers (chutes) underneath. They carry bulk materials, such as coal and ore. Closed hoppers are used to haul materials that need to be protected from the weather, such as corn, wheat, sand, salt, and fertilizer. ***Tank cars*** are large tanks on wheels. They transport liquids. ***Transport cars*** are flatcars with side rails. They are mainly used for transporting new automobiles and trucks from manufacturing plants to car dealers. A ***caboose*** is the last car on the freight train. It is used to house the train crew.

Most freight trains are made up of a combination of cars carrying different cargo. The cars may have a number of different stops before they reach their final destinations. A unit train is not like a typical freight train. ***Unit trains*** carry only one type of cargo, and all of its cars are alike. See **Figure 17-24.** A unit train goes to the same destination trip after trip.

Figure 17-21. The modern tractor makes plowing, planting, cultivating, and harvesting much more efficient. (Howard Bud Smith)

Figure 17-22. The Humvee military vehicle can be configured many different ways to perform different tasks. This one is being used to guard a traffic checkpoint. (U.S. Army)

It is more economical to haul large amounts of cargo by railroad than by highway. Carrying freight by railroad also has fewer restrictions on the weight and size of the cargo, as mandated by the Federal Railroad Administration (FRA). Highway regulations place limits on the weight and length of the loads trucks can carry.

Freight train: Several freight cars joined together and pulled by an engine or locomotive.

Boxcar: A boxlike freight car with doors on both sides.

Figure 17-23. Some common types of railway cars. A—A boxcar has large doors and a high roof to transport general cargo. B—A hopper car is designed to haul loose bulk, such as coal and grain. C—A flatcar is used to carry large machines or to "piggyback" semitrailers. (Burlington Northern Santa Fe)

A

B

C

Flatcar: A freight car that is a sturdy platform on wheels. It carries steel, lumber, truck trailers, containers, and heavy equipment.

Gondola: A freight car that has high or low sides with no tops. It transports loose material.

Hopper car: A freight car with chutes underneath. It carries bulk materials.

Tank car: A freight car that is a large tank on wheels for transporting liquids.

Transport car: A flatcar with side rails. It is mainly used for transporting new vehicles from manufacturing plants to car dealers.

Caboose: The last car on a freight train. It is used to house the train crew.

Figure 17-24. Unit trains transport only one product, such as coal.

Passenger trains

Passenger trains are trains that transport people. This form of transportation causes little traffic congestion and is safer than some other forms of transportation. There are two main types of passenger trains: long-distance rail and mass transit rail.

Long-distance rail transportation is used when passengers want to travel between distant cities on a train. One advantage of this type of transportation is that the passengers can sit back and relax on the trip because they do not have to drive the vehicle. The railroad is also often a more direct route to the destination than the highway. Long-distance passenger trains are not as convenient, however, as automobiles. Riding a train reduces schedule flexibility. Amtrak is the largest long-distance passenger train in the United States. It transports people from city to city across the United States. See **Figure 17-25.**

Mass transit rail systems are a form of rail transportation that can carry many people at one time. They transport people shorter distances, often between work and home. There are three main types of mass transit rail systems: light-rail, heavy-rail, and high-speed rail. Light-rail systems can be used in intercity (between two cities) or intracity (within a city) transportation. Heavy-rail systems typically carry more passengers and have larger trains than light-rail systems. A ***subway*** is a heavy-rail train that runs on a rail below the earth's surface. Another heavy-rail vehicle is an ***elevated train***, which is a rail system that runs above the city streets. See **Figure 17-26.** A ***monorail*** is a train that runs on a single rail. High-speed rail systems are used to connect two or more cities together. The fastest electric, high-speed train is the French TGV (*train à grande vitesse*, or high-speed train), which hit a top speed of 322 mph. See **Figure 17-27**. The newest development in high-speed rail systems is the application of electromagnetism. ***Magnetic levitation (maglev)*** trains use powerful magnets to hover above and propel down the track. See **Figure 17-28.**

Unit train: A train that carries only one type of cargo.

Mass transit rail: A form of rail transportation that can carry many people at one time.

Figure 17-25. Amtrak operates trains for long-distance travel across the United States. It replaced passenger operations run by individual railroads. (Amtrak)

Figure 17-26. The "El" is Chicago's elevated railway system, which encircles the downtown area one story above the street. Although these systems can be used with light-rail vehicles, it is more common to use heavy-rail trains. Note the massive beams supporting the rail structure.

Figure 17-27. France's electric TGV (*train à grande vitesse*, or high-speed train), the world's fastest, operates at speeds well in excess of 200 miles per hour (mph). High-speed rail tracks must be built further apart than traditional tracks because of the turbulence when trains pass each other. They must also have fences on each side of the track to ensure that animals and debris do not enter the track area.

Subway: A heavy-rail train that runs on a rail below the earth's surface.

Elevated train: A heavy-rail system that runs above the city streets.

Monorail: A train that runs on a single rail.

Magnetic levitation (maglev): A train that uses powerful magnets to hover above and propel down a track.

Figure 17-28. Magnetic levitation (maglev) trains transport passengers at high speeds in a number of countries. This train is on the outskirts of Shanghai, China. Since maglev trains do not actually touch the track, there is no friction to slow them. Several maglev rail systems are planned for the United States. (Transrapid International, Inc., GmbH)

Underwater rail systems

The Channel Tunnel project, also called the "Chunnel," is an underwater rail system linking Britain and France. For nearly 180 years, the idea of tunneling under the English Channel has been contemplated several times. This rail system consists of three tubes. Two tubes, 25′ in diameter, are linked at several points to a third tube about 16′ (4.8 m) in diameter. The smaller tube serves as a service tunnel. The two larger tunnels serve as the rail lines. One is for a westbound train, and the other is for an eastbound train. The third, center, tunnel is a service tunnel. The tubes, or tunnels, are 150 meters below sea level. Tunnel workers had to bore through chalk marl, which is the lowest layer of chalk under the channel. The tunnels were excavated using a boring machine. See **Figure 17-29.** The trains are able to carry passengers and cargo under the English Channel in 20 minutes. Electric locomotives drive these trains. See **Figure 17-30.**

Pipeline Transportation

The most efficient way to transport water from your house to a garden 100′ away would be to use a hose. This same efficiency would apply when transporting any product by pipeline. See **Figure 17-31.**

Centuries ago, pipelines were developed from bamboo. Water was transported through the hollow bamboo. Water-carrying pipelines were also made out of logs. The logs were hollowed out and fitted together. Being somewhat porous, the bamboo and the wood could not withstand the pressure of the water. Thus, by the late 1800s, iron pipelines were

Figure 17-29. Tunnel boring beneath the English Channel. A—Working 328′(100 m) below sea level, a boring machine is excavating a tunnel 25′ (7.6 m) in diameter. B—The rail has been installed for one of the tunnels. (Eurotunnel)

A

B

being constructed. Today, pipelines are made from steel or plastic and vary in diameter. One of the major pipelines you may be familiar with was constructed both above and below ground. This installation, known as the Trans-Alaska Pipeline, was constructed in 1977. It transports oil 775 miles across Alaska.

Pipelines are used to transport such products as oil, natural gas, water, coal, and gravel. Some of these substances are fluid and flow easily through a pipe. Others need to be mixed into a liquid solution called slurry, in order to be transported by a pipeline. ***Slurry*** is a mixture of a ground solid, such as coal, along with a liquid, such as water.

Pipeline construction

Construction of pipelines requires careful planning. Once a company decides to construct a pipeline, the route is first planned. The company will then contact the owners of the land where the pipeline will be constructed. Once an agreement is reached with the landowners, the path the pipeline will follow is cleared. The pipes are then laid out along the path. The digging or trenching of the earth begins. See **Figure 17-32.** Pipe is laid in the trench and welded together. The welds are examined visually and inspected using X-ray

Figure 17-30. The Channel Tunnel train pulls cars loaded with cargo and passengers. (Eurotunnel)

Figure 17-31. Pipelines stretch across many miles and carry several different kinds of materials. (The Coastal Corporation)

Figure 17-32. It takes many workers and machines to lay a pipeline. (The Coastal Corporation)

machines to ensure there will be no leaks. If the pipe passes the inspection, the trench is then refilled with earth. At this point, the pipeline is out of sight, and the earth is returned to its original condition.

Pipeline operation

A pipeline is a system of different types and sizes of pipes. See **Figure 17-33.** Each pipe has a specific task in the system. At the beginning of the process, the product to be transported is collected and stored. The pipes used at this stage are known as flow lines and ***gathering lines***. The product is pumped from its source to a processing facility or storage tank using the flow or gathering lines. These lines can be from 2–12″ in diameter. From the processing and storage facilities, the products are transported to the main line using feeder lines. The feeder lines are larger in diameter, up to 20″, because they typically carry more cargo than the flow and gathering lines. A number of feeder lines are pumped into the main lines, known as ***transmission lines***. These lines are the pipes that transport the cargo over great distances. The pipes are often buried and can be as large as 48″ in diameter. The transmission lines are routed to different places, depending on the product the pipeline serves. Once the products have reached their

Slurry: A mixture of a ground solid and a liquid.

Gathering line: A pipe in a pipeline system in which the product to be transported is collected and stored.

Figure 17-33. An oil pipeline system moves crude oil to the refinery and then moves petroleum products to storage or to transport systems that will move them wherever they are to be used.

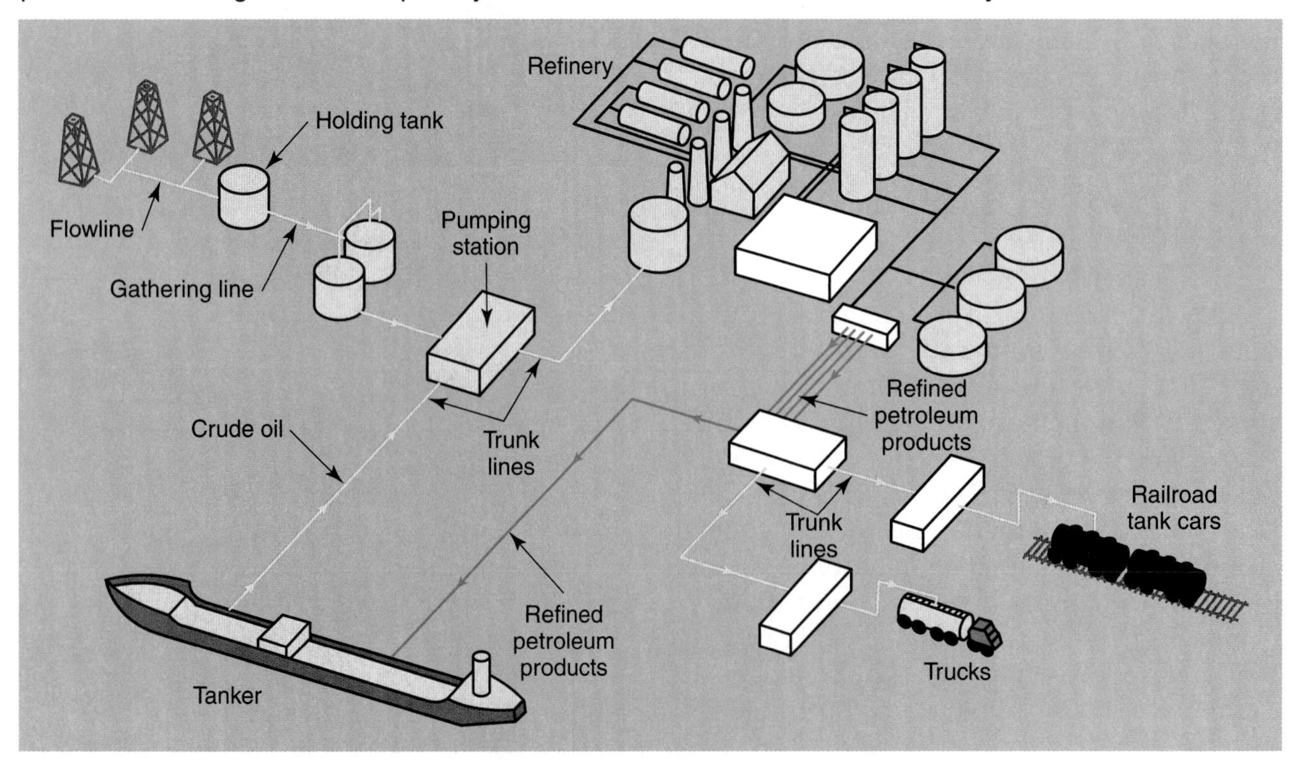

distribution centers, smaller pipelines, known as distribution lines, are used to transport the products to their final destinations.

To prevent pipeline clogging, a barrel-shaped brush, known as a ***pig***, routinely cleans the line. The pig can be blown through the line or pushed through with moving cargo. Smart pigs, or electronic pigs, can be used to determine the condition of the pipelines. They are able to detect cracks and leaks from inside the pipeline.

Several different products can be transported through the same pipeline. Products are simply pumped into the pipeline in separate batches. A batch of gasoline may be followed by diesel fuel or kerosene. This is called a ***batch sequence***. Once the batches reach the terminals, they are separated by weight with a computerized device called a ***gravitometer***. Each product is pumped into its respective holding tanks and is ready for distribution.

Transmission line: A pipe in a pipeline system that transports cargo over great distances.

Pig: A barrel-shaped brush used to clean pipelines.

Batch sequence: An order of succession of the quantity of material prepared for one operation.

Gravitometer: A computerized device that separates products by weight.

On-Site Transportation

Highway, rail, and pipeline transportation typically deal with moving people or cargo a significant distance. The distance may be across town or even across the country or continent. There is a great amount of transportation, however, that occurs on a much smaller scale. Many times, people and cargo simply need to be transported up one story in a building or to the other side of a warehouse. This type of transportation is known as on-site transportation. It can be divided into two types: material handling and people moving.

Material-handling devices are vehicles used to transport cargo within buildings and complexes. They can be as simple as hand trucks or as complex as cranes and conveyor systems. Hand trucks are *L*-shaped vehicles with two wheels. Materials are loaded on the truck, and the truck is moved using human power. Conveyors are common material-handling devices. They are often used to transport products in manufacturing settings. Belt conveyors and roller conveyors are used to transport products down an assembly line. Trolley conveyors are still another method used to transport. A trolley conveyor moves overhead on a cable, as it lifts and transports products. Cargo is often placed on pallets to help in the handling of the products. The pallets are wooden platforms with openings that allow the arms of a forklift underneath. Forklifts are vehicles that can lift pallets and move them to new locations within a warehouse or building. See **Figure 17-34.**

Figure 17-34. A forklift can move heavy pallet loads of goods from place to place and store them in stacks or in special shelving-type racks. Operator skill is important to prevent damage to the materials or storage racks.

Figure 17-35. An escalator is a people transporter designed to move people between floors in buildings. It is a type of on-site transportation.

The most popular people-moving vehicle, the elevator, was invented as a material-handling device. Lifting devices have been used to move materials for thousands of years. These devices evolved into material-moving elevators. It was not until Elisha Otis invented the safety brake in 1854 that the elevator was used to move people. Today, elevators are common people movers in public buildings and even some homes. Other people movers include escalators and moving sidewalks. See **Figure 17-35.** Both of these vehicles can be found in public areas, such as shopping malls and airports.

Recreational Land Transportation

Land transportation can also be used for fun and exercise. Vehicles used for these purposes are in the recreational category of land transportation. Bicycles may be the most popular recreational land vehicles. There are four main types of bicycles: road, mountain, dirt, and recumbent. Recumbent bikes are the newest and most unique bicycles. Riders of recumbent bicycles sit in a more "seated" position, with the pedals in front of the rider. Other recreational land vehicles include mopeds, unicycles, dirt bikes, snowmobiles, skateboards, and scooters. See **Figure 17-36.** Keep in mind that the universal systems model is at work in recreational transportation. You will find inputs, processes, outputs, and feedback.

Figure 17-36. Recreational land transportation vehicles. A—Segway personal transporters can be used on or off paved surfaces and operate for several hours on a single electrical charging. (Segway Corporation) B—A recumbent bicycle is a relaxed form of transportation. C—Snowmobiles are popular for recreational use in cold climates. (Polaris)

A

B

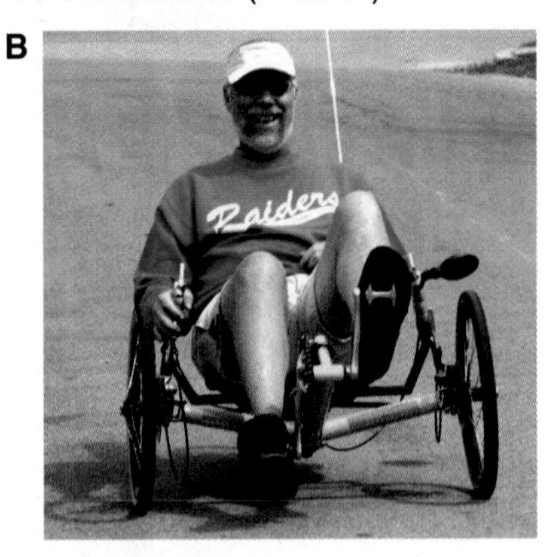

C

Summary

Land transportation is a key element in our lives. Without it, we would not be as advanced as we are today. Transportation modes on land are highways, railways, pipelines, on-site, and recreational. The wheel was the beginning of many technological advancements in land transportation. It helped lead to the invention of the bicycle, car, truck, and bus.

The development of railroads assisted in the settlement of the West. Much freight is transported by train. In the inner city, mass transit is highly effective. Light-rail trains, subways, elevated trains, and monorails are all used daily in metropolitan areas to transport hundreds of people.

Pipelines are a unique form of transporting cargo. Liquid materials, such as oil, kerosene, and water, move by pipeline over several hundreds of miles. Solid bulk material, such as coal, copper, and gravel, can also be transported by pipeline when mixed with a liquid to form a slurry. Other transporting devices located in buildings help move people and products. Elevators, escalators, moving sidewalks, and conveyors are a few of these devices.

Key Words

All the following words have been used in this chapter. Do you know their meanings?

batch sequence
boxcar
caboose
coupe
elevated train
fixed pathway
flatcar
freight train
gathering line
gondola
gravitometer
heavy equipment
hopper car
magnetic levitation (maglev)
mass transit rail
monorail
nonfixed pathway
pig
railroad
rolling stock
sledge
slurry
stationary pathway
straight truck
subway
tank car
transmission line
transport car
unit train

Test Your Knowledge

Write your answers on a separate sheet of paper. Do not write in this book.

1. In what way has the wheel benefited land transportation?
2. *True or False?* Chariots were the first land vehicles.
3. *True or False?* Powered land transportation has always existed.
4. *True or False?* Roadways have been around for thousands of years.

5. *True or False?* Rolling stock is a name referring to any wheeled vehicle.
6. Describe how diesel-electric locomotives work.
7. List and describe two types of routes for land transportation vehicles.

Matching questions: For Questions 8 through 12, match the words on the left with the correct term on the right.

8. Roads.
9. Railroads.
10. Conveyors.
11. Sidewalks.
12. Pipelines.

A. Fixed pathways.
B. Nonfixed pathways.
C. Stationary pathways.

13. State the five modes of land transportation.
14. How has the invention of the automobile made life easier?

Matching questions: For Questions 15 through 18, match the phrases on the left with the correct term on the right.

15. Usually small, "sporty" cars.
16. Typically a four-door car.
17. Raised like a pickup, with features of a sedan.
18. Can be configured for cargo or passengers.

A. Coupe.
B. Sedan.
C. Sport-utility vehicle (SUV).
D. Van.

19. Name the three main types of trucks.
20. How is a straight truck different from a tractor-trailer?
21. *True or False?* The Federal Motor Carrier Safety Administration (FMCSA) oversees pipeline transportation.
22. Write four examples of highway land transportation vehicles.
23. Cite two types of street motorcycles.
24. A train car, with high or low sides and an open top, used for transporting material such as scrap metal, iron, and stone is called a:
    A. boxcar.
    B. gondola.
    C. flatcar.
25. What is a unit train?
26. *True or False?* Mass transit rail systems carry few people.

27. An elevated train is a rail system:
    A. below the city.
    B. through the city.
    C. above the city streets.
28. Define *slurry*.
29. What device is used to unclog a clogged pipeline?
30. *True or False?* Material-handling devices are used to transport cargo within a building or complex.

Activities

1. Research the development of an early land transport vehicle and write a report on its development and use. Suggest what its impact was on society. As part of the report, suggest why it would or would not answer modern-day transportation needs.
2. Build an appearance model (one that looks like the original, but does not operate) of the vehicle you researched for Activity 1.
3. Imagine your community 50 years from now has all automobile traffic banned within the limits of the community. (If you do not live in a town or city, select a neighboring community.) Only delivery trucks and garbage trucks can enter the city. Design a system that will transport people to and from the community limits and determine which would be least polluting.

The Velaro high-speed train, shown in this design drawing, has all its drive components distributed along the bottom of the train cars, making the entire length of the train available as passenger space. These trains will go into service in Germany and Spain and are expected to operate at speeds in excess of 200 miles per hour (mph). (Siemens)

18

Land Vehicular Systems

Basic Concepts

- Cite the types of propulsion systems used in land vehicles.
- State how drive systems function.
- Identify types of guidance systems.
- List ways in which land vehicles are controlled.
- Define and describe the components of a suspension system.
- Identify the function of automobile structural systems.
- Name the support systems of land transportation.

Intermediate Concepts

- Describe the differences between series and parallel hybrid systems.
- Explain the operation of an automatic transmission.

Advanced Concepts

- Demonstrate how magnetic levitation (maglev) systems operate.

Land transportation requires vehicles such as automobiles, trucks, locomotives, and railcars to function. These vehicles and all other land vehicles have components from each of the vehicular systems. These systems include propulsion, guidance, control, suspension, structural, and support systems.

Propulsion Systems

Land vehicles, like all forms of transportation, must be propelled in order to move from one place to another. This propulsion can occur in several ways. The first way is through the use of an internal combustion engine. The second method of propulsion is electricity. The third is a hybrid of both the internal combustion engine and electricity.

Internal Combustion Engines

Internal combustion engines are the most commonly used type of engine in automobiles, trucks, and recreational vehicles. They are a type of heat engine. The engine functions by igniting a fuel to produce hot gases. The hot gases push either a piston or rotor to create mechanical energy. Mechanical energy is used to propel the land vehicle. In internal combustion engines, the hot gases are produced inside of a combustion chamber, or cylinder. There are three main types of internal combustion engines: gasoline piston engines, rotary engines, and diesel engines.

Gasoline piston engine: An engine found in automobiles, motorcycles, lawn mowers, and snowmobiles. These engines have five main components: a cylinder, a piston, a spark plug, a crankshaft, and fuel.

Gasoline piston engines

Gasoline piston engines can be found in automobiles, motorcycles, lawn mowers, and snowmobiles. The theory of the gasoline piston engine existed well before an efficient example of the engine was built. Sadi Carnot, in 1823, and Alphonse Beau de Rochas, in 1862, both published articles describing the function of a gasoline piston engine. It was not until 1876, however, that Nikolaus Otto built the first efficient gasoline piston engine.

Figure 18-1. The five major components of a gasoline piston engine are the cylinder, piston, spark plug, crankshaft, and fuel.

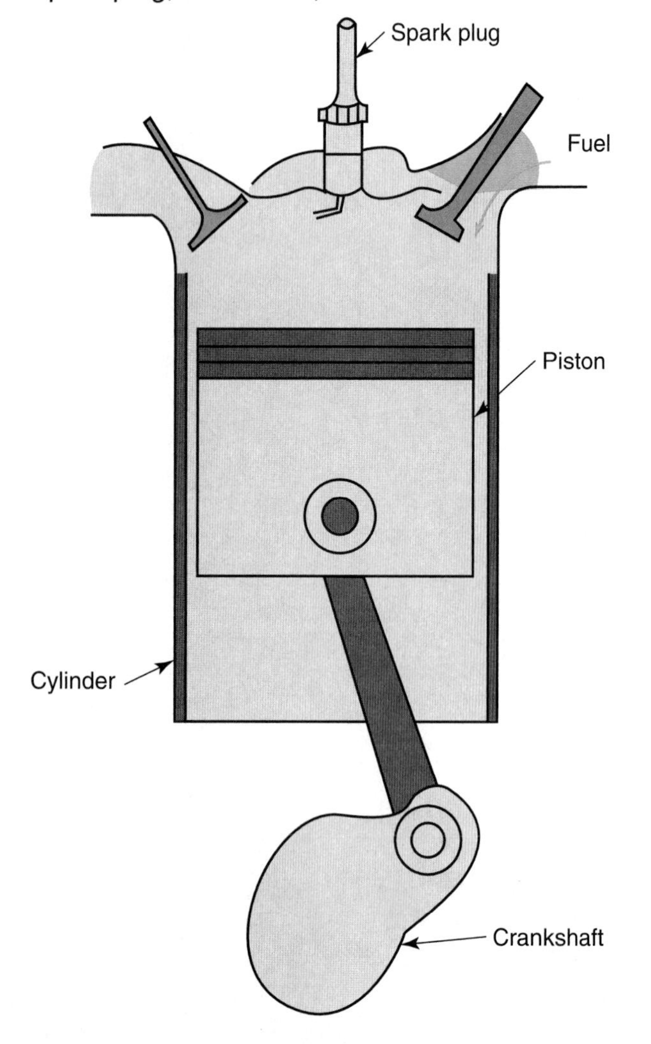

The gasoline piston engine has five main components: a cylinder, a piston, a spark plug, a crankshaft, and fuel. See **Figure 18-1.** The cylinder is an enclosed chamber with two ports. One allows fuel and air in, and the other allows exhaust to escape. The main moving part inside the cylinder is the piston. Fuel is allowed to enter the cylinder, and the piston compresses the air and fuel mixture. The compressed fuel is then ignited by a spark plug. The piston is driven down, which creates the mechanical energy. The reciprocal (up and down) motion of the piston is used to turn the crankshaft. The crankshaft is the device that transmits the power of the engine to the transmission. The function of the transmission will be discussed later.

The two ways in which gasoline piston engines operate are based on the number of strokes it takes to complete one combustion cycle. The two types of engines are known as two-stroke and four-stroke engines. The operation and advantages of each are described in greater detail in Chapter 14.

Gasoline piston engines are classified by the quantity and alignment of the cylinders. There are three main cylinder configurations. See **Figure 18-2.** The first is the in-line configuration. This configuration has all cylinders in a straight line. The second style is the flat, or opposed, engine. In this type of engine, there are two rows of cylinders arranged

Figure 18-2. Engine blocks have three different cylinder configurations. The number of cylinders varies, but it is always in multiples of two. A—In-line. B—V. C—Opposed.

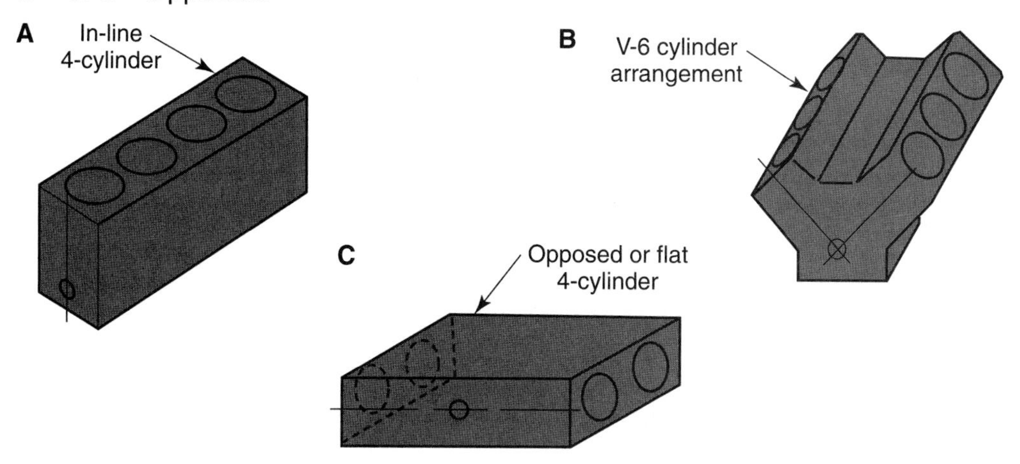

horizontally. The two banks of cylinders are on opposite sides of the crankshaft. The pistons are connected to the crankshaft and push outward from the center. The last engine configuration is the V-engine. V-engines are similar to flat engines, in that they have two rows of cylinders. The difference is that, in V-engines, the cylinders are situated to form a *V*, with the crankshaft at the bottom, instead of the cylinders being directly opposed. The number of cylinders is also listed when describing a gasoline piston engine. For example, an engine with a *V* configuration and six cylinders is called a V6. The number of cylinders and the layout of the engine affect the amount of power produced. Different configurations are used in different situations.

Rotary engines

Rotary engines are internal combustion engines that use rotors in place of pistons. They are usually called Wankel rotary engines. The name comes from their German inventor, Dr. Felix Wankel. The engine is small, lightweight, powerful, and smooth running. The smoothness comes from the use of rotary, rather than reciprocal (up and down), movement of the piston. The rotor turns inside a specially shaped combustion chamber. It lets fuel and air in, compresses it, burns it, and then releases the exhaust all in one smooth rotation. The design of rotary engines decreases the amount of moving parts used in piston engines.

Rotary engine: An internal combustion engine that uses rotors in place of pistons. Also called Wankel rotary engines.

Because rotors replace the pistons, we classify these engines differently. Whereas we refer to four, six, or eight cylinders in traditional piston engines, we call out the number of rotors in the Wankel engine. Generally, one-rotor Wankel engines are used in smaller applications, such as snowmobiles. Two-rotor Wankel engines are used for automobiles. Experimental engines with more than two rotors have been developed, but they are not yet commonly used.

Figure 18-3 shows the parts of a Wankel rotary engine. The special shape of the housing is very critical. It is somewhat like a large figure eight. This shape is called an ***epitrochoidal curve***. It allows the tips of the rotor to ride against the housing as the rotor revolves.

Epitrochoidal curve: The special shape of a Wankel rotary engine housing, somewhat like a large figure eight.

Figure 18-3. A Wankel engine uses spinning rotors instead of pistons to convert heat energy to power. Note the shape of the combustion chamber. The rotor has the shape of a slightly rounded equilateral triangle. It is attached to an eccentric (off-center) shaft, which comes through the center of the housing. Special replaceable tips maintain a tight seal between the tips and the housing. If the seal is not kept tight, the engine will have a great loss of power.

Figure 18-4. The combustion cycle of the Wankel engine. Each lobe (point) of the rotor goes through four phases in one revolution. The intake, compression, power, and exhaust stages all take place in one revolution of the rotor.

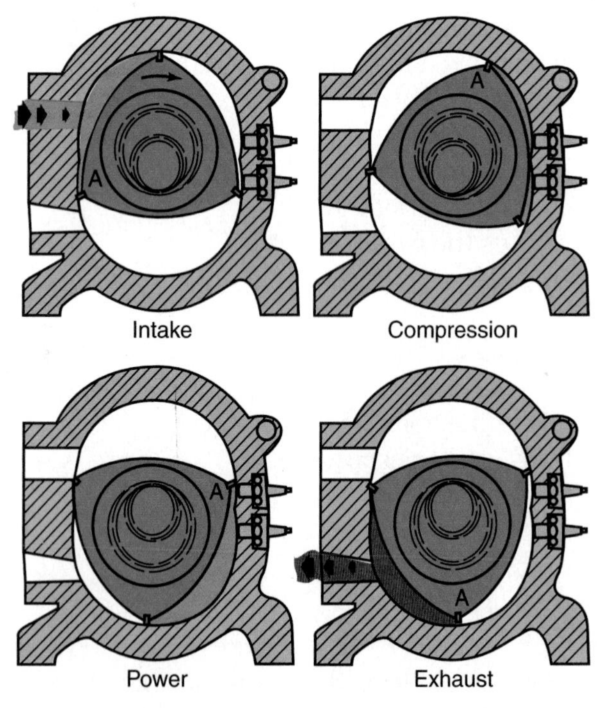

Intake and exhaust ports are located in the epitrochoidal housing. They are placed so they take advantage of the expanding and contracting spaces made by the spinning rotor. See **Figure 18-4.**

Rotary engines are typically lighter and do not cause as much vibration as gasoline piston engines. They are not, however, commonly used in automobiles. Rotary engines use more fuel than conventional engines. They are also greater pollutants than gasoline piston engines. Mazda has redesigned the original rotary engine, however, and promised that the company has improved the drawbacks to the engine. See **Figure 18-5.**

Diesel engines

Diesel engines are internal combustion engines that use heat and pressure to ignite their fuel. Rudolph Diesel, a German automotive engineer, developed them. These engines follow the same basic power cycle as the other internal combustion engines we have discussed so far (intake, compression, power, and exhaust).

They are different from gasoline engines in several ways. For one thing, they have no spark plugs. They rely on the extreme heat of the compressed air to supply the ignition. When diesel fuel is injected into the combustion chamber, the hot compressed air causes the fuel to burn with explosive force. Because the diesel engine relies on heated compressed air for ignition, it is called a ***compression-ignition engine***. During the compression stage, diesel engines only compress air, instead of an air and fuel mixture, like in gasoline engines. The fuel is not added until the compression is complete. See **Figure 18-6.** This enables diesel engines to be more efficient. Medium-duty and heavy-duty trucks, as well as many train locomotives, use diesel engines because of their efficiency.

Another point of difference is the high compression ratio of the diesel, when compared to gasoline engine ratios. The compression ratio is the difference between the volume of the combustion chamber when the piston is at bottom dead center (BDC) and the volume at top dead center (TDC). See **Figure 18-7.**

Still another difference between diesel engines and gasoline engines is diesel engines

Diesel engine: An internal combustion engine that uses heat and pressure to ignite its fuel.

Compression-ignition engine: An engine that relies on heated compressed air for ignition.

Figure 18-5. A redesigned rotary engine was introduced on the Mazda RX-8. (Mazda)

Figure 18-6. The operating cycle of the diesel engine. A—The piston compresses the air, greatly increasing the temperature of the air. B—When diesel fuel is injected into the cylinder, the heat causes combustion, driving the piston downward to provide power.

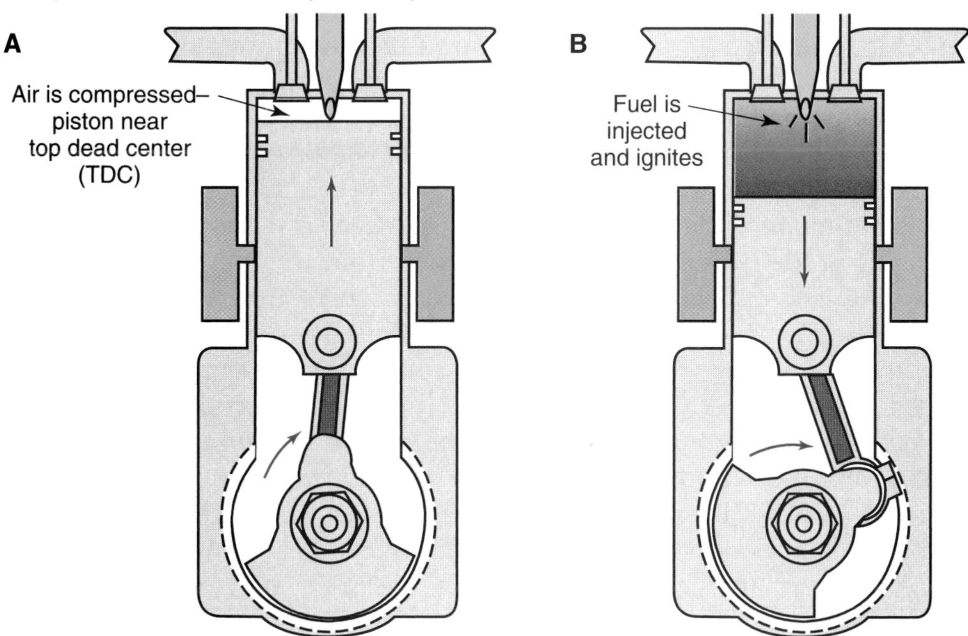

are built to be much stronger. This added strength is necessary because of the stress the higher compression ratio places on the engine parts, especially the pistons, rods, and crankshaft. The additional weight added to diesel engines makes their use on small automobiles uncommon.

Figure 18-7. Diesel engines use a much higher compression ratio than gasoline engines. If the pistons are both at bottom dead center (BDC) in their cylinders, the volume of the cylinder in the diesel engine is greater than that in the gasoline engine. This allows a greater compression ratio. Compression ratios for diesel engines are typically 16:1–21:1, while those of gas engines are usually around 9:1–10:1. The higher compression ratio is responsible for generating the temperature needed to ignite the fuel when it is injected into the cylinder.

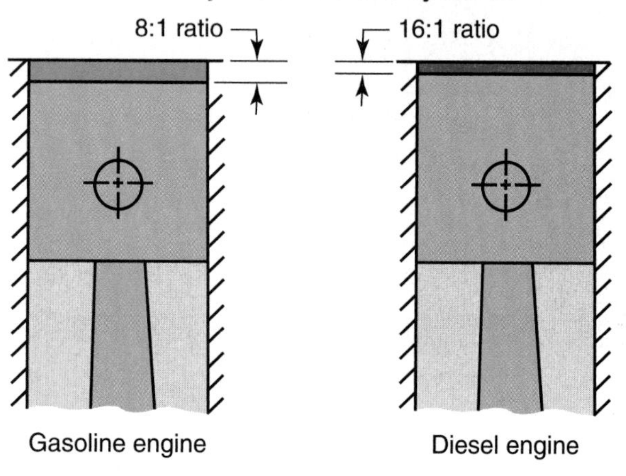

Electrical Propulsion

The second major type of land vehicle propulsion is electrical propulsion. Transportation vehicles that use electrical propulsion systems usually use an electric motor. The types of motors studied in Chapter 13 are similar to the ones used on vehicles. The motors need to be designed according to their applications. Mass transportation systems, such as electric buses and trains, need very strong motors. Motors used in golf carts do not need to be as powerful.

All vehicles propelled by electric motors must have a source of electricity to power the motors. There are two main ways electrically propelled vehicles get their energy. One is through direct connection to a source of electricity, and the other is using stored electricity.

Direct electric vehicles

A ***direct electric vehicle*** requires a connection to electricity. This is often hard to accomplish. Imagine having an automobile that has to be plugged into a wall to operate. It would take a pretty long cord to get you to school. There are direct electric vehicles, however, that you have probably seen before. Think about bumper cars at an amusement park. Bumper cars are electrically propelled vehicles that are connected to a power source. They do not carry batteries because it would be too expensive and they would have to be recharged often. Instead, the cars have long poles that connect to an electrically charged grid in the ceiling above the cars. The grid supplies the current for the electric motors in the cars to operate.

Direct electric vehicle: A vehicle that requires a connection to electricity.

This is similar to the system used by some light-rail systems. These electrically propelled vehicles get their power from overhead lines. See **Figure 18-8.** The vehicles can travel only where there are electric lines. If they become disconnected to the lines, they lose all power. Electric current is sent through the overhead lines, where special spring-tensioned rods or collapsible frames on the top of the vehicle pick the current up. The current is sent through these extensions to the motors inside the vehicle.

Other transportation vehicles, such as subway trains, receive electric current through a third ***rail***, or a long piece of steel that has an *I*-shaped cross section, located between the two main tracks. The third rail is similar to the two rails that guide the wheels of the train, except it is electrified to provide power to the train. See **Figure 18-9.** The third rail system is used in many subway systems. It requires less space above the train than the overhead line system. Third rail systems are, however, potentially very dangerous. If the rail is touched, it could be deadly.

Rail: A long piece of steel that has an *I*-shaped cross section.

The overhead lines and the third rail are powered by a power plant with substations along the way. The substations lower the voltage from the main power plant and ensure that the entire system is powered. The

Figure 18-8. Overhead lines are the power source for electric trains used in many light-rail systems, such as this one in Houston, Texas. (Siemens)

trains must keep in contact with the lines or rail to be propelled. The electrical current the train receives is then used by the onboard electric motors to propel the vehicle.

Indirect electric vehicles

In many situations, it is not possible to connect electric vehicles with a direct power source. In these cases, batteries are used to power the motors. All batteries have a specific storage capacity, rated in amp-hours, and they must be recharged when they have lost their stored power. There are several ways to recharge batteries. The first is to simply plug the battery into a power source. This is common in golf carts and other small electric vehicles. This, however, would be hard to do if you were traveling in an electric car on a cross-country trip. For this reason, several types of electricity-generating devices have been used in vehicles to create electricity that can either be directly used by the vehicle or stored in batteries for future use.

Solar propulsion

One method of supplying electric vehicles with power and recharging batteries is through the collection of solar energy. Vehicles that

Figure 18-9. Subway trains and some other rail vehicles receive their power through a special third rail. This rail runs next to or between the rails guiding the trains.

Figure 18-10. The Sunraycer is a solar-powered vehicle built to compete in a "solar challenge" race in Australia. Good design and lightweight materials helped it win. (GM-Hughes)

collect solar energy utilize photovoltaic cells to convert light energy to electrical energy. This energy is put into storage batteries, where it can be used to run electric motors.

Solar propulsion has not been popular in land transportation vehicles. The main reason is that solar cells are not highly efficient. Vehicles that rely solely on solar power have to be completely covered with photovoltaic cells. They must also be designed to be lightweight and aerodynamic, and they cannot carry many passengers.

The greatest use of solar-powered vehicles is in experimental races. See **Figure 18-10.** The GM Sunraycer is a solar-powered vehicle designed and built to participate in a "solar challenge" race in Australia. The Sunraycer proved to be a good design, as it won the race. It made good use of lightweight materials and efficiently harnessed the sun's energy.

Solar propulsion: A propulsion system that relies on the sun's energy.

Fuel cell: A device that utilizes a chemical reaction between hydrogen and oxygen to produce electricity.

Proton exchange membrane (PEM): A common type of fuel cell that works by passing hydrogen through one end and oxygen through the other.

Fuel cell propulsion

Fuel cells are a technology also used to power electric motors in automobiles. They have been used for years in the space program, but they have only recently been designed for use in automobiles. The most common type of fuel cells used in automobiles is the ***proton exchange membrane (PEM)*** fuel cell. See **Figure 18-11.** PEM fuel cells generate electricity through a chemical process. They work by passing hydrogen through one end and oxygen through the other. An anode on the hydrogen side causes the hydrogen to break into protons and electrons.

Figure 18-11. Operation of a proton exchange membrane (PEM) fuel cell is shown in this diagram. The only outputs of a fuel cell are energy and water. (DaimlerChrysler)

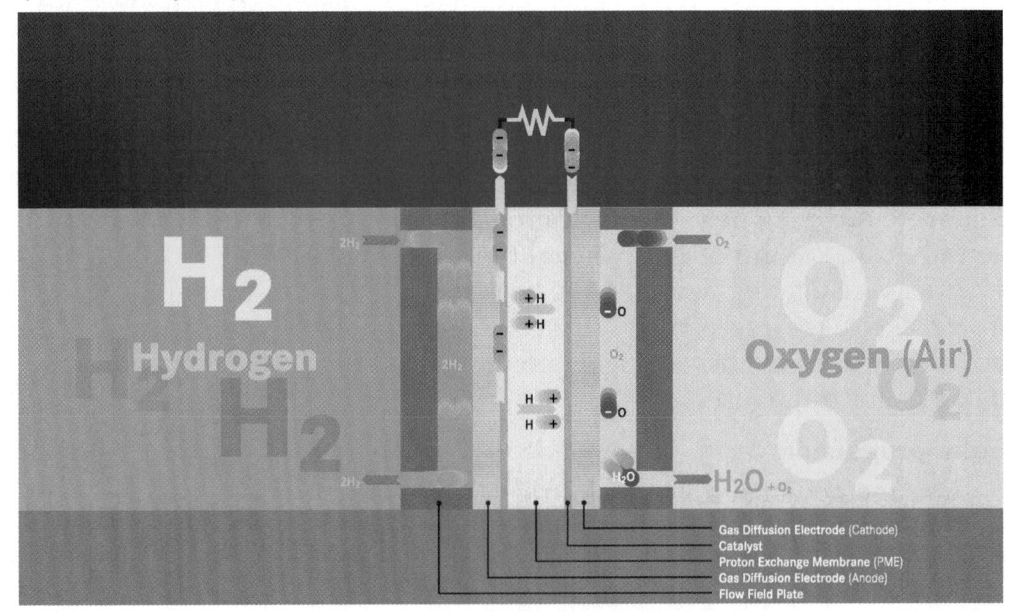

The protons then pass through the membrane in the center, on the way to the cathode side of the cell. This generates electricity. The remaining hydrogen molecules are combined with the oxygen and create water. Heat and water are the only by-products of the use of fuel cells.

The electricity created is then sent to the motors for direct use or to the batteries for storage. One problem with fuel cells is the need for hydrogen. It is easy to go to a gas station and fill a car with gasoline. These types of systems do not exist for hydrogen. One potential solution to this problem is the use of fuel reformers. Reformers are able to separate the hydrogen molecules from other forms of fuel. For example, a fuel cell vehicle with a reformer could fill up with methanol or natural gas. See **Figure 18-12.** The reformer would then send only the hydrogen molecules into the fuel cell. This does create pollutants, but still not at the level of a gasoline engine.

Guideway: A railway of magnetic levitation (maglev) vehicles.

Linear motor: An induction motor that has been flattened out.

Hybrid system: A system that uses both an internal combustion engine and an electric motor.

Electromagnetic propulsion

One other type of propulsion system that uses electricity is the electromagnetic propulsion system. These systems are quite different than the other electric propulsion systems. They are generally called *magnetic levitation (maglev)*. These vehicles are propelled using the magnetic field generated by electromagnets. Electromagnets are devices that use electricity to create magnetic fields. The magnetic field is temporary and only present when electricity flows through the wire.

The ***guideways***, or railways, of maglev vehicles are lined with electromagnets. The magnetic fields of the electromagnets come into contact with linear motors on the maglev trains. ***Linear motors*** are basically induction motors that have been flattened out. The linear motor mounted on the train acts as the stator in a typical motor. The electromagnets on the guideway repel the linear motor on the train, which pushes it down the track, or guideway. See **Figure 18-13.** The types of guideways will be discussed later in this chapter.

Figure 18-12. The first public service station offering both gasoline and hydrogen opened in Washington, D.C., in late 2004. (Shell Energy)

Figure 18-13. Linear magnetic motors on a magnetic levitation (maglev) train work with strong magnetic fields in the guideway to move the vehicle forward or backward. This train is pulling into a station in Shanghai, China. (Transrapid International, Inc., GmbH)

Hybrid Propulsion

Some systems do not use only an internal combustion engine or electric motor. They use both. These systems are known as ***hybrid systems*** because they are a combination of two different systems. Hybrid systems are able to take advantage of the positive aspects of both internal combustion engines and electric motors, without experiencing as many of the negatives. The two most popular hybrid systems are diesel-electric propulsion and gasoline-electric propulsion.

Diesel-electric propulsion

Diesel-electric propulsion: A system in which large diesel engines turn electric generators that power a locomotive's wheels.

The diesel engines you have studied earlier are most commonly used in trucks. They may, however, also be used to create electricity. These systems are known as ***diesel-electric propulsion*** systems. Diesel-electric propulsion systems are most commonly found in train engines. See **Figure 18-14.**

Gasoline-electric hybrid

Gasoline-electric hybrid: A type of propulsion system used in automobiles and small trucks. These systems are configured as either series hybrids or parallel hybrids.

Gasoline-electric hybrid systems are used in automobiles and small trucks. There are two main configurations in gasoline-electric hybrid systems. The first type is the series configuration. A ***series hybrid*** is like a series circuit in electricity. All the components are connected in a line. The gasoline engine is used to power a generator, which converts the engine's mechanical energy to electrical energy. The energy is then stored in batteries and used to power electric motors at the two front wheels. In these systems, the engines can be smaller than in a typical automobile. This is because the engine is only needed to power the generator. There is no link from the engine to the wheels.

The second type of configuration is parallel hybrid. ***Parallel hybrids*** use both the gasoline engine and electric motors to turn the wheels. The electric motors are used at speeds of under 15 to 25 miles per hour (mph), depending on the vehicle. This puts an end to engines idling at stoplights. Once cruising speeds of over 15–25 mph are reached, the gasoline engine is turned on to maintain the increased speed. The electric motors are used in the cruising stage when an additional burst of propulsion is needed, such as when passing cars on a highway. While the gasoline engine is in use, some of the power is used to operate a generator, and the rest recharges the batteries. An onboard computer controls the switching between the engine and motors. One other feature of gasoline-electric hybrids is the concept of regenerative braking. ***Regenerative braking*** is a process that transforms the car's kinetic energy into electrical energy by using the electric motor as a generator during braking. The electrical energy helps to power the batteries, which allows the gasoline engine to run even less.

The main advantage of the gasoline-electric hybrid system is fuel efficiency. Vehicles with hybrid engines also produce less harmful emissions. They are much quieter at low speeds. Hybrid systems are typically more expensive, however, than similar vehicles with traditional gasoline engines. As more hybrid systems are

Figure 18-14. Diesel-electric locomotives use large diesel engines to turn electric generators that power the locomotive's wheels. The majority of freight trains today have diesel-electric propulsion systems. Some of these train locomotives using diesel-electric propulsion can generate over 6000 horsepower (hp). The electricity that the engine—usually a large V12 diesel engine—generates could provide enough electricity to power a neighborhood of homes. (Norfolk Southern Corporation)

designed and built, the prices of these vehicles will fall. The original car manufacturers that created hybrid vehicles were Toyota, Honda, and Ford. Many of the other manufacturers will soon have hybrid vehicles as well.

Transmitting Power

The propulsion systems in land vehicles provide the power to move the vehicles. They must be connected to the wheels of the vehicle, however, in order for the vehicle to actually move. The drive system provides the transfer of power from the engine to the wheels.

The components of the ***drive system***, or power train, allow the propulsion power to be sent throughout the vehicle. Even bicycles have drive systems. See **Figure 18-15.**

Figure 18-15. A bicycle has a drive system consisting of pedals attached to a large gear (sprocket). A chain transmits power from the sprocket to drive gears on the rear wheel. This is a simple form of drive system. (Howard Bud Smith)

In automobiles, the drive systems are used to transfer the motion of the engine's crankshaft into the power that moves the vehicles. Automobiles send this constant power through transmissions, driveshafts, differentials, axles, and wheels to create movement of the vehicle. See **Figure 18-16.**

The first job of the drive system is to multiply the amount of torque the engine produces. In a drive system, the original torque is produced by the engine and present at the crankshaft. The amount of torque an engine produces is only enough to move a vehicle on a level surface and at a moderate speed.

Figure 18-16. Drive systems on automobiles have many parts that transfer power from the engine to the drive wheels. Some automobiles use the rear wheels as drive wheels, as shown here. Most current cars, however, use front-wheel drive. A few models use all four wheels.

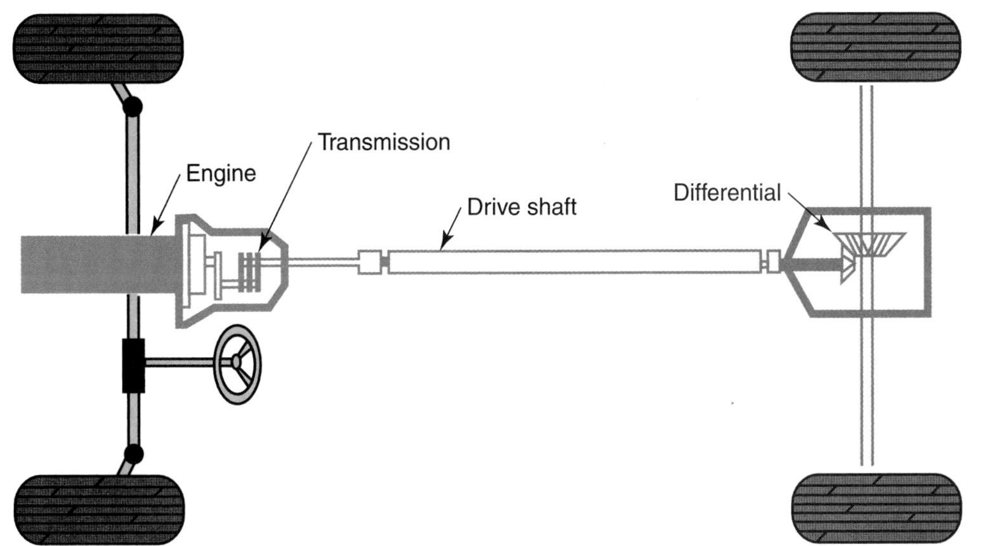

Series hybrid: A gasoline-electric hybrid system configuration in which all the components are connected in a line.

Parallel hybrid: A gasoline-electric hybrid system configuration that uses both the gasoline engine and electric motors to turn the wheels.

Regenerative braking: A process that transforms a car's kinetic energy into electrical energy by using the electric motor as a generator during braking.

Drive system: A system used to transfer the motion of the engine's crankshaft into the power that moves the vehicle.

Figure 18-17. Gears are used to transmit power. Those used in transmissions are made of high-quality steel and are designed to be tough and durable. By changing gears, the transmission is able to change the amount of torque and allow the engine to operate at its most efficient speed. A—Spur gears. B—Helical gears. These are preferred because of their great strength and smooth operation.

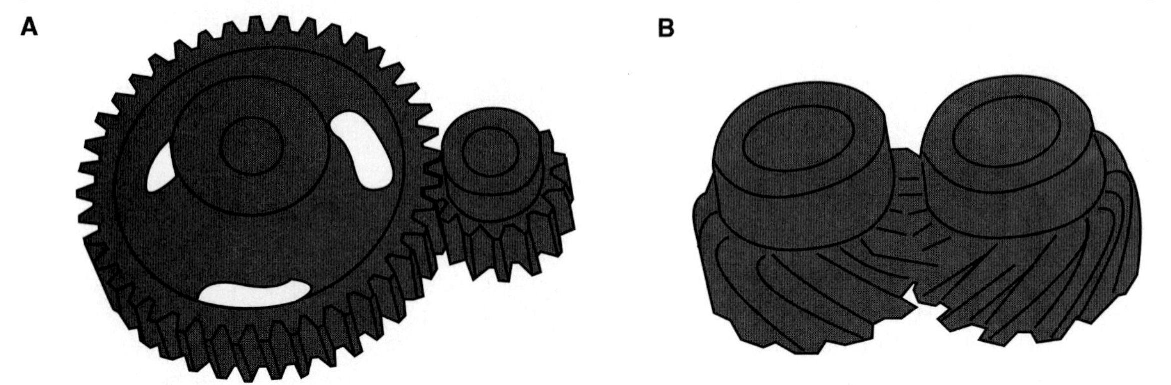

Transmission system: A device that provides for multiplying, dividing, or reversing the mechanical power and torque coming from the engine.

Manual transmission: A transmission that is totally controlled by the operator of the vehicle.

Transmission systems

Transmission systems are the devices that provide for multiplying, dividing, or reversing the mechanical power and torque coming from the engine. Two basic types of transmission gears use different arrangements of gear teeth. See **Figure 18-17.**

Manual transmissions

Manual transmissions are totally controlled by the operator of the vehicle. A lever called a stick shift controls and engages different combinations of gears. See **Figure 18-18.** In a typical five-speed manual transmission, the gears can be aligned in six different ways (five forward gears and one reverse). The gear alignments have different gear ratios. A ***gear ratio*** describes the change in the amount of torque. See **Figure 18-19.** For example, in first gear, a transmission typically has a gear ratio of 3:1. This means that, for every three revolutions of the input shaft, the output shaft makes one revolution. This slows the output shaft down and provides the torque needed to move the vehicle from a stopped position. As the vehicle gains speed, less torque is needed, and more revolutions of the output shaft are desired. Fourth gear, for example, may be a ratio of 1:1, meaning the crankshaft of the engine is spinning at the same speed as the output shaft of the transmission.

In a manual transmission, a clutch is used to disconnect the engine from the transmission. See **Figure 18-20.** Clutches allow us to do the following:

- Start the vehicle moving smoothly.
- Shift gears.
- Disengage the engine from the drive train, allowing the car to stand still with the power source still running.

Figure 18-18. A cutaway of a five-speed manual transmission. The "stick shift" lever is on the left. The clutch is contained in the "bell" housing on the right side of this drawing, opposite the stick-shift end.

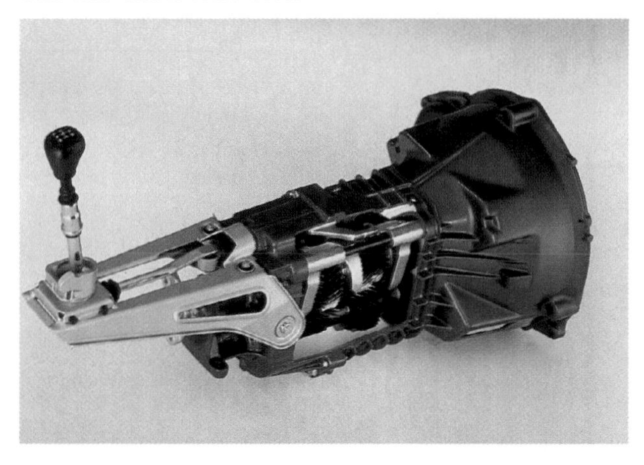

Automatic transmissions

Automatic transmissions also change the torque between the power and drive systems. They are different from manual transmissions, however, in many ways. Automatic transmissions are not dependent on the vehicle operator for control. Instead of the operator having to depress a clutch pedal and shift gears, the automatic transmission does the work itself. These transmissions use a torque converter instead of a clutch.

The ***torque converter*** is a type of fluid coupler. It uses fluid to transfer power. See **Figure 18-21.**

Automatic transmissions are also different from manual transmissions because automatic transmissions can produce an unlimited number of gear ratios. Producing a multitude of gear ratios is accomplished by using ***planetary gear sets***. See **Figure 18-22.** A planetary gear is actually composed of several gears.

Figure 18-19. How a pair of gears multiplies force. Think of the larger (driven) gear as a lever with the fulcrum at the hub, and then think of the small (drive) gear as the force moving the larger gear.

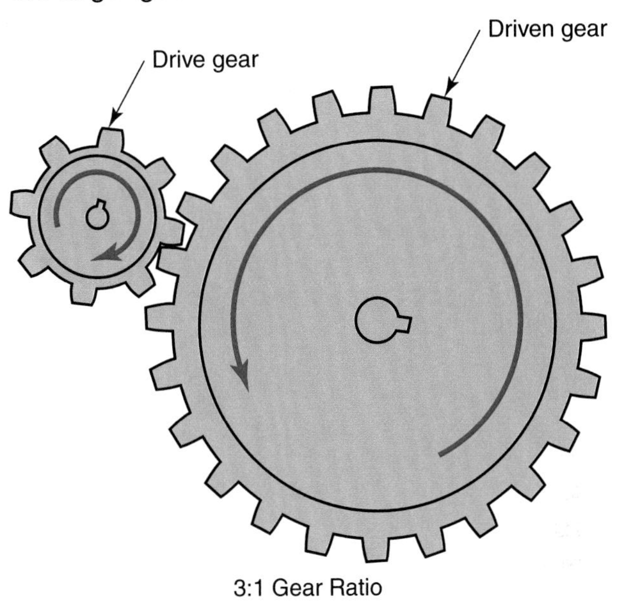

Driveshafts and axles

The output shaft of either an automatic or manual transmission is connected to either a transaxle or a driveshaft. See **Figure 18-23.** In front-wheel drive automobiles, the power is sent from the transmission to the transaxle and then to a differential. A ***differential*** takes rotational power from one source, the transaxle, and transfers it to two

Gear ratio: A ratio describing the change in the amount of torque.

Figure 18-20. A friction clutch is used in manual transmissions. This clutch uses a friction disk attached to a pressure plate on the shaft between the transmission and engine. A—The vehicle operator disengages the clutch by pressing on a foot pedal. When the clutch is disengaged, the pressure plate does not spin. The series of levers and springs that make up the clutch mechanism move the friction disk away from the flywheel. No power is transmitted between the power system and the drive system. B—When the clutch is engaged, the pressure plate is pressed against the spinning flywheel, transmitting engine power to the vehicle's driveshaft, and the vehicle moves.

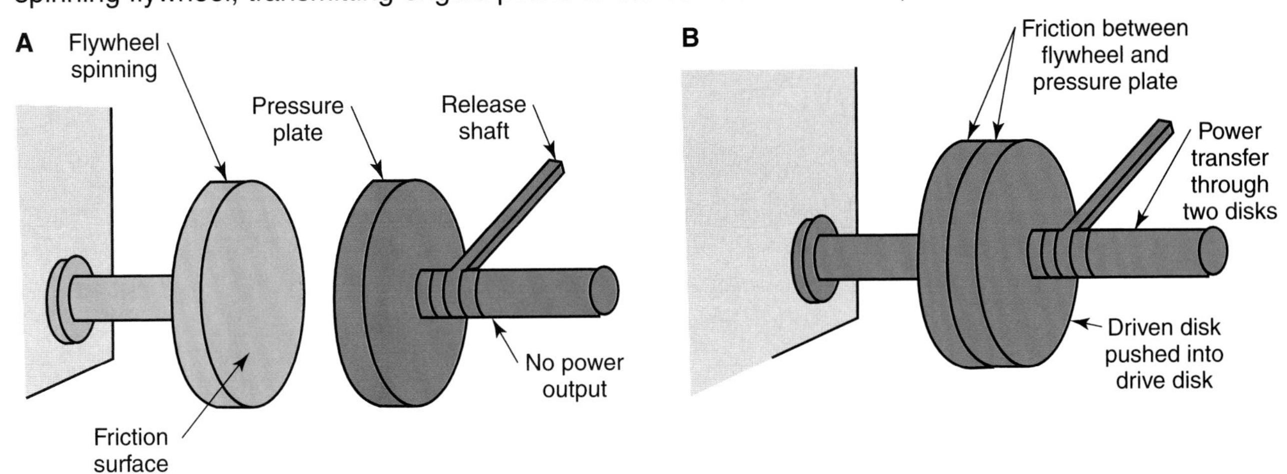

Automatic transmission: A transmission that is not dependent on the vehicle operator for control. It uses a torque converter instead of a clutch.

Torque converter: A fluid coupler used in automatic transmissions. It uses fluid to transfer power.

Planetary gear set: Several gears combined together to provide an unlimited number of gear ratios.

Differential: A component that takes rotational power from one source, the transaxle, and transfers it to two axles. It also allows the wheels to spin at different speeds, which helps make turning easier.

Figure 18-21. The torque converter is a fluid coupler used in automatic transmissions. The outer housing of the converter is attached to the flywheel and crankshaft of the engine. The inner section of the converter is a turbine that rotates within the housing and is attached to the input shaft of the transmission. The inside of the housing and turbine has blades that face each other, and the housing is partially filled with oil. As the driver accelerates the engine, the housing of the torque converter spins with the crankshaft. Oil inside the converter is set into motion and transmits power from the engine to the driven wheels. When the driver comes to a stop and the engine idles, the converter does not spin fast enough to turn the turbine and transmission shaft. This essentially disconnects the engine and transmission. In this manner, the converter serves the same function as the clutch. (DaimlerChrysler)

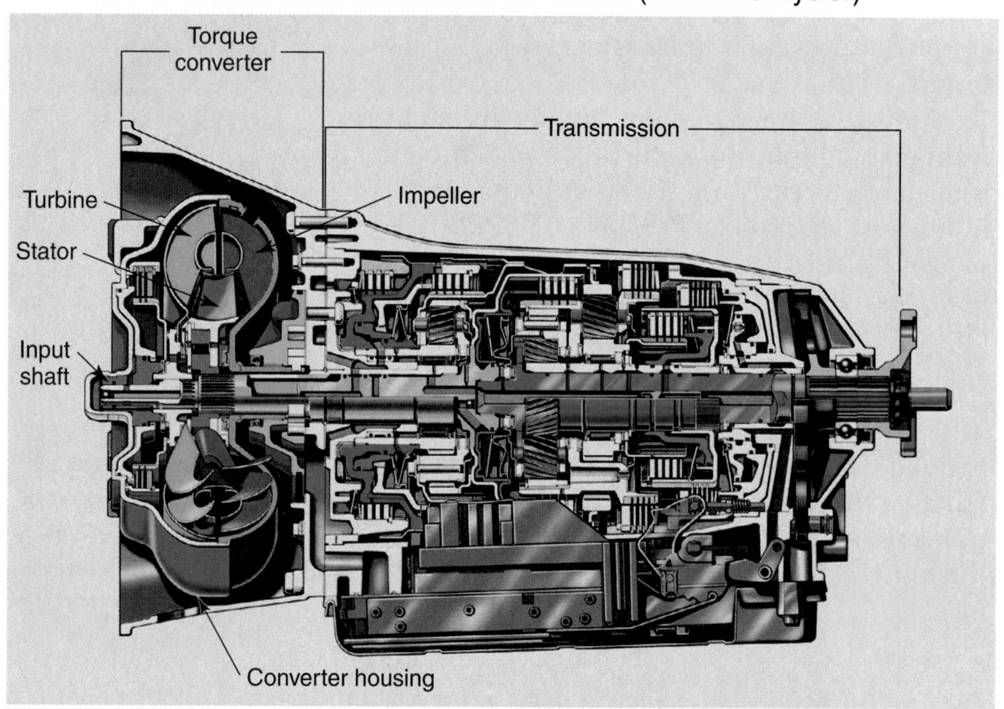

axles. It also allows the wheels to spin at different speeds, which helps make turning easier. In rear-wheel drive vehicles, the transmission is connected to a driveshaft. The driveshaft is a long rod that transfers the rotational power from the transmission to a rear differential.

In four-wheel drive vehicles, the power from the transmission must be sent to all four wheels. Power is typically sent to a transfer case. The transfer case then sends the power to the front and rear differentials.

Guidance Systems

The guidance of land vehicles includes knowing where you are and where you are going. This is referred to as navigation. There is a number of different navigational aids for land transportation vehicles, including maps, signs, and electronic navigational systems.

Maps

In land transportation, the most commonly used maps are ***road maps***. See **Figure 18-24.** These maps are often sold in collections, known as ***atlases***. Road maps use different symbols to denote different types of roads, landforms, and structures. These symbols can be found in the legend located on the map.

Traditionally, maps have been printed on paper and updated yearly. Printed maps are still very common. Electronic maps, however, are gaining in popularity as well. They can come in the form of software programs, Internet Web sites, or digital videodisks (DVDs). These resources make it very easy to plan a trip. Many electronic maps allow the user to type in the addresses of the departure and arrival locations. The programs generate a map showing the best route to follow, as well as the total distance and time of the trip. These electronic maps can then be printed out or downloaded to a disk or handheld device.

Figure 18-22. Planetary gear sets in an automatic transmission can provide a multitude of gear ratios. The ring gear is the outermost gear, and the sun gear is placed in the center. Several planetary gears connect the ring and the sun gear. By engaging and disengaging the planetary gears using bands and clutches, the transmission is able to create different gear ratios. This is done automatically once the vehicle operator has placed the vehicle in the drive gear and depressed the accelerator.

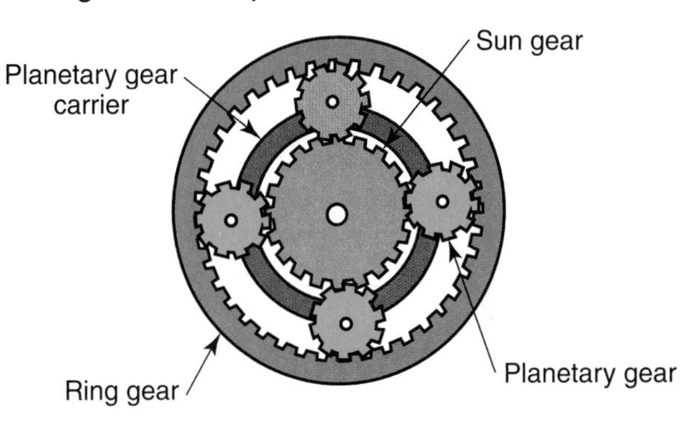

Road map: A map that uses different symbols to denote different types of roads, landforms, and structures.

Atlas: A collection of road maps.

Figure 18-23. On front-wheel-drive vehicles, the transaxle transmits power from the engine to the wheels through a differential. (Mercedes-Benz)

Curricular Connection

Science: Automobiles and Friction

Friction is a force that opposes or resists motion. In automobiles, friction can be used to start or stop vehicular motion. Vehicles come to a stop using the friction generated by applying the brakes. When brakes are applied, brake pads apply force to the rotor, causing enough friction to stop the car. The friction caused by applying the brakes produces heat, which is not used as power, but instead, is typically vented.

The clutch in a vehicle with a manual transmission uses friction to transmit power from the engine to the transmission. Located between the engine and the transmission, the friction is generated between clutch plates and the flywheel. Force is applied to the clutch disc, and that friction transfers power, which is then transmitted to the transmission. By disengaging the clutch, the clutch disc is released from this pressure, which stops the transmission from receiving this type of power.

Torque converters are used in vehicles with automatic transmission. They use a method of hydraulic coupling to transmit power to the transmission. This serves the same purpose as a clutch in a manual transmission. Friction still comes into play, however, when shifting gears. In the hydraulic coupling design, the clutch pack receives fluid pressure, causing the transmission to be powered.

Figure 18-24. Road maps show highways, expressways, local streets, and such features as rivers and lakes. They may also show schools, shopping centers, and other large features along roads. State and national road maps show all the major highways between cities in a state or country. City street maps are much more detailed and focus on the streets and alleys in a small town or large city. Street maps show local parks, schools, hospitals, and points of interest.

Navigation Systems

Maps are an essential part of the newest navigation technology, onboard navigation systems. ***Onboard navigation systems*** include a liquid-crystal display (LCD) screen in the dash that is linked to an electronic map. See **Figure 18-25.**

The entire system is connected to a global positioning system (GPS) and can display your current location on the map. If the destination has been entered, the navigation system can keep the driver informed, either on the display screen or verbally through the speakers, on how to get to the destination. Advanced systems can tell the driver which way to turn as he approaches an intersection. These systems improve the safety of land transportation because the driver does not have to take his eyes off of the road and flip through a map.

Signage

While operating a land vehicle, the most common type of guidance information used is signage. ***Signage*** is the information transmitted

Figure 18-25. By "locking on" to a series of navigational satellites, a global positioning system (GPS) receiver can show the vehicle's location within a few feet. If the destination has been entered, the navigation system can keep the driver informed, either on the display screen or verbally through the speakers, on how to get to the destination. Advanced systems can tell the driver which way to turn as he approaches an intersection. These systems improve the safety of land transportation because the driver does not have to take his eyes off of the road and flip through a map. GPS units are becoming a standard feature of some automobile models. This in-dash unit is part of the instrument panel. (Lincoln-Mercury)

Onboard navigation system: A system that includes a liquid-crystal display (LCD) screen in the dash that is linked to an electronic map.

Signage: The information transmitted by the use of signs.

by the use of signs. See **Figure 18-26.** Regulatory signs inform drivers about what they must or must not do. Warning signs, such as "railroad crossing" and "narrow road," communicate hazards lying ahead. Information signs give drivers knowledge of directions and distances of nearby cities, as well as street names.

One other type of signage used on both roads and railroads is the signal. Signals are combinations of lights that give information to the driver or operator. See **Figure 18-27.**

Electronic Guidance Systems

There are several electronic systems used on selected land transportation vehicles to help the driver guide the vehicle. Several of the systems are used on automobiles to assist in

Figure 18-26. Traffic is controlled and directed by various signs and signals, such as this electronic, travel-time display sign on an expressway. Electronic message boards are used in many urban areas to communicate road conditions and construction information. These signs can be updated to include current information. (Wisconsin Department of Transportation)

Technology Link

Communication: Traffic Lights and Signals

In many areas of our lives, visual communication systems are used to guide and direct our actions. On the road, traffic signs help to guide drivers and control the flow of traffic. These signs are communication devices that aid transportation systems. Traffic lights are dynamic, and often, they can respond to feedback they receive. In order to best control traffic, the lights are part of a system that processes information and then communicates the output to drivers.

The signals are sent from push buttons at the crosswalk, from sensors embedded in the road, from timers, or sometimes even by radio frequency from emergency vehicles. The outputs may include either three lights (red, yellow, and green) or one flashing light (yellow or red). The processor is the computer controlling the sequencing of the lights. It receives information from the control devices and selects the appropriate action, turning each light either on or off. The control devices can be manual devices, timers, or detectors. Manual devices are handheld controllers used by police officers to direct traffic. Timers are automated devices set to trigger an electrical output at a specific sequence. Detectors are the control devices used to work along with the current traffic situation.

parallel parking and include sensors and in-dash displays. Other guidance systems are used to maintain safe distances between other cars on the road. Several car manufacturers have designed ***adaptive cruise control systems*** that use lasers or radio detecting and ranging (radar) to determine the distance of the closest car on the road. The systems then monitor the conditions and adjust the speed of the car appropriately. Similar systems use sensors to detect oncoming objects and alert the driver of potential accidents.

Figure 18-27. Block signals on a railroad line warn engineers whether to proceed normally (green light), proceed with caution (yellow light), or stop (red light). In this case, the train is heading away from the station. The red signal lights warn following trains to stop until this train enters the next block, or control section.

Control Systems

Land vehicles require more than just propulsion. They must also be able to be controlled. The speed and direction of the vehicle are both able to be controlled in land vehicles. This ensures that the correct destinations are reached at the right times.

Changing Speed

Acceleration means the changing of a vehicle's speed so the vehicle moves faster. The method of acceleration depends on the type of propulsion system used by the vehicle. Those vehicles using internal combustion engines accelerate by forcing more fuel into the engine through a throttle system. This system includes a carburetor or fuel injectors. See **Figure 18-28.** Giving the engine more fuel produces more revolutions per minute (rpm) from the engine. The drive components receive the extra power and use it to produce an increase in speed.

In an electric vehicle, acceleration is signaled by a potentiometer. Potentiometers are variable resistors that change the resistance of the electric circuit. A dimmer switch is an example of a potentiometer. In electric vehicles, the gas pedal is a potentiometer. The harder it is depressed, the more power is sent to the motors, which increases the speed. Vehicles that use electromagnetic levitation for propulsion, like the maglev system, are locked into a "wave" that travels through coils in a guideway. As the operator increases the frequency of the wave, he increases the speed of the vehicle. See **Figure 18-29.**

To be safe, vehicles that are able to speed up must be able to slow down. ***Deceleration*** is the slowing down or braking of a vehicle. Manufacturers design vehicles that can decelerate gradually. Gradual deceleration provides safety for the passengers and cargo onboard. Many types of braking systems have been developed for use on a variety of land vehicles.

Hydraulic braking systems

The ***hydraulic braking systems*** used on modern wheeled vehicles link a master cylinder to a brake pedal and to one or two brake cylinders at each wheel. The cylinders are connected to the master cylinder with steel tubing. The tubing carries and contains the flow of hydraulic fluid through the system. See **Figure 18-30.** When the vehicle operator steps on the brake pedal, the master cylinder is activated. Force produced at the master cylinder is transmitted to each wheel cylinder by the hydraulic fluid. The wheel cylinders expand to activate the vehicle's brakes. In this way, each wheel receives an equal amount of force to decelerate the car

Adaptive cruise control system: An electronic guidance system that uses lasers or radio detecting and ranging (radar) to determine the distance of the closest car on the road.

Acceleration: The changing of a vehicle's speed so the vehicle moves faster.

Deceleration: The slowing down or braking of a vehicle.

Hydraulic braking system: A braking system that links a master cylinder to a brake pedal and to one or two brake cylinders at each wheel.

Figure 18-28. Acceleration in an internal combustion vehicle is achieved by providing more fuel to the engine. A—A simplified cutaway view of a carburetor. B—The operation of a fuel injection system.

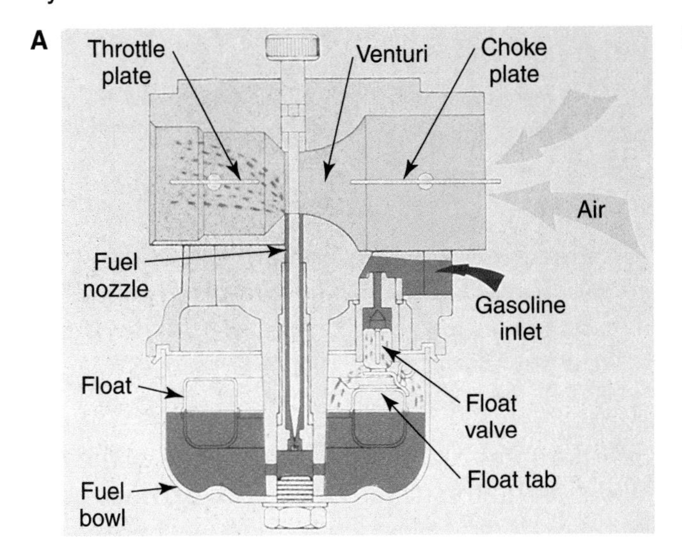

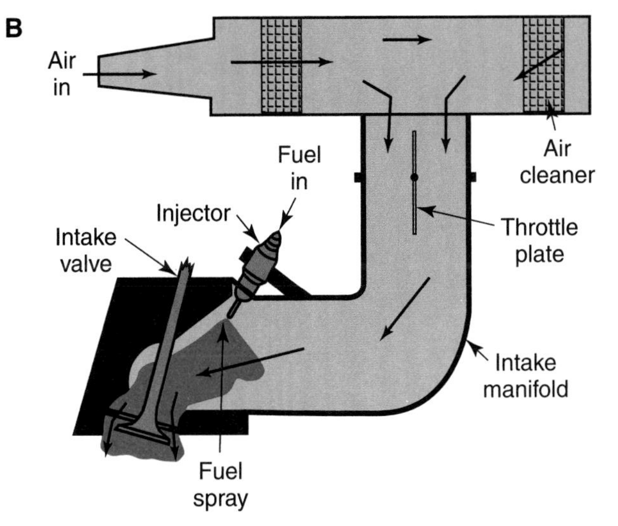

Figure 18-29. In a magnetic levitation (maglev) system, acceleration is achieved by increasing the frequency of the "wave" traveling through the guideway coils. (Transrapid International, Inc., GmbH)

Figure 18-30. Force applied to the brake pedal of a vehicle is transmitted to fluid in the master cylinder and brake lines. Fluid pressure in wheel cylinders forces the brake shoes and pads against rotating wheel components (brake drums or rotors). This reduces vehicle speed through friction.

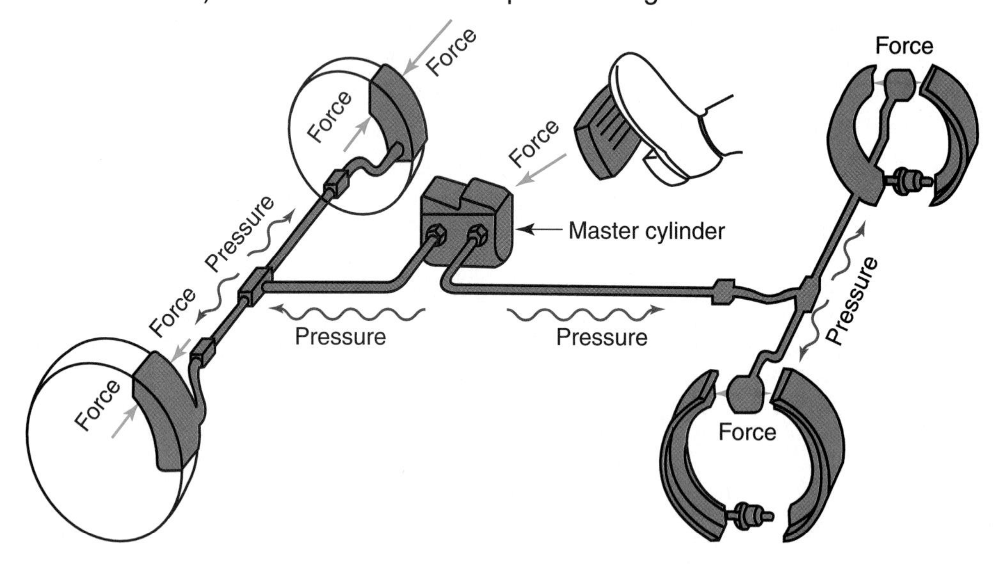

evenly. Refer to Chapter 10 for a review of the operation of hydraulic cylinders. There are various types of mechanical brakes powered by the individual wheel cylinders. Drum brakes and disc brakes are the devices that use friction to physically slow the vehicle.

Drum brake: A brake device that uses two brake shoes shaped to fit the inside of the brake drum.

Drum brakes

The ***drum brake*** uses two brake shoes shaped to fit the inside of the brake drum. The shoes are held close to the inside of the brake drum, but they are not touching it. The brake drum is attached to the wheel of the vehicle and rotates freely around the brake shoes while the vehicle is moving. As the brake pedal is depressed, the hydraulic cylinders activate the wheel cylinders. They push the brake shoes outward against the rotating brake drum. The two surfaces rub against each other. The friction from this contact slows the vehicle.

Disc brakes

Disc brakes are another commonly used brake device. Instead of a drum, a steel disc called a rotor is mounted to the wheel assembly so it spins freely. See **Figure 18-31.** A brake caliper straddles the rotating disc, but it does not touch the disc. When the disc brake is activated through the hydraulic brake system, the caliper is squeezed against both sides of the disc. The friction of the rubbing surfaces slows the disc. This, in turn, slows the wheel.

Disc brake: A brake that makes use of a rotating disc or rotor on the wheel.

Power brake: A brake system that adds a vacuum control valve between the brake pedal and the master cylinder.

Antilock braking system (ABS): A computer-controlled braking system that improves control of the vehicle in certain braking situations.

Air brake: A railroad car brake that uses compressed air to operate the brake cylinder.

Power brakes

Power brakes are found on all modern automobiles. This brake system adds a vacuum control valve between the brake pedal and the master cylinder. When the brake is pushed, the vacuum unit uses the vacuum created by the propulsion system to activate the master cylinder. The balance between vacuum pressure and atmospheric pressure is controlled in the unit. As a result, vacuum exerts the major force on the master cylinder. This allows the driver to use less foot pressure. At the same time, the brake pedal may be positioned more comfortably.

Antilock braking systems (ABSs)

Antilock braking systems (ABSs) are computer-controlled braking systems that improve the control of the vehicle in certain braking situations. An ABS uses sensors at each wheel to monitor the speed at which the wheel is turning. If one of the wheels is operating at a completely different speed while braking, the computer connected to the sensors can detect the difference. The only reason for a wheel to rotate at a very different speed would be if the tire lost traction during braking. If this occurs, the ABS controller is able to send braking pulses to the skidding wheel to help maintain the same speed as the other wheels. An ABS helps the driver control the vehicle while braking and enables the vehicle to stop faster. These systems are only used in extreme conditions when the computer senses the skidding is occurring.

Figure 18-31. Disc brakes, as shown in this phantom view, make use of a rotating disc or rotor on the wheel. Brake pads on pivoting calipers are forced against the rotor to slow and stop the vehicle. (GM-Cadillac Motor Car Division)

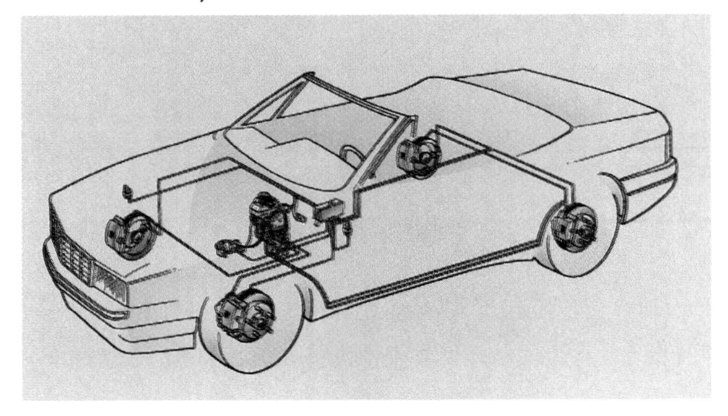

Pneumatic brakes

Railroad cars use pneumatic brakes to decelerate. Pneumatic brakes, or ***air brakes***, use compressed air to operate the brake cylinder. The brake cylinder is connected to a brake shoe that is pushed against the wheel to brake. Air brake systems can be computer controlled to provide the specific amount of pressure needed for each situation. For example, more pressure is needed when the train is loaded with cargo than when it is empty. If a train car becomes detached from the locomotive, the air brakes would automatically be applied to stop the car.

Controlling Direction

It is not hard to see that, if there were no way of steering a vehicle, the vehicle would be unsafe. Drivers and other vehicle operators would have no way of avoiding obstacles. Additionally, vehicles would be almost useless. While they could move easily, they would never arrive at the desired destination.

Land vehicles on fixed pathways

Degree of freedom: Any of a limited number of ways in which a body may move.

Railroad trains, maglev vehicles, monorails, vehicles that travel through tubes and pipes, escalators, elevators, and moving sidewalks all have one ***degree of freedom***. See **Figure 18-32.** This means they can only move forward or backward, in or on their guideways. The control of direction is mostly done automatically by the arrangement of the vehicle and guideway. Wherever the guideway goes, the vehicle must go.

Trains, for example, must follow the railroad tracks. The wheels have a flange on the inner side to help keep the train on the track. They are also ground at a slight angle and fit on the tracks so there is a gap between the wheel flange and the inside of the tracks. See **Figure 18-33.** When the train approaches a turn in the track, the whole vehicle shifts over slightly, so the outside wheel has to turn more revolutions than the inside wheel. This helps the vehicle to follow the track. Just as important, it prevents the wheels from skidding or derailing. Skid prevention also means less wear on the metal parts of the vehicle and track.

Land vehicles on nonfixed pathways

There are three basic ways to control the direction of nonfixed pathway vehicles: front steer, rear steer, and all-wheel steer. Most vehicles rely on front steering for control. See **Figure 18-34.** Both front wheels act instantly with the movement of a steering wheel.

Figure 18-32. This factory towline system has one degree of freedom. A—Towline carts move materials along fixed pathways to various parts of the factory. B—The towline connects the car to a tow chain recessed below floor level. (SI Handling Systems, Inc.)

A

B

Figure 18-33. A flange on a railroad wheel holds the wheel on the rail. A slight angle on the flange helps the wheel follow curves in the track.

Figure 18-34. Front-wheel steering is the method used on most land vehicles, such as this tractor. (Deere & Company)

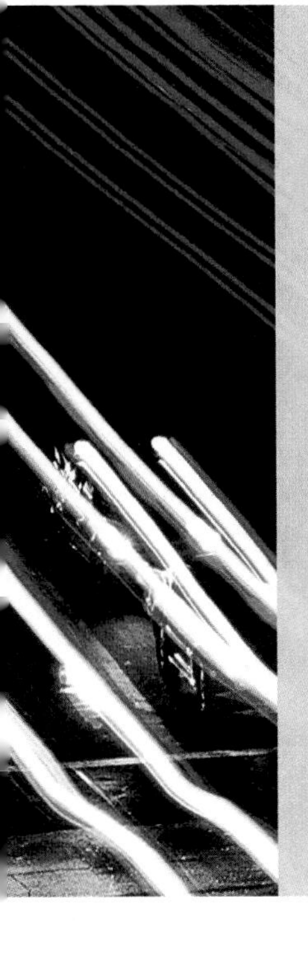

Career Connection

Locomotive Engineers

The railroad industry utilizes many types of workers. Rail yard engineers, yardmasters, switch operators, conductors, and yard laborers are all required in order for trains to operate. One of the highest paid rail occupations is the locomotive engineer.

Locomotive engineers operate trains. They begin each trip by checking the various systems on the train. The engineers are then given orders to leave the train yard. From that point on, their primary role is to monitor the status of the locomotive and train cars. They keep watch on the speed and control the throttle, as well as other system gauges. Locomotive engineers also follow all the signs along the railway and make sure the train is operating in a safe manner.

They typically work up through the ranks in a train yard. While the only initial school requirement is a high school diploma, they almost always have extensive work experience in several rail occupations before becoming an engineer. Once they are selected to become engineers, they complete an engineer training program. In the program, they spend time in the classroom, in simulators, and on a locomotive. They must also pass physical exams of their health and vision. The average salary for a locomotive engineer is just over $23 per hour. Most engineers are members of the Brotherhood of Locomotive Engineers.

Figure 18-35. Rear-steer is used on certain vehicles, such as this forklift, to permit a tight turning radius.

Forklifts, some street-cleaning machines, and other small vehicles that need to have a tight turning radius are designed with rear-wheel steering. See **Figure 18-35.** These vehicles can turn around in a very small space.

Recent improvements in the control of automobiles and trucks have led to ***all-wheel steering***. All-wheel steering is popular in larger vehicles because it increases control. Fire engines, for example, often use all-wheel steering because they need to maneuver into tight spots at the scene of a fire. All-wheel steering systems use the steering wheel to operate the front and rear wheels at the same time. There are two ways all-wheel steering systems can operate. See **Figure 18-36.** When the front and rear tires are steered in opposite directions, it is known as coordinated, or negative, steering. ***Coordinated steering*** aids in parking and decreases the turning radius of the vehicle. The opposite method of all-wheel steering is crab, or positive, mode, when all tires are turned in the same direction. ***Crab mode*** is useful in changing lanes and adding stability to the vehicle. All-wheel drive is electronically controlled, and an electronic controller selects the method of turning.

All-wheel steering: A steering system that uses the steering wheel to operate the front and rear wheels at the same time.

Coordinated steering: Negative steering, in which the front and rear tires are steered in opposite directions.

There are several types of mechanical ways that automobile and truck wheels are physically turned. One of the more common ways in small cars is by using a ***rack and pinion*** system. Racks and pinions are both types of gears. Rack gears are flat gears, and pinion gears are round. When used together, rack and pinion gears change rotary motion into linear motion. See **Figure 18-37.**

The pinion gear is mounted to the end of the steering wheel column. The teeth of the pinion gear and the rack gear mesh together. As the steering wheel is turned, the pinion gear moves the rack gear in the opposite direction. Both ends of the rack gear are connected to tie rods. The tie rods connect to the wheels. As the rack gear is moved in one direction, the wheels are turned in the opposite direction.

Figure 18-36. The two types of all-wheel steering are coordinated steering and crab steering.

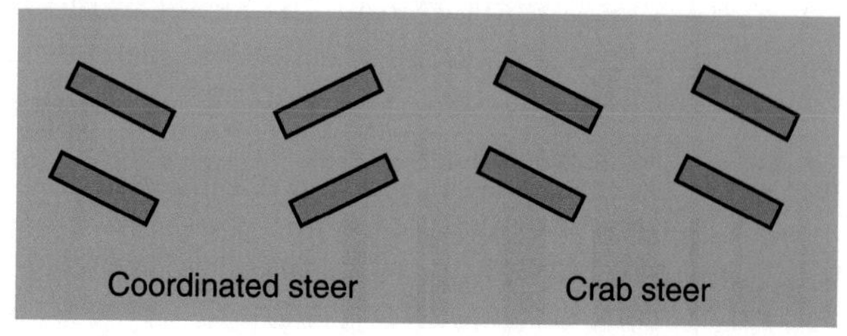

Car manufacturers are working on a new method of steering that does not use rack and pinion gears. It actually does not even use a steering column. This new method of steering is known as *drive-by-wire*. It is an electronic system that uses sensors in the steering wheel. The sensors can detect the rotation of the steering wheel and send that information to the steering system at the wheels. Electric motors at the wheels are then activated to turn the wheels.

Figure 18-37. Rack-and-pinion steering uses a round gear (pinion) and a flat gear (rack) to change rotary motion to linear motion.

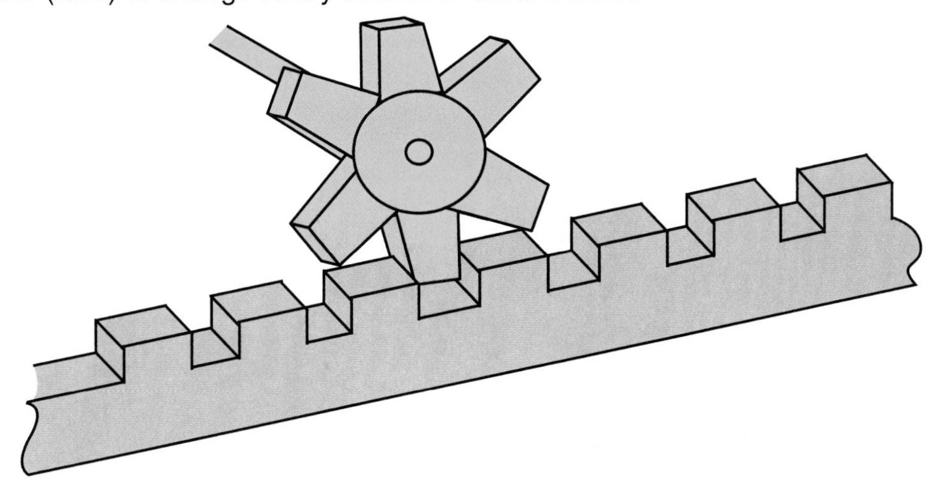

Crab mode: Positive mode, in which all tires are turned in the same direction.

Rack and pinion: Types of gears. Rack gears are flat gears, and pinion gears are round.

Articulated frame steering

Some vehicles designed for mass transit, farm operation, and construction have an ***articulated*** (hinged) section in the middle of the vehicle. See **Figure 18-38.** Steering is then controlled by swiveling the trailing wheels along with the front wheels. This system allows a small turning radius on very long vehicles.

Articulated: Hinged.

Tracked land vehicles

Tanks, bulldozers, some logging equipment, and other off-road vehicles use tracks. The tracks improve traction on difficult terrain. To change the direction of such a vehicle, the operator must vary the speed of one of the tracks. By decelerating one track and accelerating the other, she can pivot the vehicle. See **Figure 18-39.** Tracked vehicles are able to turn around in one spot by moving one track forward and one track backward.

Suspension Systems

Suspension systems include the components that link the vehicle to the guideway or path. In land vehicles, suspension systems can include tires, wheels, shocks, springs, and even magnets. Automobiles, trains, and maglev trains share some of the same suspension system components, but they each have unique pieces as well.

Automobile Suspension Systems

Automobile technology is at the forefront of suspension system design. Systems for many land vehicles have used the basic designs used for car suspensions. Trucks use the same components, but they are stronger

Figure 18-38. Hinged, or articulated, sections of this bus allow a shorter turning radius, an advantage in city mass transit. This very large bus has two hinged sections. (French Technology Press Office)

Figure 18-39. A tracked vehicle, such as this huge mining shovel, changes direction by altering the speed and direction of the two tracks. (P&H Construction Equipment)

Pneumatic tire: A tire filled with air.

Contact patch: The part of the tire actually touching the road as the tire spins.

Bias tire: A tire made of many layers of fabric, called plies.

Steel-belted radial tire: A tire constructed with wide strips of steel mesh called belts. It improves durability, wear resistance, and gas mileage.

and more heavy-duty. Even golf carts and tractors have borrowed components from car suspension technology. The main suspension components are the tires.

Pneumatic tires

Pneumatic tires (air-filled tires) are an important part of suspension systems. They provide the contact points between the vehicle and the surface on which it travels. See **Figure 18-40.** Tires actually have two important functions:

- Providing a cushioning effect as the vehicle travels over bumps and ruts. The air inside the tire compresses, and the sidewalls help to absorb the shock.
- Providing traction. This enables the car to "grip" the road surface for greater safety and better handling. The ***contact patch*** is the part of the tire actually touching the road as the tire spins.

The structure of a tire is constructed of various rubber compounds and fabric. A typical ***bias tire*** has many layers of fabric, called plies. The cords of each ply run in a different direction.

Steel-belted radial tires are the most common type of tire today. See **Figure 18-41.** These tires provide better traction and a longer tread life and permit a softer ride than other tires.

Figure 18-40. Air-filled tires are mounted on a wheel called a rim. The assembly is then bolted to the axle of the vehicle. Each modern tire can support 50 times its own weight when it is properly inflated. Air holds 90% of the weight. The structure of the tire supports the other 10%. Keep in mind, however, that the tire must also contain and withstand the pressure of the air it holds. This large tire is being mounted on a bus. (Greyhound)

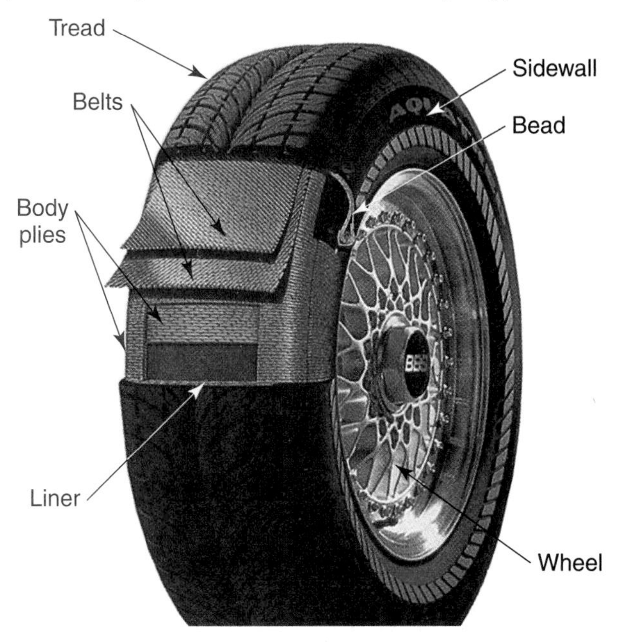

Figure 18-41. Wide strips of steel mesh called belts are key construction elements in a radial tire. They improve durability, wear resistance, and gas mileage. The plies on radial tires are placed so the cords run perpendicular to the centerline of the tread. Steel-belted radial tires also provide a softer ride than other tire types. (The Goodyear Tire & Rubber Company)

There is a thin rubber coating, or liner, on the inside of the tire, as well as a thick rubber exterior covering. The tread pattern is formed on the exterior of the tire. See **Figure 18-42.**

Springs

Springs are devices that are able to temporarily store energy and then use the energy. The springs used on automobiles are relatively large and heavy-duty. They need to support a heavy weight.

Spring: A device that is able to temporarily store energy and then use the energy.

One type of automotive spring is called a ***coil spring***. It is made by taking long steel rods and forming them around a cylinder into a helix.

Figure 18-42. Tread designs vary depending on their purpose. A—An all-season performance tire. B—An all-weather highway tire. C—An off-road tire. (The Goodyear Tire & Rubber Company)

A

B

C

Tech Extension

Reading the Tire Code

All tires have several sets of markings. See **Figure 18-A.** In this drawing, *P* signifies the type of tire and stands for "passenger." *LT* stands for "light truck." If no letter is used, the tire is meant for use in large trucks, motorcycles, tractors, or other land vehicles. *195* is the width of the tire in millimeters. *75* is the aspect ratio, which is the height of the tire, compared to the width. This means the height of the sidewall is 75% of the width. *R* signifies the type of tire construction and stands for "radial." *B* is used for bias constructed tires. *14* is the diameter, in inches, of the rim the tire is designed to fit.

Other markings include load and speed ratings. Load ratings are the amount of weight the tire can support. Speed ratings use letters to describe the maximum speed the tire can withstand. There are also quality markings. Tread-wear is graded on a scale on which 100 is average. Traction and temperature are rated on an A, B, and C scale.

Figure 18-A. Tire information, such as its pressure rating, size, construction type, and maximum inflation, are molded into the sidewall.

Coil spring: An automotive spring made by taking long steel rods and forming them around a cylinder into a helix. It is taken off the cylinder and heat-treated. The spring then retains its shape and gives the proper tension.

The springs are taken off the cylinder and heat-treated. They then retain their shape and give the proper tension. See **Figure 18-43.**

Another type of automotive spring is the ***leaf spring***. It is made of a series of steel strips, each one shorter than the next. A bolt that runs through the centers of each strip holds the leaves together. See **Figure 18-44.**

Torsion bars are another commonly used "spring" on modern land vehicles, such as trucks and automobiles. These bars do not resemble normal springs. They look more like metal rods. See **Figure 18-45.**

When a spring compresses and then rebounds, it will spring past its normal position. As it rebounds, it will compress past its normal position again. This is called ***spring oscillation***.

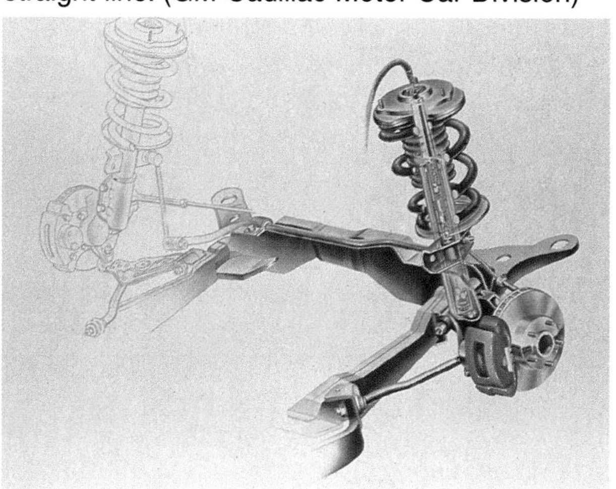

Figure 18-43. Coil springs are placed between the vehicle frame and axle to support the weight of the body. When the wheel goes over a bump, the spring compresses and stores potential energy. As the wheel moves off the bump, the spring uses the energy to expand and push the wheel back onto the road. This way, the wheel and spring move over the bump, but the vehicle frame can keep moving in a straight line. (GM-Cadillac Motor Car Division)

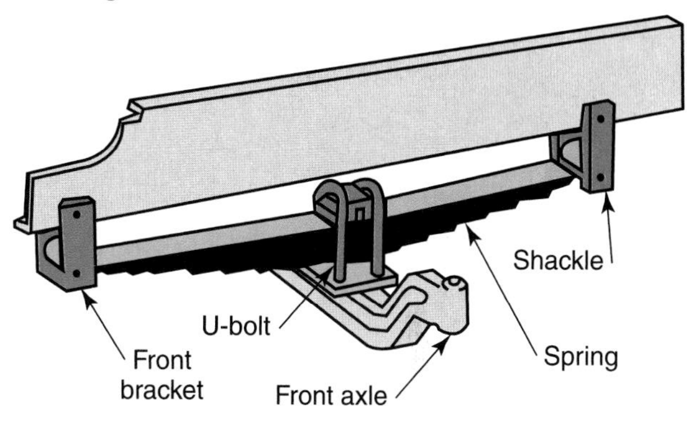

Figure 18-44. Leaf springs attach to the axle and two points on the vehicle frame. As a wheel hits a rut or bump, the leaf spring flexes up and down, storing potential energy. By returning to its original shape, the spring forces the axle up or down, to force the wheel firmly onto the road surface. This lets the wheel go over the obstacle, while the vehicle continues to travel in a straight line.

Leaf spring: An automotive spring made of a series of steel strips, each one shorter than the next.

Torsion bar: A metal rod used as a spring on modern land vehicles.

Shock absorbers

Shock absorbers absorb a road's unevenness so it is not transferred to the vehicle structure. The job of a shock absorber is to control spring oscillation. This system consists of shock absorbers connected between the car frame and the axle. There is one for each spring, usually one at each wheel. When the spring compresses and tries to oscillate, the shock will resist this motion. See **Figure 18-46.**

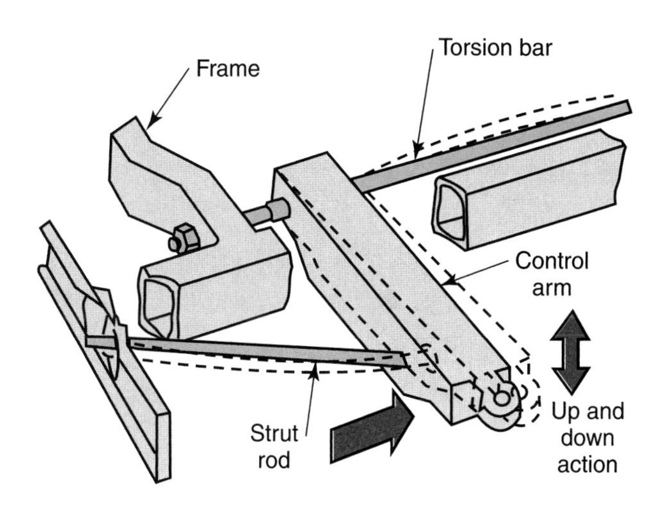

Figure 18-45. Torsion bar suspension uses the tension, or flex, in a steel bar to support the load of the vehicle. One end of the bar is attached to the wheel assembly through a lever arm. The other end is attached to a cross member of the car frame. When the wheel hits a bump or rut, the lever arm will swing up or down. The torsion bar will resist this movement and force the wheel assembly back to its normal position. Torsion bars are more stable than coil springs because they will not let the car sway from side to side as much.

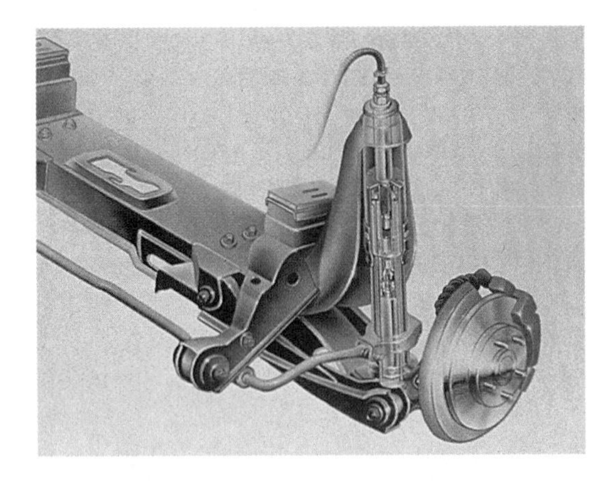

Figure 18-46. Shock absorbers smooth the vehicle's ride on bumpy roads by controlling the bounce from coil and leaf springs. A shock absorber looks like two pipes, one fitting inside the other. One end is attached to a piston rod that moves inside the other pipe. The piston on the end of the rod is made so it creates a seal between itself and the inside of the shock. On the piston, there is a two-way valve arrangement or simply ports (holes). When the shock is compressed, the hydraulic fluid in it is put under pressure too. The fluid is allowed to escape slowly through the valve or ports at a fixed rate, and so, the oscillation of the shock is slowed. This cutaway view shows the interior of a typical telescoping shock absorber. (GM-Cadillac Motor Car Division)

Spring oscillation: The compression and rebounding of a spring, in which the spring will spring past its normal position.

Shock absorber: A suspension system component that absorbs a road's unevenness so it is not transferred to the vehicle structure.

Stabilizer bars

A ***stabilizer bar***, or sway bar, is a long steel rod mounted between the two front wheel assemblies. See **Figure 18-47.** This bar helps to keep the vehicle from leaning out too far when the vehicle is going around corners.

Train Suspension Systems

Electric and diesel-electric trains use suspension components similar to those in automobiles. Wheels, springs, and shock absorbers are all used in trains. The entire suspension system of a train is contained on a ***bogie***, or truck. See **Figure 18-48.** Most train cars have two bogies, one at the front and one in the rear. The bogies are attached to the bottom of the train cars on swivels so the bogies are able to turn with the tracks. They consist of steel frames to which the suspension components are attached. Each bogie contains two axles, with a total of four steel wheels. Coil springs are used on a bogie to support the weight of the train car. Shock absorbers are used to ensure a smooth ride. Some bogies are equipped with air suspension systems that help create an even better ride. In most locomotives and electric trains, the bogies also carry the propulsion motors. The increased weight actually helps the suspension system create a smooth ride.

Figure 18-47. A stabilizer (indicated by the arrow) keeps the vehicle from leaning to the outside on tight turns. When the centrifugal force caused by the turn causes the vehicle's body to dip on the outside and lift on the inside, the bar is twisted. The stiffness of the bar resists this twisting and helps to keep the car level. This makes for a more comfortable ride. It also makes the vehicle easier to control. (Saturn Corporation)

Magnetic Levitation (Maglev) Suspension Systems

Magnetic levitation (maglev) trains have an entirely different suspension system. They are not suspended by shocks, tires, or springs. Magnetic forces suspend them in air. There are two methods of using maglev as a suspension system. The first method is by repulsion. In repulsion systems, the train levitates above the track. The magnets lining the track and those along the bottom of the train have the same polarity. Since magnets with the same polarity repel, the train floats above the track. The other method of suspension acts in the opposite manner. In attraction systems, the train wraps around the guideway. Magnets line the underneath of the guideway and the top of the lip of the train. See **Figure 18-49.** The two sets of magnets have opposite polarity and are attracted to each other. The magnetic fields are controlled so the two magnets come close to touching, but never actually touch. Both of these suspension methods create a gap between the guideway and the train that enables the train to move almost without any friction.

Figure 18-48. A train bogie holds two sets of wheels and is attached to the bottom of the car with a swivel. Note the heavy coil springs that help support the weight of the train.

Stabilizer bar: A long steel rod mounted between the two front wheel assemblies to keep the vehicle from leaning out too far when the vehicle is going around corners.

Bogie: A part of the suspension system of a train. It is attached to the bottom of the train car on a swivel and consists of a steel frame to which the suspension components are attached. Each bogie contains two axles, with a total of four steel wheels.

Figure 18-49. Magnetic forces suspend a magnetic levitation (maglev) train. In this diagram of an attraction maglev system, the magnets on the train and the magnets on the guideway are of opposite polarity and are attracted to each other. The tiny gap between the two sets of magnets allows the train to be propelled with virtually no friction. Linear motors built into each side of the train work with matching magnets in the guideway to move the train forward or backward. (Transrapid International, Inc., GmbH)

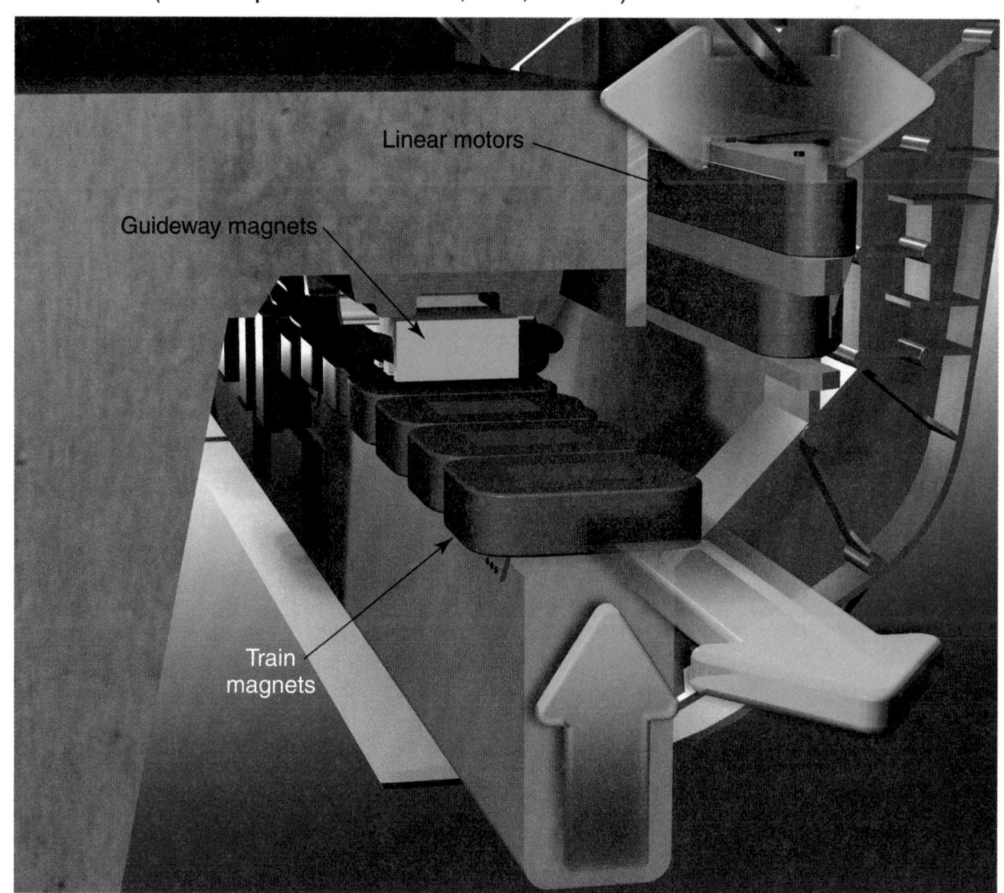

Structural Systems

Structural systems in land vehicles typically serve several purposes. They allow for a structure that contains the rest of the vehicle. These systems also form the exterior of the vehicle.

Automobile Structural Systems

Body: The enclosed part of a vehicle.

Chassis: The frame of a vehicle.

Unibody construction: A type of automobile construction that combines the body and frame in one unit.

Convertible: An automobile that has a removable roof.

The main parts of an automobile structure are the ***body*** (the exterior of the vehicle) and the ***chassis*** (the frame). The two parts historically have been made separately and then joined with bolts. Rubber blocks (spacers) are usually inserted between the body and frame to reduce vibrations from the engine and the road surface.

Unibody construction is very common on many automobiles today. See **Figure 18-50.** It is also called integral frame, or unit body, construction. This type of structure combines the body and frame in one unit. Suspension parts may be attached directly to the unibody. Sometimes, a partial frame may be attached. Partial frames are made to carry the weight of the propulsion system. They also withstand the stress put on the suspension system.

This type of construction is almost universal for all subcompact, compact, and midsize automobiles. Manufacturers, however, have kept the frame and body construction, known as body-on-frame, for many larger cars. The conventional style is better at reducing road noise.

Automobile bodies

Automobile bodies may be placed in classes by the number of doors or the type of roof. Two-door and four-door cars are clearly different and easily recognized. The roofs of many automobiles are solid sheet metal. Usually, six columns—three on each side—support them. This arrangement provides relative safety if the vehicle should roll over. One method of roof support eliminated the middle column on each side. The roof was supported only at each corner. This style improved visibility, but the roof crushed easily in rollover accidents. Many modern cars have gone back to the six-support system for safety. Designers pay special attention to placement and shape of the middle column to improve visibility.

Figure 18-50. In unibody construction, the frame and body are combined in a single unit. (GM-Cadillac Motor Car Division)

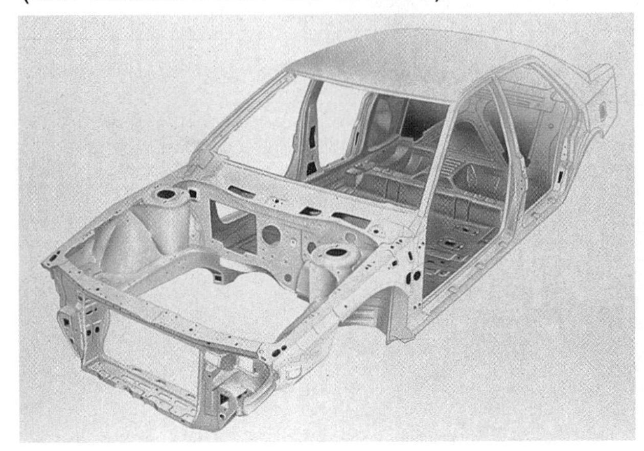

Convertibles are automobiles that have a removeable roof. The tops may be made of fabric, steel, or aluminum. Fabric tops fold down behind the passenger compartment for storage. Convertibles with metal roofs have special latches that secure them onto the body. These cars are popular, but they also have obvious safety disadvantages.

Most car bodies are made of sheet steel. Many times, special additives help the steel withstand manufacturing processes. Body parts are usually pressed, or stamped, into shape with large hydraulic presses. See **Figure 18-51.**

Some designers and manufacturers are using aluminum and fiberglass reinforced plastic for body parts. These materials are lighter and allow vehicles to have higher fuel efficiencies. The manufacture of body parts from plastic is relatively easy. One drawback, however, is plastic does not withstand impact as well as steel. Designers must keep this in mind when designing the total automobile. They do not want to reduce the level of safety for buyers of their cars.

Figure 18-51. Automotive body parts are typically stamped from sheet steel. (SI Handling Systems, Inc.)

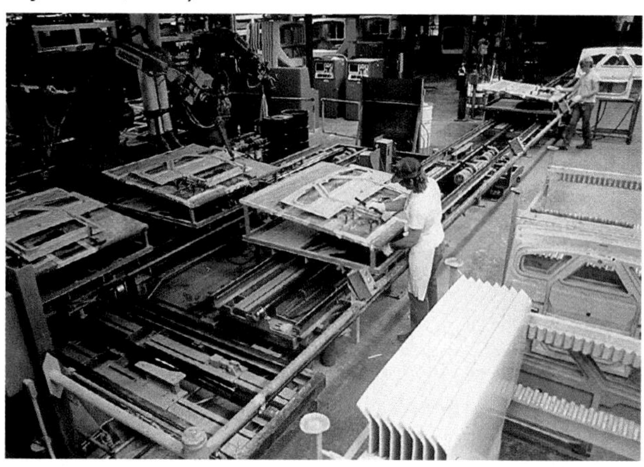

Automobile frames

Automobile chassis are the main "skeletons" of the vehicles. They are made to support and hold together the vehicle. As was explained earlier, frames may be separate from or an integral (inseparable) part of a unibody construction.

Conventional automotive frames are made of steel. The metal sheets are pressed or rolled into a box or channel shape. The frame usually consists of two long pieces running the length of the car. Steel cross members join them. The parts are usually welded, but they may also be cold riveted. Either way, the frame members must be securely joined so they provide a rigid structure.

Truck Structural Systems

Highway cargo transportation requires the use of special vehicles. Trucks are the vehicles of the trade. There are many types. The basic vehicular systems are the same for many trucks. There are different designs, however, to suit the type of cargo the truck will be hauling. See **Figure 18-52.** Single-unit trucks have one-piece structures. They are relatively short so they can maneuver easily in tight places.

Tractor-trailers are the large two-piece vehicles you see on highways. The tractor is the workhorse with a propulsion system capable of producing a great amount of horsepower (hp). These vehicles usually have more than 12 speeds in their transmissions. This enables them to start out while pulling heavy loads and to shift up to a more efficient gear when hauling on the open road and over modern expressways.

Trailers are made in various shapes and sizes. Depending on the type of cargo, bulk or break-bulk, trailers will take on many different appearances. (***Bulk cargo*** is loose material, such as grain or oil. ***Break-bulk cargo*** is a single unit or cartons of freight.) See **Figure 18-53.**

Bulk cargo: Loose material, such as grain or oil.

Break-bulk cargo: A single unit or cartons of freight.

Rail Vehicle Structural Systems

Rail vehicles have taken on many shapes, according to their purposes. Ones used for passengers are obviously different than those used for cargo. The basic framework or structure of all rail cars is steel. Most passenger cars are covered with aluminum alloy to save weight. Some passenger trains use a fiberglass-reinforced plastic body. Railcar builders are starting to use this material more because of its manufacturing ease and good appearance.

Figure 18-52. Truck bodies come in different designs, often determined by the type of cargo they will carry.

Figure 18-53. These trailers represent different types of cargo haulers found on the highways. A—A flatbed. (Fruehauf Trailer Operations) B—A closed (moving) van. C—A loose-cargo hauler. D—A tanker. (Fruehauf Trailer Operations)

A

B

C

D

Passenger trains have a variety of cars with flexible coverings between them. See **Figure 18-54.** The coverings between cars permit passenger movement between the cars. At the same time, they offer protection from the environment.

The railcars in passenger trains average about 85′ in length and are fitted with many luxuries. Coach cars are set up for comfortable passenger travel, with rows of upholstered, adjustable seats. Windows surround these cars so passengers may have full view of the countryside. Sleeping cars are divided into private berths, or bedrooms. Windows have shades to make sleeping easier and more comfortable. Dining cars have tables set for comfortable, and sometimes luxurious, dining. Other cars are set up as coffee shops and lunch counters.

Cargo railcars are designed to be functional, not comfortable. They are usually constructed of steel for strength and durability. The structures of cargo haulers are varied to suit the types of cargo they will be hauling. Cargo haulers are designed to carry weights from 50 to 150 tons. See **Figure 18-55.**

The four basic types of freight cars are the open-top gondola car, boxcar, flatcar, and tank car. Variations include refrigeration units and special designs for material handling. For example, the structures (boxes) of some gondola cars can be tipped so the contents can be dumped. Tank cars may be fitted with pumps, and boxcars may have extralarge doors so forklifts can move in and out easily. Special railcars called auto conveyors may have two or three levels. Sheet metal coverings protect the autos.

Designers have found a few advantages to creating special railcars. One is that when a railcar is designed for a particular product, it can be made so the product can be stacked on the car more efficiently. This allows each car to carry more. It also reduces the need for special dunnage. ***Dunnage*** includes the straps, blocks, and special rigging needed to securely fasten freight to a vehicle.

Figure 18-54. Passenger trains include various types of cars, such as coaches, sleepers, baggage, and dining. Flexible, accordion-type shields between cars allow passengers to safely and comfortably move from car to car.

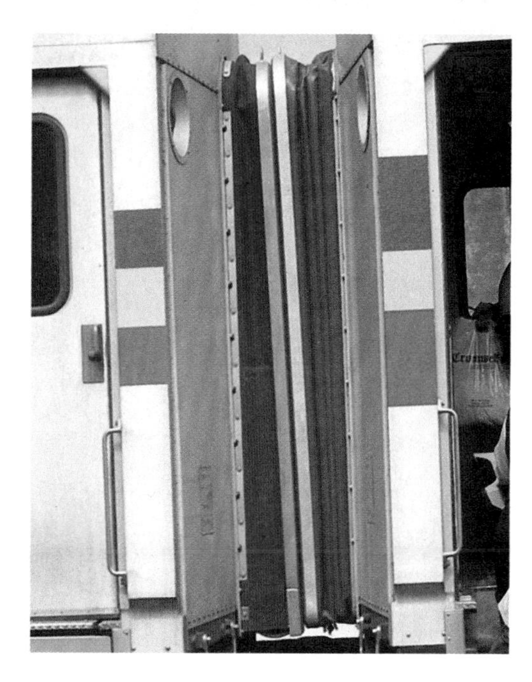

Dunnage: The straps, blocks, and special rigging needed to securely fasten freight to a vehicle.

Support Systems

Support systems are those structures that provide services to land transportation passengers and cargo. They also help to maintain the vehicles and provide pathways for transportation. Land transportation support systems can be divided into five categories, or types, of facilities. The categories are the following:

- **Related construction.** This includes the structures on which vehicles travel.
- **Passenger facilities.** These are the buildings and facilities that provide comfort and services to passengers.

Figure 18-55. Rail cars for cargo are built for strength and designed for various types of materials. A—A boxcar with large, sliding doors for loading and unloading. B—A tank car for liquid hauling. Usually, tank cars are dedicated to a specific type of liquid cargo. C—A gondola is an open-topped car that can be used to carry various types of loads. D—A center-beam flatcar designed to haul stacked lumber. E—A hopper car has unloading chutes at the bottom and typically carries dry granular cargo, such as sand or grain.

A

B

C

D

E

- **Cargo facilities.** These are the buildings and facilities that provide loading, unloading, and storage for various types of cargo.
- **Vehicle maintenance.** This includes those facilities designed to maintain and repair vehicular systems.
- **Other support systems.** These include any other systems needed for safe transportation, such as life support and rescue operations, communications systems, and regulatory agencies.

Roads and Highways

Roadbed: The foundation supporting the road surface and vehicles.

Roads and highways are constructed structures that should be included in the study of transportation technology. Improved construction designs and techniques can benefit both construction and transportation technology. Roads and highways begin with the clearing of land to make way for a ***roadbed***. This is the foundation supporting the surface and vehicles. Once the route is chosen, surveyed, and staked, hills and valleys are "moved" to level the roadbed. Construction workers on road-building equipment make cuts and fills. See **Figure 18-56.** ***Cuts*** remove excess earth from hills. This earth is usually moved to low spots, or valleys, along the route where fills are needed. A ***fill*** is the addition of material, such as rock and soil, to build up low-lying areas. By cutting and filling, construction engineers form a relatively level roadbed that will provide a safe and comfortable path for vehicle travel.

Figure 18-56. Road construction frequently involves moving large quantities of earth and rock to make the road as level as possible. This bulldozer is working in a cut, where earth has been removed. (Caterpillar, Inc.)

The soil in the roadbed is then compacted and covered with a gravel or stone subbase. This provides a strong foundation for the road surface. The road surface is usually made of asphalt or concrete. These materials make a hard, durable driving surface. Often, the road receives a surface texture that gives vehicles extra traction, especially in wet weather.

Cut: The removal of excess earth from hills.

Fill: The addition of material, such as rock and soil, to build up low-lying areas.

Once the road is built, shoulders are made along its sides to provide emergency and stopping lanes. Guidance systems in the form of road signs, traffic lights, and highway markers are then placed. All this leads to a finished support system essential for car and truck travel.

The United States has an excellent system of interstate highways that began to develop in the 1950s. These highways have smooth and wide surfaces. Many of them have four lanes, two in each direction, to promote safety and speed. The highways connect cities, usually by the shortest route possible.

Traffic planners have a major role in the design of highway and road networks. They have come up with many ways to make these systems safer and more efficient. Beltways are constructed around larger cities so all traffic does not have to go through the center of town.

Cloverleafs are used where two highways intersect. See **Figure 18-57.** They have been so named because the ramps that connect the highways are looped and look somewhat like the leaves of a clover plant. One highway actually goes over the other on a bridge, or overpass. At the ends of the ramps are acceleration or deceleration lanes that run along the edge of the highway. They make it possible to leave and enter each highway while maintaining a safe rate of speed.

Traffic planners have improved on the cloverleaf design by extending the lanes where vehicles enter and exit the main highways. Where an on-ramp and an off-ramp are close to each other, highway engineers have added extra safety lanes separate from the main highway. This provides a safer area for drivers to enter and exit the highway.

The network of roads and highways in the United States includes many types of roads. Each type of road has different specifications. For example, a country road may only have to be 15′ wide and can be constructed with gravel. On the other hand, highways must be constructed as described above and must be over three times as wide as a

Grade: The percentage of the change in height every 100′ of track.

Rail bed: Several layers of stone designed to spread the weight of the trains evenly over the compacted surface.

Figure 18-57. Cloverleaf intersections are used to bring together two major roadways without stoplights and speed the safe movement of vehicles from one road to the other. This photo was taken with a long time exposure at night, resulting in red "trails" from the taillights of moving vehicles.

country road. All the country roads, city streets, state highways, and interstate highways work together to form a system of roads that connects virtually any starting point with any destination.

Railroads

Railroad construction has some similarities to the construction of highways. First the route is planned, and then the land is obtained. Because locomotives are made to pull heavy loads and ride on smooth iron wheels, they cannot climb hills like highway vehicles can. For this reason, cuts and fills must be made accurately. There can be no steep grades anywhere on the line. The ***grade*** is the percentage of the change in height every 100′ of track. Most grades are kept below 1.5%, which would be 1.5′ change in height over a 100′ section of track. Trains are most efficient when the track is flat (0% grade) and straight. There can be no sharp turns on a rail bed. Trains are not designed to make sharp turns. The tracks must be laid so there are gradual, large-radius turns. See **Figure 18-58.** Large-radius turns are often less than 5°, while tight turns are around 10°.

Figure 18-58. Because trains are not able to make sharp turns, tracks must be laid in wide, gradual curves. Trains move along this industrial siding at low speeds, so the curves can be sharper than those where trains move at high speeds. (Howard Bud Smith)

After the land is cleared and the soil compacted by earth-moving equipment, the ***rail bed*** is put down. This is made up of several layers of stone designed to spread the weight of the trains evenly over the compacted surface. The first is a layer of crushed stone called the ***subballast***. A thick layer of ***ballast*** is then added. It is composed of larger stones. The cross-ties, usually made of wood, are then laid on top of the ballast. Their purpose is to secure the rails that are later fastened to them. More ballast is added to fill the spaces between and around the ties. The ballast holds the cross-ties in place.

Subballast: Crushed stone that forms the first layer of a rail bed.

Ballast: A layer of large stones placed on top of the subballast in a rail bed.

Bridge: A structure that spans a waterway, a ravine, or another barrier.

Rail construction crews, in the next step, lay the track on top of the ties. A track consists of two parallel rails. The wheels of a train ride on the rails. The distance between them must be kept at a constant width (gauge). This keeps the train from derailing (jumping off the track). The standard gauge for railroads in North America is 4′ 8 1/2″ wide. Rails are long pieces of steel that have an *I*-shaped cross section. See **Figure 18-59.** Modern railroad construction techniques use continuous rails instead of jointed ones. Long sections of rail are welded together so the ride is smoother and quieter.

Maglev Guideways

Maglev trains run on a different type of railway than typical locomotives do. The guideways are elevated and are either shaped like a *T* or *U*. See **Figure 18-60.** In the *T*-shaped guideway systems, the train is actually wrapped around the guideway. The maglev trains used in *U*-shaped guideways are contained within the walls of the guideway. Both systems are elevated and built on concrete piers. Maglev guideways take up much less space and are not nearly as destructive as roads or rail beds. Since the tracks are elevated, there is no need to cut and fill the land. Instead, the lengths of the piers are changed to accommodate the landscape.

Bridges

Bridges are those all-important structures that span waterways, ravines, and other barriers. They have been around for a long time. Over time, bridges have undergone many changes to make them easier, safer, and more efficient to build.

The beam bridge is the oldest type. This type is currently made of steel and concrete. Beam bridges are commonly seen connecting short spans in cloverleaf interchanges or overpasses. A variation of the beam bridge is the truss bridge. Truss bridges were often used as railroad bridges.

Arch bridges are commonly seen connecting roads over long spans, such as over rivers. The road deck may either be built on top of the arch or suspended from the arch. Cantilever bridges are noted for their strength. See **Figure 18-61.** These bridges are able to span longer lengths than beam and arch bridges. Because of their strength and

Figure 18-59. Railroad rails are long strips of steel formed into an *I* shape. Sections once were bolted together, but today, they are most often welded into a continuous length of track.

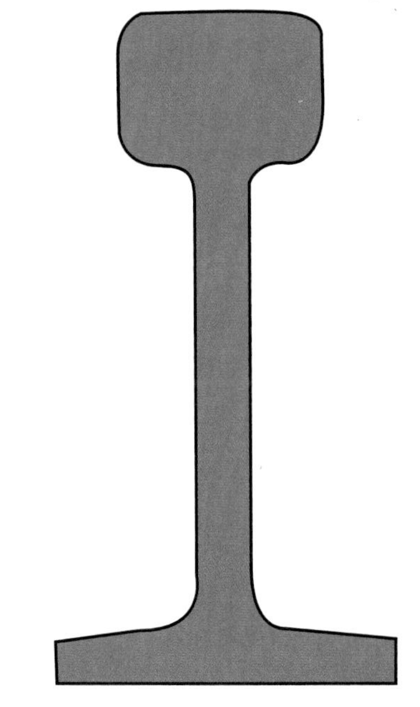

Figure 18-60. Two types of magnetic levitation (maglev) guideways. A—A *U*-shaped guideway encloses the train and is used with repulsion-type suspension systems. B—A *T*-shaped guideway is used with attraction-type suspension systems. The train rides atop the guideway. The lower section on each side of the train wraps around the guideway.

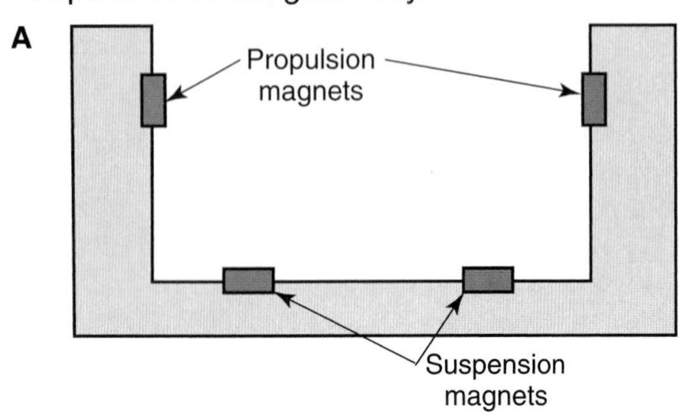

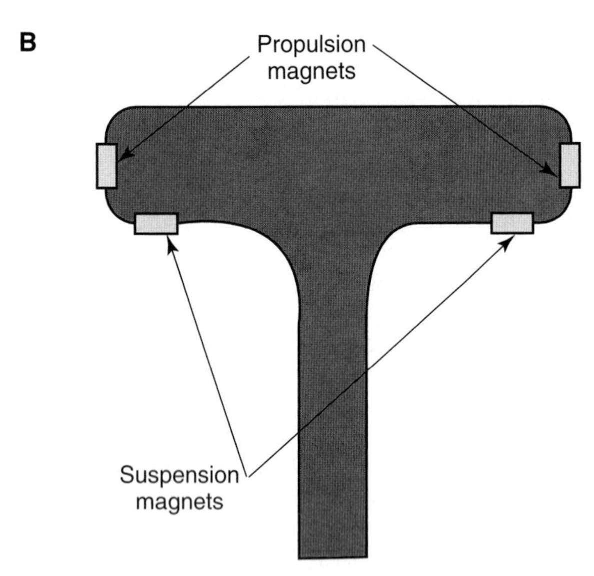

Figure 18-61. Cantilever bridges are strong in relation to material weight and can span long distances. They are often used as railroad bridges, although this example is a highway bridge.

spanning ability, they are often used as railroad bridges over large rivers. These bridges are basically made of two first-class levers. Their ends are secured firmly to the ground so the middle will not fail when heavily loaded. The large concrete and steel supports provide the fulcrum for each lever.

Of all the bridge designs, suspension bridges are capable of spanning the longest distances. The Golden Gate Bridge in California is a suspension bridge. The road deck is suspended with smaller cables from strong main cables running from one end of the bridge to the other. These cables are anchored firmly to the ground at both ends. Two towers provide support for the main cables. The height of the towers is directly related to the length of the bridge. Taller towers are needed as the span increases.

Tunnels

Where road builders have encountered mountains, they have borrowed from mining technology. ***Tunnels*** are dug or bored so straight, level paths can be maintained for vehicle travel. They are often more cost-efficient than building roads over or around mountains. See **Figure 18-62.** A tunnel-boring machine (TBM) can bore through soil and soft rock with ease. The largest TBMs are 9 1/2 yards (about 9 m) high and about the length of two football fields. Operators control them with the aid of computers and television monitors.

Tunnels are also used under bays and other bodies of water. Instead of digging the tunnel underwater, prefabricated tubes are often sunk in the water. Once the tubes are anchored and connected, they are pumped out and ready for completion. This type of tunnel technology is now commonly used.

The Chesapeake Bay Bridge Tunnel is an ambitious project that was undertaken on the East Coast of the United States. It is a series of bridges and tunnels allowing vehicles to drive from Maryland to Virginia, over and under the Chesapeake Bay. The bridges consist of a series of beam bridges. The tunnels were constructed using the sunken tube method.

Tunnel: A covered passageway.

Figure 18-62. Tunnel-boring machines (TBMs) easily burrow through soil and rock. This machine was used to bore tunnels between England and France, beneath the English Channel. (British Information Services)

The Eurotunnel links Great Britain with the European continent. It was lined with curved segments of reinforced concrete behind the TBM. Cement grout was used to seal and strengthen the joints between the segments.

Passenger Facilities

Transportation companies that move passengers need special buildings and spaces for travelers. Bus stations, airport terminals, and train stations provide comfort and services for people who are traveling. See **Figure 18-63.** Here, passengers can buy tickets and wait in sheltered areas.

Figure 18-63. Bus companies, railroads, and airlines need passenger facilities. This passenger gate at the Baltimore-Washington International Airport provides a check-in desk, a waiting area with seating, and access to the ramp used to load and unload passengers for an aircraft.

Many times, goods and services are offered, as well. Passengers can eat, shop, have their shoes shined, or read a newspaper while waiting for their rides. Planners of passenger terminals consider the special needs of people who have disabilities. These facilities must be built so everyone can use them.

Safety must be built into structures. Waiting platforms are special areas where passengers can stand before boarding their vehicle. Here, there must not be any danger of being in the path of vehicles that are arriving or leaving. Usually, these areas are out of the way of regular traffic flow. They are located so the vehicles can enter and exit easily.

Many successful businesses have been founded on providing services to motorists. As car transportation grew in popularity, people were able to travel farther from home. Families and businesspeople began to take road trips that lasted several days. This led to the creation of motels. Motels are given their name because they were first referred to as motor hotels. Besides a place to sleep, motels offer travelers other services. Relaxation by a pool or in front of a television awaits weary travelers. Today, you can probably name a number of major motel chains.

Rest stops are another type of transportation support system. Many states build structures along their major highways so travelers can stop and rest. The structures may include parklike surroundings for comfortable walks. Picnic tables are often placed so meals can be eaten in quiet, peaceful surroundings. Rest rooms and information centers are also part of highway rest stops.

Truck stops are a type of physical facility for over-the-road (OTR) cargo haulers. These facilities offer services for both vehicles and operators. The services include fuel and maintenance services, restaurants, and motels. Some facilities even have special truck-washing services.

Cargo Facilities

When you realize everything in your house, including the materials of which it is made, was carried on a truck, you have an idea of the scope of cargo hauled over the road. It should be no surprise then that trucking companies need special cargo facilities. Cargo terminals are not as comfortable as passenger terminals. These facilities do, however, have well-designed spaces for storage and movement of freight. Some sections may even be refrigerated for perishables, such as meats, fruits, and vegetables.

Loading dock: A cargo facility built next to large paved areas for trucks to back up to.

Depending on the type of vehicles they are made to serve, cargo facilities have different designs. For cargo arriving on trucks, ***loading docks*** are built next to large paved areas. Trucks can back up to the loading docks. See **Figure 18-64.** The docks are built at a height that permits forklifts easy access into the trucks.

Cargo terminals that handle rail freight must be built near rail lines. Trains must be able to pull up next to the platforms. Cargo terminals also must have elevated loading docks so forklifts can easily move on and off the railcars. Some rail cargo facilities have pits located in the ground between the tracks. Cars are positioned over the pits. Bulk (loose) cargo can be dumped from the bottoms of the cars. Conveyors carry the cargo out of the pit and move it into storage tanks or bins.

Figure 18-64. Loading docks are designed to match the floor levels of the truck and building, so cargo can easily be moved into or out of the truck. This truck is backing down a ramp built to match the two floor levels.

Vehicle Maintenance Facilities

Proper maintenance helps vehicles last longer and give better service. Many people enjoy working on their own cars. Private car owners can perform regular maintenance themselves. For convenience, however, some owners take their vehicles to commercial service facilities. Fast oil changes, "lube jobs," and tune-ups are available so car owners can maintain their vehicles without doing the work themselves. See **Figure 18-65.**

There are other types of maintenance that occasionally must be performed on automobiles. Engines are mechanical devices prone to wear and breakdowns. Other vehicular systems, such as the electrical, fluid, or mechanical system, could fail at any time. No matter how well a vehicle is designed, there are always things that can go wrong. When a car does not work properly, the problem must be found and fixed.

Early car engines could be fixed by almost anyone with mechanical ability. Modern auto engines, however, are complex pieces of machinery. It takes special training and equipment to diagnose and repair problems.

Figure 18-65. Many businesses perform maintenance tasks on automobiles and trucks.

Automotive technicians are people who have that training, and they work on other people's vehicles. Large businesses and small shops all over the country employ them. Mechanics are capable of servicing every part of an automobile. Along with their repair businesses, they provide a very important support service for transportation technology. See **Figure 18-66.**

Passenger bus services hire mechanics to service their vehicles. Bus companies usually own and operate their own maintenance facilities. There, they can concentrate on keeping every part of their vehicles in top condition. Cleaning crews also wash the buses inside and out. All this is aimed at providing safe, comfortable service for paying passengers.

For large vehicles, such as trains, special support facilities need to be set up. Mechanics are able to work on the locomotives, but sometimes other parts of trains break down. When this happens, cranes are needed to lift the heavy vehicles so they can be worked on.

When metal train wheels become worn, they often need to be reground. Special grinders built into the tracks at maintenance areas do this. They regrind the small angle on train wheels so the wheels can go around curves easily. These grinders also true the wheels so they ride smoothly on the rails. All this can be done without lifting the railcar off its tracks.

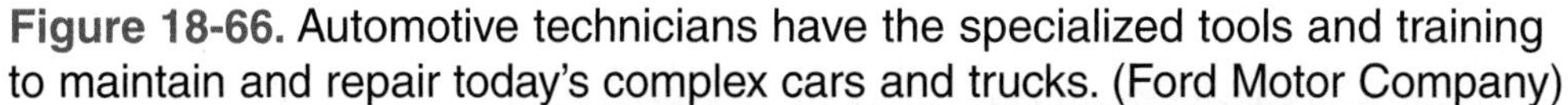

Figure 18-66. Automotive technicians have the specialized tools and training to maintain and repair today's complex cars and trucks. (Ford Motor Company)

Summary

Land vehicles are propelled by internal combustion engines, electrical propulsion, or a hybrid of both. The main types of internal combustion engines include the gasoline piston, rotary, and diesel engines. Electrical propulsion systems include direct electric, indirect electric, and electromagnetic propulsion. The two types of hybrid systems are gasoline-electric and diesel-electric.

The propulsion system provides the power, and the drive train sends the power throughout the vehicle. The main component of the drive train is the transmission, which multiplies the torque of the engine. This is a part of the control system. Brakes are also part of the control system and provide the stopping power using friction.

Land vehicle guidance is assisted with the use of road maps. These maps show the location of roads, cities, and points of interest. Onboard navigation systems also help to guide drivers. Vehicular guidance systems include adaptive cruise controls and parking assist systems.

The suspension systems of land vehicles rely heavily on the tires. Tires provide both cushioning and traction. Along with the tires, springs, shock absorbers, and stabilizer bars provide for a smooth ride in nearly all land vehicles. Magnetic levitation (maglev) trains have a suspension system that relies on the support of magnetic fields.

The structure of land vehicles is comprised of a body and frame. In many automobiles, these two pieces are one, which is known as unibody construction. Unibody construction saves weight, which can increase fuel economy. The structure of vehicles can come in many forms, depending on the passengers or cargo the vehicles are designed to carry. Trucks, trailers, and train cars are specially designed with different types of cargo in mind.

Lastly, a number of facilities support land transportation. Pathways, such as roads and railways, are constructed for land vehicles. Passenger and cargo facilities provide places to unload and load vehicles. Support systems also include vehicle maintenance facilities. All the systems work together and make land transportation as safe and efficient as possible.

Key Words

All the following words have been used in this chapter. Do you know their meanings?

acceleration
adaptive cruise control system
air brake
all-wheel steering
antilock braking system (ABS)
articulated
atlas
automatic transmission
ballast
bias tire
body
bogie
break-bulk cargo
bridge
bulk cargo
chassis
coil spring
compression-ignition engine
contact patch
convertible
coordinated steering
crab mode
cut
deceleration
degree of freedom
diesel-electric propulsion
diesel engine
differential
direct electric vehicle
disc brake
drive system
drum brake
dunnage
epitrochoidal curve
fill
fuel cell
gasoline-electric hybrid
gasoline piston engine
gear ratio
grade
guideway
hybrid system
hydraulic braking system
leaf spring
linear motor
loading dock
manual transmission
onboard navigation system
parallel hybrid
planetary gear set
pneumatic tire
power brake
proton exchange membrane (PEM)
rack and pinion
rail
rail bed
regenerative braking
roadbed
road map
rotary engine
series hybrid
shock absorber
signage
solar propulsion
spring
spring oscillation
stabilizer bar
steel-belted radial tire
subballast
torque converter
torsion bar
transmission system
tunnel
unibody construction

Test Your Knowledge

Write your answers on a separate sheet of paper. Do not write in this book.

1. *True or False?* Internal combustion engines are a type of heat engine.
2. List and describe three of the main components of a gasoline engine.
3. Describe how a rotary engine works.
4. *True or False?* Rotary engines are more commonly used in automobiles than piston engines.

5. Explain the main difference between gasoline and diesel engines.
6. *True or False?* The compression ratio of a diesel engine is greater than that of a gasoline piston engine.
7. Why are direct electric vehicles not always feasible?
8. Bumper cars are an example of a(n) _____ vehicle.
   A. indirect electric
   B. direct electric
   C. diesel
   D. solar
9. *True or False?* The two by-products of fuel cells are heat and water.
10. *True or False?* Linear motors are round induction motors.
11. Summarize the difference between parallel and series hybrid systems.
12. Write two or three sentences describing how the parts of the drive system function to transmit power from the engine to the wheels.
13. State the purpose of a transmission system.
14. A _____ is a manual device that separates the transmission from the engine.
    A. torque converter
    B. differential
    C. clutch
    D. driveshaft
15. Discuss in two or three sentences how an automatic transmission operates.
16. Cite several ways manual and automatic transmissions are different.
17. *True or False?* Torque converters use the movement of fluid to transmit power.
18. The most common type of map used in land transportation is the _____ map.
    A. political
    B. road
    C. topographical
    D. geological
19. *True or False?* The map legend lists the symbols and markers used in the map.
20. Recall and describe the three types of roadway signs.
21. Describe the function of adaptive cruise control systems.
22. *True or False?* Carburetors create a fuel-air mixture necessary for combustion within the engine.
23. *True or False?* Magnetic levitation (maglev) vehicles are accelerated by increasing the flow of fuel to the gasoline engines.
24. Paraphrase how drum and disc brakes operate.
25. Identify several rear-wheel steered vehicles.

26. Name the two types of all-wheel steering.
27. Write the two purposes of tires.
28. *True or False?* Steel-belted radial tires are the most common type of tire today.
29. Interpret what the marking "LT 225/75 R 15" means.
30. What is the difference between springs and shock absorbers?
31. *True or False?* In a repulsion maglev system, the polarities of the magnets are opposite.
32. *True or False?* The chassis is the body of an automobile.
33. What is meant by the term *unibody construction*?
34. *True or False?* Vehicles that haul bulk and break-bulk cargo use the same types of trailers.
35. List several types of support systems for rail vehicles.

Activities

1. Build a working model of an indirect electric vehicle.
2. Create a model of a maglev vehicle.
3. Create a display showing how a hybrid system operates.

19 Water Transportation Systems

Basic Concepts

- Define *buoyancy*.
- Identify several water vehicles.

Intermediate Concepts

- Discuss the different routes of water transportation.
- Describe how a boat floats.
- Explain the modes of water transportation.

Advanced Concepts

- Research and determine major sea-lanes used in transoceanic transportation.
- Calculate buoyancy and displacement.

Water transportation has always been an important method of transporting people and goods. It is also the most efficient way of moving goods across the oceans. Water transportation has been in development for many years.

The History of Water Transportation

Over 70% of the earth's surface is covered with water. See **Figure 19-1.** Water has made it possible to transport people and cargo over greater distances than by land transportation. ***Waterways*** are the bodies of water in which vessels travel. A ***vessel*** is another term for a water vehicle. Typically, ***boats*** are water vehicles less than 100′ in length. ***Ships*** are any craft over 100′ in length. Drawings and remains of boats have been found that date back over 5000 years. Not all historians agree, however, that this was the beginning of water transportation. Some believe water transportation was used over 50,000 years ago, when the Aborigines settled in Australia. The spread of civilization to new continents and remote

Figure 19-1. This picture of earth, viewed from space, shows the large proportion of water to land. The continents are patterned with dots of lakes and lines of rivers. Many of these are waterways that have been used in the development of transportation. (National Aeronautics and Space Administration)

parts of the world was only possible by water transportation. Ships and water transportation have been essential to the exploration, settling, and development of our world.

The building of boats and ships has been dependent on the technology that was available to the builder. In the beginning, logs were used to transport people and goods by water. Rafts and dugout canoes were among the first water transporting vehicles. See **Figure 19-2.** Eventually, the early shipbuilders turned to new methods of construction. In Europe and Asia, the builders created wooden frames. Animal skins were then stretched over the wooden frames of the boats. This extended the lives of the boats for a longer journey. Skins were also sewn together, inflated, and used like rubber rafts. Instead of using skins, the Native Americans used bark to construct canoes, similar to the way fiberglass is used today. These boats were powered by hand, using sticks and, later, oars and paddles.

The technology and methods of boatbuilding continued to advance. Boats became larger and able to travel greater distances. These became known as ships. Early ships were constructed of wooden planks sealed

Waterway: A body of water in which vessels travel.

Vessel: A water vehicle.

Boat: A water vehicle less than 100′ in length.

Ship: A water vehicle over 100′ in length.

Figure 19-2. Dugout canoes are made from logs that have been hollowed out. These canoes could carry several people and a small amount of cargo. This photo, taken in about 1910, shows a member of the Nez Perce tribe using a dugout canoe in the northwestern United States. (Library of Congress)

with tar or pitch. Newer ships built by the Phoenicians, Egyptians, and Romans were powered by both human and wind power. The human power was produced by many people rowing large oars. Sail power was first used around 3500 BC. Over time, sails became the sole source of power for ships. See **Figure 19-3.** In the fifteenth through the seventeenth centuries, the development of vessels brought about further improvements. Four- and five-masted ships were built so more sails could be added to propel larger, heavier ships. Some could carry 1000 tons of cargo. The use of sail power remained the popular form of propulsion until into the nineteenth century.

By the late 1700s, sailing ships were being constructed of iron and, later, steel, rather than wood. Along with the iron ships came another new power source. In 1783, a Frenchman, Claude de Jouffroy d'Abbans, added a steam engine to his boat, the *Pyroscaphe.* John Finch was the first American to power a steamboat in the United States. The first successful steamboat, however, was the *Clermont,* built by Robert Fulton in 1807. The *Clermont* used a steam engine to power a paddlewheel. See **Figure 19-4.** The early steam engines were very expensive to operate and not very reliable. Advances in technology, however, led to the development of the steam turbine. The steam turbine was efficient and reliable. Its advantages put an end to sail power for all but recreational boating. The steam turbine was a system that used a steam engine to turn a propeller. In 1839, the first propeller-driven ship was put into service. The use of propellers opened up a large amount of room that had been used for either sails or the paddlewheel. This meant ships could carry more passengers and cargo. At this time, ships also became larger, and the first ocean liners

Figure 19-3. This large sailing vessel is the clipper ship *Three Brothers*, which was 328′ long and 48′ wide. The large sails on the ship harnessed the power of the wind. Large wooded spars called masts supported the sails. The ship is shown in an 1875 color lithograph under full sail, with 30 sails set. (Library of Congress)

Figure 19-4. Steam engines were developed long before other types of heat engines and were extensively used to propel boats on the Mississippi and other major rivers in the 1800s. Today, old stern-wheelers, such as the *Delta Queen*, are used as cruise ships for vacationers on inland waterways. In addition to restored vessels, such as the *Delta Queen*, new boats with the appearance and functions of old steamboats have been built for cruising. The *Mississippi Queen*, in the background, is one such recreated steamboat. (Delta Queen Steamboat Company)

were built. Ocean liners are huge ships built to transport passengers around the world. Some of these ships are very luxurious. This type of ship was used for passenger travel from the early 1900s to around 1950. The number of passengers greatly declined in the 1950s, due to the speed and availability of air transportation.

Today, the shipping of cargo is the main use of water transportation. See **Figure 19-5.** Water transportation plays a part in nearly all U.S. imports and exports. The cargo ships today still use propellers. They are

Figure 19-5. Cargo ships carry billions of tons of raw materials and finished goods each year to ports around the globe. Many are built to carry general cargo. Others are specially designed for standardized shipping containers. Still others are huge floating tanks for oil, gasoline, and other liquids.

no longer, however, steam powered. Diesel engines are used in ships to turn the propellers. Gasoline engines, as well as human and sail power, are only used today in small recreational boats. Even nuclear power is used in some vessels. There are many kinds of vessels in the water transportation system.

Water Routes

Water transportation vehicles cannot operate without water. The main purpose of a waterway is to allow vessels to travel the safest and most efficient route from one port to another. The waters on these routes must be ***navigable***. They must be deep enough and wide enough for the boat or ship to travel through. See **Figure 19-6.** The vessels must be able to float through freely and avoid dangerous situations. If some of the waters are not navigable, serious accidents can occur. The vessel can hit large rocks and sink or run aground in shallow waters. There are two major types of waterways. They are known as sea-lanes and inland waterways.

Navigable: Deep and wide enough for a boat or ship to travel through.

Sea-Lanes

Ships and other vessels rarely collide at sea because they take regular routes when traveling across the ocean. These routes are known as ***sea-lanes***, or trade routes. The use of established sea-lanes allows vessels to travel across the waters, while avoiding other marine traffic. The sea-lanes are not marked like roads. Instead, they are shown on navigation maps and charts. A vessel must be navigated to stay in the proper sea-lane. Navigation of a vessel is simply the guidance of it. Vessels follow the sea-lanes with the aid of navigation equipment, such as compasses,

Sea-lane: A regular route taken by ships and other vessels when traveling across the ocean. Also called trade routes.

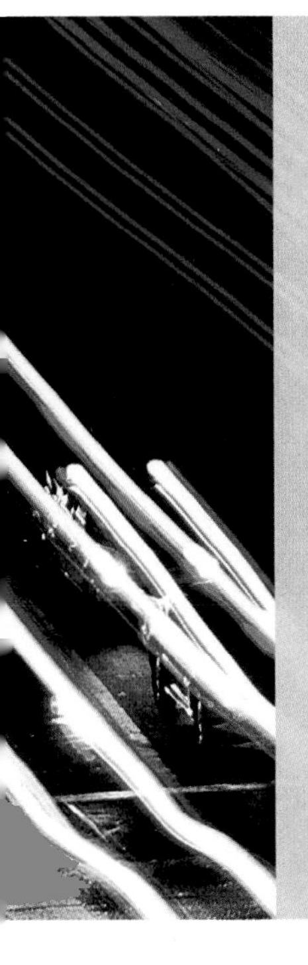

Career Connection

Merchant Marine Sailors

Commercial ships are used to transport goods within the nation and throughout the world. Those operating in the United States or outside the United States, but under the American flag, are known as the merchant marine. The people working on the ships are referred to as merchant mariners. The largest group of merchant mariners is sailors.

Sailors are used on ships to perform maintenance and keep the ship in operating condition. They are also used to keep watch and check water and ship conditions. Sailors report to either the captain or the ship's mate. They often work in 4- or 6-hour shifts, 24 hours a day and 7 days a week. Sailors are usually hired for one voyage at a time, although the voyage may last several months. When they return to port, they must find their next voyage. This means there is little job security, but this occupation allows for much flexibility.

Beginning sailors are required to have only a high school diploma. Many participate in training sessions, however, at union schools. Sailors must receive merchant marine documentation for the U.S. Coast Guard. With work experience, sailors can advance to the rank of able seaman and receive certification from the Coast Guard. Sailors typically begin employment near minimum wage. The median salary for sailors, however, is near $14 per hour.

Figure 19-6. Canoes and other small craft can safely navigate narrow, shallow waterways, such as this river. Large cargo ships, however, must have deep water and sufficient room for maneuvering.

sextants, radio detecting and ranging (radar), computers, satellites, and charts. Ship navigators also rely on various physical landmarks, such as lighthouses, stars, and buoys. Guiding a vessel is not an easy task. The navigator must be very familiar with the waterway, as well as with the vessel. Sea-lanes exist for most routes connecting foreign countries. All the countries affected by the ship's travel have a voice in determining the route of the sea-lane. Surprisingly, there are only a few routes all ships use to travel to and from the major ports of the world. When ships transport cargo between two ports on the same continent, they often use coastal sea-lanes. These sea-lanes are within 20 miles of the coast.

Inland Waterways

Inland waterway: A route taken on canals, rivers, and lakes.

Sea-lanes are routes used on the ocean. There is also a great amount of water travel, however, on other bodies of water. The routes taken on canals, rivers, and lakes are known as ***inland waterways***. Land surrounds all these waters. Inland waterways offer much guidance to a vessel. The vessels have limited routes on which to travel. Inland waterways are used for recreation, as well as for moving cargo.

Purposes and Principles of Water Transportation

Water transportation is just as important as any other form of transportation. It is used to transport both people and cargo. Cargo transported by a vessel includes oil, grain, heavy equipment, iron ore, and many other types of material. Transporting of bulk commodities is a great advantage of water transportation. Water transportation is more fuel efficient, economical, and inexpensive than other forms of transportation when carrying large amounts of cargo. It has contributed greatly to the expansion of our country. We still depend on it to a large extent.

If you have ever seen a large tanker transporting oil or a cruise ship carrying people, you may have wondered how these vessels stay afloat. See **Figure 19-7.** Archimedes first explained the principle that describes the concept of flotation. He stated that when an object is submerged in a liquid, the upward force on the object is equal to the weight of the displaced fluid. This principle explains what is known as buoyancy. ***Buoyancy*** is the upward force the water exerts on objects placed in it. See **Figure 19-8.**

The second half of Archimedes's theory describes displacement. Displacement is the weight of the water the object pushes aside. According to Archimedes's principle, when the amount of buoyancy (upward force) is equal to or greater than the amount of displacement (water pushed aside), as the object is put into the water, it will float. See **Figure 19-9.** On the other hand, if the object displaces less then the upward force, it will sink. Another way to think about displacement is to imagine placing a water vessel into wet concrete. When the ship is removed, the shape of the vessel will remain. If you fill the shape with water and weigh it, you will find the amount of displacement. See **Figure 19-10.**

Buoyancy: The upward force water exerts on objects placed in it.

Figure 19-7. This 300,000-ton tanker is made of steel and carries a heavy cargo of crude oil, but it remains afloat because its weight is less than the weight of the displaced water. (Shell Oil Company)

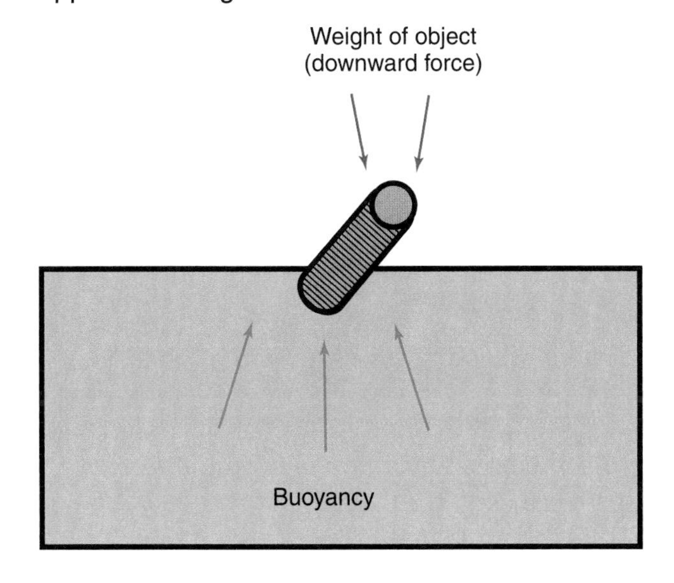

Figure 19-8. Buoyancy is the reason objects will float in water. It is the power of a fluid to exert an upward force on a body placed in it. Buoyancy is the principle that makes objects appear to be lighter underwater.

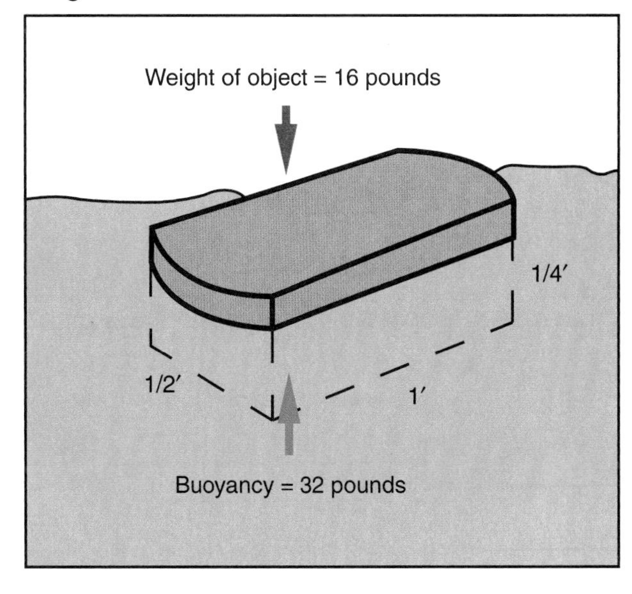

Figure 19-9. Metal will float in saltwater because the buoyancy is greater than the weight of the metal.

Curricular Connection

Math: Buoyancy Calculations

When an object is placed in water, the water exerts an upward force on the object. This force is equal to the weight of the water the object displaces. In other words, the force is equal to the weight of the water that would occupy the volume of the object beneath the surface of the water.

The density of water is 62.4 pounds per cubic foot (pcf). This means that a cube of water 1′ long on each side weighs 62 lbs. If you know the weight of a boat, you can calculate the volume of water that needs to be displaced in order for the boat to float.

For example, a fully loaded barge weighs 10,000 tons (or 20 million pounds). To calculate the volume of water that needs to be displaced for the barge to float, use the equation of density:

Density = Weight / Volume

Solve the equation algebraically for volume:

Volume = Weight / Density

Substitute the weight of the barge and the density of water:

Volume = 20,000,000 lbs. / 62.4 pcf
= 320,513 ft^3

If you know the dimensions of the barge, you can also calculate how deep the barge must sink in order to float. Assume the barge is 300′ long and 90′ wide. The volume of a rectangular box is length × width × height. Therefore, you can use the following formula:

Volume of displaced water = Barge length × Barge width × Depth of water

Rearranging the equation to solve for Depth of water:

Depth of water = Volume of displaced water / (Barge length × Barge width)
= 320,513 ft^3 / (300′ × 90′)
= 11.9′

When fully loaded, the barge must sink 11.9′ deep to displace enough water to equal its weight. If sides of the barge are 20′ deep, the barge will float. If however, the sides are only 10′ deep, the barge cannot displace enough water to support its weight, and it will sink.

Figure 19-10. Boat hulls are designed for great buoyancy.

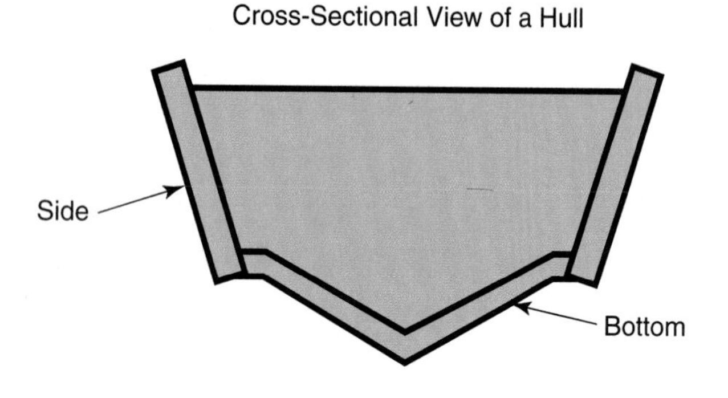

Modes of Water Transportation

There are many types of water vehicles used to serve several purposes. Several types of vessels have been in existence since the days of Phoenician and Roman shipbuilders. These ships were used for the transportation of people, the movement of cargo, and war. The same types of ships still exist today. Passenger vessels are typically used to transport people recreationally. Cargo-moving vessels transport goods within and between countries. The armed forces use military vessels in many applications. One of the newest water vehicles is the specialty craft. Specialty vessels are built to handle unique situations, such as breaking ice in the polar regions or examining wreckage at the bottom of the ocean. See **Figure 19-11.** Water transportation vehicles travel in one of two modes. See **Figure 19-12.**

- **Inland water transportation.** This is transporting people or cargo on inland waterways, including rivers, canals, and lakes.
- **Transoceanic water transportation.** This is transporting people or cargo across an ocean.

Canal: A channel constructed to connect two bodies of water.

Figure 19-11. An icebreaker is a specialty vessel designed to perform a specific task. The U.S. Coast Guard cutter *Polar Sea* has a heavily reinforced bow to break through thick ice. (U.S. Coast Guard)

Inland Waterway Systems

Inland waterways include any body of water within a landmass. Two important North American inland waterways are the Saint Lawrence Seaway and the Great Lakes. See **Figure 19-13.** Often, inland waterways are connected by canals. A ***canal*** is a channel constructed to connect two bodies of water. Various kinds of vehicles are designed for use on inland waterways.

Figure 19-12. These are the vehicles used in the two modes of water transportation.

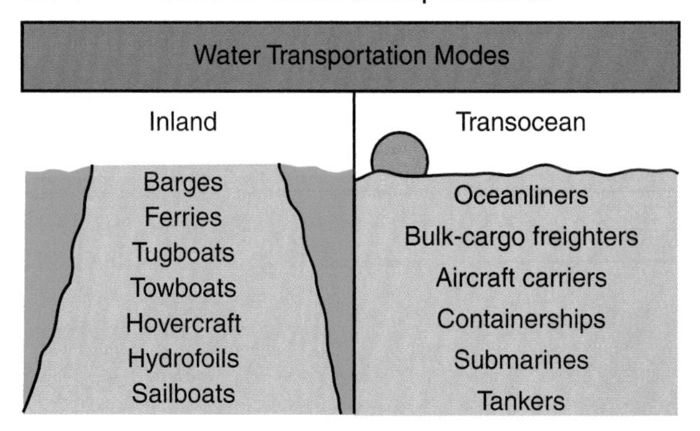

Water Transportation Modes
Inland
Transocean
Barges
Ferries
Tugboats
Towboats
Hovercraft
Hydrofoils
Sailboats
Oceanliners
Bulk-cargo freighters
Aircraft carriers
Containerships
Submarines
Tankers

Figure 19-13. The Saint Lawrence Seaway connects the Great Lakes with the Atlantic Ocean. The waterway allows oceangoing cargo ships to serve ports as far west as Duluth, at the western tip of Lake Superior. Port locations are shown as dots on the map.

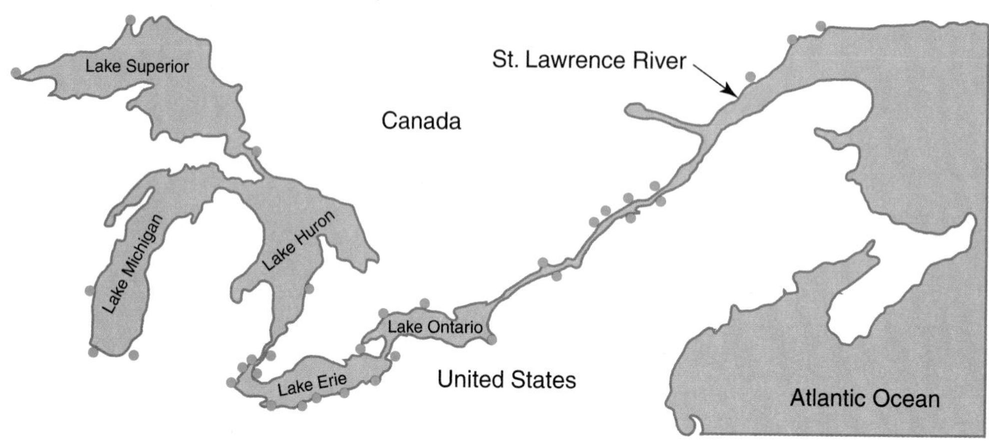

Technology Link

Construction: The Panama Canal

At times, transportation systems require structures to be built to increase the efficiency of moving goods and people. In some cases, these structures are considered to be amazing feats of engineering marvel. The Panama Canal is one such construction project. It has been regarded as one of the great engineering achievements in the history of the world. This canal is also an enormous resource in the global transportation system. Nearly all ships that cross South America use the Panama Canal. Few ships would choose to sail around South America, adding an additional 8,000 miles to the trip, rather than cut across it. It takes between 8 and 10 hours to cross the canal and requires a toll. The average toll is $45,000, which may seem like an enormous amount of money, but compared to the cost and time it takes to travel around South America, it is economical for ships to pay the toll.

The French started the construction of the canal in the late nineteenth century. They, however, quickly went bankrupt. The United States, several years later, took over the project and completed it in 1914. The construction of the canal required an infrastructure to be built to accommodate the workers and the machinery. Houses and villages with bakeries and other services were constructed to house the 19,000 construction workers who built the canal. The Panama Railroad was also built to handle the movement of the heavy construction equipment. During construction, 239 million cubic yards of earth was moved. This is enough dirt to fill the Metrodome in Minneapolis, Minnesota four times.

The finished Panama Canal is 50 miles long and is 85′ over sea level at its highest point. Ships go through a series of three locks throughout the passage from the Atlantic to the Pacific Oceans. The locks are used to raise or lower the ships to the height of the water on the opposite side of each lock. Ships that pass through the canal must be less than 965′ in length and 106′ in width. Unfortunately, many of the supertankers and large containerships being built today are much larger than the canal will permit. Due to this problem, there is much discussion about either widening the canal or creating an entirely new canal.

Barges

Barges are flat ships with blunt ends that carry very heavy loads of cargo. See **Figure 19-14.** A barge can carry up to five times its own weight. About three-fourths of all the cargo carried by water is transported by barge. A barge is not a very attractive vessel. It is not very fast either. Barges are, however, very safe. The smoothness of their rides and the amount of weight they can carry have no equals. Barges carry liquids, solids, and gases. The most common types of barges are the open hopper, covered dry cargo, liquid cargo, and deck. Such barges can carry coal, ore, oil, and grain.

Figure 19-14. Most barges have no engines. Some other vessel or series of vessels must, therefore, push or pull them. A towboat is pushing these barges. (National Park Service, Natchez Trace Parkway)

Towboats and tugboats

Towboats and tugboats are not often thought of as different from each other. They are, however, very different. A ***towboat*** is designed to push barges. See **Figure 19-15.** A towboat has a wide, flat front end to allow more surface area for pushing barges. ***Tugboats***, on the other hand, are designed to pull barges. They are very powerful. Tugboats are also used to pull ocean liners in and out of ports and help dock and undock other oceangoing vessels.

Towboat: A vessel designed to push barges.

Tugboat: A vessel designed to pull barges.

Hydrofoils

Hydrofoils operate on inland and coastal waters. See **Figure 19-16.** A ***hydrofoil*** is a passenger-transporting vessel, similar to a plane and ship put together. Hydrofoils have been used to transport people across big lakes and up and down rivers, channels, and canals. A hydrofoil has

Figure 19-15. Towboats, such as this one, are used to push barges along the inland waterways. Note the heavy pushing blocks at the front (bow) of the towboat.

Figure 19-16. Hydrofoils lift watercraft out of the water, much like wings lift planes into the air. In either case, power is needed to move the craft forward at high speed. (Turbo Power, United Technologies Corporation)

Hydrofoil: A passenger-transporting vessel, similar to a plane and ship put together. It operates on inland and coastal waters.

wings called foils. It develops its lift from the buoyancy of the water, just as an airplane receives lift from the air. As the watercraft reaches high speeds, the foils lift it out of the water. The boat is then sailing along at high speeds, just skimming the surface of the water.

Hovercraft

A hovercraft is a vessel that rides on a cushion of air. Hovercraft are also referred to as ***air-cushion vehicles***. Air pressure allows the vessel to hover (remain suspended) in the air a few feet above the water or land. The vessel makes no contact with the water or land. It just rides along on a cushion of air. See **Figure 19-17.** Gas turbine engines drive large, high-speed fans that bring in air. The air is then forced down around the base on all sides. This forms the cushion of air. Hovercraft are used for several different applications, including military, rescue, and recreation.

Figure 19-17. The basic design of the hovercraft. A cushion of air keeps the hull suspended above the surface.

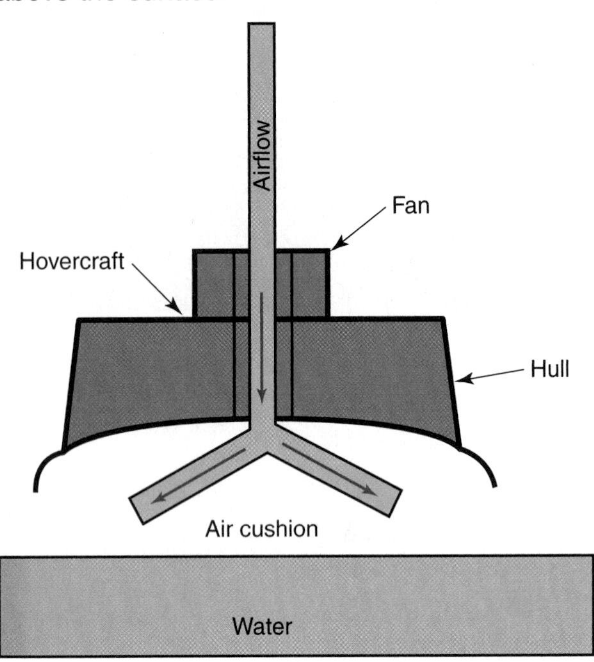

Ferries

Ferries are vessels that move people and vehicles across narrow or small bodies of water. See **Figure 19-18.** Ferries are usually used along coastal waters and on inland waterways. They transport passengers to and from islands and across rivers. One of the most popular is the Staten Island Ferry in New York Harbor. Another important ferry travels between the mainland and Vancouver Island in British Columbia, Canada. A roll-on/roll-off (RO/RO) vessel is the type of ferry used to transport vehicles. These boats allow passengers to drive their cars onboard. Larger RO/RO vessels have several levels, much like a parking garage. These usually operate in areas where time can be saved by transporting automobiles across the water instead of driving around the coastal areas.

Figure 19-18. This ferry carries passengers and vehicles between Victoria, Vancouver Island, Canada and Port Angeles, Washington.

Commercial fishing boats

Commercial fishing boats are used in inland and coastal waters. The type of fishing boat used depends heavily on the type of fish being sought. The three main types of fishing boats are trawlers, seiners, and liners. The main differences between the three are the type of capturing methods they use. Trawlers use trawl nets, which are cone-shaped nets dragged behind the boat. The fish swim into the net. Seiners use seine nets, which are placed around the area and then closed on the fish. Liners are boats in which fish are caught by hook and line. The lines can be operated mechanically or by fishermen. These boats have storage tanks onboard to carry the fish they catch.

Cruisers

Cruisers are boats that can be used for both inland and transoceanic pleasure trips. See **Figure 19-19.** Cruisers are divided into three classes: Class 1 includes boats 16–26′ in length, Class 2 includes boats 26–40′, and Class 3 includes boats 40–65′. Class 2 and 3 cruisers can be used to travel around the globe. Cruisers can be powered by engines or sails and are used for pleasure cruising in the coastal, inland, and ocean waters. These boats are sometimes referred to as yachts. Yachts have cabins for sleeping, navigation areas, and small kitchens. Large cruisers can have enough space for up to 12 people.

Recreational water transporting vehicles

Other forms of water vehicles are often found in smaller lakes and rivers. Such vehicles are most often used for recreational purposes. Sailing and windsurfing are very big sports along the coastal areas and northern lakes. Sailboats are common water vehicles. Jet skiing has become a popular pastime on the water. Pontoon boats, speedboats, canoes, rafts, and paddleboats are some other examples of water vehicles most often used for leisure. See **Figure 19-20.**

Figure 19-19. Cruisers and yachts are larger water recreation vehicles. Some are large enough to cross oceans. Sails or engines can power yachts. These sailing yachts are berthed at Annapolis, Maryland, on the Chesapeake Bay.

Transoceanic Waterway Systems

Another mode of water transportation is transoceanic waterway systems. ***Transoceanic*** means traveling across the ocean. Many vessels are designed for transoceanic travel. Some of the vessels carry cargo, and others carry people. As mentioned earlier, vessels traveling across the ocean follow sea-lanes (established routes). They stay within the designated sea-lanes by being navigated. Some transoceanic vessels are ocean liners, freighters, tankers, containerships, aircraft carriers, and submarines.

Ocean liners

Ocean liners are basically for luxury use. Cruise ships and ocean liners can carry thousands of passengers. Most people travel on ocean liners for a relaxing vacation. Ocean liners are very large and often have several decks. Decks may include bedrooms, swimming pools, restaurants, and game rooms. One of the most famous ocean liners is the *Queen Elizabeth II* from England. There are not many ocean liners in use today. Cruise ships

Air-cushion vehicle: A vessel that rides on a cushion of air. Also called a hovercraft.

Ferry: A vessel that moves people and vehicles across narrow or small bodies of water.

Transoceanic: Traveling across the ocean.

Figure 19-20. Recreational boats are also a form of water transportation. A—Windsurfing involves a surfboard fitted with a sail. (Howard Bud Smith) B—Sailboats are popular on all types of waterways. (Howard Bud Smith) C—Pontoon boats are used most often on rivers and small lakes.

A

B

C

are much more common. They are smaller. Some are a quarter of the size of an ocean liner. Cruise ships can dock at smaller ports around the world. They are primarily used for vacation purposes, rather than for transporting people to a final destination.

Bulk-cargo freighters

Bulk-cargo freighters are ships designed to carry very large quantities of cargo. Freighters usually have a series of holds, or storage areas, below the main deck. See **Figure 19-21.** Several different kinds of cargo can be carried, each in a different hold. Freighters carry such goods as coal, ore, oil, grain, sugar, cotton, and cement. The cargo is referred to as dry bulk or liquid bulk. An ore, bulk-cargo (OBO) freighter carries coal, grain, ore, and oil. An advantage to such a vessel is its ability to carry several types of cargo at one time. As it unloads a hold at one port, it can fill that hold up with something else and continue on its route.

Tankers

A tanker is a vessel designed to carry liquids. It has tank-shaped holds (sections) for carrying oil, petroleum products, chemicals, wine, and even molasses. The most common tanker is that for transporting oil. A tanker is constructed of two or three long tanks, which are divided into sections. By dividing the tanks in sections, the liquid will not move so much during transport. This makes the ship easier to handle and less likely to capsize. Because the cargo on a tanker is liquid, large pumps and hoses are used for loading and unloading. Large ocean tankers can carry up to 2 million barrels of oil.

Containerships

Containerships are very large and carry cargo stored in containers before loading. See **Figure 19-22.** If you took several milk cartons and stacked them in a rowboat, you would have the same concept, on a much smaller scale, as a containership. The containers are airtight, permanent, reusable, watertight, and fitted with at least one door on the end. Some containers are made of steel, but most are made of aluminum because it is a lighter metal that does not rust. The containers vary in length. Some are the size of a semitrailer. Containerships save time and money in the loading and unloading of cargo.

Figure 19-21. A bulk-cargo carrier has a series of holds that look like containers. These carriers are also called ore, bulk-cargo (OBO) ships because they carry ore, bulk-cargo, and oil.

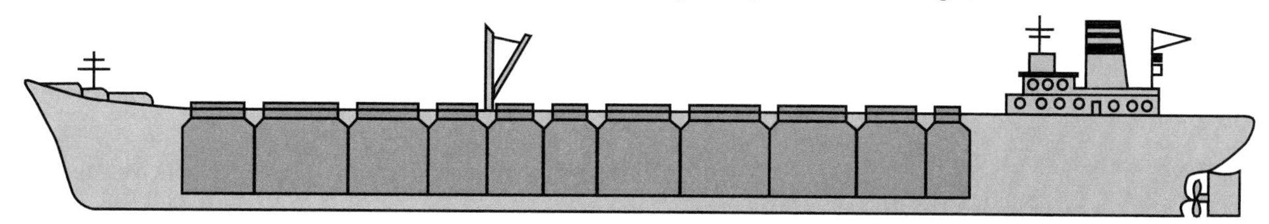

Figure 19-22. Containerships carry standardized metal boxes, called containers, which make cargo storage and handling more efficient. Containers are both carried in the ship's hold and stacked on its deck.

Military craft

Military water vehicles help to protect the nation and its interests around the world. The largest military craft is the aircraft carrier. An aircraft carrier is a very large ship with a padlike deck. Aircraft carriers carry fighter jets for the navy. The large deck allows the jets to take off and land. See **Figure 19-23.** The *Nimitz* class, which is the

Figure 19-23. An aircraft carrier is a "floating airport," with a large deck that provides room for planes to take off and land. Planes are stored and maintained on a large hangar deck below the flight deck. This is the aircraft carrier *Harry S. Truman.* (U.S. Navy)

Figure 19-24. Nuclear-powered submarines, such as the *USS Key West*, can remain submerged for months at a time. The tall structure is known as the conning tower, or "sail." (U.S. Navy)

largest type in the U.S. Navy, is over 1000′ long (over three football fields) and weighs 97,000 tons, fully loaded. These carriers can carry 85 aircraft and a total of 5680 people. Two other types of U.S. warships include cruisers and destroyers. Destroyers are smaller and faster than cruisers. Both can be used for surface combat, as well as antisubmarine and antiaircraft attacks. The military also has amphibious vehicles for landing troops and combat vehicles on land.

Submarines

Submarines are vessels that can submerge and travel underwater. See **Figure 19-24.** These vessels are used to explore the ocean and for military purposes. The first American submarine was the *Turtle*, built by David Bushnell in 1775. The *Turtle* was designed to be used in the American Revolution and could hold one person. Today, military submarines can hold 140 sailors and are over 370′ long. Many military submarines are nuclear powered and can stay underwater for months at a time. Military purposes, however, are not the only uses for submarines. Sight-seeing tours and amusement parks use submarines to allow visitors a view of the ocean.

Submarine: A vessel that can submerge and travel underwater.

Submersibles

Submersibles are small craft designed to explore shipwrecks and used for charting the ocean. See **Figure 19-25.** Submersibles use ballast tanks to submerse and rise. These vessels are equipped with specialized tools to complete the jobs they are given. For example, a submersible exploring a sunken vessel would be equipped with lights, a video camera, and a robotic arm to pick up objects.

Figure 19-25. Submersible craft are used for a variety of tasks underwater, such as charting the sea bottom, underwater archaeology, and marine biology studies. Most are unmanned, although some larger submersibles carry one or more crew members and function as small submarines. This unmanned submersible, being launched from a U.S. Navy minesweeper, is used to neutralize floating mines. (U.S. Navy)

Other transoceanic vessels

Icebreakers are a type of vessel used to clear frozen waters. The front, or bow, of an icebreaker is specially designed to be rammed on top of the ice. The weight of the vessel or ship then causes the ice to break. These ships are used to open Arctic sea-lanes. Lighter aboard ship (LASH) vessels are large ships that can carry both containers and barges. The ships are unique because they are equipped with an onboard crane so they can load and unload themselves.

Summary

Transporting people and cargo by water has been around for thousands of years. The water transportation system is an efficient form of transportation. Vessels travel in one of the two modes of water transportation: on inland waterways or across the ocean. Traveling across the ocean, a vessel follows a pathway, or route, called a sea-lane. Sea-lanes are comparable to highways on land.

Vessels can carry passengers or cargo. Passenger vessels on inland waterways include ferries, hovercraft, and hydrofoils. Transoceanic passenger vessels are ocean liners and cruise ships. The largest use of water transportation is for transporting cargo. Ships such as tankers, freighters, and containerships can carry large amounts of cargo at a lower cost than most other forms of shipping.

Buoyancy and displacement are important principles and concepts of water transportation. The upward force of water is buoyancy. Displacement is the amount of water the vessel moves. If the buoyancy is equal to or greater than the displacement of an object, the object will stay afloat.

Key Words

All the following words have been used in this chapter. Do you know their meanings?

air-cushion vehicle
boat
buoyancy
canal
ferry
hydrofoil
inland waterway
navigable
sea-lane
ship
submarine
towboat
transoceanic
tugboat
vessel
waterway

Test Your Knowledge

Write your answers on a separate sheet of paper. Do not write in this book.

1. *True or False?* Water transportation was invented 250 years ago.
2. Write the definition of *waterway*.
3. Cite the definition of the word *vessel*.
4. The _____ was the first steamship.
   A. *Monitor*
   B. *Pyroscaphe*
   C. *Turtle*
   D. USS *Indianapolis*

5. *True or False?* Steam power eventually replaced sail power as a main source of power in vessels.
6. *True or False?* Transporting cargo is the main use of water transportation today.
7. What is the purpose of a sea-lane?
8. State the definition of *buoyancy*.
9. Paraphrase the two principles that keep a ship afloat.
10. If you throw an object weighing 20 lbs. into the water and the weight of the water it displaces is 40 lbs., will the object sink or float? Explain.
11. List the two modes of water transportation.
12. *True or False?* A lake is an example of a transoceanic waterway.
13. What major engineering feat allowed 8000 miles to be saved when traveling from New York to Los Angeles?
14. A(n) _____ is used to move people over small bodies of water.
    A. towboat
    B. ocean liner
    C. ferry
    D. containership
15. An ocean liner is known for transporting _____.
    A. passengers
    B. cargo
    C. both A. and B.
16. Name three types of vessels under each mode listed in question 11.
17. Recall and describe two military vessels.

Activities

1. Take part in a boat hull design competition. The object is to design and build a boat hull for maximum buoyancy. As a group, determine material limitations. Test the boats by loading them with pebbles or other materials until they sink. Weigh the pebble load to determine the winner.
2. Obtain a world map and locate the major ocean-shipping routes.

20 Water Vehicular Systems

Basic Concepts

- Define the four types of propulsion.
- Identify tools used to find direction, speed, time, and location in water transportation.
- Cite ways in which vessels are controlled.
- State the functions of hydrofoils and hovercraft.
- List several nautical terms and parts of ships.
- Name the support systems of water transportation.

Intermediate Concepts

- Chart a course on a nautical chart.

Advanced Concepts

- Calculate the hull speed of different sized vessels.
- Design hulls for different uses of vessels.

All water vehicles are designed for specific purposes. These vehicles may be used for recreation, to move cargo, or to transport people. Whatever the vehicle's purpose, all water vehicles are moved by propulsion systems, directed by guidance systems, steered by control systems, kept afloat by suspension systems, comprised of structural systems, and docked at support systems.

Propulsion Systems

Like all forms of transportation, water vessels require a propulsion system to move. The earliest forms of propulsion were the paddle and oar, followed by the sail. The most recent additions to water propulsion are the propeller and water jet. Each of these types of propulsion has some type of contact with either the water or wind that enables the vessel to be pushed or pulled across the water.

Paddle: An implement used to propel a boat using human power.

Sail: An extent of fabric by means of which wind is used to propel a ship through water.

Paddles and Oars

Paddles were the first type of propulsion used to move boats. Early rafts and canoes were put into motion by a flattened log under human power. For many years, human power was used to row canoes, boats, and even ships. Large ships built 2500 years ago were propelled by over 150 people rowing with large oars. The first boats to use engines were still also propelled by paddles. The steam engines in paddleboats were used to turn a wheel of paddles. The contact between the paddles and water is what propelled the boats. Today, paddles and oars are still used in small recreational watercraft. See **Figure 20-1.**

Figure 20-1. Oars and paddles are used today to propel and guide small recreational watercraft, such as rowboats, canoes, and this kayak.

Figure 20-2. A sailboat has triangular sails that can change positions to move the vessel in the desired direction. (Adrienne Levatino)

Sails

A ***sail*** is a piece of fabric used to convert wind energy into a source of propulsion. Most sails are triangular and able to rotate into the wind. See **Figure 20-2.** Sails can be used to propel boats in any direction, except within 45° from the direction of the wind. This is because sails actually function as an airfoil (much like an airplane wing or race car spoiler). When the wind hits the sail, just enough air goes into the sail to give it the rounded shape. The rest of the air travels on either side. The air moving across the sail is deflected in one direction. This deflection generates motion roughly perpendicular to the bottom edge of the sail. Depending on the direction of the wind, it may tend to pull the sailboat sideways. To counteract this effect, sailboats have a keel, or centerboard, which is placed along the bottom of the boat.

The mainsail and jib are the types of sails used to drive small sailboats. The ***mainsail*** is the larger of the two and is connected to the mast and a boom. It can be rotated by allowing slack on either side in the ropes, or ***sheets***, in nautical terms. The ***jib*** is a smaller sail and is connected to the mast and the bow. The ***bow*** is the very front of a vessel. The jib can be pulled tight on either side of the boat, depending on the angle of the wind. ***Spinnaker*** sails are used at the front of the boat when traveling downwind. These sails are typically large and colorful and resemble a section of a hot air balloon. See **Figure 20-3.**

Propellers

A ***propeller*** is a rotating blade that produces thrust. Thrust is the propeller's reaction to the water that pushes the boat forward. There are several types of propellers used in vessels. See **Figure 20-4.** An important part of a propeller is its pitch. The ***pitch*** is the angle of the blade. The greater the pitch is, the faster the propeller will travel. If the propeller pitch is too great or too little, however, the propeller is inefficient and travels slower. The propeller cannot turn by itself, so it requires a power source. Internal combustion engines, diesel engines, turboelectric generators, gas turbines, and even nuclear reactors are used to power propellers.

Outboard engines are most common on fishing boats and small motorboats. In this type of system, the power source and propeller are one piece. Outboard engines are attached to the ***stern*** (back) of the boat. See **Figure 20-5.** ***Inboard/outboard engines*** are used in mid-sized recreational boats. These systems have larger engines than outboard engines mounted inside the boat, with a propeller similar to an outboard engine's propeller. ***Inboard engines*** are used on most vessels over 36′ in length. These systems have a power source mounted inside the ship attached to a propeller shaft.

On large ships, engines are used to generate and store electricity. Military submarines, for example, use a system to generate power for electric motors. This inboard propulsion system, known as a ***nuclear turbine engine,*** uses a nuclear reactor to heat water. The water becomes steam and turns a turbine engine, which creates electricity. The electricity is used to power the submarine and turn the propellers. Excess electricity is stored in batteries for later use. This allows the submarine to turn the turbine engine off, remain quiet, and still have electricity and propulsion.

Mainsail: The larger of the two sails on a sailboat. It is connected to the mast and a boom.

Sheet: A rope used on a sailboat.

Jib: The smaller sail on a sailboat that is connected to the mast and the bow.

Bow: The very front of a vessel.

Spinnaker: A sail used at the front of a boat when traveling downwind.

Figure 20-3. The three basic types of sails are visible in this photograph. The sailboat in the foreground has both a mainsail and a jib. The blue-striped sail on the boat in the background is a balloon spinnaker.

Figure 20-4. The three-bladed propeller is most common for both large and small boats. This massive unit, called a *pod prop*, has propellers on both ends and can be swiveled in any direction. A matching unit is just visible on the other side of the ship's keel. The two units can be used to steer and maneuver a large ship in any direction, eliminating the need for a rudder. (Siemens)

Propeller: A rotating blade that produces thrust.

Pitch: The angle of a propeller's blade.

The advantage of nuclear power over diesel or gas is that one load of uranium fuel can last over a year, allowing the submarine to remain submerged without having to refuel.

Water Jets

Water jets are the newest form of marine propulsion. They use an inboard engine to turn an impeller. The impeller draws in water and expels it through a nozzle. The water forced out of the jet propels the craft. Water jets are extremely popular for personal watercraft, such as Jet Ski® watercraft, because they do not contain dangerous propellers. They are used in ferries and hydrofoils because of their higher top speed. Water jets, in the form of thrusters, are also used on large ships to help the ships move in and out of docks.

Figure 20-5. An outboard motor is a widely used propulsion method for recreational watercraft ranging from small fishing boats to speedboats. The power source is an internal combustion engine that turns a shaft directly connected to the propeller. When in use, the entire unit is rotated to steer the boat. The motor can be removed when the boat is not in use. (Bayliner Marine Corporation)

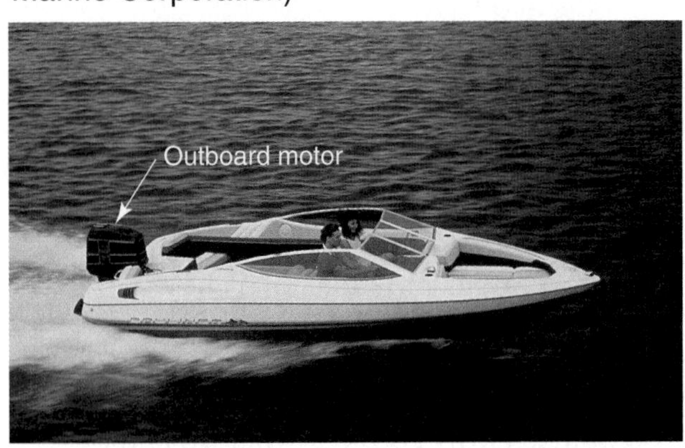

Figure 20-6. This compass is designed for use in automobiles. Instead of a simple magnetized needle that rotates on a pivot, it has a dome that rotates freely in a liquid. As the vehicle changes direction, the north-pointing magnet in the compass causes the dome to rotate. The direction the vehicle is traveling (southwest, in this case) is aligned with a red bar.

Guidance Systems

Boats and ships are able to be propelled in various directions and do not use fixed paths, like automobiles do. Shipping routes and sealanes help to keep ocean travel in set routes. These are not as easy to identify, however, as highways are. Boat and ship navigators must use navigational tools to find both their current location and their destination. The captain must then track the navigation variables: direction, speed, and time.

Direction-Finding Tools

Direction-finding devices have been used in ship navigation for hundreds of years. Several of these technologies are actually still used as references. ***Compasses*** are simple devices for determining which direction is north. The compass was an early navigational device developed in Europe and Asia in the 1100s. This technology was created soon after it became known that lodestone, a type of magnetite that has polarity, would point north when allowed to rotate freely. The first compasses were simply pieces of lodestone floating in water. Today, simple compasses use a free-floating, magnetized needle that spins on an axis. See **Figure 20-6.** The needle will always point north (unless it is placed near a magnetic field). Compasses are used to determine both heading and bearing. The direction the boat is pointed at a given time is the *heading*. The ***bearing*** is the desired direction of travel. See **Figure 20-7.** Many ships and boats also use radio direction-finding equipment. Electronic equipment aboard the vehicle receives transmissions sent by radio transmitters (beacons). The navigator adjusts the antenna until its signal is locked onto the direction of the incoming signals. The heading of the vessel can then be adjusted according to the position of the beacon. The locations of beacons are identified on nautical charts.

Outboard engine: The most common motor used on fishing boats and small motorboats. Its power source and propeller are one piece.

Stern: The back of a boat.

Inboard/outboard engine: An engine used in mid-sized recreational boats. These systems have larger engines than outboard motors mounted inside the boat.

Speed Tools

Speed indicators aboard marine transportation vehicles are called ***logs***. Historically, wooden logs were used as the first speed indicators in sailing. Logs were tied to a rope and dropped overboard. The rope had a series of knots tied in it every 47′ 3″. Once the log was in the water, a sailor would count the number of rope knots that passed the back of the ship in 28 seconds. If the log carried eight knots past the back of the boat, the boat was traveling at 8 knots (nautical miles per hour). A ***nautical mile*** is roughly equal to 1 minute, or 1/60 of a degree, of latitude around the earth. It is equal to about 1.15 statute miles. The measurement of nautical miles was first used in sailing. Today, mechanical devices have taken the place of the logs thrown overboard. The speed logs currently in use are made up of two electronic transducers and a digital monitor. The transducers are placed in the ***hull***, or the body of the boat or ship, one in front and the other in the rear. The rear transducer sends a signal to the front. The speed of the vessel determines the amount of time it takes to receive the signal. This is then displayed on the digital display.

Time Tools

Speed logs often have a setting that will display the current time or even act as a stopwatch. A timepiece of some sort is an essential tool for navigation of a vessel. It can be helpful in determining location. If the ship has been traveling at the same speed and in the same direction, it is easy for the captain to determine how far it is from shore. It is also helpful, especially for a novice sailor, to know what time it is. With a timepiece, he can get back into the harbor before nighttime, so he can use visual navigation aids.

Location Tools

Nautical charts are maps that show coastal waters, rivers, and other marine areas. These charts are specially designed to show information for navigating waterways. They are marked with special symbols that represent things such as depths of water, channel markers, buoys, underwater phone and electric lines, and wrecks. Coastal features that can aid

Figure 20-7. Heading and bearing may be different. Turning the ship until it is pointed in the desired direction (bearing) is known as "changing the heading."

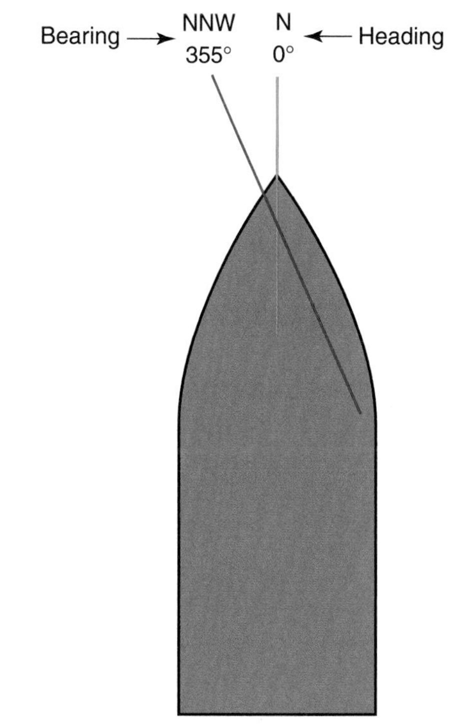

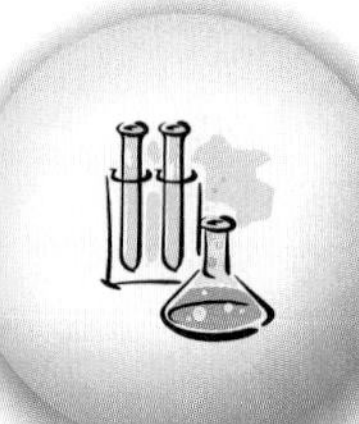

Curricular Connection

Science: Relative Velocity

Imagine you are riding in a car traveling 65 miles per hour (mph). Sitting in the back seat, you toss a ball up in the air and then catch it. The ball is in the air for a total of 1 second.

From your point of view (referred to as *frame of reference*) in the car, the ball moves straight up and down. If your friend was standing along the side of the road watching this through the car window, however, she would see the ball travel over 95′ in the air. This is the distance that you, the ball, and the car travel in 1 second at 65 mph. Thus, you are actually throwing the ball nearly 100′ when you toss it in the car!

In science, this concept is called relative velocity. Relative velocity states that the speed of an object is relative to the frame of reference from which it is viewed. This is an important concept for boats traveling on rivers. In a river, water flows downstream. A boat's propulsion system determines the boat's speed, relative to the water. For example, imagine you are rowing a rowboat in a river. The water in the river is flowing at a rate of 3 mph. You may be able to row a rowboat at a constant speed of 5 mph, relative to the water. If your friend was watching you from the riverbank, your rowboat would be moving 8 mph if you were heading downstream (the boat's 5 mph, plus the river's 3 mph). If you were heading upstream, your boat would be traveling 2 mph, relative to your friend (the boat's 5 mph, minus the river's 3 mph). The concept of relative velocity serves as the foundation for Albert Einstein's theory of relativity and perhaps physics' most well-known equation, $E = mc^2$.

Inboard engine: An engine used on most vessels over 36′ in length. Its power source is mounted inside the ship attached to a propeller shaft.

Nuclear turbine engine: An inboard propulsion system that uses a nuclear reactor to heat water.

Water jet: The newest form of marine propulsion. It uses an inboard engine to turn an impeller.

in navigation also appear on nautical charts. These include lighthouses, church steeples, and water towers. Nautical charts also show latitude and longitude lines, and some show the grid pattern used with the Loran-C navigation system. Ships use charts on the open sea and in waterways to plot their own positions and, possibly, those of other craft in the area. See **Figure 20-8.**

Electronic navigation tools

There are a number of electronic devices used to aid marine navigation. ***Radio detecting and ranging (radar)*** systems contain a transmitter, a receiver, and a display. The transmitter sends radio waves in a circle around the vessel. The receiver picks up the waves that have bounced off of an object and are returning back. The results are sent to the display, and the operator is able to view a 360° view of the objects around the vessel. Radar is especially helpful in bad weather conditions and in avoiding other vessels. ***Loran-C*** is a long-range navigation system. The location is displayed as a series of numbers that can be plotted on a nautical chart with Loran-C lines. Loran-C determines location by figuring the difference in time it takes two base stations to send timed signals to the receiver. The global positioning system (GPS) has quickly become a useful tool in marine navigation. It is extremely accurate and can be used in all weather conditions. This system consists of 24 satellites that circle the earth in a specific configuration. Both GPS and Loran-C receivers allow the navigator to input way points (destinations). Once the way points are input, the navigation receivers provide the bearing the vessel must follow to reach the destination.

Visual navigation aids

Navigational tools are helpful while in route from one place to another. When entering and exiting a harbor, port, or waterway, however, it is easier to use visual aids. A system of different colored buoys, known as the U.S. Aids to Navigation System (US ATONS), has been designed to ensure the meanings of the buoys are consistent. ***Buoys*** are painted markers anchored in a body of water to guide water vehicles. See **Figure 20-9.** Their colors have significant meanings to boat pilots. Red buoys mark the right side of a channel as the boat is coming into port. Green buoys mark the left side. Buoys painted in red and white bands mark safe water. Orange diamonds painted on buoys mark dangerous water and should be avoided. Some buoys have messages marked on them so their meanings are clear. Many buoys have lights so they can be seen easily at night. Navigators can tell what each lighted buoy means by the color of its light and the length of its flashes.

Control Systems

The control systems of water vessels perform two functions. The first is to the steer the boat or ship from side to side. The second function is to raise or lower the vessel.

Steering Systems

Steering vessels from side to side can be done by turning the propulsion units. For example, in water jet–propelled vehicles, the operators change the position of the jet nozzles to change the direction of thrust. The sterns of jet-propelled marine vehicles will move in the opposite direction of the thrust. See **Figure 20-10.** Boats with outboard engines are steered this way. When the engine is rotated, the stern of the boat is pushed in the opposite direction. The sailor is usually able to turn the engine or water jet about 45° in either direction.

Compass: A simple device for determining which direction is north.

Heading: The direction a boat is pointed.

Bearing: The desired direction of travel.

Log: A speed indicator aboard a marine transportation vehicle.

Nautical mile: Roughly 1 minute, or 1/60 of a degree, of latitude around the earth. It is equal to about 1.15 statute miles.

Hull: The body of a boat or ship.

Nautical chart: A map that shows coastal waters, rivers, and other marine areas. It is designed to show information for navigating waterways.

Figure 20-8. A ship's navigators plot courses and positions on a nautical chart. (U.S. Navy)

Figure 20-9. A red buoy such as this one marks the right side of a channel leading into a port. The opposite side of the channel is marked with green buoys.

Figure 20-10. This Jet Ski® watercraft is steered by changing the position of the jet propulsion nozzles.

In all other craft, including large ocean-going vessels, steering is done by the use of a rudder. ***Rudders*** are hinged vertical surfaces on water vehicles. They are usually located near the output of the propulsion source. Ships with more than one propeller typically have an equal number of rudders. The rudders act to change the direction of water pressure against the vessel. Because of the change in pressure, the vessel's heading is changed. Rudders control the stern of most marine vehicles. This is similar to the way rear-steered land vehicles are controlled. Rudders can either be balanced or unbalanced. An unbalanced rudder is the simpler of the two and is usually found on small sailboats. See **Figure 20-11.** In small sailboats, rudders can be turned by hand. In large ships, however, the rudders may weigh thousands of pounds. In mid- to large-sized ships, gears, hydraulics, or pneumatics turn the rudders.

Air-cushioned marine vehicles have the rudders behind the propulsion fans that move air. The rudders or vanes use the airflow to change the vehicle's direction of travel. See **Figure 20-12.**

Radio detecting and ranging (radar): An electronic navigation tool that contains a transmitter, a receiver, and a display.

Loran-C: A long-range navigation system.

Buoy: A painted marker anchored in a body of water to guide water vehicles.

Diving and Rising Systems

Submarines and submersibles operate at many depths under the surface of the water. They have very specialized control systems that allow them to dive and rise. Basically, they can increase or decrease their buoyancy or weight to control their depth.

Submersibles often carry weights that pull them down to the desired depth. They have specialized compartments that fill with water to equalize the pressure as they are descending. Once the submersible is ready to ascend, it releases the weights and allows the water to drain on the way to the surface.

Figure 20-11. Balanced or unbalanced rudders are used on different types of boats. This is an unbalanced rudder on a medium-sized sailboat. It consists of the rudder, which is usually shaped like the letter *D*; a tiller, which is the steering rod the sailor uses to turn the rudder; and the stock, which is a vertical shaft connecting the rudder with the tiller. Balanced rudders have the same configuration, except there is an additional small rudder on the opposite side of the stock. The additional rudder helps push the main rudder into position.

Submarines dive and rise a little differently than submersibles. They are constructed with inner and outer hulls. See **Figure 20-13.** To dive, the submarine allows the space between the two hulls, or the ballast area, to be filled with water. This creates negative buoyancy, and the submarine dives. Once a submarine is underwater, it uses propellers, wings, and rudders to maneuver. In order to surface, the submarine uses compressed air to pump the water out from the ballast area. The submarine again becomes buoyant, rises to the surface, and floats.

Figure 20-12. Large vanes or rudders positioned behind the propulsion fans steer airboats used in swampy areas. (U.S. Navy)

Suspension Systems

Hulls serve as the suspension systems of marine vehicles. They must keep the vessel afloat and stabilize the vessel in various water and weather conditions. Hulls stay afloat by moving, or displacing, the same

Rudder: A hinged vertical surface on a water vehicle. It acts to change the direction of water pressure against the vessel. Rudders control the stern of most marine vehicles.

Figure 20-13. The submarine dive process involves flooding the space between the submarine's hulls (ballast tanks) to create negative buoyancy. To rise to the surface, compressed air is used to "blow the tanks," forcing out the water and creating positive buoyancy.

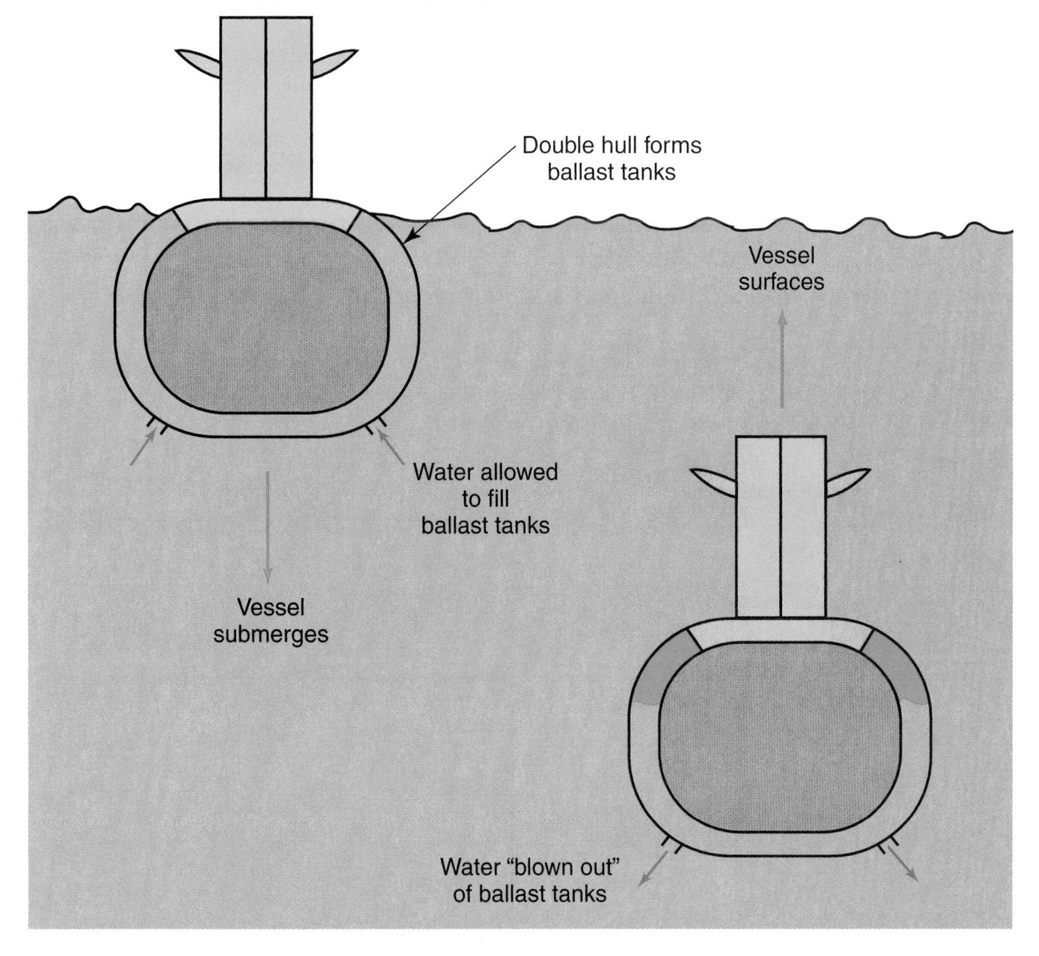

amount of water as the ship weighs. They must be designed to displace the additional weight of the people or cargo when the vessels are loaded. If the additional weight is not factored into the design, the ship will capsize as the weight of the ship becomes greater than the weight of the displaced water.

Full displacement hull: A hull that sits low in the water and has the greatest draft. It is very economical and efficient.

Draft: The distance from the waterline to the bottom of the boat.

Waterline: The location at which the water stops along the side of the hull.

Hull speed: The top speed at which hulls become inefficient and dangerous. It is figured by measuring the length of the boat at the waterline, finding the square root of the length, and multiplying by 1.34.

There are three types of hulls used on all types of watercraft, from speedboats to ocean liners. The intended use of the vessel determines whether the hull is a full displacement, semiplaning, or planing hull. At rest and at very low speeds, all hulls act as ***full displacement hulls***. Displacement hulls sit low in the water. They have the greatest draft. ***Draft*** is the distance from the waterline to the bottom of the boat. The ***waterline*** is the location at which the water stops along the side of the hull. See **Figure 20-14.** The advantage of displacement hulls is that they are very economical and efficient. A cruising vessel, or yacht, with a displacement hull may be able to obtain around 7 nautical miles to the gallon of fuel. Displacement hulls also handle better and require less power than the other two types of hulls. Because full displacement hulls have so much draft, however, they make a great amount of contact with the water, which causes friction. Due to this friction and the way in which the waves are formed under the bow of the ship, these are the slowest hulls. In fact, these hulls have a top speed at which they become inefficient and dangerous, known as ***hull speed***. Hull speed can be figured by measuring the length of the boat at the waterline, finding the square root of the length, and multiplying by 1.34. For example, a boat with a 26′ waterline would be able to travel at just under 7 nautical miles per hour (knots). You would take the square root of 26, which is 5.10, and then multiply that by 1.34 to get the answer of 6.83.

The ***planing hull*** is the exact opposite of the displacement hull. See **Figure 20-15.** When the vessel is being driven, the planing hull actually rides on top of the water. There is no maximum hull speed with planing hulls. A lot of power is required, however, to keep planing hulls on top of the water. Fuel efficiency may be as low as 1 mile per gallon. Planing hulls are also much harder to handle and much rougher in heavy waves.

Figure 20-14. A ship's draft is directly related to the speed and amount of force needed to move it through the water. A greater draft results in greater displacement and more friction.

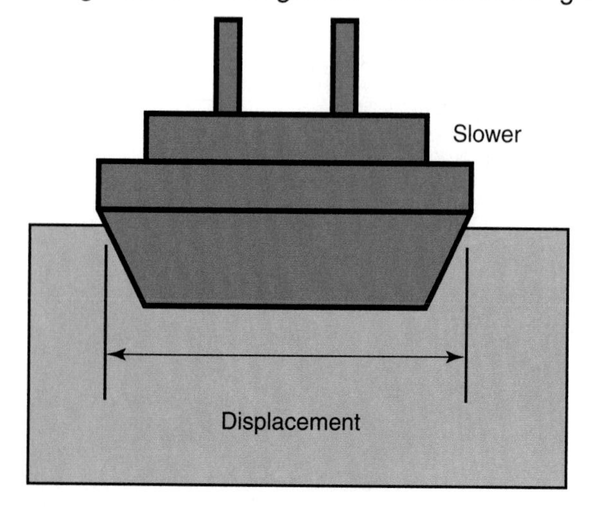

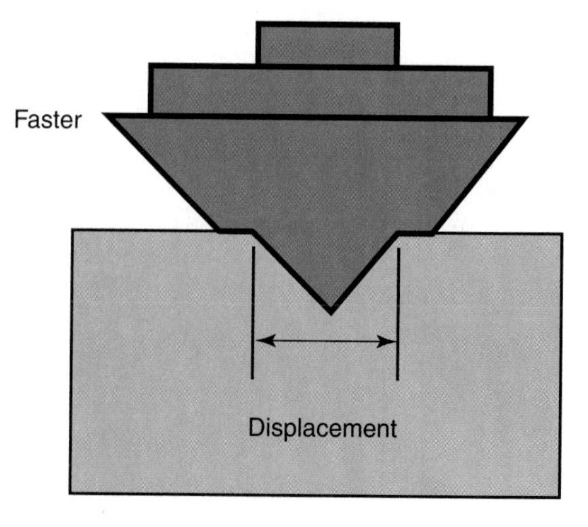

Figure 20-15. A boat with a planing hull lifts almost completely out of the water when it is "on plane." This makes the planing hull faster than the full displacement hull and the semiplaning hull. Vessels with planing hulls can travel two to three times faster than those with displacement hulls.

Planing hull: A hull that rides on top of the water. It has no maximum hull speed, but fuel efficiency is low, and it is hard to handle and rough in heavy waves.

Semiplaning hull: A hull in which the stern of the boat remains in the water, like a displacement hull, and the bow is raised on top of the water, like a planing hull.

Semiplaning hulls may be the best of both worlds. Vessels designed with semiplaning hulls can attain higher speeds than those with full displacement hulls. They can also get better gas mileage than vessels with planing hulls. This is possible because the stern of the boat remains in the water, like a displacement hull, and the bow is raised on top of the water, like a planing hull.

Hulls can take on different shapes. See **Figure 20-16.** The following are the five basic hull shapes:

- **Round hull.** These are displacement hulls and are easily moved through water. They are the least stable of the group, as they tend to roll in the water. For this reason, most round hull boats use a keel or centerboard to add stability.
- **Flat hull.** These are generally planing hulls and provide more stability because of the surface area that comes in contact with the water. Their stability can be compared to a book resting on a table. They do not roll easily.

Figure 20-16. The five basic hull shapes. The shapes allow for various degrees of performance and stability.

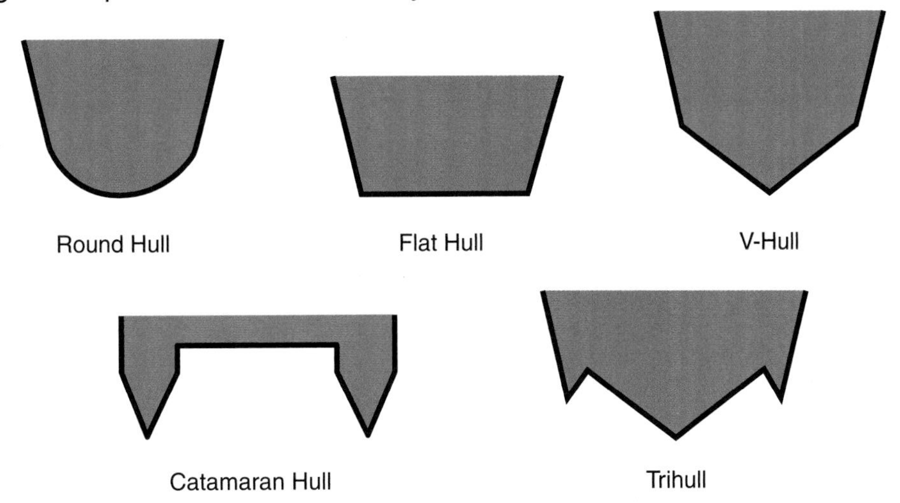

Catamaran: A stable type of boat that has two hulls in the water.

Bilge keel: An extension protruding downward from the centerline of a boat. If the boat starts to lean, it acts as a hydrofoil and pushes against the water in the opposite direction.

- **V hull.** This design is similar to the round hull, except its undersides are flatter. Because of this, it is generally more stable than the round hull design. The pointed bow of boats leads to a partial V shape in many hulls.
- ***Catamaran.*** Boats with this type of hull are some of the most stable. They essentially have two hulls in the water. The hulls are placed far apart so the boat is difficult to roll.
- **Trihull.** Boats with this type of hull are some of the most stable. They have three hulls in the water. The hulls are side by side to increase stability.

Antirolling Devices

Stability is an important aspect of suspension systems. Vehicles must be able to remain upright while in operation. Unstable vehicles compromise the safety of passengers and cargo. Ships are subjected to large waves on the ocean. These waves can cause the ship to roll, leading to an uncomfortable ride. They may even cause the ship to capsize.

Bilge keels are extensions protruding downward from the centerline of a boat. See **Figure 20-17.** If the boat starts to lean, the keel acts as a hydrofoil and pushes against the water in the opposite direction. Antirolling, or passive, stabilizer tanks are U-shaped tanks, partially filled with water, located inside the hull of a ship. As the ship rolls to one side, the water in the tank will flow to the low side, causing the ship to roll back in the opposite direction. Activated ***fin stabilizers*** are fins located on the sides of a ship, below the waterline. Fin stabilizers basically act the same as bilge keels. When the ship rolls to one side, the increased surface area offers resistance to keep the ship upright.

Figure 20-17. Bilge keels are used on sailboats to keep them from tipping over. These keels increase the amount of surface area below the water surface.

Hydrofoils

The suspension systems of hydrofoils are completely different than those of other marine vehicles. Although hydrofoils are currently not in heavy use, their design is an important part of nautical suspension. Hydrofoils use airfoil-shaped devices to hold them up in the water. The main hulls of these vehicles rise out of the water when the vehicle is in operation. See **Figure 20-18.** Hydrofoil action is similar to an airfoil's action. The basic difference in a hydrofoil is that the fluid environment is water, not air. The foil creates a pressure difference between the top and bottom of the wing. When the boat is in motion, the foil deflects the oncoming water. As the water is deflected, the wing is forced upward in the water.

Figure 20-18. Hydrofoils are produced in two designs. The v-foil is very stable, but it is not suitable for rough water conditions. The fully submerged foil provides a smoother ride in rough water.

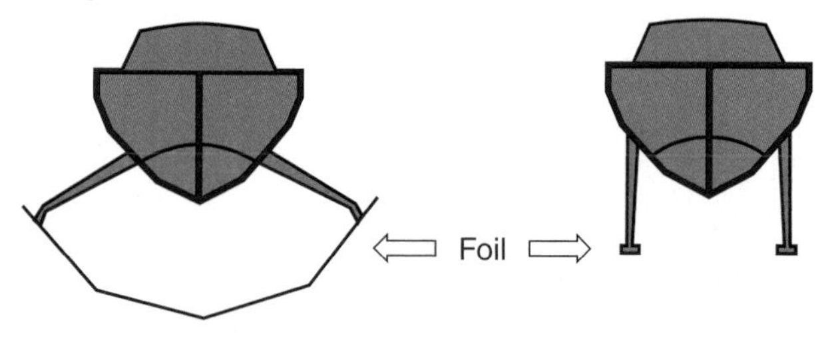

The wings are attached to the vehicle with long struts, so they push the main hull of the vehicle out of the water. The struts are lowered once the boat reaches a given speed. The foils, or wings, then lift the boat out of the water. They are always totally submerged. The foils are controlled, using a computer, to pivot the wings up and down on their struts. This increases or decreases the angle of attack and directly affects lift. The use of sensors and computer technology automates the process of control.

Air Cushions

Air-cushion vehicles, sometimes known as ***hovercraft*** or ground-effect machines, are designed to ride on a cushion of air the vehicle generates. This type of suspension system allows travel over land or water. Flexible skirts at the bottom of the vehicle let the vehicle pass over obstacles without jarring itself.

These vehicles generate the cushion of air with powerful fans. The fans draw air from the top or sides of the vehicle and force it out the bottom. Two types of hovercraft configurations are the plenum chamber and the annular jet. The plenum chamber is the simpler of the two. See **Figure 20-19.** Air is essentially pumped straight through the craft to produce lift. The annular jet configuration directs the air so the air comes from the sides of the vehicle's bottom. The air is directed inward. The cushion of air the annular jet creates is stronger than that which the plenum chamber creates. It is also created with less energy.

Structural Systems

In all vessels, the very front is known as the bow. The rear is known as the stern. The areas in-between are divided into three zones: forward, amidships, and aft. The left of the ship is the ***port*** side. The right is the ***starboard*** side. See **Figure 20-20.** The measurements of the hull are known as the length overall (LOA) and beam. ***Length overall (LOA)*** is measured from the tip of the bow to the stern. Beam is the width of the ship at its widest point. Internal components used to strengthen the hull are known as ***bulkheads***. The top edge of the hull is named the ***gunwale***, or gunnel.

The hull and other structural members of marine vessels are made from a number of materials. Traditionally, wood was used. Wood has natural buoyancy, is easy to work, and is readily available. Typical wooden boat construction requires a set of ribs built up around a keel. The keel is a frame member that runs the length of the boat on its centerline. Attached to this skeleton is wooden planking. The planking is cut and formed so it follows the contours of the skeleton.

Modern boat technology makes wide use of fiberglass-reinforced plastic, aluminum, and other lightweight alloys. Small, fiberglass boats usually do not need to be built around a central structure. Manufacturing processes

Fin stabilizer: A fin located on the side of a ship, below the waterline. When the ship rolls to one side, the increased surface area offers resistance to keep the ship upright.

Hovercraft: An air-cushion vehicle designed to ride on a cushion of air the vehicle generates. This type of suspension system allows travel over land or water.

Port: The left side of a ship.

Starboard: The right side of a ship.

Length overall (LOA): The measurement from the tip of the bow to the stern.

Bulkhead: An internal component used to strengthen the hull.

Figure 20-19. In the plenum chamber design used for air-cushion vehicles, a fan pumps air down through the vehicle to provide an air cushion. The vehicle rides above the water or ground surface on this cushion.

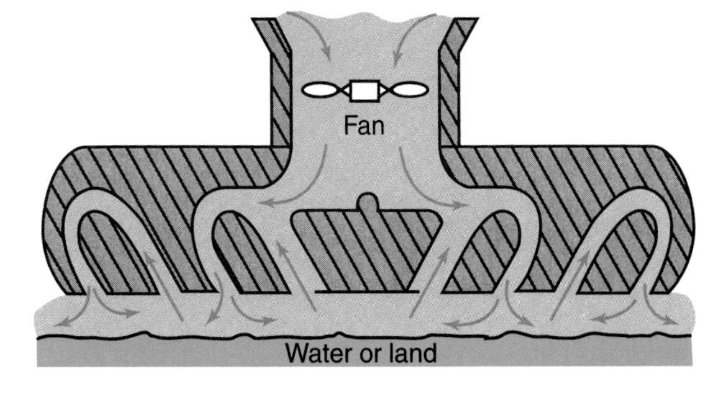

Gunwale: The top edge of the hull. Also called the gunnel.

Classification society: A society that sets and enforces the construction standards for shipbuilding.

Figure 20-20. A boat has its own special set of terms. A—The front of the boat is the bow. The rear is the stern. The left side (when facing the bow) is the port side, and the right side is the starboard side. B—When you are standing on the deck in the middle of the boat, you are amidships. If you walk toward the bow, you are going forward, and if you turn and walk toward the stern, you are heading aft.

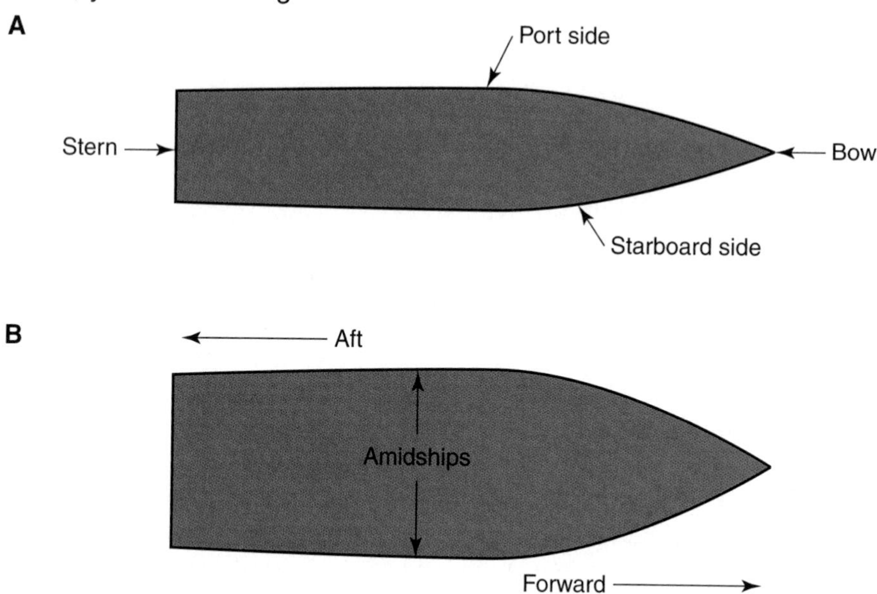

Figure 20-21. In this boat manufacturing plant, fiberglass hulls are shown in various stages of layup. (OMC)

allow a relatively strong shell to be made in a mold. See **Figure 20-21.** Attached to this shell are the propulsion and control systems. Other features that add to the comfort and pleasure are also built onto these structures. See **Figure 20-22.**

Ship Structural Systems

Oceangoing ships are made of metal and metal alloys. The main structural hull and supporting bulkheads are steel. Other structures on the vessels, such as the crew's living areas, passenger areas, and control room areas, may be constructed of lighter materials, such as aluminum alloys.

Shipbuilders must follow a rigid set of rules and regulations for construction of their vessels. Almost every shipbuilding country has its own ***classification society***. These societies set and enforce the construction standards. A prominent classification society is Lloyd's Register of Shipping in the United Kingdom. The American Bureau of Shipping is another large classification society. These organizations send representatives who watch the

Figure 20-22. Fiberglass hulls combine strength with light weight, making them ideal for racing.

construction of ships to ensure that quality materials, equipment, and construction techniques are used. They then determine the classification of each vessel. This rigid system ensures high standards for the shipbuilding industry.

In the design and construction of ships, naval architects must be aware of the major hazards the ships face. When an iceberg or some other underwater obstacle tears open a hull, stranding occurs. The use of a double hull design usually controls damage from this type of problem. A second hull is built inside the main hull. The space between hulls can also be used as a stabilizer (antirolling) tank or as added flotation. Collision

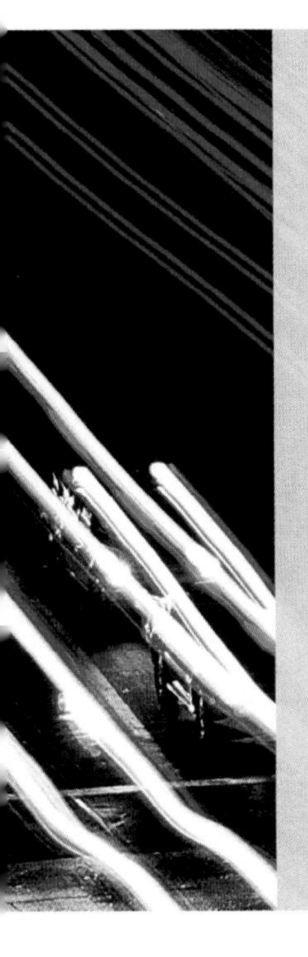

Career Connection

Marine Industry Metal Fabricators

Metal fabricators ensure that engineering details are completed and materials are on hand to fabricate a wide variety of aluminum products for the recreational and commercial marine industry. They fabricate everything from custom rails for boats to standard production equipment found on some of the most popular commercial and recreational boats on the water today. Creating marine hardware can be both challenging and fun.

Like any job, this career has both positive and negative points. One of the most desirable aspects of this job is the personal satisfaction you can feel from seeing your products in use for both work and pleasure. Some disadvantages of this job, however, are the varying demands of the production schedule and the variations in weather conditions while working in an open-air shop.

A high school diploma is essential for a marine industry metal fabricator. Beyond that, vocational training or attending a technical college would be helpful. Experience in any aspect of metal fabrication, such as machining or welding, would be useful. A business degree or business experience would also be beneficial, but neither is required. The entry-level position in this field is an assistant. After gaining some experience, you can become a welder or fabricator, and eventually you may become a production manager.

Technology Link

Manufacturing: Shipbuilding

When you think of the manufacturing processes used to build vehicles, the first thing to come to mind may be the mass production lines used to build automobiles. Not all vehicles, however, are mass-produced. Ships, for example, are not mass-produced in an assembly line. The manufacturing process of shipbuilding is a very customized and time-consuming process. Ships are simply too large and customized to be manufactured using mass production techniques.

The facilities used to manufacture ships are known as shipyards. Shipyards contain several large buildings, including design offices, machining buildings, structural assembly buildings, and a shipway, or dry dock. Once the ship is designed to meet the customer's requirements, the designs are sent to the machining buildings. Most of the cutting and bending of steel required to manufacture the ship is completed using computer-controlled equipment. The fabricated pieces are moved to a structural assembly building, where sections of the ship are assembled. Relative to the size of the entire ship, ships are built in small blocks. The blocks can be manufactured and assembled inside climate-controlled buildings. Once the blocks are completed, they are moved to a dry-dock area. The dry dock is an open-air structure adjacent to the ocean. Water can either be pumped into or out of the dry dock. Final assembly of the ship begins without water in the dock. The ship is assembled from the ground up and welded together. Once the hull of the ship is completed, water is pumped into the dry dock, and the ship begins to float. When the manufacturing is completed, the ship is moved out of the dry dock and tested in open water.

A majority of the ships built for the transportation industry are built in China and Japan. Both the United States and England, however, have several leading global shipbuilding companies. All shipyards across the globe are modernizing, in an effort to manufacture ships more efficiently and to stay ahead of the competition.

accidents may result in part of both layers of the hull being damaged at sea. In such cases, part of the ship becomes flooded. Bulkheads are used to divide the hull into different watertight sections. These are designed in such a way that only the damaged section floods. See **Figure 20-23.**

Figure 20-23. Watertight bulkheads are keeping this ship afloat, even though its bow was punctured in a collision, allowing water to flood some compartments. In such situations, the vessel is usually able to stay afloat and get to a port, where the hull can be repaired. All ships are fitted with collision bulkheads at the front. Areas where propulsion systems are located are also enclosed with bulkheads so this equipment is protected in the event of an accident. (U.S. Coast Guard)

Submarine Structural Systems

As you know, submarines are vehicles designed and built for underwater use. See **Figure 20-24.** During the study of fluid power, you learned that water exerts pressure. This pressure is directly related to the depth of water. The deeper a submarine travels, the more pressure it must be able to withstand. For this reason, submarines use a strengthened steel pressure hull.

Figure 20-24. A submarine hull is designed and built to withstand tremendous pressure from the surrounding water when it submerges. (U.S. Navy)

Usually, submarine hulls are doubled, similar to those of oceangoing ships. The void between hulls is used for fuel storage and as a water ballast. Because many submarines are designed and built for the military, the double hull technique has an added advantage. Antisubmarine weapons tend to damage the outer hull, while the inner hull remains intact.

Projecting from the top of the submarine, usually in the center of the vessel, is the ***conning tower***. This is where the ship's periscopes, radio antennas, and radio detecting and ranging (radar) equipment are located. Snorkel tubes are also part of the sail structure. They let the vehicle take in fresh air without actually surfacing.

Conning tower: Part of a submarine that projects from the top, usually in the center of the vessel. It is where the ship's periscopes, radio antennas, and radio detecting and ranging (radar) equipment are located.

Support Systems

Water transportation is made possible through its support facilities. Support facilities aid in keeping vessels maintained and operational. Vessels need places to be repaired, refueled, loaded, and unloaded. Harbors, docks, ports, locks, and terminals are all support systems. Without them, ships could not operate effectively.

Harbor: A point along the coast where the water is deep enough for the vessel to come very close to shore.

Harbors

A ***harbor*** is a point along the coast where the water is deep enough for the vessel to come very close to shore. See **Figure 20-25.** The main purpose of a harbor is to get the vessels in close to land. A ***harbormaster*** controls the flow of traffic in and out of the port. If the channel heading into a port is particularly dangerous, the harbormaster may be required to board a ship and assist in the navigation of the ship until it reaches safe waters. A small chase boat then picks up the harbormaster and returns to the port.

Figure 20-25. A modern deepwater harbor, which permits large vessels to tie up directly to a dock for loading or unloading. If a ship's draft is too great to tie up at the dock, it must anchor in deeper water. Smaller vessels called *lighters* are used to shuttle cargo between the dock and the anchored ship.

Ports

A port is a place where vessels load and unload cargo or passengers. There are many ports located along seacoasts, lakefronts, and rivers. Ports also have means for fueling and repair.

Harbormaster: An officer who controls the flow of traffic in and out of a port.

Docks

Located at the port are docks. A dock is an area totally closed in by piers. Ships are usually not docked long at a port. They must get loaded or unloaded and continue on their journeys to other ports.

Lock: A chamber-like facility constructed in a canal between two different water levels. It is made up of gates, pumps, and filling and draining valves.

Locks

Locks are used in the inland waterways. A ***lock*** is a chamberlike facility constructed in a canal between two different water levels. It is made up of gates, pumps, and filling and draining valves. See **Figure 20-26.** As the vessel enters the upper level, the lock chamber is already filled to the same level by the filling valve. When the gates close behind the vessel, the drain valve opens to lower the water level. The water is now level with the lower level. The gate will open, and the vessel will be on its way out.

Terminal: A physical facility or building used to load and unload passengers and cargo.

Terminals

Terminals are physical facilities, or buildings. They are needed to load and unload passengers and cargo. In passenger terminals, there are restaurants, waiting areas, and shops of various kinds. In cargo terminals, cargo is stored and later loaded onto another vessel or reshipped by rail, air, or highway transport.

Figure 20-26. Operation of a lock used to transfer vessels between higher and lower levels on a waterway, such as a canal. Locks work similar to the way elevators work. Whereas elevators move people from one floor to another, a lock moves a vessel from one water level to another.

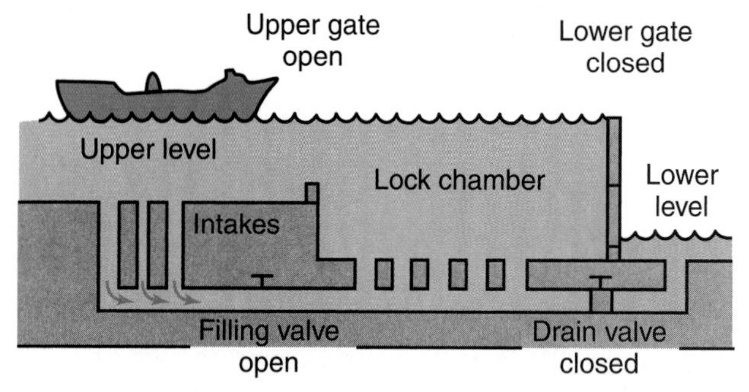

1. The lock chamber is filled to the same level as the upper level. The upper gate opens, and the vessel enters the lock chamber.

Upper gate closed
Lower gate closed
Filling valve closed
Drain valve open

2. The upper gate is closed. The water in the lock chamber is allowed to drain.

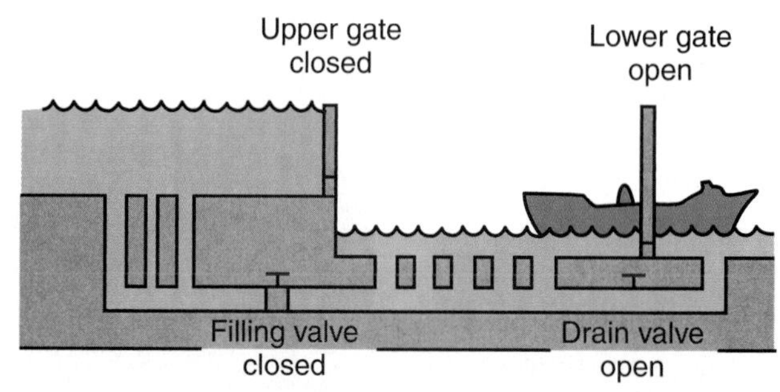

3. When the water drains to the same level as the lower level, the lower gate opens, and the vessel moves out of the lock.

Summary

Water vehicles, like all vehicles, must have systems of propulsion, guidance, control, suspension, structure, and support. Propulsion systems include paddles, sails, propellers, and water jets. Small vessels, especially recreational types, typically use sails, small engines, and water jets. Large ships use diesel power to turn propellers. Vessels use guidance systems to navigate the waterways and oceans. Being able to navigate a ship or boat requires an understanding of nautical charts and buoy markings. The control systems of vessels rely on rudders to steer ships and boats. The hull of a vessel acts as the suspension system and, when properly designed, allows the boat or ship to float. Vessels also include a structural system comprised of bulkheads and other members that keep the vessel strong. Lastly, support facilities, such as harbors, ports, and docks, aid vessels.

Key Words

All the following words have been used in this chapter. Do you know their meanings?

bearing
bilge keel
bow
bulkhead
buoy
catamaran
classification society
compass
conning tower
draft
fin stabilizer
full displacement hull
gunwale
harbor
harbormaster
heading
hovercraft
hull
hull speed
inboard engine
inboard/outboard engine
jib
length overall (LOA)
lock
log
Loran-C
mainsail
nautical chart
nautical mile
nuclear turbine engine
outboard engine
paddle
pitch
planing hull
port
propeller
radio detecting and ranging (radar)
rudder
sail
semiplaning hull
sheet
spinnaker
starboard
stern
terminal
water jet
waterline

Test Your Knowledge

Write your answers on a separate sheet of paper. Do not write in this book.

1. List the four methods of propulsion used in water transportation.
2. The _____ is the small sail connected to the mast on a sailboat.
   A. spinnaker
   B. jib
   C. mainsail
   D. trifold
3. *True or False?* Inboard engines are the most popular type of engine for small fishing boats.
4. What type of propulsion do most military submarines use?
5. Why do submarines use the type of propulsion mentioned in Question 4?
6. Discuss the difference between heading and bearing.
7. Give examples of five elements that would be included on a nautical chart.
8. The _____ uses a constellation of 24 satellites to determine a vessel's location.
   A. Loran-C system
   B. global positioning system (GPS)
   C. radio detecting and ranging (radar) system
   D. nautical charting system
9. *True or False?* Rudders are used to control ships.
10. *True or False?* Outboard engines are able to rotate 360°.
11. Describe how a submarine dives.
12. *True or False?* Full displacement hulls ride on the top of the water.
13. *True or False?* A nautical mile is longer than a statute mile.
14. Determine the hull speed of a 15′ boat.
15. Recall and describe two hull shapes.
16. Sketch the type of hull used in barges. Explain why this type is used for this boat design.
17. A _____ rises out of the water with the use of foils when it reaches top speed.
    A. hydrofoil
    B. submarine
    C. hovercraft
    D. containership

Matching questions: For Questions 18 through 21, match the words on the left with the correct term on the right.

18. Front.
19. Back.
20. Left.
21. Right.

A. Port.
B. Stern.
C. Bow.
D. Starboard.

22. The beam of a ship is its _____.
    A. length
    B. height
    C. depth
    D. width
23. *True or False?* Bulkheads are used to divide the hull into separate sections.
24. Name two examples of classification societies.
25. If a ship is returning to port, what color buoy will be on port side?
26. Write a brief description of how a lock functions.

Activities

1. Using the library or school resource center, research the history of boat propulsion. Select one system and explain how it operates. Prepare a report to the class on your findings.
2. Build a working model of a boat that can be propelled either by the release of stored energy (potential energy) or energy in motion (kinetic energy).
3. Using a container, such as a fish tank, tubing, and a small water pump system, build a working model of a lock system for raising a model boat from one water level to another.

A biometric system being field-tested in a German airport identifies airline passengers by their fingerprints, allowing more rapid movement through airport security systems. The fingerprint data is converted to a two-dimensional code of dots that is printed on the boarding pass. (Siemens)

21

Air Transportation Systems

Basic Concepts

- Define *aircraft*.
- Identify aviation services.

Intermediate Concepts

- Describe how airways are used to keep the air safe.

Advanced Concepts

- Compare lighter-than-air craft to heavier-than-air craft.
- Calculate lift in balloons.

Air transportation has evolved in just over 100 years from flying a few feet to flying around the world. Air transportation systems rely on people and vehicles. Many of the people in these systems are pilots, maintenance personnel, airport employees, and air vehicle designers. The vehicles used are known as aircraft. An ***aircraft*** is any vehicle that transports people or cargo through the air.

The History of Air Transportation

Air transportation is a relatively new form of transportation. Both land and water transportation have been around for thousands of years. Air transportation, however, is only a few hundred years old. Many people throughout history have looked to the sky with a desire to fly. Leonardo da Vinci, for example, created sketches of gliders and helicopters over five hundred years ago. ***Gliders*** are aircraft with stable wings, but no power source. ***Helicopters*** are aircraft with rotating wings. Many attempts at flying were made through the early years of transportation development. The first developers of air transportation were people who were imaginative, courageous, and creative. Many of these inventors looked at nature to

Aircraft: A vehicle that transports through the air.

Glider: An aircraft with stable wings, but no power source.

Helicopter: An aircraft with rotating wings.

gather clues as to how human flight could be possible. The Montgolfier brothers, for example, watched smoke rise through a chimney and knew, if they could trap the smoke, they could rise along with it. As a test, they filled a bag with smoke and hot air. The bag rose, as they predicted. Using this concept, they developed a balloon made out of paper and filled it with smoke and heated air. The balloon rose 6000′ and traveled 7500′ before it deflated and fell to the ground. In 1783, the Montgolfier brothers designed and built the first passenger hot air balloon. The balloon made a 5-mile flight over Paris, France. On its flight, the balloon was manned by two men, and it stayed up for 23 minutes. The development of the hot air balloon gave humans a glimpse of what was possible. This limited success with a form of flight attracted more research and development by creative and courageous people.

Aerodynamics: The study of the motion of air and how it reacts to objects passing through it.

In the early 1800s, George Cayley began to investigate the principles that would make flight possible. His studies became the basis of the new study of aerodynamics. ***Aerodynamics*** is the study of the motion of air and how it reacts to objects passing through it. Cayley's studies led to the building of many models of gliders, including a full-scale glider that could be piloted. This glider became the first human-piloted glider. Several men, including Otto Lilienthal, Samuel Langley, and Octave Chanute, continued the study of gliders and aviation through the 1800s. Lilienthal designed a number of both single- and double-winged gliders. See **Figure 21-1.** Langley, a Smithsonian director, was the first to add a gasoline engine to a glider. His tests of small models were successful. He was never able, however, to successfully fly a full-size, piloted, powered glider. Chanute developed moveable wings to add to the control of the gliders. All these pioneers were very adamant about sharing their work and test results with others. They published books and made their work available to others, including Orville and Wilbur Wright.

Figure 21-1. Otto Lilienthal with one of his double-winged gliders. His work most closely represents the hang gliders of today. One of the Wright brothers took this photo in 1895. (Library of Congress)

About a century after the first balloon flight, the Wright brothers chose to enter the aviation world. See **Figure 21-2.** They were both experienced mechanics. The brothers gathered as much information as they could about the work of the pioneers before them and set out to build the first manned, powered flying machine. They began by building a series of kites, small wings, and gliders to investigate the principles of flight. The Wright brothers did much of their building at their bicycle shop in Dayton, Ohio, and their testing was done at Kitty Hawk, North Carolina. By 1903, they had developed ways to control the movements of an aircraft, or a vehicle that transports through the air. The way these movements are controlled today is very similar to the way they were controlled in the Wrights' gliders and flyers. In 1903, the Wright brothers added an engine that provided power to two propellers at the rear of the wing, known as pushers. They named the plane *Flyer*. *Flyer* was the first successful ***airplane***, or fixed-wing aircraft kept in flight by an engine or other power source. The Wrights' airplane was airborne for 12 seconds and covered 120′. By the end of that day, the Wrights had increased the time and distance of flight to 59 seconds and 859′.

By the 1920s, companies such as Boeing, Douglas, and Lockheed were developing new and innovative airplane designs. The original biplane, an airplane with two wings on top of each other, had been discarded for the monoplane, a single-wing design. Several of these companies and others began transporting mail and then moved into the business of transporting people by airplane. By the 1940s, air travel was becoming a common method of travel. Planes were carrying passengers all over the world. See **Figure 21-3.** Airplanes, both large and small, were using piston-propeller engines. These engines used pistons that turned a crankshaft attached to the propellers. This was the common type of propulsion until the jet engine was invented in the late 1940s. Today, the jet engine has made it possible to break the sound barrier and easily travel around the world.

Figure 21-2. The world's first powered, sustained, and controlled flight took place on December 17, 1903. Orville Wright is at the controls of the *Flyer* in this historic photo that recorded the first flight. He and the *Flyer* were airborne for 12 seconds and covered 120′. By the end of that day, they had increased the time and distance to 859′ in 59 seconds. Due to patent concerns, the airplane was kept out of the public eye until 1908, when the Wright brothers made frequent appearances, showing their invention. (Library of Congress)

Airplane: A fixed-wing aircraft kept in flight by an engine or other power source.

Figure 21-3. Charles Lindbergh flew this Ryan monoplane, *Spirit of St. Louis*, from the United States to France in 1927. It was the first solo (single-person) flight across the Atlantic Ocean. The flight lasted over 33 hours. (National Aeronautics and Space Administration)

Routes

Just as water transportation has no physical highways on the water, there are no physical highways in the air. There are, however, airways. ***Airway*** is a general term for a path or route airplanes follow. The airways are designated, regulated, and controlled by the Federal Aviation Administration (FAA). They are set up in a pattern that looks very similar to a highway system. See **Figure 21-4.** The intersections of the airways are placed above very-high-frequency omnidirectional radio range (VOR) devices. VOR devices are transmitters that supply airplanes with location information.

Airway: A path or route airplanes follow.

Air transportation paths can be divided into two different types: airways and jet routes. Specifically, airways cover the area from ground level to 18,000′ above sea level. Airways serve small aircraft on short routes. ***Jet routes*** are positioned from 18,000′ up to 75,000′. They are reserved for large commercial jets and airliners. Both of these types of lanes are 8 nautical miles wide. For safety, planes must be kept apart from one another. This is done through maintaining distance, or airspace, from one another. The airways and jet routes are divided into 1000′ layers, according to their distance from earth. The layers assist the pilots in maintaining the proper distance from one another. For example, in the jet route layers, three of the layers are at 22,000′, 23,000′, and 24,000′. All even-numbered layers are for planes flying west (from 180°, south, to 359°, just shy of north). All odd-numbered layers are for planes flying east (from 0°, north, to 179°, just shy of south). See **Figure 21-5.** Aircraft must maintain either 1000′ or 2000′, depending on the lane, of vertical distance between each other. They must also leave more than 10 minutes of travel time between one another. The air layers are designed to keep the skies safe from accidents.

Jet route: An air transportation path reserved for large commercial jets and airliners.

Figure 21-4. Airways are "highways in the sky" intersecting above very-high-frequency omnidirectional radio range (VOR) transmitters that provide location information. The circled area is the Rockford, Illinois VOR transmitter.

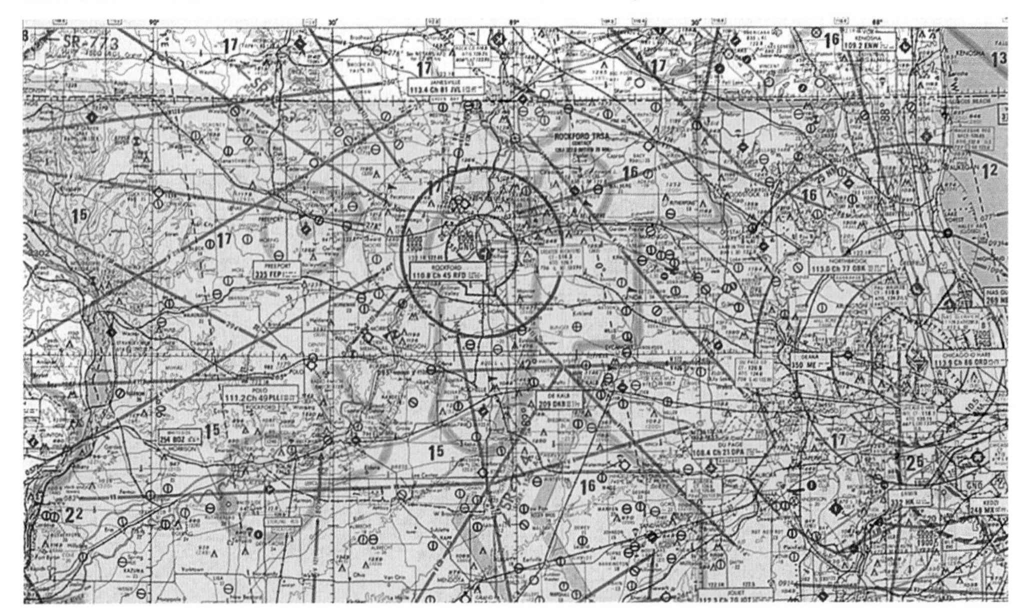

The air is also divided into classes of airspace. The different types of airspace dictate the types of planes that can enter. For example, Class B airspace surrounds major airports. To enter Class B airspace, the pilot must gain permission from the control tower. He must also have a private pilot license. Other airspaces have different requirements. Airplane pilots need to be well versed on the regulations for flying an aircraft. There is other airspace, some of which surrounds military bases, which is restricted to only military personnel. Even the airspace above and around sporting arenas is often restricted during events.

Figure 21-5. Airways divide the sky into layers so air collisions are avoided. Craft are assigned layers according to their direction of flight and type of aircraft. In all even-numbered layers, aircraft fly west. Aircraft fly east in all odd-numbered layers.

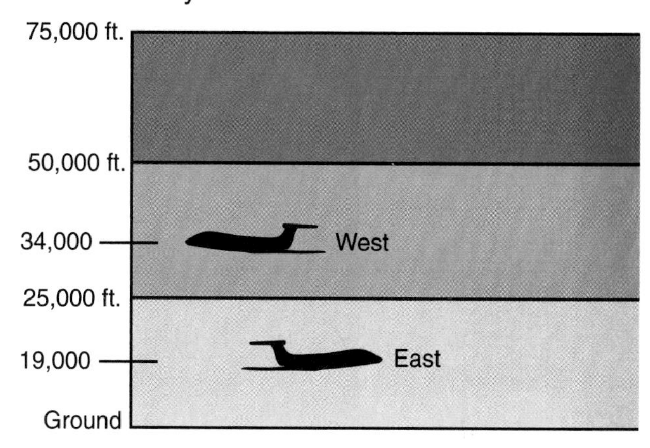

Modes of Air Transportation

An aircraft is a vehicle designed for navigation in the air. It is supported by the air against its surfaces. Lighter-than-air craft are known as balloons and airships. They rise and float. Wind, in the case of a balloon, is the only means of propulsion. Heavier-than-air craft are airplanes, helicopters, and gliders. This type of craft requires power to maintain its speed, thus, creating lift. Lift keeps the craft in the air as long as it maintains sufficient airspeed.

Lighter-Than-Air Craft

Once the subject of much experimentation, lighter-than-air craft are more energy efficient than heavier-than-air craft. Heavier-than-air craft require large quantities of energy to take off and keep them in the air. Lighter-than-air craft are held aloft by their captive gases. The gases used for lighter-than-air vehicles are selected because they are less dense than air. The density, or mass divided by volume, of air at sea level is 1.23 kg/m^3. This measurement varies according to altitude and temperature. The gases used must be less dense than 1.23 kg/m^3. The two gases typically used are helium and hot air. These gases are safe to use, are less dense than air, and can be used in several different lighter-than-air vehicles.

Balloons

The invention of balloons dates back to the 1700s, when people were curious about the upper atmosphere. The early experimental balloons were very similar to and operated on the same principles as the balloons used today. Balloons have little structure. The envelope (balloon portion) is filled with gas or hot air. Balloons used for recreation are filled with hot air. See **Figure 21-6.**

A balloon floats in the air using the same principle a boat uses to float in water. Just as objects in water have buoyancy, objects in the air have lift. Lift is the upward pressure equal to or greater than the air the object displaces. A balloon or any other object that floats in the air must weigh less than the air that has been displaced. In a hot air balloon, the load

Figure 21-6. Propane burners heat air to inflate and provide lift for a balloon. Heating the air lowers its density, making the air inside the envelope lighter than the air around it.

normally carried is the weight of the passenger basket, the burner and propane tanks, and the passengers. Hot air balloons are lowered by allowing the air to cool and become the same density as the air outside the balloon. They rely on the movement of the wind to propel them from place to place.

Balloons have been used for a number of different purposes throughout history. See **Figure 21-7.** They have periodically been called on for military duties all over the world. Balloons have also been used to measure and observe weather conditions.

Airships

In the late 1800s and early 1900s, large lighter-than-air ships were being built. They were designed to carry cargo and passengers around the world. This type of lighter-than-air ship is known as an airship. An airship is also known as a ***dirigible***, a French word meaning "steerable." See **Figure 21-8**. Airships have rudders and elevators, which are used to control the direction and altitude. They also have engines, which are used to move the airships. The balloons of airships are filled with gas, unlike hot air balloons, which are usually filled with air. Early airships were filled with hydrogen. Hydrogen is the least dense gas on earth. This meant the airships could carry a large amount of weight. These

Dirigible: An airship that has rudders and elevators, which are used to control the direction and altitude. It also has an engine.

Figure 21-7. Balloons are mainly used today for recreational purposes.

Curricular Connection

Math: Lift

It is relatively easy to figure how much hot air is needed to lift the load of a balloon. Every 1 ft^3 of hot air, at 100°C (212°F), can lift approximately .015 lbs. of load. So, if a balloon is needed to carry a load of 600 lbs., it needs to hold at least 40,000 ft^3. To figure this, divide the weight of the load (600 lbs.) by the weight 1 ft^3 of hot air can lift (.015 lbs.). The formula is $V = W / \text{lift}$.

V = Volume of the balloon
W = Weight of the load
Lift = amount of weight lifted per cubic foot (hot air = .015 lbs.)

This formula can also be used to determine how much weight a balloon can lift. For this calculation, the formula is $W = V \times \text{lift}$. If you have a balloon envelope that can hold 120,000 ft^3 of hot air, you would use the formula to figure the weight the balloon can lift.

$W = 120{,}000\ ft^3 \times .015$ lbs.
$W = 1800$ lbs.

airships could carry over 10 times the load a balloon of the same size filled with air could. Hydrogen, however, is very combustible and burns very rapidly when it is ignited. The two main types of airships are rigid and nonrigid.

A rigid airship has a metal frame surrounding the balloon and holding it in place. It can be built very large. The greatest of the old dirigible airships was the *Hindenburg*. The *Hindenburg* was docking in New Jersey after a trans-Atlantic flight, when the hydrogen gas caught fire. It burst into flames and was destroyed. The concerns over the outcome of the *Hindenburg* put an end to the commercial use of rigid airships for good, as it seemed, until recently. There are several companies in the early construction processes of new rigid airships. The main uses of the new airships will be to transport cargo to or from places that are hard to reach by train or truck. These airships will also be useful in hauling products and machinery that are too large to haul in other modes of transportation. Helium will be used, rather than hydrogen, to fill the balloons of the airships. It can carry 1 kg of load for each cubic meter of gas. This is four times more load than air can handle. So, if an airship has the same size balloon as a hot air balloon that can carry 600 kg, the airship will be able to carry 2400 kg.

Figure 21-8. An example of an early airship, or dirigible. The French dirigible *Clement Bayard* was photographed around 1900. Note the framework, with the open cabin and steering rudder suspended below the balloon. By the 1930s, large airships with enclosed cabins that could carry 100 people on long trips were being built. (Library of Congress)

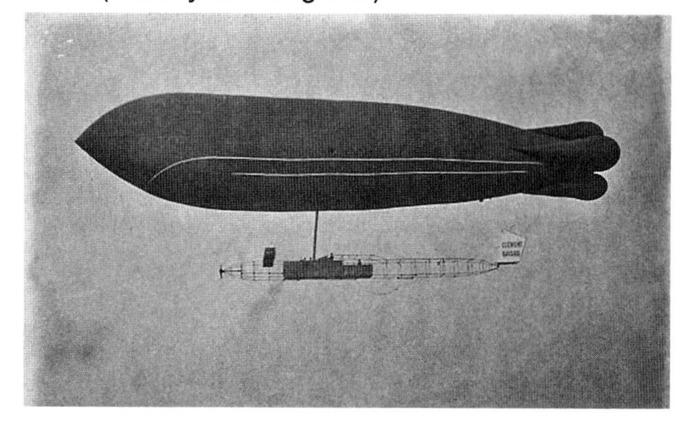

Figure 21-9. The most famous of modern blimps is this one, which advertises a brand of tires. (The Goodyear Tire & Rubber Company)

Nonrigid airships are similar to hot air balloons in that they collapse when they are not filled with a gas. These airships are more commonly called ***blimps***. They use helium instead of hydrogen or air to provide lift. Helium is a heavier gas than hydrogen, but it is much safer because it does not burn. Blimps have been used in the past for military surveillance. Today, however, they are mainly used to provide a platform for observation cameras and aerial views. They are also used for advertising and for some cargo lifting. See **Figure 21-9.**

Heavier-Than-Air Craft

Heavier-than-air craft are far more numerous than lighter-than-air vehicles. Although more energy must be expended to keep them in the air, they are usually much easier to control than lighter-than-air craft. The heavier-than-air craft include gliders, planes, and helicopters.

Blimp: A nonrigid airship that collapses when not filled with a gas. It uses helium to provide lift.

Gliders

George Cayley envisioned and developed a fixed-wing aircraft. The craft was made of light wood and stable wings, but it had no power source. It was known as a glider. Gliders can still be seen in the air today. They are used for both recreation and training. Because they are heavier-than-air craft, they have to overcome their weight before they can fly. Airplanes and helicopters use their engines to generate thrust. ***Thrust*** is the force that moves the aircraft through the air. Gliders, however, have no engines. In order for a glider to fly, it must be pulled behind another aircraft until it generates enough speed to maintain lift. See **Figure 21-10.**

Thrust: The force produced by the propulsion system that moves an aircraft through the air.

Figure 21-10. Gliders resemble small airplanes. These aircraft are built from strong, yet lightweight, materials. A glider has no engine, but it stays aloft by riding rising currents of warm air. To begin flying, the glider must be towed into the air by a powered aircraft. (DG Flugzeugbau, GmbH)

Airplanes

Since 1900, the airplane has progressed very rapidly in design and construction. See **Figure 21-11.** Planes have advanced in design from the Wright brothers' two-wing plane (biplane), which traveled roughly 7 miles per hour (mph); to airplanes that travel at Mach 1, the speed of sound; to supersonic transports (SSTs), which can travel at speeds of 1550 mph (2494 km/h). There are many supersonic aircraft in the military, but only one type was used to transport passengers. See **Figure 21-12.**

Airplanes, like all types of air vehicles, have four forces that act on the aircraft at all times. See **Figure 21-13.** Lift is the upward force that an airplane's wings produce to keep it in the air. It acts directly against ***gravity***, a natural force that tries to pull the plane to the ground. Thrust is the force produced by the plane's propulsion system. It is opposed by ***drag***, which is the force resisting forward motion of the aircraft.

There are many types of airplanes used for business, personal, sporting, agricultural, and commercial activities. Some small private planes have only two seats and one propeller. Other airplanes are used for long-distance commercial flights. These planes, known as

Figure 21-11. Aircraft design and construction have advanced rapidly in approximately a century. National Air Transport (NAT), a predecessor of United Airlines, operated the airmail plane at the top in the early 1900s. Airlines flying long routes operate large aircraft, such as the Airbus A380, which can carry hundreds of passengers. (United Airlines, ©Airbus 2005 photo by C. Brinkmann)

Figure 21-12. A supersonic transport (SST) travels over twice the speed of sound, which is known as Mach 2. The Concorde was the only supersonic airliner in commercial service. It was used to carry passengers from New York to Europe in 3 1/2 hours. British Airways and Air France built and operated a total of 20 Concorde aircraft. The planes flew for nearly 30 years and have all been retired from service.

Gravity: A natural force that tries to pull a plane to the ground.

Drag: The force resisting forward motion of an aircraft.

Curricular Connection

Social Studies: The History of the Concorde

The Concorde was the most significant commercial supersonic transport (SST) for decades. Its achievements in speed and design have been a prominent part of aviation development. The British and the French developed the Concorde. This collaboration did not build the only commercial SST. The United States considered building a commercial aircraft that could pass the sound barrier, but it was decided the effort would not be profitable. Meanwhile, Russia created the aircraft known as the Tu-144. The Tu-144 reached Mach 2 in 1969. It stopped carrying passengers after 1973, however, when a crash resulted in several deaths. The Tu-144 served as a transport for airmail for some time after the crash. Today, a Tu-144 is being used to further supersonic research.

Since the early 1970s, the Concorde has been the only commercial transport of its kind. The first prototype was built as early as 1962. The Concorde did not break the sound barrier until 1970. While the fleet enjoyed several years of successful air travel, it did have complications at times. Both British and French Concorde aircraft were temporarily grounded in 2000, when one of them crashed shortly after takeoff from Paris. The cause was not related to the aircraft itself, and service was resumed. In October of 2003, the entire fleet was grounded again, but research and developments continue.

airliners, have seating for hundreds of people. Some planes, known as seaplanes or amphibians, are designed to land on water. Other aircraft are designed to take off and land on short runways. These planes are called short takeoff and vertical landing (STO/VL) planes. See **Figure 21-14.** Other planes are jumbo jets used to transport cargo and passengers. Aircraft are also used for aerial views, photography, crop dusting, and advertising businesses by flying banners.

Figure 21-13. The four forces that act on an airplane in flight. When the forces of lift and thrust are greater than gravity and drag, the plane will fly.

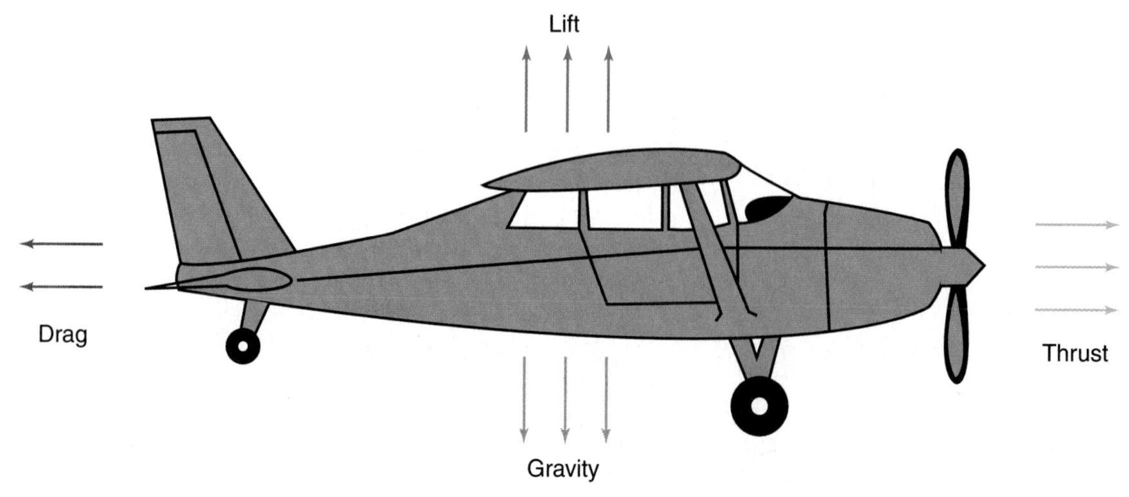

Figure 21-14. Short takeoff and vertical landing (STO/VL) aircraft. A—A Harrier jump jet uses swiveling jet nozzles to take off, land, and fly forward. B—The Osprey aircraft relies on swiveling turboprop engines, rather than jets, for takeoff, landing, and flight. (U.S. Navy)

A

B

Helicopters

Helicopters are used to transport people and cargo to places that are hard to reach by other transportation vehicles. See **Figure 21-15.** A unique quality of the helicopter is that it can take off and land in vertical flight. See **Figure 21-16.** Helicopters can hover in the air. They can also change direction of flight very quickly. Helicopters fly a little differently than airplanes do. A helicopter has rotating wings to make it fly, whereas a plane has stationary wings. The rotating wings, also known as ***rotor blades***, provide lift in the same way a wing does on a plane. Most helicopters have two types of rotors: a main rotor and an auxiliary rotor, or tail rotor. The main rotor is mounted above the cockpit and generates the lift needed to fly. The auxiliary rotor is located on the tail and keeps the helicopter from spinning along with the main rotor.

Rotor blade: A rotating wing on a helicopter that provides lift.

Figure 21-15. Helicopters play roles that are difficult or impossible for conventional aircraft. A—Medical evacuation from remote locations. (U.S. Coast Guard) B—Suppressing brush and forest fires in difficult-to-reach areas. (Sikorsky Helicopter)

A

B

Figure 21-16. The helicopter's ability to take off and land from a tiny pad is vital to transporting crews to and from ocean oil platforms. (Sikorsky Helicopter)

Technology Link

Medicine: Emergency Vehicles

The use of transportation technology allows for much faster movement of people and goods than is possible without vehicles. In some cases, the speed of transportation technology may mean life or death. Most modes of transportation have at least one type of vehicle used in cases of medical emergencies. In land transportation, these vehicles are ambulances. Ambulances are basically small emergency rooms on wheels. They are staffed by emergency medical technicians (EMTs) and are equipped with emergency medical supplies. In many cases, the EMTs are able to provide medical care while in route to a hospital.

In air transportation, there are several types of emergency vehicles. Helicopters are often used to transport patients in critical condition to large or specialty hospitals. Private companies or nonprofit organizations can provide these helicopter services. Flight nurses and, often, emergency doctors staff the helicopters. Emergency airplanes are often used to transport patients awaiting major surgical procedures, such as transplants. Angel Flight America is a nonprofit organization that has a corps of volunteer private pilots who fly these types of assignments across the United States. Medical airplanes are also used to transport injured travelers back to the United States or to their local hospitals. The use of transportation for medical purposes and the equipping of vehicles with medical technology save lives daily and provide for an efficient response to medical emergencies.

Recreational vehicles

Some types of air transportation vehicles are used only for recreation or sport. Hang gliding is done mainly for recreation. Paraplaning, parachuting, and ballooning are done for either sport or recreation. See **Figure 21-17.**

Figure 21-17. Hang gliding is a type of air transport done for recreation.

Aviation Services

Aviation describes all air transportation activities. There are three categories of aviation. They are the following:

- General aviation.
- Commercial aviation.
- Military aviation.

General Aviation

General aviation consists of privately owned planes used for a wide variety of tasks. It usually includes the use of smaller aircraft, as opposed to the very large aircraft

Career Connection

Airplane Pilots

Without pilots, airplanes and the airline industry would not perform any transportation functions. Pilots are the people who operate and fly aircraft. There are three main types of pilots: private, commercial, and airline. Private pilots are pilots who have a license allowing them to fly small aircraft for personal reasons. These pilots do not fly planes as an occupation. Conversely, commercial and airline pilots do make their livings from piloting aircraft. Commercial pilots are found in many different areas and perform many different functions, including crop dusting, flying sightseeing tours, fighting forest fires, and flying helicopters for news and police organizations. Airline pilots operate airplanes for regional, national, and international airlines.

Pilots are responsible for many activities, besides just flying the aircraft. They plan their flights, check the aircraft and instruments, review weather charts, and complete paperwork. Actually, many pilots spend nearly half of their working time completing nonflying duties. The Federal Aviation Administration (FAA) monitors the hours pilots work to ensure the pilots are not overworked and can remain alert.

The FAA also regulates the licensing of pilots. Commercial pilots must obtain a commercial pilot's license with an instrument rating. This ensures the pilots can fly in all types of weather and visibility. Airline pilots must have a airline transport pilot's license, which requires 1500 hours of flying experience (1250 more hours than the commercial pilot's license). These licenses can be obtained from flight training schools and some colleges and universities. Once hired, many airlines and commercial operations require the pilots to complete additional training. The average yearly salary is near $48,000 for a commercial pilot and $110,000 for an airline pilot.

General aviation: Privately owned aircraft used for recreational, business, and community-oriented tasks.

commercial aviation uses. General aviation is used to transport fewer people over short distances. Some services that general aviation offers are to farmers, the community, businesspeople, and individuals. See **Figure 21-18.** To farmers, general aviation performs such tasks as planting, spraying, and fertilizing crops. To individuals, general aviation provides a form of recreation and personal transportation. For businesses, general aviation offers fast and efficient transportation and communication. To the community, general aviation offers mail services, fire fighting, aerial mapping, and photography.

The aircraft used in general aviation have a wide range of sizes. Some are small single-engine craft, and some are small jet engine luxury craft. The most common craft flown in general aviation is the single-engine aircraft. Many individuals own their own planes. These airplanes can take off and land on small runways at small airports.

Commercial Aviation

Commercial aviation: Scheduled airline flights that provide passenger and cargo transportation, using aircraft that carry from dozens to hundreds of passengers.

Commuter airline service: The transport of people from several small airports to a major airport in a major city.

Regional airline service: The transport from small airports to major airports within a specific region.

Domestic airline service: The transport by way of air to and from major airports within a country.

All scheduled airline flights are examples of ***commercial aviation***. See **Figure 21-19.** ***Commuter airline service*** transports people from several small airports to a major airport in a major city. ***Regional airline service*** involves the transport from small airports to major airports within a specific region. ***Domestic airline service*** is the transport by way of air to and from major airports within a country. ***International airline service*** is a service that provides travel between countries.

Military Aviation

Military aviation consists of air activity performed by the armed forces. See **Figure 21-20.** The aircraft used in military aviation are designed to function in specific roles. These roles fall into six different

Figure 21-18. The term *general aviation* refers to privately owned aircraft used for recreational, business, and community-oriented tasks. Small, single-engine aircraft, such as this one, are typical of general aviation. (Cessna Aircraft Company)

Figure 21-19. Commercial aviation provides passenger and cargo transportation, using aircraft that carry from dozens to hundreds of passengers. This artist's conception shows the Airbus A380, which will carry more than 500 passengers on long-distance routes. The first production model of the A380 was completed in January 2005. (Airbus)

Figure 21-20. Military aircraft are used in many roles, from offensive operations (fighters and bombers) to in-air refueling, cargo transport, and surveillance. This Airborne Warning and Control System (AWACS) plane is used for surveillance work. It is a modified Boeing 707 aircraft. (The Boeing Company)

categories: surveillance, cargo, tanker, bomber, fighter, and attack. When looking at some aircraft, it is easy to see the function they serve. For example, surveillance planes often have large disks on the top that house electronic equipment used to monitor events on the ground. Cargo planes are usually large planes used to haul equipment, supplies, and troops. The tanker planes are used to refuel other aircraft in flight. See **Figure 21-21.** Bombers are typically large and relatively slow airplanes that carry cruise

International airline service: A service that provides travel between countries.

Military aviation: Air activity performed by the armed forces.

Figure 21-21. Large flying tanker aircraft can refuel fighter aircraft in flight. This view from the tanker shows the fuel boom connected to the aircraft being refueled. The close proximity of the two aircraft leaves little margin for error. (U.S. Navy)

missiles and guide bombs. Fighter and attack aircraft are often similar looking. Fighters are used for air combat with other aircraft. Attack aircraft are used for the ground support of troops. The newest advance in military aviation is the design of stealth aircraft. These aircraft perform one of the functions above, but they are designed in a manner that makes it difficult for the opposition to monitor and track them.

Tech Extension

Stealth Technology

The idea of stealth technology is to avoid the enemy's detection of an aircraft. This can be done in several ways. The first method of designing a stealth aircraft is to create a plane that radio detecting and ranging (radar) cannot detect. The second method is to create a plane that cannot be detected using infrared, heat sensing, devices.

Adding features to aircraft that would decrease the radar detection has been done for over fifty years. It was not until the 1970s, however, that the details of how radar worked were completely understood. Once radar was understood, several planes have been completely designed to minimize the reflection of radar. Aircraft such as the F-117 and the B-2 are very large aircraft, but they appear to be only the size of a bird on a radar screen. See **Figure 21-A.** These aircraft are extremely flat and contain very few curved edges. The windows are covered with a nonreflecting coating. The skin of the aircraft actually absorbs the radar that hits the airplane. It has a coating that changes the radar waves to heat waves and then absorbs the heat.

Heat, as mentioned above, is the second way an aircraft can be detected. The greatest source of heat in a jet aircraft is the air forced out the back of the engine. In stealth aircraft, this air is funneled through the plane and mixed with cool air before it escapes the aircraft. Stealth technology is highly classified and continues to evolve daily. New materials and methods are continually being developed to better evade detection.

Figure 21-A. Stealth technology used for two different types of aircraft. A—An F-117 surveillance plane. B—The B-2 bomber. (U.S. Air Force)

A

B

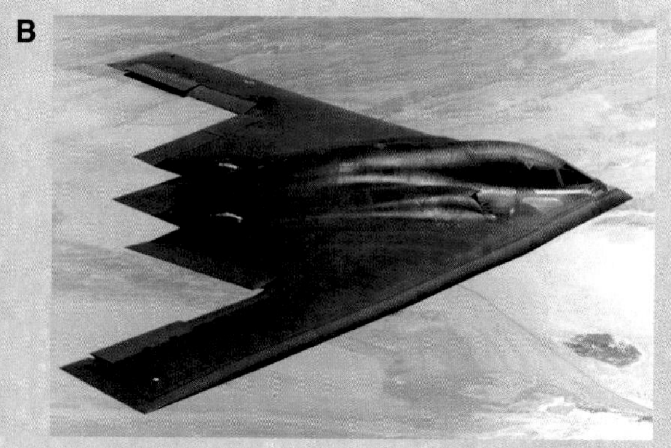

Summary

Humans have made many attempts to fly throughout the centuries. The development of air transportation has progressed rapidly since the early 1900s. The Wright brothers have been recognized throughout history for their accomplishments in airplane design. They were the first to achieve powered flight. The pace of development for air transportation has been accelerating since 1906. Inventors all over the world contributed their ideas to new designs. By 1914, flying had moved beyond being a novelty, as the military adopted the airplane.

Air transportation users have devised a set of airways. The airways are the routes airplanes travel. They are the highways of the skies. Airways are set up in layers. These air layers are designed to keep the skies safe from accidents.

Transportation through the air occurs in two different modes. There are lighter-than-air and heavier-than-air vehicles used to transport people and cargo from one place to another. Lighter-than-air vehicles include balloons, dirigibles, and blimps. Heavier-than-air vehicles include gliders, airplanes, and helicopters.

The activities that take place in air transportation are known as aviation. Aviation occurs in three categories. Privately owned aircraft are known as general aviation. Commercial aviation consists of scheduled airline businesses that make a profit on their services. Aircraft flown and used by the armed forces are part of military aviation.

Key Words

All the following words have been used in this chapter. Do you know their meanings?

aerodynamics
aircraft
airplane
airway
blimp
commercial aviation
commuter airline service
dirigible
domestic airline service
drag
general aviation
glider
gravity
helicopter
international airline service
jet route
military aviation
regional airline service
rotor blade
thrust

Test Your Knowledge

Write your answers on a separate sheet of paper. Do not write in this book.

1. *True or False?* The Montgolfier brothers flew the first hot air balloon.
2. _____ work became the basis for aerodynamics.
3. State the definition of *aircraft*.

4. _____ was the first to add a gasoline engine to a glider, but he was never able to make a successful flight.
   A. George Cayley.
   B. Otto Lilienthal.
   C. Samuel Langley.
   D. Wilbur Wright.
5. The Wright brothers' first airplane was named _____.
6. *True or False?* The propeller made it possible to break the sound barrier.
7. An airway above 18,000′ is known as a(n) _____.
8. The main reason for airways (routes) is for _____.
9. Explain how airways are used.
10. Name three heavier-than-air vehicles.
11. In order for a lighter-than-air craft to rise, the gas it contains must be less dense than _____.
12. *True or False?* A hot air balloon is a heavier-than-air vehicle.
13. If you weighed 175 lbs., how much hot air would be needed to lift you?
14. The word *dirigible* means _____ in French.
15. *True or False?* Blimps today use hydrogen as the safe alternative to helium.
16. Heavier-than-air vehicles are _____ to control than lighter-than-air vehicles.
17. Thrust is the force that moves the aircraft _____.
   A. forward
   B. backward
   C. up
   D. down
18. List four uses of an airplane.
19. _____ are used to transport people and cargo to hard to reach places.
20. Discuss the aircraft used in general aviation.
21. Fighter and attack aircraft are part of _____ aviation.

Activities

1. Construct a model of a heavier-than-air aircraft.
2. Create a display showing the history and evolution of an aircraft.
3. Write a report about the functions and authority of the Federal Aviation Administration (FAA).

22

Air Vehicular Systems

Basic Concepts

- Cite the uses of each type of jet engine.
- List common types of aircraft instruments.
- Identify the three types of stability of an aircraft.
- State how an airplane flies.
- Name the parts of an airfoil.
- Define the structural parts of an airplane.
- Cite the support systems used in air transportation.

Intermediate Concepts

- Figure gearbox ratios for different applications.
- Give examples of how moving surfaces control stability.
- Describe the control of a helicopter.

Advanced Concepts

- Plan a flight using an aeronautical chart.
- Discuss the various explanations of the creation of lift.
- Calculate the lift coefficient for various aircraft.

Aircraft require the use of several systems in order to transport people and goods. Propulsion systems provide thrust, which moves the aircraft forward. Many instruments and gauges that provide the pilot with navigation information are onboard the aircraft and are part of the guidance system. Aircraft control systems allow the pilot to steer and land the craft. Suspension systems, such as the wings, provide lift and keep the aircraft in flight. The structural system includes the trusses that keep the aircraft intact and flightworthy. These systems are all contained onboard the aircraft, while the support system is on the ground and includes airports, runways, and flight control towers.

Radial engine: An engine in which all the pistons are connected to a hub in the center of the engine.

Airfoil: The shape of a propeller blade or an airplane's wing.

Angle of attack: The angle at which a propeller blade hits the air or the upward tilt of a wing's leading edge.

Propulsion Systems

Aircraft propulsion systems are used to generate thrust. They create thrust by accelerating the oncoming air. Engines are the propulsion systems used to generate thrust for most aircraft. The only other types of propulsion systems used are the wind, in the case of hot air balloons, and human power, in the case of hang gliders and pedal-powered gliders. See **Figure 22-1.** Aircraft engines are divided into two categories: reciprocating and jet.

Reciprocating Engines

Reciprocating engines were used on the Wrights' *Flyer* and are still used today on general aviation craft. They are named for the motion the piston follows while in operation. These engines are also known as internal combustion engines because they rely on the combustion of gases inside a cylinder to create motion.

Several internal combustion engine shapes have been used in aircraft throughout aircraft development. Early aircraft used radial internal combustion engines. ***Radial engines*** are configured in a way in which all the pistons are connected to a hub in the center of the engine. See **Figure 22-2.** The pistons fan out from the center to form a circle.

Figure 22-1. The wind is the propulsion system for a hot air balloon. Air heated by a propane burner provides the lift to get the balloon into the air, but horizontal movement depends completely on the wind.

Today, the internal combustion engines used on aircraft are either in a horizontally opposed (flat) or *V* shape. In flat engines, all the cylinders share a crankshaft and are set in opposite directions from each other. In *V*-shaped engines, the cylinders form a *V* and also share the same crankshaft. Typical aircraft engines have four or six cylinders.

The reciprocating engine is used to drive propellers. The propellers provide the aircraft with thrust in the reciprocating engine systems. See **Figure 22-3.** A propeller is made up of long blades rotated around a center hub. Its blades have a shape that is very common in both air and marine transportation—the ***airfoil***. A propeller can be compared to a set of rotating airplane wings. It provides thrust by accelerating the velocity and changing the pressure of the air passing through the blades.

The amount of thrust a propeller creates is determined in part by its pitch and rotation speed. The pitch determines the angle of attack. The ***angle of attack*** is the angle at which the propeller blade hits the air. Propeller blades are twisted, so the angle of attack decreases as the distance from the center increases. Refer to Figure 22-3, noting the shape of the propeller blade. This idea was first developed by the Wright brothers and is used to make the propeller more efficient. The speed of the connected engine's crankshaft determines the rotation speed of the propeller. The crankshaft of the engine often rotates too fast, however, for the propeller to be efficient. In this case, a gearbox is used to reduce the revolutions per minute (rpm).

Propellers are very efficient propulsion systems. They are able to provide good fuel economy. The efficiency is, however, a trade-off for speed. Propeller propulsion systems have a top speed well under the speed of sound, or as referred to in aviation, ***Mach 1*** (760 miles per hour).

Jet Engines

Jet engines are able to travel at much higher rates of speed than reciprocating engines. They operate on the principle defined in ***Isaac Newton's third law of motion***. This law states that "for every action, there is an equal and opposite reaction." A simple example of the third law of motion can be conducted with a balloon. If you blow up a balloon and release it without tying the opening shut, all the air rushes out. The air that rushes out is sent in one direction, and the balloon flies in the opposite direction. The air rushing out is the action, and the balloon's movement is the reaction. See **Figure 22-4.** This release of air (and the opposite movement) is very similar to the action of a jet engine. Jet engines force hot gases from the rear of the engine. The reaction to the hot gases moves the engine forward, producing thrust.

Ramjet engines

Ramjet engines are the simplest type of all jet engines. They have no moving parts. See **Figure 22-5.** Ramjets can only operate at high speeds because they require moving air to enter the engine. They cannot be used for takeoffs and will not work at low speeds. Aircraft with ramjets are often experimental craft launched from other aircraft. Once the craft are launched, the ramjets are used to produce a great amount of thrust. This type of engine is often used on missiles and other weapons that can be fired from moving aircraft.

Figure 22-2. Radial engines consist of a series of cylinders around a central hub. The pistons are connected to the hub, causing it to rotate and spin the propeller. Because the cylinders are all at the front of the engine, the air rushing by the airplane cools the engine. Radial engines are able to produce a large amount of horsepower (hp) at relatively few revolutions per minute (rpm).

Figure 22-3. Airplane propellers are airfoils, providing thrust by accelerating and changing the pressure of the air passing over the blades. This World War II fighter aircraft is on exhibit at the Smithsonian Institution's Udvar-Hazy Air and Space Museum, just outside Washington, D.C.

Curricular Connection

Math: Gearbox Ratios

Gearboxes use ratios to express the amount of reduction they produce. A ratio is the relation between two numbers. In a gearbox, the first number of the ratio is the input value. The second number is the output value. A gearbox that slows down the revolutions has a larger number first. For example, a crankshaft may rotate at 2400 revolutions per minute (rpm), and the connected propeller may be most efficient at 1200 rpm. The input value is 2400 rpm, and the output value is 1200 rpm. This ratio would be written as 2400:1200. Expressed in the lowest value possible, this example reduces to 2:1.

If the crankshaft (input) rotates at 3825 rpm, and the desired propeller rpm is 2550 rpm, what is the ratio of the needed gearbox? The initial ratio would be 3825:2550. Divide 3825 by 2550, which equals 1.5. The ratio would be 1.5:1. Ratios should not, however, be left with decimal points. Multiply both sides by 2. The needed gearbox is 3:2.

Mach 1: The speed of sound (760 miles per hour).

Isaac Newton's third law of motion: For every action, there is an equal and opposite reaction.

Gas turbine engines

The four types of gas turbine engines are turbojet, turbofan, turboprop, and afterburning turbojet engines. See **Figure 22-6.** Each type of gas turbine engine has specific functions and designs.

Turbojet engines are used on commercial aviation vehicles. The compressor is used to increase the pressure of the air entering the engine. It has several rows of blades that spin around a central shaft. The air is compressed by the blades, known as rotors and stators, and sent to the

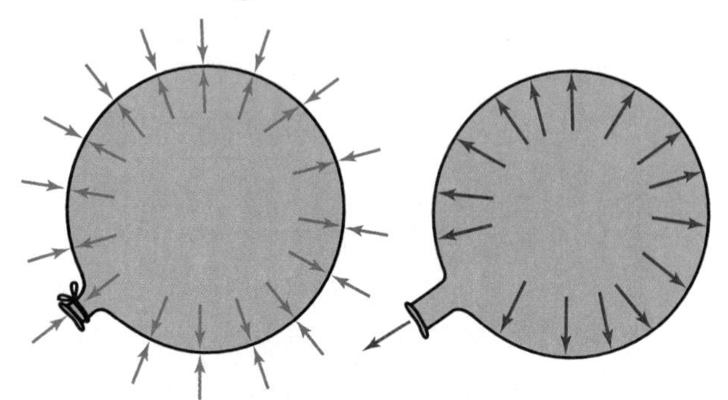

Figure 22-4. If an inflated balloon is tied shut, nothing happens. If it is left untied, the air escapes when the balloon is released. The balloon reacts by moving in the direction opposite the opening.

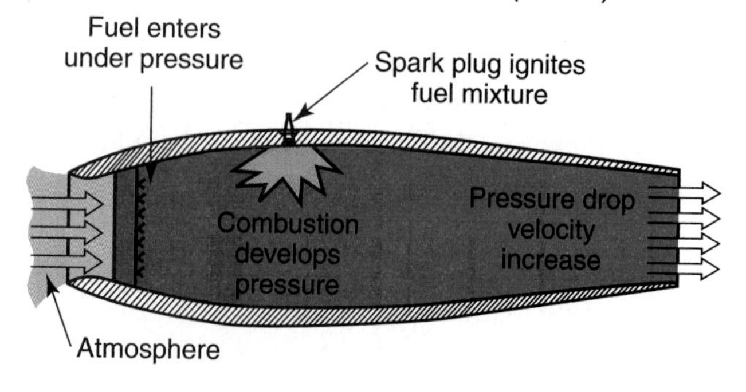

Figure 22-5. A ramjet engine can operate only when moving at high speed, since it has no moving parts and no device for drawing in air. Air enters the front of the engine as the vehicle is moving at a high rate of speed. The internal shape of the engine causes the air to be compressed. When this happens, fuel is sprayed into the combustion area so it mixes with the compressed air. This mixture is then ignited. It expands rapidly as it burns, and it is forced out the back of the engine as thrust. (Estes)

combustion chamber. It is packed into a combustion chamber in the middle of the engine. There, fuel is injected so it can mix with the compressed air. After mixing, the fuel and air are burned and forced toward the back of the engine. Once ignition has started, it becomes a continuous process inside the engine. Before the hot gases leave the rear as thrust, they pass through a turbine section. The gases spin the turbine blades at high speed. The turbine powers the compressor section and adds to the thrust the hot gases leaving the rear of the engine provide. The hot gases then escape through the exhaust nozzle at the rear.

Ramjet engine: An engine that can operate only when moving at high speed, since it has no moving parts and no device for drawing in air.

Turbojet engine: An engine used on commercial aviation vehicles.

Turbofan engines are sometimes called fan-jets or bypass engines. Turbojets produce small streams of fast air. Turbofans utilize large amounts of slower-moving air. The turbofan brings more air to the compressor section. This allows for a more efficient combustion process. The fan also forces air around the outside of the engine. This adds to the total thrust output of the engine. Turbofan engines are widely used on commercial passenger airplanes because they are efficient, as well as powerful, at low speeds.

Turboprop engines are basically turbojets that have a propeller mounted on the front. These engines use the product of the combustion to turn the propeller. Thrust from the rear of the engine is not relied on as propulsion. It is used to turn the propeller, which, in turn, moves the vehicle forward.

Figure 22-6. A—A cross section of a turbojet engine. Airflow and combustion are continuous. Notice that the engine has internal parts rotating on a shaft. Arrows show airflow from the front of the engine through the back. At the front, there is a compressor section. (Pratt and Whitney, Canada) B—The turbofan engine is a variation of the turbojet, with a fan placed in front of the compressor section. The action the engine causes supplies power, in addition to thrust. C—This cutaway view of a turboprop engine shows the gear assembly connecting the propeller shaft on the left with the turbines on the right. Notice the gearbox between the propeller shaft and the main engine shaft. (Allied Signal Aerospace, Garrett Engine Division) D—With the aid of its afterburner, a U.S. Navy F-18 Hornet fighter plane takes off from the deck of the aircraft carrier USS *Harry S. Truman*. (U.S. Navy)

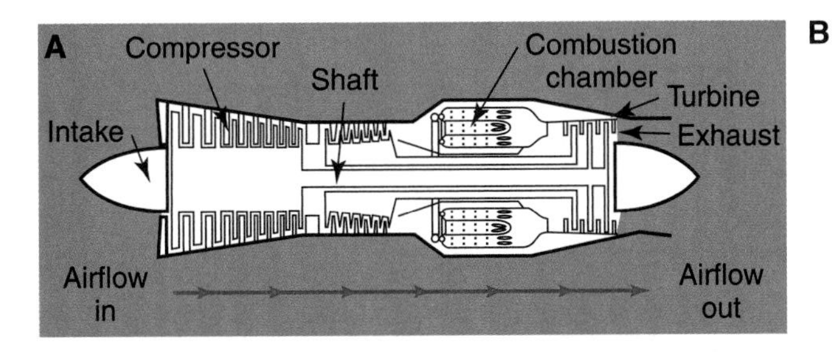

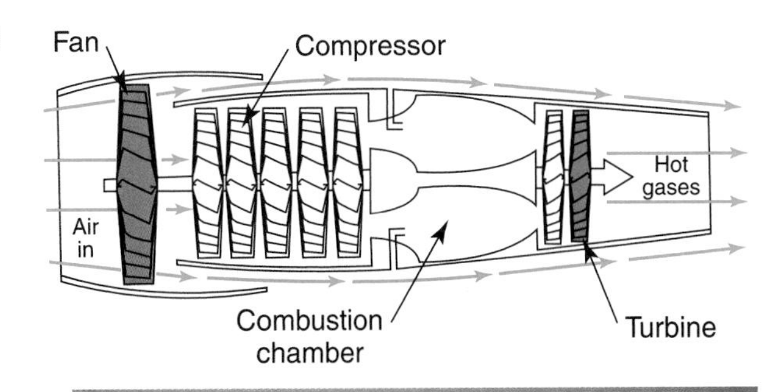

Turbofan engine: An engine widely used on commercial passenger airplanes because it is efficient, as well as powerful, at low speeds. Sometimes called a fan-jet or bypass engine.

Turboprop engine: An engine that is basically a turbojet with a propeller mounted on the front.

Afterburning turbojet engine: A turbojet or turbofan engine with an additional burner added to the nozzle.

Aeronautical chart: A chart that provides important data for airplane pilots and navigators. It is basically a topographic map on which special guidance information has been added.

The ***afterburning turbojet engine*** is a turbojet or turbofan engine with an additional burner added to the nozzle. The afterburner includes a port that allows fuel to be injected into the hot exhaust gases. When this fuel is burned, it provides additional thrust. The afterburner is, however, very inefficient. Afterburners are only used on aircraft that would, at times, need additional thrust. Common applications are on fighter jets and supersonic transports (SSTs).

Guidance Systems

Guiding an aircraft is similar to guiding a boat or ship because there are not fixed paths to follow. Pilots must use charts and instruments to plot and follow their desired course. Piloting aircraft is much different than simply driving a car and following roads. Because it is much more difficult to guide or navigate a plane, pilots must go through a great amount of training and obtain a series of licenses before they are allowed to fly. This training helps pilots learn how to use charts, instruments, and navigation systems.

Aeronautical Charts

Aeronautical charts are charts that provide important data for airplane pilots and navigators. These charts are basically topographic maps on which special guidance information has been added. Aeronautical charts show elevations of hills and mountains, as well as the locations of airports and other landing areas. They have special markings that locate prohibited areas where aircraft cannot fly legally, such as around military installations. Other areas are labeled as restricted. These are not prohibited, but they could be dangerous. Artillery ranges are marked as restricted areas.

When plotting a course for an aircraft, the pilot or navigator uses a navigation plotter. See **Figure 22-7.** It is used to measure distances, as well as directions, when placed on top of an aeronautical chart.

Instruments

The cockpits of aircraft are filled with guidance and navigation equipment. See **Figure 22-8.** The number and types of instruments required depend on the type of flying being done. Pilots follow ***visual flight rules (VFR)*** when weather conditions allow them to navigate by what they are able to see outside the cockpit. ***Instrument flight rules (IFR)*** are followed when weather conditions do not allow pilots to navigate visually. There are several essential instruments that can be found on any airplane, whether the pilot operates by IFR or VFR. See **Figure 22-9.**

Figure 22-7. A navigation plotter is used with an aeronautical chart to plot the course of an aircraft. The plotter is made from clear plastic, and it combines a protractor with a straight edge marked with different scales.

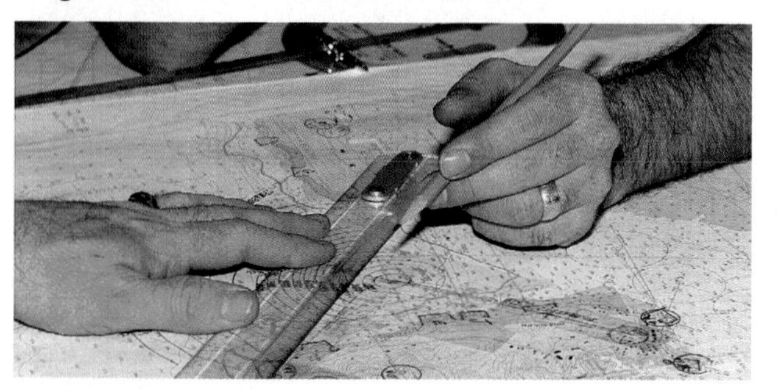

Airspeed indicators measure the difference between two pressures acting on an aircraft. As the plane starts flying, air rushes into a small tube on the outside of the aircraft. There is another tube inside the plane's ***fuselage***, or the

Figure 22-8. The cockpit of a modern airliner has nearly 1000 instruments, controls, and gauges filling almost every possible space. Smaller aircraft have many less instruments. Note that all the controls are duplicated on each side of the cockpit. The command pilot sits on the left, with the copilot on the right. Either can fly the aircraft. (Airbus)

Visual flight rules (VFR): Rules followed when weather conditions allow pilots to navigate by what they are able to see outside the cockpit.

Instrument flight rules (IFR): Rules followed when weather conditions do not allow pilots to navigate visually.

Airspeed indicator: An instrument that measures the difference between two pressures acting on an aircraft.

Fuselage: The main body of a plane.

Figure 22-9. Essential instruments found on aircraft of all sizes. This aircraft combines traditional gauges with electronic display screens. (Cessna Aircraft Company)

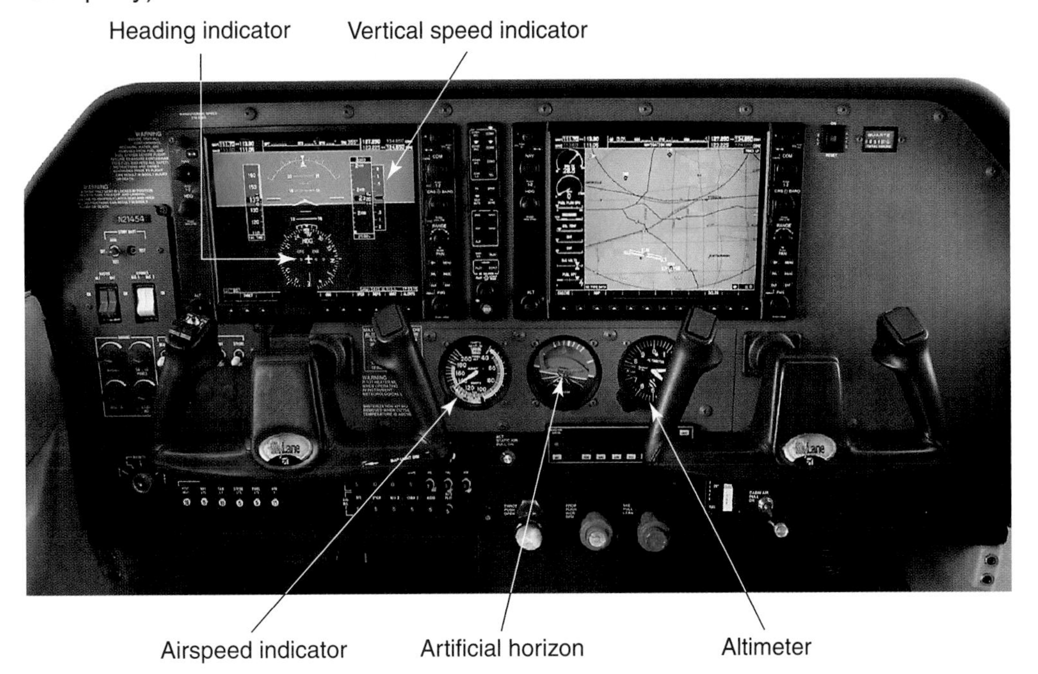

main body of the plane. Both tubes lead to the airspeed indicator. The difference in pressure between the two tubes activates a small diaphragm. The motion of the diaphragm is displayed on the face of the instrument in miles per hour (mph) and knots. The ***vertical speed indicator*** works similarly to the airspeed indicator. This instrument, however, displays the rate at which the airplane is ascending or descending. ***Altimeters*** are used to display the altitude, in feet, of the aircraft from the ground. ***Artificial horizon*** instruments, also called attitude indicators, display the amount of pitch and roll of the aircraft, compared to the horizon. The ***heading indicator*** shows the pilot the direction the plane is headed.

Vertical speed indicator: An instrument that displays the rate at which the airplane is ascending or descending.

Altimeter: An instrument used to display the altitude, in feet, of the aircraft from the ground.

Artificial horizon: An instrument, also called an attitude indicator, that displays the amount of pitch and roll of an aircraft compared to the horizon.

Heading indicator: An instrument that shows the pilot the direction the plane is headed.

Electronic Navigation Equipment

Instruments allow pilots to find their current direction, speed, attitude, and altitude. Most pilots use the instruments in combination with electronic navigation equipment to ensure efficient travel from one place to another. Electronic navigation equipment is based on the use of radio waves. Radio transmitting stations used for guidance are usually land based. They are mostly government owned and operated.

Radio waves are distinguished by their frequencies. High-frequency waves are very accurate for navigation, but they cannot travel past the horizon. Low-frequency waves are still accurate, and their signals carry for thousands of miles. The type of frequency used is determined by the function of the transmitting station. High frequencies are employed if the waves are to be sent only short-range. If the waves are to be sent over very long distances, lower frequencies are used.

Radio Direction Finding

Radio direction finding is one of the early methods for guiding airplanes and ships. Electronic equipment aboard the vehicle receives transmissions sent by radio transmitters (beacons). The navigator adjusts the antenna until it signals it is locked onto the direction of the incoming signals. The course of the vehicle can then be adjusted according to the position of the beacon.

The Very-High-Frequency Omnidirectional Radio Range (VOR) Navigation System

Very-high-frequency omnidirectional radio range (VOR) navigation was developed in the 1940s. This is a commonly used guidance system for air transportation. Each VOR station transmits a series of beams in all directions. The beams of radio waves are called ***radials***. See **Figure 22-10.** The locations of VOR stations make up typical air routes. They are easy to find using a VOR instrument. The VOR instrument shows the heading of the next VOR station. See **Figure 22-11.** Because the systems send out signals in all directions, pilots can navigate in any one of 360 verified directions toward or away from the VOR transmitter.

Very-high-frequency omnidirectional radio range (VOR) navigation: A commonly used guidance system for air transportation. Each VOR station transmits a series of beams of radio waves in all directions. A VOR instrument shows the heading of the next VOR station.

Radial: A beam of radio waves.

The Instrument Landing System (ILS)

The ***instrument landing system (ILS)*** is a system allowing pilots to land in all types of weather conditions. It uses two radio waves to mark the approach of a ***runway***, or a flat, straight path specially lit and marked

Figure 22-10. A very-high-frequency omnidirectional radio range (VOR) station sends out radio signals in all directions for aircraft to home in on. There are presently over 1000 VOR transmitting stations in the United States. The pilot selects a signal (also called a radial) and locks onto it. Each radial is different from the others so they can be kept apart and identified. The airplane has a VOR-detecting device that locks onto the signals and tells the pilot or navigator which radial it is following.

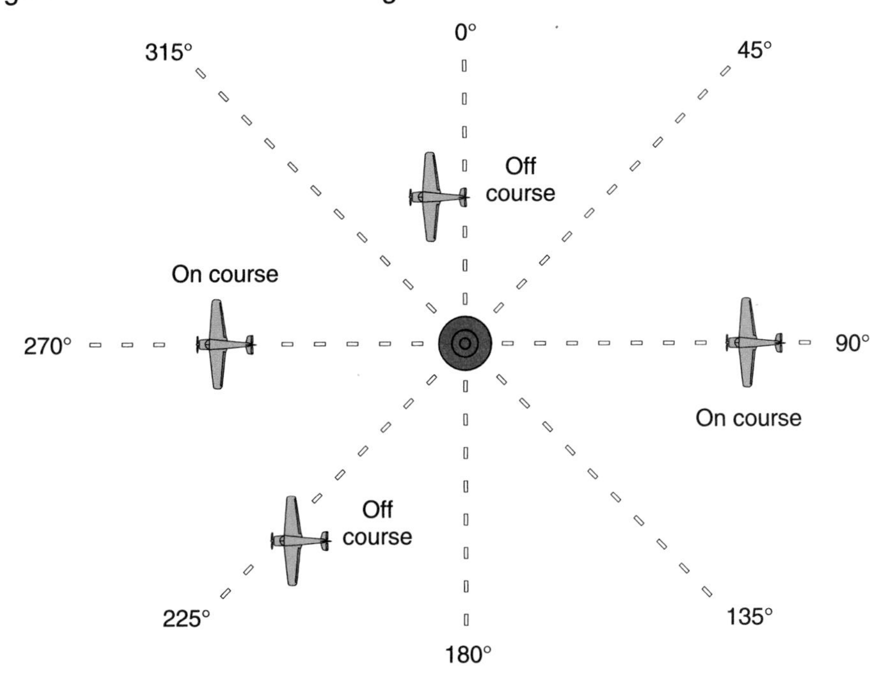

Instrument landing system (ILS): A system allowing pilots to land in all types of weather conditions. It uses two radio waves to mark the approach of a runway.

Runway: A flat, straight path specially lit and marked to aid pilots.

to aid pilots. The first wave is a localized beam marking the center of the runway. The second beam is the guide-slope beam that designates the path to be followed by the approaching aircraft. There are also three beacons indicating the beginning, middle, and end of the approach. The first is anywhere from 4 to 7 miles from the runway. The final beacon is near the end of the runway. The aircraft is equipped with an ILS indicator showing the location of the plane within the two beams and three beacons. In bad weather conditions, the ILS can be relied on to land the plane.

Global Positioning Systems (GPSs)

It is estimated that VORs and ILSs will eventually be phased out, due to the advances in satellite technology. Global positioning systems (GPSs) will ultimately be used for navigation and landing of aircraft. They use a collection of satellites that work together to locate any position on or above earth. There are two types of GPSs available to aircraft

Figure 22-11. The very-high-frequency omnidirectional radio range (VOR) instrument on the cockpit control panel. The course selector turns the compass card so the desired heading is at the top, or 12 o'clock position. The course deviation indicator (CDI) points at the desired heading when the plane is on course. Deflection to the left or right indicates the plane has gone off course. This instrument can also show the plane's position relative to two stations.

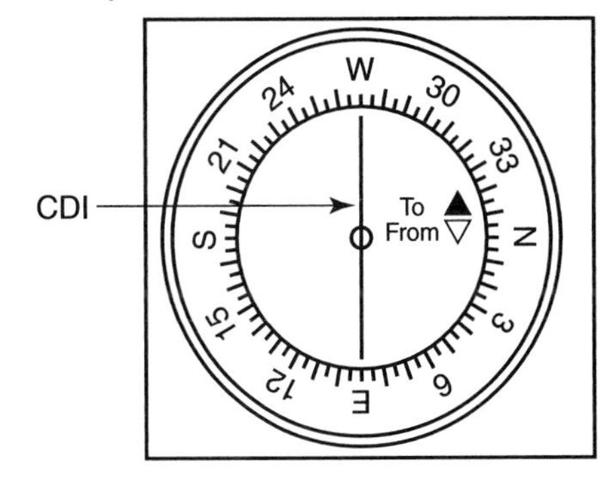

pilots. GPS receivers can be purchased as handheld or in-dash systems. The in-dash systems are installed in the aircraft and are typically more sophisticated than the handheld receivers. See **Figure 22-12.**

GPSs can be used to navigate from one point to another by placing way points. A ***way point*** is the location of the destination. Once the way point is set, the receiver can calculate the information needed to keep the aircraft on the correct heading. With the use of the Wide Area Augmentation System (WAAS) and the Local Area Augmentation System (LAAS), which make GPSs more accurate, GPSs will eventually be used to land aircraft.

Way point: The location of a destination in a global positioning system (GPS).

Attitude: The position and orientation, or directional control, of an aircraft.

Control surface: A moveable airfoil designed to change the attitude of an aircraft.

Envelope: The inside of a hot air balloon.

Control Systems

Aircraft are controlled in two ways. First, the attitude of the aircraft must be controlled. The ***attitude*** is the position and orientation, or directional control, of the aircraft. This is controlled by the use of control surfaces on the aircraft. ***Control surfaces*** are moveable flaps that deflect wind and help to steer the aircraft. Second, the speed of the aircraft must be controlled. Speed can be controlled in several different ways.

Lighter-Than-Air Vehicle Control Systems

The amount of control a pilot has of a lighter-than-air vehicle depends entirely on the type of vehicle. Hot air balloon pilots can only control the vertical flight of the balloon. Changing the temperature of the air inside the balloon controls the altitude of the hot air balloon. To raise the balloon, the pilot turns on the burner and heats air inside the balloon, or ***envelope***. See **Figure 22-13.** To lower the balloon, the pilot allows the air to cool or

Figure 22-12. An in-dash global positioning system (GPS) receiver. Chart displays pinpoint the aircraft's position and can be changed in scale to show wider or more detailed views. In-dash systems can be integrated with weather maps, terrain charts, and aeronautical charts. (Cessna Aircraft Company)

Figure 22-13. Heated air from a propane burner is directed into the envelope of the hot air balloon, providing lift, which causes the balloon to rise. To descend, the pilot can vent some heated air from the envelope.

opens a vent at the top of the balloon. The direction and speed the hot air balloon travels are not as easily controlled by the pilot. Balloon pilots are at the mercy of the wind. Pilots must raise or lower the hot air balloon to take advantage of wind currents and directions.

Blimps and dirigibles are lighter-than-air vehicles that are able to control both vertical and horizontal dimensions of flight. See **Figure 22-14.** These airships use both horizontal and vertical stabilizers to steer the craft. These ***stabilizers*** are similar to rudders used on boats. By manipulating the rudder, pilots control the flow of air. Airflow, in turn, influences the direction of travel. By raising or lowering the horizontal stabilizer, a blimp ascends or descends. Turning the vertical stabilizer right or left steers the rear of the airship sideways in either direction. The components of the propulsion systems can control some lighter-than-air vehicles. Thrust in a different direction reorients the position of the craft. Propulsion fans driven by engines swivel in order to swing the blimp in the desired direction. See **Figure 22-15.**

Heavier-Than-Air Vehicle Control Systems

Different control systems control the different types of heavier-than-air vehicles. Airplanes rely on surfaces to steer the aircraft. Instead of surfaces, helicopters use two rotating blades to control the movement of the aircraft.

Airplane control systems

Airplanes are heavier-than-air vehicles capable of great speed and control. See **Figure 22-16.** In order for an airplane to be controlled, the vehicle must be stable. There are three types of stability of great concern to all pilots: directional, lateral, and longitudinal stability. Each of these types of stability is concerned with one axis of the aircraft. See **Figure 22-17.** The axes are all 90° from each other, like the X, Y, and Z axes you may have studied in a mathematics course. Each axis is stabilized by a ***stationary surface***, which is a surface that is fixed in one position, such as an airplane wing, and can be controlled by a moveable control surface. See **Figure 22-18.**

Figure 22-14. Horizontal and vertical stabilizers on a blimp function like rudders on a boat, turning the vehicle or making it rise or descend. (The Goodyear Tire & Rubber Company)

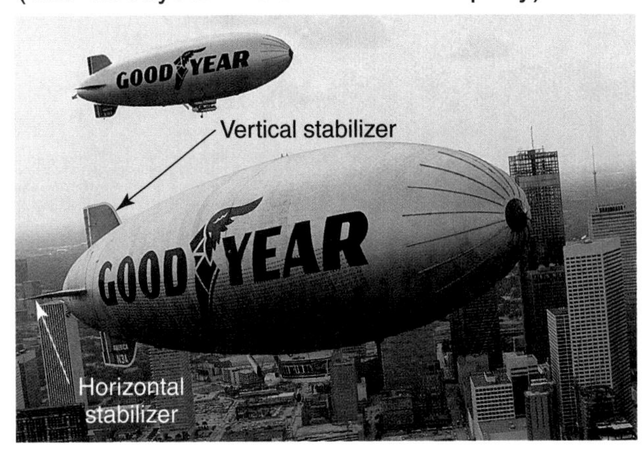

Figure 22-15. Propulsion fans mounted on the blimp's control cabin provide forward (or backward) motion and also can be swiveled to provide directional control. (The Goodyear Tire & Rubber Company)

Figure 22-16. Airplanes control three degrees of freedom and can be precisely maneuvered at high speed, as in this example of precision formation flying. (U.S. Air Force)

Directional stability allows the airplane to fly in a straight line. The horizontal stabilizer, or tail fin, is used to provide directional stability. While the plane is in motion, wind currents that push the tail of the aircraft off course can affect the directional stability. This side to side motion is known as ***yaw***. A rudder placed at the back of the tail fin controls yaw. See **Figure 22-19.**

Ailerons are moveable control surfaces placed on the main wing. The wing of the plane is the stationary stabilizer for lateral stability. ***Lateral stability*** is used to overcome the tendency of the plane's wings to dip on either side. This motion is known as ***roll***. When the ***control stick***, or the airplane lever operating the elevators and ailerons with the power to guide the aircraft, is pushed right, the right aileron is angled up, and the left aileron is angled down. This causes the plane to roll to the right.

Longitudinal stability is the ability to fly the airplane without the nose moving up or down. The ***horizontal tailplane*** is used to achieve longitudinal stability. ***Elevators*** are the moveable surfaces placed on the tailplane to control changes in this stability. Movement in this direction is known as pitch. ***Pitch*** is the up and down movement of the nose of the aircraft.

A turn, or bank, in an aircraft is known as a ***coordinated turn*** because all three control surfaces must work together. The rudder orients the nose of the plane. The ailerons roll the plane into the turn. The elevators are used to maintain altitude.

The speed of most aircraft is controlled by changing the amount of thrust the engine produces. Large aircraft have additional control surfaces to slow themselves down during landing. These control surfaces are known as ***spoilers***. When the spoilers are raised, they help to slow the aircraft down by increasing the amount of drag.

Stabilizer: A device, similar to a rudder on a boat, used by a pilot to steer an airship. It controls the flow of air.

Stationary surface: An immobile, exterior boundary of an object.

Directional stability: The ability to fly an airplane in a straight line.

Yaw: The side-to-side motion of an aircraft.

Figure 22-17. Yaw, pitch, and roll are the three motions affecting stability of an aircraft. (Estes)

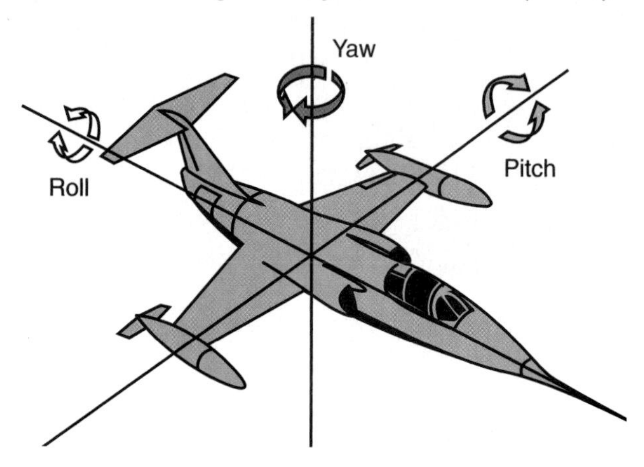

Figure 22-18. The ailerons, elevators, and rudder are moveable surfaces that help control the flight of an aircraft. (Estes)

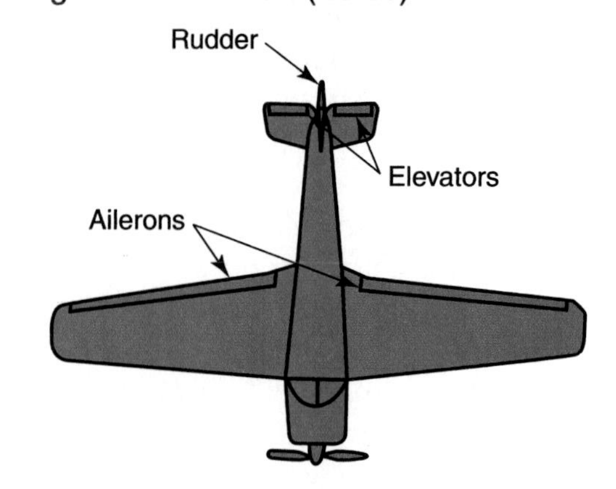

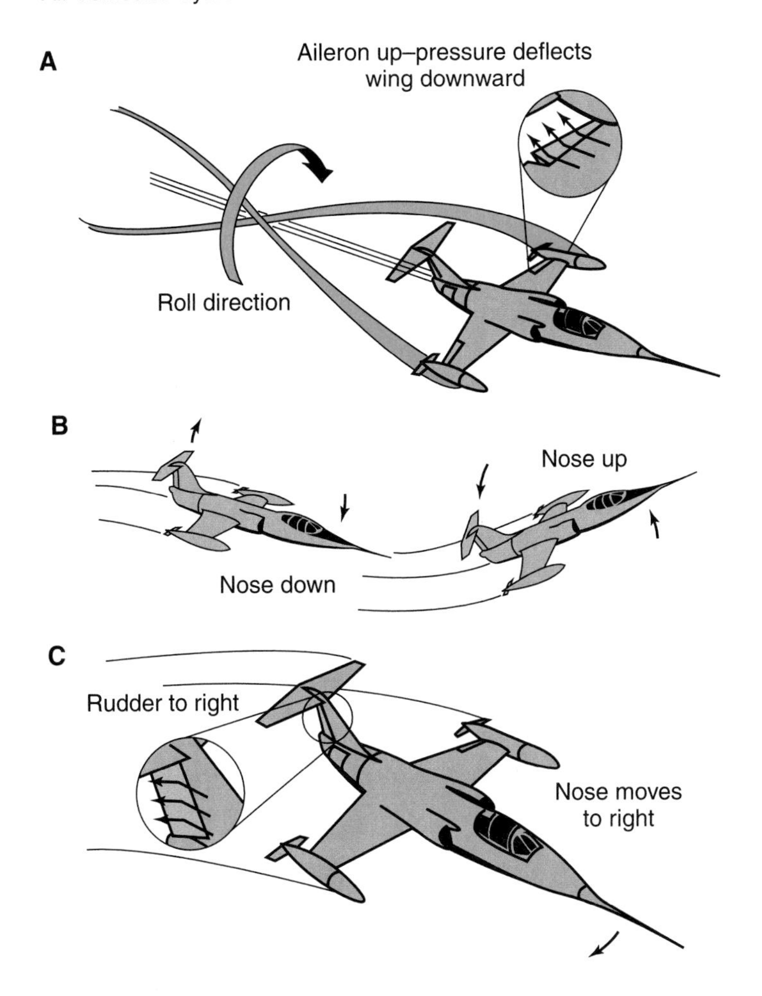

Figure 22-19. Functions of the moveable control surfaces. A—Ailerons. These surfaces can be raised or lowered to bank the plane in either direction. Moving the control stick of the airplane from right to left controls roll. B—Elevators. Moving the control stick forward or backward controls the aircraft pitch. Moving the stick backward raises the elevator and angles the nose of the plane upward. This allows the airplane to climb. C—Rudder. By turning the rudder using rudder pedals, the pilot can correct the yaw and direct the nose of the plane in the right direction. The rudder is only used to correct yaw, however, and cannot be used to turn the aircraft itself. (Estes)

Helicopter control systems

Helicopters are rotary-wing aircraft that can move in different directions with greater speed and agility than airplanes. They are more maneuverable. A ***swash plate*** located on the main rotor makes control possible. See **Figure 22-20.** Two hand levers, known as the collective pitch control lever and the cyclic control stick, control the swash plate. See **Figure 22-21.** The ***collective pitch control lever*** directs the swash plate to change both blades evenly. This allows the helicopter to ascend or descend. The ***cyclic control stick*** changes only one side of the swash plate. This changes the pitch of the blades on only one side of the rotor. The pilot can steer in any direction. By controlling cyclic pitch, pilots determine the vehicle's direction of flight forward, backward, and laterally (to the left or right).

To turn the helicopter, the pilot uses the directional control pedals at his feet. The directional pedals control the thrust of the tail rotor. In straight flight, the thrust of the tail rotor is equal to that of the main rotors. When turning, the pilot either increases or decreases the amount of tail rotor thrust to spin the helicopter in either direction.

Aileron: A moveable control surface placed on the main wing of a plane.

Lateral stability: The ability to overcome the tendency of a plane's wings to dip on either side.

Roll: The tendency of a plane's wings to dip on either side.

Control stick: The airplane lever operating the elevators and ailerons with the power to guide the aircraft.

Longitudinal stability: The ability to fly the airplane without the nose moving up or down.

Horizontal tailplane: The tail surfaces of an airplane that are parallel to the horizon, including the stabilizer and the elevator.

Elevator: The moveable surface placed on the tailplane to control changes in longitudinal stability.

Pitch: The up-and-down movement of the nose of an aircraft.

Coordinated turn: A turn, or bank, in an aircraft in which all three control surfaces work together.

Spoiler: A control surface on a large aircraft designed to slow the aircraft down during landing by increasing the amount of drag.

Swash plate: Part of a helicopter that controls the pitch of the blades on the main rotor, controlling the aircraft's horizontal and vertical movements.

Collective pitch control lever: A helicopter control that directs the swash plate to change both blades evenly, allowing the aircraft to ascend or descend.

Figure 22-20. The swash plate (see inset) controls the pitch of the blades on the main rotor of the helicopter, controlling the aircraft's horizontal and vertical movements. (Sikorsky Helicopter)

Figure 22-21. The collective pitch control lever controls upward and downward movement of the helicopter. The cyclic control stick is used to steer the aircraft. (Sikorsky Helicopter)

Suspension Systems

Airplanes must have suspension systems so they can take off, land, and maneuver on the ground. Usually, airplanes have wheels. The wheels are called landing gear.

Small aircraft usually have their landing gear arranged in one of two ways. See **Figure 22-22.** In the conventional arrangement, the single wheel is at the tail of the plane, and the two main wheels are under the cockpit.

Figure 22-22. Landing gear arrangements. A—A taildragger system with two main wheels under the wings and a small wheel beneath the tail. This arrangement was one of the first ways wheels were positioned onto planes. B—Tricycle gear, with a nose wheel and two main wheels. This arrangement keeps the plane more or less level when it is at rest and gives the pilot better visibility when taxiing and taking off.

A

B

A plane with this arrangement is known as a taildragger. It has a disadvantage in the fact that the pilot has poor visibility when the plane is on the ground. This is because the fuselage of the plane tends to point upward when at rest. This also makes takeoffs more difficult and even requires special training from the Federal Aviation Administration (FAA). In the tricycle-style landing gear arrangement, the two main wheels are placed just behind the cockpit area. The single wheel is in the plane's nose, usually under the engine area.

Cyclic control stick: A helicopter control that changes only one side of the swash plate. This changes the pitch of the blades on only one side of the rotor.

Large commercial passenger planes use the tricycle-style landing gear. Instead of three wheels, they have multiple sets of wheels. See **Figure 22-23.**

Wheels are attached to the aircraft with struts. Struts are designed so they absorb the shock of the airplane touching the ground. They are made differently, depending on the size of the vehicle they must support. Small planes can use one-piece metal struts. Large planes use hydraulic shock-absorbing devices similar to the ones used on automobile suspensions. Of course, the shock absorbers on large planes must be stronger than those used on cars.

Some airplanes are designed to land on water. Their suspension systems use pontoons instead of wheels. See **Figure 22-24.** Some planes can land on solid ground or water. These use wheels attached to the insides of the pontoons. Planes used on snow and ice can be fitted with skis.

The components of airplanes that enable flight are also parts of the suspension system. Airplanes are suspended by air flowing over and under specially shaped wings. Even though the function of wings as part of the suspension system is not easy to understand, the wings are just as important as wheels on a car. The main force we are concerned about when studying suspension is lift. The shape of the wing, speed of the airplane, and angle of attack affect lift.

Figure 22-23. To support the heavy weight of large airliners during takeoff and landing, multiple sets of wheels are used. (Airbus)

Figure 22-24. Floats called pontoons are used, instead of wheels, to allow aircraft to land on water. The pontoons are hollow cylinders, so they float. This type of aircraft is used extensively to provide transportation to remote areas of Alaska and Canada. (Howard Bud Smith)

The shape of a plane's wing is called an airfoil. See **Figure 22-25.** The front of the airfoil is called the ***leading edge***. The point at the back of the airfoil where the top and bottom meet is known as the ***trailing edge***. The line drawn between the leading and trailing edges is the ***chord line***. If you were to divide the airfoil in half between the top and bottom, the resulting curve is the mean camber line. The distance between the top of the mean camber line and the chord line is the ***camber***. The length, or span, of the wing is also an important variable. The ***span*** is the length of the wings, from tip to tip. With the span and chord line, the aspect ratio of the wing can be figured. The ***aspect ratio*** is the span divided by the chord line (aspect ratio = span/chord). For wings that do not have a constant chord line length—for example, wings that get smaller at the tips—a different formula is used. In these cases, the aspect ratio is equal to the span, squared, divided by the area of the wing. The chord line, mean camber line, span, and camber of the airfoil affect the amount of lift possible.

Leading edge: The front of an airfoil.

Trailing edge: The point at the back of an airfoil where the top and bottom meet.

Chord line: The line drawn between the leading and trailing edges of an airfoil.

Camber: The distance between the top of the mean camber line and the chord line of an airfoil.

Span: The length of a plane's wings, from tip to tip.

The way in which airfoils actually generate lift is an area of great discussion. There are two popular theories of lift: the Bernoulli theory and the Newtonian theory. Both theories were originally meant to describe phenomena, rather than lift. They have both been used, however, to describe how lift occurs. Each theory has both correct and incorrect aspects.

The Bernoulli theory states that the air traveling over the airfoil is forced to travel farther than the air traveling below. See **Figure 22-26.** This generates a higher speed of airflow across the top, which creates an area of low pressure. The area of lower pressure then lifts the airfoil up. This popular theory is incorrect for several reasons. First, not all airfoils are longer on the top. Some have either the same or less distance on the top of the airfoil than the bottom of the airfoil. Secondly, the air across the longer section of the airfoil travels too fast for this theory to be accurate. One part of the theory is, however, correct. There is an area of lower pressure created above the airfoil. It is not, however, created due to a difference in air velocity.

Figure 22-25. The major parts of an airfoil.

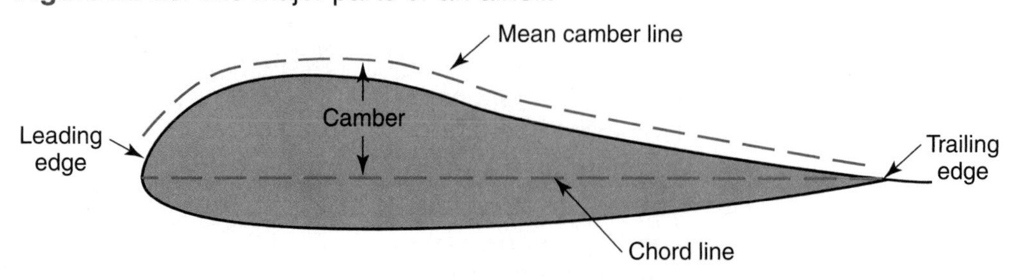

Figure 22-26. The Bernoulli theory of lift.

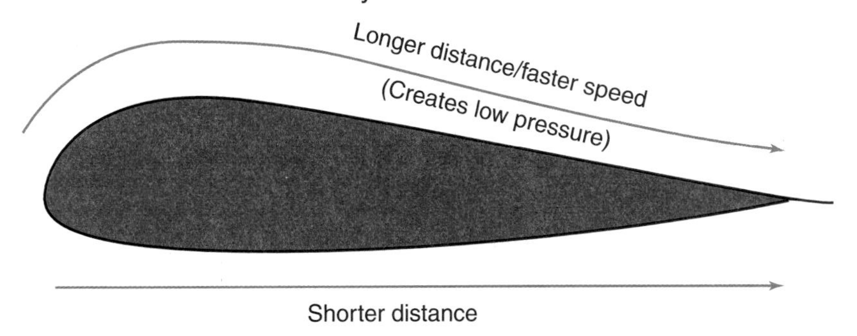

Aspect ratio: The span divided by the chord line.

The Newtonian theory is based on the ideas of Newton's third law of motion. The idea of this theory is that, as the air particles hit the bottom of the airfoil, the particles are deflected downward. See **Figure 22-27.** This deflection then causes the airfoil to react in the opposite direction, generating lift. This theory is also incorrect because it assumes that the top of the airfoil makes no difference in the amount of lift generated. It has been proven in tests that both the top and bottom of the airfoil affect lift.

The actual creation of lift is a complicated lesson in physics. The basic theory of lift is that lift is created by turning the flow of a gas. Both the top and bottom of the airfoil are used to change the direction of the air as it flows around the airfoil. By turning the flow of air, all parts of the airfoil work together to generate lift. The speed of the plane is also an important part of the creation of lift. Lift is a force, and force equals mass multiplied by acceleration ($F = M \times A$). Acceleration is a measurement of velocity (speed and direction) and time. Therefore, the greater the speed, the more lift generated.

The final variables in the factors of lift are the angle of incidence and angle of attack. The ***angle of incidence*** is the angle of the wing as it is attached to the aircraft. It is measured from an imaginary line running from the nose of the aircraft to the tail of the aircraft and cannot be changed once the aircraft is built. The angle of attack refers to the position of the wing as it hits the air. See **Figure 22-28.** As the front of the wing tips

Angle of incidence: The angle of the wing as it is attached to the aircraft.

Figure 22-27. The Newtonian theory of lift.

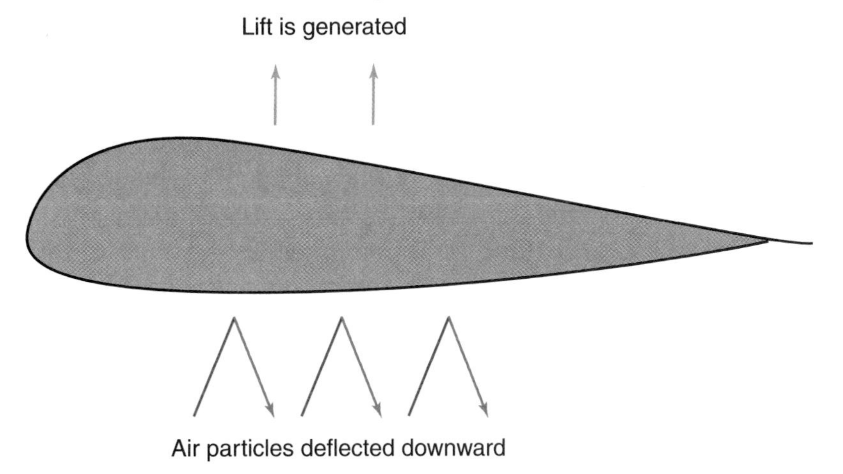

Figure 22-28. The upward tilt of the wing's leading edge is referred to as the angle of attack. As the angle of attack increases, so does the lift the wing generates.

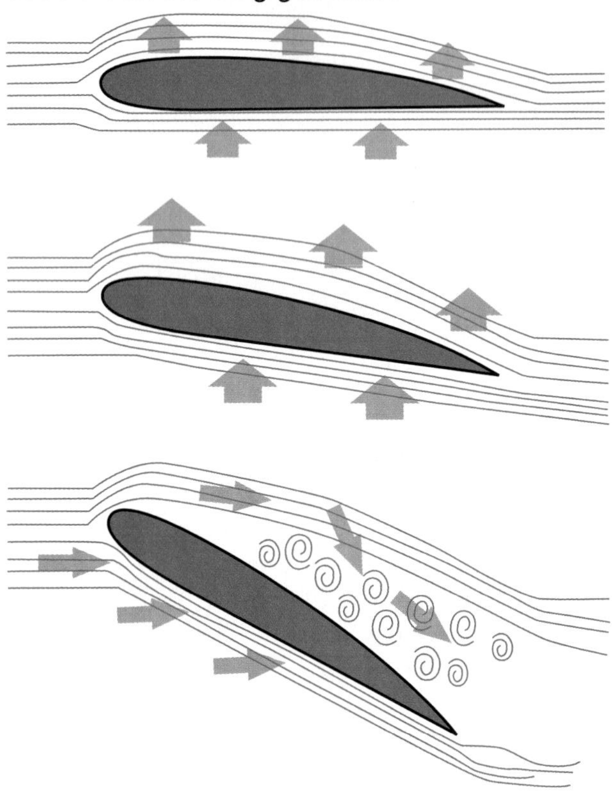

upward, the angle of attack increases. This means a great amount of the oncoming air hits directly on the bottom of the wing. This action tends to push up on the plane, forcing it higher in the air. There is a limit, however, to the angle of attack. If the angle becomes too great, the wing will quickly lose lift. This situation is known as ***stall***, and it can be very dangerous.

Stall: A situation in which the angle of attack becomes too great and the wing quickly loses lift.

Structural Systems

The structures of aircraft are substantially different in lighter-than-air and heavier-than-air vehicles. Lighter-than-air vehicles require an envelope in which they can hold either air or another gas. Heavier-than-air vehicles, typically, require a much more rigid structure.

Lighter-Than-Air Vehicle Structural Systems

Lighter-than-air vehicles, as we have seen, are designed to displace a large amount of air. Their structures need to be large in order to contain enough air or gas to rise. Each type of lighter-than-air vehicle has a slightly different structural system.

Blimps are nonrigid airships. The pressure the gas exerts maintains the shape of the vehicle. When the gas, usually helium, is let out of the blimp, the vehicle deflates and does not retain its shape. The blimp balloon, or envelope, is constructed of a strong, but lightweight, polyester or fabric. The internal structure of the balloon contains two air bags or ballonets, which can be filled with helium. Cables running within the envelope attach the cars, or gondolas. The gondolas include an area for passengers and the pilots and also have engines mounted to each side. The stabilizers are mounted to the rear of the envelope.

Hot air balloons are similar to blimps. The pressure of the hot air they contain holds their shape. Balloons are made of ripstop nylon, similar to the sails on sailboats. Steel cables running through the fabric are attached to the basket holding the fuel and occupants. See **Figure 22-29.**

An inflator fan fills the envelope with air. This gives the balloon its initial shape while it is still on the ground. When the pilot is ready and has done a safety check of all the vehicle's systems, he heats the air inside the envelope with large propane-fueled burners. As the air heats, the envelope begins to rise and stand straight up. After a few more safety checks and more heat, the balloon is ready to ascend.

Rigid airships are sometimes referred to as zeppelins, named after Count Ferdinand von Zeppelin, a German general and aeronautical designer. They have stiff frameworks that serve as hulls. The hulls retain the shape of these airships, unlike nonrigid airships. Inside the rigid airships, there are a number of airtight compartments filled with hydrogen or helium gas. The propulsion systems and stabilizers are attached to the hull structure using mounting brackets. Large, hotel-like gondolas are also part of the rigid superstructure. These airships are no longer in use.

Figure 22-29. Because it is lightweight and shock resistant, wicker has traditionally been used for the baskets of hot air balloons.

Heavier-Than-Air Vehicle Structural Systems

Heavier-than-air vehicles are designed to have large lifting capabilities. The structures of heavier-than-air vehicles can be divided between airplanes and helicopters. Both types of aircraft have specific structural needs.

Airplane structural systems

When powered flight first became a reality, airplane structures were made of wood. They had wire and cables for support and bracing. The wooden structures were covered with fabric, which gave the vehicles smooth skins. These surfaces enabled airflow to generate lift. Today, airplanes are built quite differently.

The fuselage

The fuselage is the large hollow section that holds the other parts of the plane together. In passenger aircraft, the fuselage contains a passenger cabin, a cargo hold, and a cockpit. See **Figure 22-30.** In cargo planes, the fuselage is divided into just two sections, the cargo hold and the cockpit.

The structure of the fuselage can be either constructed as a truss or shell. Early aircraft used the truss-type construction. This type of structure consists of pieces of wood or steel running the length of the plane. These are called ***longerons***. Cross-members called webs attach the longerons. The resulting truss structure is very strong and rigid, but it fills the interior with braces and wires, which cut down on useable space. Modern aircraft make use of monocoque or semimonocoque construction. ***Monocoque*** means "one shell." This type of construction results in a hollow structure with plenty of room for passengers, cargo, and equipment. See **Figure 22-31.** Monocoque construction relies on the outer covering, or skin, of the aircraft to carry most of the load. ***Semimonocoque*** construction makes use of both vertical and horizontal frame members that relieve some of the stress on the skin of the aircraft. Bulkheads, frames, and formers are the vertical members used to give greater strength. The horizontal members include longerons and stringers, and they help to keep the fuselage from bending.

Longeron: A piece of wood or steel running the length of the plane.

Monocoque: A type of fuselage construction that results in a hollow structure. A series of ringlike ribs attached to the strong metal outside covering of the plane form the skeleton of the structure.

Semimonocoque: A type of fuselage construction that makes use of both vertical and horizontal frame members, relieving some of the stress on the skin of the aircraft.

Curricular Connection

Math: Lift Coefficients

The amount of lift an airfoil will provide can be figured mathematically. In order to figure the lift, you must know the lift coefficient for the airfoil in question. The lift coefficient is a number that takes into account the shape, the angle of attack, and other features of the airfoil. It can be figured either experimentally or by using computer software. Once the lift coefficient is known, the amount of lift can be determined mathematically using the following formula:

lift = coefficient × density × 1/2 × velocity squared × wing area

or

$$L = Cl \times r \times .5 \times v^2 \times A$$

L = the weight the aircraft can carry
Cl = number determined experimentally
r = air density at the altitude to be measured
v = speed of aircraft (in feet per second)
A = the chord × the length of the wing

Here is an example:

Cl = .59
r = .00237 slug/ft^3 (density of air at sea level)
v = 250 miles per hour (mph), or 336.75 feet per second (fps)
A = 400 ft^2

$$L = 0.59 \times 0.00237 \times .5 \times (336.75)^2 \times 400$$
Lift = 31,713 lbs.

If a pilot knows any four of the variables in the above formula, she can solve for the unknown quantity. A pilot could solve for the speed of the aircraft if she knows the amount of lift, the coefficient of the airfoil, the area of the wing, and air density. For more information about lift and lift coefficients, visit the National Aeronautics and Space Administration (NASA) Glenn Research Center's on-line Beginner's Guide to Aerodynamics.

The tail section

Empennage: The tail section of a plane.

The tail section of the plane is also called the ***empennage***. See **Figure 22-32.** This section is the part that contains many of the vital control surfaces. The empennage consists of a tapered continuation of the fuselage to maintain an aerodynamic shape.

The wings

The wings are an extremely important structural part of the aircraft. The main structural members are two aluminum alloy spars. These run the length of the wing, starting at the fuselage. The airfoil shape of the wing is made by adding strong, lightweight ribs. These are spaced at regular intervals between the fuselage and the tip of the wing. The ribs are

Figure 22-30. The cockpit, passenger cabin, and cargo hold are the major sections of an airliner's fuselage. At the front of the fuselage is the cockpit. This is where the pilots and flight crew control and navigate the plane. The passenger cabin contains seats, windows, and amenities to ensure a comfortable trip. The cargo hold is typically either behind or underneath the passenger cabin. It is used to store baggage and other cargo during transport. (Airbus)

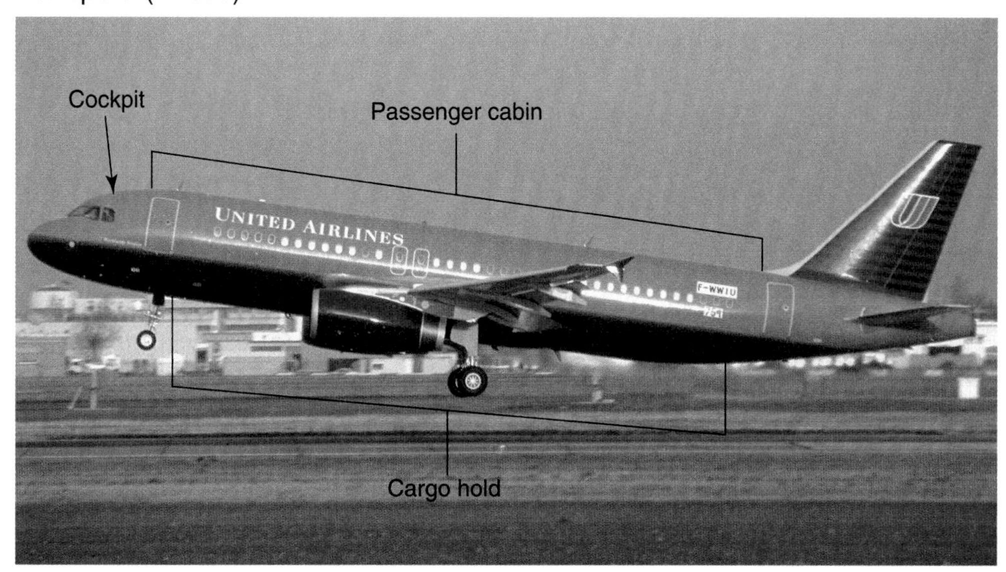

Figure 22-31. Monocoque construction results in large unobstructed cargo areas in modern aircraft. Smooth, aerodynamic shapes are also possible with monocoque construction. Instead of structural members running the length of the plane, monocoque construction relies on a series of ringlike ribs. These ribs form the skeleton of the structure. They are attached to the strong metal outside covering of the plane. (U.S. Air Force)

Figure 22-32. The empennage, or tail section, of an aircraft consists of the vertical and horizontal stabilizers. The rudder is attached to the vertical stabilizer, and the elevators are attached to the horizontal stabilizer.

Figure 22-33. This computer-aided design (CAD) drawing shows details of the wing and fuselage construction for a small plane. Note the rib construction in the wing. (Autodesk, Inc.)

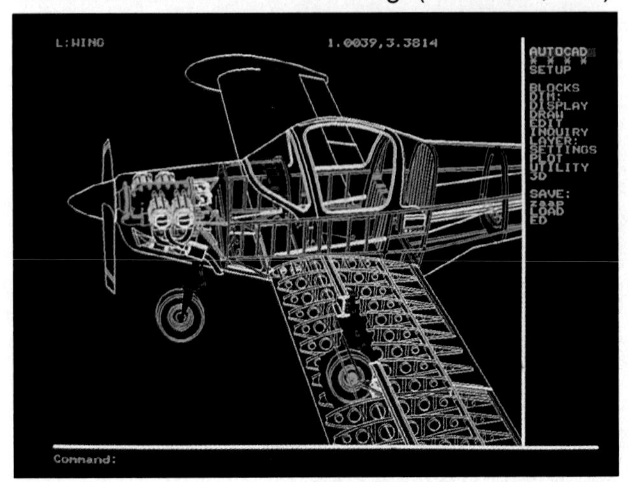

braced with cross-members to give the structure strength. This framework of spars, ribs, and braces is covered with thin, lightweight sheets of aluminum alloy. These sheets form the skin and provide the smooth aerodynamic shape essential for flight. The skin is also essential to the strength of the wing structure. Wings must be securely attached to the fuselage. See **Figure 22-33.** Usually, wings are cantilevered out from the plane. Airplane wings have taken many shapes and forms. This is the result of designers trying to improve flight characteristics of aircraft. See **Figure 22-34.**

Helicopter structural systems

Helicopter structures follow designs similar to those of airplanes. Usually, a combination of truss and semimonocoque construction methods is employed. The two main sections of the helicopter are the tail section and cabin. The cabin may be compared to the fuselage of an airplane. See **Figure 22-35.** The tail section of a helicopter is usually of the truss type. It needs to be rigid. This section usually does not, however, need to support cargo or passengers.

Figure 22-34. Different wing shapes. A—This experimental aircraft, the X-29, has a forward-swept wing. (National Aeronautics and Space Administration) B—A pivot-wing aircraft. C—Folding wing aircraft are used to save space on aircraft carrier hangar decks. (U.S. Navy)

A

B

C

Support Systems

Airports are facilities that house most of the major support systems for air transportation. They are busy places. The communication centers and control towers are important parts of this system. Their main activity is to keep track of all planes within their airspace. Controllers work in the control tower and use radio detecting and ranging (radar) screens to keep track of all the planes. See **Figure 22-36.** There are several stages of air traffic control. The air traffic control tower is located at the airport. It tracks all planes within 5 miles of the airport and handles all departures and landings. The terminal radar approach control station monitors planes between 5 and 50 miles from the airport. While planes are in route and outside of 50 miles from an airport, air route traffic control centers view their locations. Another important activity is ground support. This involves the people who guide the aircraft to the gates, service the planes, and handle baggage. All types of support systems are represented at airports. Airports are where passengers and cargo make the transition from land to air transportation.

Figure 22-35. Helicopter cabins are generally similar to the cabins of comparable-sized airplanes. They are where the passengers, pilots, and cargo are located. The fuel and propulsion systems are also part of a helicopter cabin.

Figure 22-36. A radio detecting and ranging (radar) control room in an airport control tower. Controllers rely on radar to keep track of multiple aircraft. They are also in voice contact with the pilots of the aircraft. (United Airlines)

Runways

In order to take off and land, airplanes need lengthy runways. See **Figure 22-37.** Small airports use a single runway. As air traffic increases, other runways are usually added either parallel or perpendicular to the first one. A ***taxiway*** is the name of the roadway that connects the runway to the terminal. As airplanes leave the terminal, they follow the taxiways to the runway they will use for takeoff. Runways are generally labeled based on their magnetic heading, with the last digit removed. For example, a runway positioned at 040 degrees is labeled "04," if you are heading to the northeast, and "22" (for 220 degrees), if you are heading on the same runway to the southwest.

Taxiway: A roadway that connects the runway to the terminal.

Special markings are added to runways to guide pilots of approaching and landing aircraft. Runway lighting helps guide pilots who are landing or taking off at night or in weather conditions that limit visibility. See **Figure 22-38.**

Career Connection

Air Traffic Controllers

Air transportation is a fast moving and busy part of the global transportation system. It is so busy, in fact, that some of the world's busiest airports see nearly 2500 airplanes take off from their runways each day. It is the job of air traffic controllers to direct those airplanes on the ground and in the air to make sure the aircraft maintain safe distances from each other.

There are a number of different types of air traffic controllers stationed throughout the journey of an airplane. These controllers include tower controllers, ground controllers, en route controllers, radio detecting and ranging (radar) controllers, and flight service specialists. A controller is in contact with an aircraft pilot from the time the pilot is taxiing to the runway until he lands and receives a gate location at the destination. Controllers are employed by the Federal Aviation Administration (FAA), and most work typical 40-hour work weeks.

Air traffic controllers are under high amounts of stress and must maintain deep concentration throughout their shifts. Because of this, controllers must take preemployment tests that measure their abilities to work in high-stress environments and to learn the activities of an air traffic controller. All candidates for air traffic controller jobs must have a college degree or full-time work experience. They must also attend the FAA Academy, where they receive controller training. With experience and time, controllers are able to advance through different controller positions. Average earnings for air traffic controllers are slightly over $90,000 per year.

Figure 22-37. Many airports have runways in different directions because of changeable wind patterns in the area. Airplanes usually take off and land headed into the wind. With a variety of runways, airports can always provide a relatively safe landing area. This is a diagram of Chicago's Midway Airport, which occupies a one-square-mile site.

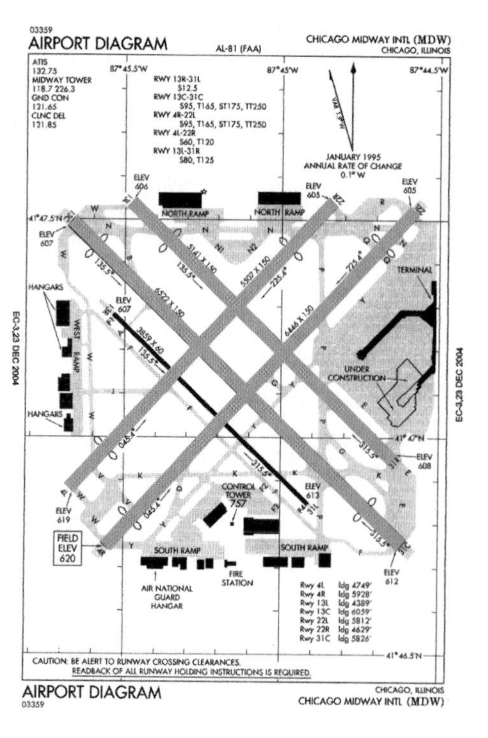

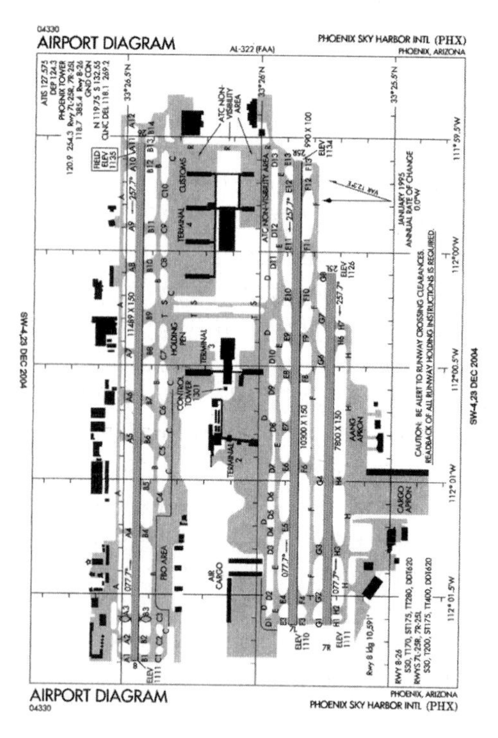

Figure 22-38. Many runways are fitted with recessed lights in the center to help at night or when weather creates poor visibility. A series of lights is also placed in rows along the sides of the runways. These lights help the pilot determine the runway width. Lights at the end of the runway blink in a sequence indicating the direction of travel.

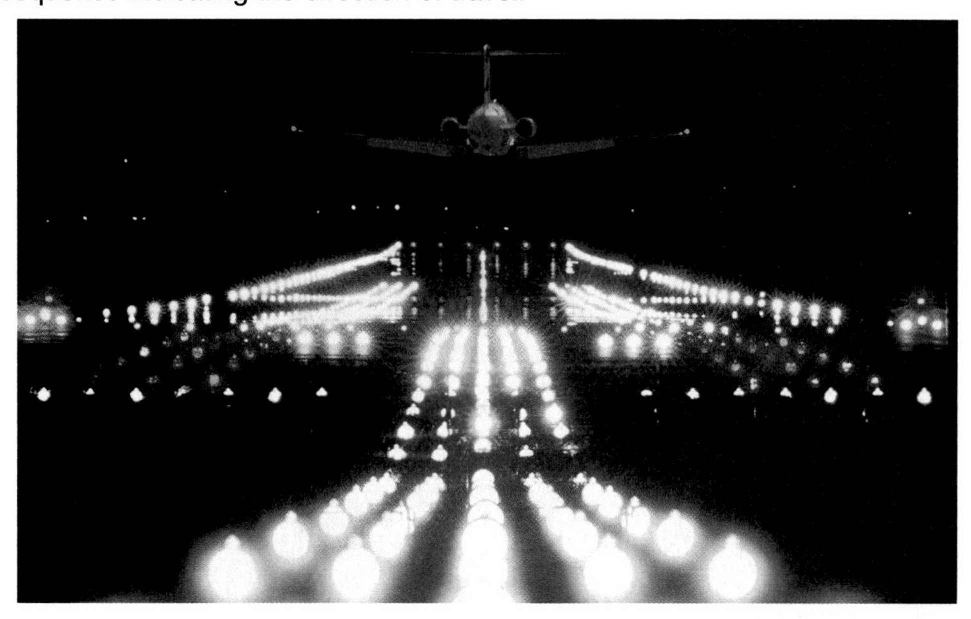

Airport Terminals

Passenger and cargo terminals are often very large at airports in major cities. See **Figure 22-39.** Passenger terminals in airports have all the services of other transportation terminals. Ticket counters, restaurants, lounges, shops, and comfortable waiting areas are available for passenger use. Separate baggage areas are in the terminal so passengers can pick up their luggage on their way to securing land transport. See **Figure 22-40.** Cargo terminals at airports are usually separate from passenger terminals. Cargo planes are directed straight to them once they land.

Aircraft Maintenance

It is extremely important that aircraft be well maintained. Proper and thorough maintenance on the ground can save lives, as well as cargo and vehicles. See **Figure 22-41.**

Inspections are a part of aircraft maintenance. The FAA requires thorough inspections on a regular basis. Because of the increasing number of accidents among aging aircraft, the FAA has increased the number of inspections. The FAA determines and enforces the inspections for all types of aircraft, commercial and private.

Various types of inspections are carried out on aircraft. See **Figure 22-42.** Because the cabins must be pressurized to fly at high altitudes, structural stress is great. Every time the vehicle takes off and gains altitude, the fuselage expands, or gets larger. When the vehicle descends to land, the fuselage contracts because of atmospheric pressure changes. This flexing is not great, by any means, but it causes metal fatigue over a long period of time. Metal fatigue results in the weakening of the material until it is

Figure 22-39. Most airport designs incorporate their facilities in the land-side–air-side layout. In this type of layout, passengers and cargo approach the airport from one side, and air traffic operates on the other. This layout permits smooth traffic flow for both land and air vehicles. Airport buildings are usually located in the center of the layout. Airport fire and rescue stations, however, are usually located away from the congested central area of the airport. In this large airport, the double roadway down the center and the semicircular parking and terminal structures are the land side. The air side consists of the runways arranged along either side of the airport; the gate areas, where planes load and unload at the terminals; and the support structures, such as cargo and maintenance facilities.

Figure 22-40. Moving baggage efficiently from check-in counters to departing aircraft and from arriving aircraft to the baggage claim areas is a major concern at large airports. This baggage handling and sorting system located in Munich, Germany uses scanners to read bar codes attached to bags and automatically route them to the correct aircraft or baggage claim area. (Siemens)

Figure 22-41. Careful maintenance of equipment is essential for safe air travel. (America West Airlines)

Figure 22-42. Inspections of aircraft engines and other systems and components are carried out on a regularly scheduled basis. These inspections are used to find faults that may cause engine, landing gear, electronic, or structural problems. The various inspections airplane support crews perform range from visual inspections to highly complex electronic inspections. (United Airlines)

prone to breaking. Should the metal fuselage break, air disasters can easily happen. Inspectors were once able to find cracks in the fuselage by locating small brown stains on the airplane. Tobacco smoke–filled air leaking from the fuselage caused the stains. Now that smoking is not allowed on most flights, more advanced methods are being used. Bright lights and cleaning solutions help inspectors see small cracks.

X-ray techniques have been developed to study internal components of aircraft. The X rays can produce images of very complex parts. The problem is that the pictures are hard to read and interpret. Misread X rays have led to crashes.

Another type of inspection uses eddy currents. A magnetic field produces a small electrical current in a piece of metal. Any cracks will disturb the flow of current. These can be monitored on a meter or screen. Cracks that are invisible to the naked eye can be detected in metal up to 5/8″ (16 mm) thick. This type of inspection is very slow, and expensive magnetic probes must be used. Different probes are used on different parts of an aircraft.

Besides the conscious and rigorous inspections, normal vehicle maintenance must be performed on aircraft. The same types of things you would do for any other vehicle, you do for an airplane. This makes sense because vehicles can last a lot longer with proper maintenance.

Technology Link

Communication: Airline Use of Internet Services

The flight segment of air transportation is usually very efficient and much faster than any other form of transportation. There are many other things, however, that go into air travel. Airport parking, ticketing lines, security checkpoints, concourses and gates, and baggage claims can all be points of stress. The airline companies realize this and have begun to use the Internet and other communication systems in order to help travelers deal with these issues.

Many airlines have Web sites that allow passengers to check the status of their flights. One of the first uses of the Internet in air transportation was the electronic ticket (e-ticket). It is now common to purchase reservations for an airline and not receive a paper ticket. The reservation is held in a computer system and accessible at the ticket counter. Before leaving for the airport, passengers are able to determine if the flight is on time or delayed. There are Web sites that track the location, speed, elevation, and estimated time of arrival of all airplanes in flight. This gives people picking up passengers at the airport an idea of when the airplane will actually land. Flight information can even be sent to mobile devices, such as personal digital assistants (PDAs) and cell phones.

Many airlines have self check-in kiosks. These kiosks allow passengers to check themselves in and receive their boarding passes. Those passengers who check baggage to be stored in baggage compartments also benefit from information technology. Baggage tags have scan bars placed on the luggage, and the passengers receive a receipt with the baggage code. Each time the baggage is moved, it is scanned. In an instance in which a piece of luggage is lost, the baggage tag can be tracked down by computer to find the last location where the baggage was scanned. While this does not always help the baggage get to the right location, the baggage can at least be tracked down.

Lastly, travelers often find themselves waiting in airport terminals for several hours, either before a flight or on a layover. For business travelers, this often means a loss of work time. Many airports have, however, installed wireless networks. These networks allow travelers to use their laptop computers or PDAs to communicate with others using e-mail or to simply surf the Internet in an effort to pass the time. While air transportation can be frustrating at times, the use of information technology has the ability to keep passengers better informed and to speed up the process of air travel.

Summary

Air vehicular systems allow both lighter-than-air and heavier-than-air vehicles to perform the act of transportation. Either the use of a reciprocating engine and a propeller or a type of jet engine generates propulsion. When an aircraft is in flight, the pilot monitors the onboard instruments and navigation equipment to make sure the aircraft has the right heading. Global positioning systems (GPSs) have begun to revolutionize how aircraft are guided and will continue to advance in the future. To keep an aircraft on course, the pilot uses aircraft control systems. Control systems in airplanes involve both stationary and moveable surfaces that change the attitude and heading of the plane. Helicopters are controlled by changing the amount of rotation in the tail rotor and the pitch of the main rotor blades. All aircraft require both suspension and structural systems. Suspension systems in airplanes rely on the wings to keep the plane in the air and on the landing gear to safely land the plane. The structural system involves the shell or system of trusses holding the plane together. Commercial airplanes contain different sections within the structural system, allowing for the pilot's cockpit, the passenger cabin, and a luggage compartment. Support systems aid all air transportation vehicles. These systems include maintenance facilities, terminals, control towers, and runways. They are most often found at airports.

Key Words

All the following words have been used in this chapter. Do you know their meanings?

aeronautical chart
afterburning turbojet engine
aileron
airfoil
airspeed indicator
altimeter
angle of attack
angle of incidence
artificial horizon
aspect ratio
attitude
camber
chord line
collective pitch control lever
control stick
control surface
coordinated turn
cyclic control stick
directional stability
elevator
empennage
envelope
fuselage
heading indicator
horizontal tailplane
instrument flight rules (IFR)
instrument landing system (ILS)
Isaac Newton's third law of motion
lateral stability
leading edge
longeron
longitudinal stability
Mach 1
monocoque
pitch
radial
radial engine
ramjet engine
roll
runway
semimonocoque
span
spoiler
stabilizer
stall
stationary surface
swash plate
taxiway
trailing edge
turbofan engine
turbojet engine
turboprop engine
vertical speed indicator
very-high-frequency omnidirectional radio range (VOR) navigation
visual flight rules (VFR)
way point
yaw

Test Your Knowledge

Write your answers on a separate sheet of paper. Do not write in this book.

1. _____ engines are configured with the pistons arranged in a circle.
2. Name the two variables that determine the amount of thrust a propeller generates.
3. What gearbox ratio is needed if the output shaft needs to turn at 1750 revolutions per minute (rpm), and the input shaft turns at 7000 rpm?
4. Jet engines operate on the principle of the _____ law of motion.
5. The simplest jet engine is the _____.
   A. turboprop
   B. ramjet
   C. afterburner
   D. turbojet
6. Explain how the turbojet engine operates.
7. *True or False?* The thrust from the rear of the turboprop engine is not the main source of propulsion for turboprop engines.
8. *True or False?* Afterburners are inefficient to operate.
9. *True or False?* The air leaving the rear of a turboprop is used to provide thrust.
10. Maps pilots use are known as _____.
11. Aeronautical charts are similar to _____ maps.
    A. political
    B. topographic
    C. road
    D. city
12. Artificial horizon instruments are used to determine _____.
    A. speed
    B. altitude
    C. pitch and roll
    D. heading
13. Summarize how the instrument landing system (ILS) operates.
14. Define the three types of stability of an aircraft.

Matching questions: For Questions 15 through 17, match the terms on the left with the correct term on the right.

15. Yaw.
16. Roll.
17. Pitch.

A. Elevator.
B. Rudder.
C. Aileron.

18. *True or False?* Propulsion systems are used to generate drag.
19. Discuss the use of the cyclic control stick in helicopter control.
20. Commercial passenger planes use _____-style landing gear.
21. Write two or three sentences explaining how an airplane flies.
22. The length from the leading edge to the trailing edge of an airfoil is the _____.
    A. chord line
    B. aspect ratio
    C. camber
    D. angle of incidence
23. Paraphrase the two explanations of the creation of lift.
24. The position of the wing as it hits the air is known as the _____.
    A. deflection angle
    B. camber
    C. chord line
    D. angle of attack
25. Determine the lift of an airfoil with the following statistics: Cl = .64, r = 0.00237 slug/ft^3, v = 300 fps, and A = 425 ft^2.
26. *True or False?* Nonrigid airships retain their shape when the air is let out.
27. The tail section is also known as the _____.
28. Identify the parts of the structural system of an airplane.

Matching questions: For Questions 29 through 31, match the phrases on the right with the correct term on the left.

29. Air traffic control tower.
30. Terminal radio detecting and ranging (radar) approach control station.
31. Route traffic control center.

A. 50 miles and further from an airport.
B. Between 5 and 50 miles from the airport.
C. Within 5 miles of the airport.

32. List three types of services available at most airports.

Activities

1. Use the FoilSim software available at the National Aeronautics and Space Administration (NASA) Web site to design and test various types of airfoils.
2. If a wind tunnel is available, test a model aircraft in the wind tunnel.
3. Construct a model of a hot air balloon and test it.
4. Build a model or display of one of the types of jet engines.
5. Create a flight plan using aeronautical charts.

23

Space Transportation Systems

Basic Concepts

- Define *spacecraft*.
- State what makes a spacecraft fly.
- Cite the definition of *orbiting*.
- Identify the two types of space transportation modes.
- List the different types of space vehicles.

Intermediate Concepts

- Describe the space environment.
- Explain the types and applications of orbits used to circle the earth.

Advanced Concepts

- Examine the impacts space technology has had on daily life.

Space transportation is the use of rockets and orbiting vehicles to explore the regions beyond the limits of the atmosphere. The space extending from 50 miles to 10,000 miles beyond the atmosphere is known as ***near space***. Beyond 10,000 miles is ***outer space***.

The History of Space Transportation

The pioneering of space travel came around the turn of the twentieth century. It was concluded that ***spacecraft***, or space vehicles, built around a large rocket would be the most effective means of escaping the earth's gravitational pull. Rockets were not necessarily a new idea. The Chinese used rockets as a weapon in warfare over 700 years ago. Eventually, rockets were used by militaries throughout the world. These rockets were effective in war, but they were incapable of launching humans into space. Three men living in different countries around the same time developed theories and models of rockets that would be capable

Near space: The space extending from 50 miles to 10,000 miles beyond the atmosphere.

Outer space: The space extending from 10,000 miles beyond the atmosphere.

Spacecraft: A vehicle used for space travel.

of space travel. These men were a Russian teacher, Konstantin E. Tsiolkovsky; an American professor, Robert H. Goddard; and a German experimentalist, Hermann Oberth.

Tsiolkovsky was the first to put his ideas on paper. In his books, *Dreams of Earth and Sky* and *Exploration of Cosmic Space by Means of Reaction Devices*, he discussed the idea of using rockets to explore the universe. In 1903, the same year as the Wright brothers' *Flyer* was flown, he first proposed the use of liquid hydrogen and oxygen as fuels for rockets. These liquid propellants are used in most rockets today. Tsiolkovsky had the practical possibilities in mind and a theory of how things would work, but he never built a rocket.

Robert Goddard, however, designed and built his own rockets. In 1926, Goddard launched the world's first liquid-fuel rocket. See **Figure 23-1.** He also demonstrated how rockets could carry scientific instruments into the upper atmosphere. His work led to hundreds of patents and paved the way for manned and unmanned space vehicles.

After World War I, Oberth was at work in Germany on the development and testing of rockets. His discoveries led to the publication of *The Rocket into Interplanetary Space*, which explained how rockets could escape the earth's gravitational pull. The discoveries of these three men became the basis for all later space transportation developments. The first successful spaceflight came in 1957.

An examination of the people involved in early rocket development and space exploration would not be complete without the mention of Wernher von Braun. Von Braun was a German rocket expert who had studied under Hermann Oberth. His work in Germany led to the development of the V-2 rocket the German army used during World War II. The V-2 rockets were capable of destroying targets over 120 miles away from their launch site. Von Braun had hoped his inventions would lead to exploration, however, and not destruction. Toward the end of World War II, he and his team of engineers surrendered to American forces. His "rocket team" developed many of the rockets that launched American satellites, probes, and even space stations into space.

Figure 23-1. Dr. Robert H. Goddard was a pioneer of American rocketry. He is shown with his first liquid-fuel rocket, which he launched in Auburn, Massachusetts, on March 16, 1926. Both the Smithsonian Institute and Charles Lindbergh financed Goddard's inventions and test flights. (National Aeronautics and Space Administration)

These rockets were not the first, however, to deliver an object into space. October 4, 1957 marked the beginning of a new age, a new dimension in transportation and exploration. The bulletins flashed around the world, carried by radio, newspapers, and television. The Soviet Union had launched the first successful artificial satellite and placed it in orbit around the earth. The satellite was called *Sputnik 1.* See **Figure 23-2.** The word *sputnik* means "traveler."

This event came as a surprise to many nations, including the United States. Reacting to a deep concern that the Soviet Union was taking the lead in the exploration of space, the U.S. federal government mobilized a program in rocket development. The response of the United States was to put the nation's first satellite, *Explorer 1*, into orbit. See **Figure 23-3.** For several years, a contest was waged between the United States and the Soviet Union. Each country was trying to outdo the other in putting new satellites into earth's orbit.

On April 12, 1961, the Soviet Union amazed the world by announcing it had just put a man into earth's orbit. At that time, the United States, feeling very much in second place because of the Soviets' accomplishments, set a

Figure 23-2. This full-scale mock-up of *Sputnik 1* was placed on display in the Soviet pavilion at the Paris air show.

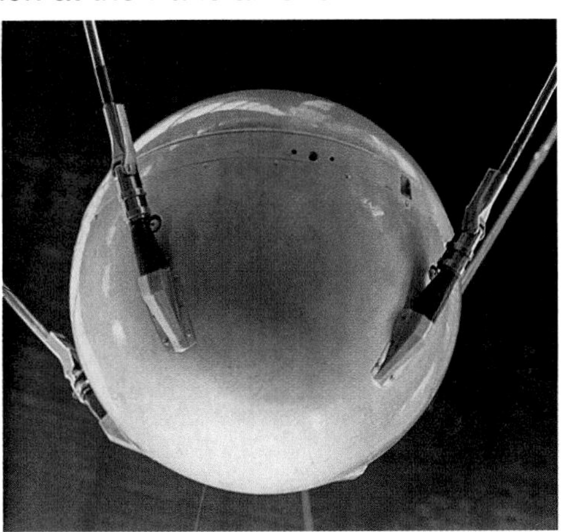

Figure 23-3. *Explorer 1*, shown in orbit in this artist's view, was less than 6′ in length. It was the first U.S. satellite and was launched in 1958, one year after Russia's *Sputnik 1*. (National Aeronautics and Space Administration)

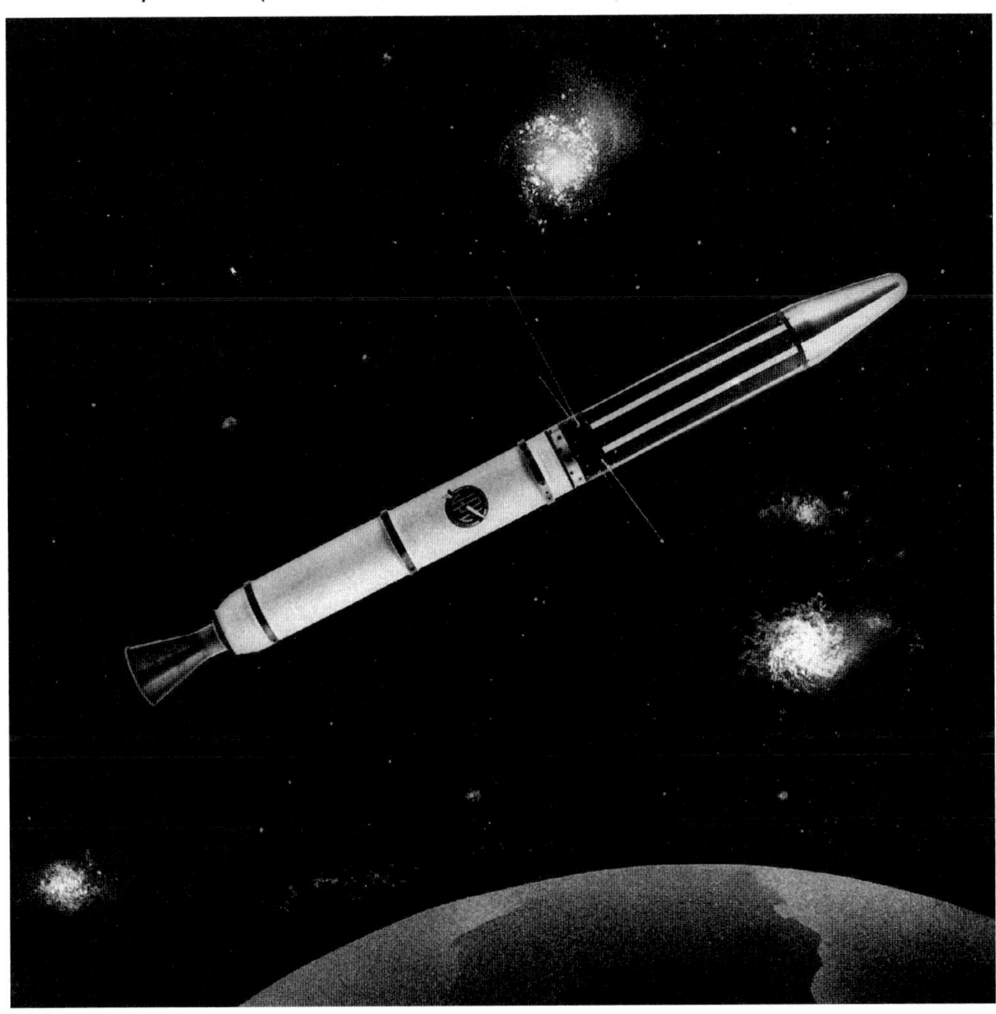

much more difficult goal: placing a human being on the moon! John F. Kennedy, then president of the United States, placed a high priority on the project and set it in motion. After some eight years of intensive planning and work, the United States launched the first manned flight to the moon. See **Figure 23-4.**

Figure 23-4. The historic moon landing in 1969. On July 20, U.S. astronaut Neil Armstrong became the first man to set foot on the moon. Armstrong was the commander of the *Apollo 11* spaceflight. The success of the spaceflight depended on new advanced technology. The technology had to enable the flight crew to complete a lunar landing and then return the space travelers safely to earth. Astronaut David Scott, commander of the mission, stood beside the U.S. flag at the landing site. The vehicle on the right served as a "space buggy," allowing the astronauts to explore a larger area than they could on foot. (National Aeronautics and Space Administration)

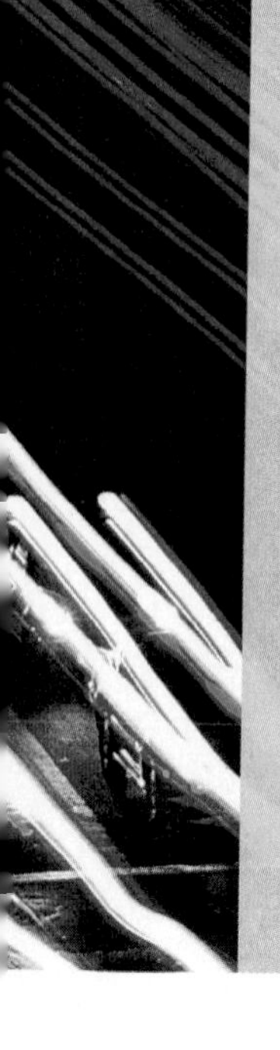

Career Connection

Astronauts

Traveling to space and conducting missions is not an average job. It makes sense then that the astronaut selection process does not target average applicants. The National Aeronautics and Space Administration (NASA) selects candidates for astronaut training every two or three years. Applicants can be either civilian or military personnel and must have at least a bachelor's degree in engineering, biological science, physical science, or mathematics. They must be in good physical condition and pass rigorous medical physicals.

Astronaut candidates can fill one of two roles: pilots or mission specialists. Pilots must have previous jet-piloting experience, most often gained in the military. They can also serve as commanders and have responsibility for the crew and mission. Mission specialists must have advanced degrees, work experience, or both to apply. Their role includes maintaining the orbiter's systems, conducting space walks, and conducting experiments in space.

Candidates receive several years of training, which includes survival training, mission simulations, and payload training. Civilian astronauts are paid on the government pay scale, earn $60,000 to $90,000 a year, and must serve for five years. Military astronauts are paid according to their rank and serve a specified tour of duty with NASA.

The moon launch was the culmination of several space projects the National Aeronautics and Space Administration (NASA) oversaw. NASA is the primary organization in charge of the U.S. space program. It was created in 1958 as a result of the Soviet launch of *Sputnik 1*. NASA was formed by combining several government agencies, including the National Advisory Committee for Aeronautics (NACA), into one organization. It has several overall goals, which include more than just traveling and exploring space. The mission of NASA includes advancing the level of understanding about the earth and universe. Another goal is to conduct research in space. An often-overlooked goal that has impacted all of us is the development and transfer of new technology. See **Figure 23-5.**

NASA is not the only space agency in the world. Many other countries either have an agency of their own or are in a partnership with other countries. The Canadian government organized the Canadian Space Agency (CSA) in 1990. Canada even developed the robotic arm for NASA's space shuttle program. Japan, Russia, and Italy also have their own space organizations. The European Space Agency (ESA) includes a number of member countries. Great Britain, Germany, France, Spain, and many other countries are members of the ESA. Today, these space agencies regularly send satellites, probes, and spacecraft past the earth's atmosphere. The space race between the Soviet Union and the United States may have ended when *Apollo 11* landed on the moon, but the technology has continued to advance. Each launch and mission adds new information about both earth and space.

The Space Environment

The atmosphere of the earth has several characteristics important to space travel. See **Figure 23-6.** The different regions of the earth's atmosphere include the troposphere, stratosphere, mesosphere, thermosphere, and exosphere.

Figure 23-5. Transferred technology resulting from the space program is widely used. Inventions such as racing suits, pacemakers, and Teflon® coating are examples of products known as spin-offs. These products were developed by the National Aeronautics and Space Administration (NASA) for the space program, but later found successful uses here on earth. Shown are three of the hundreds of devices and processes developed in this way. A—Robotic surgery. B—Hydrogen fuel cells. C—Earth Observation System (EOS) satellites use arrays of sensors to monitor global climate change. (NASA)

A

B

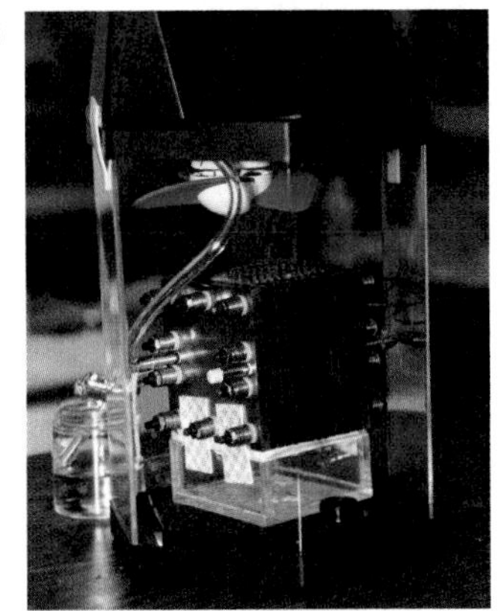

C

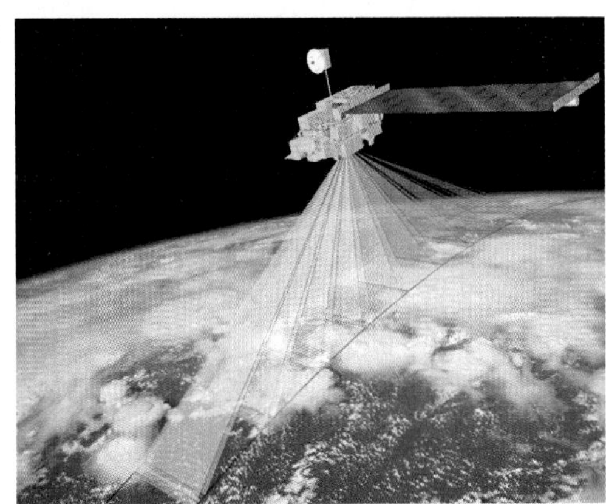

Troposphere: The closest atmospheric region to the earth. It begins at the earth's surface and stretches about 10 miles (16 km) above the earth's surface.

Stratosphere: The atmospheric region ranging from the troposphere to about 30 miles (50 km) above the earth's surface.

Atmospheric Regions

The ***troposphere*** is the closest layer to the earth. It begins at the earth's surface and stretches about 10 miles (16 km) from the earth. Most clouds and weather patterns originate in this region. The next region, the ***stratosphere***, ranges from the troposphere to about 30 miles (50 km) above the earth's surface. In this region, there is an absence of water vapor and clouds. The stratosphere also contains an ozone layer that helps absorb ultraviolet (UV) radiation from the sun. The ***mesosphere*** is located above the stratosphere and reaches about 50 miles (80 km) above the earth's surface. It is the coldest atmospheric layer. Clouds of frozen water vapor can actually exist in this region. The next region is a layer that extends from about 50 miles to about 300 miles out. This layer is known as the ***thermosphere***. In this region, the atmosphere is so thin that no sound is transmitted. The extreme outer region of the atmosphere, before getting to outer space, is a region known as the ***exosphere***. It is hard to estimate the top boundary of the exosphere because it gradually becomes outer space. Outer space is very different than the areas inside the earth's atmosphere. It contains virtually no air particles.

Figure 23-6. There are five different layers in the space environment surrounding the earth. The closest, the troposphere, is 10 miles from earth's surface. The farthest, the exosphere, is 500 miles from earth.

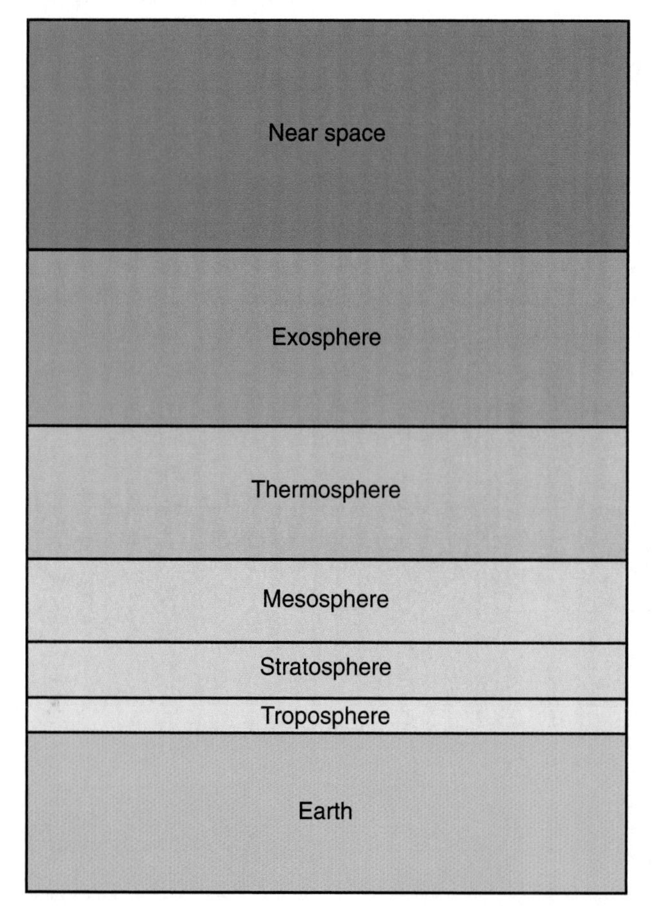

The Space Environment and Vehicle Design

Space environment characteristics affect the design and development of a spacecraft. Among the most important characteristics are the extremes of temperature and radiation levels. These have a direct effect on the materials used in the construction of space vehicles. Both temperature and radiation can damage the spacecraft in various ways. The space environment can affect the structure, instruments, and communication. Besides the effect it would have on the craft itself, there is also a concern for the health of personnel on manned spaceflights. Radiation, in particular, can seriously affect health, if precautions are not taken.

One other space environment concern is both natural and human made. Spacecraft designers must consider the possibility of the craft running into either a meteoroid or a piece of space debris. Meteoroids are the remains of comets in space. Space debris includes the human-made pieces of satellites, hardware, and rockets that have not fallen to earth. It is a major concern for spacecraft engineers and designers. The collision with an object in space could be very dangerous for a spacecraft. NASA's Orbital Debris Program Office tracks space debris and conducts research to help

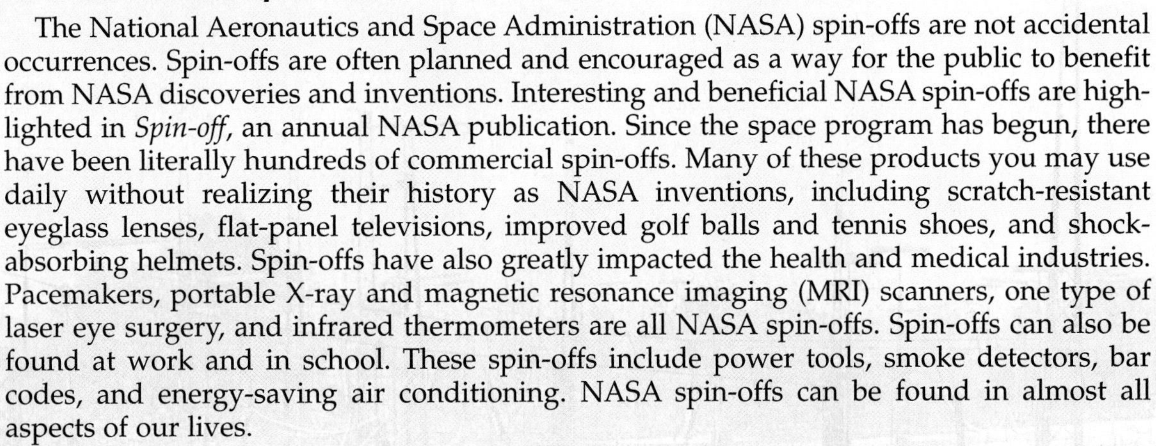

Curricular Connection

Social Studies: Spin-offs

The National Aeronautics and Space Administration (NASA) spin-offs are not accidental occurrences. Spin-offs are often planned and encouraged as a way for the public to benefit from NASA discoveries and inventions. Interesting and beneficial NASA spin-offs are highlighted in *Spin-off*, an annual NASA publication. Since the space program has begun, there have been literally hundreds of commercial spin-offs. Many of these products you may use daily without realizing their history as NASA inventions, including scratch-resistant eyeglass lenses, flat-panel televisions, improved golf balls and tennis shoes, and shock-absorbing helmets. Spin-offs have also greatly impacted the health and medical industries. Pacemakers, portable X-ray and magnetic resonance imaging (MRI) scanners, one type of laser eye surgery, and infrared thermometers are all NASA spin-offs. Spin-offs can also be found at work and in school. These spin-offs include power tools, smoke detectors, bar codes, and energy-saving air conditioning. NASA spin-offs can be found in almost all aspects of our lives.

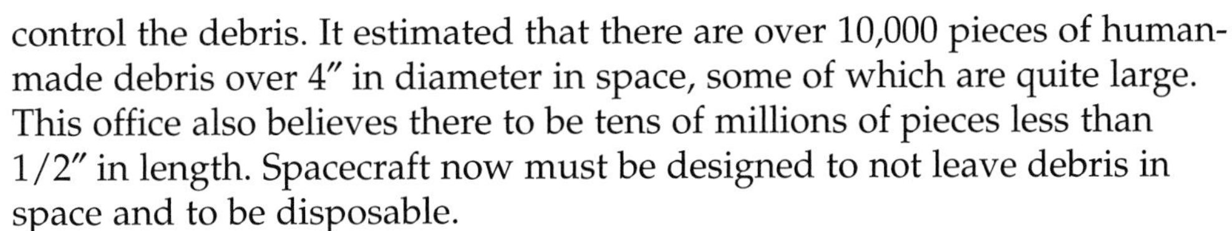

control the debris. It estimated that there are over 10,000 pieces of human-made debris over 4″ in diameter in space, some of which are quite large. This office also believes there to be tens of millions of pieces less than 1/2″ in length. Spacecraft now must be designed to not leave debris in space and to be disposable.

Another very important characteristic of space is weightlessness. ***Weightlessness*** occurs due to the forces around bodies in space. See **Figure 23-7.** Both pull and push effects take place. Gravitational forces pull the body toward the earth, and a centrifugal force pulls the body away. When gravitational forces are equal to a centrifugal force, weightlessness occurs. When you are jumping up and away from a diving board, you experience weightlessness as you reach the height of your jump.

Weightlessness has always appeared to be fun, as we see films of the astronauts in space or read about their experiences. It could be fun for a time. Weightlessness can also become frustrating as astronauts go about their daily routines. For instance, drinking liquid and taking a shower could be difficult.

Spacecraft Flight

Many inputs are needed in the launching of a spacecraft. Tools, energy, materials, money, and people are all needed to begin the process. It is the scientific principle outlined in Sir Isaac Newton's third law of motion, however, that allows the spacecraft to lift off. Newton's law states that for every action, there is an equal and opposite reaction.

Mesosphere: The atmospheric region ranging from the stratosphere to about 50 miles (80 km) above the earth's surface.

Thermosphere: The atmospheric region extending from the mesosphere to about 300 miles above the earth's surface.

Exosphere: The extreme outer region of the atmosphere, before outer space. This region is located from the thermosphere to over 5000 miles from the earth's surface.

Weightlessness: The condition occurring when gravitational forces are equal to a centrifugal force.

Launchpad: A nonflammable platform from which a rocket, launch vehicle, or guided missile can be launched.

Figure 23-7. Astronauts practice working in a weightless environment by performing tasks in a large water tank. Their suits are carefully weighted to provide neutral buoyancy. (National Aeronautics and Space Administration)

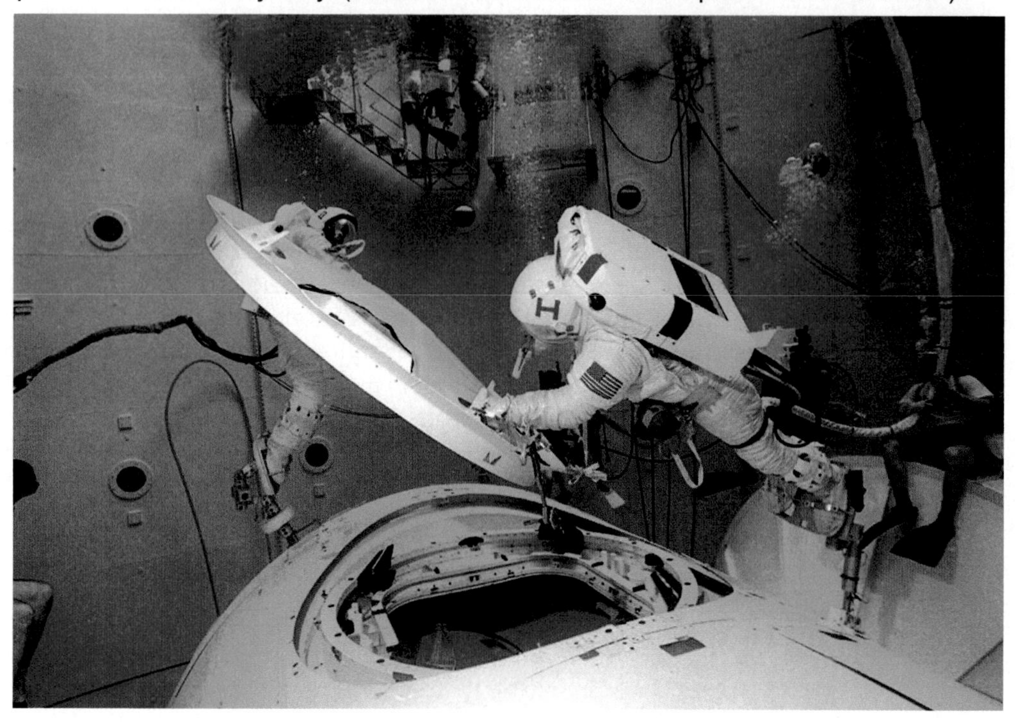

Figure 23-8. The unequal pressures inside a rocket engine propel it forward. If the rocket is aimed skyward, an upward motion is the reaction to such a force. The escaping of the gases from inside the rocket is the action. The upward movement of the rocket is the reaction. This movement is known as thrust. Thrust is a force that produces motion in a body. It is measured in pounds or newtons.

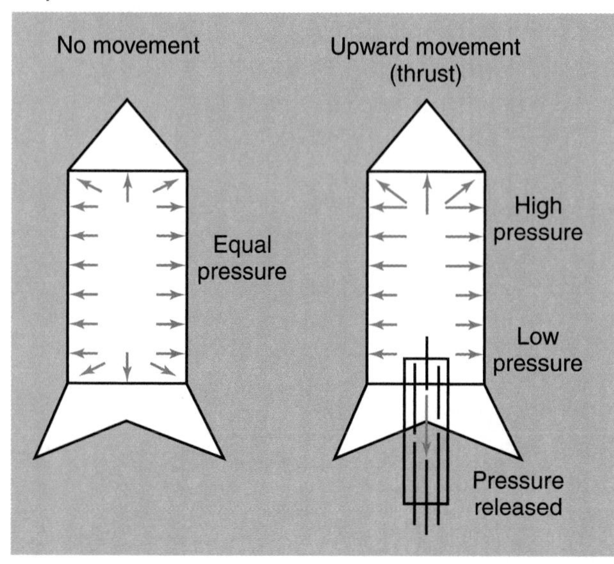

The burning of fuel creates the pressure inside a rocket engine. As the fuel burns, the pressure increases, causing a great pressure buildup within the engine. The exhaust of the engine allows for the release of pressure so there is higher pressure at the front of the rocket engine than at the tail. When the pressures within a rocket engine are unequal, the rocket will move toward the direction from which the higher pressure is exerted. See **Figure 23-8.**

In order for the spacecraft to leave the ***launchpad***, the platform designed to support the space vehicle on the ground and withstand the impact of takeoff, the reaction must be greater than the action. The thrust must be greater than the gravitational pull of earth and the weight of the vehicle. In the case of the space shuttle, the rockets must overcome a large amount of weight. The weight of the loaded shuttle, tanks, and rocket boosters at takeoff is over 4.5 million pounds. The rocket boosters and shuttle main engines both operate at takeoff and together generate over 7.8 million pounds of thrust. This amount of

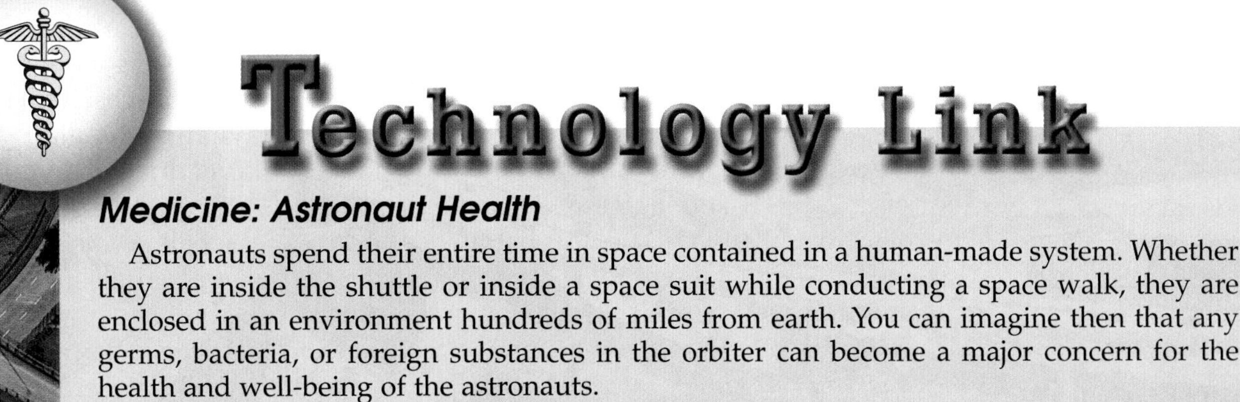

Medicine: Astronaut Health

Astronauts spend their entire time in space contained in a human-made system. Whether they are inside the shuttle or inside a space suit while conducting a space walk, they are enclosed in an environment hundreds of miles from earth. You can imagine then that any germs, bacteria, or foreign substances in the orbiter can become a major concern for the health and well-being of the astronauts.

Monitoring of the shuttle air and the astronauts' health is vitally important. The National Aeronautics and Space Administration (NASA) division of AstroBionics is responsible for creating sensors that monitor oxygen and carbon dioxide (CO_2) gas levels, heart rate, blood pressure, and other health statistics. These sensors are known as biosensors. NASA's goal is to create biosensors that can be implanted into astronauts and monitored from earth.

Physical fitness is just as important in space as it is on earth, and possibly even more important because spaceflight causes 1–2% bone loss per month in space. Astronauts who spend extended periods of time in space must remain physically active in order to reverse the effects of spaceflight. Since there is no gravity in space, however, there is very little resistance for the astronauts' muscles. NASA has developed several pieces of equipment that enable the astronauts to work out. The astronauts can lift weights with the Resistance Exercise Device (RED), run on a treadmill using the Vibration Isolation System (VIS), and ride a Cyclergometer, which is a cycle built for zero gravity.

Lastly, it is possible for astronauts to become either injured or ill while in space. The shuttle is equipped with two medical kits. One kit, the medications and bandage kit (MBK) contains pills, gauze, and bandages to be used if an astronaut is ill or has a minor injury. The other kit, the emergency medical kit (EMK) is used for more serious injuries or infections. It contains injectable medications, medical instruments, thermometers, and supplies for minor surgery. If serious problems occur, doctors are available at the Johnson Space Center and can talk the other astronauts through any procedure that might need to be completed.

thrust enables the shuttle to generate a speed of over 16,800 miles per hour (mph). Such a high speed allows the shuttle to escape the gravitational pull of the earth.

Once this upward force ceases, gravitational forces will cause it to fall back to earth. Therefore, the spacecraft needs to achieve a very high speed so it can escape the earth's gravitational pull. At this time, the gravitational forces and the centrifugal forces are equal. As a result, the spacecraft will stay in orbit. To ***orbit*** is simply to stay in a path that circles an object, which in many cases, is the earth.

Orbit: To stay in a path circling an object in space.

Orbiting

If a spacecraft increases its speed while in orbit, a greater centrifugal force will result, and the spacecraft could possibly be slung out of orbit, into space. If the spacecraft loses speed, the centrifugal force will decrease, and the spacecraft will achieve a lower orbit or reentry into the earth's

Low earth orbit (LEO): An orbit between 180 and 250 miles above the earth.

Geosynchronous (GEO) orbit: A geostationary orbit. It is often used for communication satellites, which are stationed in one spot and rotate along with the earth.

Apogee: The point in the path of an elliptical orbit farthest from the earth.

Perigee: The point in the path of an elliptical orbit closest to the earth.

atmosphere. The concept of orbiting is like swinging a ball on a string. You can demonstrate this by taking a small ball tied to a string about 3′ long and, while holding the end of the string in your hand, whirling the ball around. As you whirl the ball around slowly, with the string extended to its full length, you can understand what a satellite in orbit experiences. You have two forces acting: a centrifugal force and gravitational pull. If you decrease the speed of the ball moving in its circular path, the ball falls out of its orbit. If you increase the speed of the ball's orbit, the centrifugal force is greater, and you will probably lose control of the ball on the string. Thus, the ball will fly away from you at a rapid speed. Like a satellite, it will fly out of orbit if traveling at too great a speed.

There are several different types of orbits a spacecraft can follow. ***Low earth orbit (LEO)*** involves orbiting between 180 and 250 miles above the earth. This orbit requires the least launch energy and is the lowest in which to place a satellite. Objects in LEO are able to make a complete revolution around the earth in 90 minutes. The International Space Station (ISS), as well as many weather satellites, are located in LEO. Polar orbits are a type of LEO that orbits in a north-south direction and crosses the two poles. Since they are orbiting north to south and the earth is rotating east to west, satellites in this orbit can view the entire earth. Satellites used for imaging are often in polar orbits.

Geosynchronous (GEO) orbits, or geostationary orbits, are used for most communication satellites because they are stationed in one spot and rotate along with the earth. GEO satellites are placed in what is known as the Clarke Belt. Arthur C. Clarke calculated the distance from the earth that would be required to keep an object in one place. When objects are placed in the Clarke Belt, 22,300 miles from the earth, they make one revolution every 24 hours. Because of GEO orbits, our television satellite dishes can be aimed in one direction and do not have to be rotated.

Elliptical orbits follow an oval path. See **Figure 23-9.** The highest point of the path is the farthest away from earth. This point is called the ***apogee***. The point closest to the earth is called the ***perigee***.

Figure 23-9. Many satellites and spacecraft follow elliptical orbits. They come close to the earth for a period of time and are farther from the earth at other times. The apogee is the point farthest from earth. The perigee is the point closest to earth.

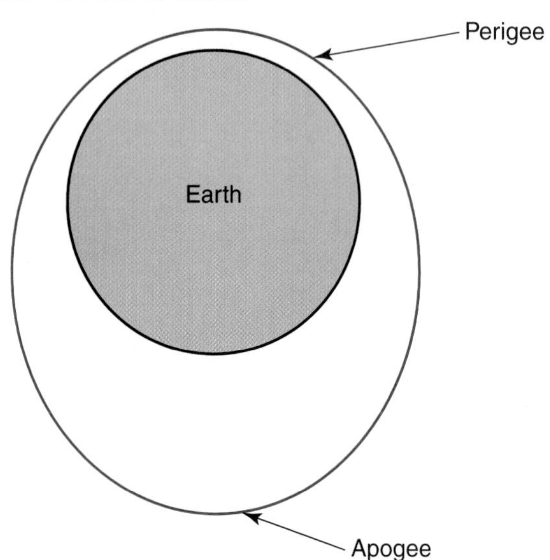

Space Vehicles

The spacecraft is the term for the space vehicle that actually travels into space. Spacecraft include sounding rockets, satellites, space probes, space shuttles, and space stations. They differ greatly in size, shape, and purpose. Some spacecraft are designed to fly by distant planets, while others are used to deliver objects into orbit. There are some used only to get into space, and there are others that are unloaded and used once they are in space. One vehicle unloaded and used in space is a jet pack. A ***jet pack*** is strapped to the back of an astronaut. Jet packs are vehicles that are not used to transport people into space, but they are essential for human transportation while people are in space. Some spacecraft are actually operated by human beings, and some are not. Therefore, the two modes of space transportation are manned and unmanned vehicles.

Jet pack: A space vehicle strapped to the back of an astronaut.

Unmanned Space Vehicles

Unmanned space vehicles include launch vehicles, satellites, space probes, and sounding rockets. They have been used for space exploration. The following paragraphs will briefly introduce you to the different unmanned spacecraft.

Unmanned space vehicle: A space vehicle not operated by human beings.

Launch vehicles

Launch vehicles are the workhorses of the space program. These vehicles are used to place other spacecraft into the atmosphere or even outer space. Whether the spacecraft is a satellite, a piece of a space station, or the space shuttle, it requires a launch vehicle to get it off the ground. The first launch vehicles were ballistic missiles that were converted to rockets. Today, the United States uses three main launch vehicles. These vehicles are the Delta, Titan, and Atlas/Centaur rockets. Each type of rocket has different capabilities and uses. Delta rockets were used to send the *Pathfinder* spacecraft to Mars. See **Figure 23-10.** The most powerful launch vehicle the United States used was the Saturn V, which is no longer in service. The Saturn V was used to send humans to the moon.

Satellites

A satellite is simply any object that orbits around another object in space. The earth is a satellite of the sun, and the moon is a satellite of the earth. Since 1957, with the launch of *Sputnik 1*, humans have been placing artificial satellites in orbit around earth. These artificial satellites are of many designs and purposes. Communication satellites bounce telephone, television, and radio waves from one transmitter on earth to another. See **Figure 23-11.** Environmental satellites are launched into space to monitor the conditions of the earth. Astronomical satellites are used for scientific research about our solar system and beyond. See **Figure 23-12.** Navigational satellites are used more frequently with the availability of global positioning system (GPS) receivers.

Figure 23-10. Currently, rockets are the only reliable method of launching spacecraft. The National Aeronautics and Space Administration (NASA) widely uses the Delta rocket as a launch vehicle. This launch, in January 2005, carried the *Deep Impact* spacecraft into space. The objective of the Deep Impact program is to collide with a comet and investigate its chemical makeup. (NASA)

Figure 23-11. Because of satellites, we can have instant communication around the world. A signal from Indianapolis, Indiana, bounced off the satellite, can be picked up almost instantaneously in Seoul, South Korea.

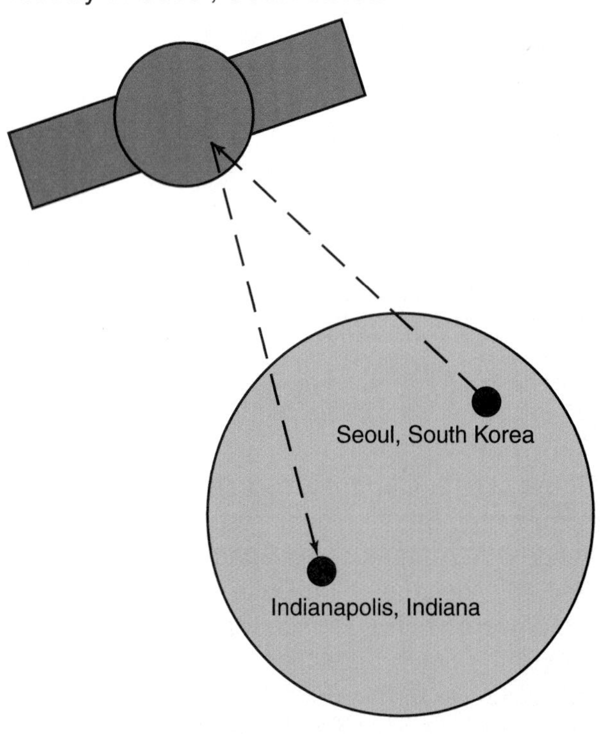

Space probes

Space probes are launched very far from the earth, where they can escape the earth's gravitational attraction. They are used for research purposes. These probes are sent to explore outer space. Probes can be designed to fly by, orbit, land, or rove. Flyby probes travel near objects, but not close enough to be affected by their gravitational pull. *Voyager 2* was a flyby probe launched in 1977. Its mission lasted until 1989, and it reached Jupiter, Saturn, Uranus, and Neptune. Orbiting probes travel to a distant planet and then orbit around the planet. See **Figure 23-13.** Landing probes are designed to touch down at their destinations. The Surveyor probes landed on the moon prior to the human landing. Roving probes are usually placed inside landing probes. See **Figure 23-14.**

Sounding rockets

Sounding rockets were the first type of spacecraft to be launched. They are able to travel above the range of aircraft. The rockets used today range from 7′ to 65′ tall. Sounding

Figure 23-12. An astronomical satellite may be launched to take pictures of the moon, other planets, or landing sites for astronauts. The most famous astronomical satellite may be the Hubble Telescope. The Hubble Space Telescope orbits the earth every 97 minutes at an altitude of 353 miles. It has provided extraordinary high-quality images of space objects light-years away from earth. (National Aeronautics and Space Administration)

Figure 23-13. The *Odyssey* spacecraft has been orbiting the planet Mars since late 2001. In this artist's conception, the spacecraft passes over the South Pole of Mars. (National Aeronautics and Space Administration's Jet Propulsion Laboratory)

Figure 23-14. Once a rover has landed on the surface, it can be driven remotely from earth. Two exploration rovers, *Spirit* and *Opportunity*, landed successfully on opposite sides of Mars in January 2004 and provided more than a year of photographic surveying and data gathering. The *Spirit* rover was used to explore Mars and collect information about the soil. This photo, taken by *Spirit*, shows the rover's robotic arm deploying a microscopic imager. The imager takes high-resolution, extreme close-up images to help scientists analyze rocks and soils on the Martian surface. (National Aeronautics and Space Administration's Jet Propulsion Laboratory)

rockets carry payloads that are ejected during flight and fall back to earth. During the fall, the payload is used to conduct experiments or gather data. Sounding rockets gather information about the sun and stars. They use electronic devices to retrieve information. Sounding rockets can measure temperatures, take photographs, and record important data. They also gather information about the sun's radiation and solar activity. Sounding rockets have collected some very valuable information about space that has aided scientists in new discoveries.

Manned Space Vehicles

Manned space vehicle: A vehicle sent to space with a crew in it.

Manned space vehicles are those vehicles sent to space with a crew in them. In the late 1950s and early 1960s, several manned spacecraft flights were tested. A space project known as Project Mercury tested several manned spaceflights. U.S. astronauts Alan Shepard and Gus Grissom were the first two Americans in space. They were in space for just 15 minutes each in 1961. In 1962, however, John Glenn became the first American to orbit the earth. His orbit was in a Project Mercury capsule named *Freedom 7*. By the end of 1963, three other astronauts had orbited the earth. When Project Mercury came to an end, it had accomplished its goals of orbiting the earth, giving Americans a chance to test their ability to function in space, and returning crew members and spacecraft safely to earth. Project Gemini was the second of the projects for manned spaceflights. Its main purpose was to continue exploration of space and resolve some other technological problems before attempting to land anyone on the moon. The *Gemini*, meaning "twin," could hold two astronauts, unlike the Mercury capsule. Project Gemini saw many space firsts, including the first American space walk. It was a stepping-stone for Project Apollo. The goal of Project Apollo was to put a man on the moon and bring him home safely. On July 20, 1969, after centuries of dreaming of setting foot on the moon, this dream came true. Three Americans, Neil Armstrong, Michael Collins, and Edwin Aldrin, were aboard *Apollo 11* as it traveled to the moon. Collins stayed above the surface of the moon in the command module. Armstrong and Aldrin landed and set foot on the lunar surface in the lunar module. See **Figure 23-15.** The Apollo capsule was the main vehicle used to transport people into space until the early 1980s.

Figure 23-15. The manned spacecraft *Apollo 11* landed on the moon in July 1969. A—The crew, left to right, Neil Armstrong, commander; Michael Collins, command module pilot; and Edwin Aldrin, Jr., lunar module pilot. They spent 21.6 hours on the moon before returning safely to earth. B—Aldrin descended the steps of the lunar module ladder in preparation for a walk on the moon. (National Aeronautics and Space Administration)

A

B

The space shuttle

In 1981, the first reusable space transportation vehicle was put in use. This vehicle is known as the space transportation system (STS), or the space shuttle. The space shuttle was developed primarily to be a reliable and reusable means of space transportation. It has been used for transporting data-gathering equipment into space. The STS has been used to transport pieces of the ISS into space, as well as to deliver crew members to the station. It is also useful in docking with space equipment in need of repair.

A space shuttle is made up of two solid rocket boosters, an external fuel tank, and an orbiter. See **Figure 23-16.** The components of the space shuttle are assembled and moved to the launch site on a mobile launchpad. Once it is ready to be launched, the countdown procedure begins. Approximately one hour after the initial countdown, the orbiter achieves orbit. See **Figure 23-17.**

A total of six orbiters have been built. The first shuttle, *Enterprise*, was built for testing and has never been in space. See **Figure 23-18.** The other five, *Columbia*, *Discovery*, *Atlantis*, *Endeavor*, and *Challenger*, have all been to space on several missions. Unfortunately, of the five orbiters built for space, only three remain. Shuttles have been used for over twenty years. During this time, there have been two catastrophic events. In 1986, after a very successful year of space travel, tragedy struck *Challenger*. *Challenger* burst into flames just seconds after liftoff. See **Figure 23-19.** The following year, 1987, the United States sent no astronauts into space. The space program was grounded, while NASA examined, researched, and retested space shuttle transportation. Shuttle flights resumed in the 1990s, with a number of the scientific missions including crew members from several nations. In 2003, after 15 years of successful operations, tragedy struck again. The space shuttle *Columbia* broke up over Texas during reentry.

Space stations

The Soviet Union launched *Salyut 1*, the world's first space station, in 1971. Two years later, on May 14, 1973, the United States

Figure 23-16. The space shuttle consists of two booster rockets, an external fuel tank, and an orbiter.

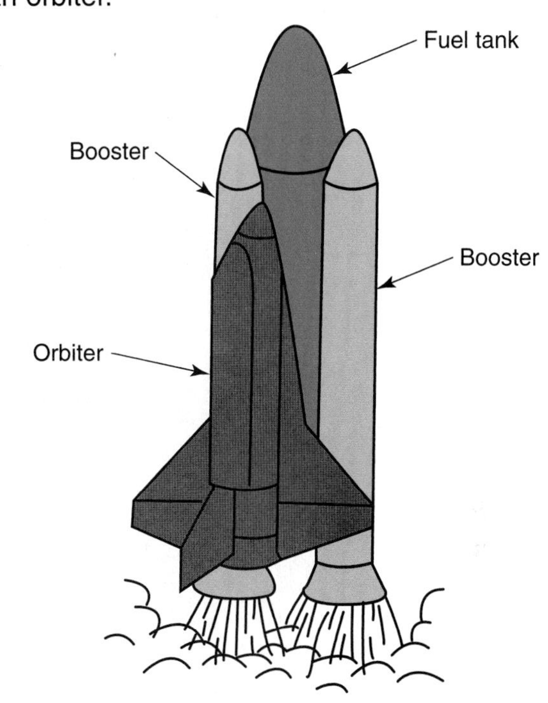

Figure 23-17. A launch of the shuttle. At liftoff, the rocket engines ignite at the same time. Two minutes after liftoff, the rockets separate from the space shuttle. A parachute opens, and the rockets slowly drop into the ocean, where tugboats collect them. Nine minutes into flight and just prior to entering orbit, the shuttle's external fuel tank separates. As it reenters the earth's atmosphere, it burns up.

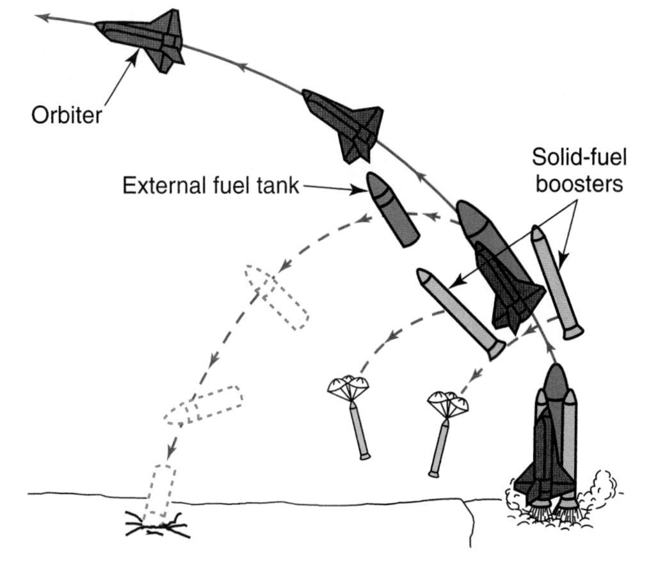

Skylab: A space station the United States launched into orbit on May 14, 1973.

International Space Station (ISS): The newest space station. It is a joint effort of 16 countries, with the United States in charge of the operations.

Figure 23-18. The shuttle *Enterprise*, built for testing, has been installed in the Smithsonian Institution's Udvar-Hazy Air and Space Museum near Washington, D.C. It is shown here being readied for exhibition. *Enterprise* is the centerpiece of the Museum's large space exploration wing.

Figure 23-19. The actual launch of the space shuttle *Challenger* on January 28, 1986. An accident 73 seconds after liftoff claimed both vehicle and crew. The nation looked on in horror, as this tragic accident killed all seven crew members. (National Aeronautics and Space Administration)

launched a space workshop and laboratory into orbit. This space station is known as ***Skylab***. Throughout the six years *Skylab* was orbiting the earth, three different crews were launched to work in its laboratories and perform experiments. The final launch was the longest mission, lasting 84 days. This mission was a great success. During this time, many studies were conducted, leading to important discoveries. Over the six years, this pioneer space station had compiled some very important and impressive statistics and records. It set the pace for future programs in space. The Soviet Union had one other very successful space station, named *Mir*. *Mir* was a major improvement over both *Salyut 1* and *Skylab*. The station was occupied continuously for over nine years. Dr. Valeri Polyakov, a Russian cosmonaut, set a new record for number of days in space. He was aboard *Mir* for 438 straight days. American astronauts visited *Mir* often to study the station. The information they gathered was essential to the development of the newest space station, the ISS. See **Figure 23-20.** The ***International Space Station (ISS)*** is a joint

Tech Extension

The X Prize

Space transportation has generally been a government-funded activity. In 1996, however, this changed. The X Prize Foundation was formed, in order to encourage private investors and companies to enter the space transportation industry. The encouragement was in the form of a $10 million prize. The prize was offered to the first team of engineers that could design and fly an efficient and safe space plane. The vehicle had to be privately financed without government aid. It had to reach the height of 62 miles above earth and return safely home. The vehicle had to then be flown into space again within 14 days after the first flight, without major modifications to the vehicle.

There were a total of over 20 teams that registered for the X Prize competition. These teams represented seven countries around the world, including the United States, the United Kingdom, Russia, and Israel. The designs varied from planes that took off on traditional runways, planes launched on rockets, and planes deployed in the air from other aircraft. On October 4, 2004, the spacecraft *SpaceShipOne* claimed the X Prize. *SpaceShipOne* was launched from the aircraft *White Knight* on September 29, 2004 and again on October 4. The spacecraft was designed by Burt Rutan and Scaled Composites and reached an altitude of 71.5 miles. This historic pair of flights may serve as the start of space tourism.

Figure 23-20. The International Space Station (ISS) is being built in stages over a period of years. The assembly of the ISS began in 1998, when astronauts attached the first two pieces, the Russian *Zarya* control module and the U.S. *Unity* connecting module, in space. This photo was taken in December 2000, after installation of a key component—the 240′-long, 38′-wide solar array (the "wings" at the top of the photo). By the end of the assembly, there will be a total of 46 U.S. and Russian spaceflights aimed at constructing the station. The station will allow for a constant presence in space. Astronauts and cosmonauts have already spent a combined 1000 straight days on the ISS. This allows for science experiments that have never before been possible. (National Aeronautics and Space Administration's Marshall Space Flight Center)

effort from 16 countries around the globe. It is three times larger then *Mir* and will include U.S., Russian, Canadian, Japanese, and ESA components. The United States is in charge of the operations of the station.

Extravehicular Mobility Units (EMUs)

Extravehicular activity (EVA): A spacewalk.

In many space missions, especially in the assembly missions to ISS, it is necessary for the astronauts to walk in space. These spacewalks are often called ***extravehicular activities (EVAs)***. Because of the extreme conditions of space, the astronauts must wear specialized suits. The suits have 13 layers of material and are equipped with a heating and cooling system. They cost over $10 million apiece. In order for the spacewalker to navigate in space, the suits are equipped with jet packs. See **Figure 23-21.** In the early 1980s, two astronauts strapped backpacks on their backs. This type of jet pack device was known as an Extravehicular Mobility Unit (EMU). The astronauts are able to move around with EMUs by releasing bursts of compressed nitrogen gas, which is shot through tiny thrusters.

Figure 23-21. A jet pack or Extravehicular Mobility Unit (EMU) allows an astronaut to move around outside the space shuttle without a tether line. EMUs became the first human spaceships. Today, a smaller device known as a Simplified Aid for EVA Rescue (SAFER) is used as a jet pack. These are smaller than EMUs and are only used in case of emergency. If a tether line were to break, the astronaut could use the SAFER to propel himself back to the ship. These two astronauts are wearing EMUs as they work outside the payload bay of the shuttle *Endeavor*. On the right is the Remote Manipulator System robot arm Canada developed for the shuttle program. (National Aeronautics and Space Administration)

Summary

Space transportation has developed over a time when humans' dreams came true. For centuries, humans have dreamed of a manned flight to the moon. This dream moved a step closer to reality in 1957, when Russia launched its first successful satellite into orbit. They named it *Sputnik 1*. That day in October added a new dimension to space transportation and exploration.

Galvanized into action, the United States soon had a space program underway to research and develop space technology. The National Aeronautics and Space Administration (NASA) then, as now, controlled the program. The United States successfully placed its first satellite in orbit in 1958 and, 11 years later, placed a manned spaceship on the moon. Since that time, NASA has launched over 100 space shuttle missions. It has also launched two space stations into orbit.

The two types of space transportation modes are manned and unmanned vehicles. Unmanned space vehicles are satellites, space probes, and sounding rockets. Manned space vehicles are space shuttles, space stations, and jet packs. Both modes of space transportation are used to research, experiment with, and explore space. They represent the future of our travels into space.

Key Words

All the following words have been used in this chapter. Do you know their meanings?

apogee
exosphere
extravehicular activity (EVA)
geosynchronous (GEO) orbit
International Space Station (ISS)
jet pack
launchpad
low earth orbit (LEO)
manned space vehicle
mesosphere
near space
orbit
outer space
perigee
Skylab
spacecraft
stratosphere
thermosphere
troposphere
unmanned space vehicle
weightlessness

Test Your Knowledge

Write your answers on a separate sheet of paper. Do not write in this book.

1. Write the definition of *spacecraft*.
2. Summarize Robert Goddard's influence on space transportation.
3. *True or False?* The National Aeronautics and Space Administration (NASA) is the only space agency in the world.
4. When is a spacecraft said to be in orbit?

Matching questions: For Questions 5 through 10, match the phrases on the left with the correct term on the right.

5. Ten miles from the earth's surface.
6. A region that contains an ozone layer.
7. A region that has clouds of frozen water vapor.
8. A region in which no sound is transmitted.
9. The outer region of the atmosphere, before outer space.
10. A region beyond the atmosphere.

A. Exosphere.
B. Mesosphere.
C. Outer space.
D. Stratosphere.
E. Thermosphere.
F. Troposphere.

11. *True or False?* Space debris is a serious concern in space travel.
12. Discuss several environmental factors of space affecting the design of spacecraft.
13. Cite the principle allowing a rocket engine to work.
14. Describe two types of orbits.
15. *True or False?* The farthest point away from the earth in the elliptical path of orbit is known as a perigee.
16. Name the two modes of space transportation.
17. _____ was the first successful satellite to orbit the earth.
18. *True or False?* Satellites are used only for television broadcasting.
19. What is the main purpose of a space probe?
20. *True or False?* The *Gemini* spacecraft was the first manned space vehicle.
21. State the purpose of the space transportation system (STS).
22. The first U.S. space station was _____.

Activities

1. Design and construct your own version of a space station. Check with your resource center for books and other materials on its design.
2. Choose a specific space mission and create a display showing the history and technology used in the spaceflight.

24

Space Vehicular Systems

Basic Concepts

- State the function of a rocket engine.
- Identify the ways navigation information is collected.
- State how manned and unmanned spacecraft are controlled.
- Name the functions of orbiter suspension systems.
- Identify the structures of various types of spacecraft.
- List several support systems for space vehicles.

Intermediate Concepts

- Describe how thrusters operate in the vacuum of space.
- Explain how the Mission Control Center (MCC) operates.

Advanced Concepts

- Relate how an ion engine operates.
- Discuss the aerodynamics of a delta wing.

The goal of spacecraft is to transport people and cargo into space. In order for that to occur, space vehicles must be designed to include several vehicular systems. Propulsion systems, typically rocket engines, are used to propel the craft into space. Guidance systems are used to locate the vehicles in space. Once the location is known, the control systems are used to guide and steer the craft. Suspension systems of spacecraft are used while in the earth's atmosphere, but once the spacecraft are in space, these systems are not of much use because of the lack of gravity. A vehicle's structural system is used to contain and protect the astronauts and payload. Personnel at several support facilities supervise and control the entire process of space transportation.

Propulsion Systems

In order for a spacecraft to operate in the environment it is designed for, space, it must first leave the earth's atmosphere. This, however, is easier said than done. In order for all 4.5 million pounds of a space shuttle to reach orbit, it must be propelled 200 miles above the earth. See **Figure 24-1.** It takes quite an engine to provide this amount of thrust.

Rocket Engines

Currently, the only type of engine able to develop this amount of thrust is the rocket engine. Rocket engines are a type of reaction engine that, as described in Chapter 23, works according to the principle of Newton's third law of motion. This law states that for every action, there is an opposite and equal reaction.

There are two main types of reaction engines: the air stream reaction engine and the rocket engine. Rocket engines are, by far, the more powerful type of reaction engine. The air stream reaction engine, or jet engine, requires an external source of air to operate. Space vehicles, however, operate in an environment without air. For this reason, the jet engine would be useless on a spacecraft. Rocket engines do not require an external source of air to operate, which makes them the perfect choice for space travel. They carry their own fuel and oxygen, known as an oxidizer. An ***oxidizer*** is a chemical substance that mixes with fuel to allow combustion. Rocket engines are classified by the type of propellant they use. ***Propellants*** are mixtures of fuel and oxidizers. There are two types of rocket engine propellants: solid fuel and liquid fuel.

Oxidizer: A chemical substance that mixes with fuel to allow combustion.

Propellant: A mixture of fuel and an oxidizer.

Figure 24-1. Tremendous power is needed to lift a space shuttle into orbit. In this view of a shuttle launch, the two solid rocket boosters (SRBs) and the three main engines on the shuttle itself are all operating, generating 7 million pounds of thrust. (National Aeronautics and Space Administration)

Curricular Connection

Science: Newton's Laws of Motion

Sir Isaac Newton presented three natural laws of motion in 1686. The first law states that an object will either remain at rest or in a state of perpetual motion in a straight line until an external force acts upon it. In other words; objects at rest stay at rest, and objects in motion stay in motion. This law is evident in outer space. Once a spacecraft is put into motion, it continues because there are no external forces. If the spacecraft were inside the earth's atmosphere, forces such as gravity and drag would slow it down.

The second law states that a force acting on an object gives the object acceleration in the same direction as the force. The magnitude is also inversely proportional to the mass of the object. This law can be written as the formula "Force = Mass x Acceleration ($F = m \times a$)." The second law of motion is used to calculate the motion of spacecraft and aircraft.

The third law states that for every action, there is an equal and opposite reaction. For example, when you are swimming, as you move your arm through the water, the water pushes you in the opposite direction. This law is observed in both the lift and thrust of spacecraft and aircraft.

Solid-fuel rocket engines

Of the two types of rocket engines, solid-fuel is the simpler one. It is so simple, in fact, it is used in model rockets. See **Figure 24-2.** The ***solid-fuel rocket*** engines contain solid propellant, a combination of fuel and oxidizers, packed into a cylindrical container. The propellant is typically placed inside the cylinder with a channel down the center. This channel serves as the combustion chamber. In the design stage of a solid-fuel rocket engine, the engineer selects the types of chemicals and the proportion of fuel and oxidizer to create a rocket engine with a specific amount of thrust. There is usually more oxidizer than fuel in the propellant mixture. When ready for use, the propellant is ignited at the bottom. As the fuel burns, thrust is produced from the bottom of the engine. See **Figure 24-3.**

Figure 24-2. Model rockets use solid-fuel engines that are miniature versions of those that power full-size launch vehicles. This rocket is being launched during a high school technology education activity. (Estes)

Liquid-fuel rocket engines

Liquid-fuel rockets are much more complex than solid-fuel rockets. They are also safer and offer control over the amount of thrust produced. See **Figure 24-4.** These engines use two separate liquids, which are ignited in a combustion chamber. The two propellants are kept in separate tanks inside the vehicle. They are pumped into a combustion chamber, where

Solid-fuel rocket: A rocket engine that contains solid propellant packed into a cylindrical container.

Figure 24-3. A model rocket in flight. The drawings on the right show the burning of the solid fuel as the flight continues. The process takes place very rapidly and continues until the fuel supply is exhausted. (Estes)

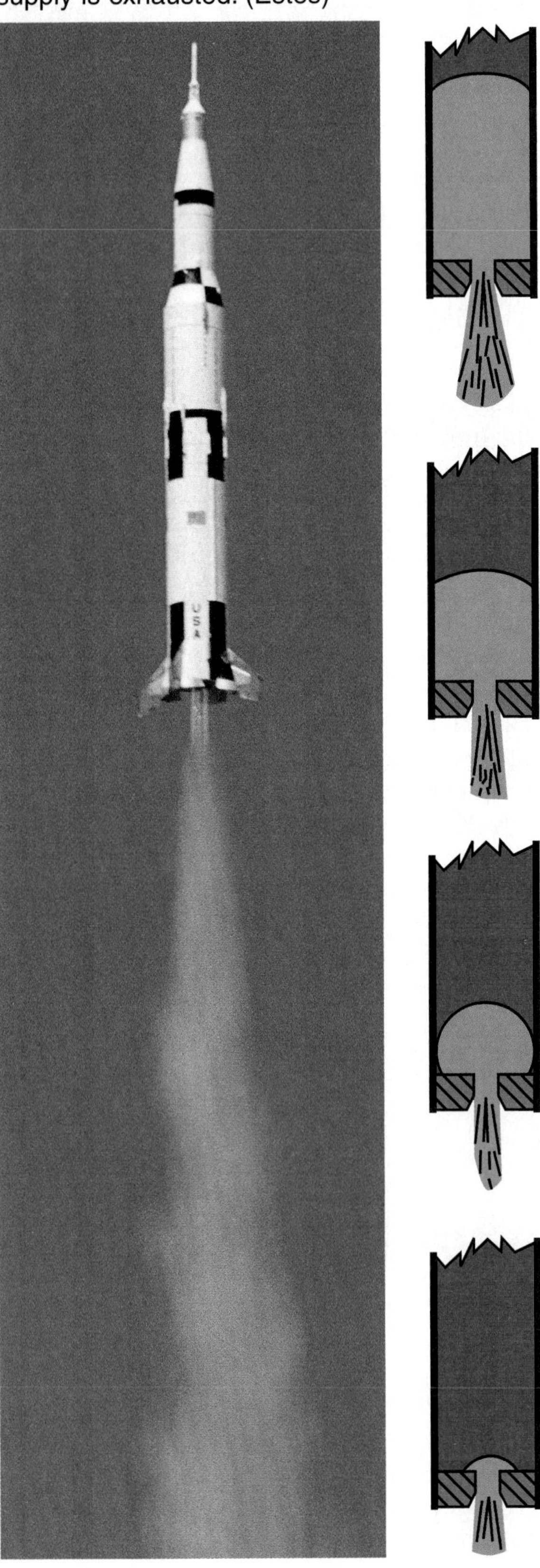

they are mixed and burned. Valves are used to control the amount of flow of each liquid fuel. See **Figure 24-5.** As propellant is burned in the combustion chamber, it leaves the bottom of the engine as thrust.

Launch Vehicle Propulsion

Launch vehicles are used to take spacecraft and satellites into orbit and outer space. They are configured in a number of ways, depending on the payload they are carrying. The launch of a small satellite, for example, may only require a single liquid-fuel rocket. If the vehicle launches a spacecraft on a mission to another planet, it requires more thrust. For example, in 2003, the ***National Aeronautics and Space Administration (NASA)***, the U.S. agency set up for research and development of space exploration, launched the *Mars Rover* onboard the *Delta II* launch vehicle. The vehicle was comprised of a liquid-fuel engine that used kerosene and liquid oxygen as the propellant and produced 200,000 pounds of thrust. This would not be enough, however, to propel the craft into outer space. So, nine solid-fuel rockets were fastened to the *Delta II*. The solid-fuel boosters, as they are called, added over 1 million pounds of thrust to the launch vehicle.

Early manned space vehicles, such as the Saturn V launch vehicle that carried the Apollo mission, relied on ***staging***. This technique places several propulsion systems on top of each other. As the first stage burns out, it is released from the vehicle. This exposes the second stage of propulsion systems. Stages are burned and released until the vehicle reaches its final orbit altitude. By using the staging method, the vehicle can get rid of unneeded mass during its flight. See **Figure 24-6.** Launch vehicles still use staging today.

Space Shuttle Propulsion

Like many launch vehicles, the space shuttle uses both solid- and liquid-fuel rockets. The shuttle has three main propulsion components: the main engines, the solid rocket boosters (SRBs), and the external tank (ET). The ***Space Shuttle Main Engines (SSMEs)*** are

Figure 24-4. Liquid-fuel rockets are more controllable and safer than solid-fuel rockets. (National Aeronautics and Space Administration)

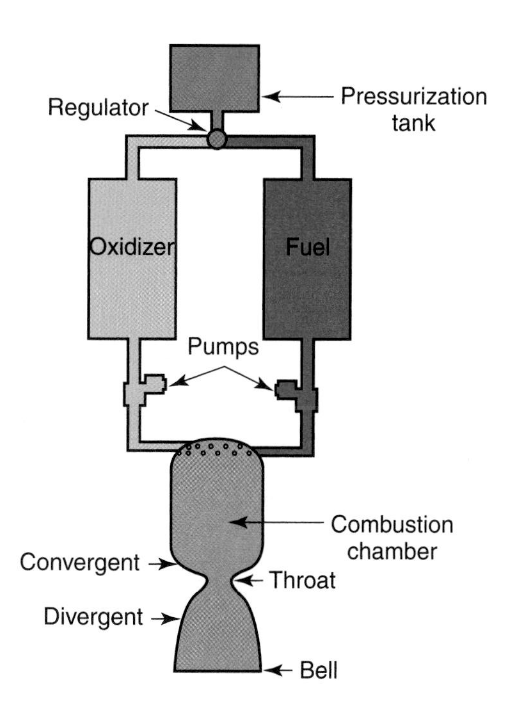

Figure 24-5. A simple liquid-propellant system for a model rocket. One liquid is the fuel burned in the combustion chamber. Fuels often used include liquid hydrogen and kerosene. The second liquid is the oxidizer, which allows the combustion to take place. The oxidizer is often in the form of liquid oxygen. The fuel and oxidizer are sprayed into the combustion chamber to ensure a good mix and efficient combustion. Most combustion chambers also have an igniter to begin the combustion process. Some fuels and oxidizers are so volatile, however, that they ignite on contact and do not require a spark. For this reason, propellants are put into the rocket shortly before they are to be used and are not stored in the rockets. (Estes)

located at the rear of the orbiter. See **Figure 24-7.** There are three engines that make up the SSME system and produce nearly 400,000 lbs. of thrust. During takeoff, the engines receive the needed fuel from the ET. When the ET is released, it burns up and breaks apart over the ocean.

Liquid-fuel rocket: A rocket engine that uses two separate liquids, which are ignited in a combustion chamber.

The ***solid rocket boosters (SRBs)***, on the other hand, are recovered and reused. The shuttle is equipped with two SRBs, one on each side of the ET. The SRBs supply the majority of the power at takeoff—3.3 million pounds of thrust each. The propellant used in the SRBs is a combination of ammonium perchlorate, aluminum, iron oxide, and mixture of chemicals used to

Figure 24-6. The Saturn V launch vehicle used in the early years of the U.S. manned spaceflight program consisted of three stages. The vehicle's first stage consisted of five liquid-fuel rocket engines and their fuel supply. It was able to produce about 7.5 million foot-pounds (10.2 million Joules) of thrust. The second stage consisted of five smaller engines capable of producing over 1 million pounds of thrust. The third stage consisted of one engine and its fuel supply. It carried the *Apollo* spacecraft to its final orbit altitude. As the fuel in each stage was exhausted, it separated and dropped away. (National Aeronautics and Space Administration)

Figure 24-7. The shuttle main engines can be stopped and started as needed. During launch, they burn at full thrust, along with the solid rocket boosters (SRBs). Launch fuel for the main engines is contained in the external fuel tank. The tank is situated on the underside of the orbiter and is the largest part of the space shuttle. The top of the tank holds liquid oxygen, while liquid hydrogen is stored in the bottom. The liquid oxygen and hydrogen are used so quickly that the tanks are empty within 8 minutes after takeoff. The hydrogen is used the quickest, at a rate of over 40,000 gallons per minute (GPM). (A garden hose flows at a rate near 8–10 GPM.) The tank's contents are exhausted during launch, and the tank drops away and burns up during reentry. (National Aeronautics and Space Administration)

Career Connection

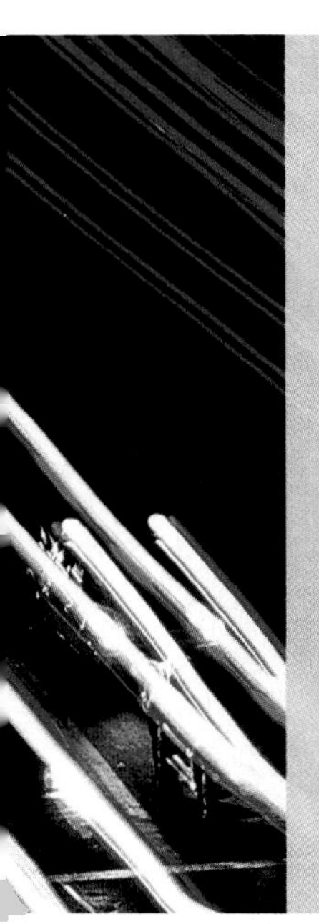

Astronautical Engineers

All technology depends on the development of new products. In air and space transportation, aerospace engineers develop new vehicles. There are two main types of aerospace engineers, aeronautical and astronautical. Aeronautical engineers design aircraft, and astronautical engineers design spacecraft. Aerospace engineers often specialize in one area of design, such as aerodynamics, control, structure, or propulsion.

Astronautical engineers not only design spacecraft, such as the shuttle and space station, but they also design satellites, space communication systems, and ballistic missiles. Research and development companies, such as Boeing, Lockheed Martin, and Raytheon, typically employ astronautical engineers. These companies provide products to the federal government and the National Aeronautics and Space Administration (NASA). Some astronautical engineers, however, work directly for NASA and the federal government.

At the very least, astronautical engineers are required to have bachelor's degrees in aerospace or astronautical engineering. Many, however, obtain master's and doctorate degrees. The average salary for an aerospace engineer is slightly over $70,000.

hold everything together. Two minutes after launch, the boosters are empty and released from the shuttle. Several small thrusters at the top and bottom of the SRB are used to ensure the boosters do not come in contact with the rest of the shuttle.

New Propulsion Technologies

Ion propulsion is a type of propulsion much different from liquid- or solid-fuel rockets. Traditional rockets use chemical reactions to accelerate gases that move the spacecraft. Ion propulsion uses the electrical charge of atoms to move vehicles. As you might remember, electrons move between atoms to balance their charges. Ion propulsion uses this movement of electrons to propel a vehicle through space. See **Figure 24-8.** Ion propulsion is not powerful enough to launch spacecraft, but it may eventually be the best method of propelling spacecraft once they are in space.

Guidance Systems

The guidance systems of spacecraft are responsible for two functions—navigation and guidance. Navigation is being able to establish the spacecraft's location in space. Guidance is determining if the vehicle is in the correct place and figuring how to change its location if it is not correct.

In order to determine if a spacecraft is in the correct location and traveling on the correct path, there first must be a flight plan. A flight plan is a detailed account of the path the spacecraft is supposed to travel. Because objects in space are always in motion, the flight must be well planned. See **Figure 24-9.**

National Aeronautics and Space Administration (NASA): The U.S. agency set up for research and development of space exploration.

Staging: A technique that places several propulsion systems on top of each other. Stages are burned and released until the vehicle reaches its final orbit altitude.

Space Shuttle Main Engine (SSME): An engine that can be stopped and started as needed. It is located at the rear of the orbiter.

Solid rocket booster (SRB): A shuttle propulsion component that is recovered and reused. It supplies the majority of the power at takeoff.

Ion propulsion: A type of propulsion that uses the electrical charge of atoms to move vehicles.

Figure 24-8. An ion propulsion engine powered *Deep Space 1*, a spacecraft launched in 1998 to visit a comet far outside our solar system. The spacecraft was retired in 2001, after surpassing all expectations. Currently, ion propulsion is slower than traditional methods, but it can last much longer. The ion engine set a new endurance record—3.5 years of continuous work. *Deep Space 1* approaches the comet 19P/Borrelly in this artist's view. (National Aeronautics and Space Administration's Jet Propulsion Laboratory)

Deep Space Network (DSN): A system with three radio antennas located on three continents. Two of the antennas determine the distance of the spacecraft from themselves. The antennas then determine the distance of a known object in space. All these distances are computed, and the location of the spacecraft in space is determined.

Navigation

Determining a spacecraft's position in space is the role of navigation systems. This is more complex than might be imagined because the entire solar system is constantly moving. Even if a spacecraft remains in one place, its position relative to everything else continues to change. In order to accurately determine the position of spacecraft, navigation systems have to determine three pieces of information: location, velocity, and attitude.

Location

In order to determine the location of a spacecraft, the distance and angle from earth must be known. The distance from earth is most often figured by bouncing a radio signal from earth off the spacecraft. The time it takes for the signal to be returned can be calculated to find the spacecraft's distance from earth. To calculate the angle of the spacecraft to the earth, a process of triangulation is used. NASA's ***Deep Space Network (DSN)***, a system with three radio antennas located on three continents around the earth, takes the necessary measurements. See **Figure 24-10.**

Figure 24-9. The *Stardust* spacecraft was launched in 1999, with the intention to collect samples from the comet Wild 2 in 2004. The flight plan had to take into account where the comet would be five years after the launch. It also had to include two revolutions around the sun prior to meeting the comet, in order to gain enough speed. The flight was successfully planned, and on January 2, 2004, *Stardust* encountered the comet Wild 2. (National Aeronautics and Space Administration)

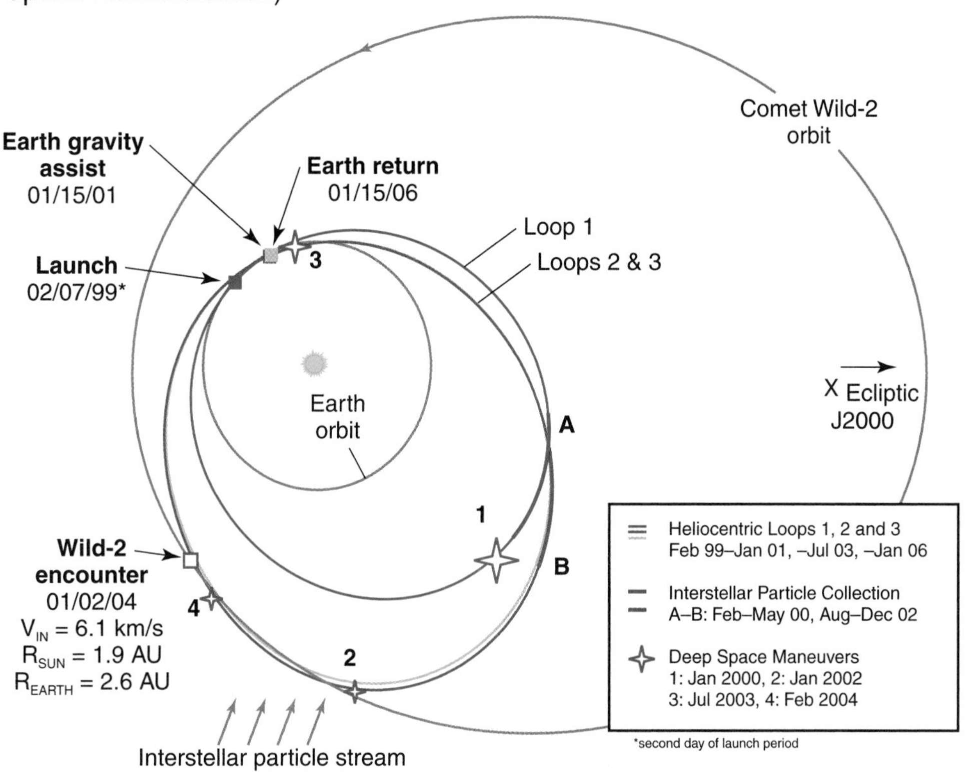

Figure 24-10. Triangulation is used to compute the location of a spacecraft or other object in space. A—The Deep Space Network (DSN) uses two of its antennas to determine the distance of the spacecraft from themselves. B—The antennas then determine the distance of a known object in space, such as a star. C—All these distances are computed, and the location of the spacecraft in space is determined. (National Aeronautics and Space Administration)

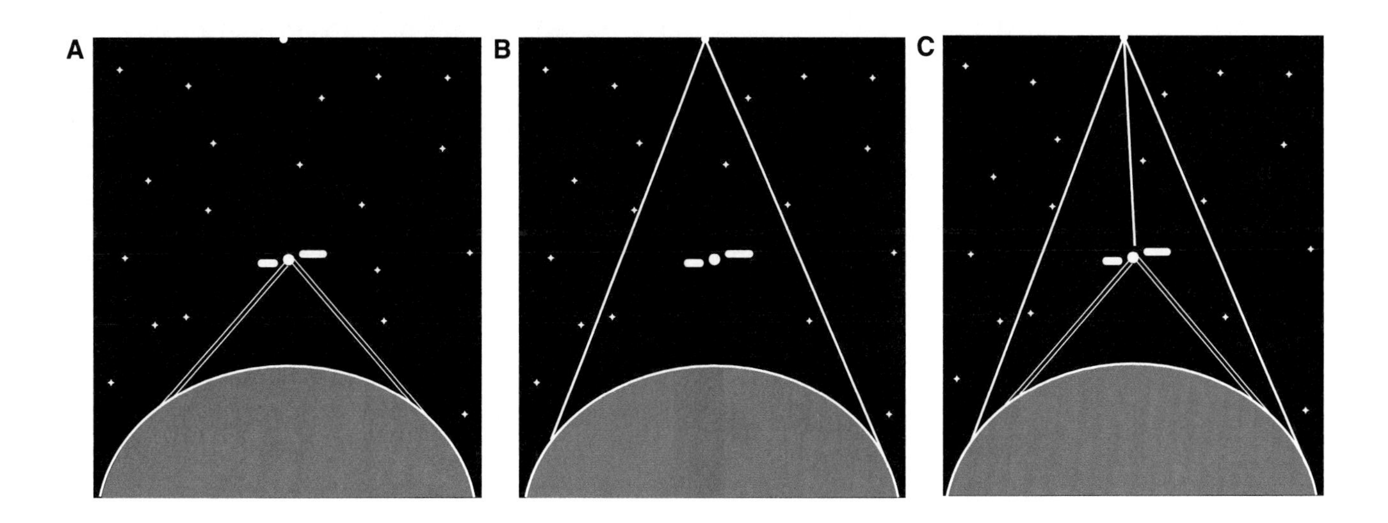

Velocity

The velocity, or speed, of spacecraft is often also computed by NASA's DSN. See **Figure 24-11.** The DSN is able to determine the speed a spacecraft is traveling by measuring the Doppler shift. This shift is based on the Doppler effect, which is the change in pitch that occurs as an object moves past an observer. For example, if you are sitting at a racetrack, the pitch of the engine gets higher as the car approaches you and decreases as the car passes. This same effect occurs in space. The radio antennas at the DSN are able to detect the change in pitch and calculate the velocity of the moving spacecraft.

Attitude

The attitude, or rotation, of a spacecraft can be determined in two ways: celestial reference and inertial reference. ***Celestial reference devices*** use the location of objects in space to figure the rotation of the spacecraft. ***Inertial reference systems*** use gyroscopes to determine the attitude of the spacecraft. Gyroscopes are spinning devices that can measure the amount of rotation from an initial setting.

Celestial reference device: A device that uses the locations of objects in space to figure the rotation of a spacecraft.

Inertial reference system: A system that uses gyroscopes to determine the attitude of a spacecraft.

Complete systems

Systems have been designed that combine the measurements of location, velocity, and attitude into integrated functions. These systems are known as Space Integrated Global Positioning System and Inertial Navigation Systems (SIGIs). A SIGI uses both a global positioning system (GPS) and inertial navigation systems (INSs) to navigate. This system is used on the International Space Station (ISS) and may be used in the future on space shuttles. A GPS is already used on shuttles and other orbiting spacecraft to quickly determine location and attitude.

Figure 24-11. Deep Space Network (DSN) antennas at the Goldstone site in California's Mojave Desert are used for communication and to measure the Doppler shift when determining the velocity of a spacecraft. (National Aeronautics and Space Administration's Jet Propulsion Laboratory)

Guidance

Knowing the location, velocity, and attitude of the spacecraft is only one piece of guiding the vehicle. The other part is determining if the spacecraft is in the right place. This is the job of guidance systems. Guidance systems, usually computers, examine the navigation information and flight plan to determine if the spacecraft is on the proper trajectory. A ***trajectory*** is the course or route of the craft. If the trajectory is incorrect, the guidance computers or personnel from ground control centers can use the spacecraft's control system to change its course. Periodic trajectory changes are usually built into a flight plan because there are many variables in space that can cause a spacecraft to fly off course.

Trajectory: The course or route of a spacecraft.

Control Systems

Having enough power to propel the spacecraft into space is only one part of space travel and exploration. The space vehicle must also be able to be controlled. A spacecraft is controlled in several different ways, depending on the type of vehicle.

Unmanned Spacecraft

Satellites and space probes are controlled much differently than space shuttles. Both are unmanned and must be completely controlled from ground control centers. The rotation, or attitude, of both of these spacecraft is extremely important. For example, imagine if a television satellite was at the wrong attitude and, instead of transmitting television programming to the earth, it sent the transmission into outer space. There would be a lot of unhappy customers. The same is true for a space probe. Imagine if millions of dollars were spent to build a probe to explore Mars and instead it headed to the moon and crashed.

To avoid those types of situations, the control systems of unmanned spacecraft are designed to be highly accurate and have many backups in place. The guidance systems are used to determine the position in which the spacecraft should be. The job of the control system is to rotate the vehicle into that position. Three types of systems are used to maneuver a satellite or space probe. The first is the use of thrusters. There are typically several thrusters pointed at 90° from one another. By using these thrusters, the satellite can rotate itself into position. Another method of rotating is using reaction wheels. ***Reaction wheels*** are large rotating wheels that can generate momentum and spin the spacecraft. Three or four wheels are typically placed at different angles to ensure the vehicle can be rotated into any position. The last type of control is known as spin stabilization. This works because if the satellite or a part of the satellite remains spinning, it will be naturally stable. This system also uses small thrusters to make any minor corrections that may be needed.

Reaction wheel: A large rotating wheel that can generate momentum and spin an unmanned spacecraft.

Manned Spacecraft

The pilots onboard manned spacecraft are able to control the vehicles. The two main types of manned spacecraft in use today include space shuttles and manned maneuvering units (MMUs). Each vehicle uses different systems to control the direction of the vehicle.

Space shuttle orbiters

Elevon: A flap located at the rear of a shuttle orbiter's wing. It is used to control pitch and roll.

Gimbal: To rotate the nozzle surrounding each of the three main engines.

There are two different control systems used on the space shuttle orbiter. The first type of control system is the use of control surfaces. Control surfaces, as you may remember from Chapter 22, are used on aircraft. They are used on space shuttles for the same purposes as in aircraft, to control the pitch, roll, and yaw. See **Figure 24-12.** Flaps at the rear of the wing, known as ***elevons***, control pitch and roll. See **Figure 24-13.** The use of a body flap located along the bottom rear of the shuttle, under the main engines, also controls the pitch. A rudder positioned on the tail fin of the orbiter controls the yaw. Switches on the instrument panel and foot pedals at both the commander's seat and the pilot's seat activate the control surfaces. These surfaces operate by deflecting air in the opposite direction of the desired movement. Therefore, once the shuttle leaves the atmosphere, the surfaces are not useful because there is no air to deflect.

Once outside the earth's atmosphere, the second control system must be used. This system uses three sets of small and large rocket engines to control the movement of the orbiter. The first set of engines used for control is the SSME. While these engines are mainly used to propel the craft into space, they can also help control direction. The nozzle surrounding each of the three main engines can be rotated, which is known as ***gimbaling***. By directing the flow of the escaping gas, the engines are able to aid in steering the space shuttle.

Figure 24-12. The space shuttle's control surfaces are used to control pitch, roll, and yaw. Pitch is movement around the X axis, roll is movement around the Y axis, and yaw is movement around the Z axis.

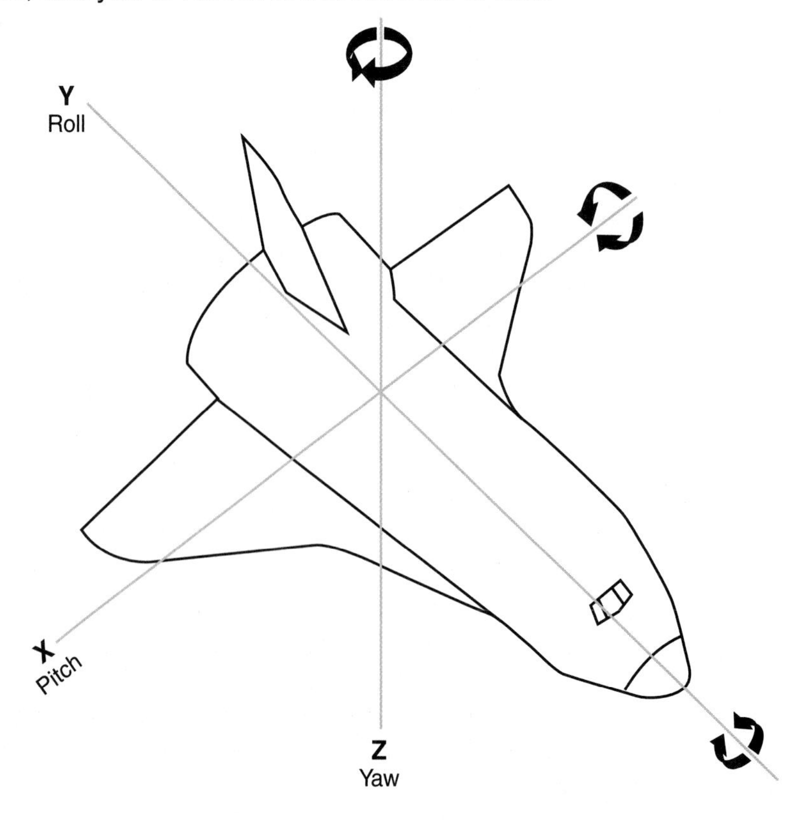

Figure 24-13. Elevons are flaps at the rear of the shuttle's wings that serve the same purposes as ailerons and elevators in a conventional aircraft. The rudder performs the same function—yaw control—on both conventional aircraft and the shuttle. (National Aeronautics and Space Administration)

The second set of engines is used to control the rotation, or attitude, of the orbiter. These are the ***orbital maneuvering system (OMS) engines***, located at the rear of the orbiter above the main engines, one on each side of the tail fin. They are used primarily to maneuver the shuttle into orbit and to slow and deorbit the craft prior to reentry. In most missions, engines are fired twice to place the shuttle into orbit and once to remove it from orbit. These engines use two propellants that ignite on contact, nitrogen tetroxide and hydrazine. Therefore, an igniter is not needed. This type of rocket is known as a ***hypergolic engine***.

The third collection of engines is known as the ***reaction control system (RCS) engines***. See **Figure 24-14.** These engines serve as thrusters that enable the shuttle to maneuver in space. The crew of the shuttle can operate the RCS engines using two types of controllers. The first is known as a rotational hand controller. This controller resembles a video game joystick. It is used to change the pitch, roll, and yaw of the shuttle. To change the pitch, the stick is pushed forward or backward, and the nose of the shuttle lowers and rises. If the controller is moved to the right or left, the shuttle banks left or right, changing the roll. To change the yaw, the controller is turned clockwise or counterclockwise to rotate the shuttle. When the controller is pushed in any direction, a signal is sent to a computer, which then turns on the required thrusters and moves the shuttle in the desired direction. The other type of controller is the translational hand controller. This resembles a knob that can be moved up and down, right and left, and in and out. The movement is sent to a computer, and thrusters are activated to move the shuttle in the desired direction. The translational hand controller only produces movements along the X, Y, or Z plane. This is helpful when the shuttle is docking with a space station or another spacecraft.

Orbital maneuvering system (OMS) engine: An engine located at the rear of the orbiter above the main engines. It is used primarily to maneuver the shuttle into orbit and to slow and deorbit the craft prior to reentry.

Hypergolic engine: A type of rocket in which the engines use two propellants that ignite on contact, nitrogen tetroxide and hydrazine. Therefore, it does not require an igniter.

Reaction control system (RCS) engine: An engine located in a cluster in either the front or rear of a shuttle. It serves as a thruster that enables the shuttle to maneuver in space.

Figure 24-14. Reaction control system (RCS) engines are located in clusters in the front and rear of the shuttle. There is a total of 14 primary and 2 secondary RCS engines in the front and 12 primary and 2 secondary RCS engines on each side of the tail fin in the rear. The primary thrusters are used for most movements, and the secondary engines are used to hold a position or for small movements. In this view, taken during a mission of the shuttle *Discovery*, the RCS engines are firing upward and to the left. (National Aeronautics and Space Administration's Johnson Space Center)

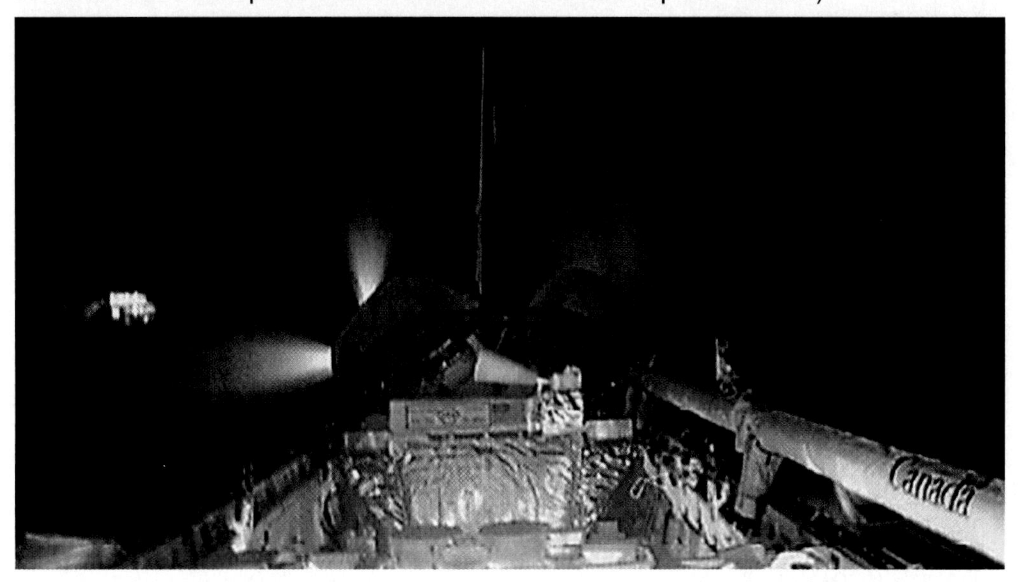

Manned maneuvering units (MMUs)

Manned maneuvering unit (MMU): A specialized space vehicle developed early in the shuttle program to allow astronauts to maneuver outside the vehicle.

Some specialized space vehicles use inert gas jets to propel them while they are in space. ***Manned maneuvering units (MMUs)***, developed by NASA, are one example. See **Figure 24-15.** These use small jets that emit nitrogen gas. The escaping gas provides the action that causes a reaction in the opposite direction.

Suspension Systems

Vehicular suspension systems are the components of a vehicle supporting it in its environment. In land transportation, the suspension system is the wheel. In air transportation, it is the wings, and in water transportation, the hull works to support the ship. In space transportation, the suspension system is a little different. Some spacecraft do not even need suspension systems. Spacecraft that travel in outer space have no need for suspension systems to keep them in the air because space is a vacuum. Anything will float in space. There is also very little, if any, friction, so aerodynamics are of minor concern.

Spacecraft that orbit the earth, such as satellites and space stations, do not require sophisticated suspension systems. Once they are placed in orbit, the centrifugal force and gravitational pull keep the craft suspended. As long as a spacecraft can keep the proper speed and distance from earth, it will continue to be suspended. This is the same reason space debris remains in orbit for years, until it finally loses speed and falls to earth.

Figure 24-15. Early in the shuttle program, the National Aeronautics and Space Administration (NASA) developed the manned maneuvering unit (MMU) to allow astronauts to maneuver outside the vehicle. When the operator wants to move in a certain direction, he activates the jet nozzle, which will produce movement in that direction. In this 1984 view, astronaut Bruce McCandless performed the first spacewalk without a tether. In recent years, the more efficient Extravehicular Mobility Units (EMUs) have replaced MMUs. (NASA)

The only time a space vehicle uses a suspension system is when it exits and reenters the earth's atmosphere. This is especially important for a space shuttle, since the shuttle carries passengers. The suspension system used on space shuttles is the orbiter wings. The wings are a unique design, known as a ***delta wing***. See **Figure 24-16.** Delta wings are triangular and relatively flat. This style of wing is not commonly used on aircraft because it is ineffective and hard to control at speeds under the speed of sound, or subsonic speeds. The two most common vehicles that use the delta wing, Concordes and space shuttles, travel at supersonic speeds. Supersonic speeds are between Mach 1 and 5, or between the speed of sound and five times the speed of sound. On reentry, space shuttles even reach ***hypersonic speeds***, which are speeds over Mach 5. Travel at these speeds makes the delta wing a good choice for space shuttles.

Delta wing: An orbiter wing used as the suspension system on space shuttles. It is triangular and relatively flat.

Hypersonic speed: Speed over Mach 5 (five times the speed of sound).

The wing of the orbiter serves two purposes. The first is to suspend the shuttle in the air as it is guided back to earth during reentry. The orbiter begins reentry at over 16,000 miles per hour (mph) and has 25 minutes to decrease its speed to 200 mph, so it can land safely. The

Reinforced carbon-carbon panel: The covering for the leading edge of an orbiter wing. It is made of a material that has several layers of carbon and can withstand extreme temperatures.

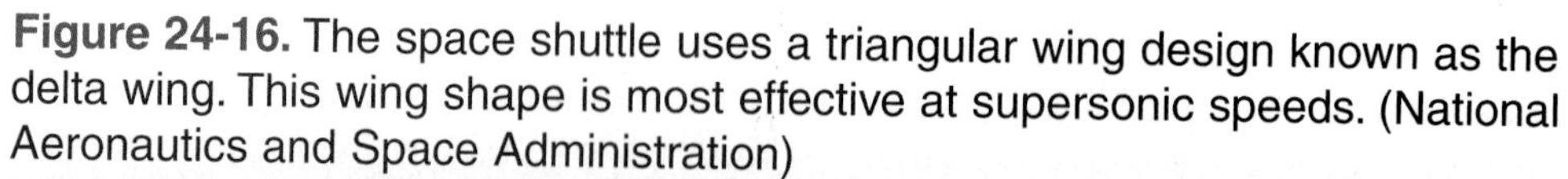

Figure 24-16. The space shuttle uses a triangular wing design known as the delta wing. This wing shape is most effective at supersonic speeds. (National Aeronautics and Space Administration)

shuttle is flown at a high angle of attack, where the nose is elevated at an angle between 25° and 40°. Flying at such an angle causes a large amount of drag, which slows the orbiter down. The orbiter also goes through a series of *S* turns to help it decrease speed. The entire reentry is computer controlled. The process of controlling an object going this fast would be very hard for a person to handle. In fact, a pilot has controlled the process only once. The landing process also requires a high amount of accuracy. The orbiter, as part of the deorbiting process, burns all the remaining fuel and is left with no functioning engines. So, it actually works like a glider and would be in trouble if it decreased its speed too quickly or not quickly enough because it has no engines to retry the landing.

The second function of the wing is to serve as a thermal protection system. The underside and leading edge of the wings are exposed to temperatures over 2000°F on reentry. To protect the shuttle, the wings are covered with two different materials. The leading edge of the wing is covered with ***reinforced carbon-carbon panels***. These panels begin as pieces of rayon cloth and are subjected to several chemical processes. The result is a material that has several layers of carbon and can withstand extreme temperatures. The underside and nose of the shuttle are covered with tiles of a different material. See **Figure 24-17.** These tiles

Figure 24-17. A technician installs high-temperature reusable surface insulation tiles on the wing of a space shuttle. Most of the tiles are 6 × 6 in and range in thickness from 1″ to 5″. When heated, the tiles glow red, but they cool so quickly that within seconds, they are cool enough to touch. (National Aeronautics and Space Administration)

are known as ***high-temperature reusable surface insulation tiles*** and are made of nearly pure silica fibers. The specially designed tiles rapidly dissipate heat and can withstand temperatures from below –200°F to over 2000°F.

High-temperature reusable surface insulation tile: A specially designed tile, made of nearly pure silica fibers, that rapidly dissipates heat and can withstand temperatures from below –200°F to over 2000°F. These tiles cover the underside and nose of shuttles.

The final space shuttle suspension device is the landing gear. See **Figure 24-18.** The pilot completes the landing of the shuttle with the use of an electronic landing system, much like the one commercial aircraft use.

Structural Systems

The builders of most space vehicles use construction techniques similar to those used on aircraft. Variations of truss and monocoque construction are relied on. Designs for spacecraft depend on many things and vary greatly. Some considerations include the vehicle's destination, whether it is to be manned or unmanned, the tasks to be performed on the mission, and conditions the vehicle will have to endure.

Unmanned Spacecraft

Unmanned spacecraft all have similar components based around a main structure. The main structure, also known as the ***bus***, can come in different shapes, sizes, and materials. Rectangular, cylindrical, and polygonal buses are all common. The size of the structure often ranges from several feet to slightly over 10′ in width. The vehicle used to launch the spacecraft constrains the size. If it is to be deployed from a space shuttle, the satellite or probe can be larger than if it is to be launched from the nose of a rocket. The bus material is most often an aluminum alloy, a

Bus: The main structure of an unmanned spacecraft.

Figure 24-18. The space shuttle uses tricycle-style landing gear, with one set of tires in the front and two sets further back, under the fuselage. The front set can be steered, while the two in back have brakes. The shuttle tires are safe to use up to 225 miles per hour (mph) and are replaced after each flight.

titanium alloy, or a magnesium alloy. Composite materials, such as Kevlar® resins, are becoming more popular, but they are often used in combination with an alloy.

The main purpose of the bus is to serve as a structure to which the spacecraft components can be attached. These components include solar panels, antennas, guidance and navigation equipment, measurement instruments, and monitoring devices. The types of components used depend on the type of mission the spacecraft is to complete. A broadcast satellite would have much different components than a space probe traveling to Jupiter.

Manned Spacecraft

Manned spacecraft require additional structural considerations. The structure must be able to hold human astronauts and scientists. The structures of both space stations and space shuttles are designed to contain and protect the human travelers. The structures, however, are still very different.

Space stations

The ISS, when complete, will be the largest space station ever constructed. The entire structure will be over 200′ long and 350′ wide. The station will have several modules in a continuous line that will serve as the living quarters and research area for the crew. See **Figure 24-19.** At the center of the space structure is the *Unity* capsule, which serves as a hub for other capsules.

Figure 24-19. The International Space Station (ISS) is constructed from a series of interconnected modules produced by the United States, Russia, and Japan. Extending from one side of *Unity* are two Russian modules, *Zarya* and *Zvezda*. Each module has solar collectors attached to provide power to the station. *Zarya* serves as the control center, and *Zvezda* is the living quarters. Connected to the other side of *Unity* is the U.S. lab *Destiny*, which serves as a science lab. Three other modules will be added to the ISS. (National Aeronautics and Space Administration)

Construction: The International Space Station (ISS)

The International Space Station (ISS) is not only the largest space station ever built and the largest international project ever completed, but it is also a combination of several types of technology. In order for the ISS to be completed, it relies on transportation, manufacturing, and construction technology. The station itself is a transportation vehicle, as it carries astronauts, equipment, and experiments around the earth. It also contains all the vehicular systems present in all vehicles.

The ISS, however, is not a typical mass-produced vehicle. The station contains many components, each of which is custom manufactured. There are several aspects of the manufacturing of the components that make it very challenging. First, the components are manufactured in phases and sent into space at different times. When the components are sent into space, it is the first time they have been connected to each other. The ISS was not built on earth first and then disassembled and sent to space for reassembly. One other challenge is that different companies build the components all over the world. To overcome these challenges, the manufacturers of the components must ensure that they build each component exactly as planned, with very high tolerances.

After the components have been manufactured and are ready to be assembled, they are sent to space. The individual pieces are assembled in what can be viewed as a construction project. The astronauts onboard the ISS serve as construction workers and assemble the components. Several cranes and robotic arms have been designed into the ISS to help the astronauts with the construction. When working on the ISS, the astronauts are tethered to the station, and their tools are tethered to their space suits to make sure the astronauts remain safe and their tools do not drift away. The ISS may seem like strictly a transportation vehicle, however, without both manufacturing and construction technologies, it could never be built.

Running perpendicular to the research modules and centered on the *Unity* module will be a truss system. The aluminum trusses, when completed, will span 300′ and contain the cooling system and solar arrays needed to maintain the ISS. The entire structure will require over 40 spaceflights and 160 space walks to complete.

Space shuttles

When you think of a space shuttle, the first image that may come to mind is a shuttle sitting on the launchpad, ready for takeoff. See **Figure 24-20.** When the shuttle is in this position, it is actually made of three distinctly different parts. It consists of two SRBs, a large ET, and a shuttle orbiter.

The SRBs are the largest solid-fuel rocket engines ever developed. See **Figure 24-21.** These boosters are also the first ones to be used on any manned space vehicle. They are constructed of a series of hardened steel rings. The booster sections are attached with high-strength steel pins. The resulting joints are sealed with rubberlike O-rings and then covered with a

Figure 24-20. A space shuttle being moved to the launchpad at the Kennedy Space Center (KSC) aboard a giant transporter. The shuttle orbiter is attached to a large external fuel tank and two solid rocket boosters (SRBs). The structure of each part is unique to the space program. (National Aeronautics and Space Administration)

Figure 24-21. During launch, the two solid rocket boosters (SRBs) create a huge amount of thrust. The three main engines of the shuttle, as seen on the right, fire at the same time as the SRBs. (National Aeronautics and Space Administration)

fiberglass tape to make a smooth, aerodynamic shape. Inside the nose cone of the SRBs are flight electronics and a parachute. The SRBs are reusable, and once they have been disconnected from the orbiter, they parachute into the ocean, where U.S. Navy vessels recover them.

The large ET is the largest part of the launch vehicle. It holds the liquid fuel the shuttle's main engines will use. There are three parts to the ET. The top section is the liquid oxygen tank. To produce an aerodynamic shape, the top of this tank is tapered to a point. The bottom section is the liquid hydrogen tank. It is the larger of the two tanks. This section also holds the mounting brackets for the SRBs and the shuttle orbiter. A collar joins the two tanks. This is the third part of the ET structure. When the ET is released from the orbiter, it descends to earth and is burned up during reentry into the atmosphere.

The orbiter is the only component of the three that enters orbit. The shuttle orbiter makes use of the construction techniques found in the aircraft industry. The fuselage, or body, of an orbiter can be divided into three sections, much like a ship. The forefuselage is the only pressurized section and contains all the operating and living quarters. It is divided into two decks. The upper deck, or ***flight deck***, contains all the flight controls. One of the latest modifications to space shuttles is in the cockpit area. See **Figure 24-22.** The midfuselage is also known as the ***payload area***. This area holds the cargo the shuttle is carrying on its mission. A robotic arm is housed in the midfuselage to help load and unload the mission cargo. See **Figure 24-23.** When astronauts exit the shuttle to spacewalk, they exit from the forefuselage to the midfuselage, through an ***air lock***. The air lock ensures that the cabin remains pressurized and provides a place for the astronauts to put on their space suits. The rear section of the shuttle is the aftfuselage. This area contains the majority of the propulsion system (the main engines, orbital maneuvering engines, and reaction control thrusters).

Flight deck: The upper deck of a space shuttle's forefuselage, which contains the flight controls.

Payload area: The midfuselage of a space shuttle, which holds the cargo.

Air lock: An area through which astronauts exit the shuttle. It ensures that the cabin remains pressurized and provides a place for the astronauts to put on their space suits.

Support Systems

As you know, space transportation is very specialized and experimental. Governments of various countries support their space exploration efforts by supplying money for research and development. In the United States, NASA is the agency set up to conduct work in this area. NASA extends its arms into all areas of aviation and space travel. It is the primary support system of space transportation.

Figure 24-22. Space shuttles are being upgraded with the installation of a "glass cockpit," in which conventional gauges have been replaced with full-color graphic display screens. These displays improve the safety and flight conditions for the pilot and commander. Similar systems are being used in modern airliners. (National Aeronautics and Space Administration)

Figure 24-23. A robotic arm aboard the space shuttle is used to unload cargo. Oftentimes, the cargo is either a satellite or a section of the International Space Station (ISS). The robotic arm is helpful in moving the objects and can be operated at the payload control center at the rear of the flight deck. In this view, the Hubble Space Telescope is being lifted out of the cargo bay and redeployed after a servicing mission. (National Aeronautics and Space Administration)

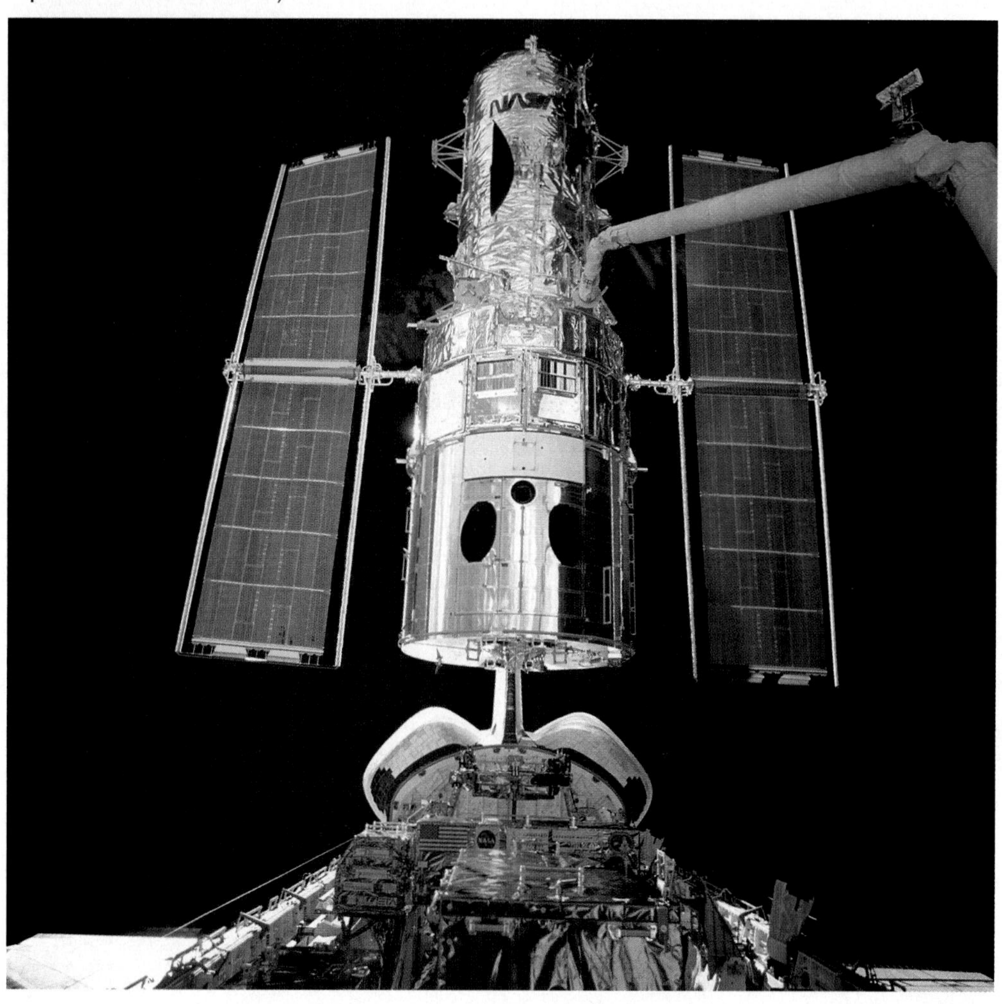

Orbiter Processing Facility (OPF): A facility that houses the orbiter from the time it lands until a week before the next scheduled launch.

Vehicle Assembly Building (VAB): A facility in which the orbiter is stood vertically and the solid rocket boosters (SRBs) and external tank (ET) are attached.

NASA has a number of space centers, research centers, and test facilities that conduct research and tests and prepare spacecraft for flight. The most well-known support centers are the Kennedy Space Center (KSC) in Florida and the Johnson Space Center (JSC) in Texas. These centers are the most highly involved with the launching and flight control of space shuttles. The shuttles are prepared and launched from the KSC. The preparation for launch is done in the Orbiter Processing Facility (OPF) and the Vehicle Assembly Building (VAB). The ***Orbiter Processing Facility (OPF)*** houses the orbiter from the time it lands until a week before the next scheduled launch. In the OPF, the orbiter is examined and reconditioned. Any problems are fixed, and needed modifications are made. From the OPF, the orbiter is taken to the ***Vehicle Assembly Building (VAB)***. Here, the orbiter is stood vertically, and the SRBs and ET are attached. See **Figure 24-24.**

The ***Mission Control Center (MCC)*** at the JSC directs the actual liftoff of the shuttle, the mission of the orbiter, and the control of the ISS. See **Figure 24-25.** All personnel work facing a large screen in the front of the room displaying the vital information and location of the spacecraft. At each desk, the engineers have equipment monitoring the specific functions of the spacecraft and astronauts. The lead support person is the ***flight director***, who is the team leader. The flight director is in charge of making decisions based on the information each engineer and officer provides. The spacecraft communicator then relays the information to the commander of the spacecraft.

The JSC also contains the space vehicle mock-up facility. This facility serves as one of the main training areas for astronauts. There are full and partial orbiters, as well as replicas of space station components, that serve different training functions. Astronauts are able to perform routine training and prepare for emergencies at this location.

Unmanned spacecraft are supported by NASA's DSN. The DSN is a system of three deep space antennas that are able to track space probes. The antennas are placed across the globe in California, Spain, and Australia so the spacecraft can be "seen" by at least two of the antennas at all times. The DSN is able to monitor and correct trajectories, as well as collect scientific data the spacecraft is sending.

The KSC, JSC, and DSN are only a few of the support facilities used in space transportation. Private companies also provide much support. These companies help to maintain and improve the existing spacecraft and research ways to enhance space travel.

Figure 24-24. In the Vehicle Assembly Building (VAB) at the Kennedy Space Center (KSC), a shuttle orbiter is being mated to the external fuel tank and solid rocket boosters (SRBs). The assembly is completed on a mobile platform, which is later driven to the launchpad. The pad is only about 4 miles away, but the trip takes 6 hours, due to the weight of the shuttle. At the launchpad, the shuttle goes through a number of inspections prior to liftoff. (National Aeronautics and Space Administration)

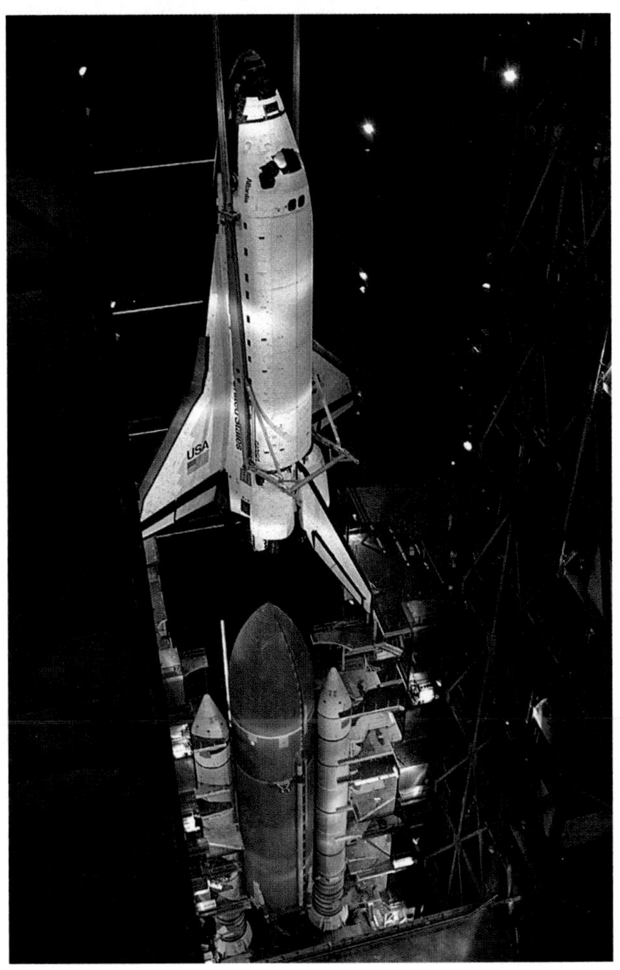

Mission Control Center (MCC): An area in the Johnson Space Center (JSC) where the liftoff of the shuttle, the mission of the orbiter, and the control of the International Space Station (ISS) is directed.

Flight director: The team leader at the Mission Control Center (MCC).

Figure 24-25. The Mission Control Center (MCC) at the Johnson Space Center (JSC) in Texas controls and monitors all U.S. spaceflights. It has two separate flight control rooms—one for the shuttle operations and the other for the space station. The rooms are laid out with many desks that have specific purposes. This view was taken during a television transmission from the Apollo 13 mission in 1970. During the launch and recovery of a space vehicle, technicians and engineers work at all the consoles. (National Aeronautics and Space Administration)

Summary

In order for space transportation vehicles to function, they must first get off the ground. This is the job of the propulsion system. All spacecraft are currently propelled into outer space using rocket engines. The rocket engines are either solid or liquid fueled and are the largest engines used in the world. Once in space, thruster or rotational devices control the spacecraft. The spacecraft are controlled in an effort to remain on the planned trajectory. The information gathered from the guidance systems helps to determine if the spacecraft is on the correct path and how much correction is needed.

For most spacecraft, other than space shuttles, suspension systems are unimportant because the spacecraft fly in the vacuum of space. For a shuttle's orbiter, the delta-style wings serve as suspension as the craft reenters the earth's atmosphere. The structure of spacecraft varies, depending of the use of the craft. The components of satellites and space probes are attached to a frame serving as the structural center. A number of space agencies and facilities across the world support space research and transportation. In the United States, the National Aeronautics and Space Administration (NASA) is the supporting agency of space transportation. Its facilities and research centers, along with private companies, research, design, build, and operate space vehicles that orbit the earth and travel to distant planets.

Key Words

All the following words have been used in this chapter. Do you know their meanings?

air lock
bus
celestial reference device
Deep Space Network (DSN)
delta wing
elevon
flight deck
flight director
gimbal
high-temperature reusable surface insulation tile
hypergolic engine
hypersonic speed
inertial reference system
ion propulsion
liquid-fuel rocket
manned maneuvering unit (MMU)
Mission Control Center (MCC)
National Aeronautics and Space Administration (NASA)
orbital maneuvering system (OMS) engine
Orbiter Processing Facility (OPF)
oxidizer
payload area
propellant
reaction control system (RCS) engine
reaction wheel
reinforced carbon-carbon panel
solid-fuel rocket
solid rocket booster (SRB)
Space Shuttle Main Engine (SSME)
staging
trajectory
Vehicle Assembly Building (VAB)

Test Your Knowledge

Write your answers on a separate sheet of paper. Do not write in this book.

1. What is the difference between an air stream reaction engine and a rocket engine?
2. Summarize how thrusters operate in the vacuum of space.
3. *True or False?* An oxidizer is mixed with fuel to create a propellant.
4. *True or False?* Liquid-fuel rockets are simpler and offer less control than solid-fuel rockets.
5. Paraphrase how a liquid-fuel rocket works.
6. What is staging?
7. *True or False?* Ion propulsion was the first type of rocket propulsion used.
8. Write two or three sentences relating how an ion engine operates.
9. How are the distance and angle of a spacecraft from earth figured?
10. Select one type of attitude measurement device and explain how it is used.
11. *True or False?* A global positioning system (GPS) cannot be used in space.
12. *True or False?* Reaction wheels use momentum to maneuver the spacecraft.
13. Rotating the nozzle of an engine to change direction is known as _____.
14. Why are reaction control system (RCS) engines necessary for controlling spacecraft?
15. How are the functions of the rotational hand controller and translational hand controller different?
16. The orbiter wing style is a(n) _____ wing.
17. Analyze the aerodynamics of a delta wing.
18. Speeds of over Mach 5 are known as _____ speeds.
19. *True or False?* The high-temperature tiles used on the orbiter retain heat for a long period of time.
20. Cite the three components that make up a space shuttle.
21. The _____ is the section of the orbiter containing the flight controls.
22. _____ is the major organization supporting space transportation in the United States.
23. Recall and describe two space transportation support facilities.
24. The lead person of flight control is the _____.

Activities

1. If your class has access to a model rocket kit, construct and launch a rocket with your instructor's assistance. This can be a class project or a team project. Prepare a report on what you did and what you observed.
2. Construct a model of a satellite and display it.

25

Intermodal Transportation and Vehicular Systems

Basic Concepts

- Define *intermodal transportation.*
- State the importance of intermodal transportation in our society.
- Identify the importance of containerization.
- List and discuss the advantages of intermodal transportation.
- Define passenger and cargo intermodal transportation.

Intermediate Concepts

- Describe legislation and government agencies that control intermodal transportation.

Advanced Concepts

- Plan an intermodal shipping route.

A mode of transportation is a method of moving people and goods. Modes of transportation exist in all the environments of transportation: land, air, water, and space. Each mode of transportation is responsible for the moving of people and goods to new locations. In reality, however, getting from one place to another often requires more than one mode of transportation. Imagine sending a package from your home to an international destination. The package would be picked up by a small truck or van and delivered to a sorting terminal. See **Figure 25-1.** From there, the package, along with hundreds of other packages, would be taken to an airport terminal. An airplane would be used to transport the package to the destination country. Once in the foreign country, the package may travel in several different vehicles before reaching the final destination. In the transit from the starting point to the destination point, the package was transported using several modes of transportation. This is an example of intermodal transportation. The

Figure 25-1. Intermodal transportation. Many packages are transported by truck from customers' homes to a sorting facility and then to an airline terminal to continue their journeys. (United Parcel Service)

prefix *inter-* comes from the Latin language and means "between." Thus, intermodal means between several modes or involving more than one mode.

Intermodal Transportation

Intermodal Surface Transportation Efficiency Act (ISTEA) of 1991: The act that created the National Commission on Intermodal Transportation (NCIT).

Intermodal transportation, or intermodalism, has been a reality in the United States since the mid-1980s. The U.S. government recognized the need for an increase in intermodalism and passed the ***Intermodal Surface Transportation Efficiency Act (ISTEA) of 1991***. The Act called for the creation of the National Commission on Intermodal Transportation (NCIT). NCIT studied the state of intermodal transportation and made suggestions to Congress through its final report in 1994. The Commission found that intermodal transportation has great benefits for the nation, including increased productivity and decreased congestion. See **Figure 25-2.** Today, the U.S. Department of Transportation's Office of Intermodalism oversees intermodal transportation in the United States.

Intermodal systems involve less material handling, which results in less damage and loss of goods. See **Figure 25-3.** This type of transportation has cut down on shipping time and costs. Intermodal transportation also reduces energy consumption.

This mode of transportation takes time to plan and organize. It depends on each mode working on a timetable. Service can easily break down if all modes do not stay on schedule. If one mode does not pick up where the other one ends, the process is slowed down. There are delays and confusion. Both timing and planning are vital to intermodal transportation. Intermodal transportation can be broken into two categories. These categories are intermodal cargo transportation and intermodal passenger transportation.

Intermodal Cargo Transportation

Intermodal cargo transportation is the movement of goods and products using two or more modes of transportation. Transporting cargo from one point to another is often done by using several different modes of transportation. For example, the shipping of grain from the United States to Russia may go something like this:

- The grain is first harvested by the farmer and brought out of the fields in trucks or wagons. See **Figure 25-4.**
- The farmer sells the grain and hauls it to a grain elevator.
- The grain is loaded into railroad cars or semitrucks and transported to a port.
- Here, the grain is loaded onto a ship that carries it across the ocean.
- Once the grain arrives in Russia, it must then be unloaded and distributed by way of land vehicles to various places of use.

There are a few different ways in which to ship cargo by intermodal transport. Since the intermodal transportation system has grown, special shipping containers and transport vehicles have been developed.

Figure 25-2. The increased use of intermodal transportation has reduced highway traffic congestion like this.

Figure 25-3. This port facility is an example of intermodal transportation. Shipping containers arriving onboard the ships are unloaded and stored before being transferred to railcars or truck chassis to continue their journey. Movement in the opposite direction also takes place. Containers arriving by rail or highway are stored and then loaded onto ships.

Intermodal cargo transportation: Transporting cargo from one point to another by using several different modes of transportation.

Figure 25-4. The first intermodal transport step following the harvest of this shelled corn is to haul it by truck to a grain elevator. The corn might travel by train, barge, and ship before it reaches its final destination.

Containerized shipping

Imagine you are traveling to a relative or friend's house across the country for a week-long vacation. Would you carry each individual piece of clothing, or would you pack all the clothes into a suitcase? Of course, the clothes would be easier to carry if they were packed together in a piece of luggage. See **Figure 25-5.** ***Containerization*** is a method of handling goods by packing many small packages, which are going to the same destination, in one large container. This method eliminates the handling of many small packages every time they change vehicles. It is easier and requires less time and people to move one container, rather than many small packages.

Technology Link

Agriculture: Intermodal Delivery of Crops

It may not surprise you that agriculture uses transportation throughout many of its processes. Tractors, grain trucks, and harvesters are commonly associated with agriculture. You may not realize, however, that in order for crops to be delivered to the market or food plant, most products are shipped through intermodal transportation. This occurs with food products, including grain, milk, and tropical produce.

When wheat is harvested, it is collected using a harvester, or combine. The combine separates the wheat into the grain and straw. The straw is sent out the back, and the grain is stored in the combine and later unloaded into an open-topped grain truck. The grain is driven to a grain elevator or silo for sale or storage. At the elevator or silo, augers and conveyors transport the grain. From the grain elevators, the grain is often loaded onto either barges or railroad cars. The grain is then transported to another elevator across the nation or world and unloaded into hoppers that feed back into an elevator. From there, it is transported by truck to the final destination.

Milk is also transported by several modes of transportation. As you may imagine, tanker trucks from the farms that milk the cows transport it to the dairy that processes the milk. It is also transported, however, within the dairy farm and dairy processing plant by pipelines. Pipes and tubes are used to connect milking machines, storage tanks, heat-treating machines, and bottling machines.

Imagine the intermodal processes that bananas, mangos, and coffee beans harvested in the tropical regions of the world must travel through to reach our supermarkets. The crops are harvested and transported using farm equipment, transported using tractor-trailers, and then shipped to the United States on ships, finally arriving at the supermarket on trucks. Without efficient intermodal transportation it would be impossible to have fresh produce in many parts of the United States year-round.

Figure 25-5. Containerized shipping is similar to using a suitcase, since it is easier to pack a number of items into one container than to transport them individually.

Containerization: A method of handling goods by packing many small packages, which are going to the same destination, in one large container.

Dry cargo container: A fully enclosed shipping container resembling a large metal box with doors either on the side or end.

The less the individual packages are handled, the less likely it is that they will be accidentally damaged or stolen. This method of packing makes intermodal transportation safer and more efficient than other modes of transportation.

Containerized shipping is an efficient method of transporting goods. The ***dry cargo container*** is the most frequently used and most commonly recognized. See **Figure 25-6.** This container is fully enclosed, resembling a large metal box with doors either on the sides or ends. Products requiring circulating air are shipped in ***ventilated containers***. These containers have openings in the sidewalls allowing air to flow around the cargo. Open top and flat rack containers are used for large equipment and bulk goods. The open top containers have removeable or retractable roofs, and the flat rack containers do not have sides or a top. ***Thermal containers*** are used to transport perishable food products. They can be insulated, refrigerated, heated, or both refrigerated and heated. The standard container size is 8′ wide, 8′ 6″ high, and 40′ long. The inside volume of a standard container is about 2385 ft^3. Planners organize the products to be shipped in the best possible configuration to minimize wasted space.

Figure 25-6. Dry goods containers, such as the one shown here being unloaded by crane, are the most common type. Several more specialized containers are also used. (Port of Long Beach)

Ventilated container: A container that has openings in the sidewalls allowing air to flow around the cargo.

Thermal container: A container used to transport perishable food products. It can be insulated, refrigerated, heated, or both refrigerated and heated.

When transported by water, the containers can be swiftly loaded onto ships with the aid of cranes. See **Figure 25-7.** Containers filled with products to be sent to the same place are transported by rail or highway. Those that have contents to be sent several different places are moved to a central distributing point. There, they are opened, and the contents are sent on to their final destinations. When using a highway mode, the container is lifted onto a frame with eight wheels. See **Figure 25-8.** A truck tractor can then pull the container. Railroad flatcars can also carry the containers. The containers can be lifted directly onto the flatcar and transported to their destination by way of rails. See **Figure 25-9.** Once they arrive at their destination terminal, the containers are unloaded onto a truck trailer and transported across the land. A container transported on a railroad flatcar is referred to as a ***Container on Flat Car (COFC).*** Transporting the containers across the ocean on barges and ships is done quite often. The ships are designed to carry these standard containers. They are referred to as ***containerships.*** See **Figure 25-10.** Another way to ship goods is transporting containers by air.

Figure 25-7. Containerized shipping. Large metal containers filled with goods are loaded onto flat railcars to be transported. (CSX Creative Services)

Figure 25-8. Often, containers are loaded onto a trailer frame and hauled away by a tractor. (CSX Creative Services)

Trailers on Flat Cars (TOFCs)

Trailer on Flat Car (TOFC) is another method of intermodal transportation. It involves trailers full of cargo carried on a rail flatcar. See **Figure 25-11.** This method of intermodal transportation is called ***piggybacking.***

Other forms of intermodal cargo transportation

Liquids and mined materials are also transported by using intermodal methods. Oil, milk, inks, chemicals, and other liquids are transported through pipelines. For instance, oil is transported to a terminal through pipelines and then pumped into a tank truck and hauled to its destination. See **Figure 25-12.** Mined materials, such as coal, ore, and various stones, are first transported by way of conveyor out of the ground. From the stationary conveyor, they are loaded into trucks or hopper railroad cars. The railroad car or truck hauls the materials to their final destination.

Intermodal Passenger Transportation

People also use intermodal transportation. Often, we use it several times a day without realizing it. If you ride your bike to your friend's house, and then you both ride to school in a car, you have just experienced intermodal transportation. Let us say you have

Figure 25-9. Containers are often loaded on flatcars for long-distance land transportation. These container flats are shown in a rail yard along with a number of other specialized freight cars.

Figure 25-10. Containerships are designed to carry the maximum number of containers, both in their holds and stacked on deck.

Container on Flat Car (COFC): A container transported on a railroad flatcar.

Containership: A ship designed to carry standard containers.

Trailer on Flat Car (TOFC): A method of intermodal transportation involving trailers full of cargo carried on a rail flatcar.

Piggyback: To carry a trailer or container on wheels on a flat railroad car.

Figure 25-11. Trailer on Flat Car (TOFC) is a widely used form of intermodal transportation. Loaded trailers are placed on flatcars for long-distance transportation and then hauled to their final destination as a tractor-trailer unit.

just won a trip to Maui, one of the Hawaiian islands. You live in Boston, Massachusetts. You would probably get to the airport by way of a bus or car. You would catch a plane to San Francisco, California. See **Figure 25-13.** Once you are in California, you have to change planes. You get on the moving sidewalk that goes through the airport, and you board the next plane. You then fly to Honolulu, Hawaii. Once you are there, you take a taxi to board a ferry, which will take you to the island of Maui. ***Intermodal passenger transportation*** is the movement of people using two or more modes of transportation. In the previous example, you experienced intermodal passenger transportation by using a combination of land, air, and water transportation.

Intermodal passenger transportation: Transporting passengers from one point to another by using several different modes of transportation.

These modes of transportation, however, are not as streamlined as cargo intermodal transportation. You have to purchase separate tickets for the bus, airplane, and ferry, and you may even have to walk a distance from the bus stop to the airport. Passenger transportation companies are beginning to see the importance of seamless intermodal travel and are fixing this problem. Due to partnerships and alliances between companies, in some places, you can buy a ticket from one company that will include all the bus, plane, light-rail, and shuttle tickets you will need to get from the starting point to the destination point. Companies are also building terminals alongside or within other transportation stations. In many cities, the intercity bus station and train station are in the same building or terminal. See **Figure 25-14.** Another common example is the placement of rental car companies and shuttle bus pickups inside or

Figure 25-12. Some tractor-trailers, called tankers, haul liquid cargo. (Freightliner Corporation)

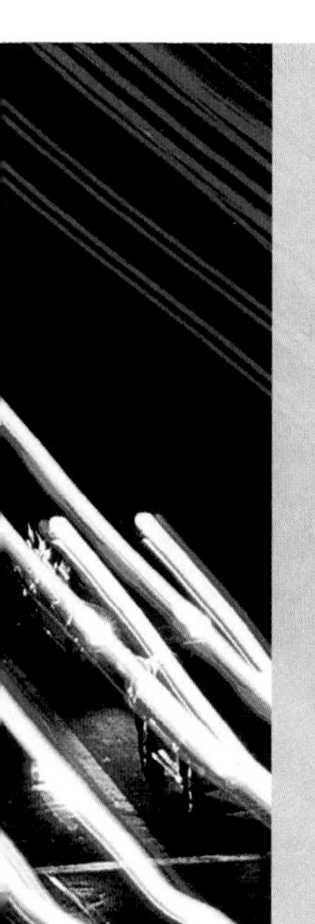

Career Connection

Pipelayers, Pipefitters, and Steamfitters

The use of pipelines requires the work of pipelayers, pipefitters, and steamfitters. These construction professionals lay, install, and fit the pipes needed to move materials within buildings and across cities. Pipelayers lay city utilities, such as water mains and sewer systems, as well as oil and gas lines. Pipefitters and steamfitters install piping used in power production and manufacturing plants.

Pipelayers, pipefitters, and steamfitters typically work 40-hour workweeks, with overtime pay if extra work is required. The nature of the work is, at times, physically demanding and, often, outdoors. Entry requirements into the profession are a high school diploma or the equivalent and being 18 years old. Training is provided through apprenticeship programs. The apprenticeship program through the United Association of Journeymen and Apprentices of the Plumbing, Pipefitting, Sprinkler Fitting Industry of the United States and Canada (UA), which is the largest pipefitting union, lasts 5 years. Each year of the apprenticeship consists of 1700–2000 hours of on-the-job training and 216 hours of classroom instruction. The classroom instruction covers topics such as drafting, mathematics, safety, and local codes. The average pipelayer makes over $13 per hour, while the highest paid makes over $25 per hour.

next to airport terminals. Intermodal transportation should be efficient and save passengers time. The closer these facilities are to each other, the more time can be saved.

Figure 25-13. Airplanes are often part of an intermodal transportation system for both freight and passengers. This aircraft is being guided to an airport gate, where the jet way will be connected to allow passengers to unload and new passengers to board.

Figure 25-14. Urban mass transit systems are examples of intermodal passenger transportation. A—This suburban transit station outside of Washington, D.C., allows travelers arriving by car or bus to continue their journey by subway. B—Passengers arriving at the station by subway can then transfer to buses or cars parked in adjacent lots.

A

B

Vehicular Systems of Intermodal Transportation

Land, water, air, and space transportation all have vehicles and vehicular systems that are unique to their environments. Intermodal transportation is different, however, in that it uses vehicles in multiple environments to transport people and goods. The vehicles selected and used determine the vehicular systems used in intermodal transportation. There are really only two vehicular systems intermodal transportation impacts—structure and support.

Structural Systems

The structures of the vehicles used in cargo intermodal transportation must be able to carry the container being transported. In land transportation, special trailers and rolling stock have been designed to carry the containers. Trailers used in highway transportation to move intermodal containers are known as chassis. A chassis is a frame that connects to the tractor, or semi, in the front and has wheels in the rear. A chassis is very similar to the underside of a typical semitrailer. The containers are placed on top of the chassis and secured in each of the four corners. Because the sizes of the containers are fairly standard, most chassis are made in standard lengths. The structure of the chassis, however, can also be made to be shortened or extended to fit various sizes of containers.

Well car: A railcar designed so the containers actually sit near or below the height of the axles, allowing two containers to be stacked on top of one another.

Railcar structures are also designed differently for intermodal transportation. Flatcars are standard railcars without walls. These are typically used for hauling trailers, as previously described. ***Well cars*** are designed so the containers actually sit near or below the height of the axles. See **Figure 25-15.** This allows two containers to be stacked on top of one another, without worrying about the height of tunnels and other obstructions along the railroad. Well cars are often linked with other well cars and share the same bogies. These cars can fit various sizes of containers.

Lastly, the structures of ships must also be designed to carry containers. Most often, containerships and barges carry intermodal containers. Containerships are large ocean-going vessels that can carry hundreds of containers. Barges are smaller and travel in inland waterways. Containers in these ships can be stacked as many as eight high. This makes the shipping of containers more cost-efficient.

Support Systems

Like the other four transportation systems, intermodal transportation also requires support systems. For example, ports and harbors are needed to load and unload ships. Unlike a port used for only one mode of transportation, however, intermodal ports must also have facilities for the loading and unloading of semitrucks and trains. Intermodal terminals must have facilities for all the modes of transportation they serve. This often makes intermodal stations very expensive to construct. The money saved from the efficiency of intermodal transportation, however, often makes up for the costs of the facilities.

Figure 25-15. Rail transport. These containers have been stacked into well-type cars to be transported partway to their final destination. (Norfolk Southern Corporation)

Curricular Connection

Social Studies: Transportation and Landforms

The earth is covered with different landforms. Oceans, rivers, plains, mountains, valleys, forests, and swamps can be found around the globe. Vehicles and vehicular support systems have been designed to move goods and people across each of the different landforms. There are, however, few vehicles that can cross all landforms. To travel long distances around the globe, intermodal transportation is used to carry passengers and cargo across different landforms.

Imagine the landforms you would cross and the vehicles you would use to travel on a vacation from Dodge City, Kansas to Venice, Italy. Starting in Dodge City, Kansas, you might drive a car across the plains to the mountains of Colorado. From the airport in Denver, Colorado, you might take an airplane across over half of the continental United States to New York, and then you would probably get on another flight to travel over the Atlantic Ocean to Rome. Once in Rome, you might choose to rent a car and drive over the Apennines Mountains until you reach Venice. In order to reach your hotel in Venice, you would have to ride a gondola through the canal system.

The landforms of the earth slowed the migration of people before many early forms of transportation were developed. Today, there are various vehicles that can be used to cross different landforms. When used together, they form an intermodal transportation system.

At intermodal ports and docks, large cranes are used to pick the containers off the ships and place them into a storage area. See **Figure 25-16.** From there, other cranes or specialized pieces of equipment are used to place the containers onto trucks or railcars. At railroad intermodal facilities, loading equipment is used to transfer the containers from the railcars to the truck chassis and vice versa. These facilities also employ people to monitor the transfer of the containers and make sure the containers are sent to the correct places on the correct mode of transportation.

Figure 25-16. Giant cranes are used at intermodal ports to load containers on ships, unload them for storage, or transfer them to other means of transportation.

Summary

Intermodal transportation is a system that uses a combination of several modes of transportation to get from one place to another. It is an efficient system of cargo transportation. Containerization is a very popular method of effective intermodal transportation because it cuts back on material handling, and it also saves on labor costs. Containers can be loaded onto ships, flat railroad cars, and airplanes. Carrying a trailer or container on wheels on a flat railroad car is referred to as piggybacking. Passengers also use intermodal transportation.

Key Words

All the following words have been used in this chapter. Do you know their meanings?

containerization
Container on Flat Car (COFC)
containership
dry cargo container
intermodal cargo transportation
intermodal passenger transportation
Intermodal Surface Transportation Efficiency Act (ISTEA) of 1991
piggyback
thermal container
Trailer on Flat Car (TOFC)
ventilated container
well car

Test Your Knowledge

Please do not write in this text. Place your answers on a separate sheet.

1. Write the definition of *intermodal transportation.*
2. *True or False?* The federal government has made no effort to recognize the need for intermodal transportation.
3. What is the importance of intermodal transportation in our society?
4. List the two categories of intermodal transportation.
5. List vehicles that may be used to move freight from Tokyo, Japan to St. Louis, Missouri.
6. Why is containerization useful?
7. Name at least two advantages of intermodal transportation.
8. *True or False?* Containerization is an efficient method of handling cargo.
9. Describe two types of containers.
10. What does *TOFC* stand for?

11. Carrying truck trailers on the back of railroad cars is known as:
    A. hauling.
    B. containerization.
    C. piggybacking.
    D. intermodal passenger transportation.
12. *True or False?* Travelers hardly ever use intermodal transportation systems.
13. Describe how you have used intermodal transportation in the past.
14. *True or False?* Intermodal harbors and ports are the same as those used for only one mode of transportation.

Activities

1. Design and build a scale model of a transportation system with two modes.
2. As a class project, develop a proposal for an intermodal transportation system to relieve traffic congestion in a major city with a population of 2 million people.

26

Energy, Power, Transportation, and the Environment

Basic Concepts

- State the various aspects of the environment that pollution from the energy, power, and transportation industries most commonly affects.
- List ways in which producing power has been harmful to the environment.
- Name ways in which present-day methods of transportation are harmful to the environment.
- Identify ways in which obtaining energy resources has been harmful to the environment.

Intermediate Concepts

- Discuss the steps that have been taken to minimize environmental impacts associated with the transportation of goods and people.
- Describe the steps that have been taken to minimize the environmental impacts of harvesting and refining energy for use.
- Give examples of the steps that have been taken to minimize environmental impacts associated with the generation of large-scale power.

Advanced Concepts

- Explain one major piece of legislation associated with environmental protection.
- Summarize advanced and futuristic concepts associated with environmental protection emerging from the energy, power, and transportation industries.

There is an ever-growing concern that activities such as large-scale power generation and the transportation of goods and people are having a profoundly adverse effect on our global environment. See **Figure 26-1.** At the local level, energy, power, and transportation technologies can have effects such as producing smog and haze, reducing the quality of

Figure 26-1. The very conveniences that make our lives easier and improve our standard of living are also harmful to the environment. A—Vehicle exhausts from cars and trucks are a major source of pollutants. B—Although air quality has been improved in recent years, many major cities still experience high levels of pollution at times.

A

B

breathing air, and polluting lakes and streams. At the global level, there is concern that the rapid consumption of so many fossil fuels over such a short period of time is resulting in worldwide climate change.

Environmental Pollution

Air pollution: The action of contaminating the mixture of gases surrounding the earth, especially with human-made waste.

Contaminants in the air, water, and land cause environmental pollution. ***Air pollution*** is the action of contaminating the mixture of gases surrounding the earth, especially with human-made waste. The greatest source of air pollution is a direct result of the burning of fossil fuels. Fuels are consumed to produce electricity, for industrial processes, and to power internal combustion engines. See **Figure 26-2.**

Figure 26-2. Air pollution can occur through natural phenomena or human activity. Fossil fuel consumption for transportation and power generation is responsible for many pollutants.

Pollutant	Natural Source	Human Source
Carbon dioxide	Decay from oceans	Wood and fossil fuel combustion
Nitrogen oxides	Lightning, bacteria in soil	High-temperature combustion
Sulfur dioxide	Decay, volcanic eruptions	Coal and oil combustion, smelting of ores
Ozone	Produced in the troposphere	Smog from auto and industrial emissions
*Particulate matter	Forest fires, volcanic eruptions, wind erosion	Waste building, road building, mining
Methane	Termites, anaerobic decay, animal waste	Combustion, natural gas leaks
Ammonia	Anaerobic decay	Sewage treatment plants
*Not a specific chemical substance, but still classified as a pollutant		

Air Pollution

Air pollution is known to have certain effects on the body. See **Figure 26-3.** As Americans began to recognize and acknowledge the significant air quality problems associated with the modernization of our nation, it became evident that steps must be taken to protect the environment. One of the most significant steps taken was the Air Pollution Control Act of 1955.

Prior to the mid-1950s, air pollution and air quality issues were the responsibility of state and local governments. In 1955, the first federal legislation controlling air pollution was passed. It was known as the Air Pollution Control Act. Congress passed the ***Clean Air Act (CAA)*** in 1963. This act set emissions standards for certain sources of pollution. Subsequent amendments to the legislation strengthened the federal role in preserving air quality. The most significant of these amendments are associated with the CAA amendments of 1990. This series of amendments included provisions for an acid rain control program, funding for state-run permit programs for the operation of many sources of air pollutants, and even funding for the retraining of displaced workers who lost their jobs as a direct result of implementation of the CAA amendments. Changes implemented in 1990 include the following:

- More stringent automobile emissions standards.
- Reformulated gasoline and alternative fuels for some of the most populated areas of the country.
- A phaseout schedule for many of the most ozone-depleting chemicals.
- The establishment of National Ambient Air Quality Standards for various types of air pollutants.
- The strengthening of the Environmental Protection Agency (EPA)'s role in oversight and authority to assess penalties for noncompliance.

Clean Air Act (CAA): The Act Congress passed in 1963 that set emissions standards for certain sources of pollution.

Figure 26-3. Sources of various pollutants and their effects on human health. Other effects are suspected as well.

Pollutant	Source	Effects
Carbon monoxide	Automobile emissions	High concentrations can cause death, low concentrations impair judgment
Lead	Smelting and manufacturing processes	Lead poisoning—impaired mental ability in younger children
Nitrogen oxides	Fossil fuel consumption	Smog, acid rain, respiratory and eye irritation
Ozone	Photochemical reactions with by-products of combustion	Respiratory irritation, reduced lung function
Particulate matter	Smoke, dust, automobile emissions, industrial processes	Breathing difficulties
Sulfur oxides	Industrial processes involving combustion	Smog, upper respiratory disease
Toxic air pollutants	Asbestos, arsenic, benzene, chemical and industrial processes	Lung disease, cancer

Nonattainment: A classification of an area in which the Environmental Protection Agency (EPA)'s minimum air quality guidelines are not met.

Water pollution: The action of contaminating the liquid that descends from the clouds as rain, especially with human-made waste.

Suspended particle: A particle in water that absorbs light and makes water cloudy.

The EPA has the authority to establish minimum air quality guidelines. It is the state, however, that implements plans to meet the air quality guidelines. Each state has to submit a state implementation plan to the EPA to ensure guidelines are being met. Those guidelines not met are referred to as ***nonattainment*** areas. Nonattainment areas are classified as marginal, moderate, serious, severe, or extreme. A plan is then submitted to the EPA, along with a timeline to bring the air quality for a particular pollutant into compliance. In the case of severe or extreme air pollution, a timeline of up to 20 years may be necessary to improve air quality to the guidelines established in 1990.

Water Pollution

Water pollution is the action of contaminating the liquid that descends from the clouds as rain, especially with human-made waste. It can affect ponds, streams, rivers, and even oceans. There are natural impurities and human activities that pollute water supplies. Naturally occurring particles that appear in water are not always pollutants. They are typically divided into three categories of particles. ***Suspended particles*** absorb light and make water cloudy. Colloidal particles require special filtration to be removed from water. Dissolved matter, including molecules and ions of various substances, is the tiniest of all particles. Human activities are often

Career Connection

Safety Professionals

The roles of safety professionals vary with the company or organization with whom they are employed. Safety professionals are prepared to identify hazards in the workplace and to implement ways to minimize those hazards. Some industries require highly specific safety expertise. Most trained safety professionals are well prepared to recognize hazards, conduct safety inspections and audits, implement fire protection plans, and ensure regulatory compliance with agencies such as the Occupational Safety and Health Administration (OSHA) and the Environmental Protection Agency (EPA). They also manage hazardous materials, improve the workplace through an understanding of ergonomics (the physiological and psychological limitations of the body), ensure environmental protection, and provide safety training to employees. Safety professionals are often responsible for conducting accident investigations, advising management by helping to establish safety objectives, maintaining an accurate record keeping system for safety-related data, and implementing and managing a comprehensive safety program.

The largest percentage of safety professionals are employed in the manufacturing and production sector of the economy. Many safety professionals are employed in the insurance industry, for governmental agencies such as OSHA or the EPA, or as independent consultants. The minimum educational requirement for a safety professional is a bachelor's degree in a safety-related field. Most safety professionals earn a Certified Safety Professional (CSP) designation by passing a professional exam administered by the Board of CSPs. Salaries for safety professionals range from about $30,000 for a safety inspector to over $150,000 for an experienced manager of a comprehensive safety program in a large industry. The average starting salary for an entry-level safety professional is approximately $40,000.

the cause of localized water pollution. For instance, industrial spills, agricultural runoff, or untreated sewage runoff can contaminate rivers. Even oceans can be subjected to large-scale pollution, due to garbage dumping and oil spills. Oceans rich in animal and plant life cover more than 70% of the earth. Protecting the earth from water pollution is essential to the long-term survival of the human species. Water is also essential to maintaining our quality of life. It takes an estimated 400 gallons of water to produce 1 gallon of gasoline. 50,000 gallons of water are needed to produce an automobile.

There are certain water pollutants specifically associated with the power generation and transportation industries. These pollutants are oils, hydrocarbons, and lead. See **Figure 26-4.** Another water pollution problem that is a direct result of burning fossil fuels is acid rain.

Acid rain forms when water vapor and certain elements combine chemically with natural and human-made pollutants in the stratosphere. Emissions, such as sulfur dioxide and nitrogen oxide, combine with water vapor in the air to form sulfuric and nitric acids. When this mixture combines with rain and falls back to earth, it can contaminate lakes, rivers, and streams. Acid rain can make these bodies of water more acidic than they would be naturally. The subsequent effects can be detrimental to plant and animal life. Acid rain is also known to leach natural aluminum from the soil. This aluminum is extremely toxic to many organisms, such as plants that exist in the water.

It is estimated that ***natural pollutants***, such as acid rain and smog created as the result of wildfires caused by lightning strikes, account for only about 10% of all the acid rain on the planet. The remaining 90% is estimated to be occurring as a result of the combustion of fossil fuels. This problem was first noticed around the beginning of the Industrial Revolution in the late eighteenth century. At that time, fish kills in ponds located close to industrial plants were reported. A ***fish kill*** results in the death of a large number of fish, due to a rapid change in the characteristics of their environment. The change could be a severe temperature change in a short period of time or a severe change in the toxicity of the water. The solution of the Industrial Revolution era was to build taller smokestacks so the waste would be dispersed over a larger area.

The ***pH*** of a solution is a measure of its acidity. The pH scale consists of a range from 0 to 14. A 7 on the pH scale represents a perfectly neutral solution. ***Acidic solutions*** have a pH rating of less than 7. A rating above 7

Natural pollutant: Something produced by nature that contaminates the environment.

Fish kill: The death of a large number of fish, due to a rapid change in the characteristics of their environment.

pH: A measure of a solution's acidity.

Acidic solution: A solution with a pH rating of less than 7.

Figure 26-4. Water pollutants associated with transportation and power generation.

Pollutant	Source	Effects
Oils and hydrocarbons	Oil spills, oil leaks, oil field run-off	Death of fish, disruption of food chain, contamination of drinking water, possible liver and kidney damage from eating contaminated fish
Lead	Some hydrocarbon fuels	Lead poisoning

Curricular Connection

Social Studies: The Exxon Valdez *and Oil Spills*

On March 24, 1989, the supertanker *Exxon Valdez* strayed from its course in Prince William Sound and slammed into rocks underneath the surface. See **Figure 26-A.** The result of the disaster was an oil spill that took more than three years to clean up. Approximately 11 million gallons of oil were spilled over an area that grew to reach hundreds of square miles. More than 1300 miles of shoreline were affected. Of those, about 200 miles were moderately to heavily oiled, with the remainder receiving a lighter coating to only trace amounts of oil. See **Figure 26-B.**

A team of more than 10,000 workers used various techniques to clean the shoreline of the oil until 1992, when the U.S. Coast Guard declared the cleanup operations complete. Despite the fact that many fish and birds were killed by the oil spill, Prince William Sound continues to yield record harvests for salmon. The *Exxon Valdez* disaster remains one of the worst accidental oil spills in American waters and cost more than $2.1 billion to clean up, according to Exxon. As of 2003, the *Exxon Valdez* spill was no longer considered to be one of the top 50 oil spills worldwide. It is still widely considered to be the worst spill, however, in terms of environmental damage, due to the pristine nature of Prince William Sound and the abundance of unique fish and wildlife found there.

One of the major lessons learned by the *Exxon Valdez* disaster was that spill prevention, response, and containment procedures were totally inadequate. Since the time of the accident, the following improvements have been made to oil shipping through Prince William Sound:

Figure 26-A. The *Exxon Valdez* tanker aground in Prince William Sound. (National Oceanic and Atmospheric Administration)

Figure 26-B. Workers cleaning the shoreline of oil from the *Exxon Valdez* oil spill. (National Oceanic and Atmospheric Administration)

Basic solution: A solution with a pH rating above 7.

represents a ***basic solution***. Pure water has a pH of 7, while rainwater and stream water are slightly acidic. Most organisms can tolerate minor acidity, but higher levels of acidity can significantly reduce the health of ecosystems. Acid rain includes precipitation with a pH lower than 5.6. Some of the most acidic rains have reported pH ratings as low as 2.4.

- All fully laden tankers are now satellite monitored by the U.S. Coast Guard from departure at the Trans-Alaska Pipeline terminal until they exit Prince William Sound.
- Two escort vessels now accompany each tanker through the entire sound.
- Special marine pilots with specific knowledge of Prince William Sound board tankers and work with the tanker crew to navigate the 25 most difficult miles of the 70-mile journey through the sound.
- Weather criteria for safe navigation have been established so ships cannot leave the terminal unless weather conditions are favorable.
- Congress has enacted legislation that will require all tankers operating in Prince William Sound to be double hulled by 2015. A double-hulled vessel is much more likely to contain a spill, in the unlikely event of an accident.
- Emergency crews practice and plan for oil spills in Prince William Sound on an annual basis. The spill scenario they train for is 12.6 million gallons.
- Skimming equipment is now in place that can skim 10 times the amount of oil from the water as the equipment available at the time of the *Exxon Valdez* spill.

The worst oil spill worldwide occurred in 1991. It was not accidental at all. During the Persian Gulf War of 1991, Saddam Hussein, dictator of Iraq, ordered 250 million gallons of crude oil dumped into the waters of the Persian Gulf. This deliberate oil spill was the magnitude of 20 times greater than the *Exxon Valdez* disaster. Hussein's army also set fire to approximately 600 oil wells in Kuwait, creating an environmental disaster of epic proportion. See **Figure 26-C.** Using private contractors, the well fires were extinguished in about one year.

Figure 26-C. Oil well fires in Kuwait during the 1991 Persian Gulf War. (National Oceanic and Atmospheric Administration)

Other Forms of Pollution

Air and water pollution are the most obvious forms of pollution. There are, however, many other more subtle forms of pollution. Some of these have been brought about as a result of energy, power, and transportation industries.

Lead poisoning: Chronic intoxication produced by the absorption of lead into the system.

Electromagnetic field (EMF): A field of electromagnetic energy produced by electricity flowing through power lines.

Light pollution: The excessive amount of light in the nighttime sky that often surrounds urban areas.

Noise pollution: Annoying or harmful noise in an environment.

Land pollution: The action of contaminating the surface of the earth and its natural resources.

Surface mining: The process of working the external layer of the earth and extracting mineral substances.

Reclamation: A technique that requires land that has been disturbed to be graded and covered with topsoil, so as to be suitable for alternate use at the conclusion of mining operations.

Lead contamination, or ***lead poisoning***, was more prevalent many years ago, when lead was a primary additive in gasoline. Breathing lead fumes could lead to lead poisoning, leaving victims with permanently diminished mental capacity. Lead was phased out as a gasoline additive and is no longer used in gasoline today.

Electromagnetic fields (EMFs) are all around us. Electricity flowing through power lines produces them. There is considerable debate about the health effects of EMFs. People who live in close proximity to high-voltage power transmission lines are of particular concern for health risks. Reputable studies have not proven that exposure to EMFs causes any significant health risks at this time. There is concern, however, that chronic exposure to EMFs may increase the risk of incidence for some specific types of cancer. The effects of chronic exposure to electricity on cell structure and development have not been thoroughly researched.

Light pollution is a term used to describe the excessive amount of light in the nighttime sky that often surrounds urban areas. The primary consequence of light pollution is that artificially created light can trick plants, birds, and animals. Migrating birds can be drawn toward the light, as they confuse it for daylight on the horizon. Plants can bloom prematurely if exposed to too much artificial light. Nocturnal animals living near an urban area can also become confused between day and night.

Noise pollution is a problem commonly associated with some forms of power generation and many forms of transportation. Regarding the installation of a wind farm to generate electricity, noise can be a primary concern. Geothermal power plants are also known to be very noisy. Concern about noise pollution has resulted in the modifications of airplane approach ways to airports and even aircraft engine design.

Land pollution is a form of pollution that occurs when land is harvested (typically surface mined) to the extent that it renders the remaining land unusable. It is a particular problem in the fossil fuel energy industry, especially regarding the mining of coal. ***Surface mining***, often referred to as *strip mining*, refers to the process of extracting natural resources, such as coal, from the earth by digging the coal out from the surface, as opposed to extracting it from underground mines. Years ago, strip or surface mining plots were simply left ungraded and unusable for alternate purposes at the conclusion of a mining operation. Legislation now mandates that strip mining tracts undergo ***reclamation*** at the conclusion of mining operations. This technique requires the land that has been disturbed to be graded and covered with topsoil so as to be suitable for alternate use at the conclusion of mining operations.

Thermal pollution is a form of pollution most commonly associated with power plants. Historically, it has been easier and cheaper to simply expel waste heat, a by-product of the combustion process, into the atmosphere. Many power plants were constructed in the 1950s and 1960s. At that time, little attention was paid to the effects on aquatic life. Today, as a result of more stringent regulations, power plant cooling systems must be designed to take the local environmental characteristics of lakes, rivers, and streams into account. For instance, let us say an environmental impact study determines that the discharge water from a proposed power plant will significantly warm the water in a river near the location where the power plant is to be built. Such a change would be drastic enough to change the ecosystem of the river. Therefore, the water would have to be cooled to within a few degrees of the temperature of the river, prior to discharge. This is typically accomplished with the use of cooling towers, which are responsible for transferring the excess heat from the discharge water to the surrounding air, prior to letting the water flow back into the river. See **Figure 26-5.**

Thermal pollution: A form of pollution most commonly associated with power plants, in which expelled waste heat adversely affects aquatic life.

Choosing Environmentally Friendly Energy, Power, and Transportation Sources

There are many environmentally friendly alternatives to using conventional energy, power, and transportation sources. These alternatives, however, are often not as economical or convenient as traditional sources. For instance, let us say you live in a suburb and work in a city less than 10 miles away. You typically drive to work, but you would like to consider some environmentally friendly transportation alternatives. Riding a bicycle is the most economical and environmentally friendly alternative (in addition to being the healthiest), but it is also the least convenient. A compromise might be to consider commuting with other people from the neighborhood who also work in the city, thereby using only one vehicle, instead of three or four. Another popular and economical alternative is to use mass transit, such as a subway or bus system. This method of

Figure 26-5. A typical thermal electric generating station, shown in schematic form. Large power plants require tremendous amounts of cooling water. The cooling water, shown in color, flows from the intake through the condenser to remove heat from the steam in the turbine loop. The water is returned to its source at a higher temperature, which can affect fish and other forms of aquatic life. A system must be designed to accommodate a specific site in a way that is not harmful to the environment.

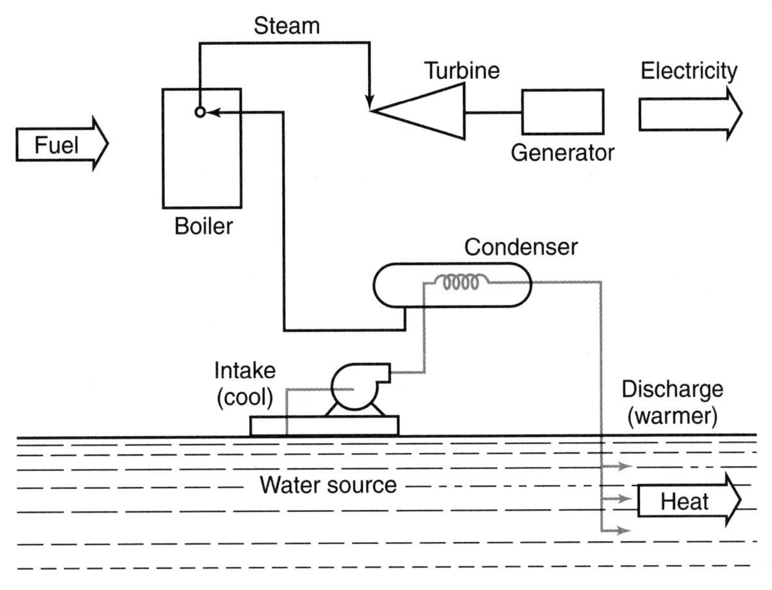

transportation is more environmentally friendly than driving yourself, and it is also more economical. It might be slightly less convenient than driving yourself, but this might be an inconvenience with which you can live.

Ultimately, factors such as convenience, economics, and available technology influence decisions about energy, power, and transportation that affect the environment. Such technological trade-offs are common—we make them all the time. In the summer, when your house is hot, you have several options. The most environmentally friendly option is to open the windows, but this can be inconvenient and might not work without a breeze. The least environmentally friendly is to use air conditioning. It is also the most expensive choice, but it would require only the flick of a switch. A good compromise might be to use a fan. Fans require much less electricity to run than air conditioning and are equally easy to use. Again, this might be an environmentally friendly option with which you can live.

The most environmentally friendly options are usually selected when they are also the most cost-effective and convenient. A hybrid vehicle typically offers similar performance to that of a regular automobile, but it has much better gas mileage. Hybrid vehicles are more environmentally friendly than regular automobiles. Although they currently cost more to purchase than regular automobiles, the cost will come down as more people express interest in purchasing them. The high initial cost is also offset by lower fuel costs, due to the good gas mileage. Future hybrid vehicles may truly represent a technology that is not only the most environmentally friendly alternative to owning a regular car, but also the most economical and convenient.

Technology Link

Agriculture: Environmentally Friendly Uses of Energy

There is no question that our supplies of fossil fuels will eventually run out. For this reason, we need to develop alternate forms of energy. It is important that any form of energy we use harms the environment as little as possible.

One renewable source of energy that is gaining popularity is using waste from cows to produce electricity. The development of methane digesters has made it possible for a dairy farm to produce enough energy to operate the farm and fuel the family car! Using material created by the farm to operate the farm greatly reduces energy costs, and the methane digester also reduces the amount of harmful gases released into the atmosphere.

A methane digester is located in a lagoon near the farm, and it captures naturally occurring gas from the cow waste and converts it into electricity. Creating energy from dairy biogas is a good solution because it turns waste into something usable, while decreasing air and water pollution. The economic and environmental benefits the methane digester provides are very promising. Digesters will be more common on farms in the future, and they will help to make farms self-sufficient and less dependent on other sources of energy.

Reducing Pollution in the Energy, Power, and Transportation Industries

Much pollution is directly attributable to the power generation and transportation industries. It makes sense to focus on what these industries are doing to reduce pollution. The worldwide motor vehicle fleet is now estimated to be approaching 700 million vehicles. These vehicles produce significant pollution, particularly in urban areas. The automobile is now responsible for about 33% of all crude oil consumption worldwide. Driving automobiles is now widely regarded as the single most polluting human activity on earth. As a result, North America, Europe, and Japan have developed significant pollution control standards for automotive vehicles. These standards have resulted in tremendous advancements in pollution control for gasoline-powered automobiles. Some of these advancements are beginning to make their way into the small gas engine industry as well. Similar technologies are under development for diesel-powered vehicles. This is important because diesel fuel emits 30–70 times more particulate waste than gasoline fuel. Diesel is a popular fuel in the trucking industry in North America and in many third world nations.

Automobile Emissions

A strong relationship between automobile emissions and air quality was suggested as long ago as the 1940s. Eye and throat irritation and decreased visibility as the result of smog were detected in Los Angeles as early as 1943. By 1948, the California legislature established air pollution control districts that had the authority to curb emissions sources. Of course, the technology available to reduce pollution in the 1940s was very limited, in comparison to that of today. The initial efforts generally focused around reducing particle emissions flowing from industrial smokestacks, not automobiles. As a result of these efforts, visibility improved. Eye irritation and smog, however, remained. The particulate matter was not the primary cause of these effects. Further research showed that, when in the presence of sunlight, nitrogen dioxide and hydrocarbon compounds (by-products of burning fossil fuels) react to form ozone and other irritants. This discovery led to a series of emission control requirements for automobiles.

A catalytic converter is part of the exhaust system of an automotive engine. It contains a chemical that acts as a catalyst to reduce polluting emissions. The catalytic converter is one of the most effective pollution control devices for reducing internal combustion engine exhaust gases that has been invented to date.

Studies have shown that lowering the peak temperature of combustion within an engine will lower the amount of nitrous oxides emissions produced during combustion. An exhaust gas recirculation (EGR) system performs this task by routing some of the exhaust gas from the engine into the air intake of the engine. The system will reduce nitrous oxide pollutants as long as it is functioning properly.

Fossil Fuel Power Plant Emissions

Over the years, a variety of filtering techniques have evolved to improve the emissions from power plants. Large-scale power generation requires the consumption of tremendous volumes of fuel. There is some advantage, however, in producing so much power in one location. The advantage is that advanced emissions control techniques can be implemented that are not possible to use in smaller power generation applications.

The following is an overview of techniques used independently or in conjunction with one another to help reduce power plant emissions.

Baghouse: A technique used to reduce power plant emissions in which particles are trapped on filters made of cloth, paper, or similar materials. These particles are then shaken or blown from the filters down into a collection hopper.

Wet scrubber: A technique used to reduce power plant emissions in which particulates, vapors, and gases are controlled by passing the gaseous stream of emissions through a liquid solution.

- ***Baghouses.*** Particles are trapped on filters made of cloth, paper, or similar materials. These particles are then shaken or blown from the filters down into a collection hopper. Baghouses are used to control air pollutants from power plants, as well as steel mills, foundries, and other industrial furnaces. They can collect more than 98% of the particulates.
- **Electrostatic precipitators.** By use of static electricity, these precipitators attract particles in much the same way that static electricity in clothing picks up small bits of dust and lint. See **Figure 26-6.**
- ***Wet scrubbers.*** Particulates, vapors, and gases are controlled by passing the gaseous stream of emissions through a liquid solution, or "scrubber." Scrubbers are used in coal-burning power plants, asphalt and concrete plants, and a variety of other facilities that emit gases that are highly soluble in water. Wet scrubbers are often used for corrosive, acidic, or basic gas streams.

These techniques have vastly improved the quality of power plant emissions. There is, however, still more work to be done. Fossil fuel pollutants and the consumption of coal for power generation, in particular, remain the principal causes of acid rain and the greenhouse effect.

Figure 26-6. Electrostatic precipitators and other control devices help to reduce the emissions of particulate matter and gases from pollution sources, such as this steel mill. They are regarded as being 98–99% effective. Electrostatic precipitators are often used instead of baghouses when the particles are suspended in very hot gases, such as in emissions from power plants, steel and paper mills, smelters, and cement plants.

Summary

The energy, power, and transportation industries are some of the most polluting industries on earth. They are also some of the industries most responsible for improving our quality of life. Therefore, it is necessary to continue to use these technologies, while striving to minimize the environmental damage they can cause. Pollution can come in many forms. The most obvious form is air pollution. Acid rain and other forms of water pollution can be attributed to power generation and our extensive use of the automobile. There are also a variety of other forms of environmental pollution that are not as well known. These forms of pollution include lead poisoning, light pollution, noise pollution, land pollution, thermal pollution, and even possible pollution from electromagnetic fields (EMFs). Many devices have been created to curb harmful emissions from escaping into the atmosphere.

Key Words

All the following words have been used in this chapter. Do you know their meanings?

acidic solution
air pollution
baghouse
basic solution
Clean Air Act (CAA)
electromagnetic field (EMF)
fish kill
land pollution
lead poisoning
light pollution
natural pollutant
noise pollution
nonattainment
pH
reclamation
surface mining
suspended particle
thermal pollution
water pollution
wet scrubber

Test Your Knowledge

Write your answers on a separate sheet of paper. Do not write in this book.

1. *True or False?* The Clean Air Act (CAA) of 1955 was enacted at the state level.
2. *True or False?* Most acid rain can be attributed to natural causes.
3. _____ was a gasoline additive that was eventually phased out because of the adverse health effects it caused.
4. *True or False?* Electromagnetic fields (EMFs) from power lines have no proven side effects to date.
5. Write one or two sentences about how obtaining energy resources has harmed the environment.
6. *True or False?* Reclamation refers to the reconditioning of mined land for alternate use.

7. *True or False?* Thermal pollution from power plants has no serious consequences.
8. Automobiles are responsible for about _____ percent of all crude oil consumption worldwide.
9. *True or False?* Automobiles are collectively regarded as the most polluting of all human inventions.
10. Describe three common techniques used to reduce fossil fuel pollutants at power plants.
11. *True or False?* Environmental protection should be a primary concern when developing any new technology.

Activities

1. Construct a wet scrubber or electrostatic precipitator in class.
2. Review automotive emissions control devices using a working automobile and with a professional mechanic.
3. Review state and federal environmental legislation on-line.
4. Research new and emerging pollution control techniques.

27

Energy, Power, Transportation, and the Future

Basic Concepts

- List future trends in energy technology.
- Name future advances in power technology.
- Identify future trends in transportation technology.

Intermediate Concepts

- Describe how space planes differ from airplanes and space shuttles.
- Explain how fuel cells operate.

Advanced Concepts

- Discuss nanotechnology.

If you ask your grandparents and great-grandparents about the technological changes they have seen in their lifetimes, you will hear many stories. Automobiles have gone from being a replacement for horse-drawn vehicles to becoming a major force in changing people's lifestyles. See **Figure 27-1.** Airplanes have made tremendous advances in design since the first flying machine. Not all that long ago, space travel was only for dreamers and science fiction writers. Houses that were heated by wood and coal are now heated by electricity and natural gas. The number of electronic devices used in homes and businesses would have been unimaginable. Today, we hardly give it a thought to warm food in the microwave or to surf the Internet on our computers or handheld devices.

When you consider that all this has occurred within the last 100 years, it seems incredible. There has been an explosion of innovative ideas that have helped to shape energy, power, and transportation technologies. New and improved transportation vehicles and energy systems continue to affect the way we live, work, and relax. What was new and exciting even 10 years ago seems commonplace now.

Figure 27-1. Evolution of the automobile. A—The Model T Ford was a "horseless carriage" with a body style that resembled a buggy or carriage. (Ford) B—Rugged sport-utility vehicles (SUVs) can function on highways or off-road, reflecting changing lifestyles. (Subaru)

A

B

Trends of the past will continue, and inventions will continue to come from creative people. Transportation technology will continue to evolve. Improvements will be made to existing systems. New systems will emerge.

The Future of Energy Technology

Today, the main energy sources used to heat and cool homes and businesses, generate electricity, and propel vehicles are primarily based on fossil fuels. Natural gas, coal, and oil are the three main energy sources. See **Figure 27-2.** These three, combined, account for two-thirds of the electricity generated and 99% of all transportation fuels. Unfortunately, these energy sources are nonrenewable and have major impacts on the environment.

Figure 27-2. Fossil fuels, such as this coal being stockpiled at a generating plant, remain the main energy source for industrial societies. Generating facilities that use coal are often located along waterways to allow low-cost transportation of fuel by barge. (Howard Bud Smith)

Current Trends in Energy Technology

Current projections show that the demand for and use of fossil fuels will continue to rise for at least the next 20 years. In the future, it will become necessary to find alternative energy sources. In order for alternative energy sources to compete with fossil fuels, the alternative sources must be affordable and easy to use. Many consumers are not ready to make drastic changes in their energy use or pay more for energy because they do not see the value in it. There are, however, several trends and initiatives in the use of renewable and inexhaustible energy sources that may have the potential to impact the current dependence on fossil fuels.

Solar energy

Solar energy has been on the minds of energy researchers for several decades. Much research and many solar energy installations have been made across the nation. Solar energy, however, only makes up a very small portion of the energy used in the United States. The potential for solar energy to create a large impact is present, and new systems are being explored. In the future, it is very possible that solar energy will make up a large portion of the energy used to create electricity.

There are two main types of solar energy systems being heavily researched. Photovoltaic cells and active solar collectors are being studied, and technology is being created to make these systems more efficient and available at a lower cost. See **Figure 27-3.** There are also other solar applications that may see use in the future. The first is ***solar lighting***. Solar lighting systems use a collector, located on a rooftop, which sends light into the building through fiber-optic cables. These cables are used in conjunction with the lighting system inside the building. On sunny days, the fiber-optic cables can be used to supply all the light needed in the

Solar lighting: A system that uses a collector, located on a rooftop, which sends light into the building through fiber-optic cables.

Figure 27-3. Photovoltaic cells are seeing increasing use as an alternative energy source. This large array covers the roof of a solar cell manufacturing plant in the Netherlands. (Shell Energy)

building, cutting down on the building's use of electricity. Another solar system currently being discussed involves the use of solar collectors as satellites or placed on the moon. The collectors would be placed on the sunny side of the moon, gather solar energy, and then beam it down to earth in the form of microwaves. The microwaves would be collected by receivers on earth and converted into electricity.

Wind energy

Wind energy has been the largest growing energy source in the last 10 years. The Department of Energy's Wind Powering America program spurred much of the growth. This program ensures that wind energy will be a large contributor to the energy grid in the future. In the future, ***wind farms***, collections of wind turbines, will dot the American landscape and perhaps the oceans as well. See **Figure 27-4.**

Wind farm: A collection of wind turbines used to create electricity.

In order to generate electricity at a cost comparable to current methods, wind turbines must be designed to be highly efficient. The National Wind Technology Center is researching and developing new wind turbine technology. One of their main focuses is the development of low wind speed turbines. These will generate electricity at much lower wind speeds than the current wind turbines. By creating more electricity, the cost of each watt produced is decreased. In the future, these turbines may make wind energy a viable source of energy for large-scale power production.

Figure 27-4. Wind farms are increasingly important energy sources, especially in the American West. This large installation is located in Montana. Wind turbines not only help to create electricity, but they also have economic benefits. Farmers who allow wind turbines to be placed on their property receive royalty payments that help to subsidize their farming efforts. (Shell Wind Energy)

Ocean energy

The ocean is a largely untapped source of energy. There are several ways the ocean can be used to create energy, including harnessing the power of the tides and waves and using the temperature differences between surface waters and depth waters to generate electricity. These sources of energy are not being heavily researched in the United States, but they are being examined and developed in several European and Asian countries. As these countries further develop their technology, ocean energy will be used more extensively across the globe.

Bioenergy

Bioenergy, or biomass, is currently the most widely used renewable energy source. It is used to supply 3% of the U.S. energy demand. Biomass is also the most likely energy source to have the quickest impact in the future. It is energy generated by releasing the energy stored in organic materials. Biomass energy can be used for a number of purposes, including transportation fuel, electricity generation, industrial heat, and chemical production. The materials used in creating biomass energy are currently centered on agricultural crops and industrial residues. See **Figure 27-5.**

One of the future developments of biomass energy will be the creation of biorefineries. Biorefineries will function similarly to petroleum refineries, except they will convert organic materials to fuels, chemicals, and other products, rather than oil. They have the potential to help alleviate the dependence on foreign oils, create jobs, and reinforce the U.S. economy.

Figure 27-5. Agricultural crops, such as corn and the soybean plants shown here, provide most of the biomass for energy generation and vehicle fuels. In the future, however, biomass energy will be generated using agriculture and forestry residues. Eventually, the biomass industry will grow special plants and grasses to be used in the generation of bioenergy. (U.S. Department of Agriculture)

Fossil fuels

While the renewable and inexhaustible energy sources will definitely be used in the future, fossil fuels are the main source of current energy. For this reason, technologies are being developed and will continue to be developed to lessen the impacts of fossil fuels on the environment and to improve the efficiency of fossil fuels. The Department of Energy's Office of Fossil Energy oversees much of the research. One development includes the design of a future zero-emission coal power plant. The plant, known as FutureGen, will create electricity and hydrogen gas from coal, without generating pollutants. New technologies will also be created in the future to effectively locate and obtain fossil fuels from the earth. Currently, much energy is wasted exploring and collecting fossil fuels.

Conserving Energy for the Future

New and exciting energy sources are on the horizon of the future. There are also, however, many things that we can do in the present to better utilize our current energy supply. In our everyday lives, we waste energy that we could conserve with several easy steps.

Our homes are large sources of wasted energy. The primary source of wasted home energy in most homes is air leaks. These leaks allow cold air into the home, which is known as ***infiltration***. Infiltration can be reduced by sealing cracks with caulk, foam, weather stripping, and other specialty products, such as gaskets for wall outlets. Another main source of heat loss in a home is ***conduction***, or the flow of heat through the walls as a result of poor insulation. Building codes of the future will require greater insulation (R) values to help conserve energy in homes. They are also beginning to require insulation in the bottom floor of a home, and some municipalities may even require infiltration tests to ensure that the houses are tightly built to conserve energy.

Infiltration: Cold air forcing its way into a home through cracks and other penetrations.

Conduction: The transfer of heat from molecule to molecule, straight through a material or group of materials.

Technology Link

Construction: R-Values for Common Building Materials

Every building material has an R-value relating to its ability to resist heat flow by conduction. The R-value has little to do with the thickness of a given construction material. Rather, the R-value reflects the material's ability to resist heat flow by conductivity so energy conservation can be planned and heating and cooling systems can be sized appropriately. The following products are commonly used in the construction industry to provide insulation, thus conserving energy:

- *Fiberglass insulation* is the most widely used product to prevent heat loss by conduction and infiltration. It is available as rolled batt insulation and as bagged clumps of loose-fill insulation. The standard 3 1/2″ thick insulation is made to fit the stud cavity of a 2 × 4 exterior wall. The 5 1/2″ thick insulation is designed for use with 2 × 6 exterior wall studs or in the attic or floor of a home. Loose-fill insulation is generally only used in the attic of a home. The vapor barrier is a layer attached to the face of most batt insulation and is designed to face the warm, moist air to block moisture from entering the stud cavity.
- *Rigid foam board insulation* is made of phenolic foam and offers a better R-value per thickness than fiberglass insulation (up to R-10 per inch, or an equivalent of about 3″ of fiberglass insulation). It has such a good R-value that it is often used as a replacement for plywood or flake board sheathing under finished siding. Other applications include insulating the foundation of homes.
- *Blown-in batt insulation* is sometimes used to fill stud cavities of new homes. This insulation is literally sprayed into the stud cavities and smoothed down with a striker before drying in place. It offers a better R-value than fiberglass insulation. Blown-in batt insulation is, however, usually more expensive than fiberglass insulation.
- *Foam bead insulation* is sometimes used to retrofit older homes that did not have insulation installed when they were constructed. It can be pumped into the stud cavities high in the walls and allowed to settle into the cavity. This process is one of the only cost-effective means of adding some insulation without major expenditures to renovate the exterior or interior of a home.

Home lighting uses a large amount of energy that could be saved using energy conservation techniques. Most homes use incandescent lighting. These lights are typically inexpensive lightbulbs. They are much less efficient, however, than fluorescent lightbulbs. Fluorescent lightbulbs often cost 25 times more than incandescent lights, but they use 75% less energy and can last several years longer. The Department of Energy (DOE) suggests that, by replacing 25% of the incandescent lights in high-use areas of your home with fluorescent lightbulbs, you can save 50% of the energy that would have been used.

Transportation vehicles also use a great amount of energy. Typically, as the size of a vehicle becomes larger, the more fuel it uses. A large sport-utility vehicle (SUV) uses more fuel per gallon than a small sedan does. Often, families buy larger vehicles than they really need and, therefore,

When considering the appropriate R-values for construction or renovation, it is important to conform to all applicable state and local building codes. Recommended R-value for walls, ceilings, and floors in different parts of the country vary widely based on weather patterns and geography. **Figure 27-A** shows a sampling of R-values for typical building materials.

Figure 27-A. This chart shows the insulation values of common construction materials.

Material	R-Value
1/8″ vinyl floor	0.05
15-lb. building paper	0.06
1″ concrete, sand, and gravel masonry	0.08
Tile or slate	0.08
1/4″ mineral fiber	0.21
4″ concrete or stone	0.32
Vinyl siding	0.33
1/2″ plasterboard	0.50
1/2″ sheathing board (flake board)	0.50
1/2″ wood siding	0.50
1″ maple, oak, and other hardwoods	0.91
3/4″ plywood, softwood	0.93
3/4″ wood bevel siding	1.05

Material	R-Value
4″ lightweight aggregate concrete blocks	1.50
8″ concrete blocks	1.60
Carpet and fibrous pad	2.08
1″ cellular or foam glass	2.50
1 3/4″ wood door with storm door	3.12
1/2″ sheathing board (insulation board)	3.60–5.00
3 1/2″ fir, pine, and other softwoods	4.35
1″ expanded polyurethane insulation	7.00
3 1/2″ fiberglass insulation	11.00–13.00
4″ mineral batt insulation	14.00
5 1/2″ fiberglass insulation	19.00–23.00

waste energy. There are, however, ways to conserve energy with all vehicles. Keeping vehicles in good working order is one of the best ways to conserve energy. When the tires are properly inflated and the air filter is clean, the vehicle will use less energy than when the tires are low and the filter is clogged. Lastly, carpooling and decreasing the amount of short trips in automobiles can also conserve energy.

The Future of Power Technology

The future of power is centered on greater efficiency. Improved batteries and new propulsion technologies are also possible in the near future. We will look at these possibilities in the following sections.

Efficiency

Over the years, the use of energy to power products and devices has become more efficient, due to technological advances. For example, if you examine an early computer, it is easy to see that the large machine required much more electricity to do much less work than a modern computer. This is even the case in transportation. Advances in wheels, axles, bearings, aerodynamics, and a host of other aspects have made vehicles easier to move, if you compare vehicles by their weight. This trend will continue, largely due to advancements in materials science and computer control.

Materials science

Advances in materials science will allow devices, homes, buildings, and other products to require less energy for power. For example, as future materials are designed for lighting, lights will shine brighter and use less power. A compact fluorescent bulb requires two-thirds less electricity to produce the same amount of light as a standard incandescent screw-in lightbulb. Additionally, a compact fluorescent bulb will last six to ten times longer than a regular lightbulb. The same increasing efficiency is true of insulation materials. As materials are created to serve as better insulators of buildings, the buildings will require less energy to power the heating and cooling systems. Through persistent research, fiberglass insulation measuring 3.5″ in thickness now provides an R-value of 13. Only a few years ago, fiberglass insulation of the same thickness offered an R-value of only 11, and 20 years prior to that, it offered an R-value of only 9. In transportation, materials science often concentrates on lowering the weight of the materials. Generally, the less the vehicles weigh, the less power is required.

Career Connection

Science Technicians

The future of energy, power, and transportation is heavily linked with science. New advances in technology often have scientific components. A key role in the many science industries is the science technician. Science technicians normally work in either research or production roles in various industries. Agriculture, materials science, nuclear technology, biotechnology, oil and gas extraction, and metallurgy are all industries that will be important in the future of energy, power, and transportation and that employ science technicians.

Science technicians work under a scientist and prepare and conduct experiments, monitor conditions, and calculate results. Technicians must be very detailed and keep good logs of their work. Most technicians work in highly computerized labs that require a working knowledge of specific computer software. Knowledge of mathematics, good language skills, problem-solving abilities, and scientific knowledge relevant to the industry they work in are required. Technicians typically have high school diplomas and associate degrees in science or technology. They may also have certification in a specific area of science. The average salary of a science technician ranges from around $15 to $18 per hour, depending on the specific industry.

These advances will also allow electrical devices to be designed in ways that require less power. See **Figure 27-6.** In the future, nanotechnology will be used to create devices that are more powerful, but use much less electricity. ***Nanotechnology*** is the design of products and devices at a molecular level. For example, a nanosized computer processor chip would be smaller than a grain of sand and thousands of times more powerful than today's supercomputer processors. Such a chip would also require less power because the material is less resistant to the flow of electricity than the silicon used in current processor chips.

Figure 27-6. Improved technology is allowing electrical devices and systems to become smaller and more efficient. This scientist is observing production of superconducting ceramic ribbon that can carry 50 times more electrical current than a copper conductor of the same diameter. (Siemens)

Computer control

Computer control will also affect the use of power in a very positive manner. Computers and automated networks will be used more heavily in the future to monitor and control power systems. Networks will send power to areas that need it, without wasting power in other areas of the network. This type of control will be used in all areas of technology. For example, electricity grids will be better connected and controlled by computers. See **Figure 27-7.** This type of computer control is already available in automobile traction control systems. Automated networks will be developed in more depth and used in many other areas in the future.

Nanotechnology: The design of products and devices at a molecular level.

Batteries

Batteries have been a part of life since the late eighteenth century, when Alessandro Volta produced a current from zinc and silver plates. Today, they are used to power everything from household products and cellular telephones to medical devices and transportation vehicles. See **Figure 27-8.** Currently, lithium-ion batteries are the most advanced batteries. In the future, however, it is expected that several new technologies will take over the battery market.

Paper batteries

One new technology is the paper battery. ***Paper batteries*** are ultrathin dry batteries. The batteries function like larger dry batteries, but they can be printed on a number of different products, including paper products. Since the paper batteries are dry batteries, they do not contain harmful chemicals. Imagine flyers and bulletin boards that can light up or play recorded messages. This may not be far off, with paper battery technology. These ideas, however, are just the beginning of the impacts of this technology. In the future, it may be possible to power laptops or cell phones using these thin paper batteries.

Paper battery: An ultrathin dry battery that functions like larger dry batteries, but can be printed on a number of different products, including paper products.

Figure 27-7. Technicians monitor computer systems at an electrical transmission control center in China. (Siemens)

Fuel cells

Another technology that will surely change the face of batteries is the fuel cell. Fuel cells are typically discussed in regards to automobiles. There are a number of future uses, however, large and small, for fuel cells. Fuel cells, like batteries, produce electricity through a chemical reaction. The

Figure 27-8. Batteries are a vital power source for many devices in our lives. In the future, this will be no different. Batteries will be used more than ever. A—Dry cell batteries provide portable power for devices ranging from large flashlights to small personal music devices. B—The starting system of automobiles and other transportation vehicles depends on the 12-volt lead-acid battery that can be discharged and recharged many times. (Exide)

A

B

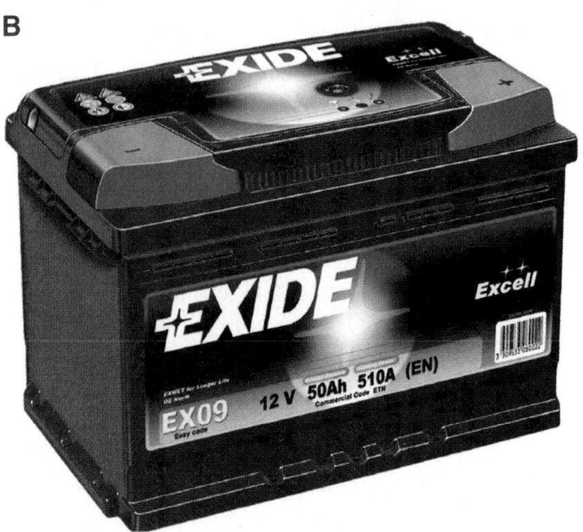

reaction, however, is quite different. Fuel cells operate by passing hydrogen atoms through a catalyst. The catalyst frees the electrons from the hydrogen. The remainder of the hydrogen atom passes through a membrane to the positive side of the cell. The electrons, however, must pass through an electrical circuit to reach the positive side. This creates electricity.

Fuel cells, unlike batteries, never lose their power. As long as fuel cells have a supply of hydrogen, they will produce electricity. Supplying the hydrogen is the biggest challenge today and in the future. The fuel cells work best and create no pollution when pure hydrogen is used. Pure hydrogen is hard to store, however, and there is not a network of hydrogen stations like there are gasoline stations. Fuel cells can also be designed to operate on fuels rich in hydrogen, such as natural gas and methane. See **Figure 27-9.**

Once issues of size, cost, and availability are worked out, fuel cells may be a substantial source of power. Fuel cells can be used to power everything from handheld devices to automobiles to homes. Many people believe that, in the future, fuel cells may be used as neighborhood power plants to supplement the power received by larger power stations.

Figure 27-9. Mass transit vehicles and automobiles using efficient and pollution-free fuel cells are achieving growing popularity. Iceland's public transportation system operates fuel cell buses, such as the one shown here refueling its hydrogen supply at a service station. In this type of system, a reformer is used to convert the fuels to hydrogen prior to entry into the fuel cell. This kind of system is currently more practical than one that uses pure hydrogen, but it increases the size, complexity, and cost of the fuel cell system. As fuel cell use grows, more and more hydrogen refueling stations will be available. (Shell Energy)

Propulsion Technologies

Providing power for transportation vehicles is one of the largest uses of energy. There will be several technologies in the future that will help to decrease the use of transportation energy. Hybrid vehicles will decrease the use of fossil fuels to generate power. They use both electric motors and gasoline engines to power vehicles. See **Figure 27-10.** Fuel cells will also transform the power used to propel vehicles. In fuel cell cars, the electricity the cells generate will power the electric motors that propel the vehicle.

The power used in space vehicles will also be much different in the future. Space vehicles are limited to the amount of power they can carry onboard because of size and weight restrictions. In the future, space vehicles may use a technology to propel the spacecraft that does not require onboard power. ***Solar sails*** may be this future technology. They are large sails that operate on the principle that light exerts a small force on the objects it touches. The solar sails would be used to collect light energy. The more light they collect, the greater the "push" they receive. It is believed that space vehicles using solar sails may be able to generate a great amount of speed and would be able to reach distant planets much faster than by current methods. Solar sails may also be beneficial because the power used to move the vehicle would be free and readily available.

Solar sail: A large sail that operates on the principle that light exerts a small force on the objects it touches. It will be used to collect light energy to propel spacecraft.

Figure 27-10. Hybrid vehicles, such as the one shown in this ghost view, are powered by both a gasoline engine and electric motors. Depending on the driving situation, the vehicle automatically switches from the gasoline engine to electric motors to a combination of the two. This can cut down on the amount of gasoline the vehicle uses by up to 50%. (Ford Motor Company)

The Future of Transportation Technology

Transportation systems have been evolving and changing since early humans made sleds and rafts. In the future, transportation technology will advance in ways that will make the vehicles of today seem as primitive as those early sleds and rafts. There will be advances in each environment of transportation, as well as trends in all vehicles.

Tech Extension

Futuristic Large-Scale Power Generation

One futuristic application of solar energy envisions a huge solar collector assembled in space that could receive sunlight almost 24 hours a day. The sunlight would also be much stronger than what we receive on earth, since it would not have to penetrate earth's atmosphere. The energy would then be sent by microwave to an antenna located on earth. The antenna could require a five-mile diameter area. The microwaves would then be converted back to electricity.

Curricular Connection

Social Studies: Future Cities

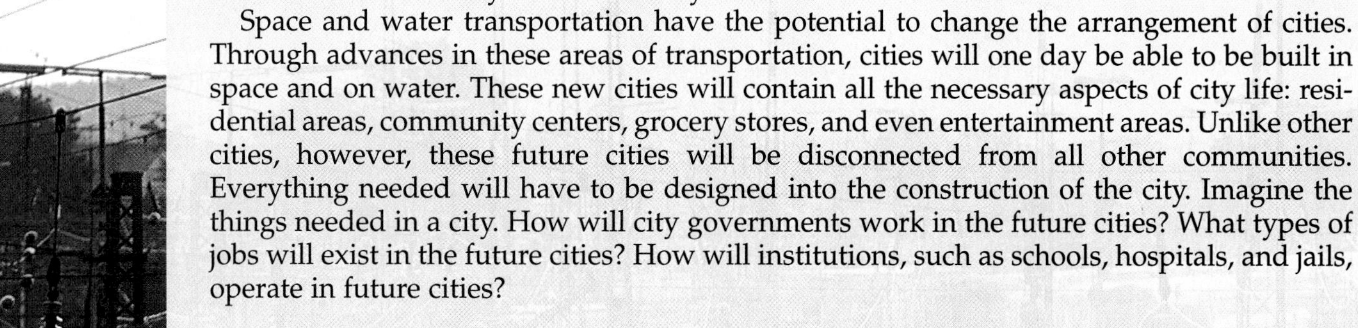

As transportation has evolved, so have towns and cities. As automobiles became more affordable, cities became more accessible. People began to move outside of city centers into suburban areas. Suburbs grew, and new businesses were established to serve the new residential areas. It is not hard to imagine that, in the future, transportation will again change the structure of the city as we currently know it.

Space and water transportation have the potential to change the arrangement of cities. Through advances in these areas of transportation, cities will one day be able to be built in space and on water. These new cities will contain all the necessary aspects of city life: residential areas, community centers, grocery stores, and even entertainment areas. Unlike other cities, however, these future cities will be disconnected from all other communities. Everything needed will have to be designed into the construction of the city. Imagine the things needed in a city. How will city governments work in the future cities? What types of jobs will exist in the future cities? How will institutions, such as schools, hospitals, and jails, operate in future cities?

Trends in Transportation

There are several future trends that will guide the development of technology in all transportation environments. Most vehicle operators and passengers are concerned with several aspects of their vehicles. First, they want their vehicles to be safe for all passengers. Accidents do occur in all environments and modes of transportation, and passengers want to know that their vehicles are as safe as possible. Second, operators are concerned with fuel efficiency. When gas prices rise, efficiency becomes even more important because driving a vehicle with low fuel efficiency means a greater amount of money is spent at the gas pump. These two trends, safety and efficiency, will guide the development of technology in all transportation environments.

Safety

Safety is naturally an important factor for the traveling consumer. It is stressed in commercial transportation. The travel industry wants passengers to return to use its services. Vehicles in the future will be designed to better handle accidents. In transportation, this is known as ***crashworthiness***. When vehicles are crashworthy, it is believed that they can handle and provide safety during an accident. Private vehicle owners also know the need for safe transportation. Changes are being made so travel will be safer for the private car owner. In the future, safety will be better tested using more sensitive crash test dummies. See **Figure 27-11.**

Crashworthiness: The ability to handle and provide safety during an accident.

Efficiency

We also need to increase energy efficiency in transportation. This will conserve the earth's limited supply of fossil fuels. There is a need to reduce our dependence on these precious, limited resources. Vehicle

Figure 27-11. Crash test dummies simulate the effects of vehicle impacts on human drivers and passengers. Dummies in the future will be better designed to react like humans would in crashes. In this photo, a technician is preparing dummies for a crash test. (Ford Motor Company)

designers will continue to experiment with new types of energy-efficient propulsion systems. Though still using petroleum fuels, new engines are more efficient than those just a few years ago. Many vehicles are built much larger, however, than they were in the past. When the more efficient engines have to propel larger vehicles, the benefits are lost. A change in the outlook of drivers is also needed. Fuel economy must be a priority for those buying new vehicles, in order for major changes to take place.

Future Land Transportation

The U.S. Department of Transportation has created a division that examines and studies how to improve vehicles for the future. This program, known as the Intelligent Vehicle Initiative, has the goal of improving the safety of all highway vehicles in all types of driving conditions. Its main objectives are improving vehicle warning systems so accidents can be avoided, enhancing the vision of drivers using technology, and creating more stable vehicles that are safer to drive and can better withstand accidents. See **Figure 27-12.** Technologies that enhance these areas will be present in the highway vehicles of the future.

The Department of Transportation is also concerned with the highway infrastructure of the future. The highway ***infrastructure*** is the roadways, bridges, ramps, and other structures that make up the highway and

Infrastructure: The roadways, bridges, ramps, and other structures that make up the highway and roadway system.

Figure 27-12. "Heads-up" vehicle information systems, similar to those used on military fighter aircraft, are being developed for use in automobiles. The systems display vehicle speed, vehicle condition, and navigation as a projected image in the driver's line of sight, eliminating the need for the driver to take her eyes off the road to check instruments on the dashboard. Global positioning systems (GPSs) to aid in vehicle navigation are being offered as original equipment on some production models. Key information from the GPS display, to the right of the steering wheel, will be included in the "heads-up" display. (Siemens)

roadway system. In the future, the infrastructure will be designed to take advantage of new communication, data processing, navigation, and computer technologies. One example of future infrastructure is the creation of a smart road. ***Smart roads*** will communicate with vehicles and help to control the speed, braking, and steering of the vehicles. This will help to minimize the congestion and accidents on highways. The design and construction of smart roads will require automobile and truck manufacturers and road builders to work together to design a feasible system for the vehicle and roadway.

Smart road: A component of future highway infrastructure that will communicate with vehicles and help to control the speed, braking, and steering of the vehicles.

Small Aircraft Transportation System (SATS): An organization formed by the National Aeronautics and Space Administration (NASA) and the Federal Aviation Administration (FAA) to research a solution to the overcrowding of major airports.

Future Water Transportation

Water transportation currently carries roughly 95% of the freight the United States imports and exports. The volume of this freight is expected to triple in the next 20 years. The infrastructure of water transportation facilities is not equipped to handle the increased traffic. Many of the facilities must be improved to handle increased traffic. This will include dredging current channels and creating more channels to harbors. Locks and canals must be modernized and improved to handle new ships. See **Figure 27-13.** Ships and terminals will have to use a greater amount of computer technology to enhance the efficiency of docking, loading, and unloading. The unloading of freight will also become more computerized. Containers will use digital devices to communicate with robotic machines, in order to be properly unloaded and placed in storage or on intermodal vehicles for further transport.

The vessels used in water transportation will also change in the future. One future change will be in the propulsion of water vessels. Many believe that fuel cells being developed in space and land transportation may also serve a purpose in water transportation. Another possibility in the future of water vessels is the creation of floating cities. For years, designers have been creating plans of large ships that could serve as entire cities.

Figure 27-13. Locks on the Panama Canal and many other waterways must be expanded in the near future to permit use by the larger and more efficient ships now being launched.

Future Air Transportation

The future of air transportation, like other forms of transportation, looks to be crowded. Many airports are currently faced with overcrowding and long delays. An organization formed by the National Aeronautics and Space Administration (NASA) and the Federal Aviation Administration (FAA) is researching a potential future solution to this problem, called the ***Small Aircraft Transportation System (SATS)***. SATS's solution to the overcrowding of major airports is to better utilize smaller regional and rural airports. One of the main ideas in this initiative is to create a system of air taxis. The air taxis will be smaller general aviation aircraft, as well as business jets. Along with air taxis, SATS is also examining ways to make flying easier and more accessible to people. New technologies are being developed and will continue to develop in the future that may potentially make flying a plane as easy as driving a car. These technologies will monitor and track the current position of the aircraft and also the position of nearby aircraft. This information will be displayed on a screen in the digital cockpit. See **Figure 27-14.**

In the future, it may be common to build airports on islands in the sea. These large airports will serve as hubs for international flights. The island airports will allow airlines to use larger aircraft because the noise created would not affect people living near the airport. Since it will be many years until offshore airports are designed, built, and used, several airplane manufacturers are currently trying to design new aircraft that are faster and create much less noise. A current problem in air transportation is, when an airplane breaks the sound barrier, it creates a sonic boom that can be heard on land. If commercial airliners all created this sonic boom every time they passed Mach 1, it would be highly distracting, if not deafening, to people near the airports. Being able to fly at speeds of over Mach 1 would be extremely beneficial for commercial airline companies, however, because they could transport people to their destinations faster. See **Figure 27-15.**

Figure 27-14. Digital displays have replaced many analog gauges and displays in the cockpits of both commercial airliners and smaller aircraft. This is the pilot-side instrument panel on a medium-range jet airliner. (Airbus)

Figure 27-15. Although no supersonic airliners remain in service today, the concept remains attractive because of the ability to move passengers more rapidly between distant locations. This is a concept drawing of a supersonic transport (SST). No major aircraft manufacturer, however, is currently developing such an airplane. (The Boeing Company)

Future Space Transportation

The largest wave of the future of space transportation is space tourism. Space tourism is the ability for anyone to travel to space and experience what it is like to be outside the earth's atmosphere. In order for space tourism to be functional, new vehicles must be designed, built, and tested. These new vehicles—several are currently under development—are a combination of airplanes and space shuttles. The idea behind the ***space planes*** is that they will be reusable and more cost-efficient than space shuttles. See **Figure 27-16.** The first step in the development of space tourism is for the space planes to safely take passengers to suborbital heights. Suborbital flight occurs above 62 miles from the earth, but at speeds not fast enough to enter orbit. These flights will allow passengers to view the earth from above the atmosphere for a short period of time.

Another method of reaching space is also being examined. NASA and other space organizations have studied the idea of a space elevator. A ***space elevator*** will, in essence, be a long cablelike structure extending from the earth to space. Elevator cars will be attached to the cable and used to transport people to and from space.

Space plane: A combination of an airplane and a space shuttle that will be reusable and more cost-efficient than space shuttles.

Space elevator: A long cablelike structure with cars attached, extending from the earth to space, that could be used to transport people to and from space.

Figure 27-16. Space planes are reusable passenger and cargo vehicles that will be more cost efficient to produce and operate than the current National Aeronautics and Space Administration (NASA) space shuttle. *SpaceShipOne*, shown landing in California after a recent test flight, is a privately developed space plane that has successfully flown beyond earth's atmosphere. (Scaled Composites)

Summary

The future of energy, power, and transportation promises to bring exciting changes. There will be changes and improvements to existing systems and totally new, innovative designs. Advances in energy technology will provide viable alternatives to the use of fossil fuels as the main source of energy. Biomass, hydrogen, solar, and wind energies seem to be the most promising energy sources in the future. The future of power systems will be based around increased efficiency, new batteries, and advanced propulsion systems. Fuel cells will surely be one of the largest innovations in power technology. The future of transportation will see improvement in vehicles and infrastructure in all transportation environments. The safety, efficiency, and speed of transportation are sure to increase in the future. Consider all the advancements we have made in the last 100 years. Imagine what you will see in your lifetime!

Key Words

All the following words have been used in this chapter. Do you know their meanings?

conduction
crashworthiness
infiltration
infrastructure
nanotechnology
paper battery
Small Aircraft Transportation System (SATS)
smart road
solar lighting
solar sail
space elevator
space plane
wind farm

Test Your Knowledge

Write your answers on a separate sheet of paper. Do not write in this book.

1. Today, energy for heating, cooling, electricity, and propulsion are generated using _____, _____, and _____.
2. Give examples of two possible future solar energy technologies.
3. Discuss the purpose behind the design of low wind speed turbines.
4. State three uses for biomass energy.
5. Summarize how materials science will affect power technology.
6. Explain nanotechnology.
7. Computers will be used to _____ and _____ power systems.
8. Identify two possible future advances in batteries.
9. Supplying the _____ is the largest challenge in fuel cell technology.

10. Explain how solar sails are powered.
11. _____ and _____ are two trends that affect all transportation systems.
12. Write the goal of the Intelligent Vehicle Initiative.
13. _____, _____, and _____ are examples of highway infrastructure.
14. Describe one possible future change in water transportation.
15. Discuss the concept of a digital cockpit.
16. Airports in the future may be built in the _____.
17. Space _____ will be the largest focus in future space transportation.
18. Identify how space planes are different from airplanes and space shuttles.
19. Summarize the differences between suborbital and orbital flights.

Activities

1. Research information on a future energy, power, or transportation technology. Prepare a written report of the results of your research.
2. Imagine what transportation might be like in the year 2050 and write a scenario about traveling from one city to another 250 miles away.
3. Design a model of a future vehicle and prepare sketches of it.
4. Choose a topic or technology in energy, power, or transportation and conduct research that will allow you to create a forecast about the use of that technology in 50 years. Report your findings.

Technical Terms

A

Acceleration: The rate of change in velocity, such as the increase of a vehicle's speed.

Accelerator: The part of a vehicle that controls speed by changing the amount of power the propulsion system generates.

Accumulator: A device that stores hydraulic liquid under pressure, sometimes even when the hydraulic pump is not running.

AC generator: A device that converts mechanical energy into electrical energy by using electromagnetic induction.

Acid rain: Rain contaminated by the by-products of combustion, such as carbon dioxide (CO_2), nitrous oxides (NO_x), and sulfur oxides (SO_x), which condense in our atmosphere.

Active solar energy collection: A type of system that uses circulating pumps and fans to collect and distribute heat.

Actual mechanical advantage (AMA): The ratio of the increase of force or distance by a machine, including energy lost through friction.

Actuator: A device that converts fluid power to mechanical power in both hydraulic and pneumatic systems.

Adaptive cruise control system: An electronic guidance system that uses lasers or radio detecting and ranging (radar) to determine the distance of the closest car on the road and controls the vehicle's speed accordingly.

Advantage-gaining device: A device that modifies the effort and rate characteristics of power in order to achieve a goal.

Aerodynamics: The study of the motion of air and how it reacts to objects passing through it.

Aeronautical chart: A chart that provides important data for airplane pilots and navigators. It is similar to a topographic map on which special guidance information has been added.

Afterburning turbojet engine: A turbojet or turbofan engine with an additional burner added to the nozzle.

Agricultural technology: Systems that produce outputs by growing plants and animals. The outputs are typically foods and fibers.

Aileron: A moveable control surface placed on the main wing of a plane used to control turning and banking.

Air brake: A brake that uses compressed air to operate the brake cylinder often used on railroad cars.

Air compressor: A device that converts mechanical power into pneumatic power, creating the necessary pressure to make the system work.

Aircraft: A vehicle that transports through the air.

Air-cushion vehicle: A vessel that rides on a cushion of air. Also called a hovercraft.

Air exchanges (AXs) per hour: A measurement of the amount of air from within a heated structure that escapes and is replaced by fresh air at the outdoor temperature in one hour.

Airfoil: The shape of a propeller blade or an airplane's wing.

Air lock: An area through which astronauts exit the shuttle. It ensures that the cabin remains pressurized and provides a place for the astronauts to put on their space suits.

Airplane: A fixed-wing aircraft kept in flight by an engine or other power source.

Air pollution: The action of contaminating the mixture of gases surrounding the earth, especially with man-made waste.

Airspeed indicator: An instrument that measures the difference between two pressures acting on an aircraft to calculate the speed of the aircraft.

Air transportation: A transportation system using either lighter-than-air or heavier-than-air modes of transportation, including airplanes, hot air balloons, airships, hang gliders, military fighter jets, and helicopters.

Airway: A path or route airplanes follow.

All-wheel steering: A steering system that uses the steering wheel to operate the front and rear wheels at the same time.

Alternating current (AC): A type of current in which electrons flow first in one direction and then reverse and flow in the other direction.

Alternator: See *AC generator*.

Altimeter: An instrument used to display the altitude, in feet, of the aircraft from sea level.

Amperage: The rate at which electrons or coulombs move through a conductor.

Ampere: In electrical power, the measurement for rate of flow.

Amplifier: A device that receives electrical waves and increases the power of the signal, so it has greater strength, before sending it on to the speaker.

Anaerobic digestion: Decay without the use of oxygen.

Analog signal: An input signal used to transmit variable data, such as percentage of light, loudness of a sound, or weight of an object.

Analog-to-digital converter (A/D converter): An electronic circuit that converts the analog information sent from sensors to digital representations a computer can understand.

AND logic: A form of control logic in which input from two or more devices is required before an action can take place.

Angle of attack: The angle at which a propeller blade hits the air or the upward tilt of a wing's leading edge.

Angle of incidence: The angle of the wing as it is attached to the aircraft.

Anthracite coal: Coal that is hard and brittle. It appears shiny on the surface and has a high carbon content. Because it burns cleaner than other forms of coal, it is often used for home heating. It does not have as much energy content per volume as bituminous coal.

Antilock braking system (ABS): A computer-controlled braking system that improves control of the vehicle in certain braking situations.

Apogee: The point in the path of an elliptical orbit farthest from the earth.

Aquifer: An underground rock formation that acts as a reservoir for large quantities of water. It is used to store gas from pipelines.

Armature: A series of wires around a metal core in which electricity is induced, which is free to rotate.

Articulated: Hinged.

Artificial horizon: An instrument, also called an attitude indicator, that displays the amount of pitch and roll of an aircraft compared to the horizon.

Aspect ratio: The span of an aircraft wing divided by the chord line of the airfoil.

Astable application: An integrated circuit (IC) application that provides continuous pulses at timed intervals.

Atlas: A collection of maps.

Atom: The "building block" of everything we know of on earth. Atoms are made up of protons, neutrons, and electrons.

Atomized: Broken into small droplets.

Attitude: The position and orientation of an aircraft.

Automatic control: The level of control technology achieved by using sensors and other automatically functioning devices, such as timers, to turn things on or off without human intervention.

Automatic transmission: A transmission that is not dependent on the vehicle operator for control. It uses a torque converter instead of a clutch.

Azimuth path: The sun's movement from east to west.

B

Bag filter: A device that works like a bag on a vacuum cleaner, trapping all solid particles in the waste stream prior to the hot waste gases exiting through the smokestack of a power plant.

Baghouse: A technique used to reduce power plant emissions in which particles are trapped on filters made of cloth, paper, or similar materials. These particles are then shaken or blown from the filters down into a collection hopper.

Ballast: a: A layer of large stones placed on top of the subballast in a rail bed. **b:** A heavy substance used to improve stability in the hull of a boat.

Barrage: A permanent or floating dam that seals a natural bay in order to use tidal power.

Batch sequence: A technique used in pipeline transportation in which different materials are transported in succession and separated when they are received.

Battery: A common device that produces direct current (DC).

Bearing: a: A specially shaped piece of metal used to support shafts and reduce friction between metal parts as they move past or revolve around each other. **b:** The desired direction of travel.

Bias tire: A tire made of many layers of fabric, called plies.

Bilge keel: An extension protruding downward from the centerline of a boat. If the boat starts to lean, the keel acts as a hydrofoil and pushes against the water in the opposite direction.

Bimetallic coil: The part of a thermostat formed from strips of two metals with differing coefficients of expansion.

Binary-coded decimal (BCD): A numbering system that represents values using only the digits *0* and *1* (*false* and *true*).

Bioconversion: The process that produces energy from the waste products of our society.

Biomass: Waste products that can be used in bioconversion.

Bipolar transistor: A transistor with three junction points: an emitter, a base, and a collector.

Bituminous coal: Coal with a high carbon content. It is denser and blacker than most other forms of coal. This type of coal is principally used for the production of electricity.

Blimp: A nonrigid airship that collapses when not filled with a gas. It uses helium to provide lift.

Boat: A water vehicle less than 100′ in length.

Body: The enclosed part of a vehicle.

Bogie: A part of the suspension system of a train. It is attached to the bottom of the train car on a swivel and consists of a steel frame to which the suspension components are attached. Each bogie contains two axles, with a total of four steel wheels.

Boiling water reactor (BWR): A type of fission reactor in which water surrounds the nuclear fuel core within the reactor. Control rods sit between the fuel rods and absorb stray neutrons. When the control rods are retracted, the fission process begins to occur, and a tremendous amount of heat is produced. This heat converts the surrounding water to steam. The force of the expanding steam spins a turbine to produce electricity.

Bore: The diameter of a cylinder.

Bottom dead center (BDC): As low in the cylinder as the top of the piston can go.

Bow: The very front of a vessel.

Boxcar: A boxlike freight car with doors on both sides.

Brake horsepower (bhp): The amount of power available at the rear of the engine under normal conditions.

Break-bulk cargo: A single unit or cartons of freight.

Breeder reacting: Creating nuclear fuel from a substance that is not fissionable.

Bridge: A structure that spans a waterway, a ravine, or another barrier.

British thermal unit (Btu): A basic unit of measurement for heat energy. It is the amount of heat necessary to raise one pound of water one degree Fahrenheit.

Brush contactor: The part of a direct current (DC) motor that supplies electricity to the armature.

Bulk cargo: Loose material, such as grain or oil.

Bulkhead: An internal component used to strengthen the hull or outer shell of a vehicle.

Buoy: A painted marker anchored in a body of water to guide water vehicles.

Buoyancy: The upward force water exerts on objects placed in it.

Burst pressure: The pressure at which a conductor will fail by rupturing.

Bus: The main structure of an unmanned spacecraft.

C

Caboose: The last car on a freight train. It is used to house the train crew.

Cam-activated valve: An automatically operated valve. When the cam is triggered, the valve shifts its position, allowing another action to occur.

Camber: The distance between the top of the mean camber line and the chord line of an airfoil.

Cam lobe: A curved projection of the rotating piece in a mechanical linkage. Cams are used to open and close the valves in engines.

Camshaft: A shaft to which a cam is fastened. In a gasoline engine, the camshaft controls the movement of the intake and exhaust valves.

Canal: A channel constructed to connect two bodies of water.

Capacitor: A device that has the ability to store an electrical charge. Unlike batteries, capacitors can store and discharge electricity very quickly.

Carbon-zinc battery: A type of primary cell in which the carbon is the positive electrode and the zinc is the negative electrode.

Catamaran: A stable type of boat that has two hulls in the water.

Celestial reference device: A device that uses the locations of objects in space to figure the rotation of a spacecraft.

Cell: A common device for storing electrical power. A cell converts chemical energy to electrical energy.

Centrifugal force: The energy that makes objects fly outward when spinning around.

Centrifugal pump: A pump that uses centrifugal force to move fluids in a system.

Chassis: The frame of a vehicle.

Check valve: A valve needed when using reciprocating pumps to keep fluid from moving backward in the system.

Chemical energy: The potential energy locked within a substance.

Choke: A platelike device that varies the amount of air that can enter the carburetor.

Chord line: The line drawn between the leading and trailing edges of an airfoil.

Circuit breaker: A restorable device that breaks the circuit if too much electrical current passes through it.

Classification society: A society that sets and enforces the construction standards for shipbuilding.

Clean Air Act (CAA): The Act Congress passed in 1963 that set emissions standards for certain sources of pollution.

Closed circuit: A properly functioning circuit in which all loads are energized.

Closed loop solar collection: A type of system in which a collection medium is used to collect the heat and transfer it to water or air for end use.

Closed loop system: A control system that considers the output of a system and makes adjustments based on that output.

Clutch: A mechanical device that connects the power source to the rest of the machine.

Coefficient of expansion: The rate at which a metal expands and contracts as temperature rises or falls.

Coil spring: An automotive spring made by taking long steel rods and forming them around a cylinder into a helix. It is taken off the cylinder and heat-treated. The spring then retains its shape and gives the proper tension.

Collective pitch control lever: A helicopter control that directs the swash plate to change both blades evenly, allowing the aircraft to ascend or descend.

Combustion: The process of burning.

Combustion chamber: An enclosed space in which burning takes place.

Commercial aviation: Scheduled airline flights that provide passenger and cargo transportation, using aircraft that carry from dozens to hundreds of passengers.

Communication technology: Systems associated with the dissemination of information and ideas. Products include schematics, advertisements, Web pages, and media messages.

Commutator: A circular conductive strip connected to the end of an armature loop.

Commuter airline service: The transport of people from several small airports to a major airport in a major city.

Compass: A magnetic device used to determine direction.

Complementary metal oxide semiconductor (CMOS): A type of integrated circuit (IC) that typically works with voltages up to 16 V and draws very little current.

Complex machine: A machine that uses more than one simple machine to accomplish its tasks.

Compound gear cluster: More than two gears arranged in such a way as to gain significant mechanical advantage of force or speed.

Compound parabolic collector: A combination of a flat-plate collector and a parabolic collector, offering advantages of both types. It is stationary mounted and does not need to track the sun. It is also a linear-concentrating collector, offering a greater collection ratio than a flat-plate collector.

Compressibility: The extent to which any substance can be packed down into a smaller size or volume.

Compression ignition engine: An engine that relies on heated compressed air for ignition.

Compression ratio: The mathematical relationship between the volume available in the cylinder with the piston at bottom dead center (BDC) and the volume available in the cylinder with the piston at top dead center (TDC).

Compression stroke: An upward movement of the piston and connecting rod assembly.

Compression test: A test performed with a pressure gauge covering the spark plug hole. Once the gauge is installed, the engine is rotated as it would be for starting. If the pressure that would build up during compression is significantly less than the manufacturer's specification, a leak is likely occurring.

Condensing unit: In solar cooling operations, the unit where refrigerant is condensed from a gas back into a liquid and gives up its heat. The condenser is located outside the refrigeration area and transfers heat out of the area to another environment.

Conduction: The transfer of heat from molecule to molecule, straight through a material or group of materials.

Conductor: A material made of atoms that transfer electrons easily.

Connecting rod: An engine component that connects the piston with the crankshaft.

Conning tower: Part of a submarine that projects from the top, usually in the center of the vessel. It is where the ship's periscopes, radio antennas, and radio detecting and ranging (radar) equipment are located.

Construction technology: Systems associated with the creation of structures for residential, commercial, industrial, and civil use.

Contact patch: The part of the tire actually touching the road as the tire spins.

Containerization: A method of handling goods by packing many smaller packages, which are going to the same destination, in one large container.

Container on Flat Car (COFC): A container transported on a railroad flatcar.

Containership: A ship designed to carry standard containers.

Continuity: The continuous flow through a component or an entire circuit.

Continuity checker: A device used to check for continuous electron flow throughout an electrical circuit or through circuit components.

Control rod: Part of a fission reactor that sits between the fuel rods and absorbs stray neutrons. When the control rods are retracted, the fission process begins to occur.

Control stick: The airplane lever operating the elevators and ailerons with the power to guide the aircraft.

Control surface: A moveable airfoil designed to change the attitude of an aircraft.

Control system: A system necessary to control the power within a system.

Convection: The natural movement of a heated gas in an upward direction.

Convective loop:. The loop created when convection carries heated air from an isolated gain solar collector up to a structure, and cold air return vents allow cooler air to return to the collector. It does not require any circulating pumps or fans.

Conversion method: A necessary process to convert energy so some type of work is produced.

Convertible: An automobile that has a removable roof.

Cooling fin: A projecting rib on an engine cylinder that moderates heat.

Cooling subsystem: The system responsible for keeping the engine operating within an efficient temperature range.

Coordinated steering: Negative steering, in which the front and rear tires are steered in opposite directions.

Coordinated turn: A turn, or bank, in an aircraft in which all three control surfaces work together.

Coulomb: A unit of electrical charge equal to the amount of electricity transported by 1 ampere in 1 second.

Counting function: A computer subroutine that indicates by units so as to find the total number of units involved.

Coupe: A small, usually two-door, car.

Crab mode: Positive mode steering, in which all tires are turned in the same direction.

Crankpin journal: An offset on the crankshaft that converts the downward movement of the piston and connecting rod into rotary motion.

Crankshaft: An engine component that converts the reciprocating motion of the piston and rod assembly into rotary motion.

Crashworthiness: The ability of a vehicle to handle and provide safety during an accident.

Creosote: A tarlike substance that can build up on the walls of a chimney when wood is burned.

Crude oil: Oil in its natural state.

Current: The flow of electrons in a conductor.

Current flow: The rate at which electrons move, or amperage.

Cut: The removal of excess earth from hills.

Cycle: One complete performance of a periodic process.

Cycle timer: A device that turns a load on and off on a continuous cycle.

Cyclic control stick: A helicopter control that changes only one side of the swash plate. This changes the pitch of the blades on only one side of the rotor, which enables the helicopter to turn.

Cylinder: A hole bored in the block of an engine that directs the piston during movement.

D

DC generator: A device that relies on the principle of electromagnetic induction to create direct current (DC). It consists of an armature loop mounted on a shaft that rotates. On two sides of the armature loop are magnets. A north pole and a south pole are directed toward the armature loop. As the armature loop rotates between the magnetic fields, a current is produced.

Deceleration: The slowing down or braking of a vehicle.

Deep mining: A mining operation that uses shafts and special machinery to remove the coal from deep below the earth's surface.

Deep Space Network (DSN): A system with three radio antennas located on three continents. Two of the antennas determine the distance of the spacecraft from themselves. The antennas then determine the distance of a known object in space. All these distances are computed, and the location of the spacecraft in space is determined.

Degree of freedom: Any of a limited number of ways in which a body may move.

Delta wing: An orbiter wing used as the suspension system on space shuttles. It is triangular and relatively flat.

Diaphragm: The part of a dynamic microphone that is a thin membrane designed to receive sound waves in the air and vibrate accordingly.

Diesel-electric propulsion: A system in which large diesel engines turn electric generators that power a locomotive's wheels.

Diesel engine: An internal combustion engine that uses heat and pressure to ignite its fuel.

Differential: A component that takes rotational power from one source, the transaxle, and transfers it to two axles. It also allows the wheels to spin at different speeds, which helps make turning easier.

Digital signal: An input signal that has only two possible states—the sensor either sends a signal or does not send a signal.

Diode: A device that allows for the one-way flow of electricity. It can also perform a switching function, in that electricity cannot flow through it until it reaches a certain voltage.

Direct converter: A device that changes one form of energy or power into another form of energy or power in one step.

Direct current (DC): A type of current in which electrons move only in one direction.

Direct electric vehicle: A vehicle that requires a connection to electricity.

Direct gain approach: The most common type of passive solar collection. In this type of system, there are no significant architectural provisions made to collect solar energy, other than facing windows in the proper direction.

Directional-control valve: A valve used to control which path fluid takes in a circuit in a fluid power system.

Directional stability: The ability to fly an airplane in a straight line.

Dirigible: An airship that has rudders and elevators, which are used to control the direction and altitude. It also has an engine.

Disc brake: A brake that makes use of a rotating disc or rotor on the wheel.

Displacement: a: A vector quantity that includes both distance and direction. **b:** The amount of water or air pushed aside by either a water vehicle or lighter-than-air aircraft.

Domestic airline service: The transport by way of air to and from major airports within a country.

Double-acting cylinder: A cylinder that uses the force of a fluid to move the piston in both directions.

Double-pole, double-throw (DPDT) switch: A switch used in conjunction with two single-pole, double-throw (SPDT) switches, allowing one or more loads to be controlled from three or more locations.

Draft: The distance from the waterline to the bottom of the boat.

Drag: The force resisting forward motion of an aircraft.

Drive system: A system used to transfer the motion of the engine's crankshaft into the power that moves the vehicle.

Drum brake: A brake device that uses two brake shoes shaped to fit the inside of the brake drum.

Dry cargo container: A fully enclosed shipping container resembling a large metal box with doors either on the side or end.

Dunnage: The straps, blocks, and special rigging needed to securely fasten freight to a vehicle.

Dynamic coil microphone: A common type of microphone with a voice coil attached to the diaphragm.

E

Efficiency: The extent to which an energy form is usefully converted into another form of energy.

Effort: The force behind movement in a power system.

Electrical circuit: A power source, a load, and conductors connected together so electrical current flows in a complete path.

Electrical energy: The energy associated with the flow of electrons.

Electrical subsystem: In engines, the subsystem that produces the current that fires the spark plug.

Electrical system: A power system that uses electrical energy to do work.

Electrode: A terminal on a cell or battery.

Electrolysis: The process of separating the hydrogen-oxygen bond in water using an electrical current.

Electrolyte: A liquid or paste that surrounds and touches electrodes, causing a chemical reaction between the electrodes and electrolyte, which produces an electrical current.

Electromagnet: A conductor wrapped around an iron core. The two ends of the conductor are attached to a power source. When current passes through the conductor, the iron core becomes magnetized.

Electromagnetic field (EMF): A field of electromagnetic energy produced by electricity flowing through power lines.

Electromagnetic induction: The production of electricity in conductors with the use of magnets.

Electron: A negatively charged atomic particle.

Electronic circuit: A group of electronic components arranged on a circuit board so that they work together to perform a specific function.

Electronic device: A device made of one or more electronic circuits.

Electron theory: Electrons flow from a negative point to a positive point.

Electrostatic precipitator: A device that works by positively charging waste particles and attracting them to a negatively charged electrode. The solid particles are then washed off the electrode and collected.

Elevated train: A heavy-rail system that runs above the city streets.

Elevator: The moveable surface placed on the tailplane to control changes in longitudinal stability.

Elevon: A flap located at the rear of a shuttle orbiter's wing. It is used to control pitch and roll.

Embargo: Restriction of trade for political means.

Empennage: The tail section of a plane.

Energy: The ability to do work.

Energy and power technology: Systems that gather energy and convert it to useful power for the use and benefit of society. Products include electricity for household use and the engines that produce mechanical power for automobiles.

Energy conservation: Making better use of the available supplies of energy.

Energy consumption: The use of energy resources.

Energy conversion: The changing of one form of energy into another.

Energy converter: A device that changes a form of energy into a useful form of power.

Energy inverter: A device that inverts a form of power back into a form of energy.

Energy source: A force that has the capacity to do work.

Engine: A device that converts heat energy into mechanical energy.

Engine ring: A circular band for sealing and limiting the amount of oil that makes its way into the combustion chamber.

Entropy: A measure of the unavailable energy in a closed system.

Envelope: The shell of a hot air balloon.

Environmental Protection Agency (EPA): An organization that supports the monitoring of acid rain in the United States.

Epitrochoidal curve: The special shape of a Wankel rotary engine housing, somewhat like a large figure eight.

Ethanol: Ethyl alcohol, sometimes referred to as grain alcohol.

Evaporative emissions control: A device used to regulate vaporized substances discharged into the air.

Evaporator: A freezing unit used in solar cooling operations. Pure liquid refrigerant absorbs the heat from the surrounding area and turns back into a gas in the evaporator. The evaporator is located within a large room or building insulated from exterior heat for refrigeration.

Exhaust stroke: The final movement in the four-cycle process—an upward stroke of the piston.

Exosphere: The extreme outer region of the atmosphere, before outer space. This region is located from the thermosphere to over 5000 miles from the earth's surface.

External combustion engine: The steam engine used a little over 100 years ago as the major source of power for vehicles and many other applications requiring mechanical power production. Heat is produced outside the cylinder in an external combustion engine.

Extravehicular activity (EVA): A spacewalk.

F

False signal: No signal being sent, such as when a switch remains in the "off" position or a thermostat does not reach its set point. Also referred to as a low signal.

Feedback: Information on how a system is performing or has performed.

Feeler gauge: A thin strip of metal machined to a specific thickness.

Fermentation: The decomposition of carbohydrates found in plants with the production of carbon dioxide (CO_2) and acids.

Ferry: A vessel that moves people and vehicles across narrow or small bodies of water.

Field-effect transistor: A switching device often used because it can carry much more current than a bipolar transistor. It has three junction points: a gate, a drain, and a source. Sometimes referred to as a metal-oxide-semiconductor field-effect transistor (MOSFET) or a junction field-effect transistor (JFET).

Fill: The addition of material, such as rock and soil, to build up low-lying areas.

Filter, regulator, lubricator (FRL) device: A device that controls pressure coming from the compressor, filters harmful moisture from the air, and adds lubrication to the air so the wear on pneumatic equipment is minimized.

Fin stabilizer: A fin located on the side of a ship, below the waterline. When the ship rolls to one side, the increased surface area offers resistance to keep the ship upright.

First-class lever: A lever that has the fulcrum positioned between the input force and the load.

Fish lift: A giant elevator-like device installed on dammed rivers to help fish return upstream during spawning season.

Fixed pathway: A pathway that has a fixed route.

Flatcar: A freight car that is a sturdy platform on wheels. It carries steel, lumber, truck trailers, containers, and heavy equipment.

Flat-plate collector: The most common type of active solar collector. It is typically stationary, mounted on a rooftop, facing south in the northern hemisphere. This collector has the ability to collect heat from diffuse sunlight, even on cloudy days.

Flight deck: The upper deck of a space shuttle's forefuselage, which contains the flight controls.

Flight director: The team leader at the Mission Control Center (MCC).

Flow-control valve: A valve used to start or stop the flow of fluid in a system.

Flowmeter: A device that measures the rate of flow in fluid power systems.

Fluid motor: A device that converts fluid power into rotary mechanical motion.

Fluid system: A power system that uses the energy created by liquids and gases to do work.

Fluorescence: The process of converting one form of light to another.

Flux: The lines of force on a magnet.

Fly ash: A solid waste by-product produced by burning coal.

Flywheel: A heavy wheel for opposing and moderating any fluctuation of speed in the machinery with which it revolves.

Foot-pound (ft.-lb.): The amount of force necessary to move a 1-lb. load a distance of 1′.

Force: Effort, in mechanical power.

Forward bias: The flow of electricity occurring through a standard diode when electricity of the proper polarity is applied.

Four-stroke cycle engine: An engine that requires four movements of the piston in its cylinder to complete a full cycle.

Four-way valve: A common type of spool valve that allows both the pressure and return lines to reverse themselves when the valve is triggered, thus allowing the end effectors to be driven in the opposite direction.

Fractionating tower: A large tower in which crude oil is separated into various products.

Freight train: Several freight cars joined together and pulled by an engine or locomotive.

Frequency: The number of cycles in a given time interval, usually one second.

Frequency converter: A device that changes one frequency of radiant energy directly into another frequency of radiant energy.

Frictional horsepower (fhp): The amount of hp necessary to overcome the internal friction of an engine and other forms of frictional loss.

Fuel-air charge: Small droplets of liquid fuel mixed with air.

Fuel cell: A device that utilizes a chemical reaction between hydrogen and oxygen to produce electricity.

Fuel subsystem: The subsystem responsible for creating the fuel-air mixture used to power the engine and delivering that charge to the combustion chamber, based on how much fuel-air charge the governing system allows.

Fulcrum: The fixed point around which a lever rotates.

Full displacement hull: A hull that sits low in the water and has the greatest draft. It is very economical and efficient.

Fuse: A filament that breaks the circuit if too much electrical current passes through it. It prevents damage to the rest of the circuit in the event of an overload.

Fuselage: The main body of a plane.

Gallons per minute (GPM): One of the most common measurements of rate of flow in a fluid system.

Gasohol: An automobile fuel made from grains and gasoline, typically 90% gasoline and 10% ethyl alcohol.

Gasoline-electric hybrid: A type of propulsion system used in automobiles and small trucks. These systems are configured as either series hybrids or parallel hybrids.

Gasoline piston engine: An engine found in automobiles, motorcycles, lawn mowers, and snowmobiles. These engines have five main components: a cylinder, a piston, a spark plug, a crankshaft, and fuel.

Gathering line: A pipe in a pipeline system in which the product to be transported is collected and stored.

Gear: A metal wheel with small notches cut into its rim.

Gear pump: A pump in which two gears are positioned so they mesh with each other inside a housing. The fluid caught between the gear teeth is forced out of the pump housing at a higher pressure.

Gear ratio: A ratio describing the change in the amount of torque.

General aviation: Privately owned aircraft used for recreational, business, and community-oriented tasks.

Geosynchronous (GEO) orbit: A geostationary orbit. It is often used for communication satellites, which are stationed in one spot and rotate along with the earth.

Geothermal energy: Heat from the earth.

Gimbal: A device that can be rotated in many directions. These are used on the space shuttle orbiter to rotate the nozzles surrounding each of the three main engines.

Glider: An aircraft with stable wings, but no power source.

Global positioning system (GPS): A satellite-based navigation system.

Global warming: An increase in the average temperature of the earth's atmosphere, possibly resulting in the melting of ice caps, which could alter shorelines, change weather patterns, and alter agricultural productivity.

Gondola: A freight car that has high or low sides with no tops. It transports loose material.

Governing mechanism: A mechanism that determines how much fuel-air mixture needs to enter the combustion chamber, based on the speed of the engine.

Governing subsystem: The engine subsystem designed to keep an engine running at a desired speed, regardless of the load applied to the engine.

Grade: The percentage of the change in height every 100′ of a railway or roadway.

Gravitometer: A computerized device that separates products by weight.

Gravity: A natural force that has a downward effect on an aircraft.

Greenhouse effect: The situation caused by a layer of greenhouse gases surrounding our planet, produced by the burning of fossil fuels. This layer does not allow the heat produced by the sun to escape the earth's atmosphere as easily as it once did.

Ground Fault Circuit Interrupter (GFCI): A device that can trip to open a circuit in case of an imbalance between the feed and return legs. It is designed to protect people and so is more sensitive than a circuit breaker or fuse and provides a quick reaction time.

Guidance system: A system that provides the information required to make a vehicle follow a particular path or perform a certain task.

Guideway: A railway for magnetic levitation (maglev) vehicles.

Gunwale: The top edge of the hull. Also called the gunnel.

H

Half-life: The time it takes for half the atoms present in an unstable element to transform into a new element.

Harbor: A point along the coast where the water is deep enough for the vessel to come very close to shore.

Harbormaster: An officer who controls the flow of traffic in and out of a port.

Heading: The direction a vehicle is pointed.

Heading indicator: An instrument that shows the pilot the direction the plane is headed.

Heat energy: Energy with a longer wavelength than light energy. It is generally not visible to the eye, but it can be measured in terms of temperature. Also referred to as *infrared energy*.

Heating season (HS): A period of the year associated with making the outside air temperature warm.

Heating unit: The equivalent of 100,000 Btu.

Heat pump: In geothermal energy, an earth-coupled heat pump is an application for residential heating and cooling. A system of these pumps consists of pipes buried in the shallow ground and located near a home. Water flows through the pipes and conducts heat from the ground for transfer into the home.

Heavier-than-air: An aircraft of greater weight than the air displaced.

Heavy equipment: A large and powerful vehicle used for reasons such as moving earth, farming fields, and conducting warfare.

Helicopter: An aircraft with rotating wings.

Hertz: A unit of frequency equal to one cycle per second.

High-temperature reusable surface insulation tile: A specially designed tile, made of nearly pure silica fibers, that rapidly dissipates heat and can withstand temperatures from below –200°F to over 2000°F. These tiles cover the underside and nose of shuttles.

Hopper car: A freight car with chutes underneath. It carries bulk materials.

Horizontal tailplane: The tail surfaces of an airplane that are parallel to the horizon, including the stabilizer and the elevator.

Horsepower (hp): The standard measuring unit of power equal to the energy needed to lift 33,000 lbs. 1′ in 1 minute. 1 hp of electricity is equivalent to 746 watts.

Hovercraft: An air-cushion vehicle designed to ride on a cushion of air the vehicle generates. This type of suspension system allows travel over land or water.

Hull: The body of a boat or ship.

Hull speed: The top speed at which hulls become inefficient and dangerous. It is figured by measuring the length of the boat at the waterline, finding the square root of the length, and multiplying by 1.34.

Hybrid system: In vehicles, a system that uses both an internal combustion engine and an electric motor.

Hydraulic: A type of fluid power system that uses a liquid, such as oil, to transmit and control power.

Hydraulic braking system: A braking system that links a master cylinder to a brake pedal and to one or two brake cylinders at each wheel.

Hydraulic pump: A pump that supplies and transmits the pressure needed to operate a hydraulic power system. It converts mechanical energy into fluid power, creating the necessary flow in a system.

Hydraulic system: A system that controls and transmits energy through the use of liquids.

Hydroelectric energy: The use of flowing water from waterfalls and dams to produce electricity.

Hydrofoil: A passenger-transporting vessel, similar to a plane and ship put together. It operates on inland and coastal waters.

Hydrogen: The first and simplest element on the periodic table. It is one of the most common elements in the galaxy.

Hydrogen sulfide gas: A gas that smells like rotten eggs, found within the steam of geothermal energy plants.

Hypergolic engine: A type of rocket in which the engines use two propellants that ignite on contact, nitrogen tetroxide and hydrazine. Therefore, it does not require an igniter.

Hypersonic speed: Speed over Mach 5 (five times the speed of sound).

I

Ideal mechanical advantage (IMA): A ratio of the forces or the distances involved in a mechanism. It assumes 100% efficiency.

Idle: The condition an engine will run under when it is warmed up to temperature and not under load.

Idle bypass circuit: A small passageway that allows some fuel-air mixture to escape around the throttle plate and keep the engine running, even when the throttle is closed.

Idler gear: See *Intermediate gear*.

Impeller: A device that has many small blades mounted on a shaft. It spins inside its housing, drawing liquid through the inlet port from a reservoir.

Inboard engine: An engine used on most vessels over 36′ in length. Its power source is mounted inside the ship attached to a propeller shaft.

Inboard/outboard engine: An engine used in mid-sized recreational boats. These systems have larger engines than outboard motors. The motor on an inboard/outboard engine is mounted inside the boat.

Inclined plane: A simple machine that makes use of sloping surfaces.

Indicated horsepower (ihp): The maximum potential hp produced by an engine under ideal conditions.

Indirect converter: A device that changes one form of energy or power into another form of energy or power using several intermediate steps.

Indirect gain approach: A type of passive solar collection that makes use of a storage medium to store heat for later use. The storage medium may be rocks, concrete, or water.

Induced: Made to flow.

Inertia: The tendency of a body in motion to remain in motion.

Inertial reference system: A system that uses gyroscopes to determine the attitude of a spacecraft.

Inexhaustible energy source: An energy source that will never run out.

Infiltration: Cold air forcing its way into a home through cracks and other penetrations.

Infrastructure: The roadways, bridges, ramps, and other structures that make up the highway and roadway system.

Inland waterway: A route taken on canals, rivers, and lakes.

Input: A resource needed to begin a system and make it operate.

Instrument flight rules (IFR): Rules followed when weather conditions do not allow pilots to navigate visually.

Instrument landing system (ILS): A system allowing pilots to land in all types of weather conditions. It uses two radio waves to mark the approach of a runway.

Insulation: The state of being separated from conducting bodies by means of nonconductors so as to prevent transfer of heat.

Insulator: A material made of atoms that do not transfer electrons easily.

Intake stroke: The downward stroke of the piston that begins the process of producing power.

Integrated circuit (IC): A collection of electronic circuits, undistinguishable to the naked eye, etched into a thin layer of silicon and installed into a plastic or ceramic housing.

Intelligent control: The highest level of control technology. It uses machines and programming techniques capable of solving complex problems without human intervention. The technology emulates human thought processes using sophisticated software that makes use of artificial intelligence principles.

Intermediate gear: A gear between a driver gear and a driven gear.

Intermodal: A transportation system that combines various modes of transportation to move people and products.

Intermodal cargo transportation: Transporting cargo from one point to another by using several different modes of transportation.

Intermodal passenger transportation: Transporting passengers from one point to another by using several different modes of transportation.

Intermodal Surface Transportation Efficiency Act (ISTEA) of 1991: The Act that created the National Commission on Intermodal Transportation (NCIT).

Intermodal transportation: A transportation system that uses more than one environment or mode.

Internal combustion engine: An engine that produces heat inside the cylinder containing the piston.

International airline service: A service that provides travel between countries.

International Space Station (ISS): The newest space station. It is a joint effort of 16 countries, with the United States in charge of the operations.

Ion propulsion: A type of propulsion that uses the electrical charge of atoms to move vehicles.

Isaac Newton's third law of motion: For every action, there is an equal and opposite reaction.

Isolated gain approach: A type of passive solar collection in which the solar collector is isolated from the structure to be heated. The collector is typically located next to or beneath the home, and the system relies on convection to carry heated air up to the structure. Cold air return vents typically allow cooler air to return to the collector.

Isotope: One of two or more atoms with the same number of protons but with different numbers of neutrons.

J

Jet engine: An engine that uses spinning turbines to compress air and propel an aircraft.

Jet pack: A space vehicle strapped to the back of an astronaut.

Jet route: An air transportation path reserved for large commercial jets and airliners.

Jib: The smaller sail on a sailboat that is connected to the mast and the bow.

K

Kerogen: An oily substance contained by 40–50 million-year-old sedimentary rock.

Key: On an engine, a small metal piece that holds the flywheel in a properly aligned position on the crankshaft for spark to occur.

Kilowatt-hour (kWh): Wattage multiplied by the number of hours the wattage is used and then divided by 1000. It is the unit by which customers are billed for electricity usage.

Kinetic energy: Energy in motion.

Kyoto Protocol: Targets set in 1997 by countries wishing to increase capacity while reducing carbon dioxide (CO_2) emissions.

L

Laminar flow: The smooth flow of liquids in stratified layers that may not be visible to the eye.

Land pollution: The action of contaminating the surface of the earth and its natural resources.

Land reclamation: The restoration of land to a usable condition after strip mining has taken place.

Land transportation: A transportation system using vehicles on land, including subways, buses, trains, trucks, bicycles, and motorcycles.

Latching: A function used to start a machine at the touch of a button and have it remain running indefinitely.

Latching relay: The process of using momentary contact switches and a magnetic motor starter to start a device. This ensures that, if the power goes off during operation, the device cannot turn back on at an inopportune time. The only way to start the device is to push the start button again. The start button triggers a magnetic contactor that closes and remains latched until the stop button is depressed or the power is interrupted.

Lateral stability: The ability to overcome the tendency of a plane's wings to dip on either side.

Launchpad: A nonflammable platform from which a rocket, launch vehicle, or guided missile can be launched.

Lead-acid battery: A common secondary cell used in automobiles. It is a combination of several cells. This type of battery includes a series of positive and negative metal plates in a weak sulfuric acid electrolyte.

Leading edge: The front of an airfoil.

Lead poisoning: Chronic intoxication produced by the absorption of lead into the system.

Leaf spring: An automotive spring made of a series of steel strips, each one shorter than the next.

Length overall (LOA): The measurement from the tip of the bow to the stern of a water vessel.

Lever: A rigid bar that rotates around one fixed point.

Lift: The upward force that keeps aircraft in the air.

Light energy: Energy visible to the eye.

Lighter-than-air: Of less weight than the air displaced.

Light pollution: The excessive amount of light in the nighttime sky that often surrounds urban areas.

Lignite coal: A type of coal containing some woody decomposition that can be recognized as peat, but it has a higher energy content than peat. It contains a large amount of moisture and is brownish in color, as opposed to the black substance typically recognized as coal.

Linear-concentrating parabolic collector: An active solar collector that tracks the sun's movement, generating temperatures of hundreds of degrees Fahrenheit on a clear day. Mounted along the focal line is often a single collection tube through which water flows. This type collects much more energy than a flat-plate collector, in much shorter periods of time, while occupying much less space.

Linear motor: An induction motor that has been flattened out.

Line diagram: A pictorial representation of a control circuit.

Line of force: A theoretical line running between the poles on the outside of a magnet.

Liquefied natural gas (LNG): Gas that has been placed under pressure at very low temperatures, allowing transportation by railroad tankers, truck tankers, and ships.

Liquid-fuel rocket: A rocket engine that uses two separate liquids, which are ignited in a combustion chamber.

Load: An output that is the final goal of the power system. It is the work done by the system.

Loading dock: A cargo facility built next to large paved areas for trucks to back up to.

Lock: A chamberlike facility constructed in a canal between two different water levels. It is made up of gates, pumps, and filling and draining valves.

Log: A speed indicator aboard a marine transportation vehicle.

Longeron: A piece of wood or steel running the length of an airplane.

Longitudinal stability: The ability to fly the airplane without the nose moving up or down.

Loran-C: A long-range navigation system.

Low earth orbit (LEO): An orbit between 180 and 250 miles above the earth.

Lubrication subsystem: The oil distribution mechanism, the various oil seals, and the lubricating oil itself.

M

Mach 1: The speed of sound (760 miles per hour).

Machine: A device used to manage mechanical power.

Magma: Molten rock located miles beneath the earth's surface.

Magnet: A material attracted to any metal containing iron.

Magnetic levitation (maglev): A train that uses powerful magnets to hover above or below and propel down a track.

Magnetohydrodynamic (MHD) generator: An advanced, highly efficient direct conversion device that generates electricity from fossil fuels.

Mainsail: The larger of the two sails on a sailboat. It is connected to the mast and a boom.

Management process: The "behind the scenes" part of an energy, power, or transportation system necessary to plan, organize, and control the system.

Manned maneuvering unit (MMU): A specialized space vehicle developed early in the shuttle program to allow astronauts to maneuver outside the vehicle.

Manned space vehicle: A vehicle sent to space with a crew in it.

Mantle: Rock that conducts heat coming from the earth's core.

Manual control: The simplest type of control technology, requiring human input in order to function.

Manual transmission: A transmission that is totally controlled by the operator of the vehicle. A shift lever controls and engages different combinations of gears.

Manufacturing technology: Systems that transform raw materials into useful products in a central location. These products must be marketed, transported, and distributed for end use.

Margin thickness: The thin edge at the top of a valve that wears away, due to friction, over the course of an engine's life.

Mass transit rail: A form of rail transportation that can carry many people at one time.

Measuring device: A device required in power and transportation systems that provides a source of feedback to monitor how well the system is functioning.

Mechanical advantage: An increased force and the benefits of that increase created by using a machine to transmit force.

Mechanical efficiency: In engines, the percentage of power developed in the cylinder compared to the power actually delivered to the crankshaft.

Mechanical energy: Energy produced by mechanical devices, such as gears, pulleys, levers, or more complex devices, such as internal combustion engines.

Mechanical subsystem: In engines, the subsystem that converts the force of the expanding gases during combustion into mechanical power, delivering the power to the crankshaft.

Mechanical system: A power system that uses mechanical energy to do work.

Medical technology: Systems used to maintain health and treat injuries and illnesses. The end product of medical technologies is a healthier society.

Memory: The ability of a control circuit to remember the last command it received.

Mesosphere: The atmospheric region ranging from the stratosphere to about 50 miles (80 km) above the earth's surface.

Methane digester: A vessel that converts shredded organic materials into methane gas, which can be used for heating, used for power generation, or purified and stored for distribution.

Methanol: A clean-burning liquid used as fuel to power vehicles. It can be made from nonrenewable sources of energy, such as coal, or from renewable sources of energy, such as wood, plants, and manure. Methanol produces more energy than ethanol, per volume, and burns more slowly than gasoline.

Methyl alcohol: See *Methanol.*

Micrometer: A basic precision measuring instrument used to check for wear points on engine parts.

Military aviation: Air activity performed by the armed forces.

Mission Control Center (MCC): An area in the Johnson Space Center (JSC) where the liftoff of the shuttle, the mission of the orbiter, and the control of the International Space Station (ISS) is directed.

Momentary contact switch: A switch that makes or breaks a circuit based on the input or touch of the switch. When the switch is not depressed, the switch returns to its default status.

Monocoque: A type of fuselage construction that results in a hollow structure. A series of ringlike ribs attached to the strong metal outside covering of the plane form the skeleton of the structure.

Monorail: A train that runs on a single rail.

Monostable application: An integrated circuit (IC) application that turns something on or off for a specific period of time.

Motor: A device that produces mechanical energy by converting electrical energy.

Multigrade: Many positions in a scale of qualities, such as multiweight oil that has winter and summer viscosity ratings.

N

Nacelle: An enclosure that houses the gearbox, the generator, and a variety of equipment necessary to keep a wind turbine properly positioned into the wind and spinning at a safe speed.

NAND logic: A combination of NOT logic and AND logic.

Nanotechnology: The design of products and devices at a molecular level.

National Aeronautics and Space Administration (NASA): The U.S. agency set up for research and development of space exploration.

Natural gas: Gas usually found within the vicinity of petroleum reserves.

Natural pollutant: Something produced by nature that contaminates the environment.

Nautical chart: A map that shows coastal waters, rivers, and other marine areas. It is designed to show information for navigating waterways.

Nautical mile: Roughly 1 minute, or 1/60 of a degree, of latitude around the earth. It is equal to about 1.15 statute miles.

Navigable: Deep and wide enough for a boat or ship to travel through.

Navigation: The act of guiding a vehicle.

Near space: The space extending from 50 miles to 10,000 miles beyond the atmosphere.

Neutron: An uncharged atomic particle.

Noise pollution: Annoying or harmful noise in an environment.

Nomograph: In fluid power, a chart helpful in determining the inside diameter of conductors or for estimating flow or velocity, if two of the three variables are known.

Nonattainment: A classification of an area in which the Environmental Protection Agency (EPA)'s minimum air quality guidelines are not met.

Nonfixed pathway: A pathway in which a vehicle has freedom to move in various directions.

Nonrenewable energy source: A resource that cannot be replaced once used.

NOR logic: A combination of NOT logic and OR logic.

NOT logic: A form of control logic in which an output signal is provided *unless* there is an input signal.

Nuclear energy: The power of the atom.

Nuclear fission. The process of splitting a larger atom to produce two smaller atoms and a tremendous amount of energy.

Nuclear fusion: The combining of two nuclei into a larger nucleus. The large nucleus weighs less than the two smaller nuclei that formed it. The result of this process yields a large energy release.

Nuclear turbine engine: An inboard propulsion system that uses a nuclear reactor to heat water. The steam produced by the heated water is used to power a turbine.

Nuclear Waste Fund: A multibillion-dollar fund used for the development of a permanent nuclear waste disposal site.

Nuclear Waste Policy Act: An Act passed by Congress in 1982 promising that the federal government is to take nuclear waste from the utilities for permanent storage.

Nucleus: The center portion of an atom containing the protons and neutrons.

O

Ohm's law: Voltage (E) can be determined by multiplying current (I) by resistance (R).

Oil dipper: A device that splashes the moving parts of the engine with oil when the engine is running.

Oil shale: 40–50 million-year-old sedimentary rock containing an oily substance.

Onboard navigation system: A system that includes a liquid-crystal display (LCD) screen in the dash that is linked to an electronic map and a global positioning system (GPS) receiver.

On-delay function: A common timing function typically created to provide a margin of safety from the time the start button on a machine is pushed until the machine actually begins running.

Open circuit: A circuit or part of a circuit that is not energized.

Open loop solar collection: Systems in which the heated water or air is directly distributed for use.

Open loop system: A control circuit in which the system output has no effect on the control.

Operational amplifier (op-amp): An integrated circuit (IC) that has the ability to take in an alternating current (AC) or direct current (DC) signal and amplify the output by as much as 100,000 times the input.

Orbit: To stay in a path circling an object in space.

Orbital maneuvering system (OMS) engine: An engine located at the rear of the orbiter above the main engines. It is used primarily to maneuver the shuttle into orbit and to slow and deorbit the craft prior to reentry.

Orbiter Processing Facility (OPF): A facility that houses the orbiter from the time it lands until a week before the next scheduled launch.

Organization of Petroleum Exporting Countries (OPEC): A group of nations committed to the strength and success of the oil market. These countries export oil to consumer nations.

OR logic: A form of control logic in which input must be received from either one or more devices before output will occur.

Outboard engine: The most common motor used on fishing boats and small motorboats. Its power source and propeller are one piece located off the back of the boat.

Outer space: The space extending from 10,000 miles beyond the atmosphere.

Output: The end result of a system.

Output device: A small load that makes things happen, such as a relay, an indicator light, a horn, a bell, a solenoid-controlled valve, a heating element, and a small motor.

Overburden: When strip mining, the soil remaining after the topsoil is removed.

Overshot waterwheel: A waterwheel that relies on an elevation change and makes use of the weight of the water, in addition to the water's force.

Oxidation: A union between fuel and oxygen.

Oxidizer: A chemical substance that mixes with fuel to allow combustion.

P

Paddle: An implement used to propel a boat using human power.

Paper battery: An ultrathin dry battery that functions like larger dry batteries, but can be printed on a number of different products, including paper products.

Parabolic dish collector: An active solar collector that is point focusing. It has a tremendous collection ratio, permitting large point-focusing collectors to produce temperatures of thousands of degrees Fahrenheit. This type of collector tracks the height of the sun in the sky, as well as the azimuth path.

Parallel hybrid: A gasoline-electric hybrid system configuration that uses both the gasoline engine and electric motors to turn the wheels.

Pascal's law: When there is nothing but liquid in a container, compression applied to any part of the container is distributed equally in all directions. The initial pressure multiplies as it is distributed through a container with a larger diameter.

Passive solar energy collection: Systems that do not make use of any externally powered, moving parts, such as circulation pumps, to move heated water or air.

Payload area: The midfuselage of a space shuttle, which holds the cargo.

Peat: The first step in the formation of coal. It is formed from water and the decomposition of organic materials.

Penstock: A dam tunnel through which stored water rushes to a water turbine.

Perigee: The point in the path of an elliptical orbit closest to the earth.

Petroleum: Oil.

Phase change: A change in a substance's state (liquid, solid, or gas).

Photoelectric effect: The emission of free electrons when photons strike certain metals.

Photosynthesis: The process by which carbohydrates are compounded from carbon dioxide (CO_2) and water in the presence of sunlight and chlorophyll.

Photovoltaic cell: A cell that converts sunlight directly into electricity. This occurs when positively charged photons strike the cell and displace electrons from the valence shell of the material making up the cell. Sometimes referred to as a solar cell.

Pig: A barrel-shaped brush used to clean pipelines.

Piggyback: To carry a trailer or container on wheels on a flat railroad car.

Pin-out: A device that shows all the pins on an integrated circuit (IC) and may indicate their purpose.

Piston: A cylindrical engine component that slides back and forth in the cylinder when propelled by the force of combustion.

Piston engine: An engine that produces heat by compressing and igniting a mixture of air and fuel.

Pitch: a: The angle of a propeller's blade.
b: The up-and-down movement of the nose of an aircraft.

Planetary gear set: Several gears combined together to provide a wide range of gear ratios.

Planing hull: A hull that rides on top of the water. It has no maximum hull speed, but fuel efficiency is low, and it is hard to handle and rough in heavy waves.

Plutonium 239 (P-239): A fissionable fuel created from uranium 238 (U238) by a breeder reactor.

Pneumatic: A type of fluid power system that uses a gas, such as air, to transmit and control power.

Pneumatic tire: A tire filled with air.

Polarity: The type of charge an atomic particle has.

Port: The left side of a ship.

Positive crankcase ventilation (PCV) valve: A valve that vents blowby gases from the crankcase back into the engine for reburning.

Potential energy: Energy waiting to happen.

Potentiometer: A switch that can provide variable motion control. It can vary the resistance within the switch, which affects both the current and voltage flowing out of the switch.

Pounds per square inch (psi): A unit used to measure pressure.

Pounds per square inch absolute (psia): A type of pressure gauge that accounts for atmospheric pressure in its measurement.

Pounds per square inch gauge (psig): A pressure instrument that determines the approximate pressure developed in a fluid circuit but does not take atmospheric pressure into account.

Power: The rate at which work is performed or energy is expended.

Power brake: A brake system that adds a vacuum control valve between the brake pedal and the master cylinder.

Power converter: A conversion device used to change one form of power to another.

Power stroke: The stroke in which mechanical movement is transferred from the piston to the connecting rod and then to the crankshaft.

Power system: A system in which energy is harnessed, converted, transmitted, and controlled to perform useful work.

Practical efficiency: A measurement of how efficiently an engine uses its fuel supply.

Pressure: Effort, in fluid power.

Pressure gauge: A device that measures the difference between the pressure within a fluid circuit and the pressure in the surrounding atmosphere.

Pressure-reducing valve: A valve that reduces the pressure within a fluid circuit to levels suitable for use.

Pressure-regulating valve: A valve that controls pressure coming from the compressor.

Pressure-relief valve: A valve placed in a fluid power system to make sure the pressure does not get too high.

Pressure tank: A tank that stores air under pressure in pneumatic systems.

Pressurized water reactor (PWR): A reactor that works similarly to a boiling water reactor (BWR), except it makes use of a heat exchanger known as a steam generator. A PWR can operate at higher pressures and temperatures than a BWR. Unlike the BWR, the steam generator in a PWR allows the turbine to remain free of radioactive contamination.

Primary cell: A type of cell that cannot be recharged.

Primary coil: The coil attached to the input side of the iron core of a transformer.

Primary loop: The part of a pressurized water reactor (PWR) in which the water is heated. It surrounds the reactor core.

Principle of magnetism: Like forces repel one another, and opposing magnetic forces are attracted to one another.

Process: The portion of the technological systems model in which the desired goal is in sight. The inputs are being changed in this step.

Production process: The "on the scene" part of a transportation system, including receiving, holding, loading, moving, unloading, storing, and delivering.

Programmable control: The level of control technology that typically uses a dedicated microprocessor or computer as the brains of the system.

Programmable logic controller (PLC): A microcomputer designed exclusively for control purposes.

Proliferation: The use of by-products of nuclear power for the production of nuclear weapons.

Prony brake: A device used to measure the effort produced by a twisting or turning force. It is based on the principle that if an opposite force equals the effort produced by a spinning object, movement will cease.

Propellant: A mixture of fuel and an oxidizer.

Propeller: A rotating blade that produces thrust.

Propulsion system: The components of a vehicular system that produce the power needed to move a vehicle.

Protection device: A device that shields components in power circuitry from excessive effort or excessive rate of flow.

Proton: A positively charged atomic particle.

Proton exchange membrane (PEM): A common type of fuel cell that works by passing hydrogen through one end and oxygen through the other.

Proximity sensor: A device that responds to physical closeness and transmits a resulting impulse.

Pulley: A solid disc that rotates around a center axis. It usually has a groove around the outside edge that allows ropes or belts to easily ride around them.

Push-button make/break (PBMB) switch: A switch that retains its status until pressed again.

Pyrolysis: The process of separating the hydrogen-oxygen bond in water using heat.

Q

Quad: An accepted abbreviation for 1 quadrillion Btu.

R

Rack and pinion: Types of gears. Rack gears are flat gears, and pinion gears are round.

Radial: A beam of radio waves.

Radial engine: An engine in which all the pistons are connected to a hub in the center of the engine.

Radiant energy: Energy transferred by radiation, including infrared rays, visible light rays, ultraviolet (UV) rays, X rays, radio waves, and gamma rays.

Radiation: Energy radiated in the form of waves or particles.

Radiator: A heat exchanger that transfers the heat from a liquid to the surrounding environment.

Radioactivity: A property of some atoms, such as those of uranium decay, in which they give off atomic particles. The particles emitted are harmful to humans and other living things.

Radio detecting and ranging (radar): An electronic navigation tool that contains a transmitter, a receiver, and a display.

Rail: A long piece of steel that has an *I*-shaped cross section.

Rail bed: Several layers of stone designed to spread the weight of the trains evenly over the compacted surface.

Railroad: A permanent road made of a line of tracks fixed to wooden or concrete ties.

Ramjet engine: An engine that can operate only when moving at high speed, since it has no moving parts and no device for drawing in air.

Rate: The characteristic of power that expresses a certain quantity per unit of time.

Rated horsepower (rhp): About 80% of the engine's brake horsepower (bhp) capability.

Reaction control system (RCS) engine: An engine located in a cluster in either the front or rear of a shuttle orbiter. It serves as a thruster that enables the shuttle to maneuver in space.

Reaction wheel: A large rotating wheel that can generate momentum and spin an unmanned spacecraft.

Reciprocating pump: A type of pump with a piston that moves back and forth in a cylinder to move hydraulic fluid.

Reclamation: A technique that requires land that has been disturbed to be graded and covered with topsoil, so as to be suitable for alternate use at the conclusion of mining operations.

Recursive function: The ability to make something happen over and over again.

Regenerative braking: A process that transforms a car's kinetic energy into electrical energy by using the electric motor as a generator during braking.

Regional airline service: The transport from small airports to major airports within a specific region.

Reinforced carbon-carbon panel: The covering for the leading edge of an orbiter wing. It is made of a material that has several layers of carbon and can withstand extreme temperatures.

Relay: An electromechanical switching device.

Renewable energy source: A resource that can be replaced when needed.

Repeater: A specialized amplifier used in a telephone system. It receives electronic communication signals and sends out corresponding amplified signals.

Reservoir: A container to hold liquids to be used again in a system.

Resistance: Opposition to the flow of current.

Resistor: A device that resists the flow of electricity.

Revolutions per minute (rpm): A measurement of movement in a rotational mechanical system.

Rich mixture: A fuel mixture with more fuel vapor than normal.

Ring: The path followed by electrons in an atom.

Roadbed: The foundation supporting the road surface and vehicles.

Road map: A map that uses different symbols to denote different types of roads, landforms, and structures.

Rocket engine: An engine that produces thrust by expelling hot gases from a rear nozzle.

Roll: The tendency of a plane's wings to dip on either side.

Rolling stock: A railroad car pulled by a locomotive.

Rotary engine: An internal combustion engine that uses rotors in place of pistons. Also called Wankel rotary engines.

Rotor: The spinning coil of an alternating current (AC) motor that is connected to the motor shaft.

Rotor blade: A rotating wing on a helicopter that provides lift.

Rudder: A hinged vertical surface on a water vehicle. It acts to change the direction of water pressure against the vessel. Rudders control the stern of most marine vehicles.

Runway: A flat, straight path specially lit and marked to aid pilots.

Run winding: A winding that powers a motor once it is up to speed.

S

Sail: An extent of fabric by means of which wind is used to propel a ship through water.

Scalar quantity: A physical quantity specified by the magnitude of the quantity and expressed by a number or unit.

Schematic drawing: A drawing that traces the path electron flow will take in an electrical or electronic circuit. Symbols are included that represent the components in the circuit.

Science: The body of knowledge related to the natural world and its phenomena.

Scientific method: Methodology that pursues new knowledge by the collection of data through observation and experimentation to test a hypothesis.

Screw: A simple machine consisting of a very long inclined plane wrapped around a shaft.

Sea-lane: A regular route taken by ships and other vessels when traveling across the ocean. Also called trade routes.

Sealing contact: A contact in a motor starter that is part of the control circuit. When these contacts close, they maintain memory, bypassing the start button and feeding power to the coil of the motor starter.

Secondary cell: A type of cell that can be discharged and recharged many times.

Secondary coil: The coil attached to the output side of the iron core of a transformer.

Secondary loop: The part of a pressurized water reactor (PWR) in which steam is created.

Second-class lever: A lever that has the load placed between the fulcrum and the input force.

Semiconductor: A material that is both a conductor and an insulator.

Semimonocoque: A type of fuselage construction that makes use of both vertical and horizontal frame members, relieving some of the stress on the skin of the aircraft.

Semiplaning hull: A hull in which the stern of the boat remains in the water, like a displacement hull, and the bow is raised on top of the water, like a planing hull.

Series hybrid: A gasoline-electric hybrid system configuration in which all the components are connected in a line.

Set point: A predetermined output level at which a closed loop system makes an adjustment.

Shaft: A long, cylindrical piece of metal used to transfer mechanical energy in many types of machines.

Sheet: A rope used on a sailboat.

Ship: A water vehicle over 100′ in length.

Shipping cask: A container designed to ship spent fuel from one facility to another.

Shock absorber: A suspension system component that absorbs a road's unevenness so it is not transferred to the vehicle structure.

Short circuit: A circuit in which the load is bypassed and the hot wire comes directly into contact with the return leg or with something grounded.

Signage: The information transmitted by the use of signs.

Simple machine: A lever, a pulley, a wheel and axle, an inclined plane, a screw, or a wedge.

Single-acting cylinder: A cylinder that uses the force of a fluid to move the piston in one direction.

Single-phase induction motor: A type of alternating current (AC) motor that only has one set of run windings.

Single-pole, double-throw (SPDT) switch: A switch used to control a load from two different locations.

Single-pole, single-throw (SPST) switch: A switch that makes or breaks one set of contacts to turn a load on and off.

Skylab: A space station the United States launched into orbit on May 14, 1973.

Sledge: An early example of the modern sled, built using logs.

Slip ring: See *Commutator*.

Slow-blow fuse: A fuse designed to withstand the initial surge of current associated with starting a motor when it is at rest.

Slurry: A mixture of a ground solid and a liquid transported in a pipeline.

Small Aircraft Transportation System (SATS): An organization formed by the National Aeronautics and Space Administration (NASA) and the Federal Aviation Administration (FAA) to research a solution to the overcrowding of major airports.

Smart road: A component of future highway infrastructure that will communicate with vehicles and help to control the speed, braking, and steering of the vehicles.

Society of Automotive Engineers (SAE): An organized group of individuals working together because of common training in the branch of engineering relating to self-propelled vehicles.

Socket: In electronics, a base into which an integrated circuit (IC) is installed, allowing the IC to be connected with other components in a circuit.

Solar lighting: A system that uses a collector, located on a rooftop, which sends light into the building through fiber-optic cables.

Solar propulsion: A propulsion system that relies on the sun's energy.

Solar sail: A large sail that operates on the principle that light exerts a small force on the objects it touches. It will be used to collect light energy to propel spacecraft.

Solenoid: A device that converts electricity into mechanical movement.

Solenoid-operated valve: A valve shifted by an electrical signal.

Solid-fuel rocket: A rocket engine that contains solid propellant packed into a cylindrical container.

Solid rocket booster (SRB): A shuttle propulsion component that is recovered and reused. It supplies the majority of the power at takeoff.

Solid-state: A type of device that can perform a switching function without any physical moving parts.

Sound: Vibration traveling through matter in the form of a wave motion.

Spacecraft: A vehicle used for space travel.

Space elevator: A long cablelike structure with cars attached, extending from the earth to space, that, in theory, could be used to transport people to and from space.

Space plane: A combination of an airplane and a space shuttle that will be reusable and more cost-efficient than space shuttles.

Space Shuttle Main Engine (SSME): An engine that can be stopped and started as needed. It is located at the rear of the orbiter.

Space transportation: A transportation system in which people and cargo are moved within near space and into outer space.

Span: The length of a plane's wings, from tip to tip.

Spark plug: A part that fits into the cylinder head of an internal combustion engine and carries two electrodes separated by an air gap across which the current from the ignition system discharges to form the spark for combustion.

Spark plug feeler gauge: A tool used to properly gap a spark plug.

Spark test: A test performed to determine if the magneto is producing appropriate spark and if the spark plug is firing.

Spinnaker: A sail used at the front of a boat when traveling downwind.

Spoiler: A control surface on a large aircraft designed to slow the aircraft down during landing by increasing the amount of drag.

Spool valve: Another name for a directional-control valve. On early models of directional-control valves, the interiors resembled spools that hold thread.

Spring: A device that is able to temporarily store energy and then use the energy.

Spring oscillation: The compression and rebounding of a spring, in which the spring will spring past its normal position.

Stabilizer: A device, similar to a rudder on a boat, used by a pilot to steer an airship.

Stabilizer bar: A long steel rod mounted between the two front wheel assemblies to keep the vehicle from leaning out too far when the vehicle is going around corners.

Staging: A technique that places several propulsion systems on top of each other. As the first stage burns out, it is released from the vehicle. This exposes the second stage of propulsion systems. Stages are burned and released until the vehicle reaches its final orbit altitude.

Stall: A situation in aircraft in which the angle of attack becomes too great and the wing quickly loses lift.

Starboard: The right side of a ship.

Start winding: A special winding required by a single-phase induction motor because its single set of run windings do not create enough magnetic attraction and repulsion to move the rotor past the neutral point of rotation.

Stationary pathway: A transporting path in which the structure does not move.

Stationary surface: An immobile, exterior boundary of an object.

Stator: The part of an alternating current (AC) motor that remains stationary and does not rotate. It is typically connected to the housing of the motor and the motor frame.

Steel-belted radial tire: A tire constructed with wide strips of steel mesh called belts. It improves durability, wear resistance, and gas mileage.

Step-down transformer: A transformer used to decrease voltage supplied to a circuit.

Step-up transformer: A transformer used to increase voltage supplied to a circuit.

Stern: The back of a boat.

Stoichiometric ratio: The optimal air-to-fuel ratio for reducing emissions.

Storage medium: A device that is necessary when power must be stored for use at a later point in time.

Straight truck: A truck that has one frame that connects both the front and rear axles.

Stratosphere: The atmospheric region ranging from the troposphere to about 30 miles (50 km) above the earth's surface.

Strip mining: The mining or removing of coal close to the earth's surface. It is done mainly with the use of large pieces of machinery, such as mechanical shovels and bulldozers. Also called *surface mining.*

Stroke: The movement of the piston from the bottom limit of its travel to the top limit (or vice versa).

Structural system: The parts of vehicles that hold other vehicular systems and the loads they will carry.

Subballast: Crushed stone that forms the first layer of a rail bed.

Subbituminous coal: Coal that contains greater energy content per volume than lignite and is frequently used as an industrial fuel, for space heating, and for the generation of electricity. This coal is dull black in color and is categorized as a "soft" coal.

Submarine: A vessel that can submerge and travel underwater.

Subway: A heavy-rail train that runs on a rail in tunnels below the earth's surface.

Support system: The external operations and facilities that maintain transportation systems.

Surface mining: The process of working the external layer of the earth and extracting mineral substances.

Suspended particle: A particle in water that absorbs light and makes water cloudy.

Suspension system: The vehicle system that supports or suspends the vehicle in or on its given environment, providing a method to smooth the ride for passengers and cargo.

Swash plate: Part of a helicopter that controls the pitch of the blades on the main rotor, controlling the aircraft's horizontal and vertical movements.

System: A combination of related parts that work together to accomplish a desired result.

T

Tank car: A freight car that is a large tank on wheels for transporting liquids.

Tar sand: A source of crude oil. The sand is mined and mixed with hot water or steam to extract the thick oil known as bitumen.

Taxiway: An airport roadway that connects the runway to the terminal.

Technological method: Method of problem solving that yields new products through a process of researching, testing, and refining.

Technological system: The inputs, processes, and outputs designed to meet human needs and wants.

Technology: The body of knowledge related to the human-made world. The technological world includes human-made products and their impacts.

Terminal: A physical facility or building used to load and unload passengers and cargo.

Test pressure: The maximum pressure that a conductor is designed to withstand.

Therm: The equivalent of approximately 100 cf (cubic feet) of natural gas, or 100,000 Btu.

Thermal container: A container used to transport perishable food products. It can be insulated, refrigerated, heated, or both refrigerated and heated.

Thermal efficiency: A measurement of how much heat is actually used to drive the piston downward.

Thermal pollution: A form of pollution most commonly associated with power plants, in which expelled waste heat adversely affects aquatic life.

Thermocouple: A transducer consisting of two different metals joined end-to-end to produce a loop.

Thermographic imaging: A technology that allows heat loss to show up in the form of various colors when viewed through a heat-sensitive lens.

Thermosphere: The atmospheric region extending from the mesosphere to about 300 miles above the earth's surface.

Thermostat: A temperature-controlled flow valve.

Third-class lever: A lever that has the input force positioned between the fulcrum and the load.

Three Mile Island (TMI) accident: A nuclear disaster occurring in 1979 near Harrisburg, Pennsylvania. The Unit 2 reactor reached excessively high temperatures through a series of faulty readings and operator errors. Eventually, a small piece of the reactor core melted, rendering the reactor unusable before the situation was brought under control.

Three-phase induction motor: An alternating current (AC) motor constructed using three sets of stator windings.

Throttle: A platelike device on an internal combustion engine, located in back of the venturi, that regulates the amount of fuel-air mixture entering the carburetor.

Thrust: The force produced by the propulsion system that moves an aircraft through the air.

Tidal fence: A barrier intended to use the power of tides to generate electricity.

Timing function: A computer subroutine that observes and records the elapsed time of a process.

Top dead center (TDC): As high in the cylinder as the top of the piston can go.

Torque: Effort, in rotary mechanical power.

Torque converter: A fluid coupler used in automatic transmissions. It uses fluid to transfer power.

Torsion bar: A metal rod used as a spring on modern land vehicles.

Towboat: A vessel designed to push barges.

Trade-off: A situation in which a technological development solves one problem, only to create other problems.

Trailer on Flat Car (TOFC): A method of intermodal transportation involving trailers full of cargo carried on a rail flatcar.

Trailing edge: The point at the back of an airfoil where the top and bottom meet.

Trajectory: The course or route of a spacecraft.

Transducer: A device for converting one form of energy into another. It is often used to measure quantities in a system and convert them to a proportional unit displayed on a meter or scale.

Transformer: A device used to increase or decrease voltage supplied to a circuit.

Transistor: A solid-state electronic device similar to an electron tube.

Transistor-transistor logic (TTL): A type of integrated circuit (IC) that works on low voltage, typically 5 V or less.

Transmission line: A pipe in a pipeline system that transports cargo over great distances.

Transmission path: A means of transporting energy from the point where it was generated to the point where it will produce work.

Transmission system: A device that provides for multiplying, dividing, or reversing the mechanical power and torque coming from the engine.

Transoceanic: Traveling across the ocean.

Transportation: The movement of people or products from one place to another.

Transportation system: An organized process of relocating people and cargo using the various modes of transportation.

Transportation system input: A resource needed in order to begin and maintain the use of the system, including people, capital, knowledge, material, energy, and finances.

Transportation system output: The relocation of people or cargo.

Transportation system process: An action that converts inputs into desired outputs.

Transportation technology: Systems designed to move people and products from one place to another.

Transport car: A flatcar with side rails. It is mainly used for transporting new vehicles from manufacturing plants to car dealers.

Trombe wall: A common application of the indirect gain approach. The Trombe wall is heated throughout the day. The storage medium takes a long time to heat up, but it gives off its heat slowly throughout the night, requiring less use of conventional energy.

Troposphere: The closest atmospheric region to the earth. It begins at the earth's surface and stretches about 10 miles (16 km) above the earth's surface.

True signal: A positive signal sent when a switch is moved to "on" or a sensor reaches its set point. Also referred to as a high signal.

Truth table: A graphic method of representing the possible results from inputs.

Tugboat: A vessel designed to pull barges.

Tunnel: A covered passageway.

Turbofan engine: An engine widely used on commercial passenger airplanes because it is efficient, as well as powerful, at low speeds. Sometimes called a fan-jet or bypass engine.

Turbojet engine: An engine used on commercial aviation vehicles.

Turboprop engine: An engine that is similar to a turbojet with a propeller mounted on the front.

Turbulence: The opposition to flow that occurs within a liquid itself.

Two-stroke cycle engine: An engine in which every upward stroke is a compression stroke and every downward stroke is a power stroke.

U

Undershot waterwheel: A waterwheel that does not require a significant elevation change and primarily makes use of the force of flowing water.

Unibody construction: A type of automobile construction that combines the body and frame in one unit.

Unit train: A train that carries only one type of cargo.

Universal joint: A joint that allows connected shafts to spin freely, while permitting a change in direction.

Unlatching: A function used to turn a machine off.

Unmanned space vehicle: A space vehicle not operated by human beings.

Uranium 235 (U235): An element whose atoms can be split more easily than most others, making it suitable for refining into nuclear fuel.

Uranium 238 (U238): A type of uranium that is a nonfissionable element. About 99% of all uranium mined is this type.

V

Valence ring: The outermost ring of electrons in an atom.

Valve: A part of an air compressor at the top of the cylinder that lets air pass in only one direction.

Valve lifter: A lifter that transfers power from the cam lobe to the valve.

Variable-flow restrictor: See *Flow-control valve.*

Vector quantity: A quantity that has both magnitude and direction.

Vehicle Assembly Building (VAB): A facility in which the orbiter is stood vertically and the solid rocket boosters (SRBs) and external tank (ET) are attached.

Vehicular system: A collection of separate systems that allow the machine to move through its environment safely and efficiently.

Velocity: A vector quantity that includes speed and direction.

Ventilated container: A container that has openings in the sidewalls allowing air to flow around the cargo.

Venturi: The narrow, restricting section of the carburetor, where air speed increases and drafts the fuel vapor along with it into the combustion chamber.

Vertical speed indicator: An instrument that displays the rate at which the airplane is ascending or descending.

Very-high-frequency omnidirectional radio range (VOR) navigation: A commonly used guidance system for air transportation. Each VOR station transmits a series of beams of radio waves in all directions. A VOR instrument shows the heading of the next VOR station.

Vessel: A water vehicle.

Viscosity: A measurement of internal friction, or the resistance of a fluid to flow.

Visual flight rules (VFR): Rules followed when weather conditions allow pilots to navigate by what they are able to see outside the cockpit.

Voice coil: The wire in a dynamic coil microphone that is attached to the diaphragm and moves back and forth as the diaphragm vibrates.

Voltage: In electricity, the effort behind the movement of electrons.

Volumetric efficiency: A measurement of how well the engine breathes.

W

Waste-to-energy plant: A plant that shreds and burns waste. The heat energy is then used to produce electricity or for industrial processes.

Water jacket: A space machined into the block of an engine. In a liquid cooling system, a water and antifreeze solution is pumped through the water jackets surrounding each cylinder.

Water jet: The newest form of marine propulsion. It uses an inboard engine to turn an impeller.

Waterline: The location at which the water stops along the side of the hull.

Water pollution: The action of contaminating the liquid that descends from the clouds as rain, especially with human-made waste.

Water transportation: A transportation system in which people and cargo are moved on bodies of water. Vessels used include ships, sailboats, barges, tugboats, and submarines.

Waterway: A body of water in which vessels travel.

Watt: A measurement of power in electrical power systems.

Wattage: A measurement of electrical power calculated by multiplying voltage and amperage.

Watt's law: Power equals effort multiplied by rate. This law describes the relationship among voltage, amperage, and wattage.

Wavelength: One frequency of radiant energy.

Way point: The location of a destination in a global positioning system (GPS).

Wedge: A simple machine based on the principle of the inclined plane.

Weightlessness: The condition occurring when gravitational forces are equal to a centrifugal force.

Well car: A railcar designed so the containers actually sit near or below the height of the axles, allowing two containers to be stacked on top of one another.

Wet scrubber: A technique used to reduce power plant emissions in which particulates, vapors, and gases are controlled by passing the gaseous stream of emissions through a liquid solution.

Wheel and axle: A large-diameter wheel and its small-diameter axle are attached to each other to move as one unit.

Wind farm: A collection of wind turbines used to create electricity.

Wind turbine: A propeller driven by the wind and connected to a generator. The wind makes it turn the generator, which produces electricity.

Wind velocity profile: Data characterizing the number of expected hours of a given wind speed for a particular location.

Wiper: The electrical connection on a potentiometer that can change position along the resistive material, based on the position of the knob or slider.

Work: The application of force that moves an object a certain distance.

Working pressure: The normal operating pressure for which a conductor is designed.

X-ray tube: A device that converts electricity into x-radiation. It produces a stream of negatively charged electrons that strike a tungsten filament, causing the tungsten to emit X rays.

Yaw: The side-to-side motion of an aircraft.

Yucca Mountain storage facility: A government-owned facility in southern Nevada that is a planned site for permanent storage of nuclear waste. The waste would be stored in stable rock formations deep within the earth's surface.

Z

Zener diode: A diode that conducts current in the reverse-biased direction. It blocks current until the voltage exceeds a certain level, and then it allows the current to flow. Once the zener diode allows current to flow, it is able to maintain a steady voltage.

Zenith path: The height of the sun in the sky when measured vertically from the horizon line.

Index

A

D

E

F

M

P

T

U

V

W

X

Y

Z